# ANNUAL REVIEW OF ENTOMOLOGY

# ANNUAL REVIEW OF ENTOMOLOGY

## VOLUME 34, 1989

THOMAS E. MITTLER, *Editor*

University of California, Berkeley

FRANK J. RADOVSKY, *Associate Editor*

North Carolina State Museum of Natural Sciences, Raleigh

VINCENT H. RESH, *Associate Editor*

University of California, Berkeley

ANNUAL REVIEWS INC.    4139 EL CAMINO WAY    P.O. BOX 10139    PALO ALTO, CALIFORNIA 94303-0897

R| ANNUAL REVIEWS INC.
Palo Alto, California, USA

*International Standard Serial Number: 0066–4170*
*International Standard Book Number: 0–8243–0134-X*
*Library of Congress Catalog Card Number: A56–5750*

Annual Review and publication titles are registered trademarks of Annual Reviews
Inc.

∞   The paper used in this publication meets the minimum requirements of American National Standard for Information Sciences—Permanence of Paper for Printed
Library Materials, ANSI Z39.48-1984.

Annual Reviews Inc. and the Editors of its publications assume no responsibility
for the statements expressed by the contributors to this *Review*.

Typesetting by Kachina Typesetting Inc., Tempe, Arizona; John Olson, President
Typesetting coordinator, Janis Hoffman

PRINTED AND BOUND IN THE UNITED STATES OF AMERICA

# PREFACE

The Livraria Kosmos, in one of the fashionable downtown streets of Rio de Janeiro, was in the late 1950s a meeting place for professionals, aspiring scientists, and intellectuals of all sorts. Run by a dealer of old European tradition, this bookstore was the only one in town to carry the latest English, French, German, and Italian titles in the sciences and the arts. Kosmos was my favorite stop whenever I was in the neighborhood, although another powerful enticement in the area was Café Simpatia, where they prepared, and still do, a delightful coconut juice, a perfect thirst quencher for the scorching Rio summers.

Entomological books were rare in Rio at that time, and one needed access to the libraries of the Instituto Oswaldo Cruz or the Escola Nacional de Agronomia, both inconveniently located away from city center, to obtain some of the basic texts. Brazilian titles were also scarce: Costa Lima's *Insetos do Brasil*, Foratini's *Entomologia Médica*, and Jacintho Guérin's *Cóleopteros do Brasil* were the best known, and even these were not generally available. The shelves of Kosmos contained a handful of entomological titles that often remained unsold for months, and I always had time to save the money needed to buy such books as Johannson and Butt's *Embryology of Insects and Miriapods*, Snodgrass's *Insect Morphology*, and Wigglesworth's *Principles of Insect Physiology*. I remember that for many months I admired a copy of Bodenheimer's *Geschichte der Entomologie*, but neither my means nor my German was sufficient for that opus.

As a young, rather self-taught entomologist with very limited resources, I often flirted with books for a long time before I could afford to buy them, and I wanted to make sure that each title was really worth the price. It was during one of my frequent visits to Kosmos in June of 1958 that I saw for the first time Volume 1 of the *Annual Review of Entomology* (1956). I grabbed the book and was immediately fascinated by the concept. I could now have a full library on the most current entomological topics under a single cover. This time I did not hesitate among the various options of recent imports; I bought that volume right away and made sure to ask Mr. Eisner, the bookdealer, to obtain the already published Volumes 2 and 3 and all future issues for me.

I have kept up my collection of the *Annual Review of Entomology* ever since, and it is the most valuable resource in my personal library, for myself and for my students. Never in my wildest fantasies did I imagine myself

(*continued*)   v

involved in the production of this publication. And yet 30 years later here I am—two-time author and a member of its editorial committee. The association with the committee has been one of the most rewarding of my career. My memories of Kosmos and my problems obtaining the literature of my chosen profession in a foreign country intensify my admiration for an enterprise that has managed to keep up with the changing nature of science and to maintain a fresh approach to the reviewing process.

This volume of the *Annual Review of Entomology* is dedicated to the Entomological Society of America, which is celebrating its centennial this year. In 1953 a committee of the Entomological Society of America examined the problem of providing adequate reviews of the entomological literature. It was recommended that the needs of entomology would best be met by a review publication of the type published by Annual Reviews Inc. The first volume of the *Annual Review of Entomology* appeared in 1956, and for the past 35 years the Entomological Society of America and its members have had a close and mutually rewarding relationship with the *Annual Review of Entomology*.

The contributions of our authors, the conscientious efforts of our Production Editor, Ms. Andrea Perlis, and the high quality of work of our compositor and printer are greatly acknowledged.

MARCOS KOGAN
FOR THE EDITORIAL COMMITTEE

Annual Review of Entomology
Volume 34, 1989

# CONTENTS

viii    CONTENTS    (*continued*)

# OTHER REVIEWS OF ENTOMOLOGICAL INTEREST

From the *Annual Review of Ecology and Systematics*, Volume 19 (1988)

From the *Annual Review of Microbiology*, Volume 42 (1988)

ANNUAL REVIEWS INC. is a nonprofit scientific publisher established to promote the advancement of the sciences. Beginning in 1932 with the *Annual Review of Biochemistry,* the Company has pursued as its principal function the publication of high quality, reasonably priced *Annual Review* volumes. The volumes are organized by Editors and Editorial Committees who invite qualified authors to contribute critical articles reviewing significant developments within each major discipline. The Editor-in-Chief invites those interested in serving as future Editorial Committee members to communicate directly with him. Annual Reviews Inc. is administered by a Board of Directors, whose members serve without compensation.

For the convenience of readers, a detachable order form/envelope is bound into the back of this volume.

*Ann. Rev. Entomol. 1989. 34:1–16*

# THE HISTORY OF THE VEDALIA BEETLE IMPORTATION TO CALIFORNIA AND ITS IMPACT ON THE DEVELOPMENT OF BIOLOGICAL CONTROL

## L. E. Caltagirone

Division of Biological Control, University of California, Albany, California 94706

## R. L. Doutt

1781 Glen Oaks Drive, Santa Barbara, California 93108

## INTRODUCTION

The suppression of the cottonycushion scale, *Icerya purchasi,* by the vedalia beetle, *Rodolia cardinalis,* in California citrus groves in 1889 is unparalleled in the annals of entomology for its drama, human interest, political ramifications, and continuing significance.

This spectacular demonstration of applied ecology established the practicality of biological control, which was singled out as the preferred pest control method of the Entomological Society of America in 1964 (22). During the past 100 years biological control has been widely, repeatedly, and thoroughly tested. So this review of the vedalia beetle is entirely appropriate as a celebration of the centennial of biological control.

Biological control has proved superbly successful in an array of diverse situations. It has succeeded in both temperate and tropical climates; on continents, subcontinents, and islands; and in both northern and southern latitudes. It has suppressed target pests of great taxonomic diversity. It has worked well in depauperate biotas and in those of great complexity. It has been a boom to primitive agriculture and has been economically successful in

1

0066-4170/89/0101-0001$02.00

modern, progressive, and intensified agriculture. It has been used where chemical control measures were not feasible, and it has been used harmoniously with pesticide applications. Most significantly, it has worked wherever farmers have persevered in its use (8, 9, 19, 20, 39).

## HISTORY OF THE VEDALIA BEETLE IMPORTATION TO CALIFORNIA

The initial demonstration of applied biological control sprang from a special set of circumstances. By the middle of the nineteenth century the leading American agricultural entomologists, Asa Fitch in New York and Benjamin Walsh in Illinois, were well aware of the foreign origin of this country's most devastating pests (23, 34, 44, 63). Since both men clearly understood the role of entomophagous species in the suppression of host or prey populations, they suggested alleviating the problems caused by these alien pests by importing their natural enemies (18, 24, 63). These men were not only early and especially articulate proponents of the biological control of insect pests; they also suggested the control of weeds by using phytophagous insects (18, 63).

Benjamin Walsh was the mentor of Charles Valentine Riley, who came to this country from England at the age of 17. At the age of 25 Riley was appointed State Entomologist of Missouri. He was a restless man, ambitious to build up a large organization. He was a great schemer, striving constantly to make his work appear important (34). His aggressive personality was bound to make enemies, and some people who were not especially fond of Riley referred to him as "the General" (44).

However, as an entomologist Riley had become such a national figure by 1878 that at the age of 35 he was appointed Entomologist for the United States Department of Agriculture and Chief of the Division of Entomology.

To understand the circumstances that set the stage for the famous vedalia project it is necessary mentally to enter a time warp and go 100 years into the past.

Perhaps nowhere in the world is the history of agriculture so well documented from its inception as in California. This is the case because there was virtually no agriculture in the state before 1850. The Indian tribes were not agricultural. The Spaniards did not appear upon the scene as landholders until the last of the eighteenth century, and they did little more than tend small mission gardens and perfunctorily run some cattle on the hillsides. Thus agriculture in California really did not begin until after the gold rush, when settlers in large numbers remained to till the soil (21). These were inherently a pioneering breed faced with new conditions and not guided by precedent or bound by rigid and ancient customs. They were inventive entrepreneurs and from the outset experimented with crops, irrigation systems, and new methods of pest control. They were progressive, imaginative, and fiercely competitive. These qualities were essential to economic survival because the major markets for their agricultural products were 3000 miles to the east across a nation that had no transcontinental railroad until 1869. These were the farmers who were so impressed with the fact that their major pests were of foreign origin that they enacted the first quarantine legislation in the Western Hemisphere (54) fully a quarter of a century before the federal laws were passed. These were the farmers who as practical entomologists without technical assistance developed the poison bait method of grasshopper control. These were the men who invented the fumigation method for controlling pests on citrus, who experimented with sprays, and who enthusiastically supported the first significant demonstration on classical biological control (18, 69).

In April 1887 Charles Valentine Riley addressed the Fruit Growers' Convention at Riverside, California, a receptive and politically powerful audience. The citrus industry, which held great promise as a major economic force in California, was just in its infancy, but the crop was being devastated by an almost incredibly dense infestation of *Icerya purchasi*. The growers desperately needed help, so they had invited Riley to address their convention. He did not disappoint his audience. He told them that *I. purchasi* was from Australasia, although he was not sure whether it was Australia or New Zealand. He believed that it had been accidentally introduced to California at Menlo Park on acacia in 1868. Finally, he said (17):

> It had doubtless occurred to many of you that it would be very desirable to introduce from Australia such parasites as serve to keep this fluted scale in check in its native land . . . This state—yes, even Los Angeles County—could well afford to appropriate a couple of thousand dollars for no other purpose than the sending of an expert to Australia to devote some months to the study of these parasites there and to their artificial introduction here. But the agent must be an expert entomologist and his selection should be left to some competent authority.
>
> I would not hesitate, as United States Entomologist, to send some one there with the consent of the Commissioner of Agriculture were the means for the purpose at my command; but unfortunately, the mere suggestion that I wanted $1,500 or $2,000 for such a purpose would be more apt to cause laughter and ridicule on the part of the average committee in Congress than serious and earnest consideration, and the action of the last Congress has rendered such work impossible by limiting investigation to the United States.

This last comment referred to a limitation that was part of the appropriations bill for the Department of Agriculture for that year, which prohibited foreign trips by employees and was purposely inserted in the bill to stop the frequent European junkets taken by Riley (17).

As a result of Riley's address and persuasion the convention adopted a resolution favoring the idea of sending someone to Australia to seek out natural foes of *I. purchasi* and to bring them to California.

To implement this resolution the growers put pressure on their Washington representatives. In a letter to Representative Felton, Norman J. Colman, the Commissioner of Agriculture, pointed out that while travel by the Division of Entomology was restricted to the United States, there was to be an International Exposition in Melbourne in which the United States Government was to take part. Colman (11a) said "This exposition in many ways would further the investigation referred to in your memorial." This opportunity to exploit the appropriation for the Melbourne Exposition was quickly grasped. The Honorable Frank McCoppin of San Francisco was made US Commissioner to the Exposition, and he arranged with other members of his commission to set aside $2000 to pay the expenses of an entomologist who was to accompany them, ostensibly to represent the State Department but actually to collect enemies of *I. purchasi* (17).

When Riley addressed the Fruit Growers' Convention he had two field agents working in California. One was Albert Koebele, a naturalized German immigrant, who was stationed in Alameda to investigate the history and habits of California insects. The other was Daniel W. Coquillet, a native of Illinois, who was in Los Angeles to work on cottonycushion scale. Riley selected Koebele to accompany the commission, although the California growers had suggested Coquillet. According to his letters, Riley himself had intended to follow Koebele, but his trip never materialized. Koebele sailed for Australia on August 25, 1888.

In October 1888 Koebele (quoted in 51) wrote Riley, "So far my work has been much more successful than I expected." He had found the dipterous parasite *Cryptochetum*[1] *iceryae* in large numbers and three predaceous larvae, one a *Chrysopa* species and the other two "larvae of a coccinella."

The peril of making predictions about imported entomophagous species before they are tested is illustrated in Riley's reply (quoted in 17) to Koebele that "the sending of the coccinellids is of course desirable but I think we have much more to hope from the *Cryptochaetum*." The coccinellids were of course the vedalia beetles.

Through Koebele's efforts a total of approximately 12,000 *C. iceryae* individuals were sent to California. Koebele first found vedalia beetles feeding upon *I. purchasi* in a North Adelaide garden on October 15, 1888. He sent only 129 vedalia specimens to Coquillet, who was assigned by Riley to receive and culture the species in Los Angeles; Coquillet received 28 beetles on November 30, 1888; 44 specimens on December 29; and 57 specimens on January 24, 1889 (13).

The beetles, as received, were placed under a tent on an *I. purchasi*–infested orange tree on the J. W. Wolfskill ranch in Los Angeles. This tent (Figure 1) thus became the first insectary in California. Here the beetles were

---

[1]The spelling *Cryptochetum* is correct (59). *Cryptochaetum* has been kept in this text when it was so used by the authors quoted.

allowed to breed undisturbed, and on April 12 one side of the tent was removed to allow the ladybirds to spread to adjoining trees. At this time Coquillet began sending out colonies of the vedalia to various parts of the state. By June 12 he had distributed 10,555 of the voracious ladybirds to 228 different orchardists.

On July 2, 1889 a citrus grower, J. R. Dobbins (quoted in 17), reported:

> The Vedalia has multiplied in numbers and spread so rapidly that every one of my 3,200 orchard trees is literally swarming with them . . . People are coming here daily, and by placing infested branches upon the ground beneath my trees for two hours, can secure colonies of thousands of Vedalia, which are there in countless numbers seeking food. Over 50,000 have been taken away to other orchards during the past week, and there are millions still remaining . . . I feel positive from my own experience, that the entire valley will be practically free from *Icerya* before the advent of the New Year.

A. Scott Chapman, who a year earlier had stated that infestations of *I. purchasi* were forcing him to abandon fruit growing, reported on October 18, 1889 that the vedalia had cleared the scale from 150 acres of his land. The

*Figure 1*   The first 129 Vedalia beetles imported to California were released in this screened tree in the Wolfskill Orchard, Los Angeles (51).

shipments of oranges from Los Angeles County jumped in one year from 700 to 2000 carloads. The vedalia was hailed as a miracle of entomology. The cost of the project was about $1500 (23).

## IMPACT OF THE VEDALIA PROJECT ON THE DEVELOPMENT OF BIOLOGICAL CONTROL

In retrospect, one could not have purposely designed a more dramatic, effective, public demonstration of biological control. An entire industry was being ravaged by a scale pest that was particularly conspicuous because of its huge white egg masses. The farmers themselves collected and placed on their trees the strikingly marked orange and black vedalia beetles. The voracious feeding activities of the beetles were visible and impressively dynamic, and the beneficial results were instantly apparent.

### Vedalia and Cryptochetum iceryae as Controlling Agents

Smith (56) has listed (a) dispersal ability, (b) rapidity of population increase, and (c) searching ability as features that make it possible for a parasite or a predator to reduce the numbers of a pest effectively. Similarly, Hagen et al (31) indicate that some of the essential attributes for successful pest control by a predator are (a) multivoltinism, (b) narrow prey specificity, (c) adult longevity, and (d) high prey-searching efficiency. The vedalia beetle, *Rodolia cardinalis,* possesses most of these attributes (49, 61). According to Smith (57), however, the characteristics of *R. cardinalis* are not the only ones that explain the success of the predator. The limited dispersal of *I. purchasi,* which tends to be strongly colonial, favors the predator.

*Icerya purchasi* has a high reproductive capacity (each female produces 325–450 eggs) (49), is predominantly hermaphroditic and capable of self-fertilization (4, 36–38), and has three generations per year under most California conditions. These characteristics enable the scale to establish thriving colonies in a short time starting from very few individuals.

The reproductive capacity of *R. cardinalis* is 150–190 eggs per female, and the number of generations varies from 12 in the hot, dry inland areas of southern California to eight in the cooler coastal areas (49).

Less favorable conditions for vedalia, such as cooler temperatures, are favorable for the cryptochetid *Cryptochetum iceryae*. This parasite has a reproductive capacity of some 50 individuals per female and nine generations per year. It can exploit various stages of the host, from the early stages, which barely provide enough nutrients for the development of an individual parasite, to large, fully developed females, which can support a maximum of 11 parasites. *C. iceryae* also has excellent powers of dispersal and host location. These attributes enable this parasite to exploit *I. purchasi* even when this host reaches very low population levels.

The interaction between the cottonycushion scale and vedalia on favorable plants such as *Citrus, Pittosporum,* and *Acacia* species results in a reduction of the scale to very low levels, causing a drastic reduction of the predator's own food supply (an "intrinsic" food shortage) (2). In areas where conditions are favorable for both *R. cardinalis* and *C. iceryae,* interspecific competition can occur; the outcome is determined by the conditions at the time of competition. If there is an abundance of scale, competition will be minimal. Otherwise, if vedalia arrives first at a scale colony it will control it before *C. iceryae* gets a foothold; if the latter species colonizes the patch first, there will be a number of parasitized scales (those containing mature fly larvae or pupae) that will be unacceptable to the beetles. In California there are areas in which both natural enemies can exist simultaneously, maintaining the scale population at reduced levels. In these areas each natural enemy could probably control the scale as effectively by itself (61). When both species coexist, each creates an intrinsic food shortage for itself and an extrinsic food shortage for the other (2). Quezada & DeBach (49) consider these two species ecological homologs and consequently claimed that they could not coexist permanently. They explained those cases in southern California where both vedalia and *C. iceryae* coexist during certain periods of the year by postulating that repeated colonizations take place: Vedalia arrives from the warmer desert areas and *C. iceryae* from the cooler coastal areas (49). However, these entomophages coexist in Bermuda (6) without the benefit of repeated immigration.

Today, 100 years after their introduction to California, these two natural enemies still regulate the populations of cottonycushion scale at levels that are extremely low, not only on cultivated *Citrus* spp., but also on numerous ornamentals (e.g. *Pittosporum* spp., *Acacia* spp.). Populations of the scale occasionally increase; historically scale increases have been due to resurgence after application of pesticides toxic to vedalia (8, 16, 45, 46, 52, 58). These increases last only for a very short time before either or both natural enemies locate the population and quickly bring it down again to low levels.

In California the role of each of the natural enemies is determined by the environmental conditions of the area where the interaction takes place. In the interior areas (e.g. southern desert, San Joaquin Valley) the coccinellid seems to be the dominating regulatory factor, while in the coastal areas the fly is the more important mortality factor. In the intermediate areas the scale is exploited by one or the other natural enemy depending on the prevailing environmental conditions (49). Quezada & DeBach (49) concluded that the distribution of these natural enemies in southern California is the result of competitive displacement; *R. cardinalis* has displaced *C. iceryae* in the interior, while the reverse is true in the coastal areas.

The regulation of cottonycushion scale by vedalia, *C. iceryae,* or both together has not been demonstrated by well-designed experiments. It is generally accepted that to determine the effect of an exotic or indigenous natural enemy on the population of a pest it is necessary to quantify pest densities in areas with and without natural enemies. It is also necessary to evaluate the population densities both before and after release of the natural enemies (42). In the *I. purchasi–R. cardinalis–C. iceryae* system, (*a*) in most cases, shortly after the natural enemy has been introduced the scale has been reduced to low population levels; (*b*) whenever the natural enemies have been eliminated or drastically reduced (e.g. by inclement weather or chemical pesticides) the population of the scale has increased shortly thereafter; and (*c*) whenever the decimated natural enemies have been allowed to recolonize, the scale has quickly returned to its formerly low levels. The performance of vedalia and *C. iceryae* in California and elsewhere during the past 100 years is "unquestionable proof of long-term control by natural enemies" (42). To object to this conclusion because it is not based on well-designed experiments would be "quibbling over details" (60).

Understandably, as a result of the 1888–1889 success the California farmers and agricultural officials became ardent and enthusiastic supporters and advocates of biological control. They were eager to apply the method more widely. However, Riley thought he was not given his due credit for the vedalia introduction, while Koebele was praised throughout the state and given a gold watch and diamond earrings for his wife. Disagreements between Riley and the California officials became so intense and acrimonious that Riley recalled his agents from the state. Coquillet returned to Washington, but Koebele resigned to work for the territory of Hawaii (17).

Riley retired from his position as Chief of the Division of Entomology in 1894. His successor was L. O. Howard, who had been Riley's loyal assistant for 13 years. It is therefore not surprising that he held Riley's viewpoints about California. Howard wrote that persons with strange beliefs migrate to California and that southern California is the home of all heterodoxies (34). He was annoyed by the enthusiasm for biological control and the frank criticisms of the entomological force in Washington voiced by the influential and articulate California officials. Howard actually believed that the vedalia project and subsequent enthusiasm for biological control caused the state to make no advances in the fight against insects for many years.

## Effect of Host Plant on Effectiveness of Vedalia

*I. purchasi* is not effectively controlled where the climate is unfavorable to vedalia, as in the foothills of the Pyrenees and the Alps (29), when host plants render the scale unsuitable for *R. cardinalis*, or in areas where the climatic conditions are not favorable for *C. iceryae*. Some of these plants are the Spanish broom, *Spartium junceum*, in Italy, France, Peru, and Israel (7, 33, 48, 66; Svi Mendel, personal communication); *Plumbago zeylandica* in Malta

(7); Scotch broom, *Cytisus scoparius,* in California (8, 25); and maple in California (25). When *I. purchasi* feeds on *S. junceum* it sequesters alkaloids that render the scale unacceptable as prey for *R. cardinalis* (48). In Peru the unsuitability of *I. purchasi* growing on these plants is somewhat different (66). *R. cardinalis* was imported to Peru in 1932 to the Huánuco area, where the predator completely controlled the pest on citrus. It was distributed to other areas, where it has been equally successful. In the Huánuco area the scale also infests *S. junceum, Ambrosia artemisiaefolia, Baccharis* spp., and *Caesalpinia tinctoria.* Vedalia does not completely control the scale on these plants, where scales and predators persist for long periods. These areas are permanent sources of vedalia that will disperse into new infestations of scale on citrus and bring them under control. This suggests that in Peru the scale on these plants becomes partially unsuitable, preventing the beetle from fully realizing its potential to reduce the prey populations.

## Infatuation With Ladybird Beetles

The astounding success of vedalia in California was soon duplicated in other infestations of *I. purchasi* as colonies of the ladybird were moved from California to other areas (Figure 2). Again, in most cases the triumph of vedalia was spectacular (10). This success quickly led to an increased interest

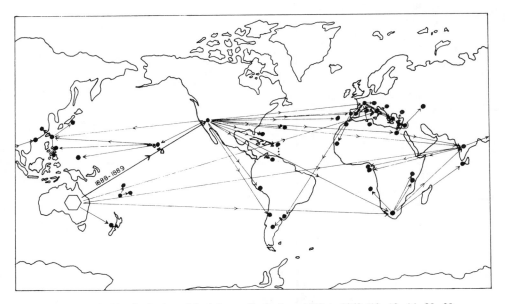

*Figure 2*   World colonization of *Rodolia cardinalis* from 1889 to 1968 (10, 13, 14, 28, 29, 39–41, 50, 64).

in exploratory surveys for natural enemies of pests in other areas (50, 55, 68) and to a worldwide infatuation with coccinellids. Lounsbury (41) scornfully referred to this attitude as the "ladybird fantasy." Species of ladybirds, mostly from Australia, were haphazardly introduced into many areas of South Africa, and virtually all failed to provide control. For example, tens of thousands of *Hippodamia convergens,* a species native to California, failed to become established.

The enthusiasm for coccinellids was particularly noticeable in the years immediately after the California success, the beginning of the "intermediate period" in the history of biological control (55). This was a period in which "enthusiasm for biological control was unrestrained among orchardists, self-trained entomologists, and horticultural officials in California" (12). From late 1891 to the middle of 1892 Koebele sent to California from Australia, New Zealand, New Caledonia, and Fiji some 40,000 specimens of coccinellids comprising some 40 species. These were introduced against scale and mealybugs on citrus (30, 53). Only four of the species became established, but only one, vedalia, was a successful controlling agent. So widespread was this enthusiasm to control scales with natural enemies that some growers left their orchards unsprayed. Commenting on this attitude, Howard (34) stated: "It is safe to say that a large share of the loss through insects suffered by California from 1888 until, let us say, 1908 was due to this prejudicial and badly based policy."

During the period 1895–1902 many coccinellids were moved from Eastern Australia to Western Australia. Thereafter the importations, many of coccinellids, were made from abroad. Only a few of them became established (67). In all these introductions the essential requirements and basic biology of the various species of coccinellids were ignored or unknown. This period of great interest in coccinellids lasted until the 1920s (32).

A more positive effect on Australia's perspective on biological control was stimulated by Koebele's successful trips and the subsequent visits of a number of workers in biological control, such as G. Compere, R. C. L. Perkins, F. Muir, H. Compere, C. E. Pemberton, E. J. Vosler, and S. E. Flanders. As the ladybird "craze" died down it was clear that a sounder approach to biological control was needed. L. O. Howard, who was always critical of the enthusiasm engendered in California by the vedalia project, wrote that biological control work in the state would have been stopped by the federal government had California not appointed Harry Scott Smith to take charge of the parasite work for the state in 1912.

Smith was made an official collaborator of the United States Department of Agriculture, and his subsequent biological control efforts were thus to some extent in cooperation with the federal government (34).

## Vedalia Successes Elsewhere

The spectacular success of vedalia and *C. iceryae* in California resulted in an understandable interest in biological control of the cottonycushion scale wherever it had become a pest. In a number of cases the California experience was relived. In South Africa, then the Colony of the Cape of Good Hope, *I. purchasi* was widespread, attacking a number of plants, especially citrus and roses. When news of the successful vedalia project in California reached the Cape, T. A. Louw, a member of the Cape parliament, was sent to California to import a breeding stock of the beetles. "Louw landed at Cape Town on 29 January 1892, bringing a supply of vedalia, which was in part liberated at Stellenbosch and in part used as breeding stock at Fernwood. The outcome of introducing the vedalia was spectacular. Within two years Australian bug was subjugated far and wide" (41).

In Peru *I. purchasi* was detected in 1932 in Huánuco and in nine other localities. Shortly thereafter the vedalia beetle was imported from the United States, and a colony was established in Huánuco. After only a few months the beetle population had increased and considerable control of the pest was achieved. Vedalia was colonized in the other areas where the scale was present, and in each of them the pest was eliminated after a short period (65).

Graf-Marín & Cortés-Peña (27) attributed the development of biological control in Chile to, among other reasons, a remarkable development of biological control abroad, trips to the United States by Chilean plant protection authorities, visits to Chile by foreign specialists in biological control (e.g. H. Compere, F. Silvestri), similarity in climate and other ecological characteristics between the agricultural areas of Chile and those of California where biological control had become very important, and the alarming increase in Chile of pests such as *I. purchasi* and *Pseudococcus fragilis* (=*gahani*). The vedalia beetle was imported to Chile on July 16, 1931 by A. Graf-Marín. On his way back to Chile after completing his doctorate at Cornell University, Graf-Marín stopped at the University of California, Riverside, where he obtained colonies of *R. cardinalis* and *C. iceryae*. Very few beetles arrived alive but they were placed in a caged orange tree heavily infested with the scale. From this tree the beetles were distributed to other locations. In most of the infested areas the ladybird beetle completely controlled the pest (26).

The effectiveness of the predaceous vedalia and its companion parasitoid fly against the cottonycushion scale in the overwhelming majority of cases undoubtedly had a decisive influence on the implementation of biological control programs the world over.

# GENETIC VARIABILITY OF COLONIZED VEDALIA

When procuring and colonizing exotic natural enemies for use in biological control, sufficiently large numbers of individuals should be collected from the diverse sites in which a population or groups of populations occur so as to ensure that the genetic diversity in the area searched is well represented. The aim is to avoid genetic bottlenecks that might impede colonization. Maximum diversity is desirable in the founding colony because establishment of an exotic species is dependent on the evolution of a type adapted to the local conditions (43). When the field-collected founder material is not abundant, the risk of a genetic bottleneck effect in the resulting laboratory or insectary population is exacerbated. This is because the effective breeding population, defined as the weighted average of the number and degree to which individuals participate in the reproductive effort (62), in a laboratory is considerably smaller than the whole population.

In some of the successful colonizations of *R. cardinalis* the released specimens represented a restricted sample of the population(s) from which they were collected. Presumably the genetic pool represented was in each case a restricted one. The specimens released in California in 1889 consisted of the progeny of the 129 individuals received in the first three shipments from Australia (see history section above) and 385 received in February and March 1889 and released directly in orange groves (14). These individuals were the genetic ancestors of colonies that were subsequently distributed to many areas of the world from California (Figure 2).

The initial effort to introduce vedalia to France was made by P. Marchal in 1912. He was able to obtain from Italy only eight individuals, three of which survived the trip. They were the progenitors of the first vedalia beetles reared in France. Additional material was subsequently obtained from Italy, Portugal, and the United States. The first shipment of vedalia from the United States to Italy consisted of 36 beetles; Italy obtained other shipments from Portugal. In 1920 this predator was introduced to Sri Lanka (then Ceylon), where *I. purchasi* was attacking *Acacia decurrens;* the scale was "much reduced by the predator" (38a). The parent stock consisted of 50 adults (50).

Some 200 individuals of *R. cardinalis* from California were released in a citrus grove in Phoenix, Arizona in October 1926 in an attempt to control an outbreak of cottonycushion scale. In spite of the adverse weather, enough individuals survived to produce abundant progeny by early spring. The scale was completely controlled by the middle of the summer (3). Bennett (5) reported that a heavy infestation of *I. purchasi* in St. Kitts, West Indies, was brought under control after less than 500 *R. cardinalis* individuals were introduced.

The successful colonization of vedalia in places with climates very different from that of California and the areas in Australia where the founder material was collected suggests the enormous adaptability of the descendants of the original colonizers brought to California. Beetles from California were colonized, directly or through other areas, without additional material from other sources, in Egypt, Cyprus, the Soviet Union, Portugal, Puerto Rico, Venezuela, Peru, Chile, Hawaii, the Philippines, Guam, Uruguay, Argentina, Taiwan, and Palau. Vedalia is among the very few species of natural enemies that have colonized climatically diverse areas through the progeny of a few individuals.

## VEDALIA, THE EXCEPTION THAT PROVES THE RULE

It is extremely difficult to assess the impact of the vedalia beetle on the implementation of other biological control programs. From Clausen's worldwide review (9) a few interesting conclusions can be reached. During the period 1889–1921 a total of 75 programs of biological control of insects and mites were initiated around the world. During the period 1922–1968 232 new projects were initiated; only 108 would have been expected had the rate been the same as in the previous period. This difference is even more marked if the number of introductions of natural enemies in each project is considered. From 1889 to 1921 there were 281 importations, while from 1922 to 1968 there were 2136 (5.3 times more than the 404 expected). The figures on importations do not include the 224 natural enemies of the gypsy moth (79 species) imported during the period 1963–1977 (11, 15). We presume that after the lull of the 1910s, possibly caused by the ladybird beetle disenchantment, and as the knowledge of the impact of natural enemies on pests increased, biological control regained recognition as an important component of pest-control programs.

The search for generalizations applicable to natural phenomena is one of the primary goals of the scientific endeavor. Biological control is basically applied population ecology; thus there must be general principles that if well understood would make the application of this pest control strategy more predictable. The consistently successful performance of vedalia during the past 100 years as a colonizer and a regulator of its prey populations should make this predator a model from which general principles could be readily learned. However, when some of the generally accepted features of successful biological control projects are measured against those involving vedalia, one reaches the disturbing conclusion that the undisputed prime example of a successful biological control agent is an exception in several aspects.

Of the arthropod natural enemies used in biological control listed by Laing & Hamai (39), 295 of 353 are parasitoids. *Rodolia cardinalis* is a predator. Coccinellidae are not very good colonizers, as strongly suggested by the fact that only four species became established of the 40 that were imported during the "ladybird fantasy" period. *R. cardinalis* is one of those four species. In biological control, the numbers of specimens released in the target areas should be sufficiently large to increase the probability of establishment. Although no one has satisfactorily defined "sufficiently large," there is no argument against this guideline. However, *R. cardinalis* proved to be an extremely successful colonizer, even when the founder population consisted of a few individuals and the genetic diversity of the released material was comparatively restricted.

Although the spectacular suppression of cottonycushion scale by the vedalia beetle may be a fortuitous exception to the general ineffectiveness of coccinellids as classical biological control agents, its is a milestone in applied entomology. The search for effective pest control measures continues, but in the long pursuit of these goals no single achievement has more thoroughly, soundly, and significantly established a major pest control tactic than the vedalia project. All subsequent projects, programs, advances, and refinements in the theory and practice of biological control have sprung from this single event.

ACKNOWLEDGMENTS

We thank K. S. Hagen for his help in procuring reference material.

*Literature Cited*

1. Deleted in proof
2. Andrewartha, H. G., Birch, L. C. 1984. *The Ecological Web. More on the Distribution and Abundance of Animals.* Chicago: Univ. Chicago Press. 506 pp.
3. Arizona Commision of Agriculture and Horticulture. 1927. The cottony cushion scale and the vedalia beetle in Arizona. *XVII & XVIII Annu. Rep.*, pp. 35–36. Phoenix, Ariz: Ariz. Comm. Agric. Hortic.
4. Bacci, G. 1965. *Sex Determination.* Oxford, UK: Pergamon. 306 pp.
5. Bennett, F. D. 1971. Some recent successes in the field of biological control in the West Indies. *Rev. Peru. Entomol.* 14:369–73
6. Bennett, F. D., Hughes, I. W. 1959. Biological control of insect pests in Bermuda. *Bull. Entomol. Res.* 50:423–36
7. Borg, P. 1930. Entomological notes. *Rev. Appl. Entomol. Ser. A* 18:261–62
8. Clausen, C. P. 1956. *Biological Control of Insect Pests in the Continental United States. US Dep. Agric. Tech. Bull.* Vol. 1139. 151 pp.
9. Clausen, C. P., ed. 1978. *Introduced Parasites and Predators of Arthropod Pests: A World Review. US Dep. Agric. Agric. Handb.* Vol. 480. 545 pp.
10. Clausen, C. P. 1978. Margarodidae. See Ref. 9, pp. 132–36
11. Clausen, C. P. 1978. Limantriidae. See Ref. 9, pp. 195–204
11a. Colman, N. J. 1888. Letter to Rep. Felton. *Pac. Rural Press,* Feb. 11
12. Compere, H. 1969. Changing trends and objectives in biological control. *Proc. 1st Int. Citrus Symp., Riverside, Calif., 1968,* 2:755–64. Riverside, Calif: Univ. Calif. Riverside
13. Coquillet, D. W. 1889. The imported Australian ladybird. *Insect Life* 2:70–74
14. Coquillet, D. W. 1890. Report on various methods for destroying the red scale of California. *Dep. Agric. Bull.* 22:9
15. Coulson, J. R. 1981. Recent history: 1961–77. In *The Gypsy Moth: Research Toward Integrated Pest Management. US Dep. Agric. For. Serv. Tech. Bull.* 1584:302–10
16. DeBach, P., Bartlett, B. R. 1951. Effects of insecticides on biological con-

trol of insect pests of citrus. *J. Econ. Entomol.* 44:372–83

17. Doutt, R. L. 1958. Vice, virtue and the vedalia. *Bull. Entomol. Soc. Am.* 4:119–23

18. Doutt, R. L. 1964. The historical development of biological control. In *Biological Control of Insect Pests and Weeds,* ed. P. DeBach, pp. 21–42. London: Chapman & Hall. 844 pp.

19. Doutt, R. L. 1967. Biological control. In *Pest Control. Biological, Physical, and Selected Chemical Methods,* ed. W. W. Kilgore, R. L. Doutt, pp. 3–30. New York: Academic. 477 pp.

20. Doutt, R. L. 1972. Biological control: parasites and predators. In *Pest Control Strategies for the Future,* ed. Agric. Board Div. Biol. Agric., Natl. Res. Counc., pp. 288–97. Washington, DC: Natl. Acad. Sci. 376 pp.

21. Ebeling, W. 1979. *The Fruited Plain. The Story of American Agriculture.* Berkeley, Calif: Univ. Calif. Press. 433 pp.

22. Entomological Society of America. 1964. Statement on pesticides. *Bull. Entomol. Soc. Am.* 10:18

23. Essig, E. O. 1931. *A History of Entomology.* New York: Macmillan. 1029 pp.

24. Fitch, A. 1856. *Sixth, Seventh, Eighth and Ninth Reports on the Noxious, Beneficial and Other Insects of the State of New York.* Albany, NY: NY State. 259 pp.

25. Flanders, S. E. 1940. Environmental resistance to the establishment of parasitic Hymenoptera. *Ann. Entomol. Soc. Am.* 33:245–53

26. González, R. H., Rojas, S. 1966. Estudio analítico del control biológico de plagas agrícolas en Chile. *Agric. Tec. Santiago* 26:133–47

27. Graf-Marín, A., Cortés-Peña, R. 1941. Introduction de hiperparásitos en Chile: resumen de las importaciones hechas y de sus resultados. *Proc. 6th Pac. Sci. Congr., Berkeley, Stanford, San Francisco, 1939,* 4:351–57. Berkeley, Calif: Univ. Calif. Press

28. Greathead, D. J. 1971. *A Review of Biological Control in the Ethiopian Region. Tech. Commun. 5.* Farnham Royal, Slough, England: Commonw. Agric. Bur., Commonw. Inst. Biol. Control. 162 pp.

29. Greathead, D. J. 1976. *A Review of Biological Control in Western and Southern Europe. Tech. Commun. 7.* Farnham Royal, Slough, England: Commonw. Agric. Bur., Commonw. Inst. Biol. Control. 182 pp.

30. Hagen, K. S. 1974. The significance of predaceous Coccinellidae in biological and integrated control of insects. *Entomophaga* 7:25–44

31. Hagen, K. S., Bombosch, S., McMurtry, J. A. 1976. The biology and impact of predators. In *Theory and Practice of Biological Control,* ed. C. B. Huffaker, P. S. Messenger, pp. 93–142. New York: Academic. 788 pp.

32. Hagen, K. S., Franz, J. M. 1973. A history of biological control. In *A History of Entomology,* ed. R. F. Smith, T. E. Mittler, C. N. Smith, pp. 433–76. Palo Alto, Calif: Ann. Rev.

33. Hodek, I. 1966. Food ecology of aphidophagous Coccinellidae (review). In *Ecology of Aphidophagous Insects,* ed. I. Hodek, pp. 23–30. Prague: Czech. Acad. Sci. 360 pp.

34. Howard, L. O. 1930. A history of applied entomology. *Smithson. Misc. Collect.* 84:1–564

35. Deleted in proof

36. Hughes-Schrader, S. 1925. Cytology of hermaphroditism in *Icerya purchasi. Z. Zellforsch. Mikrosk. Anat.* 2:264–92

37. Hughes-Schrader, S. 1927. Origin and differentiation of the male and female germ cells in the hermaphrodite of *Icerya purchasi. Z. Zellforsch. Mikrosk. Anat.* 6:509–40

38. Hughes-Schrader, S. 1930. Contribution to the life history of the icerine coccids, with special reference to parthenogenesis and hermaphroditism. *Ann. Entomol. Soc. Am.* 23:359–80

38a. Hutson, J. C. 1920. The fluted scale and the vedalia beetle. *Trop. Agric. Mag. Ceylon Agric. Soc.* 55:95–97

39. Laing, J. E., Hamai, J. 1976. Biological control of insect pests and weeds by imported parasites, predators and pathogens. See Ref. 31, pp. 685–743

40. Leonard, M. D. 1932. The cottony cushion scale in Puerto Rico. *J. Econ. Entomol.* 25:1103–7

41. Lounsbury, C. P. 1940. The pioneering period of economic entomology in South Africa. *J. Entomol. Soc. South. Afr.* 3:9–29

42. Luck, R. F., Shepard, B. M., Kenmore, P. E. 1988. Experimental methods for evaluating arthropod natural enemies. *Ann. Rev. Entomol.* 33:367–91

43. Mackauer, M. 1976. Genetic problems in the production of biological control agents. *Ann. Rev. Entomol.* 21:369–85

44. Mallis, A. 1971. *American Entomologists.* New Brunswick, NJ: Rutgers Univ. 549 pp.

45. Metcalf, R. L. 1975. Insecticides in pest management. In *Introduction to Insect Pest Management,* ed. R. L. Metcalf,

W. Luckman, pp. 235–73. New York: Wiley. 587 pp.

46. Metcalf, R. L. 1980. Changing role of insecticides in crop protection. *Ann. Rev. Entomol.* 25:219–56

47. Deleted in proof

48. Poutiers, R. 1930. Sur le comportement de *Novius cardinalis* vis-a-vis de certains alcaloides. *C. R. Soc. Biol.* 103:1023–25

49. Quezada, J. R., DeBach, P. 1973. Bioecological studies of the cottony cushion scale, *Icerya purchasi* Mask., and its natural enemies, *Rodolia cardinalis* Mul. and *Cryptochaetum iceryae* Will., in southern California. *Hilgardia* 41:631–88

50. Rao, V. P., Ghani, M. A., Sankaran, T., Mathur, K. C. 1971. *A Review of the Biological Control of Insects and Other Pests in South-East Asia and the Pacific Region. Tech. Commun.* 6. Farnham Royal, Slough, England: Commonw. Agric. Bur. Commonw. Inst. Biol. Control. 149 pp.

51. Riley, C. V. 1889. The fluted scale. In *Report of the Entomologist, Charles V. Riley, for the Year 1888,* pp. 80–93. Washington, DC: GPO

52. Rivnay, E. 1964. The influence of man on insect ecology in arid zones. *Ann. Rev. Entomol.* 9:41–62

53. Rosen, D., DeBach, P. 1978. Diaspididae. See Ref. 9, pp. 78–128

54. Ryan, H. J. 1969. *Plant Quarantines in California. A Committee Report.* Berkeley, Calif: Univ. Calif. Div. Agric. Sci. 251 pp.

55. Simmons, F. J., Franz, J. M., Sailer, R. I. 1976. History of biological control. See Ref. 31, pp. 17–39

56. Smith, H. S. 1935. The role of biotic factors in the determination of population densities. *J. Econ. Entomol.* 28:873–98

57. Smith, H. S. 1939. Insect populations in relation to biological control. *Ecol. Monogr.* 9:311–20

58. Stehr, F. W. 1975. Parasitoids and predators in pest management. See Ref. 45, pp. 147–88

59. Stone, A., Sabrosky, C. W., Wirth, W. W., Foote, R. H., Coulson, J. R. 1965. *A Catalog of the Diptera of America North of Mexico, US Dep. Agric. Agric. Handb.* Vol. 276. 1696 pp.

60. Taylor, R. J. 1984. *Predation.* New York: Chapman & Hall. 166 pp.

61. Thorpe, W. H. 1930. The biology, post embryonic development and economic importance of *Cryptochaetum iceryae* (Diptera, Agromyzidae) parasitic on *Icerya purchasi* (Coccidae, Monophlebini). *Proc. Zool. Soc. London* 60:929–71

62. Unruh, T. R., White, W., Gonzalez, D., Gordh, G., Luck, R. F. 1983. Heterozygosity and effective size in laboratory populations of *Aphidius ervi* (Hym.: Aphidiidae). *Entomophaga* 28:245–58

63. Walsh, B. D. 1866. Imported insects: the gooseberry sawfly. *Pract. Entomol.* 1:119–25

64. Waterhouse, D. F., Norris, K. R. 1987. *Biological Control. Pacific Prospects.* Melbourne, Australia: Inkata. 454 pp.

65. Wille, J. E. 1941. Resumen de las diferentes labores ejecutadas en el Peru para combatir insectos daninos por el "metodo biologico." See Ref. 27, pp. 369–71

66. Wille, J. E. 1952. *Entomología Agrícola del Peru.* Lima, Peru: Junta Sanid. Veg. Dir. Gen. Agric., Min. Agric. 543 pp. 2nd ed.

67. Wilson, F. 1960. *A Review of the Biological Control of Insects and Weeds in Australia and Australian New Guinea. Tech. Commun.* 1. Farnham Royal, Bucks, England: Commonw. Agric. Bur., Commonw. Inst. Biol. Control. 102 pp.

68. Wilson, F., Huffaker, C. B. 1976. The philosophy, scope, and importance of biological control. See Ref. 31, pp. 3–15

69. Woglum, R. S. 1912. Hydrocyanic-acid gas fumigation in California—fumigation of citrus trees. *US Dep. Agric. Bur. Entomol. Bull.* 90:1–81

Ann. Rev. Entomol. 1989. 34:17–52

# THE ECOLOGY OF *HELIOTHIS* SPECIES IN RELATION TO AGROECOSYSTEMS[1]

## Gary P. Fitt

CSIRO Division of Entomology, P.O. Box 59, Narrabri, New South Wales, Australia 2390

## INTRODUCTION

Members of the noctuid genus *Heliothis*[2] are agricultural pests of worldwide significance, but of the 80 or so species currently recognized (171), only *H. armigera*, *H. zea*, and *H. virescens* have achieved major pest status. Apart from these, *H. punctigera*, *H. assulta*, *H. peltigera*, and *H. viriplaca* (syn. *H. dipsaceae*) are also pests of some crops, though each has either a more restricted geographic distribution *(H. punctigera, H. viriplaca)* or host range *(H. assulta, H. peltigera)* than the major pest species. This review is limited to *H. armigera*, *H. zea*, *H. virescens*, and *H. punctigera*. The other species, though of local significance, do not consistently achieve damaging populations and have only been subject to superficial ecological study (10, 27, 67, 105, 135, 179, 192). Other minor pest species occur in the African genus *Heliocheilus* but are not dealt with here (see 105a).

As key pests of several agricultural and horticultural crops, the four species I have singled out for review have attracted an enormous volume of research work. Consequently, the literature on their ecology, biology and management is vast. Rather than attempt a comprehensive review of all facets of the biology and ecology of *Heliothis* spp. in agricultural systems, I concentrate on

---

[1]For a detailed review of the biological control of *Heliothis* spp., see the following chapter in this volume.

[2]The generic status of *Helicoverpa*, first applied to *H. zea*, *H. armigera*, *H. punctigera*, and *H. assulta* by Hardwick (66), is receiving increasing support on taxonomic grounds (105a). However, since this nomenclature is not yet widespread in the ecological literature, the broader traditional interpretation of *Heliothis* is retained here.

17

0066-4170/89/0101-0017$02.00

those characteristics of their ecology that have allowed them to achieve major pest status. The general biology of *Heliothis* spp. is typical of the Noctuidae and has been described in many publications (66, 84, 113, 117, 127, 194). There have also been a number of recent reviews dealing directly with particular species or aspects of the ecology of *Heliothis* spp. (42, 84, 92, 129, 134, 164, 184, 194). Much of the older work on the *Heliothis* spp. remains excellent in its ecological insight (24, 66, 81, 99, 113, 117, 127), and is worth perusal to set the massive current effort against these pests in historical perspective. Indeed Reed & Pawar (135) believe much recent research on species of *Heliothis* has done little more than rediscover what was reported in the early 1900s.

## DISTRIBUTION AND PEST STATUS

*Heliothis armigera* has one of the widest distributions of any agricultural pest, occurring throughout Africa, the Middle East, southern Europe, India, central and southeastern Asia, eastern and northern Australia, New Zealand, and many eastern Pacific Islands (27, 28–28c, 66, 194). Other *Heliothis* spp. are more restricted (Table 1). *Heliothis zea* and *H. virescens* occur in both North and South America with permanent populations in most areas between latitudes 40° N and 40° S. *Heliothis punctigera* is endemic to Australia and occurs throughout that continent.

The four major pest species are highly polyphagous and collectively attack a wide range of food, fiber, oil, and fodder crops (Table 1) as well as many horticultural and ornamental crops. Relative to the areas grown, cotton, soybeans, tobacco, and the pulses (chick-pea and pigeon pea), which are high value crops or staple foods, account for most of the economic loss due to *Heliothis* spp. Cotton, tobacco, sweet corn, and many horticultural crops (e.g. tomatoes, cut flowers) receive a disproportionate amount of the pesti-

**Table 1** *Heliothis* species of major agricultural importance, their distributions, and principal cultivated host plants

| Species | Distribution | Principal crop hosts | Ref. |
|---|---|---|---|
| *Heliothis zea* | North and South America | maize, sorghum, cotton, tomato, sunflower, soybean | 66, 99, 113, 150 |
| *Heliothis armigera* | Africa, southern Europe, Asia, Southeast Asia, Australia, eastern Pacific | maize, sorghum, sunflower, cotton, tobacco, soybean, pulses, safflower, rapeseed, groundnuts | 27, 117, 134, 173, 181, 194 |
| *Heliothis punctigera* | Australia | cotton, sunflower, lucerne, soybean, chick-pea, safflower | 27, 181, 194 |
| *Heliothis virescens* | North and South America | tobacco, cotton, tomato, sunflower, soybean | 99, 113, 150 |

cides applied (73) because of their low damage tolerances. Economic losses, both from direct yield reduction and from the cost of chemicals, application, and scouting required to control them, may be considerable. Annual estimates of the cost of damage include over $1 billion in the United States due to *H. zea* and *H. virescens* on all crops (84), $300 million in India by *H. armigera* on legumes (135), and $25 million in Australia due to *H. armigera* and *H. punctigera* on many crops (187). The level of damage to other crops varies greatly throughout the world and among species, making generalization difficult. *Heliothis armigera* often requires control on sunflower in Kenya (91), as it does on sorghum in some years in Australia (176), but these are unusual cases.

The control of *Heliothis* spp. on most crops still relies heavily on insecticides. Insecticide resistance often reduces the efficacy of chemical applications, necessitating the more frequent use of harder chemicals to achieve control. This is particularly true in intensive production of high value crops such as cotton, and it has led to the abandonment of cotton growing in some areas (13, 73). The costs of resistance are difficult to estimate, and are rarely incorporated in economic analyses of pest management strategies. With fewer new insecticides being developed it appears likely that resistance will remain a problem for many cropping systems, despite the introduction of resistance management programs in some countries (48, 122).

Some species have not always been major pests. The elevation of *H. virescens* to major pest status in the southeastern and western United States coincided with the introduction of synthetic organic pesticides (13) and has been ascribed to the disruption of natural enemy populations by insecticides. However, other changes in the cropping ecosystem may have contributed. Brazzel et al (16) believed an increase in the acreage of crimson and white clovers, both suitable spring hosts, favored population increase in the first generation, and that cotton became the main host thereafter. However, this scenario of a *Heliothis* species as an induced pest does not hold for any of the other species. In other parts of the world *H. armigera, H. zea,* and *H. punctigera* were considered major pests before the widespread use of pesticides and remain so even in areas where insecticide use is minimal (73, 86, 127, 181). In the absence of pesticides, natural control of *Heliothis* spp. is insufficient to avoid economic damage to high value crops in the United States (56, 96), Australia (82), Sudan (12), Thailand (148), southern Africa (17), and many areas of South America (153).

Why are these few species such damaging agricultural pests? One reason is the feeding preference of larvae for plant structures that are high in nitrogen (66), principally the reproductive structures and growing points of their hosts (e.g. cotton buds and bolls, corn ears, tobacco buds, and sorghum heads), enabling the insects to influence yield directly. Another reason is the high value of some crops (cotton, tobacco, sweet corn, tomato) and consequent low damage thresholds.

More broadly, the pest status of the *Heliothis* spp. derives from a suite of four physiological, behavioral, and ecological characteristics that enable them to survive in unstable habitats and in turn to colonize and exploit agricultural systems successfully: polyphagy, high mobility, high fecundity, and a facultative diapause. Below I show how each elevates *Heliothis* spp. to major pest status in many cropping systems. I then examine the exploitation of agroecosystems by *Heliothis* spp. in the light of these characteristics, emphasizing the importance of movement in the dynamics of their populations and the consequent need for a regional approach to research and management.

## CHARACTERISTICS ADAPTING *HELIOTHIS* SPECIES TO AGROECOSYSTEMS

### Polyphagy

All *Heliothis* spp. of agricultural importance are highly polyphagous, attacking a diverse array of plant species in many plant families. Worldwide, *H. armigera* has been recorded from at least 60 cultivated and 67 wild host plants (135). Collectively, *H. armigera* and *H. punctigera* in Australia have been recorded from 161 plant species in 49 families (194). Lists of cultivated and uncultivated hosts for different species are available for Africa (117), the Americas (66, 99, 113, 158), Australia (194), New Zealand (168), and India (10). Predominant host families include Asteraceae, Fabaceae, Leguminaceae, Malvaceae, Poaceae, and Solanaceae (117, 194).

The importance of polyphagy to the population dynamics and pest status of *Heliothis* spp. is threefold. Firstly, populations may develop simultaneously on a number of hosts within a region. Secondly, populations may develop continuously during suitable periods by exploiting a succession of different cultivated and uncultivated hosts through the season. Thirdly, populations can persist at low density in seemingly unsuitable areas, since females have a high probability of locating a host able to sustain larval development. Although there is much variation in the suitability of host plants for larval survival and development and subsequent adult fecundity (59, 81, 111), polyphagy affords *Heliothis* spp. a great potential for population persistence and increase.

Polyphagy requires both physiological mechanisms to allow larvae to deal with diverse plant chemistry and flexibility in host selection behavior by adult females to allow oviposition on a wide range of plants. Larvae develop through five to seven instars; six is usual (51, 111, 175). Larval survival and rate of development are dependent on temperature and diet (19, 20, 51, 62, 111, 143, 151, 173, 175). Like most generalist feeders, *Heliothis* larvae have well developed detoxification systems (15, 26), but early instars are susceptible to many allelochemicals in crop plants (106, 136, 186). Host incompatibility due to allelochemicals has not been adequately defined in studies of larval mortality, but it may explain much of the high hatchling mortality often recorded.

Host selection behavior of females is not well understood, but a range of physical cues (e.g. leaf pubescence, leaf shape, plant height) and chemical cues (volatiles, nectar, surface chemicals) are involved (2, 44–46, 130, 150). Eggs [0.5–0.6 mm in diameter (66)] are laid singly or in groups of two or three on or in the vicinity of young growing points or buds, the preferred feeding sites of larvae (41, 46, 150). Individual females oviposit on many different plants. The most consistent pattern in host selection is a strong preference by all species for the flowering stage of their hosts (83, 116, 127, 144, 150, 181). The potential and realized host ranges of *Heliothis* spp. are not identical. Females do not oviposit on all plants that are suitable for larval development, nor are all plants accepted for oviposition suitable as hosts (150, 194). Parsons (116) suggested that in the absence of flowering hosts, females lay indiscriminately, sometimes on plants not usually selected as hosts.

Although all are polyphagous, the pest species display preferences for particular hosts (78, 83, 144, 150). The expression of these preferences depends heavily on the temporal and spatial availability of hosts at the preferred stage of development. All species readily attack legumes, but *H. zea* and *H. armigera* attack maize and grain sorghum preferentially over most other crop hosts (113, 117, 144, 166). By contrast, these graminaceous hosts are not attacked by *H. virescens* and *H. punctigera*. In Australia *H. punctigera* appears to be restricted to dicotyledonous hosts, while *H. armigera* occurs on both dicots and monocots (181, 194). In the southeastern United States *H. virescens* attacks tobacco whenever available and is most abundant on cotton and other hosts once the tobacco crop becomes unattractive. Cotton, a crop considered particularly susceptible to damage by *Heliothis* spp., is not a preferred host of any species (44, 78, 83). In many areas cotton is heavily attacked only after alternative hosts have senesced.

## Mobility

The ability to undertake extensive local and interregional movements is the second major factor leading to the success of *Heliothis* spp. as pests. The ephemeral nature of annual crop and uncultivated habitats requires that they have the mobility, on both local and regional scales, to cope with a changing spatial and temporal mosaic of hosts. Noctuids adopt two strategies to cope with the seasonality of their habitats: spatial redistribution by migration and diapause through periods of cold or drought (42, 43, 52). In contrast to some noctuids that are obligate migrants, e.g. *Spodoptera exempta* (52), *Heliothis* spp. are facultative migrants; they migrate in response to poor local conditions for reproduction (shortage of adult nectar sources or larval hosts) and the passage of weather systems conducive to such movements. The weather conditions on the night of emigration largely determine the types of movement undertaken and the destination of migrants. Thus moths emerging from the same area on different nights may display differing patterns of movement. There is considerable variation among the pest species in the relative importance of migratory movements. Farrow & Daly (42) ranked the species in decreasing order of migratory activity as *H. punctigera* > *H. zea* > *H. virescens* > *H. armigera*. However, local movements within crops and between nearby alternative crop and wild hosts are also important in the seasonal dynamics of these pests, especially in diverse cropping systems where feeding and oviposition sites may be continuously available.

NOCTURNAL FLIGHT BEHAVIOR    Although some activity may occur during the day (194), most is restricted to nighttime. The recent application of specialized night vision technology has permitted quantification of general activity within crops, local movements across crop boundaries, and responses to monitoring traps (100; R. A. Farrow & G. P. Fitt, unpublished results). On the first night after eclosion adults undertake short flights (< 200 m), and they may feed before settling, usually in the same habitat (100a). On subsequent nights adult activity usually commences at dusk, when feeding and local redistribution begins. A well-defined period of adult takeoff from emergence sites may occur at this time, lasting little more than 30 min (35, 149). Some moths are active within crops, while others move locally, usually traveling downwind within the flight boundary layer (36, 87, 149, 174; V. A. Drake & R. A. Farrow, unpublished observations). Some moths display a characteristic vertical takeoff flight, which carries them above the flight boundary layer and enables them to undertake migratory movement in upper wind systems.

For moths engaged in local movements, takeoff is followed by a period of activity lasting 1–2 hr, during which the adults move, feed, and oviposit. This is followed by a quiet period that lasts until about midnight, when mating activity begins. Mating continues until 3:00–4:00 AM (100, 174). During this time females are inactive, releasing pheromone from near the tops of plants, while males engage in characteristic high-speed, directed flights in search of pheromone plumes (R. A. Farrow & G. P. Fitt, unpublished observations).

TYPES OF MOVEMENT    Farrow & Daly (42) defined three categories of movement by *Heliothis* spp.: short-range, long-range, and migratory movements; each involves different patterns of behavior. All are important in the colonization and exploitation of agroecosystems, but distinctions between the categories are not well defined, and other classifications based on the displacement achieved during movement are possible (34). Researchers are only beginning to understand the relative importance of different types of movement in the redistribution of populations.

Short-range or trivial movements occur within or just above the host canopy and involve appetitive behaviors such as feeding, oviposition, mating, and sheltering. These movements continue for 1–2 hr after dusk and are generally confined within habitats over distances of 100–1000 m.

Long-range movements (1–10 km) involve less-frequent responses to external stimuli and occur above the canopy at heights of up to 10 m (36, 42), where the insect can control its flight speed and orientation in relation to the prevailing wind. Nevertheless such flights are usually downwind, and the scale of these local movements is dependent on flight duration. Long-range flights include movements between crops, between feeding and oviposition sites, and between emergence and local oviposition sites and could therefore be classed as migratory. In agricultural areas, long-range movements between crops are well established (131, 164–166, 173, 174), although intercrop movement is more often inferred from patterns of population development, emergence, and appearance on successive hosts than directly observed and quantified.

Migratory movement occurs above the flight boundary layer, where moths take advantage of synoptic-scale wind systems at altitudes of up to 1–2 km. These flights may continue for several hours, making possible downwind displacements of hundreds of kilometers (35, 37). The distinctive behavioral feature of migration is the suppression of appetitive responses to external stimuli that might otherwise arrest the insect (90).

As yet the physiological and environmental cues that initiate local or migratory movements in *Heliothis* spp. have not been clearly defined. Flight-mill studies suggested that long flights in *H. armigera* are induced by poor larval and adult nutrition, which delay reproductive maturity (63). However, poor larval nutrition is not a factor in the migration of *H. zea* from maize-growing areas of Mexico into the southern United States (see below). These moths emigrate in response to a widespread shortage of breeding habitats after the usually synchronous corn crop has senesced. In Mexico in 1987 corn was extensively replanted, creating a patchwork of crops at different stages of development, and no emigration was reported, presumably because emerging moths found suitable local breeding sites (R. A. Farrow, personal communication). Similarly, mass migrations of *H. punctigera* from inland areas of Australia follow the widespread senescence of the winter annuals (43). Moths then follow the progressive growth of spring annuals in southern and eastern Australia, where most of the extensive cropping areas are located.

In the Sudan, Topper (172) found few *H. armigera* individuals flying above 5 m, and Schaefer (149) suggested that less than 10% of an emerging population undertakes migratory movements above the flight boundary layer.

In general, we have little information on (*a*) the proportion of any emerging population of *Heliothis* spp. that may undertake migratory as opposed to local movements, (*b*) how this proportion is affected by weather conditions or other factors, and (*c*) how and when newly emerged adults assess local conditions prior to moving. Considerably more research is warranted on these questions.

EVIDENCE FOR MIGRATORY MOVEMENT    Until recently much evidence for migratory movements of *Heliothis* spp. has been circumstantial (see 42 and references therein), based on unexplained appearances of moths in areas where they cannot overwinter, in numbers that could not have been locally produced, or before local emergence from diapause. Evidence of moths flying at altitudes above the flight boundary layer, consistent with migratory movement, has been obtained for *H. zea, H. armigera,* and *H. punctigera* using tower-mounted light traps and kite- and plane-mounted nets (42). A number of mark-recapture experiments have documented movements by individual moths over distances ranging from 25 to 160 km (74, 132, 156). Hendrix et al (75) identified pollen attached to trapped *H. zea* males and were thus able to document movements of 750–1000 km from southern Texas to south-central Arkansas. The pollen was derived from geographically restricted plants *(Pithecellobium* and *Calliandra)* not found in the trapping area.

The occurrence of *H. zea* in north-central states of the United States is attributed to migration from southern states, since often few or none of these insects survive the winter (66). In the southern states there is increasing evidence of regular spring migrations of *H. zea* from large source areas of maize and sorghum in northern Mexico (70, 131). *Heliothis virescens* may also move northward in the spring (131). Adults of *H. zea* produced on corn in northern Mexico and the lower Rio Grande valley of Texas in early summer (May and early June) may provide a source of moths moving north, northwest, or northeast to other parts of Texas and the Delta States. The appearance of moths in traps in these areas before local emergence and the occurrence of weather systems that would be suitable for the proposed movements lend support to this hypothesis. Asynchrony between peak trap captures of males and emergence from diapause has been observed elsewhere in the southern United States (70, 102, 147, 160), again suggesting immigration. However, the regularity of these influxes and their importance to local dynamics in other areas are not clear. In North Carolina, seasonal dynamics of both *H. zea* and *H. virescens* seem explicable in terms of local population processes (166), whereas Mueller et al (109) concluded that in Mississippi and Arkansas locally overwintering *H. zea* cannot account for infestations on early spring weed hosts, which suggests that regular immigrations occur.

In some areas of Australia the phenology of *H. punctigera* is explicable only by invoking substantial migratory movement. Movements of *H. punc-*

*tigera* from the mainland to Tasmania and within the eastern half of Australia are probably regular events (35, 37, 42), and migrations of 2000 km to New Zealand (42) demonstrate this species' potential for movement. In South Australia, *H. punctigera* is rare during autumn, when there are few larval hosts, but reappears each spring. Cullen & Browning (29) were unable to explain this phenology in terms of diapause. Similarly, overwintering populations of *H. punctigera* are extremely difficult to locate in the irrigated cropping areas of northern New South Wales and southeastern Queensland (188; G. P. Fitt & J. C. Daly, unpublished observations). Diapausing populations consist almost entirely of *H. armigera,* yet each spring *H. punctigera* reappears in trap catches in large numbers. Migration from extensive winter breeding grounds in inland Australia is implicated as the source of these immigrants (43; P. Gregg, M. P. Zalucki, G. P. Fitt & P. Twine, unpublished observations).

*Heliothis armigera* appears to be more sedentary than *H. punctigera* in Australia (42, 108, 181). It is almost certainly a facultative migrant (31), but it is more closely associated with cropping areas than *H. punctigera* (43, 194) and may rarely experience the widespread deterioration of breeding habitat needed to induce regular mass movements. Nevertheless, the absence of significant genetic differentiation between widely separated subpopulations of *H. armigera* (31) suggests that gene flow occurs over much of eastern Australia.

Elsewhere, there is evidence of substantial wind-assisted movement by *H. armigera.* Pedgley (118) showed that *H. armigera* migrates up to 1000 km to reach Britain and other parts of Europe from sources in southern Europe and North Africa. Patterns of egg-laying by *H. armigera* on cotton in the Sudan Gezira also suggest considerable mobility within the irrigation area (65). In central India there is circumstantial evidence of migration of *H. armigera* on southeasterly winds from coastal cropping areas in the spring (7, 119).

Much of the evidence for long distance movements of *Heliothis* spp. suggests migration from low latitudes to higher latitudes in summer, usually on warm winds preceding cold fronts (37, 190). Without a return migration to warmer overwintering areas this might appear disadvantageous, since it leads to an accumulation of insects in areas where they cannot overwinter successfully. Evidence for return migrations in autumn is scanty. In the southern United States persistent southerly winds blow for much of the spring and summer and are replaced by persistent northerly flows in autumn (131, 190). Pair et al (115) provided circumstantial evidence for an autumn return migration of *H. zea* to lower latitudes. Even in the absence of return migrations, poleward migration is not necessarily maladaptive if it is seen as part of a general strategy for survival in ephemeral habitats (43).

## Diapause

WINTER DIAPAUSE    The third key element in the life cycle of all the pest species of *Heliothis* spp. is the ability to enter a facultative diapause as pupae. In subtropical and temperate parts of their range most *Heliothis* spp. appear to pass the winter in diapause, though regular movements between tropical and higher latitudes (see above) supplement or reestablish populations of some species. Diapause defines the seasonal occurrence of *Heliothis* spp. in many areas and contributes to their pest status by maintaining local populations during periods when hosts are unavailable or conditions are not conducive to reproduction and population survival. Diapausing pupae are more tolerant of cold, dry conditions than are nondiapausing pupae (39, 145).

A facultative diapause ensures that widespread species are able to respond to differing environmental conditions and prospects for reproduction and survival in different parts of their geographic range. In *H. armigera, H. zea,* and probably other species the incidence of diapause increases with increasing latitude. In the tropics, populations breed continuously (24, 131), although a small proportion (2–4%) of pupae may diapause (64, 133). In subtropical and temperate regions most but not all individuals diapause (29, 66, 88, 113, 145, 189). In these areas some nondiapausing individuals emerge late in the autumn and die without reproducing. Most, however, emerge the following spring, usually before the diapausing portion of the population (189). This pattern allows the species to exploit favorable conditions that occasionally occur early in the season.

In the field, diapause is induced during late-instar larval and prepupal development by the shortening days and decreasing temperatures of autumn, which provide cues of impending winter conditions unfavorable for reproduction. Photopheriods of 11.5–12.5 hr, accompanied by low or decreasing mean temperatures of 19–23°C, are optimal for diapause induction (29, 64, 66, 117, 141, 145, 183). A simple photoperiod-dependent sigmoid relationship was adequate to explain diapause initiation of *H. zea* and *H. virescens* in North Carolina (35°46' N) (165). In this area *H. zea* entered diapause about one month earlier than *H. virescens,* and diapause induction was spread over a longer period. This early initiation of diapause in *H. zea* was significant in allowing large overwintering populations to develop on late-sown maize. High temperatures (26–29°C) during the prepupal or early pupal stages negate the effects of diapause-inducing conditions experienced earlier in development (18, 64, 145). The importance of adult experience in inducing diapause in their progeny is less clear (29, 64, 141). The requirement for decreasing daylengths during development and the countering effects of high temperature presumably ensure that diapause is not induced in spring when photoperiods are similar to those in autumn but temperatures are generally rising (but see below).

After diapause initiation, pupae undergo a period of diapause development at low temperatures before becoming responsive to increasing spring temperatures, which terminate diapause and allow the resumption of morphogenesis (64, 101, 145, 189). Temperature thresholds for the resumption of development [17°C for *H. armigera* (189); 18°C for *H. zea* (101)] are higher than for development of nondiapausing pupae (12–13.5°C). Emergence patterns of natural diapausing pupal populations have been described for all species (88, 102, 117, 145, 189; Table 2), but simple models for predicting the emergence of adults in spring are available only for *H. armigera* in northern New South Wales (30, 189). In most areas of the United States *H. virescens* emerges in the spring about 10 days earlier than *H. zea* (60, 101, 102, 113, 128).

One consequence of diapause is that overwintering pupae attain a uniform physiological state before morphogenesis resumes. Consequently, the emergence of pupae formed over a long period becomes concentrated, and the date of emergence is not related to the date of pupation (102, 189). In all species emergence from diapause occurs over a period of 3–6 wk, with females emerging slightly earlier than males. Kay (88) described a consistently bimodal emergence pattern of *H. armigera* in southeastern Queensland (27°35' S); some adults emerged in late September and October (early spring), and a second, larger emergence occurred in late November and early December. Wilson et al (189) attributed the early peak of emergence to nondiapausing individuals. Whatever the explanation, the extended period of emergence of most species ensures that the entire overwintering population does not encounter unfavorable conditions for reproduction, such as unseasonal weather or lack of synchrony with spring hosts.

The incidence and intensity of diapause in *Heliothis* spp. has shown much variation in different experiments, which reflects in part the variation of results obtained under differing conditions and possibly also the existence of genetic variation in response to diapause-inducing stimuli (66, 77, 80, 128, 145). The genetic basis of differences in diapause development in *H. zea* has been demonstrated by selection of strains with differing responses to diapause-inducing conditions (77, 88). Parent-offspring heritabilities of up to 0.77 were obtained (88). Whether tne geographic variation in diapause response observed in various species is genetically based or simply reflects variation in the inducing stimuli can only be studied by exposing populations from a wide latitudinal range to a standard set of controlled conditions in the laboratory. No such study has been conducted.

SUMMER DIAPAUSE    Summer or drought-responsive diapause has also been documented in some species. This phenomenon was first noted in populations of *H. fletcheri* in the Sudan Gezira, which enter diapause as the dry season advances and may remain in diapause for as long as 9.5 mo, until the following rainy season (64). Hackett & Gatehouse (64) also observed the maintenance at high temperatures of a diapause induced at low temperature in *H. armigera* in the Sudan Gezira; 36% of diapause pupae transferred from 22 to 35°C remained in diapause for at least 80 days and broke diapause after a transfer back to 26°C, which simulated the drop in soil temperatures associated with the June rains that follow the dry season in the Sudan (November–June). This mechanism may allow a proportion of pupae to survive in diapause through the hot, dry season when there are few hosts within the Gezira.

A true summer diapause induced by exposure of larvae to high temperatures alone (above 32°C) has now been demonstrated in *H. virescens* (21) and is suspected in some populations of *H. punctigera* in Australia (D. A. H. Murray, personal communication). Daily exposure to 43°C for 8 hr (similar to summer conditions in the cropping areas of Arizona and California) induced summer diapause in 96% of male and 59% of female *H. virescens* pupae, which delayed emergence by several months until the autumn. Male pupae not in diapause were sterilized by exposure to extreme high temperatures (21, 76), whereas diapausing males retained their fertility (21). Summer diapause thus serves to enhance the survival of pupae in extreme soil conditions, ensure the reproductive viability of the resulting adults, particularly the males, and bridge an unfavorable host period (76).

DIAPAUSE STRATEGY OF *HELIOTHIS PUNCTIGERA*    *Heliothis punctigera* is endemic to Australia and has evolved a strategy of diapause and movement well suited to a highly unpredictable environment. Of all the species, *H. punctigera* is the most mobile (42) and has the most complex diapause strategy (29, 189; D. A. H. Murray, personal communication). In subtropical and temperate areas of eastern Australia many pupae enter a facultative winter diapause, but some survive the winter without diapausing, emerging before the bulk of the diapausing population. A portion of the first generation of *H. punctigera* may then enter a spring diapause in October (29; D. A. H. Murray, personal communication), when photoperiods correspond to those that are most effective in inducing diapause in the autumn, and when soil temperatures remain low after winter. Adults emerge from these pupae during early summer (D. A. H. Murray, personal communication), when the second generation normally emerges. This strategy is adaptive in the unpredictable environment found in Australia, in ensuring that a proportion of the population will survive unsuitable late spring conditions. Some *H. punctigera* individuals may also enter a summer diapause during January, which delays their emergence until late summer or early autumn. This complex diapause behavior, combined with the mobility of *H. punctigera* and its wide range of hosts (both cultivated and uncultivated), allows it to cope with highly varying local conditions and greatly complicates its dynamics in cropping areas. At any one time the effective gene pool, which is large (31), may well consist of many subpopulations on different cultivated and uncultivated hosts, some breeding year round, others in varying types of diapause, all linked by at times extensive movements. The composition of populations colonizing cultivated hosts may include elements of recent local origin, locally produced individuals that had entered diapause some time before, and long-distance immigrants whose numbers are affected by seasonal climatic patterns far from the cropping areas.

## Fecundity

The fourth major characteristic contributing to the pest status of *Heliothis* spp. is high fecundity, which combined with a short generation time, gives these species a high capacity for population increase. High fecundity is common to many noctuids (e.g. *Spodoptera, Mythimna*), but in contrast to many of these, females of the *Heliothis* spp. deposit eggs singly rather than in batches. Up to 3000 eggs have been recorded from single *H. armigera* (133) and *H. zea* (127) females, but most estimates of fecundity, over the reproductive lifetime of about 8–10 days, range between 1000 and 1500 eggs per female (51). The prereproductive period varies from 2 to 5 days, depending largely on temperature (40, 81). There is some genetic variation in the onset of pheromone calling in *H. armigera* (97), related perhaps to differing propensities for migration and hence delayed reproduction. Fecundity is influenced by temperature, humidity, and larval and adult nutrition (2, 51, 81, 111, 185) and is positively though weakly linked to body size (117, 133). The highly variable estimates of fecundity obtained by different workers reflect differences in the way adults are maintained in the laboratory. Life and fecundity tables constructed in the laboratory for cohorts reared on different hosts demonstrate the potential for high rates of increase and give some measure of the influence of diet on the intrinsic rate of increase (e.g. 11). However, the calculated statistics (e.g. net reproductive rate, $R_0$) are applicable only to populations with a stable age distribution, which populations of *Heliothis* spp. are unlikely to achieve. In all species, prolonged exposure to temperatures above 35°C reduces adult survival, fertility, and fecundity (51, 76, 175). There are no estimates of realized fecundity in the field, and it is not clear how well laboratory estimates apply. Modelers have assumed field fecundities ranging from 500 to 3000 eggs per female (72, 96, 165), modified by host and temperature.

# POPULATION DYNAMICS

## Seasonal Phenology

The seasonal dynamics of all the pest species throughout their ranges are too diverse to cover in detail (but see 99, 113, 117, 133, 144, 173, 181). The number of generations possible each year is directly influenced by temperature, host sequence, and host suitability. Seasonal abundance is also influenced by these factors. Rainfall indirectly influences seasonal abundance by affecting the abundance and suitability of host plants. Where hosts are available in the tropics, *Heliothis* spp. may breed continuously, completing a generation in as little as 28–30 days and passing through 10–11 generations per year (24, 131). In most subtropical and temperate cropping systems three to five generations are more usual. Discrete generations are not always clear, as considerable overlap occurs after the first or second spring generations. The sequence of host plants available and utilized during the breeding season determines in large part the damaging potential of the moth populations. The importance of uncultivated hosts in early-season dynamics is discussed below. In many areas *Heliothis* spp. do not become pests on cultivated crops until the second, third, or even fourth generation (113, 158, 165, 181).

Hartstack et al (69) observed a consistent pattern of progressive population increase during the season. They found that the size and timing of early-season light trap catches of *H. zea* and *H. virescens* in central Texas could be used to predict the seasonal dynamics thereafter. This observation formed the basis of the MOTHZV program (72). Population dynamics of *Heliothis* spp. are less predictable elsewhere. In Australia, *H. punctigera* invariably appears in large numbers in cropping areas in early spring, but population change thereafter is highly variable, depending on the interaction of climate and host-plant availability (47, 181).

Some workers (112) have suggested that generation cycles of *Heliothis* spp. are synchronized by lunar cycles, independently of host plant phenology and temperature. A reduction in general adult activity and egg-laying has been observed in response to the full moon (112, 120). This hypothesis was supported in Texas in midsummer, when generation lengths of *H. zea* matched those of lunar cycles (112); however, there is no evidence of lunar control of populations in spring and autumn, when generation times are considerably longer. Hartstack et al (69) also observed depressed light trap catches and oviposition activity associated with the moon during July in Texas, which they claimed as support for the above hypothesis. However, they disregarded patterns in their own data from earlier in the season, when peak trap catches and egg counts coincided with full moon. In southern Africa, although light trap catches were often depressed by the full moon, there was no depression of egg numbers, which suggests that adult activity was unaffected (55).

## Mortality and Population Regulation

In many cropping systems species of *Heliothis* spp. suffer considerable mortality through the actions of weather, predators, and parasites and through direct effects of the host plant on survival and growth. Much of this mortality is concentrated during the egg stage and early instars (56, 93, 178, 180), resulting in a typical type III survivorship curve (mortality decreasing with age). However, Hogg & Nordheim (79) reported a type I curve (mortality rate increasing with age) on cotton in one generation and a type II curve (mortality rate constant with age) in the next generation. Lifetime mortality has rarely been quantified and partitioned among the various mortality agents, and few quantitative life tables have been produced for field populations (79, 170, 178). Extremes of temperature and humidity and other weather factors (e.g. wind) are thought to be responsible for mortality of eggs, larvae, and pupae of most species (117, 126). Effects of abiotic factors are often underestimated, and the general importance of natural mortality, particularly due to predators and parasites, in the seasonal abundance of *Heliothis* spp. is poorly understood.

Some life tables produced for *H. armigera* on various crops in India (e.g. 9) suggest extremely high generation survival (0.50–0.87). By their design these studies omitted abiotic mortality factors and may severely underestimate egg and first-instar mortality. Moreover, the estimated rates of increase between successive field generations ignore the effect of immigration and emigration on population change and are therefore meaningless.

Most cropping systems support a diversity of beneficial arthropods, and studies of their impact are numerous (56, 73, 93, 94). Lists of key predators and parasites of *Heliothis* spp. are available for many areas (24, 56, 95, 117, 142, 161), but comparisons of the degree of control exercised by them in different areas are complicated by differences in methodology. The impact of predators and parasites has been presented in terms of consumption rates, percentage reduction in larval density, or percentage of different stages parasitized, but rarely in terms of their effect on seasonal patterns of abundance.

The role of predators and parasites in the dynamics of *Heliothis* spp. should be viewed from both local and regional perspectives. They may suppress pest populations within a particular field to below damaging levels but have only a small effect on the size of the population in that region. Alternatively, beneficials may reduce the regional population size of the pest. Evidence of an impact at the local level is found in some (1, 94) though not all (96) cotton systems in the United States and in some areas of Central America (153), where naturally occurring predators and parasitoids frequently suppress populations below economic thresholds if undisturbed by insecticides (95). Much of the evidence for a regulatory function of predators and parasitoids stems from so-called pest resurgences following the disruption of beneficial communities by insecticides (153). However, despite numerous studies identifying key beneficial organisms in the United States and efforts to incorporate them into pest thresholds (1), Goodenough et al (56) concluded that quantitative studies of their impact remain inadequate and that beneficial organisms cannot yet be exploited in economic pest management. The same is true of most other areas where *Heliothis* spp. are pests. In Australian cotton, predators of *H. armigera* and *H. punctigera* cannot be relied upon for control, since they never approach the densities of 0.5–1.0 predators per egg or larva suggested as necessary for control (1, 85). The effectiveness of parasitoids other than those attacking the egg stage [*Trichogramma* spp. (92)] or early instars *(Microplitis)* is generally slight in limiting damage caused by larvae because, although they may reduce the rate of feeding, many do not kill the host until the final-instar larva or pupa, after considerable damage has been done.

Much research has been devoted to the development of techniques for the augmentation of natural enemies for the management of *Heliothis* spp. (92–94). Augmentation of the egg parasitoid *Trichogramma* spp. is a favored approach (92), but does not appear feasible for intensive systems at this stage (95). Parasitism may reach high levels (50–70%), but it may vary in intensity from inversely density dependent to density independent as the spatial pattern of eggs on host plants varies (107), and it does not necessarily cause irreplaceable mortality. The conservation of natural populations of parasitoids and predators through the use of efficient sampling and action thresholds, selective pesticides, maintenance of noncrop refugia, and strip cropping (23, 85) has also been advocated for many systems. Some of these features should be incorporated in all production systems where *Heliothis* spp. are pests. However, until the efficacy of beneficials is quantified, their potential is unlikely to be utilized efficiently.

Evidence for a regulatory function of beneficial organisms at the regional level is sparse for phytophagous insects in general, but especially for *Heliothis* spp. (33, 154, 177). The mobility and high rate of increase of *Heliothis* spp. permit rapid colonization and population growth in new habitats, which overwhelms the capacity of natural enemies to respond functionally and numerically (124). Asynchrony in colonization of crops by *Heliothis* spp. and their natural enemies is one of the major factors limiting the effectiveness of natural control. This factor seems most obvious in systems where population development is disrupted by a cold or dry season, such that the pest population and its predators become disjointed in time. This is the case in most intensive cropping areas in temperate and subtropical regions. In a tropical area (Uganda), Coaker (24) concluded that *H. armigera* was not a major pest of cotton because breeding occurred year round on several hosts, allowing predators and parasites to stabilize the pest at low levels. Similar arguments for the role of beneficial organisms in regulating abundance of *Heliothis* spp. have been advanced for the Canete Valley of Peru (5) and some isolated cotton growing areas of Colombia (I. de Polania, personal communication). Whether these conclusions are generally applicable to tropical systems seems doubtful. As with many other r-type pests, it seems probable that regional abundance of *Heliothis* spp. is determined more by climatic (abiotic) factors, which act directly on the insect or indirectly through effects on host plant abundance and quality, than by biotic factors (154, 155, 177).

## Sources and Mortality of Overwintering Pupal Populations

The size of overwintering pupal populations, their location and survival, and their contribution to the development of populations the following season are not well established for most areas where *Heliothis* spp. are pests. The contribution of larval populations feeding on different hosts to the total overwintering population is highly variable, depending on when hosts are suitable for larval development relative to the time when diapause is induced. In several areas crops produce a substantial proportion of the overwintering generation (158; Table 2), but in some areas uncultivated hosts may also contribute substantially. The relative importance of contributions from culti-vated and uncultivated hosts has rarely been quantified.

Survival of overwintering pupae is undoubtedly the best studied in cropping areas of the United States. *Heliothis zea* is able to overwinter in diapause as far north as 45° N in the western and eastern United States and up to 40° N in the central region (66). In most areas and in most seasons overwintering survival is low (<5%; Table 2) owing to mortality of larvae prior to pupation, mortality of pupae, and mortality of adults trapped in the pupal chamber after eclosion. Cultivation of crop residues during winter undoubtedly removes a substantial proportion of the overwintering generation, but the proportion of the total population removed in this way is not known. Mortality within undisturbed pupal populations is due to low temperatures, exposure to mois-ture at subfreezing temperatures, soil compaction, flooding, parasitism, pre-dation, and disease (22, 38, 39, 50, 58, 78, 152, 193). The interaction of temperature and soil moisture seems most important; pupal survival is much lower in cold, moist soil than in cold, dry soil (38, 147). Rummel et al (147) reported overwintering survival of up to 25% in dry winters, even when ambient temperatures fell for short periods to $-17.0°C$. Mortality of moths in their emergence tunnels after eclosion can also be significant, generally removing 10–20% of the surviving population (22, 152, 159; D. A. H. Murray, personal communication).

The relevance of many measures of overwintering survival based on the use of artifically produced pupal populations, often enclosed within containers or beneath cages, is questionable. Mortality due to predation and parasitism may often be underestimated, but mortality due to temperature extremes may be magnified because vegetation cover is often excluded; hence survival under weedy fallows and crop residues may be higher than that measured.

Although overwintering survival is generally low, there may be high carryover in undisturbed populations during mild winters. Substantial survival

**Table 2** Dates of initiation and termination of diapause, estimates of overwintering survival, and principal hosts producing the overwintering generation for species of *Heliothis* in different parts of their ranges

| Species | Area | Diapause initiation | Spring emergence | Range in % survival (years of data) | Major sources of overwintering generation | Examples of possible densities (per ha)[a] | Ref. |
|---|---|---|---|---|---|---|---|
| **United States** | | | | | | | |
| *H. zea* | Southern Texas | Sept–Nov | April | — | Cotton | 6,400 a | 58, 158 |
| *H. virescens* | | Sept–Nov | March | — | Cotton | 11,700 a | 158 |
| *H. zea* | Central Texas | Late Oct–Nov | April/May | 12–25 (2) | | — | |
| *H. virescens* | | Late Oct–Nov | April/May | <1 (2) | — | 13,000 b | 38, 102 |
| *H. zea* | Texas high plains | Late Sept–Oct | Mid April/May | 2–25 (4) | Maize, cotton | — | 147 |
| *H. zea* | Georgia | Sept–Oct | Late March/June | 23–39 (4) | — | | 158 |
| *H. virescens* | | Sept–Oct | Late March/May | — | | | |
| *H. zea* | Georgia | Sept–Oct | March/April | 42.5 (1) | | 7,750 c | 60 |
| *H. virescens* | | Sept–Oct | March/April | 25.7 (1) | | | |
| *H. zea* | South Carolina | Mid Aug–Sept | May | significant (1) | Cotton, various un-cultivated hosts | 7,760 a | 140, 158 |

| | | | | | | | | |
|---|---|---|---|---|---|---|---|---|
| *H. zea* | North Carolina | Aug–Mid Sept | May/June | 2–28 (3) | Maize | 4,117 | d | 22, 58, 113, 158 |
| *H. virescens* | North Carolina | Aug–Mid Sept | April/May | | Soybeans, uncultivated hosts | 40,000 | e | 158 |
| *H. zea* | Mississippi | Late Aug–Sept | Late April/May | 0.4–5.0 (3) | Cotton, soybeans | | | 160 |
| *H. virescens* | | Late Aug–Sept | Late April/May | 4 (2) | Weeds | | | 158, 159 |
| *H. zea* | Arkansas | Late Aug–Sept | Late April/May | 0.4–1.4 (2) | Cotton | | | 152, 158 |
| *H. zea* | Oklahoma | Late Aug–Sept | Late April/May | 0.0–0.05 (2) | Lucerne | 450 | f | 193 |
| Australia | | | | | | | | |
| *H. armigera* | Northern NSW | Late March–April | Oct | 50.0 (1) | Cotton | 20,000 | a | 188, b |
| *H. punctigera* | | April ? | Late Sept/Oct | — | | | | |
| Africa | | | | | | | | |
| *H. armigera* | Botswana | May | Sept | 46–48 (1) | — | — | | 145 |
| *H. armigera* | South Africa/ Rhodesia | April–May | Sept | — | Cotton, maize | | | 117 |

[a]Adults emerging from naturally populations under a, cotton; b, pigeon pea; c, dolichos; d, maize; e, soybean; f, lucerne.
[b]G. P. Fitt, unpublished.

of both *H. zea* (42.5%) and *H. virescens* (25.7%) has been reported in natural populations under *Dolichos* sp. in Georgia (60). Both species also had high survival rates under pigeon pea in central Texas (102).

Estimates of overwintering mortality from other parts of the world are few. Wilson (188) claimed that overwintering survival of *H. armigera* was high in the Namoi valley of New South Wales; 45–50% survival was reported in a natural population of 45,000 pupae per hectare of the same species under cotton residues in one year (G. P. Fitt, unpublished observations). Conditions at soil level in this area (30° S) are rarely as extreme as in most cropping areas of the United States. Frosts occur regularly, but soil temperatures at 5 cm never fall below 0°C. The same is true of most Australian cropping areas where *Heliothis* spp. are abundant.

The importance of overwintering survival to subsequent population dynamics is poorly understood in most areas. There is disagreement over the relative importance of overwintering survival of local populations and immigration from the south in the southern United States (discussed earlier). Claims that *H. zea* does not overwinter in the United States and that all populations are refounded by immigration (123) are incorrect (Table 2). The relative importance of the two factors undoubtedly varies between areas and years. Certainly some individuals do overwinter successfully in much of the southeastern United States. Although the size of this first generation may be low, this widely dispersed, low-density population has a high potential rate of increase on abundant spring hosts (110). Survival of only 1–2% over winter may be sufficient to allow populations to reach damaging levels on crops by the third or fourth generation, even without immigration. Laster et al (98), using mark-recapture techniques, estimated early spring adult densities of *H. virescens* that had presumably survived from the overwintered population in the Mississippi delta. Densities of 5.7 and 20.9 moths per hectare were recorded in a 148-km² study area during 1982 and 1983. Given the high fecundity of *Heliothis* spp., these densities are significant.

## Modeling

Detailed population models have been constructed only for *Heliothis* spp. in some cropping systems in the United States (72, 165), though work is underway in Australia (G. P. Fitt & G. Hamilton, unpublished; M. P. Zalucki, I. Titmarsh, D. A. H. Murray & P. Twine, unpublished) and India (A. King, personal communication). HELSIM-2 for *H. zea* (165) and MOTHZV for *H. zea* and *H. virescens* (72) represent ambitious approaches to modeling complex life systems. Both are applicable only to the systems where they were developed (North Carolina and Texas, respectively), which reflects the dominating role of the ecosystem in the population dynamics of these species. Both attempt to model only within-season dynamics, having as initial conditions emergence from diapause (HELSIM) or early-season trap catches (MOTHZV). Neither accounts for long-distance movement. Some simple models of between-season dynamics have been produced (96), but these assume a direct numerical relationship between the last generation of one season and the first generation of the next season, with no immigration. Although both HELSIM and MOTHZV make a number of assumptions about adult movement, they nevertheless successfully predict the timing if not the size of peaks of adult or egg populations in particular areas. They have been used for research (89, 162) as well as management (71). These models represent an excellent basis for further development, which will benefit from improved understanding of local and migratory movement, host-plant location, overwintering mortality, and the relationship between host-plant suitability and immature survival.

# EXPLOITATION OF AGROECOSYSTEMS BY *HELIOTHIS* SPECIES

Many of the pest problems of modern agricultural systems are attributed to a reduction in species diversity caused by growing crops in monoculture (121, 182), which makes plants more apparent to their consumers. It has been argued that diversity should be increased in agroecosystems, at either the regional or field level, in order to reduce pest problems (4, 139). However, in the case of *Heliothis* spp. the evidence strongly suggests that populations are more abundant in diverse cropping systems (3, 4, 167). The suite of characteristics outlined earlier enables *Heliothis* spp. to exploit diverse areas that provide a succession of cultivated and uncultivated hosts; this adaptation allows rapid rates of increase and the potential for devastating attacks on susceptible crops. High mobility allows rapid colonization of suitable habitats, and diapause ensures the persistence of populations in many parts of the breeding range. In many temperate and subtropical areas the typical pattern of heliothine population development prior to the development of agriculture may have been a major spring increase on an extensive and reasonably predictable spring flush of uncultivated plants, followed by widespread dispersal and the subsequent maintenance of small populations in localized favorable summer and overwintering habitats. The diversification of summer cropping and particularly the development of irrigation schemes in relatively arid areas have greatly favored *Heliothis* spp. by providing a succession of suitable hosts (67, 73, 173). This is true for many polyphagous pests (177); indeed, Way (182) claimed that "fundamentally most of our pests occur because there is too much diversity in the form of alternate food and refuges."

Although *Heliothis* spp. may consistently infest and damage sorghum, sunflowers, maize, lucerne, various legumes, and other field crops, they are rarely controlled on these crops because the economic thresholds for damage are higher (e.g. 176) than on more susceptible crops. Moreover, although beneficial organisms undoubtedly reduce population densities in some crops (e.g. sorghum, sunflower), in others (e.g. maize) developing larvae are protected from much natural mortality. These crops thus have the potential to produce enormous populations of adults. Densities well above 10,000 pupae per hectare have been recorded under maize, sunflower, sorghum, pigeon pea, and adzuki bean in Australia (G. P. Fitt & G. Hamilton, unpublished observations), with maximum densities of 70,000 and 280,000 pupae per hectare recorded under maize and adzuki beans, respectively. Maize crops in the United States may regularly produce 40,000–50,000 *H. zea* adults per hectare (131, 166). Most of these hosts are suitable for population development for only one generation before senescing, whereas high-value crops such as cotton and tobacco may be susceptible for several months. These alternative hosts thus act as a source of *Heliothis* spp., while the susceptible, heavily sprayed crops act as sinks for these recruits.

The problem depends not only on the identity of alternative hosts but also on their phenology with respect to susceptible crops. This point is most clearly illustrated by the role of maize, the preferred host of *H. armigera* and *H. zea*, in the dynamics of these two species throughout the world. Maize is highly attractive (e.g. more so than cotton) to both species only briefly during silking (though many eggs may be laid on vegetative maize). The crop may act as a sink or a source of moths depending on the timing of production in relation to flowering of other hosts (117). In much of the southern United States maize is the predominant early-season host of *H. zea* (113, 131, 166, 167), and the abundance of *H. zea* on subsequent crops is related to the area and timing of the maize crop (14). In North Carolina over 90% of the first two generations of *H. zea* occurs in maize (166). The crop in this area acts initially as a diversionary host, but it is also a source of large numbers of moths, which move to cotton and soybeans late in the season after maize has senesced (99, 113, 166, 167). Stinner (162) estimated that the removal of maize from the cropping system in North Carolina would reduce densities of *H. zea* on tobacco, cotton, and soybeans by 92–97%.

Other examples of problems with *H. zea* or *H. armigera* on cotton due to the local production of maize can be found on the high plains of Texas (14, 146) and throughout southern and western Africa (117, 133), Central America (153), India (7), and Thailand (148) where maize is grown as a staple food crop. The increasing importance of *H. armigera* in Australia, Sudan, India, and the Middle East can be attributed to increased diversity of cropping systems, the introduction of suitable hosts spanning periods when previously few hosts were available, and the general increase in productivity through the use of fertilizer; these changes have produced environments more conducive to population growth.

In the Sudan Gezira, the emergence of *H. armigera* as a major pest of cotton has been attributed to (*a*) the introducion of alternative hosts (groundnuts, sorghum), which support large populations before cotton (86, 173), (*b*) the earlier sowing of cotton (86), and (*c*) the introduction of more attractive acala cotton types in place of barbadense (12). *Heliothis armigera* is a pest on cotton in the Gezira up to late October (173), even though the crop does not reach maximum flowering, its most attractive stage, until November. The timing of the attack is determined by the emergence of moths from groundnuts (173). There are similar examples involving the movement of *Heliothis* spp. to cotton from source crops of chick-pea in the Middle East (67) and sorghum in Texas (125) and India (A. King, personal communication).

In Australia, hosts such as sorghum, sunflower, and small areas of maize probably produce many of the *Heliothis* moths that move to cotton throughout the season (181; G. P. Fitt, unpublished observations). Further diversification of the traditional summer cropping areas where cotton is grown is now underway, with the introduction of chick-pea as a winter and spring crop and pigeon pea as an autumn crop, both highly suitable hosts for *H. punctigera* and *H. armigera*.

Sprayed crops may also be major sources of adult *Heliothis* spp. Despite spraying, significant densities of larvae (up to 1 larva/m$^2$) may survive. In some areas (e.g. northern New South Wales and Texas) sprayed crops such as cotton are grown in virtual monoculture, and alternatives such as maize, in which larval densities may be spectacularly high, cover a relatively insignificant area. The seasonal increase and decline in levels of pyrethroid resistance in populations of *H. armigera* in Australian cotton areas (48) indicate that adults produced in cotton contribute significantly to the late-season populations infesting that crop.

## Importance of Uncultivated Hosts

*Heliothis* spp. are most apparent as high-density damaging populations on cultivated crops. However, widely dispersed, usually low-density populations also occur on native or naturalized plants. These small, unapparent populations are important in maintaining populations when and where suitable crop hosts are unavailable. Populations may also reach high densities on uncultivited hosts (43, 158), but it is difficult to quantify the contribution of these populations to the seasonal dynamics of *Heliothis* spp. on crop hosts (158). Nevertheless, in many areas uncultivated hosts are important in the initial buildup of the first spring generation, before crop hosts are widely available (99, 110, 117, 133, 158, 161, 181, 188). Despite the often high mortality of eggs and larvae on uncultivated hosts (99, 161), the high fecundity of subsequent adults allows rapid multiplication of the populations (110). Because the size of populations developing on spring hosts may determine the scale of populations on crops for the remainder of the season (158), strategies of area-wide management have been proposed (68, 96, 109; and below).

Additionally, in intensive cropping systems uncultivated hosts, along with unsprayed crops, may act as refugia for pesticide-susceptible genotypes, which can dilute the selection for resistance in sprayed crops.

## ECOLOGICAL IMPLICATIONS FOR THE MANAGEMENT OF *HELIOTHIS* SPECIES

The management of *Heliothis* spp. and chemical control in various crops, including the implementation of integrated pest management, have been covered extensively elsewhere (13, 49, 73, 134, 138). Here I concentrate on other methods of management that require an ecological understanding of *Heliothis* spp. I stress the importance of a regional understanding of population ecology in management.

### Cultural Control

Many agronomic practices have been suggested for the management of *Heliothis* spp. in various cropping systems, including manipulation of crop planting dates, stubble cultivation and destruction of crop residues, the use of closed seasons, and destruction or manipulation of alternate hosts (5, 14, 73, 103, 117, 146). These practices achieve some degree of pest control either by destroying overwintering populations in soil or residues or by providing a host-free period during which populations are greatly reduced. Overwintering populations beneath crop residues offer scope for management if a significant proportion of the diapausing pupae can be destroyed through cultivation (50, 103). The effectiveness of this control is dependent on the diversity of alternative hosts, on the agronomic practices applied in other parts of the system, and on the role of immigrants in local dynamics. For example, the introduction of conservation tillage practices in Alabama, which might have increased overwintering survival, had little impact on the dynamics of *H. zea* and *H. virescens* (53), perhaps because early-season populations were initiated by recruits from outside that cropping system. Except on the Texas high plains, where early plantings of short-season determinate cotton varieties usually limit the damage caused by *H. zea* (73), few specific cultural measures have been adopted for management.

### Area-Wide Suppression

Chemical control of *Heliothis* spp. is usually applied on a farm-to-farm and crop-to-crop basis when populations exceed economic thresholds. An alternative to this "uncoordinated defensive strategy" (96) is to develop coordinated areawide systems to suppress pest numbers to low levels throughout an entire ecosystem. Area-wide programs based on synchronous application of insecticides to large areas of cotton covering many farms have been adopted in parts of Arkansas since 1978 (25, 114) and have brought reductions in pesticide applications for *H. zea* and *H. virescens* and increased returns to growers (25). The appropriateness of this strategy depends largely on the movement behavior of the pest (6).

Knipling & Stadelbacher (96) suggested a strategy for management of *H. zea* and *H. virescens* in the Mississippi delta region based on area-wide reduction of the first spring generation using a range of techniques: mass release of parasites or predators, application of pathogens, autocidal techniques, and destruction of spring hosts. According to their simple population model, a 90% reduction in reproduction of the first generation, which is small and localized, would reduce the level of damage to susceptible cotton and soybeans over the following season below economic levels. Harris & Phillips (68) tested one technique for area-wide suppression by mowing spring hosts in a 6500-ha area of Arkansas. Although larval populations in weeds were reduced by 50%, the effects on subsequent infestations in cotton and soybean were highly variable, perhaps because the treated and control areas were too small to overcome immigration from outside. Despite the inconclusiveness of this test and some flaws in the theory, related to the origin of spring populations and the impact of controls on beneficial populations (109), the strategy of area-wide suppression suggested by Knipling & Stadelbacher (96) may be economically and technically feasible for that agroecosystem. The particular conditions of the delta region do not pertain to other agroecosystems, however. In Australia early spring populations of *H. punctigera* occur on widespread native host plants within and outside the cropping areas, where the possibilities for suppression are limited. Nevertheless, the concept of area-wide suppression of *Heliothis* spp., where and when the population is small and localized at some time in the year, deserves careful consideration and much further research.

## Trap Cropping

In some cropping systems different crops act inadvertently as trap crops for *Heliothis* spp. at particular times of the season (see above). The manipulation of cropping systems to incorporate trap crops as diversionary hosts or to provide refuges for beneficial organisms that will later colonize susceptible crops has often been suggested (23, 85, 117) but has rarely been applied successfully. Increased field-level diversity is not easily accommodated in large-scale intensive cropping systems (see the earlier discussion of the role of maize). Most potential trap crops are only briefly more attractive than the crops they are intended to protect, so sequential plantings are necessary. Moreover, unless the trap crop can be destroyed promptly, it may act as a concentrated source of moths. Several studies have shown species of *Heliothis* to be more abundant in intercropped systems (3).

## Pesticide Resistance

*Heliothis virescens, H. armigera,* and to a lesser extent *H. zea* have demonstrated a propensity to develop resistance to a range of pesticides in many parts of the world (32, 61, 104, 148, 157, 191), which has led to the collapse of cropping systems in some areas (13, 73). Most recently, resistance to synthetic pyrethroids has been documented in *H. armigera* in Australia (61), Thailand, Turkey (148), and India (A. B. S. King, personal communication) and in *H. virescens* in the United States (104) and Colombia (I. de Polania, personal communication). The development of resistance is influenced by genetic, ecological, behavioral, and agronomic factors (32, 54, 57), which affect the proportion of the total population selected with insecticides and the selection pressure exerted on sprayed populations.

The importance of ecological factors in the development of resistance is well illustrated in Australia, where *H. punctigera* occurs with *H. armigera* in the same cropping systems and often on the same crops. Although both species have been subject to intense selection in cotton, *H. punctigera* has not developed resistance to any chemical. *H. punctigera* is highly mobile, has an extensive geographic distribution, and occurs on many native and introduced uncultivated hosts in addition to its many crop hosts. Thus this species maintains adequate populations in unsprayed refugia to dilute selection acting in the cropping areas. By contrast, *H. armigera* has a more limited range of uncultivated hosts and is usually most abundant in cropping areas (43, 194). Consequently a higher proportion of the *H. armigera* population is regularly exposed to selection. This proportion is increased at certain times of the growing season and in drought years, when much of the population may be concentrated in sprayed crops such as cotton (48).

In intensive cropping areas of the southeastern United States, *H. virescens* is largely restricted to sprayed crops of cotton, soybean, or tobacco through much of the season. Overwintering populations are also derived largely from these crop hosts. Thus there is a direct genetic relationship between populations in successive seasons. *Heliothis zea* historically has become resistant to several organochlorine and organophosphate insecticides (157), but has yet to develop resistance to pyrethroids. *Heliothis zea* maintains large populations on extensive unsprayed crops (maize, sorghum) throughout the South and Southeast; thus the proportion of the population exposed to selection is reduced. Moreover the regular influx of immigrants from Mexico, discussed earlier, may oppose selection for resistance in the cropping areas.

These examples suggest that resistance to pesticides is not an inevitable consequence of their use. Strategies for the controlled use of pesticides, based on ecological, behavioral, and toxicological understanding of the pest(s), have now been implemented in Australia (48), the United States (122), and Thailand (148), though their success is uncertain.

## Host-Plant Resistance

The development of crop cultivars that are resistant to, or tolerant of, feeding damage has great potential in the regional management of *Heliothis* spp. (73, 89). Many crops display characters that can be exploited by breeders to reduce attractiveness to ovipositing adults or suitability for larvae (89, 169, 186).

The value of host-plant resistance (HPR) depends on the type of resistance, the behavior of the pest, and the diversity of the cropping system. The importance of a regional understanding of population dynamics to the application of HPR is demonstrated in the simulations of Kennedy et al (89). They showed that in a multicrop system in North Carolina, increasing antixenotic resistance (nonpreference) of maize to *H. zea* was likely to increase infestation of other crops in the system; the increase could be up to 700% in the case of cotton. On the other hand, increasing antibiosis in maize would result in a decrease in infestation of all crops. By contrast, increased antibiosis of cotton would not benefit cotton greatly, because the pest populations do not originate on that crop, but immigrate from other hosts.

## FUTURE DEVELOPMENTS

Despite the wealth of research, *Heliothis* spp. remain major, recurring pests requiring considerable control, and there is little prospect of change in the immediate future. One reason for this may be that research has too often focused on the pest species as problems in single crops, partly no doubt because of funding restrictions. Consequently, the link between hosts within cropping systems and the regional aspect of population dynamics has often been obscured (135, 158, 194). The polyphagous, highly mobile nature of *Heliothis* spp. dictates that management be applied on an area-wide or regional level (6, 163, 194). In the past 20 years regional approaches to both research and management of *Heliothis* spp. have emerged, but there are many areas where our knowledge remains inadequate.

The potential importance of long-distance movement to the spread of insecticide resistance throughout populations and to the dynamics of *Heliothis* spp. populations is clear. However, its importance in the dynamics of any species in any cropping system has not yet been quantified. We are far from being able to propose management strategies that take account of, or cope with, large-scale immigration, perhaps because in many systems migration events do not occur regularly in a predictable fashion. This lack of a quantitative understanding of movement is not surprising. Studies of insect movement have usually covered only brief periods during a season, and although they have provided much insight into the characteristics of migratory flight, they give little idea of the regularity or intensity with which migration occurs. The coordinated multidisciplinary research effort required to monitor all aspects of local population dynamics (131, 163) and flight activity (e.g. 100), including the phenology and abundance of immigrants (37, 131, 149) and their impact on the numerical and genetic structure of populations (31), will be expensive, but it is much more likely to produce worthwhile results than a fragmented approach.

Area-wide suppression using chemicals appears viable in some systems (25, 114). There also appears to be much potential to manipulate the spatial, temporal, or genetic structure of agroecosystems (89, 167) to reduce the damaging effects of *Heliothis* species. These approaches could include the development of host-plant resistance in various crops, manipulation of crop composition or timing, the use of trap crops, the manipulation of early-season hosts to reduce breeding of the first generation, cultural measures such as cultivation to destroy overwintering populations, and augmentation of natural enemies. At present we do not sufficiently understand the regional ecology of *Heliothis* spp. to recommend major changes, which would undoubtedly disturb significant sections of the agricultural community. Moreover, the economics of some of these approaches may mitigate against their successful adoption. The potential for managing *Heliothis* spp. by the use of trap crops or other techniques of behavioral manipulation involving attractants and deterrents [e.g. the "push-pull" strategy (137)] is far from clear and also requires further research.

Perhaps the major problem facing pest managers is that where different parts of the agricultural community are involved in different activities, management options are constrained by the cropping system. Even with adequate understanding of the regional dynamics of the pests, it may be difficult to alter cropping systems sufficiently to reduce attacks on susceptible crops. Stinner's analysis (162) suggests that the removal of maize from North Carolina would be advantageous for growers of other crops, but this is unlikely to happen. In intensive cropping areas of Australia, the crops that serve as sources for *H. armigera* and *H. punctigera* and the highly susceptible crops, such as cotton or tobacco, are not usually grown by the same producers. In much of Africa and Asia peasant farmers must grow food crops, which are often sources of *Heliothis* spp., and cash crops, which are attacked by them. Notwithstanding the likelihood of long-distance migrations of moths into cotton areas, the economic disruption necessary to alter cropping patterns greatly would negate any efforts in this direction. In free-market economies there is little prospect for control over activities at this scale.

ACKNOWLEDGMENTS

I am grateful to Myron Zalucki, Brian Fletcher, Alistair Drake, Joanne Daly, Roger Farrow, Graeme Hamilton, Norm Thomson, Brian Hearn, and Neil Forrester for their constructive comments on earlier drafts of the manuscript, which greatly enhanced the clarity of my presentation. Daphne Johnston deserves special thanks for her efforts in producing the final manuscript. Carol Murray (CSIRO Black Mountain Library) assisted greatly with the bibliography and provision of references. My research on *Heliothis* spp. has been generously supported by the Australian Cotton Research Council.

*Literature Cited*

1. Ables, J. R., Goodenough, J. L., Hartstack, A. W., Ridgway, R.L. 1983. Entomophagous arthropods. In *Cotton Insect Management With Special Reference to the Boll Weevil*, ed. R. L. Ridgway, E. P. Lloyd, W. H. Cross, pp. 103–28. Washington, DC: US Dep. Agric., Agric. Res. Serv.
2. Adjei-Maafo, I. K., Wilson, L. T. 1983. Association of cotton nectar production with *Heliothis punctigera* (Lepidoptera: Noctuidae) oviposition. *Environ. Entomol.* 12:1166–70
3. Altieri, M. A., Liebman, M. 1987. Insect, weed and plant disease management in multiple cropping systems. In *Multiple Cropping Systems*, ed. C. A. Francis, pp. 183–218. New York: Macmillan
4. Andow, D. 1983. The extent of monoculture and its effects on insect pest populations with particular reference to wheat and cotton. *Agric. Ecosyst. Environ.* 9:25–35
5. Barducci, T. D. 1972. Ecological consequences of pesticides used for the control of cotton insects in Canete Valley, Peru. In *The Careless Technology. Ecology and International Development*, ed. M. Taghi Farvar, J. P. Milton, pp. 423–38. New York: Nat. Hist.
6. Bellows, T. S. 1987. Regional management strategies in stochastic systems. *Bull. Entomol. Soc. Am.* 33:151–54
7. Bhatnagar, V. S., Davies, J. C. 1981. Pest management in intercrop subsistence farming. In *Proc. Int. Workshop Intercropping 1979, Hyderabad, India*, pp. 249–57. Hyderabad, India: ICRISAT
8. Deleted in proof
9. Bilapate, G. G. 1981. Investigations on *Heliothis armigera* (Hübner) in Marathwada. XXIII. Key mortality factors on cotton, pigeonpea and chickpea. *Proc. Indian Natl. Sci. Acad. Part B* 47:637–46
10. Bilapate, G. G., 1984. *Heliothis* complex in India—a review. *Agric. Rev. London* 5:13–26
11. Bilapate, G. G., Raodeo, A. K., Pawar,

V. M. 1980. Investigations on *Heliothis armigera* Hübner in Marathwada. V. Life fecundity tables on sunflower and maize. *Proc. Indian Natl. Sci. Acad. Part B* 46:652–58
12. Bindra, O. S. 1985. Relation of cotton cultivars to the cotton-pest problem in the Sudan Gezira. *Euphytica* 34:849–56
13. Bottrell, D. G., Adkisson, P. L. 1977. Cotton insect pest management. *Ann. Rev. Entomol.* 22:541–82
14. Bradley, J. R., Herzog, G. A., Roach, S. H., Stinner, R. E., Terry, L. I. 1986. Cultural control in southeastern USA cropping systems. See Ref. 84, pp. 22–27
15. Brattsten, L. B. 1987. Metabolic insecticide resistance in boll weevil compared to those in a resistance-prone species. *Pestic. Biochem. Physiol.* 27:1–12
16. Brazzel, J. R., Newsom, L. D., Roussel, J. S., Lincoln, C., Williams, F. J., et al. 1953. Bollworm and tobacco budworm as cotton pests in Louisiana and Arkansas. *La. Agric. Exp. Stn. Tech. Bull.* Vol. 482. 47 pp.
17. Brettel, J. H. 1983. Strategies for cotton bollworm control in Zimbabwe. *Zimbabwe Agric. J.* 80:105–8
18. Browning, T. O. 1979. Timing of the action of photoperiod and temperature on events leading to diapause and development in pupae of *Heliothis punctigera* (Lepidoptera: Noctuidae). *J. Exp. Biol.* 83:261–69
19. Butler, G. D. 1976. Bollworm: development in relation to temperature and larval food. *Environ. Entomol.* 5:520–22
20. Butler, G. D., Hamilton, A. G. 1976. Development of *Heliothis virescens* in relation to constant temperature. *Environ. Entomol.* 5:759–60
21. Butler, G. D., Wilson, L. T., Henneberry, T. J. 1985. *Heliothis virescens* (Lepidoptera: Noctuidae): initiation of summer diapause. *J. Econ. Entomol.* 78:320–24
22. Caron, R. E., Bradley, J. R., Pleasants, R. H., Rabb, R. L., Stinner, R. E. 1978. Overwintering survival of *Heliothis zea* produced on late planted

corn in North Carolina. *Environ. Entomol.* 7:193–96

23. Cate, J. R. 1985. Cotton: status and current limitations to biological control in Texas and Arkansas. In *Biological Control in Agricultural IPM Systems,* ed. M. A. Hoy, D. C. Herzog, pp. 537–56. New York: Academic

24. Coaker, T. H. 1959. Investigations on *Heliothis armigera* in Uganda. *Bull. Entomol. Res.* 50:487–506

25. Cochran, M. J., Nicholson, W. F., Parvin, D. W., Baskin, R., Phillips, J. R. 1985. An assessment of the Arkansas experience with bollworm management communities: evaluated from three perspectives. *Proc. Beltwide Cotton Prod. Res. Conf. 1985, New Orleans, La.,* pp. 160–62. Memphis, Tenn: Natl. Cotton Counc. Am.

26. Collins, P. J., Hooper, G. H. S. 1984. The microsomal mixed function oxidase system of *Heliothis punctiger* Wallengren and *H. armiger* (Hübner) (Lepidoptera: Noctuidae). *Comp. Biochem. Physiol. B* 77:849–55

27. Common, I. F. B. 1953. The Australian species of *Heliothis* (Lepidoptera: Noctuidae) and their pest status. *Aust. J. Zool.* 1:319–44

28. Commonwealth Institute of Entomology. 1967. *Distribution Maps of Pests,* Ser. A, No. 238, *H. virescens.* London: Commonw. Inst. Entomol.

28a. Commonwealth Institute of Entomology. 1967. *Distribution Maps of Pests,* Ser. A, No. 239, *H. zea.* London: Commonw. Inst. Entomol.

28b. Commonwealth Institute of Entomology. 1968. *Distribution Maps of Pests,* Ser. A, No. 15, *H. armigera* (Revised). London: Commonw. Inst. Entomol.

28c. Commonwealth Institute of Entomology. 1969. *Distribution Maps of Pests,* Ser. A, No. 263, *H. punctigera.* London: Commonw. Inst. Entomol.

29. Cullen, J. M., Browning, T. O. 1978. The influence of photoperiod and temperature on the induction of diapause in pupae of *Heliothis punctigera. J. Insect. Physiol.* 24:595–601

30. Cunningham, R. B., Lewis, T., Wilson, A. G. L. 1981. Biothermal development: a model for predicting the distribution of emergence times for diapausing *Heliothis armigera. Appl. Stat.* 30:132–40

31. Daly, J. C., Gregg, P. 1985. Genetic variation in *Heliothis* in Australia: species identification and gene flow in the two pest species *H. armigera* (Hübner) and *H. punctigera* Wallengren (Lepidoptera: Noctuidae). *Bull. Entomol. Res.* 75:169–84

32. Daly, J. C., Murray, D. 1988. Evolution of resistance to pyrethroids in *Heliothis armigera* (Hübner) (Lepidoptera: Noctuidae) in Australia. *J. Econ. Entomol.* In press

33. Dempster, J. P. 1983. The natural control of populations of butterflies and moths. *Biol. Rev.* 58:461–81

34. Drake, V. A. 1988. Methods for studying adult movement in *Heliothis.* In *Resource Book on Heliothis Research Techniques,* ed. M. P. Zalucki. Brisbane, Australia: Queensl. Dep. Primary Ind. In press

35. Drake, V. A., Farrow, R. A. 1985. A radar and aerial trapping study of an early spring migration of moths (Lepidoptera) in inland New South Wales. *Aust. J. Ecol.* 11:223–35

36. Drake, V. A., Farrow, R. A. 1988. The influence of atmospheric structure and motions on insect migration. *Ann. Rev. Entomol.* 33:183–210

37. Drake, V. A., Helm, K. F., Readshaw, J. L., Reid, D. G. 1981. Insect migration across Bass Strait during spring: a radar study. *Bull. Entomol. Res.* 71:449–66

38. Eger, J. E., Sterling W. L., Hartstack, A. W. 1983. Winter survival of *Heliothis virescens* and *Heliothis zea* (Lepidoptera: Noctuidae) in College Station, Texas. *Environ. Entomol.* 12:970–75

39. Eger, J. E., Witz, J. A., Hartstack, A. W., Sterling, W. L. 1982. Survival of pupae of *Heliothis virescens* and *Heliothis zea* (Lepidoptera: Noctuidae) at low temperatures. *Can. Entomol.* 114:289–301

40. Ellington, J. J., El-Sokkari, A. 1986. A measure of the fecundity, ovipositional behavior, and mortality of the bollworm, *Heliothis zea* (Boddie) in the laboratory. *Southwest. Entomol.* 11:177–93

41. Farrer, R. R., Bradley, J. R. 1985. Within-plant distribution of *Heliothis* spp. (Lepidoptera: Noctuidae) eggs and larvae on cotton in North Carolina. *Environ. Entomol.* 14:205–9

42. Farrow, R. A., Daly, J. C. 1987. Long-range movements as an adaptive strategy in the genus *Heliothis* (Lepidoptera: Noctuidae): a review of its occurrence and detection in four pest species. *Aust. J. Zool.* 35:1–24

43. Farrow, R. A., McDonald, G. 1988. Migration strategies and outbreaks of noctuid pests in Australia. *Insect Sci. Appl.* In press

44. Firempong, S. K. 1987. *Some factors affecting host plant selection by* Heliothis armigera *(Hübner) (Lepidoptera: Noctuidae).* PhD thesis. Univ. Queensland, Brisbane, Australia

45. Fitt, G. P. 1986. Host selection in *Heliothis*. In *Proc*. Heliothis *Ecol. Workshop 1985, Brisbane*, pp. 47–59. Brisbane: Queensl. Dep. Primary Ind.

46. Fitt, G. P. 1987. Ovipositional responses of *Heliothis* spp. to host plant variation in cotton *(Gossypium hirsutum)*. In *Insects-Plants. Proc. 6th Int. Symp. Insect-Plant Relat., Pau, France, 1986*, pp. 289–94. Dordrecht, the Netherlands: Junk

47. Fitt, G. P., Zalucki, M. P., Twine, P. 1989. Temporal and spatial patterns in pheromone trap catches of *Helicoverpa* sp. in cotton growing areas of Australia. *Bull. Entomol. Res*. In press

48. Forrester, N. W., Cahill, M. 1987. Management of insecticide resistance in *Heliothis armigera* (Hübner) in Australia. In *Biological and Chemical Approaches to Combating Resistance to Xenobiotics*, ed. M. G. Ford, D. W. Holloman, B. P. S. Khambay, R. M. Sawicki, pp. 127–37. Amsterdam: Elsevier

49. Frisbie, R. E., Adkisson, P. L. 1985. IPM: definitions and current status in USA agriculture. See Ref. 23, pp. 41–50

50. Fye, R. E. 1978. Pupation preferences of bollworms, tobacco budworms and beet armyworms and impact on mortality resulting from cultivation of irrigated cotton. *J. Econ. Entomol.* 71:570–72

51. Fye, R. E., McAda, W. C. 1972. Laboratory studies on the development, longevity and fecundity of six lepidopterous pests of cotton in Arizona. *US Dep. Agric. Agric. Res. Serv. Tech. Bull.* Vol. 1454. 72 pp.

52. Gatehouse, A. G. 1986. Migration in the African armyworm *Spodoptera exempta:* genetic determination of migratory capacity and a new synthesis. In *Insect Flight, Dispersal and Migration*, ed. W. Danthanarayana, pp. 128–44. Heidelberg, West Germany: Springer-Verlag

53. Gaylor, M. J., Foster, R. 1987. Cotton pest management in the southeastern United States as influenced by conservation tillage practices. In *Arthropods in Conservation Tillage Systems. Misc. Publ. Entomol. Soc. Am.* 65:29–34

54. Georghiou, G. P., Taylor, C. E. 1977. Genetic and biological factors influencing the evolution of insecticide resistance. *J. Econ. Entomol.* 70:319–23

55. Gledhill, J. A. 1982. Progress and problems in *Heliothis* management in tropical southern Africa. See Ref. 134, pp. 375–84

56. Goodenough, J. L., Gaylor, J. J., Harris, V. E., Mueller, T. F., Heiss, J., et al. 1986. Efficacy of entomophagous arthropods. See Ref. 84, pp. 75–91

57. Gould, F. 1984. The role of behavior in the evolution of insect adaptation to insecticides and resistant host plants. *Bull. Entomol. Soc. Am.* 30:34–40

58. Graham, H. M., Fife, L. C. 1972. Overwintering of *Heliothis* spp. in the Lower Rio Grande Valley, Texas. *J. Econ. Entomol.* 65:708–11

59. Gross, H. R., Young, J. R. 1977. Comparative development and fecundity of corn earworm reared on selected wild and cultivated early-season hosts common to the southeastern USA. *Ann. Entomol. Soc. Am.* 70:63–65

60. Gross, H. R., Young, J. R., Forbes, I. 1975. *Heliothis* spp.: seasonal occurrence and overwintering on the hyacinth bean plant in Tifton, Georgia. *J. Econ. Entomol.* 68:169–70

61. Gunning, R. V., Easton, L. R., Greenup, L. R., Edge, V. E. 1984. Pyrethroid resistance in *Heliothis armiger* (Hübner) (Lepidoptera: Noctuidae) in Australia. *J. Econ. Entomol.* 77:1283–87

62. Hackett, D., Gatehouse, A. G. 1978. Larval development of *Heliothis armigera* on groundnuts, sorghum and cotton in the Sudan Gezira. In *Ciba-Geigy 3rd Sem. Strategy Cotton Pest Control Sudan, Basle*, pp. 4–16. Basle, Switzerland: Ciba-Geigy

63. Hackett, D., Gatehouse, A. G. 1982. Studies on the biology of *Heliothis* spp. in Sudan. See Ref. 134, pp. 29–38

64. Hackett, D., Gatehouse, A. G. 1982. Diapause in *Heliothis armigera* (Hübner) and *H. fletcheri* (Hardwick) (Lepidoptera: Noctuidae) in the Sudan Gezira. *Bull. Entomol. Res.* 72:409–22

65. Haggis, M. J. 1981. Spatial and temporal changes in the distribution of eggs by *Heliothis armiger* (Hübner) (Lepidoptera: Noctuidae) on cotton in the Sudan Gezira. *Bull. Entomol. Res.* 71:183–93

66. Hardwick, D. F. 1965. *The Corn Earworm Complex. Mem. Entomol. Soc. Can.* Vol. 40. 247 pp.

67. Hariri, G. 1982. The problems and prospects of *Heliothis* management in southwest Asia. See Ref. 134, pp. 363–74

68. Harris, V. E., Phillips, J. R. 1986. Mowing spring host plants as a population management technique for *Heliothis* spp. *J. Agric. Entomol.* 3:125–34

69. Hartstack, A. W., Hollingsworth, J. P., Ridgway, R. L., Coppedge, J. R. 1973. A population dynamics study of the boll-

worm and the tobacco budworm with light traps. *Environ. Entomol.* 2:244–52

70. Hartstack, A. W., Lopez, J. D., Muller, R. A., Sterling, W. L., King, E. G., et al. 1982. Evidence of long range migration of *Heliothis zea* into Texas and Arkansas. *Southwest. Entomol.* 7:188–201

71. Hartstack, A. W., Witz, J. A. 1983. Models for cotton insect pest management. See Ref. 1, pp. 359–81

72. Hartstack, A. W., Witz, J. A., Hollingsworth, J. P., Ridgway, R. L., Lopez, J. D. 1976. MothZV-2: a computer simulation of *Heliothis zea* and *Heliothis virescens* population dynamics. User manual. *US Dep. Agric. Agric. Res. Serv. S-127,* Washington DC. 55 pp.

73. Hearn, A. B., Fitt, G. P. 1988. Cotton cropping systems. In *Field-Crop Ecosystems,* ed. C. Pearson. Amsterdam: Elsevier. In press

74. Hendricks, D. E., Graham, H. M., Raulston, J. R. 1973. Dispersal of sterile tobacco budworms from release points in northeastern Mexico and southern Texas. *Environ. Entomol.* 2:1085–88

75. Hendrix, W. H., Mueller, T. H., Phillips, J. R., Davis, O. K. 1987. Pollen as an indicator of long-distance movement of *Heliothis zea* (Lepidoptera: Noctuidae). *Environ. Entomol.* 61:1148–51

76. Henneberry, T. J., Butler, G. D. 1986. Effects of high temperature on tobacco budworm (Lepidoptera: Noctuidae): reproduction, diapause, and spermatocyst development. *J. Econ. Entomol.* 79:410–13

77. Herzog, G. A., Phillips, J. R. 1976. Selection for a diapause strain of the bollworm, *Heliothis zea.* *J. Hered.* 67:173–75

78. Hillhouse, T. L., Pitre, H. N. 1976. Oviposition by *Heliothis* on soybeans and cotton. *J. Econ. Entomol.* 69:144–46

79. Hogg, D. B., Nordheim, E. V. 1983. Age-specific analysis of *Heliothis* spp. populations on cotton. *Res. Popul. Ecol.* 25:280–97

80. Holtzer, T. O., Bradley, J. R., Rabb, R. L. 1976. Geographic and genetic variation in time required for emergence of diapausing *Heliothis zea.* *Ann. Entomol. Soc. Am.* 69:261–65

81. Isely, D. 1935. Relation of hosts to abundance of cotton bollworm. *Arkansas Agric. Exp. Stn. Bull.* Vol. 320. 30 pp.

82. Ives, P. M., Wilson, L. T., Cull, P. O., Palmer, W. A., Haywood, C., et al. 1984. Development and field usage of SIRATAC: an Australian computer-based pest management system for cotton. *Prot. Ecol.* 6:1–21

83. Johnson, M. W., Stinner, R. E., Rabb, R. L. 1975. Ovipositional response of *Heliothis zea* to its major hosts in North Carolina. *Environ. Entomol.* 4:291–97

84. Johnson, S. J., King, E. G., Bradley, J. R., eds. 1986. *Theory and tactics of* Heliothis *population management. I. Cultural and biological control. South. Coop. Ser. Bull.* Vol. 316. 161 pp.

85. Johnson, S. J., Pitre, H. N., Powell, J. E., Sterling, W. L. 1986. Control of *Heliothis* spp. through conservation and importation of natural enemies. See Ref. 84, pp. 132–54

86. Joyce, R. J. V. 1978. The strategy of cotton pest control in the Sudan Gezira. See Ref. 62, pp. 85–103

87. Joyce, R. J. V. 1983. Aerial transport of pests and pest outbreaks. *EPPO Bull.* 13:111–19

88. Kay, I. R. 1982. The incidence and duration of pupal diapause in *Heliothis armiger* (Hübner) (Lepidoptera: Noctuidae) in southeast Queensland. *J. Aust. Entomol. Soc.* 21:263–66

89. Kennedy, G. G., Gould, F., Deponti, O. M., Stinner, R. E. 1987. Ecological, agricultural, genetic, and commercial considerations in the deployment of insect-resistant germplasm. *Environ. Entomol.* 16:327–38

90. Kennedy, J. S. 1986. Migration, behavioral and ecological. In *Migration: Mechanisms and Adaptive Significance. Contrib. Mar. Sci.* 27(Suppl.): 5–26

91. Khaemba, B. M., Mutinga, M. J. 1982. Insect pests of sunflower (*Helianthus annuus* L.) in Kenya. *Insect Sci. Appl.* 3:281–86

92. King, E. G. 1989. Potential for the biological control of *Heliothis* species. *Ann. Rev. Entomol.* 34:53–75

93. King, E. G., Bull, D. L., Bouse, L. F., Phillips, J. R., eds. 1985. *Biological Control of Bollworm and Tobacco Budworm in Cotton by Augmentative Releases of* Trichogramma. *Southwest. Entomol. Suppl.* Vol. 8. 172 pp.

94. King, E. G., Hopper, K. R., Powell, J. E. 1985. Analysis of systems for biological control of crop arthropod pests in the USA by augmentation of predators and parasites. See Ref. 23, pp. 201–25

95. King, E. G., Powell, J. E., Smith, J. W. 1982. Prospects for utilisation of parasites and predators for management of *Heliothis* spp. See Ref. 134, pp. 103–22

96. Knipling, E. F., Stadelbacher, E. A. 1983. The rationale of area-wide management of *Heliothis* (Lepidoptera: Noc-

tuidae). populations. *Bull. Entomol. Soc. Am.* 29:29–37

97. Kou, R., Chow, Y. 1987. Calling behavior of the cotton bollworm, *Heliothis armigera* (Lepidoptera: Noctuidae). *Ann. Entomol Soc. Am.* 80:490–93

98. Laster, M. L., Kitten, W. F., Knipling, E. F., Martin, D. F., Schneider, J. C., et al. 1987. Estimates of overwintered population density and adult survival rates for *Heliothis virescens* (Lepidoptera: Noctuidae) in the Mississippi delta. *Environ. Entomol.* 16:1076–81

99. Lincoln, C., Phillips, J. R., Whitcomb, W. H., Dowell, G. C., Boyer, W. P., et al. 1967. The bollworm–tobacco budworm problem in Arkansas and Louisiana. *Arkansas Agric. Exp. Stn. Bull.* Vol. 720. 66 pp.

100. Lingren, P. D., Sparks, A. N., Raulston, J. R. 1982. The potential contribution of moth behaviour research to *Heliothis* management. See Ref. 134, pp. 39–47

100a. Lingren, P. D., Warner, W. B., Raulston, J. R., Kehat, M., Henneberry, T. J., et al. 1988. Observations on the emergence of adults from natural populations of corn earworm, *Heliothis zea* (Boddie) (Lepidoptera: Noctuidae). *Environ. Entomol.* 17:254–58

101. Lopez, J. D., Hartstack, A. W. 1985. Comparison of diapause development in *Heliothis zea* and *H. virescens* (Lepidoptera: Noctuidae). *Ann. Entomol. Soc. Am.* 78:415–22

102. Lopez, J. D., Hartstack, A. W., Beach, R. 1984. Comparative pattern of emergence of *Heliothis zea* and *H. virescens* (Lepidoptera: Noctuidae) from overwintering pupae. *J. Econ. Entomol.* 77:1421–26

103. Luttrell, R. G., Phillips, J. R., Pfrimmer, T. R. 1986. Cultural control in mid-south USA cropping systems. See Ref. 84, pp. 28–37

104. Luttrell, R. G., Roush, R. T., Ali, A., Mink, J. S., Reid, M. R., et al. 1987. Pyrethroid resistance in field populations of *Heliothis virescens* (Lepidoptera: Noctuidae) in Mississippi in 1986. *J. Econ. Entomol.* 80:985–89

105. Manjunath, T. M., Patel, R. C., Yadav, D. N. 1976. Observations on *Heliothis peltigera* (Schiff.) (Lep., Noctuidae) and its natural enemies in Anand (Gujarat State, India.) *Proc. Indian Acad. Sci. Sect. B* 83:(2):55–65

105a. Matthews, M. 1987. *The classification of Heliothinae (Noctuidae).* PhD thesis. Br. Mus. & Kings Coll., London. 253 pp.

106. Montandon, R., Stipanovic, R. D., Williams, H. J., Sterling, W. L., Vinson, S. B. 1987. Nutritional indices and excretion of gossypol by *Alabama argillacae* (Hübner) and *Heliothis virescens* (F) (Lepidoptera: Noctuidae) fed glanded and glandless cotyledonary cotton leaves. *J. Econ. Entomol.* 80:32–36

107. Morrison, G., Lewis, W. J., Nordlund, D. A. 1980. Spatial differences in *Heliothis zea* egg density and the intensity of parasitism by *Trichogramma* spp.: an experimental analysis. *Environ. Entomol.* 9:79–85

108. Morton, R., Tuart, L. D., Wardhaugh, K. G. 1981. The analysis and standardisation of light-trap catches of *Heliothis armiger* (Hübner) and *H. punctiger* Wallengren (Lepidoptera: Noctuidae). *Bull. Entomol. Res.* 71:207–25

109. Mueller, T. F., Harris, V. F., Phillips J. R. 1984. Theory of *Heliothis* (Lepidoptera: Noctuidae) management through reduction of the first spring generation: a critique. *Environ. Entomol.* 13:625–34

110. Mueller, T. F., Phillips, J. R. 1983. Population dynamics of *Heliothis* spp. in spring weed hosts in southeastern Arkansas: survivorship and stage-specific parasitism. *Environ. Entomol.* 12:1846–50

111. Nadgauda, D., Pitre, H. 1983. Development, fecundity, and longevity of the tobacco budworm (Lepidoptera: Noctuidae) fed soybean, cotton, and artificial diet at three temperatures. *Environ. Entomol.* 12:582–86

112. Nemec, S. J. 1971. Effects of lunar phases on light-trap collections and populations of bollworm moths. *J. Econ. Entomol.* 64:860–64

113. Neunzig, H. H. 1969. The biology of the tobacco budworm and the corn earworm in North Carolina, with particular reference to tobacco as a host. *NC Agric. Exp. Stn. Tech. Bull.* Vol. 196. 76 pp.

114. Nicholson, W. F., Phillips, J. R., Bernhardt, J. L., Marx, D. B., Slosser, J. E. 1984. Sampling in an area-wide bollworm management community. *Southwest. Entomol. Suppl.* 6:17–22

115. Pair, S. D., Raulston, J. R., Rummel, D. R., Westbrook, J. K., Wolf, W. W., et al. 1987. Development and production of corn earworm and fall armyworm in the Texas High Plains: evidence for reverse fall migration. *Southwest. Entomol.* 12:89–99

116. Parsons, F. S. 1940. Investigations on the cotton bollworm, *Heliothis armigera,* Hübn. Part III. Relationships between oviposition and the flowering

curves of food-plants. *Bull. Entomol. Res.* 31:147–77

117. Pearson, E. O. 1958. *The Insect Pests of Cotton in Tropical Africa.* London: Commonw. Inst. Entomol. 355 pp.

118. Pedgley, D. E. 1985. Windborne migration of *Heliothis armigera* (Hübner) (Lepidoptera: Noctuidae) to the British Isles. *Entomol. Gaz.* 36:15–20

119. Pedgley, D. E., Tucker, M. R., Pawar, C. S. 1988. Windborne migration of *Heliothis armigera* (Hubner) (Lepidoptera: Noctuidae) in India. *Insect Sci. Appl.* 8:599–604

120. Persson, B. 1974. Diel distribution of oviposition in *Agrotis ipsilon* (Hufn), *Agrotis munda* (Walk.), and *Heliothis armigera* (Hbn.), (Lep. Noctuidae), in relation to temperature and moonlight. *Entomol. Scand.* 5:196–208

121. Pimentel, D. 1977. The ecological basis of insect pest, pathogen and weed problems. In *Origins of Pest, Parasite, Disease and Weed Problems,* ed. J. M. Cherrit, G. R. Sagar, pp. 3–34. Oxford: Blackwell

122. Plapp, F. W. 1987. Managing resistance to synthetic pyrethroids in the tobacco budworm. *Proc. Beltwide Cotton Prod. Res. Conf. 1987,* pp. 224–26. Memphis, Tenn: Natl. Cotton Counc. Am.

123. Plapp, F. W., McWhorter, G. M., Vance, W. H. 1987. Monitoring for pyrethroid resistance in the tobacco budworm in Texas—1986. See Ref. 122, pp. 324–25

124. Price, P. W. 1981. Relevance of ecological concepts to practical biological control. In *Biological Control in Crop Production, Beltsville Symp. Agric. Res. No. 5,* ed. G. C. Papavizas, pp. 3–19. Totowa, NJ: Allanheld, Osmun

125. Puterka, G. J., Slosser, J. E., Price, J. R. 1985. Parasites of *Heliothis* spp. (Lepidoptera: Noctuidae): parasitism and seasonal occurrence for host crops in the Texas Rolling Plains. *Environ. Entomol.* 14:441–46

126. Qayyum, A., Zalucki, M. P. 1987. Effects of high temperature on survival of eggs of *Heliothis armigera* (Hübner) and *H. punctigera* Wallengren (Lepidoptera: Noctuidae). *J. Aust. Entomol. Soc.* 26:295–98

127. Quaintance, A. L., Brues, C. T. 1905. *The Cotton Bollworm.* US Dep. Agric. Tech. Bull. Vol. 50. 155 pp.

128. Rabb, R. L., Bradley, J. R., Stinner, R. E., Pleasants, R. H., Pearce, L. 1975. Diapause in North Carolina strains of *Heliothis zea* and *H. virescens. J. Ga. Entomol. Soc.* 10:191–98

129. Rabb, R. L., Kennedy, G. G., eds. 1979. *Movement of Highly Mobile Insects: Concepts & Methodology in Research.* Raleigh NC: Univ. Graphics

130. Ramaswamy, S. B., Ma, W. K., Baker, G. T. 1987. Sensory cues and receptors for oviposition by *Heliothis virescens. Entomol. Exp. Appl.* 43:159–68

131. Raulston, J. R., Pair, S. D., Martinez, F. A., Westbrook, J. K., Sparks, A. N., et al. 1986. Ecological studies indicating the migration of *Heliothis zea, Spodoptera frugiperda* and *Heliothis virescens* from northeastern Mexico and Texas. See Ref. 52, pp. 204–20

132. Raulston, J. R., Wolf, W. W., Lingren, P. D., Sparks, A. N. 1982. Migration as a factor in *Heliothis* management. See Ref. 134, pp. 61–74

133. Reed, W. 1965. *Heliothis armigera* (Hb.) (Noctuidae) in western Tanganyika. I. Biology, with special reference to the pupal stage. *Bull. Entomol. Res.* 56:117–25

134. Reed, W., Kumble, V., eds. 1982. *Proc. Int. Workshop* Heliothis *Management, Patancheru, India, 1981.* Patancheru, India: ICRISAT. 418 pp.

135. Reed, W., Pawar, C. S. 1982. *Heliothis:* a global problem. See Ref. 134, pp. 9–14

136. Reese, J. C., Chan, B. G., Waiss, A. C. 1982. Effects of condensed tannins, maysin (corn) and pinitol (soybeans) on *Heliothis zea* growth and development. *J. Chem. Ecol.* 8:1429–36

137. Rice, M. 1986. Semiochemicals and sensory manipulation strategies for behavioural management of *Heliothis* spp. See Ref. 45, pp. 27–46

138. Ridgway, R. L., Bell, A. A., Veech, J. A., Chandler, J. M. 1984. Cotton protection practices in the USA and the world. In *Cotton,* ed. R. J. Kohel, C. F. Lewis, pp. 266–365. Madison, Wis: Am. Soc. Agron.

139. Risch, S. J., Andow, D., Altieri, M. A. 1983. Agroecosystem diversity and pest control: data, tentative conclusions and new research directions. *Environ. Entomol.* 12:625–29

140. Roach, S. H. 1981. Emergence of overwintered *Heliothis* spp. moths from three different tillage systems. *Environ. Entomol.* 10:817–18

141. Roach, S. H., Adkisson, P. L. 1970. Role of photoperiod and temperature in the induction of pupal diapause in the bollworm, *Heliothis zea. J. Insect Physiol.* 16:1591–97

142. Room, P. M. 1979. Parasites and predators of *Heliothis* spp. (Lepidoptera: Noctuidae) in the Namoi Valley, New

South Wales. *J. Aust. Entomol. Soc.* 18:223–28

143. Room, P. M. 1983. Calculations of temperature-driven development by *Heliothis* spp. (Lepidoptera: Noctuidae) in the Namoi Valley, New South Wales. *J. Aust. Entomol. Soc.* 22:211–15

144. Roome, R. E. 1975. Activity of adult *Heliothis armigera* (Hb) (Lepidoptera: Noctuidae) with reference to the flowering of sorghum and maize in Botswana. *Bull. Entomol. Res.* 65:523–30

145. Roome, R. E. 1979. Pupal diapause in *Heliothis armigera* (Hübner) (Lepidoptera: Noctuidae) in Botswana: its regulation by environmental factors. *Bull. Entomol. Res.* 69:149–60

146. Rummel, D. R., Leser, J. F., Slosser, J. E., Puterka, G. J., Neeb, C. W., et al. 1986. Cultural control of *Heliothis* spp. in southwestern USA cropping systems. See Ref. 84, pp. 38–53

147. Rummel, D. R., Neece, K. C., Arnold, M. D., Lee, B. A. 1986. Overwintering survival and spring emergence of *Heliothis zea* (Boddie) in the Texas Southern High Plains. *Southwest. Entomol.* 11:1–9

148. Sawicki, R. M., Denholm, I. 1987. Management of resistance to pesticides in cotton pests. *Trop. Pest Manage.* 33:262–72

149. Schaefer, G. W. 1976. Radar observations on insect flight. In *Insect Flight*, ed. R. C. Rainey, pp. 157–97. Oxford: Blackwell

150. Schneider, J. C., Benedict, J. H., Gould, F., Meredith, W. R., Schuster, M. F., et al. 1986. Interaction of *Heliothis* with its host plants. See Ref. 84, pp. 3–21

151. Sharpe, P. J. H., Schoolfield, R. M., Butler, G. D. 1981. Distribution model of *Heliothis zea* (Lepidoptera: Noctuidae) development times. *Can. Entomol.* 113:845–56

152. Slosser, J. E., Phillips, J. R., Herzog, G. A., Reynolds, C. R. 1975. Overwintering survival and spring emergence of the bollworm in Arkansas. *Environ. Entomol.* 4:1015–24

153. Smith, R. F., Reynolds, H. T. 1972. Effects of manipulation of cotton agroecosystems on insect pest populations. See Ref. 5, pp. 373–406

154. Southwood, T. R. E. 1977. The relevance of population dynamic theory to pest status. See Ref. 121, pp. 35–56

155. Southwood, T. R. E., Comins, H. N. 1976. A synoptic population model. *J. Anim. Ecol.* 45:949–65

156. Sparks, A. N. 1979. An introduction to the status, current knowledge and research on movement of selected Lepidoptera in southeastern United States. See Ref. 129, pp. 382–85

157. Sparks, T. C. 1981. Development of insecticide resistance in *Heliothis zea* and *Heliothis virescens* in North America. *Bull. Entomol. Soc. Am.* 27:186–92

158. Stadelbacher, E. A., Graham, H. M., Harris, V. E., Lopez, J. D., Phillips, J. R., et al. 1986. *Heliothis* populations and wild host plants in the southern USA. See Ref. 84, pp. 54–74

159. Stadelbacher, E. A., Martin, D. F. 1980. Fall diapause, winter mortality and spring emergence of the tobacco budworm in the delta of Mississippi. *Environ. Entomol.* 9:553–56

160. Stadelbacher, E. A., Pfrimmer, T. R. 1972. Winter survival of the bollworm at Stoneville, Mississippi. *J. Econ. Entomol.* 65:1030–34

161. Stadelbacher, E. A., Powell, J. E., King, E. G. 1984. Parasitism of *Heliothis zea* and *H. virescens* (Lepidoptera: Noctuidae) larvae in wild and cultivated host plants in the delta of Mississippi. *Environ. Entomol.* 13:1167–72

162. Stinner, R. E. 1975. Models as aids in predicting trends in *Heliothis* spp. populations on cotton and need for their control. *Proc. Beltwide Cotton Prod. Res. Conf. 1975, New Orleans, La.*, pp. 164–65. Memphis, Tenn: Natl. Cotton Counc. Am.

163. Stinner, R. E. 1979. Biological monitoring essentials in studying wide-area moth movement. See Ref. 129, pp. 199–211

164. Stinner, R. E., Barfield, C. S., Stimac, J. L., Dohse, L. 1983. Dispersal and migration of insect pests. *Ann. Rev. Entomol.* 28:319–35

165. Stinner, R. E., Rabb, R. L., Bradley, J. R. 1974. Population dynamics of *Heliothis zea* (Boddie) and *H. virescens* (F.) in North Carolina: a simulation model. *Environ. Entomol.* 3:163–68

166. Stinner, R. E., Rabb, R. L., Bradley, J. R. 1977. Natural factors operating in the population dynamics of *Heliothis zea* in North Carolina. *Proc. 15th Int. Congr. Entomol., Washington, DC*, pp. 622–42

167. Stinner, R. E., Regniere, J., Wilson, K. 1982. Differential effects of agroecosystem structure on dynamics of three soybean herbivores. *Environ. Entomol.* 11:538–43

168. Thanee, N. 1987. *Oviposition preferences, larval feeding preferences and larval food quality of Heliothis armigera*. PhD thesis. Massey Univ., Palmerston North, NZ

169. Thomson, N. J., Lee, J. A. 1980. Insect

resistance in cotton: a review and prospectus for Australia. *J. Aust. Inst. Agric. Sci.* 46:75–86

170. Titmarsh, I. 1985. Population dynamics of *Heliothis* spp. on tobacco in far North Queensland. MSc thesis. James Cook Univ., Townsville, Australia

171. Todd, E. L. 1978. A checklist of species of *Heliothis* Oschenheimer (Lepidoptera: Noctuidae). *Proc. Entomol. Soc. Wash.* 80:1–14

172. Topper, C. P. 1981. *The behaviour and population dynamics of* Heliothis armigera *(Hb) (Lepidoptera: Noctuidae) in the Sudan Gezira.* PhD thesis. Cranfield Inst. Technol., Bedford, UK. 245 pp.

173. Topper, C. P. 1987. The dynamics of the adult population of *Heliothis armigera* (Hübner) (Lepidoptera: Noctuidae) within the Sudan Gezira in relation to cropping pattern and pest control on cotton. *Bull. Entomol. Res.* 77:525–39

174. Topper, C. P. 1987. Nocturnal behaviour of adults of *Heliothis armigera* (Hübner) (Lepidoptera: Noctuidae) in the Sudan Gezira and pest control implications *Bull. Entomol. Res.* 77:541–54

175. Twine, P. H. 1978. Effect of temperature on the development of larvae and pupae of the corn earworm, *Heliothis armigera* (Hübner) (Lepidoptera: Noctuidae). *Queensl. J. Agric. Anim. Sci.* 35:23–38

176. Twine, P. H., Kay, I. 1982. A determination of an economic injury level of *Heliothis armigera* (Hübner) in sorghum for southeast Queensland. See Ref. 134, pp. 189–96

177. van Emden, H. F., Williams, G. F. 1974. Insect stability and diversity in agro-ecosystems. *Ann. Rev. Entomol.* 19:455–76

178. Vargas, R., Nishida, T. 1960. Life table of the corn earworm, *Heliothis zea* (Boddie), in sweet corn in Hawaii. *Proc. Hawaii. Entomol. Soc.* 23:301–7

179. Vermeulen, J. B. 1973. The occurrence and abundance of *Heliothis assulta afra* (Hardwick) in light traps in South Africa (Lepidoptera: Noctuidae). *J. Entomol. Soc. South. Afr.* 36:309–10

180. Wallach, D. 1984. An algorithm for the short-term prediction of insect populations. *Prot. Ecol.* 6:181–98

181. Wardhaugh, K. G., Room, P. M., Greenup, L. R. 1980. The incidence of *Heliothis armigera* (Hübner) and *H. punctigera* Wallengren (Lepidoptera: Noctuidae) on cotton and other host plants in the Namoi Valley of New South Wales. *Bull. Entomol. Res.* 70:113–31

182. Way, M. J. 1977. Pest and disease status in mixed stands vs monocultures: the relevance of ecosystem stability. See Ref. 121, pp. 127–38

183. Wellso, S. G., Adkisson, P. L. 1966. A long-day short-day effect in the photoperiodic control of the pupal diapause of the bollworm, *Heliothis zea* (Boddie). *J. Insect Physiol.* 12:1455–65

184. Widmer, M. W., Schofield, P. 1983. Heliothis *Dispersal and Migration. TDRI Inf. Serv. Annot. Bibliog. No. 2.* London: Trop. Dev. Res. Inst. 41 pp.

185. Willers, J. L., Schneider, J. C., Ramaswamy, S. B. 1987. Fecundity, longevity and caloric patterns in female *Heliothis virescens:* changes with age due to flight and supplemental carbohydrate. *J. Insect Physiol.* 33:803–8

186. Williams, W. G., Kennedy, G. G., Yamamoto, R. T., Thacker, J. D., Bordner, J. 1980. 2-Tridecanone: a naturally occurring insecticide from the wild tomato, *Lycopersicon hirsutum* f. *glabratum. Science* 207:8888–89

187. Wilson, A. G. L. 1982. Past and future *Heliothis* management in Australia. See Ref. 134, pp. 343–54

188. Wilson, A. G. L. 1983. Abundance and mortality of overwintering *Heliothis* spp. *J. Aust. Entomol. Soc.* 22:191–99

189. Wilson, A. G. L., Lewis, T., Cunningham, R. B. 1979. Overwintering and spring emergence of *Heliothis armigera* (Hübner) (Lepidoptera: Noctuidae) in the Namoi Valley, New South Wales. *Bull. Entomol. Res.* 69:97–109

190. Wolf, W. W., Sparks, A. N., Pair, S. D. Westbrook, J. K. 1986. Radar observations and collections of insects in the Gulf of Mexico. See Ref. 52, pp. 221–34

191. Wolfenbarger, D. A., Bodegas, V. P. R., Flores, G. R. 1981. Development of resistance in *Heliothis* spp. in the Americas, Australia, Africa, and Asia. *Bull. Entomol. Soc. Am.* 27:181–85

192. Yathom, S. 1971. Distribution and flight period of Heliothidinae species in Israel in 1959–69. *Israel J. Agric. Res.* 2:51–61

193. Young, J. H., Price, R. G. 1977. Overwintering of *Heliothis zea* in southwestern Oklahoma. *Environ. Entomol.* 6:627–28

194. Zalucki, M. P., Daglish, G., Firempong, S., Twine, P. 1986. The biology and ecology of *Heliothis armigera* (Hübner) and *H. punctigera* Wallengren (Lepidoptera: Noctuidae) in Australia: What do we know? *Aust. J. Zool.* 34:779–814

*Ann. Rev. Entomol. 1989. 34:53–75*

# POTENTIAL FOR BIOLOGICAL CONTROL OF *HELIOTHIS* SPECIES[1,2]

*E. G. King and R. J. Coleman*

Subtropical Agricultural Research Laboratory, USDA/ARS, Weslaco, Texas 78596

## INTRODUCTION

Synthetic insecticides are often used to suppress *Heliothis* populations because of their availability, portability, and potential for quick intervention and prevention of serious plant damage by larvae. However, *H. armigera, H. zea,* and *H. virescens* rapidly develop resistance to insecticides (33, 154). Moreover, chemicals raise other concerns such as environmental safety, production efficiency, pest resurgence due to natural enemy mortality, and energy conservation. Consequently, growers may be unable to rely solely on synthetic chemical control for *Heliothis* suppression. In fact, the concept of integrated pest management (IPM), i.e. movement away from reliance upon a single control procedure toward the use of all available management methods, emerged as many insecticides ceased to be effective for controlling insect populations such as *H. virescens* (tobacco budworm) (6).

A generally accepted component of IPM strategy is the conservation and maximum use of naturally occurring biological control agents (natural enemies). For purposes of this review, we define biological control as the management of biological control agents (predators, parasites, and microbial organisms) and their products to reduce pest population densities and their effects. This definition does not include genetic or autocidal approaches, host-plant resistance, or cultural control, nor does it include physical or chemical control except where these approaches supplement or affect natural enemies. We summarize research and the status of the use of predators, parasites, and pathogens to control *Heliothis* species. Pathways for further exploiting the potential of this vital IPM component are suggested.

---

[1]The US Government has the right to retain a nonexclusive, royalty-free license in and to any copyright covering this paper.

[2]For a detailed review of the ecology of *Heliothis* spp. in relation to agricultural systems, see the preceding chapter in this volume.

## HELIOTHIS PEST SPECIES

### Systematics

The genus *Heliothis* consists of more than 75 species or subspecies (145) and includes some of the most important insect pests of crops in the world. *Heliothis armigera* (corn earworm) and *H. zea* (bollworm, corn earworm, tomato fruitworm) are perhaps the most infamous of the *Heliothis* species. Until 1953 they were considered the same species (20), and interspecific matings under laboratory conditions may result in fertile progeny (80).

In 1965, Hardwick (44) concluded that members of the corn earworm complex were morphologically distinct from *Heliothis* species and not closely related phylogenetically; thus he proposed the generic name *Helicoverpa* for this complex. Hardwick's conclusions have not been widely accepted by entomologists because of the potential inconvenience and possible confusion related to changing the genus name applied to members of the corn earworm complex. In addition, the gender status of *Heliothis* has been questioned (145). A referendum of a 1981 international workshop resulted in a recommendation to the International Commission on Zoological Nomenclature that the well-established *Heliothis* spp. names, including *armigera, peltigera,* and *punctigera,* should continue to be used (114). Other more recent reviews (62, 71, 134) have retained the use of the generic name *Heliothis* as well as the original orthography of the species epithets. For purposes of this review, *Heliothis* is used as the nominate genus and includes *Heliothis* and the often recognized *Helicoverpa* as subgenera (see 77). Differences among *Heliothis* species are important in biological control. For example, *H. virescens* and *H. subflexa* are susceptible hosts for the larval parasite *Cardiochiles nigriceps,* but eggs and larvae of this parasite are encapsulated by hemocytes of *H. zea* larvae (84). Whatever the specific or generic confines, we must seek to understand those general principles that pertain to *Heliothis* as a taxon, yet we must also deal with the uniqueness of each member.

### Distribution

Ranging between latitudes 45° N and 45° S, *H. armigera* and *H. zea* are cosmopolitan species of the Old World and New World, respectively (44). *Heliothis armigera* is a major pest of cotton, tobacco, and other crops in southern Europe (Spain, Portugal, Turkey, and Greece), southern Asia (China, India, and Thailand), southern and eastern Africa, and Australia. *Heliothis punctigera* is restricted to Australia, where it is a major pest of cotton. *Heliothis zea* and *H. virescens* are major pests of a wide range of crops in the Americas and the Caribbean (78). Other pest species of lesser importance or more restricted distribution include *H. gelotopoeon* in southern

South America, *H. viriplaca* in southern Europe, *H. assulta* in southern Asia, Africa, and Australia, *H. peltigera* in southern Asia, and *H. nubigera* and *H. fletcheri* in Africa (44, 45).

## Hosts and Economic Importance

*Heliothis armigera, H. punctigera, H. zea,* and *H. virescens* have been recorded as feeding on over 100 different plant species each (141, 159). Field crops attacked by *Heliothis* spp. are some of the world's most important, including corn, cotton, sorghum, soybeans, vegetables, pulses, wheat, and tobacco. Pulse crops, which are the primary source of protein for a predominately vegetarian human populace in India, are severely attacked by *H. armigera* (124). *Heliothis assulta* and other species attack tobacco in southeastern Asia and Africa. *Heliothis viriplaca* and *H. peltigera* attack various crops in southwestern Asia (45), and *H. fletcheri* attacks sesame and sorghum in the Sahel of Africa (40). *Heliothis gelotopoeon* attacks cotton, corn, tobacco, soybean, flax, and other crops in southern South America (151).

In the United States, *H. zea* and *H. virescens* may cause over one billion dollars damage to crops annually despite insecticide applications costing another $250 million per year (2). Damage losses to chick-pea and pigeon pea alone in India are estimated to exceed $300 million annually (124). Estimated losses due to *Heliothis* spp. in Australia were $23.5 million annually (124). Estimated losses due to *Heliothis* damage and control have increased by 35% ($16 million to $25 million) in Queensland alone since 1979 (146).

## Biology

Generally, *Heliothis* spp. are fruit feeders, though crops such as tobacco and lettuce undergo most damage to leaves. Consequently, they are in direct competition with humans for food and fiber. Structures fed on by *Heliothis* spp. are either rendered unusable or greatly reduced in quality, and feeding often facilitates infection by pathogenic organisms. Moreover, once larvae enter the fruit, e.g. corn, they are protected from predators and parasites (122). Thus, the number of larvae tolerated is often so low that biological control agents must be supplemented with other forms of control. However, in field corn, insecticide applications for *H. zea* or *H. armigera* usually are not economically feasible. Thus with larvae protected from natural enemy attack, *H. zea* or *H. armigera* populations survive at such high levels that other crops are threatened. *Heliothis armigera, H. assulta, H. punctigera, H. zea,* and *H. virescens* are multivoltine with facultative diapause, highly fecund, and capable of moving long distances as adults. Thus they can rapidly exploit host crops, particularly monocultures, even in advance of natural enemies. Another factor contributing to *Heliothis* pest status is the relatively large size and quick development of the insects, which mature from egg to adult in less than 30 days; consequently, food is consumed at a high rate.

# DISTRIBUTION AND ABUNDANCE OF NATURAL ENEMIES

## Entomophagous Arthropods

The worldwide distribution and abundance of natural enemies of *Heliothis* spp. have been reviewed (71). *Microplitis rufiventris, Bracon brevicornis, Chelonus inanitus, Zele* spp., and *Cotesia* (=*Apanteles*) spp. are the predominant parasites attacking *Heliothis* spp. in Egypt (53). The predominant entomophagous arthropods of *Heliothis* spp. in Australia have been grouped into egg parasites (*Telenomus* sp. and *Trichogramma* spp.), larval parasites (*Microplitis* sp.), and pupal parasites (*Heteropelma scaposum* and *Carcelia noctuae*) (146). Numerous predator species were noted in host crops attacked by *Heliothis* populations; *Nabis capsiformis, Geocoris lubra, Deraeocoris signatus,* and spiders were emphasized (146). Ma & Ding (93) listed 35 predators and parasites of *H. armigera* in China. Specifically mentioned were the predators *Orius similis, Misumenops tricuspidatus, Erigonidium graminicola, Chrysopa sinica,* and *Nabis sinoferus* and the parasites *Trichogramma* sp. and *Campoletis chlorideae* (93).

Manjunath et al (94) cited 77 parasites attacking *H. armigera,* 13 attacking *H. assulta,* and 16 attacking *H. peltigera* in India. The most important hymenopterous parasites were *Trichogramma chilonis, C. chlorideae,* and *Eriborus* sp., while the most important tachinids were *Carcelia illota, Goniophthalmus halli,* and *Palexorista laxa.* In addition, 33 arthropod predator species were cited, as well as the nematodes *Ovomermis albicans* and *Hexamermis* spp. A difference in habitat preference by the parasites was noted: Hymenopterous parasites occurred most frequently in cereal crops; dipterans frequented legumes; and egg parasitism by *Trichogramma* spp. was low in chick-pea and pigeon pea (94). *Campoletis chlorideae* was cited as the most important parasite of *Heliothis* spp. in Pakistan, but surveys were largely incomplete, perhaps reflecting the relatively low pest status of *Heliothis* spp. (101).

From a database of 7717 documents, 60 species of hymenopterous parasites in six families, 61 species of dipterous parasites in four families, and 142 species of predators from eight Insecta and two Arachnida orders were reported as parasites and predators of *H. virescens* and *H. zea* in North, Central, and South America (78). In the United States, the most common egg parasites were *Trichogramma* spp. and the most common larval parasites were *Cardiochiles nigriceps, Microplitis croceipes,* and *Cotesia* (=*Apanteles*) *marginiventris* as well as several species of *Campoletis* and *Hyposotor.* The tachinid *Eucelatoria bryani* attacks late instar larvae, while another tachinid, *Archytas marmoratus,* was cited as a common larval and pupal parasite. The predominant predators included members of the Coleoptera (especially *Hippodamia convergens, Collops* spp., *Coleomegilla maculata,* and *Scymnus* spp.), Hemiptera (especially *Geocoris* spp., *Orius* spp., *Nabis* spp., and *Podisus* spp.), Neuroptera (primarily *Chrysopa* spp.), and the Araneida.

Napompeth (106) cited predators and parasites of *H. armigera* and *H. assulta* from southeast Asia (Indonesia, the Philippines, Thailand, and Vietnam), but did not discuss their frequency of occurrence or level of importance. *Trichogramma* spp. were the only natural enemies cited as commonly occurring in these countries.

The natural enemies of *H. armigera, H. peltigera,* and *H. viriplaca* in western Europe include the parasites *B. brevicornis, Cotesia* (=*Apanteles*) *kazak,* and *Hyposotor didymator* (99). Levels of parasitism were low in southern and eastern Africa during the era of high insecticide usage, but earlier reports indicated that the parasites *Cotesia* (=*Apanteles*) spp., *Cardiochiles* sp., *Gonia bimaculata,* and *Palexorista* spp. were important biocontrol agents (36). Predator species noted as important were *Orius insidiosus, Chrysopa* sp., and certain ants, especially *Pheidole* sp. and *Dorylus* sp.

## Microbial Agents

The most common microbial agents of *Heliothis* spp. in the United States are the bacterium *Bacillus thuringiensis* (Bt), the fungi *Nomuraea rileyi* and *Entomophthora* spp., the protozoa *Nosema heliothidis* and *Vairimorpha necatrix,* and several viral pathogens including the baculovirus *H. zea* nuclear polyhedrosis virus (NPV) (156). High mortality of *H. zea* larvae in sorghum has also been caused by *Entomophthora aulicae* (43). Naturally occurring epizootics by pathogens, particularly *N. rileyi,* have been noted in cotton, corn, sorghum, and soybean (17, 32, 43, 135, 156). However, such epizootics typically occur when host densities are high late in the season, after the need for plant protection from *Heliothis* spp. has passed. Microbial agents of *H. armigera* reported from India include the NPV, cytoplasmic polyhedrosis virus, granulosis virus, Bt, *Beauveria bassiana, Beauveria brongniartii, Metarhizium anisopliae, Nosema* sp., and *Vairimorpha* sp. (58). Most of the microbial agents mentioned appear to be widely distributed, but major emphasis has been on propagating the NPV and Bt to augment other control measures in the field.

## USE OF BIOLOGICAL CONTROL AGENTS

Naturally occurring predators and parasites are important in regulating *Heliothis* populations (see 62, 71, 75). In the absence of insecticides, natural enemies may maintain *Heliothis* populations at subeconomic levels. Efforts to preserve and increase the effectiveness of natural enemies via environmental manipulations, particularly the judicious use of pesticides, is biological control by conservation. When natural enemy numbers are too low, they may be augmented through propagation and release or application. Introduction and establishment of additional natural enemies may also be attempted.

## Conservation

The potential effect of natural enemies, particularly predators and parasites, is often recognized in insect control guides, especially for *Heliothis* control in cotton. However, explicit instructions for incorporating them into decision-making regarding action versus nonaction are generally lacking (67). Regardless, growers cannot be expected to adopt a practice unless they can realize a positive monetary return compared to that realized from alternative technologies or no control. Other factors entering into decision-making include the development of insecticide-resistant populations as well as environmental safety. Attendant regulations governing the use of pesticides may require the adoption of alternative control approaches.

Most American cotton-insect control guides provide a listing of the predators (e.g. *Geocoris* spp., *Orius spp.,* chrysopids, coccinelids, nabids, and spiders) that may be encountered while surveying insect pest infestations. Parasitic insects are usually mentioned but not by name. Often these guides provide picture sheets to illustrate key natural enemies (e.g. 64). Some guides discuss techniques for quantifying predators (e.g. 60, 61), but only certain guides provide instructions for deciding on treatment versus no treatment based on abundance of natural enemies (e.g. 23, 107). Scholer Resource Corporation, a West German project based in the Philippines, has developed a cartoon guidebook *(No. 3: Pesticide Use in IPM)* that includes a pegboard for making such decisions for *Heliothis* control in cotton. As cotton plants are examined, a peg is moved forward if *Heliothis* eggs or larvae are found, but if predators are present the peg may stay in place. It may even be moved backward if no pest stage is present. Position of the pegs at the conclusion of the inspection dictates the action taken.

According to many of the American cotton insect control guides, predators and parasites are best conserved by delaying the application of insecticides during early season (67). A guide from South Carolina (133) perhaps states the concept most succinctly: "Do not apply organophosphates during the early squaring periods, if possible, as they may delay maturity, severely reduce natural enemies, and result in subsequent bollworm outbreaks."

Some of the best evidence available concerning the value of naturally occurring predators and parasites for suppressing *Heliothis* populations has been discovered as a consequence of first eliminating natural enemies with insecticides. History has demonstrated that insecticide overuse results in resurgence of treated insect populations to levels equal to or greater than pretreatment levels and outbreaks of pests other than those against which the insecticides were directed (112).

Induced *Heliothis* buildup where aldicarb was used in-furrow at planting or as a side-dress application in cotton at rates greater than or equal to 1.12 kg of active ingredient per hectare has often been associated with decreased predator numbers (67). According to one American review (130), 50–90% of *Heliothis* spp. eggs and larvae are normally destroyed by naturally occurring predators and parasites in cotton. Predator:egg ratios of at least 2:1 prevented economic damage to cotton by *H. virescens,* and mean predation for eggs within 24 hr after oviposition was 73% (96). The cumulative real mortality attributable to predators during the egg stage and first two larval instars was 99.3% (97). A study in Oklahoma showed that if one or more predators (*Hippodamia convergens* or *Collops* spp.) were present per 0.8 m of row, then damage to cotton was unlikely to occur (158). Reduction in populations of predators following foliar applications of insecticides for control of the cotton fleahopper, *Pseudatomoscelis seriatus,* was correlated with buildup of *Heliothis* populations and subsequent increase in damaged fruit and reduced yield (90).

Using the MOTHZV-2 model (46), survival of late-instar *Heliothis* larvae was predicted based on the effects of different densities of total predators. As the predator populations increased, the number of *Heliothis* larvae and the amount of plant damage decreased and the total production of bolls increased (1).

Most efforts to spare natural enemy populations are directed toward predaceous arthropods, perhaps because they are among the most frequently detected species in field crops owing to their abundance and broad habitat range. However, larval parasites are important in suppressing *Heliothis* populations (56, 62, 71). For example, one half of the August brood of *Heliothis* spp. larvae collected from cotton fields in Texas and Mississippi in 1904 was destroyed by the braconid *M. croceipes* (122). Though this species remained a component of the parasitic insect complex attacking *Heliothis* spp., a survey of the literature reveals very low parasitism levels (4–23%) during the 1960s and 1970s. This decline may have been a consequence of high parasite mortality associated with the extensive use of broad-spectrum insecticides (74). However, high rates of larval parasitism (30–50%), primarily by *M. croceipes,* were recorded in Arkansas cotton fields in 1981–1982 (74). This increased rate of parasitism was attributed to reduced insecticide usage in cotton and the change to more selective insecticide groups for *Heliothis* spp., e.g. methomyl at ovicidal rates, pyrethroids, and chlordimeform. Tolerance of *M. croceipes* adults to these insecticide groups was confirmed in the laboratory (10, 121). Near 100% survival of adults was observed in spray table experiments involving direct exposure of *M. croceipes* to recommended field rates of the pyrethroid fenvalerate or the carbamate thiodicarb (26). *Microplitis croceipes* is (*a*) highly susceptible to phosphorothionate-type chemicals, but relatively tolerant of phosphates; (*b*) highly susceptible to cyclodienes, but relatively and highly tolerant of toxaphene and DDT, respectively; (*c*) tolerant of oxime-type carbamate compounds; and (*d*) highly tolerant of pyrethroids (9).

The ideal insecticides to use in IPM are those ·detoxified by metabolic processes that are well developed in predators and parasites but not in target pests (118). Generally, polyphagous plant-feeding insects such as *Heliothis* spp. have a relatively high level of oxidative detoxifying enzymes, whereas predators and parasites do not. However, predators and parasites do have relatively high levels of esterase enzymes; consequently they are more efficient at hydrolyzing toxic compounds (118). Thus selective insecticides are available for use in IPM, and efforts should be made to employ those materials that control pests yet spare entomophagous arthropod populations to the extent possible.

The average economic value of predators in the Mississippi delta was calculated at $43.47 per hectare, but at only $27.91 per hectare in nondelta areas because of reductions in predator populations due to early-season insecticide applications, primarily targeted at the boll weevil, *Anthonomus grandis* (115). Another indirect measure of natural enemy value may be surmised from the decline (55%) of cotton insecticide expenditures after the boll weevil was eliminated from 8000–16,000 ha of cotton in North Carolina (16). Increased natural enemy populations contributed to the reduced *Heliothis* populations (148).

## Propagation and Release of Natural Enemies

Any effort that results in increased natural enemy density may be considered biological control by augmentation. Such an increase in numbers may be achieved by (*a*) providing supplementary food, nonviable hosts, or alternate prey, (*b*) applying behavior-modifying chemicals, (*c*) capture and redistribution, or (*d*) propagation and release. This section emphasizes the last tactic.

The goal of biological control by augmentation is suppression of insect pests and their damage to acceptable levels. Depending on the natural enemy augmented, the effective management unit selected for an augmentation program could be a single field, a geographically defined plant community, or an entire agroecosystem.

A propagation and release program should proceed in a logical manner, beginning with colonization of the *Heliothis* sp. and a natural enemy and ending with utilization of the natural enemy and economical suppression of *Heliothis*. First the natural enemy must be collected and colonized or cultured. Candidate species should be selected on the basis of definitive attributes. Ability to propagate the natural enemy is a prerequisite to studies on host– or prey–natural enemy interactions and field evaluations. Finally, economic feasibility must be assessed before the technology can be recommended and implemented.

ENTOMOPHAGOUS ARTHROPODS    Knipling (76) developed theoretical models for appraising the value of augmentative releases of predators and parasites for control of pest arthropods, including *Heliothis* spp. The technical feasibility of suppressing *Heliothis* spp. by augmenting parasite or predator populations has been demonstrated. However, results are often inconsistent, and economic feasibility has rarely been shown (75). The difficulty of mass-producing predators and parasites at a cost competitive with that of other control strategies is a major factor limiting use of the augmentation strategy (70, 73). Consequently *Trichogramma* spp. and *Chrysopa* spp. have been emphasized because they can be mass-propagated.

The use of *Trichogramma* spp. in augmentative releases to control lepidopterous pests including *Heliothis* spp. has been reviewed (126). The state of the art for controlling *Heliothis* spp. in cotton by augmentative releases of *Trichogramma pretiosum* was demonstrated in a 3-yr pilot test in the United States (68). *Trichogramma* spp. are presently the most commonly augmented entomophagous arthropods in the world (68), particularly in the Soviet Union (eight million hectares) (152) and the People's Republic of China (two million hectares) (85).

In the United States, 60% parasitism and 1-mo control of *Heliothis* sp. in cotton was achieved after three releases of 494,000 parasites per hectare per release (88). Aerial release of 124,000–247,000 parasites per hectare per release resulted in 51% egg parasitism on five Texas cotton farms (129). In another study (143), *Heliothis* larval populations in cotton were suppressed by release of *T. pretiosum* at rates of 957,000 per hectare. *Trichogramma pretiosum* releases in 1981–1982 in Arkansas and in 1983 in North Carolina resulted in increased egg parasitism in release plots as compared to untreated controls. In 1983, release plots yielded more than the controls but less than insecticide-treated plots (69). A mean egg parasitism rate of 47% by *T. pretiosum* augmented in Australia for control of *H. armigera* in cotton was insufficient to provide adequate control (147). On the other hand, releases of *Trichogramma dendrolimi* and *Trichogramma confusum* on 8000 ha of cotton during 1977–1980 in three China provinces at rates of 215,000–640,000 parasites per hectare resulted in 60–91% egg parasitism with 70–98% *Heliothis* suppression (85). Experiments in Hebei province, China (C. F. Sheng, personal communication) involved releasing *Trichogramma* six to seven times on third generation *H. armigera* and four to eight times on fourth generation insects. A total of 1.6–3 million parasites were released per hectare. This resulted in higher suppression compared to the check, and natural enemy populations were 8–20-fold higher. Other countries reportedly using *Trichogramma* for control of *Heliothis* include Colombia, Peru, and Mexico (47, 59, 66, 79, 132). The Scholer Resource Corporation has developed a booklet *(No. 4: TRICON-Farmer's Biological Pest Control)* that describes a cottage industry for rearing and releasing *Trichogramma* spp. in cotton for *Heliothis* control.

There is limited experimental evidence to support the commercial use of *Trichogramma* spp. for *Heliothis* control. Factors that influence the effectiveness of the augmented parasites include numbers released, density of the pest, species or strain of *Trichogramma* released, vigor of the released parasite, method of distribution, crop phenology, number of other natural enemies present, and proximity of insecticide use (68).

The predator *Chrysopa carnea* is widely augmented for control of arthropod pests but not for *Heliothis* control in cotton. However, good experimental data (91, 127, 128) indicate that at the release rates employed (near $10^6$ ha$^{-1}$), *Heliothis* spp. could be suppressed below threshold levels. Based on these studies, Ridgway & Jones (128) concluded that release of 123,500 *C. carnea* larvae per hectare might provide effective control of *Heliothis* spp. However, at the price of $12.50 per 1000 larvae, control costs could amount to nearly $400 per hectare (70).

Other augmentatively released predators that have suppressed *Heliothis* spp. under caged conditions include *Nabis americoferus, Geocoris punctipes*, and *Podisus maculiventris* (91, 92, 149). The predator *Jalysus spinosus* was tested under field conditions for control of *H. virescens* and *Manduca sexta* in tobacco (25). Wasp predators have also been indicated as suppressants of *Heliothis* spp. (34, 81). A study in China reported that *Polistes* spp., when introduced in colonies of about 1500 wasps per hectare, reduced *H. armigera* larval populations by 70–80% within 5–7 days after nest transfer (55). Li et al (86) also reported that *Polistes antennalis* wasps could be used to control *H. armigera* in cotton.

Larval parasites have also shown potential in suppressing *Heliothis* spp. after augmentative releases. *Campoletis sonorensis* was released in field cages in cotton at the rate of 680 wasps per day for 10 consecutive days, and 85% of the H. *virescens* larvae were parasitized over nine consecutive weeks (87). In field cage experiments in cotton, 50% parasitism of *H. virescens* larvae was observed after release of the tachinids *Eucelatoria bryani* and *Palexorista laxa* at the rate of 6175 female flies per hectare (57). Calculations based on a two-year study of naturally occurring populations of *Cardiochiles nigriceps* in cotton indicated that 80% parasitism of *H. virescens* larvae could be achieved with 988–1482 *C. nigriceps* females per hectare (83). Large-scale rearing procedures have been developed for the larval and pupal parasite *Archytas marmoratus*, and mechanically extracted maggots applied to corn reduced *H. zea* adult emergence by 56–82% (39).

Perhaps the most abundant and important larval parasite of *Heliothis* spp. in the United States is *Microplitis croceipes,* a solitary wasp that is specific for *Heliothis* spp. (74). *Microplitis croceipes* prefers to parasitize third-instar larvae. All parasitized instars move and feed less on cotton plants, which results in less fruit damage (49, 50). Release of 15,000 adults in a 4-ha cotton field resulted in 76% parasitization of *Heliothis* larvae (K. R. Hopper, J. A. Powell & E. G. King, unpublished data). Insecticides are available that may be used to spare *M. croceipes* as well as other natural enemy populations (118). A semiochemical governing host-seeking behavior has been isolated and identified (65). Also, mate-finding behavior has been described, and a sex pheromone has been discovered and extracted (27). In summation, *M. croceipes* exhibits many attributes of an effective entomophage: relatively high host search rate, specificity for *Heliothis* spp., adaptability to the environment, and tolerance of certain insecticides. Large-scale rearing procedures have been developed for conducting experiments (120), but testing for area-wide suppression of *Heliothis* spp. by augmented wasps will require mass-production capability.

Efficient and cost-effective methods for rearing entomophagous arthropods must be developed if augmentative releases are to be economically feasible. The rearing of predators and parasites necessitates the rearing of host insects, which is expensive and often complex (72). The cost of media for rearing predators and parasites would probably be minimal compared to the cost of rearing them on natural or unnatural hosts, and dietary ingredients could be monitored to assure adequate nutrition. Some progress has been made in this area. A potassium chloride and magnesium sulfate solution was demonstrated to be an excellent ovipositional stimulant for *T. pretiosum* (111), and eggs collected from the solution and transferred to an artificial diet produced adults. An artificial host egg that contained no insect derivatives supported *Trichogramma* oviposition and development (155). In another study (48), *T. pretiosum* was reared from egg to adult on artificial media. The tachinid *A. marmoratus* was induced to larviposit on acidic solutions of a protein extracted from *H. virescens* frass (110). Rearing of hymenopterous larval endoparasites of *Heliothis* spp. has proven more complex; none have been successfully reared to the adult stage on artificial diets. *Cotesia marginiventris* and *M. croceipes* have been reared on artificial media through the first instar (35). Diets for the predators *Coleomegilla maculata* (3) and *G. punctipes* (19) have been developed. Progress on artificial diets for *C. carnea* has included development of an artificial diet for rearing predator larvae (150), encapsulation of the larval diet (95), and formulation of a nutritious adult diet (42).

Another area vital to augmenting entomophagous arthropod populations is maintaining control characteristics in the insectary-reared population. Selection for individuals suited to laboratory conditions takes place within four to five generations, with a consequent decline in genetic variation (5). Components that define the overall quality of an organism include adaptability, sexual activity, host selection, and mobility (70). Most routine measurements of traits are production oriented; traits more directly related to behavioral performance such as diurnal rhythmicity, flight propensity, genetic variation, and host-finding ability are not routinely measured in augmentation programs.

Development of storage and release techniques is also essential. The shelf life of parasites and predators is brief compared to that of chemical pesticides. Also, the demand for entomophages is seasonal; *Heliothis* control is only required for brief periods in selected crops. Storage of entomophages usually involves retarding development with reduced temperatures, but viability and effectiveness may be reduced (e.g. 102, 144). Thus production costs are inflated, and the viability and performance of the product may be compromised. Mechanized methods have been developed for release of *T. pretiosum* (7) and *C. carnea* (125). The aerial release technology for *Trichogramma* is sufficiently developed for use in commercial release programs.

Provision of supplemental resources such as food to maintain the released or indigenous predator or parasite could improve entomophage performance. A product composed of the yeast *Saccharomyces fragilis* and its whey substrate sprayed on cotton increased the effectiveness of *C. carnea* predation on *H. zea* eggs and larvae (41). In addition, semiochemicals could be used to attract, arrest, retain, or stimulate natural enemies. The chemistry of substances found in moth scales and in the frass, cuticle, and salivary glands of larvae stimulates female parasites to search for *Heliothis* spp. (113). Volatiles from *H. virescens* larvae have been used to attract and increase populations of the tachinid *E. bryani* (108, 109). Preflight conditioning to the plant-host complex improved searching by *M. croceipes* (24). Volatile chemicals present in glanded varieties of cotton had a positive effect on foraging by the ichneumonid *C. sonorensis* that was not produced by glandless or Old World species of cotton (29). Additionally, the attraction of *C. sonorensis* varied with host plant species (28). Thus the plant species upon which the *Heliothis* sp. feeds may affect the response of the entomophage. Moreover, *M. croceipes* is attracted to the windborne odor of *H. virescens* larvae and frass (30), and it responds to nonvolatile kairomones produced by *Heliothis* larvae (38, 65, 82).

MICROBIAL AGENTS     There are presently only two microbial products in the United States that are registered for suppression of *Heliothis:* the *H. zea* NPV and the δ-endotoxin of Bt. Preparations of the fungus *N. rileyi* have shown some promise against lepidopterous pests, including *Heliothis* spp., but commercial development does not appear likely (18). Microbe-based formulations have great potential in IPM programs. They may be used to complement the effects of other biological control agents because of their environmental safety and pest selectivity. Moreover, they are amenable to commercial production. They can be mass-cultured, formulated, packaged, stored, and marketed much like chemical insecticides, and equipment used for application of chemical pesticides is generally suitable for application of microbial agents (31). Despite these advantages, microbial agents are not widely used for control of *Heliothis* spp. in the United States. The δ-endotoxin of Bt was registered for use against lepidopterous larvae, including *Heliothis* spp., in 1961. The *H. zea* NPV was registered as Elcar® for use against *Heliothis* spp. on cotton in 1975 by Sandoz, Inc. By about 1982 Sandoz had ceased to produce Elcar®, and Bt was rarely used for control of *Heliothis* spp.

McKinley (98) tabulated data from Burges (13) on the use of *H. zea* NPV in 150–200 field tests for control of *Heliothis* in cotton, corn, and soybeans. Control of light to moderate *Heliothis* infestations with the virus was comparable to the chemical insecticide standard, but at higher infestations control by the virus was inferior to the standard. Control was better in soybean than in the other crops. Perhaps control on cotton was inferior because the frugivorous (as opposed to defoliating) feeding behavior of *Heliothis* sp. on this crop may have resulted in the insects' ingestion of insufficient quantity of the virus. Additionally, the virus is quickly inactivated on the cotton plant owing to the interaction of ultraviolet light, high temperature, and the moderate to high alkaline conditions that sometimes occur on cotton plant surfaces (157).

Field studies have shown that applications of Bt at dosages of 3.6–7.3 × $10^9$ international units (IU) per 0.4 ha suppressed *Heliothis* larval populations and resulted in increased yield over an untreated check (8, 116, 117). However, the degree of control was generally less than that obtained using effective chemical insecticides.

Cadavers of *H. virescens* infected with *N. rileyi* were applied to soybean plots and produced an epizootic of the fungus 14 days in advance of that occurring in untreated plots (140). *Heliothis zea* populations and pod damage were reduced. Application of a *N. rileyi* conidia-based suspension was also reported to induce an epizootic (54). This preparation was pilot tested, but with variable results (18).

Microbial agents have not been as cost effective in controlling *Heliothis* spp. as the more effective chemical insecticides, particularly pyrethroids. For example, the half-life for *H. zea* NPV on cotton may be as low as 12 hr (12). Moreover, several days usually elapse before a larva dies after ingesting a lethal dose of the microbial agent. In contrast, certain chemical insecticides may have ovicidal as well as larvicidal activity, and most have longer residual activity than microbials. Nevertheless, microbial effectiveness may be enhanced by formulations that inhibit inactivation by UV light (11), addition of feeding or gustatory stimulants such as Coax® or Gustol® (4), and more precise placement of formulations on the plant (138, 139).

## Importation and Establishment

The use of importation and establishment for biological control of *Heliothis* spp. has recently been reviewed (63, 104, 105, 119). Earlier, the probability of establishing new, more effective natural enemies was considered poor except in island-type situations with a nonmigrant *Heliothis* population (21). Nevertheless, this approach must be considered a high priority because of the "long term results and [it] is preferred as no further input is required, biological or chemical, once control has been established. With other methods, regular farmer cooperation is necessary." (37).

There is a striking example of successful establishment of an effective natural enemy on *H. armigera* in New Zealand. Initial attempts to import exotic natural enemies (an ichneumonid, three tachinids, and four *Trichogramma* species) did not result in establishment (14). However, *Cotesia kazak* was imported from Europe and released in 1977, and it was recovered in 1981. Now it has spread throughout the northern half of the North Island, where it has altered the number of *H. armigera* larvae attaining damaging size.

The successful establishment of *C. kazak* was on an island with a continuous, predominantly nonmigrant population of *H. armigera* that was devoid of a natural complex of larval parasites. In the United States, at least 17 species of parasites from seven countries as well as several predators have been imported for establishment on *Heliothis* spp. (63, 119). However, none have become established, though initial recovery was reported for *Microplitis demolitor* from Australia. Introduction and release of *Trichogramma* spp., *C. chlorideae*, *C. marginiventris*, *C. kazak*, and *Hyposotor didymator* in Australia was reviewed (100). Of these species, *T. pretiosum*, *C. chlorideae*, and *C. marginiventris* have been recovered. Beginning in 1969, numerous attempts were initiated to establish new natural enemies of *Heliothis* spp. in India. Of these, *Chelonus blackburni* and *E. bryani* from the United States have reportedly been established, but their population levels are low (105). Strains of Bt and *H. zea* NPV have been exchanged from one country to another, but this has typically been for augmentative applications, not for evaluation for establishment.

Successful importation and establishment of new, more effective natural enemies is dependent on several factors, including detailed knowledge of the ecology of the *Heliothis* species targeted as the new host as well as the natural enemy complex of the donor host. There must be a climatic and ecological match for species to be introduced into a new area, the introduced natural enemy species should have a high reproductive potential and should not super- or multiparasitize, and the developmental time should be synchronized with the target host (15). Other technical problems related to selecting the natural enemy, shipping and clearing through quarantine, and release and evaluation of the organism in the field have been reviewed (119).

In 1982 the International *Heliothis* Biological Control Work Group (affiliated with the International Organization of Biological Control for Noxious Pests and Weeds) was formed. A newsletter, which provides a medium in which biological control researchers can communicate, particularly on identification and exchange of natural enemies, is distributed once or twice per year. Two workshops, "Biological Control of *Heliothis:* Increasing the Effectiveness of Natural Enemies," New Delhi, India, 1985, and "The Role of Biosystematics in Biocontrol of *Heliothis,*" Vancouver, Canada, 1988, were organized by the work group to further the exchange of natural enemies. The former workshop identified the importance of the genera *Cotesia, Bracon, Campoletis, Cardiochiles, Obelonus, Copidosoma, Hyposotor, Microplitis, Telenomus, Trichogramma,* and *Trichogrammatoidea* for exchange programs.

## PROSPECTS

Mitigation of *Heliothis* spp. as pests remains a complex issue. The genetic plasticity of *H. virescens* and *H. armigera* is apparent in their ability to detoxify many synthetic insecticides in addition to the secondary plant metabolites present in their wide array of host plant species. Field corn, wheat, forage crops, and many species of uncultivated plants are generally not managed for *Heliothis* spp. Hence, these hosts serve as refugia for *Heliothis* populations that later disperse to higher-value crops, i.e. vegetables, cotton, and tobacco. Dispersal to and exploitation of a succession of food sources occurs rapidly with adults capable of mesoscale and/or long-range movement. As a consequence of this movement, entomophagous arthropods may be unable to effect timely biological control.

Strategies for managing *Heliothis* spp. vary among crops, within and among agroecosystems, and over time. Usage patterns and modes of action of chemical insecticides (including available and newly developed ones) will continue to have primary impact on the effectiveness of natural enemies for regulating *Heliothis* populations. Generally, host-plant resistance and biological control are viewed as complementary approaches to pest suppression (123). However, evidence indicates that there are important situations in which they are not compatible. For example, the number of predatory arthropods on resistant cotton lines was reduced 68% compared with populations on a standard variety (103). Parasitism of eggs and larvae tended to decline with increasing *Heliothis* resistance in genotypes of chick-pea and pigeon pea (136, 137). In contrast, increased parasitism by *C. sonorensis* of *H. virescens* larvae feeding on resistant cotton lines was reported (89). Thus compatibility of biological control and host-plant resistance cannot be predicted, but may best be evaluated under typical field conditions and grower practices (123).

Host-plant resistance should optimally be viewed in the context of habitat manipulation to increase the effectiveness of natural enemies. This subject has been comprehensively reviewed for *Heliothis* spp. (62, 142). Interacting factors that influence natural enemy effectiveness are fertilization, canopy enhancement, chemical properties of the host plant, provision of alternative foods, polyculture, and weed management.

In view of the vast array of biotic and abiotic factors affecting *Heliothis* populations and their natural enemies, the development of computer-based decision-making technology is imperative. Ideally, intergenerational models and data bases should be developed to monitor and assess the regional occurrence and densities of *Heliothis* spp. This information would include surveillance and prediction technology to assess *Heliothis* movement among fields and across regions. The role of alternate hosts on the dynamics of *Heliothis* would also be considered. Realistically, intragenerational (short time horizon) models have more immediate use potential. Hopper & Stark (51) have developed a model that has structures for *Heliothis* feeding, the impact of natural enemies on *Heliothis* feeding and survival, cotton plant growth, mortality of the *Heliothis* sp. and entomophagous arthropods from insecticides, and the economics of insecticide inputs and returns. For input, this model requires updated information on certain insect densities and crop status. It generates output comparing the net economic returns of intervention (different insecticides and dosages) and nonaction. More recently, the model TEXCIM (W. L. Sterling, A. Hartstack & A. Dean, personal communication) has been field-tested and validated. It incorporates several submodels, including models of the cotton fleahopper, *Heliothis* spp., cotton growth, insecticides, natural enemies, and economics, which interact to predict accurately the number of pests and damaged fruit 5–10 days into the future. It is only through this type of technology that full advantage can be taken of all available information, including knowledge of natural enemy densities, each time a decision is made to treat or not treat a *Heliothis* population.

The cost of mass-rearing entomophagous arthropods of a consistently high quality presently restricts biological control by augmentative releases to the use of *Trichogramma* spp. Greater research efforts must be directed toward improving systems for in vivo rearing as well as toward the development of artificial diets and in vitro rearing procedures before economic feasibility can become a reality. In conjunction with rearing there is the opportunity to improve entomophages genetically through selection for key attributes, e.g. insecticide tolerance. Modeling provides a framework for better understanding the complex interactions among the host or prey, the augmented entomophage, and their biotic and abiotic environment. For each augmentation program, the relationship between the numbers of entomophages released and their impact on the *Heliothis* pest population as well as on the commodity protected must be assessed and documented. Behavior modification with semiochemicals may provide an additional tool for monitoring entomophage populations and for managing dispersal and prey or host finding. Improved understanding of these processes as well as of the effects of spatial heterogeneity of the target agroecosystem on the entomophage and the *Heliothis* pest should allow for more predictable performance by the entomophage. Development of the above technology should emphasize the egg-parasite *Trichogramma* spp., the braconid larval endoparasite *M. croceipes*, the braconid larval ectoparasite *B. brevicornis*, the tachinid larval parasite *E. bryani*, the tachinid larval and pupal parasite *A. marmoratus*, and the predators *Chrysopa* spp., *C. maculata*, and *Geocoris* spp.

The effectiveness of microbial agents must be increased if they are to compete in the marketplace. Through conventional and biotechnological genetic engineering it may be possible to develop and select more potent and stable strains of the *Heliothis* NPV and Bt that affect a wider range of hosts. Application of sufficient quantities of the microbial agent at a site near the feeding larva requires production research to reduce costs of the material. On the other hand, the development of unique formulation and application systems may be bypassed by genetically engineering the plant to produce toxins such as that for Bt. Tomato plants that contain this trait are now being field-tested (D. A. Fischoff, personal communication).

There are opportunities to import and establish new and more effective natural enemies of *Heliothis*, as with the establishment of *C. kazak* on *H. armigera* in New Zealand. However, sufficiently favorable conditions are rare, and under most circumstances *Heliothis* life stages are already attacked by an array of natural enemies. Thus biological control of *Heliothis* spp. in most parts of the world will necessarily continue to be considered as only one component of an overall IPM and crop-management strategy that emphasizes timely and economic production of affected commodities.

## Literature Cited

1. Ables, J. R., Goodenough, J. L., Hartstack, A. W., Ridgway, R. L. 1983. Entomophagous arthropods. See Ref. 131, pp. 103–27
2. Agricultural Research Service. 1976. *ARS Natl. Heliothis Planning Conf., New Orleans, La.* Washington, DC: US Dep. Agric. 36 pp.
3. Attalah, Y. H., Newsom, L. D. 1966. Ecological and nutritional studies on *Coleomegilla maculata* DeGeer (Coleoptera: Coccinellidae). I. The development of an artificial diet and a laboratory rearing technique. *J. Econ. Entomol.* 59:1173–79
4. Bell, M. R., Romine, C. L. 1980.

Tobacco budworm field evaluation of microbial control in cotton using *Bacillus thuringiensis* and a nuclear polyhedrosis virus with a feeding adjuvant. *J. Econ. Entomol.* 73:427–30
5. Boller, E. F. 1972. Behavioral aspects of mass rearing of insects. *Entomophaga* 17:9–25
6. Bottrell, D. E., Adkisson, P. L. 1977. Cotton insect pest management. *Ann. Rev. Entomol.* 22:451–81
7. Bouse, L. F., Morrison, R. K. 1985. Transport, storage, and release of *Trichogramma pretiosum. Southwest. Entomol.* 8:36–48 (Suppl.)
8. Bull, D. L., House, V. S., Ables, J. R.,

Morrison, R. K. 1979. Selective methods for managing insect pests of cotton. *J. Econ. Entomol.* 72:841–46

9. Bull, D. L., King, E. G., Powell, J. E. 1988. Effects and fate of selected insecticides after application to *Microplitis croceipes*. *Southwest. Entomol. Suppl.* In press

10. Bull, D. L., Pryor, N. W., King, E. G. Jr. 1987. Pharmacodynamics of different insecticides in *Microplitis croceipes* (Hymenoptera: Braconidae), a parasite of lepidopteran larvae. *J. Econ. Entomol.* 80:739–46

11. Bull, D. L., Ridgway, R. L., House, V. S., Pryor, N. W. 1976. Improved formulation of the *Heliothis* nuclear polyhedrosis virus. *J. Econ. Entomol.* 69:731–36

12. Bullock, H. R. 1967. Persistence of *Heliothis* nuclear-polyhedrosis virus on cotton foliage. *J. Invertebr. Pathol.* 9:434–36

13. Burges, H. D. 1981. *Microbial Control of Pests and Plant Diseases, 1970–1980*. New York: Academic

14. Cameron, P. J., Valentine, E. W. 1988. Importation and establishment of new natural enemies of *Heliothis* into New Zealand. See Ref. 71, In press

15. Carl, K. P. 1988. Attributes of effective natural enemies, including identification of natural enemies for introduction purposes. See Ref. 71, In press

16. Carlson, G. A., Suguiyama, L. 1985. Economic evaluation of area-wide cotton insect management: boll weevils in the southeastern United States. *NC Agric. Res. Serv. Bull. 473.* 24 pp.

17. Carner, G. R. 1980. Sampling pathogens of insect pests. In *Sampling Methods in Soybean Entomology*, ed. M. Kogan, D. C. Herzog, pp. 559–74. New York: Springer-Verlag

18. Carner, G. R., Yearian, W. C. 1988. Development and use of microbial agents for control of *Heliothis* spp. in the USA. See Ref. 71, In press

19. Cohen, A. C., Urias, N. M. 1986. Meat based artificial diets for *Geocoris punctipes*. *Southwest. Entomol.* 11:171–76

20. Common, I. F. B. 1953. The Australian species of *Heliothis* (Lepidoptera: Noctuidae) and their pest status. *Aust. J. Zool.* 1:319–44

21. Commonwealth Institute of Biological Control. 1978. Possibilities of biological control of *Heliothis armigera* and *H. zea*. *Status Pap. 13*, Commonw. Inst. Biol. Control, Slough, UK. 12 pp.

22. Coulson, J. R., ed. 1981. *Proc. Joint Am.-Soviet Conf. Use Beneficial Organ. Control Crop Pests, Washington, DC,* 1979. College Park, Md: Entomol. Soc. Am. 62 pp.

23. Drees, B. M. 1984. Management of cotton insects in south and east Texas counties. *Tex. Agric. Ext. Serv. B-1204.* 16 pp.

24. Drost, Y. C., Lewis, W. J., Zannen, P. O., Keller, M. A. 1986. Beneficial arthropod behavior mediated by airborne semiochemicals. I. Flight behavior and preflight handling of *Microplitis croceipes* (Cresson). *J. Chem. Ecol.* 12:1247–62

25. Elsey, K. D. 1975. *Jalysus spinosus:* Increased numbers produced on tobacco by early-season releases. *Tob. Sci.* 19:13–15

26. Elzen, G. W., O'Brien, P. J., Snodgrass, G. L., Powell, J. E. 1987. Susceptibility of the parasitoid *Microplitis croceipes* (Hymenoptera: Braconidae) to field rates of selected cotton insecticides. *Entomophaga* 31:545–50

27. Elzen, G. W., Powell, J. E. 1988. Mating behavior and sex pheromone response of the *Heliothis* parasitoid *Microplitis croceipes*. *Proc. Beltwide Cotton Prod. Res. Conf., New Orleans, La.* Memphis, Tenn: Natl. Cotton Counc. Am. In press

28. Elzen, G. W., Williams, H. J., Vinson, S. B. 1983. Response by the parasitoid *Campoletis sonorensis* (Hymenoptera: Ichneumonidae) to chemicals (synomones) in plants: implications for host habitat location. *Environ. Entomol.* 12:1873–77

29. Elzen, G. W., Williams, H. J., Vinson, S. B. 1986. Wind tunnel flight responses by hymenopterous parasitoid *Campoletis sonorensis* to cotton cultivars and lines. *Entomol. Exp. Appl.* 42:285–89

30. Elzen, G. W., Williams, H. J., Vinson, S. B., Powell, J. E. 1987. Comparative flight behavior of parasitoids *Campoletis sonorensis* and *Microplitis croceipes*. *Entomol. Exp. Appl.* 45:175–80

31. Falcon, L. A. 1985. Development and use of microbial insecticides. See Ref. 52, pp. 229–42

32. Gaugler, R. R., Brooks, W. M. 1975. Sublethal effects of infection by *Nosema heliothidis* in the corn earworm *Heliothis zea*. *J. Invertebr. Pathol.* 26:57–63

33. Georgheiou, G. P. 1986. The magnitude of the resistance problem. In *Pesticide Resistance: Strategies and Tactics for Management*, pp. 14–43. Washington, DC: Natl. Acad. Sci. 471 pp.

34. Gillaspy, J. E. 1979. Management of *Polistes* wasps for caterpillar predation. *Southwest. Entomol.* 4:334–52

35. Greany, P. 1984. Insect parasitoids:

Finding new opportunities for biological control. *BioScience* 34:690–96

36. Greathead, D. J., Girling, D. J. 1988. Distribution and economic importance of *Heliothis* in southern and eastern Africa including a listing and assessment of the importance of their natural enemies and host plants. See Ref. 71, In press

37. Greathead, D. J., Waage, J. K. 1983. Opportunities for biological control of agricultural pests in developing countries. *World Bank Tech. Pap. 11.* 44 pp.

38. Gross, H. R. Jr., Lewis, W. J., Jones, R. L., Nordlund, D. A. 1975. Kairomones and their use for management of entomophagous insects: III. Stimulation of *Trichogramma achaeae, T. pretiosum,* and *Microplitis croceipes* with host-seeking stimuli at time of release to improve their efficiency. *J. Chem. Ecol.* 1:431–38

39. Gross, H. R. Jr., Pair, S. D., Layton, R. C. 1985. *Archytas marmoratus* screened cage performance of mechanically extracted maggots against larval populations of *Heliothis zea* and *Spodoptera frugiperda. J. Econ. Entomol.* 78:1354–57

40. Hackett, D. S., Gatehouse, A. G. 1979. New records of *Helicoverpa fletcheri* Hardwick from the Sudan Gezira and observations on diapause in the "American bollworm". *PANS* 25:316–17

41. Hagen, K. S., Hale, R. 1974. Increasing natural enemies through use of supplementary feeding and non-target prey. In *Proc. Summer Inst. Biol. Control Plant Insects Dis.,* ed. F. G. Maxwell, F. A. Harris, pp. 170–81. Jackson, Miss: Univ. Press Miss. 647 pp.

42. Hagen, K. S., Tassan, R. L. 1970. The influence of Food Wheast® and related *Saccharomyces fragilis* yeast products on the fecundity of *Chrysopa carnea. Can. Entomol.* 102:806–11

43. Hamm, J. J. 1980. Epizootics of *Entomophthora aulicae* in lepidopterous pests of sorghum. *J. Invertebr. Pathol.* 36:60–63

44. Hardwick, D. F. 1965. *The Corn Earworm Complex. Mem. Entomol. Soc. Can.* Vol. 40. 247 pp.

45. Hariri, G. 1982. The problems and prospects of *Heliothis* management in southwest Asia. See Ref. 56, pp. 369–74

46. Hartstack, A. W., Witz, J. A. 1983. Models for cotton insect pest management. See Ref. 131, pp. 359–81

47. Hassan, S. A., ed. 1984. Trichogramma *News.* Braunschweig, FRG: Fed. Biol. Res. Cent. Agric. For. Messeweg. 27 pp.

48. Hoffman, J. D., Ignoffo, C. M., Dickerson, W. A. 1975. In vitro rearing of the endoparasitic wasp, *Trichogramma pretiosum. Ann. Entomol. Soc. Am.* 68:335–36

49. Hopper, K. R., King, E. G. 1984. Preference of *Microplitis croceipes* (Hymenoptera: Braconidae) for instars and species of *Heliothis* (Lepidoptera: Noctuidae). *Environ. Entomol.* 13:1145–50

50. Hopper, K. R., King, E. G. 1984. Feeding and movement on cotton of *Heliothis* species (Lepidoptera: Noctuidae) parasitized by *Microplitis croceipes* (Hymenoptera: Braconidae). *Environ. Entomol.* 13:1654–60

51. Hopper, K. R., Stark, S. B. 1987. A simulation model for making decisions about *Heliothis* control. *Proc. Beltwide Cotton Prod. Res. Conf., Dallas, Tex.,* pp. 286–90. Memphis, Tenn: Natl. Cotton Counc. Am.

52. Hoy, M. A., Herzog, D. C., eds. 1985. *Biological Control in Agricultural IPM Systems.* Orlando, Fla: Academic. 589 pp.

53. Ibrahim, A., Fayad, Y. 1988. Distribution and economic importance of *Heliothis* in Egypt including a listing of the importance of their natural enemies and host plants. See Ref. 71, In press

54. Ignoffo, C. M., Marston, N. L., Hostetter, D. L., Puttler, B. 1976. Natural and induced epizootics of *Nomuraea rileyi* in soybean caterpillars. *J. Invertebr. Pathol.* 27:191–98

55. Institute of Agricultural and Forestry Sciences of Shang-Chiu. 1976. A preliminary study on the bionomics of hunting wasps and their utilization in cotton insect control. *Acta Entomol. Sin.* 19:303–8

56. International Crops Research Institute for the Semi-Arid Tropics. 1982. *Proc. Int. Workshop* Heliothis *Manage., 1981.* Patancheru, India: ICRISAT. 418 pp.

57. Jackson, C. G., Bryan, D. E., Neemann, E. G., Wardecker, A. L. 1970. Results of field cage tests with parasites of *Heliothis* spp. *3rd Q. Rep. Cotton Insects Biol. Control Invest., Tucson, Arizona.* 5 pp.

58. Jayaraj, S., Rabindra, R. J., Narayanan, K. 1988. Development and use of microbial agents for control of *Heliothis* in India. See Ref. 71, In press

59. Jimenez, E. 1980. Review of some interesting developments. (3.2) Plant protection, Mexico. *Int. Organ. Biol. Control Newsl.* 15:5

60. Johnson, D. R., Kimbrough, J. J., Wall, M. L., Bonner, C. M. 1985. Cotton

insect management. *Univ. Arkansas Coop. Ext. Serv. Leaflet 52.* 2 pp.

61. Johnson, F. A., Donahoe, M. 1983. Cotton insect control. *Univ. Fla. Coop. Ext. Serv. Ext. Entomol. Rep. 61.* 13 pp.

62. Johnson, S. J., King, E. G., Bradley, J. R. Jr., eds. 1986. *Theory and Tactics of* Heliothis *Population Management. I. Cultural and Biological Control. South. Coop. Ser. Bull.* Vol. 316. 161 pp.

63. Johnson, S. J., Pitre, H. N., Powell, J. E., Sterling, W. L. 1986. Control of *Heliothis* spp. by conservation and importation of natural enemies. See Ref. 62, pp. 132–54

64. Jones, F. G., Kowalski, E. 1985. Cotton insect control for 1985. *Univ. Mo.-Columbia Ext. Div. 4252.* 4 pp. (Revised)

65. Jones, R. L., Lewis, W. J., Bowman, W. C., Beroza, M., Bierl, B. A. 1971. Host-seeking stimulant for parasite of corn earworm: isolation, identification, and synthesis. *Science* 173:842–43

66. Kamalov, K. 1982. Possible protection without insecticides. *Zashch. Rast.* 7:5–6 (Abstr.)

67. King, E. G. 1986. Insecticide use in cotton and the value of predators and parasites for managing *Heliothis. Proc. Beltwide Cotton Prod. Res. Conf., Las Vegas, Nev., 1985,* pp. 155–62. Memphis, Tenn: Natl. Cotton Counc. Am.

68. King, E. G., Bull, D. L., Bouse, L. F., Phillips, J. R. 1985. Introduction: biological control of *Heliothis* spp. in cotton by augmentative releases of *Trichogramma. Southwest. Entomol. Suppl.* 8:1–10

69. King, E. G., Coleman, R. J., Phillips, J. R., Dickerson, W. A. 1985. *Heliothis* spp. and selected natural enemy populations in cotton: a comparison of three insect control programs in Arkansas (1981–82) and North Carolina (1983). *Southwest Entomol. Suppl.* 8:71–98

70. King, E. G., Hopper, K. R., Powell, J. E. 1985. Analysis of systems for biological control of crop arthropod pests in the U.S. by augmentation of predators and parasites. See Ref. 52, pp. 201–27

71. King, E. G., Jackson, R. D. 1988. *Increasing the Effectiveness of Natural Enemies. Proc. Int. Workshop Biol. Control* Heliothis, *1985.* New Delhi: Far East. Reg. Res. Off., US Dep. Agric. In . press

72. King, E. G., Leppla, N. C., eds. 1984. *Advances and Challenges in Insect Rearing.* New Orleans, La: US Dep. Agric., Agric. Res. Serv. 306 pp.

73. King, E. G., Morrison, R. K. 1984. Some systems for production of eight entomophagous arthropods. See Ref. 72, pp. 206–22

74. King, E. G., Powell, J. E., Coleman, R. J. 1985. A high incidence of parasitism of *Heliothis* spp. (Lepidoptera: Noctuidae) larvae in cotton in southeastern Arkansas, USA. *Entomophaga* 30:419–26

75. King, E. G., Powell, J. E., Smith, J. W. 1982. Prospects for utilization of parasites and predators for the management of *Heliothis* spp. See Ref. 56, pp. 103–22

76. Knipling, E. F. 1979. *The Basic Principles of Insect Population Suppression and Management. US Dep. Agric. Agric. Handb.* Vol. 512. 623 pp.

77. Knutson, L. 1988. Systematics of *Heliothis* species and their natural enemies as a basis for biological control research. See Ref. 71, In press

78. Kogan, M., Helm, C. G., Kogan, J., Brewer, E. 1988. Distribution and economic importance of *Heliothis virescens* and *Heliothis zea* in North, Central, and South America including a listing and assessment of the importance of their natural enemies and host plants. See Ref. 71, In press

79. Kovalenkov, V. G. 1984. The biomethod in integrated protection of cotton. *Zashch. Rast.* 8:12–14 (Abstr.)

80. Laster, M. L., Goodpasture, C., King, E. G., Twine, P. 1986. Results from crossing the bollworms *Helicoverpa armigera × H. zea* in search of backcross sterility. *Proc. Beltwide Cotton Prod. Res. Conf., New Orleans, La.,* pp. 146–47. Memphis, Tenn: Natl. Cotton Counc. Am.

81. Lawson, F. R., Rabb, R. L., Guthrie, F. E., Bowery, T. G. 1961. Studies on an integrated control system for hornworms on tobacco. *J. Econ. Entomol.* 54:93–97

82. Lewis, W. J., Jones, R. L. 1971. Substance that stimulates host-seeking by *Microplitis croceipes* (Hymenoptera: Braconidae), a parasite of *Heliothis* species. *Ann. Entomol. Soc. Am.* 64:471–73

83. Lewis, W. J., Sparks, A. N., Jones, R. L., Barras, D. J. 1972. Efficiency of *Cardiochiles nigriceps* as a parasite of *Heliothis virescens* on cotton. *Environ. Entomol.* 1:468–71

84. Lewis, W. J., Vinson, S. B. 1971. Suitability of certain *Heliothis* (Lepidoptera: Noctuidae) as hosts for the parasite *Cardiochiles nigriceps. Ann. Entomol. Soc. Am.* 64:970–72

85. Li, L. 1984. Research and utilization of *Trichogramma* in China. *Proc. Chin. Acad. Sci.–US Natl. Acad. Sci. Jt. Symp. Biol. Control Insects,* ed. P. L.

Adkisson, S. Ma, pp. 204–23. Beijing: Science. 445 pp.

86. Li, T. S., Li, C. Z., Wei, J. G. 1984. Biology of *Polistes antennalis* Perez and its use in the control of lepidopterous insects in cotton fields. *Nat. Enemies Insects* 6:101–3

87. Lingren, P. D., 1977. *Campoletis sonorensis:* maintenance of a population on tobacco budworm in a field cage. *Environ. Entomol.* 6:72–76

88. Lingren, P. D., Kim, J. G. 1970. *Inundative releases of* Trichogramma *sp. for control of bollworm and tobacco budworm attacking cotton.* Presented at Ann. Meet. Entomol. Soc. Am., Miami

89. Lingren, P. D., Lukefahr, M. J., Diaz, M. Jr., Hartstack, A. W. Jr. 1978. Tobacco budworm control in caged cotton with a resistant variety, augmentative releases of *Campoletis sonorensis,* and natural control by other beneficial species. *J. Econ. Entomol.* 71:739–45

90. Lingren, P. D., Ridgway, R. L., Cowan, C. B. Jr., Davis, J. W., Watkins, W. C. 1968. Biological control of the bollworm and tobacco budworm by arthropod predators affected by insecticides. *J. Econ. Entomol.* 61:1521–25

91. Lingren, P. D., Ridgway, R. L., Jones, S. L. 1968. Consumption by several common arthropod predators of eggs and larvae of two *Heliothis* species that attack cotton. *Ann. Entomol. Soc. Am.* 61:613–18

92. Lopez, J. D. Jr., Ridgway, R. L., Pinnel, R. E. 1976. Comparative efficacy of four insect predators of the bollworm and tobacco budworm. *Environ. Entomol.* 5:1160–64

93. Ma, S., Ding, Y. 1988. Distribution and economic importance of *Heliothis armigera* in China including a list of their natural enemies. See Ref. 71, In press

94. Manjunath, T. M., Bhatnagar, V. S., Pawar, C. S., Sithanantham, S. 1988. Economic importance of *Heliothis* in India and an assessment of their natural enemies and host plants. See Ref. 71, In press

95. Martin, P. B., Ridgway, R. L., Schuetze, C. E. 1978. Physical and biological evaluations of an encapsulated diet for rearing *Chrysopa carnea*. *Fla. Entomol.* 61:145–52

96. McDaniel, S. G., Sterling, W. L. 1979. Predator determination and efficiency on *Heliothis virescens* eggs in cotton using 32P. *Environ. Entomol.* 8:1083–87

97. McDaniel, S. G., Sterling, W. L., Dean, D. A. 1981. Predators of tobacco budworm larvae in Texas cotton. *Southwest. Entomol.* 6:102–8

98. McKinley, D. J. 1982. The prospects for the use of nuclear polyhedrosis virus in *Heliothis* management. See Ref. 56, pp. 123–36

99. Meierrose, C., Araujo, J., Bues, R., Cabello, T., Mercadier, G., et al. 1988. Distribution and economic importance of *Heliothis* in western Europe including a listing and assessment of the importance of their natural enemies and host plants. See Ref. 71, In press

100. Michael, P. J. 1988. Importation and establishment of new natural enemies of *Heliothis* into Australia. See Ref. 71, In press

101. Mohyuddin, A. I. 1988. Distribution and economic importance of *Heliothis* in Pakistan including a listing and assessment of the importance of their natural enemies and host plants. See Ref. 71, In press

102. Morrison, R. K., Jones, S. L., Lopez, J. D. 1978. A unified system for the production and preparation of *Trichogramma pretiosum* for field release. *Southwest. Entomol.* 3:62–68

103. Musset, K. S., Young, J. H., Price, R. G., Morrison, R. D. 1979. Predatory arthropods and their relationship to fleahoppers on *Heliothis* resistant cotton varieties in southwestern Oklahoma. *Southwest. Entomol.* 4:35–39

104. Nagarkatti, S. 1982. The utilization of biological control in *Heliothis* management in India. See Ref. 56, pp. 159–67

105. Nagarkatti, S., Singh, S. P. 1988. Importation and establishment of new natural enemies of *Heliothis* into India. See Ref. 71, In press

106. Napompeth, B. 1988. Distribution and economic importance of *Heliothis* in southeast Asia including a listing and assessment of the importance of their natural enemies and host plants. See Ref. 71, In press

107. Neeb, C. W., Leser, J. F., Boring, E. P. III, Fuchs, T. W. 1983. Management of cotton insects in the High Plains, Rolling Plains, and Trans-Pecos areas of Texas. *Tex. Agric. Ext. Serv. B-1209.* 26 pp.

108. Nettles, W. C. Jr. 1979. *Eucelatoria* sp. females: Factors influencing response to cotton and okra plants. *Environ. Entomol.* 8:619–23

109. Nettles, W. C. Jr. 1980. Adult *Eucelatoria* sp.: Response to volatiles from cotton and okra plants and from larvae of *Heliothis virescens, Spodoptera eridania,* and *Estigmene acrea. Environ. Entomol.* 9:759–63

110. Nettles, W. C. Jr., Burks, M. L. 1975. A substance from *Heliothis virescens* larvae stimulating larviposition by females of the tachinid, *Archytas marmoratus. J. Insect Physiol.* 21:965–78

111. Nettles, W. C. Jr., Morrison, R. K., Zie, Z. N., Bull, D. L., Shenkir, C. A., et al. 1982. Synergistic action of potassium chloride and magnesium sulfate on parasitoid wasp oviposition. *Science* 218:164–66

112. Newsom, L. D., Brazzel, J. R. 1968. Pests and their control. In *Advances in Production and Utilization of Quality Cotton: Principles and Practices,* ed. F. C. Elliot, M. Hoover, W. K. Porter, Jr., pp. 367–405. Ames, Iowa: Iowa State Univ. Press. 532 pp.

113. Nordlund, D. A., Jones, R. L., Lewis, W. J., eds. 1981. *Semiochemicals: Their Role in Pest Control.* New York: Wiley. 306 pp.

114. Nye, I. W. B. 1982. The nomenclature of *Heliothis* and associated taxa (Lepidoptera: Noctuidae): past and present. See Ref. 56, pp. 3–8

115. Parvin, D. W., Smith, J. W., Cooke, F. T. 1988. Measurement of the impact of natural enemies on *Heliothis* populations. See Ref. 71, In press

116. Pfrimmer, T. R. 1979. *Heliothis* spp.: Control on cotton with pyrethroids, carbamates, organophosphates, and biological insecticides. *J. Econ. Entomol.* 72:593–98

117. Pfrimmer, T. R., Furr, R. E., Stadelbacher, E. A. 1971. Materials for control of boll weevils, bollworms, and tobacco budworms on cotton at Stoneville, Mississippi. *J. Econ. Entomol.* 64:475–78

118. Plapp, F. W. Jr., Bull, D. L. 1988. Modifying chemical control practices to preserve natural enemies. See Ref. 71, In press

119. Powell, J. E. 1988. Importation and establishment of new natural enemies of *Heliothis* into the USA. See Ref. 71, In press

120. Powell, J. E., Hartley, G. G. 1987. Rearing *Microplitis croceipes* (Hymenoptera: Braconidae) and other parasitoids of Noctuidae with multicellular host-rearing trays. *J. Econ. Entomol.* 80:968–71

121. Powell, J. E., King, E. G., Jany, C. S. 1986. Toxicity of insecticides to adult *Microplitis croceipes. J. Econ. Entomol.* 79:1343–46

122. Quaintance, A. L., Brues, C. T. 1905. *The Cotton Bollworm. US Dep. Agric. Bur. Entomol. Bull. No. 50.* 155 pp.

123. Reed, W., Lateef, S. S., Sithanantham, S. 1988. Compatibility of host-plant resistance and biological control of *Heliothis.* See Ref. 71, In press

124. Reed, W., Pawar, C. S. 1982. *Heliothis:* A global problem. See Ref. 56, pp. 9–14

125. Reeves, B. G. 1975. *Design and evaluation of facilities and equipment for mass production and field release of an insect parasite and an insect predator.* PhD thesis. Texas A & M Univ. 180 pp.

126. Ridgway, R. L., Ables, J. R., Goodpasture, C., Hartstack, A. W. Jr. 1981. *Trichogramma* and its utilization for crop protection in the U.S.A. See Ref. 22, pp. 42–48

127. Ridgway, R. L., Jones, S. L. 1968. Field-cage releases of *Chrysopa carnea* for suppression of populations of the bollworm and tobacco budworm on cotton. *J. Econ. Entomol.* 61:892–98

128. Ridgway, R. L., Jones, S. L. 1969. Inundative release of *Chrysopa carnea* for control of *Heliothis* on cotton. *J. Econ. Entomol.* 62:177–80

129. Ridgway, R. L., King, E. G., Carillo, J. L. 1977. Augmentation of natural enemies for control of plant pests in the Western Hemisphere. In *Biological Control by Augmentation of Natural Enemies,* ed. R. L. Ridgway, S. B. Vinson, pp. 379–416. New York: Plenum. 480 pp.

130. Ridgway, R. L., Lingren, P. D. 1972. Predaceous and parasitic arthropods as regulators of *Heliothis* populations. *South. Coop. Ser. Bull.* 169:48–56

131. Ridgway, R. L., Lloyd, E. P., Cross, W. H., eds. 1983. *Cotton Insect Management With Special Reference to the Boll Weevil. US Dep. Agric. Agric. Handb.* Vol. 589. 591 pp.

132. Ridgway, R. L., Morrison, R. K. 1985. Worldwide perspective on practical utilization of *Trichogramma* with special reference to control of *Heliothis* on cotton. *Southwest. Entomol. Suppl.* 8:190–98

133. Roof, M. E., Manley, D. G., Chapin, J. W. 1985. Cotton insect management. *Clemson Univ. Coop. Ext. Serv. Inf. Card 97.* 4 pp. (Revised)

134. Schneider, J. C., Hammond, A. M., Jackson, D. M., Mitchell, E. R., Roush, R. T., eds. 1987. *Theory and Tactics of Heliothis Population Management: II— Insecticidal and Insect Growth Regulator Control. South. Coop. Ser. Bull.* Vol. 329. 54 pp.

135. Schwehr, R. D., Gardner, W. A. 1982. Disease incidence in fall armyworm

and corn earworm populations attacking grain sorghum. *J. Ga. Entomol. Soc.* 17:38–46

136. Sithanantham, S., Rao, V. R., Reed, W. 1982. The influence of host-plant resistance in chickpea on parisitism of *Heliothis armigera* Hb. larvae. *Int. Chickpea Newsl.* 6:21–22

137. Sithanantham, S., Rao, V. R., Reed, W. 1983. Influence of pigeonpea resistance to *Heliothis* on the natural parasitism of *Heliothis* larvae. *Int. Pigeonpea Newsl.* 2:64–65

138. Smith, D. B., Hostetter, D. L., Ignoffo, C. M. 1977. Ground spray equipment for applying *Bacillus thuringiensis* suspension on soybeans. *J. Econ. Entomol.* 70:633–37

139. Smith, D. B., Hostetter, D. L., Ignoffo, C. M. 1978. Formulation and equipment effects on application of a viral *(Baculovirus heliothis)* insecticide. *J. Econ. Entomol.* 71:814–17

140. Sprenkel, R. K., Brooks, W. M. 1975. Artificial dissemination and epizootic initiation of *Nomuraea rileyi*, an entomogenous fungus of lepidopterous pests of soybeans. *J. Econ. Entomol.* 68:847–51

141. Stadelbacher, E. A., Graham, H. M., Harris, V. E., Lopez, J. D., Phillips, J. R., et al. 1986. *Heliothis* populations and wild host plants in the southern U.S. See Ref. 62, pp. 54–74

142. Stinner, R. E., Bradley, J. R. Jr. 1988. Habitat manipulation to increase effectiveness of predators and parasites. See Ref. 71, In press

143. Stinner, R. E., Ridgway, R. L., Coppedge, J. R., Morrison, R. K., Dickerson, W. A. 1974. Parasitism of *Heliothis* eggs after field releases of *Trichogramma pretiosum* in cotton. *Environ. Entomol.* 3:497–500

144. Stinner, R. E., Ridgway, R. L., Kinzer, R. E. 1974. Storage, manipulation of emergence, and estimation of numbers of *Trichogramma pretiosum*. *Environ. Entomol.* 3:505–7

145. Todd, E. L. 1978. A checklist of species of *Heliothis* Ochsenheimer (Lepidoptera: Noctuidae). *Proc. Entomol. Soc. Wash.* 80:1–14

146. Twine, P. H. 1988. Distribution and economic importance of *Heliothis* in Australia including a listing of the importance of their natural enemies and host plants. See Ref. 71, In press

147. Twine, P. H., Lloyd, R. J. 1982. Observations on the effect of regular releases of *Trichogramma* spp. in controlling *Heliothis* spp. and other insects in cotton. *Queensl. J. Agric. Anim. Sci.* 39:159–67

148. United States Department of Agriculture. 1981. *Biological Evaluation: Beltwide Boll Weevil/Cotton Insect Management Programs. Appendix A. SEA–AR Staff Rep.*, Sci. Educ. Adm.–Agric. Res. 142 pp.

149. van den Bosch, R., Leigh, T. F., Gonzalez, D. 1969. Cage studies on predators of the bollworm in cotton. *J. Econ. Entomol.* 62:1486–89

150. Vanderzant, E. S. 1973. Improvements in the rearing diet for *Chrysopa carnea* and the amino acid requirements for growth. *J. Econ. Entomol.* 66:336–38

151. Velasco de Stacul, M., Barral, J. M., Orfila, R. N. 1969. Taxonomia, especificidad y caracteres biologicos diferenciados del complejo de especies denominadas "oruga del capullo" del algodon, "oruga de la espiga" del maiz, "oruga del brote" del tabaco y "bolillera" del lino. *Rev. Invest. Agropec.* 6:19–68

152. Voronin, K. E., Grinberg, A. M. 1981. The current status and prospects of *Trichogramma* utilization in the USSR. See Ref. 62, pp. 49–51

153. Wilson, A. G. L. 1982. Past and future *Heliothis* management in Australia. See Ref. 56, pp. 343–54

154. Wolfenbarger, D. A., Bodegas, P. R., Flores, R. 1981. Development of resistance in *Heliothis* spp. in the Americas, Australia, Africa, and Asia. *Bull. Entomol. Soc. Am.* 27:181–85

155. Wu, Z., Qin, J., Li, P. X., Chang, Z. P., Liu, T. M. 1982. Culturing *Trichogramma dendrolimi* in vitro with artificial media devoid of insect material. *Acta Entomol. Sin.* 25:128–35

156. Yearian, W. C., Hamm, J. J., Carner, G. R. 1986. Efficacy of *Heliothis* pathogens. See Ref. 71, pp. 92–103

157. Yearian, W. C., Young, S. Y. 1974. Persistence of *Heliothis* NPV on cotton plant parts. *Environ. Entomol.* 3:1035–36

158. Young, J. H., Willson, L. J. 1984. A model to predict damage reduction to flower buds or fruit by *Heliothis* spp. in the absence or presence of two Coleoptera predators. *Southwest. Entomol.* 6:33–38

159. Zalucki, M. P., Dalglish, G., Firempong, S., Twine, P. H. 1986. The biology and ecology of *Heliothis armigera* (Hübner) and *H. punctigera* Wallengren (Lepidoptera: Noctuidae) in Australia: What do we know? *Aust. J. Zool.* 34:779–814

Ann. Rev. Entomol. 1989. 34:77–96

# NEUROTOXIC ACTIONS OF PYRETHROID INSECTICIDES

*David M. Soderlund and Jeffrey R. Bloomquist*

Department of Entomology, New York Agricultural Experiment Station, Cornell University, Geneva, New York 14456

## PERSPECTIVES AND OVERVIEW

The insecticidal properties of the pyrethrins, the naturally occurring insecticides found in pyrethrum flowers and various extracts, have been recognized for at least two centuries (17, 71, 90). A number of synthetic analogs of the pyrethrins (called pyrethroids) were made and tested from 1940 to 1970 (38, 41). A few of these compounds, such as allethrin, tetramethrin, and resmethrin, exhibited excellent insecticidal activity and were developed as commercial insecticides for household, stored-product, and veterinary use. However, all shared the environmental instability of the natural pyrethroid esters, which prevented their widespread use in agriculture. The announcement by Elliott and coworkers in 1973 (40) of permethrin, the first potent, photostable pyrethroid, renewed the interest of the worldwide agrochemical industry in this class of compounds and catalyzed intense research and development. The numerous photostable pyrethroids that emerged from these efforts have proven to be extremely effective agricultural insecticides and have also displaced earlier pyrethroids and insecticides of other classes in a variety of nonagricultural uses (38, 41, 42). As a consequence, the pyrethroids are now widely recognized as one of four or five major classes of synthetic insecticides.

Several lines of evidence point to the nervous system as the primary locus of pyrethroid action. In insects, pyrethroids rapidly produce signs of intoxication (loss of coordinated movement, periods of convulsive activity, and ultimate paralysis) indicative of an action on the nervous system (90). In mammals, the signs of pyrethroid poisoning are also indicative of an action on the nervous system, and two distinct intoxication syndromes have been

77

0066-4170/89/0101-0077$02.00

described (52, 53, 109). One, produced by pyrethrins and several early synthetic pyrethroid esters, was characterized by whole-body tremors similar to those produced by DDT. In contrast, deltamethrin produced a distinctly different syndrome, characterized by sinuous writhing convulsions (choreoathetosis) accompanied by profuse salivation. Subsequent studies with a wide variety of pyrethroids have shown that compounds such as deltamethrin, cypermethrin, and fenvalerate, which contain the $\alpha$-cyano-3-phenoxybenzyl alcohol moiety, generally produce the choreoathetosis/salivation (CS) syndrome, whereas most noncyano compounds produce the tremor (T) syndrome. Two syndromes of intoxication have also been reported in insects exposed to these groups of compounds (9, 47, 96), but the differences are less clearly defined than those observed with mammals. A general and widely used classification of pyrethroids, based on symptomology in both insects and mammals as well as on effects observed in neurophysiological assays, identifies Type I compounds (a structurally diverse group that produces the T syndrome) and Type II compounds (predominantly $\alpha$-cyano-3-phenoxybenzyl esters that produce the CS syndrome) (18, 47). These classes are not absolute, in that some pyrethroids exhibit intermediate properties (47, 96, 109). During the past decade, substantial effort has been directed at defining the neurotoxic actions of pyrethroids in ways that could account for the different poisoning syndromes produced by Type I and Type II compounds at the level of the whole animal.

The most recent coverage of the literature on pyrethroid mode of action in the *Annual Review of Entomology* appeared in 1982 (6). This chapter summarizes the literature through 1987, with emphasis on studies published since 1981. In particular, we focus on the body of literature that implicates the voltage-sensitive sodium channel of nerve membranes as the principal site of pyrethroid action and on the evidence for and against the involvement of the $\gamma$-aminobutyric acid (GABA) receptor–chloride ionophore complex as a target site for Type II pyrethroids.

## ACTIONS OF PYRETHROIDS ON NERVES

The deleterious effects of pyrethroids on normal nerve function provide an essential context in which to explore molecular mechanisms of action. The earliest electrophysiological studies of the action of pyrethrum on nerves (reviewed in 90) employed extracellular electrodes to record compound nerve action potentials in insect and crayfish ventral nerve cord preparations. Despite methodological limitations, these studies identified the principal features of pyrethrum and pyrethroid poisoning at the level of the whole nerve: the induction of repetitive discharges, measured either as an increase in spontaneous activity or as a volley of action potentials evoked by a single electrical stimulus, followed by conduction block. These findings were confirmed in the first intracellular recording studies (72, 73), in which low concentrations of allethrin prolonged the falling phase of the evoked nerve action potential in cockroach giant fiber preparations and induced repetitive discharges, whereas high concentrations of the same compound reduced the amplitude of the action potential, eventually blocking nerve conduction. These findings were also the first to implicate specifically a modification of the transient sodium conductance mediated by the voltage-sensitive sodium channel as the primary cause of the observed effects of allethrin on nerve action potentials. Subsequent studies with several Type I pyrethroids have confirmed these findings in a variety of axonal preparations (reviewed in 90). More recently, studies comparing Type I and Type II compounds in invertebrate giant axons (26, 70) and frog sciatic nerve (112) have shown that Type I compounds produce repetitive discharges similar to those described for pyrethrins and allethrin, whereas Type II compounds do not produce repetitive discharges, but instead produce stimulus-dependent nerve depolarization and block.

Intracellular recordings have also been used to address the stereospecificity of pyrethroid actions on nerves. Most pyrethroids have two or three asymmetric carbon atoms and therefore exist as mixtures of four or eight optical and geometrical isomers. In both insects (39, 42) and mammals (53), only cyclopropanecarboxylic acid esters having the $R$ absolute configuration at cyclopropane C-1 (e.g. the 1$R$,$trans$ and 1$R$,$cis$ isomers of resmethrin or permethrin) and $\alpha$-cyano-3-phenoxybenzyl esters having the $S$ absolute configuration at C-$\alpha$ (e.g. deltamethrin) are toxic. Pharmacokinetic studies in insects (103, 106) and intracerebral toxicity determinations in mice (51, 61) suggest that toxicity differences between stereoisomers reflect the stereospecificity of the neuronal target. Lund & Narahashi (70) compared several toxic/nontoxic enantiomer pairs in crayfish nerve assays and found that minimum neuroactive concentrations of the nontoxic isomers were at least 100-fold higher than those of the corresponding toxic isomers, thus confirming the stereospecificity of the pyrethroid target site in these preparations.

In addition to having well-characterized effects on nerve axons, pyrethroids also affect other neuronal elements. Sensory structures of both invertebrates and vertebrates are particularly sensitive to pyrethroids, and these preparations often respond differently from axonal preparations. In locust crural nerve sensory fibers (26), housefly larva sensory nerves (85), cockroach cercal nerves (47), the crayfish stretch receptor organ (21, 91), and the frog lateral line organ (112) pyrethroids cause very long spontaneous or elicited trains of high frequency impulses. In comparisons of the action of a variety of pyrethroid structures, two patterns of response were noted. In cockroach (47) and crayfish (21) preparations short sensory bursts were associated with compounds classified as Type I (e.g. allethrin, tetramethrin), whereas in locust (26) and frog (112) preparations prolonged sensory bursts were produced by compounds classified as Type II. Pyrethroids also alter the electrical properties of insect neurosecretory neurons. The neurohemal organ of the stick insect *Carausius morosus* has proven to be a very useful preparation for studying the electrical properties of neurosecretory neurons (84). In this preparation pyrethroids produce prolonged burst discharges similar to those seen in other neurons (81, 83, 84). Pyrethroids also alter the firing pattern and increase the firing rate of nerves innervating the corpora cardiaca of *Rhodnius prolixus* (81, 82). In both *R. prolixus* and *Locusta migratoria*, pyrethroids produce ultrastructural changes in the cells of the corpora cardiaca and cause the release of neurohormones (99–102). This effect may explain the profound diuresis that accompanies pyrethroid intoxication in some insects (19, 103).

Pyrethroids also affect synaptic transmission through an effect on presynaptic nerve terminals. Salgado et al (94, 95) demonstrated that motor nerve terminals associated with insect larval body wall muscles are depolarized by pyrethroids, which results in neurotransmitter release. These effects are measured as an initial increase in the frequency of miniature excitatory postsynaptic potentials recorded in the muscle, followed by blockage of neuromuscular transmission. Abolition of the pyrethroid effect by tetrodotoxin (TTX), a specific blocker of voltage-dependent sodium channels, provided evidence that nerve terminal depolarization was mediated by an effect of pyrethroids on the sodium channel. A recent study of pyrethroid effects on the cercal nerve–giant fiber synapse in the sixth abdominal ganglion of *Periplaneta americana* (59) demonstrated a transient facilitation of synaptic transmission due to enhanced neurotransmitter release, followed by conduction block due to neurotransmitter depletion. The former effect is analogous to that observed at neuromuscular synapses (94, 95). Pyrethroids also induce repetitive firing in frog motor nerve terminals at concentrations that are too low to alter axonic transmission (43, 92, 116).

Evidence for effects of pyrethroids on neurotransmitter-mediated responses at the postsynaptic membrane is more limited. Pyrethroids have no discernible postsynaptic effects on cholinergic receptor function at the frog neuromuscular junction (92, 116), but a recent study indicated that deltamethrin and tralomethrin may alter postsynaptic responses to acetylcholine at the cockroach cercal nerve–giant fiber synapse (59). In the latter assay system, however, postsynaptic effects require much longer to develop than those involving presynaptic depletion of neurotransmitter, so the relevance of postsynaptic effects to the pyrethroid-dependent synaptic block is unclear. The evidence for effects of Type II pyrethroids on the inhibitory GABA receptor–chloride ionophore complex is considered in a separate section below.

Although most observations of the effects of pyrethroids on excitable cells are restricted to neurons, recent studies suggest that some pyrethroids may alter the excitability of vertebrate muscle in addition to affecting the associated motor nerves. Deltamethrin and other Type II and intermediate-syndrome pyrethroids enhanced the muscle twitch response to nerve stimulation, apparently through a direct effect on the muscle (45, 46, 117). Deltamethrin also increased the force of contraction of vertebrate cardiac muscle (7, 44). A series of compounds encompassing both Type I and Type II esters prolonged muscle action potentials in perfused rat diaphragm preparations (117). Similarly, pyrethroids with Type II and intermediate symptomology produced repetitive action potentials in frog pectoralis muscle preparations that were treated with D-tubocurarine to block acetylcholine-mediated synaptic input (92). The significance of the effects of Type II compounds on muscle in terms of acute intoxication is difficult to determine. The effects on rat muscle (45, 46, 117) were measured at intravenous doses close to the $LD_{50}$ and therefore may contribute to intoxication at these doses. There has been a single report of a direct effect of fenpropathrin on locust leg muscle (26), but studies with deltamethrin and fenvalerate in housefly larvae failed to confirm this observation (94).

Finally, any consideration of the actions of pyrethroids on nerves must also include the actions of DDT. Early physiological studies of DDT and pyrethroids on invertebrate axons (reviewed in 74) and on frog peripheral nerves (reviewed in 108) demonstrated that these structurally diverse compounds produced very similar effects. The subsequent design of a series of DDT-pyrethroid hybrid insecticides, which contain elements of structure that are optimal for insecticidal activity in both the DDT and pyrethroid series in a single molecule (57), further suggests a common mode of action for these two insecticide groups. In the following sections we consider DDT analogs and pyrethroids to have similar if not identical modes of action and summarize evidence for their action at a single site.

## ACTIONS ON VOLTAGE-SENSITIVE SODIUM CHANNELS

### Biophysical Studies

The effects of pyrethroids on whole nerves provide evidence for actions at sites in the nervous system involved in the generation of nerve action potentials. In most nerves, voltage-activated sodium and potassium channels conduct, respectively, the transient inward and outward currents that form the nerve action potential. The voltage clamp method allows the observation of the currents carried by these ions under conditions where the voltage-dependent opening and closing of ion channels can be manipulated. An excellent overview of the voltage clamp method and its application to studies of insecticide action has been given by Shankland (98). Applications of the voltage clamp technique to studies of pyrethroid action on invertebrate giant axons and frog nodes of Ranvier have been summarized in several recent reviews (75–78, 110, 111, 113). Early voltage clamp experiments with allethrin showed that both the sodium and potassium conductances were altered but that effects on sodium channel inactivation were most susceptible to modification by this compound. Normally, the sodium channel becomes open (activated) and closed (inactivated) in response to a change in membrane potential with a time course of a few milliseconds. Allethrin selectively delayed inactivation, thereby prolonging the sodium currents observed during depolarizing pulses under voltage clamp conditions; it also produced a residual, slowly decaying sodium current (the "tail current") that is evident after repolarization.

More recent studies have defined the actions of a variety of pyrethroids and DDT analogs in voltage clamp preparations (70, 111, 114, 115). They have shown that the time course of the decay of the tail current following step repolarization differs depending on the structure of the insecticide assayed. Although the absolute values of the time constants for sodium tail currents can vary substantially for any given compound between assay systems, the relative time constants measured for a series of compounds in a single assay system form a consistent pattern. DDT, DDT analogs, and Type I pyrethroids generally produce rapidly decaying tail currents, whereas Type II pyrethroids typically produce prolonged tail currents. The division between these two groups is not clearly delineated, however, and examination of tail currents for a larger group of compounds revealed a continuum of time constant values (70, 111). It is of particular interest that compounds in the middle of this continuum (e.g. cyphenothrin) were identified as those that cause intermediate intoxication syndromes (109). In the only study to date of the actions of combinations of active and inactive isomers on sodium currents under voltage clamp conditions, Lund & Narahashi (69) found that the 1S isomers of tetramethrin were able to antagonize the actions of the neurotoxic 1R isomers indirectly through an action at a different binding site on the sodium channel. These observations are not obviously related to effects of isomer mixtures in toxicity bioassays or studies of pyrethroid actions on whole nerves. Thus, further exploration of the interactions between toxic and nontoxic stereoisomers appears warranted. Recently, pyrethroids have also been found to modify frog muscle sodium channels and produce tail currents of varying duration under voltage clamp conditions (65, 66).

Patch clamp recording techniques offer the opportunity to observe the currents generated by the opening and closing of individual sodium channels in small patches of cell membrane electrically isolated across the fire-polished tip of a glass micropipette (54). To date, this method has been used to examine the actions of only a few pyrethroids on sodium channels in cultured mouse N1E-115 neuroblastoma cells. Under patch clamp conditions, [1R,trans]-tetramethrin produced a population of modified sodium channels for which the kinetics of opening and sodium conductance were unaltered but the distribution of channel open times was greatly prolonged (118). A subsequent study with deltamethrin (23) confirmed the findings with tetramethrin and also showed that this compound slowed the kinetic transitions between conducting and nonconducting channel states. Prolongation of single-channel currents by fenvalerate has also been reported in a preliminary communication (58). The data from patch clamp studies do not alter the view of pyrethroid effects on sodium channels obtained under macroscopic voltage clamp conditions, but instead confirm that pyrethroids selectively alter the kinetics of sodium channel inactivation without greatly affecting other properties.

## Biochemical Studies

Although biophysical studies provide detailed insight into the consequences of sodium channel modification, they are less valuable for describing molecular interactions between insecticides and the sodium channel. They are also technically difficult in mammalian brain and insect central nervous system preparations, where toxicologically significant interactions are expected to occur. Three biochemical strategies are available to study pyrethroid interactions with sodium channels: measurement of $^{22}Na^+$ uptake into cultured cells or isolated presynaptic nerve terminal vesicles (synaptosomes) prepared from brain tissue; measurement of the specific binding of ligands known to interact with binding sites on the sodium channel; and measurement of the stimulated release of radiolabeled neurotransmitter from presynaptic nerve terminals. These methods have been employed in several recent studies to define the interactions of pyrethroids and DDT analogs with sodium channels.

The actions of pyrethroids on sodium channel–mediated sodium uptake were first defined in assays with cultured mouse neuroblastoma (N1E-115 and C9) cells (60). Pyrethroids alone failed to stimulate sodium uptake over control levels, but deltamethrin and kadethrin enhanced the stimulation of sodium uptake produced by veratridine (VTD), batrachotoxin, and dihydrograyanotoxin II, all of which activate sodium channels. However, other neurotoxic pyrethroids such as cismethrin and [1R,cis]-permethrin were ineffective as enhancers of VTD-dependent activation but inhibited the enhancing actions of kadethrin. Polypeptide neurotoxins from scorpion (Leiurus quinquestriatus) and sea anemone (Anemonia sulcata) venoms also activated sodium channels in neuroblastoma cells, and pyrethroids enhanced the stimulation of sodium uptake caused by these compounds as well. A subsequent study of neurotoxic DDT analogs and DDT-pyrethroid hybrid compounds (56) failed to detect any effects on neurotoxin-stimulated sodium uptake in this system.

Sodium flux assays have also been used to explore the interactions of pyrethroids and DDT analogs with sodium channels in mouse brain synaptosomes. Pyrethroids and DDT did not in themselves stimulate sodium uptake, but they enhanced the uptake caused by VTD (51, 104). In an isomer comparison study (51), deltamethrin and its neurotoxic noncyano analog (NRDC 157) enhanced VTD-dependent sodium uptake, but the nontoxic 1$S$ isomers were ineffective. Moreover, the 8.8-fold difference in potency between deltamethrin and NRDC 157 in this assay agrees well with the 10-fold difference in their toxicity to mice after intracerebral injection (51). Cismethrin was also effective as an enhancer of VTD-dependent uptake in synaptosomes, as were DDT and the DDT analog EDO, but DDE and the DDT-pyrethroid hybrid compound GH401 were much less effective (104). The enhancement of uptake by these compounds, some of which were inactive in assays using neuroblastoma cells (56, 60), suggests that the insecticide recognition properties of synaptosomal sodium channels differ from those in neuroblastoma cells. A more detailed study of the interactions of DDT, cismethrin, and deltamethrin with sodium channel activators (11) showed that these insecticides exert bimodal effects in sodium flux assays, enhancing sodium channel activation by VTD and batrachotoxin but inhibiting activation by aconitine. This action is unprecedented among the various types of natural and synthetic toxins known to modify sodium channel activation. The requirement for chemical activation of sodium channels to observe effects of pyrethroids in both neuroblastoma cells and synaptosomes may reflect the selective effects of pyrethroids on sodium channel inactivation identified in biophysical studies. It may also reflect the low abundance of spontaneously opening sodium channels in neuroblastoma cells and synaptosomes under the conditions used.

The allosteric interactions between insecticides and sodium channel activators observed in sodium uptake studies are also found in radioligand binding studies using mammalian brain synaptosomes and postsynaptic vesicles. Deltamethrin, the neurotoxic isomers of cypermethrin, and DDT increased the binding affinity of rat brain synaptosomal sodium channels for [$^3$H]batrachotoxinin A-20-$\alpha$-benzoate (BTX-B), a batrachotoxin analog that labels the activator site on the sodium channel (16, 87). These results are consistent with sodium uptake data (11) and suggest that the binding of BTX-B can serve as an allosteric probe of insecticide–sodium channel interactions. Attempts to define binding sites for radiolabeled pyrethroids in insect and mammalian nerve preparations have been frustrated by the extreme lipophilicity of these compounds, which results in extremely high levels of nonspecific, nonsaturable binding to membranes (22, 105). Thus, binding studies with [$^3$H]NRDC 157 in mouse brain membranes (105) defined a stereospecific site with an affinity for this ligand similar to the concentration of this compound required to produce half-maximal stimulation of sodium uptake into synaptosomes (51). However, the low levels of specific binding encountered in these studies precluded the use of this assay to explore the interactions of other pyrethroids with this site.

Assays of the release of preloaded radiolabeled neurotransmitter from mammalian synaptosomes and brain slices have also been used to characterize the effects of insecticides on presynaptic nerve terminals. Stimulation of the release of [$^3$H]GABA from preloaded guinea pig or rat brain synaptosomes by deltamethrin, cypermethrin, permethrin, and DDT was inhibited by TTX, which implicated an effect on sodium channel–dependent vesicle depolarization (32, 80). Similarly, stimulation of the release of labeled acetylcholine and dopamine from rabbit striatal slices by fenvalerate was completely inhibited by TTX and greatly reduced by the calcium channel blocker nitrendipine (36). In contrast, the stimulation of [$^3$H]norepinephrine release by resmethrin and by a series of substituted benzyl chrysanthemate esters was TTX-insensitive, which implicated an effect on some other process related to neurotransmitter release (33). Brooks & Clark (15) failed to detect any direct insecticide-stimulated release of [$^3$H]norepinephrine from continuously perfused rat brain synaptosomes. However, they detected pyrethroid-dependent enhancement of the release stimulated by a depolarizing pulse of potassium ion in the perfusate. In these studies, insecticide-dependent enhancement of neurotransmitter release was greatest with pyrethroids containing the $\alpha$-cyano-3-phenoxybenzyl moiety and was correlated with calcium uptake into synaptosomes. Since experiments employing TTX or sodium-free media were not reported, it is not clear whether the effects of pyrethroids on calcium uptake indicate a direct effect on calcium channels in the presynaptic terminal or whether they arise from indirect voltage-dependent activation of those channels through a sodium-dependent depolarization. Most of the reported effects of pyrethroids on neurotransmitter release are consistent with an action on presynaptic sodium channels, but actions on calcium channels or other processes related to neurotransmitter release cannot be ruled out.

Efforts to extend biochemical assays of pyrethroid–sodium channel interactions to insect nervous tissue have been limited by the small amounts of tissue available and by the difficulties encountered in preparing and using insect synaptosomes. Intact synaptosomal vesicles can be prepared from homogenates of fly heads and dissected insect ganglia using density gradient centrifugation methods that avoid the hypertonic sucrose solutions routinely used in mammalian synaptosome preparations (12, 34, 35). VTD stimulated the release of [$^3$H]acetylcholine from locust (13) and cockroach (79) ganglionic synaptosomes in a manner analogous to its effect in mammalian synaptosomes. Deltamethrin alone produced a weak stimulation of acetylcholine release from cockroach synaptosomes, but it strongly enhanced the release stimulated by VTD (79). This effect of deltamethrin is analogous to its enhancement of VTD-stimulated sodium uptake and potassium-stimulated neurotransmitter release in mammalian synaptosomes, which suggests that effects on neurotransmitter release from insect synaptosomes result from interactions with presynaptic sodium channels. However, efforts to demonstrate directly the stimulation of $^{22}Na^+$ uptake by VTD in these preparations have been unsuccessful (R. E. Grubs & D. M. Soderlund, unpublished results). Recently, a high-affinity binding site for [$^3$H]BTX-B that exhibits several properties expected of the activator binding site of the sodium channel has been identified in house fly head membrane preparations (D. M. Soderlund, P. M. Adams & R. E. Grubs, manuscript in preparation). However, BTX-B binding in this system is not enhanced by deltamethrin, which precludes the use of this assay to explore allosteric interactions between activators and pyrethroids in these preparations.

## ACTIONS ON THE GABA RECEPTOR–IONOPHORE COMPLEX

During the past several years considerable research effort has been directed at determining whether the GABA receptor–chloride ionophore complex of inhibitory synapses represents an important primary or secondary molecular target for Type II pyrethroids. A toxicologically relevant effect of Type II pyrethroids on the GABA receptor–chloride ionophore complex was first proposed on the basis of in vivo drug-insecticide interactions in cockroaches and mice (49). Pretreatment with diazepam, a benzodiazepine anticonvulsant known to act on the GABA receptor complex, selectively delayed the onset of intoxication of both cockroaches and mice by Type II but not Type I compounds. The pretreatment also delayed the onset of deltamethrin-induced burst discharges in cercal motor nerves. However, diazepam pretreatment produced nonselective effects on acute toxicity in mice, antagonizing the toxicity of deltamethrin (Type II) sixfold but antagonizing the toxicity of [1R,cis]-permethrin (Type I) ninefold.

Radioligand binding studies also support an effect of Type II compounds on the GABA receptor complex. Deltamethrin, but not its nontoxic $\alpha$-R-cyano epimer, inhibited [$^3$H]dihydropicrotoxinin binding to the chloride ionophore component of the rat brain GABA receptor complex (64). The development of [$^{35}$S]t-butylbicyclophosphorothionate (TBPS), an improved radioligand for the GABA-gated chloride channel (107), permitted more extensive studies of the interactions of pyrethroids with this site. Lawrence & Casida (62) demonstrated that only Type II pyrethroids were effective displacers of TBPS binding to rat brain membranes and presented a qualitative correlation between relative TBPS displacement and relative acute toxicity by intracerebral injection. Although this report claimed absolute stereospecificity for neurotoxic isomers, a subsequent study demonstrated that the nontoxic $1S,cis,\alpha S$ isomer of cypermethrin significantly inhibited TBPS binding to native and detergent-solubilized rat brain receptors (97). Other studies (28, 67) confirmed the inhibitory actions of pyrethroids on TBPS binding and showed that pyrethroids specifically affected the binding of ligands to the chloride ionophore site and not the binding of ligands that label sites for GABA or benzodiazepines. Binding sites for TBPS have also been detected in house fly thorax and abdomen (27) and cockroach ganglion (68) preparations. A binding site for a TBPS analog has been detected in house fly head preparations (86). However, Type II pyrethroids do not displace the binding of these ligands to house fly preparations (27, 86).

Although binding studies have demonstrated an interaction of Type II pyrethroids with the mammalian GABA receptor complex, functional assays have suggested that this interaction may be of limited toxicological relevance. Chloride ion flux assays designed to assess the functional coupling of GABA receptors to their associated chloride channels in mammalian brain preparations (55) permit the evaluation of the consequences of pyrethroid interactions with the TBPS binding site of the chloride channel (2, 8, 10). The most detailed study of pyrethroid effects using this assay compared the actions of deltamethrin, its nontoxic enantiomer, and its insecticidal noncyano analog, NRDC 157 (10). The inhibition of GABA-stimulated chloride uptake into mouse brain vesicles by deltamethrin was incomplete, reaching a maximum of approximately 60% inhibition at a concentration of 30 $\mu$M. The enantiomer of deltamethrin, which is at least 500-fold less toxic to mice than deltamethrin after direct injection into the brain (51), was only 10-fold less potent than deltamethrin as an inhibitor of GABA-dependent chloride uptake. As predicted from TBPS binding assays, NRDC 157, a Type I ester, was inactive as an inhibitor of chloride uptake. Although these findings confirm that pyrethroid interactions with the TBPS site can affect GABA receptor–ionophore function, they show that deltamethrin is 1000-fold less potent as an inhibitor of GABA-dependent chloride uptake than as an enhancer of sodium channel activation (51). Moreover, the incomplete stereospecificity exhibited in the inhibition of chloride uptake by deltamethrin and its enantiomer is inconsistent with the absolute stereospecificity observed both in neurotoxicity and in assays of the effects of these compounds on intact nerves and on sodium channels. Functional assays with invertebrate GABA receptor preparations similarly suggest that pyrethroids affect this target only at very high concentrations. Deltamethrin and other Type II pyrethroids antagonized the actions of GABA at the crayfish neuromuscular junction (48), but the concentrations of deltamethrin that affected muscle GABA receptors were several orders of magnitude higher than those that produced profound disruption of crayfish nerves through an effect on voltage-dependent sodium channels (20).

# ACTIONS ON OTHER NEUROCHEMICAL SYSTEMS

## Peripheral-Type Benzodiazepine Receptor

Although pyrethroids do not affect the binding of benzodiazepine radioligands to the benzodiazepine recognition site associated with the GABA receptor complex, they inhibit the binding of the convulsant benzodiazepine Ro5-4864 to a different site, which has been designated the "peripheral-type benzodiazepine receptor" (50, 63). Recently, the ability of both Type I and Type II pyrethroids to interact with this receptor site has been correlated with their ability to act as proconvulsants by reducing the threshold for the production of pentylenetetrazole-induced seizures in rats (30, 31). Both the proconvulsant and receptor-binding effects exhibit appropriate stereospecificity for neurotoxic isomers. It is of interest that maximal proconvulsant effects in vivo are obtained at doses that are well below those required to produce pyrethroid-dependent acute intoxication (31). Ro5-4864 was also reported to delay the onset of intoxication by both deltamethrin and permethrin in cockroaches (50) and to inhibit the binding of [$^3$H]flunitrazepam, a ligand for GABA receptor–associated sites in mammals, to insect nerve and muscle preparations (1, 68). Because the actions of benzodiazepines in insects are difficult to correlate with those in mammals, the relevance of these findings as evidence of interactions of pyrethroids with either the GABA receptor complex or a peripheral-type benzodiazepine receptor in insects is unclear.

## Nicotinic Acetylcholine Receptor

Pyrethroid interactions with the nicotinic acetylcholine receptor of eel electroplax have been explored using as a probe the binding of [$^3$H]perhydrohistrionicotoxin ($H_{12}$-HTX) to a site associated with the acetylcholine-gated ion channel. Pyrethrins, allethrin, resmethrin, and tetramethrin were the most potent and rapid inhibitors of $H_{12}$-HTX binding, whereas permethrin and several $\alpha$-cyano–substituted esters were less potent and exhibited slow association kinetics (3–5). These pyrethroids also inhibited $^{45}Ca^{2+}$ uptake stimulated by carbachol, an acetylcholine analog, which was presumed to represent flux through the relatively nonspecific cation channel of the acetylcholine receptor (4, 5). However, subsequent studies showed no effect of pyrethroids on $^{22}Na^+$ uptake, which suggests that the apparent effect on calcium uptake actually represented an effect on calcium binding to the receptor (37). These findings suggest that pyrethroids interact with the $H_{12}$-HTX binding site of the nicotinic acetylcholine receptor without affecting ion transport.

## ATP-Hydrolyzing Enzymes and Calcium Regulation

Adenosine triphosphate (ATP) has a central role in cell energy metabolism and as the energy source for the ion pumps that maintain ionic gradients across cell and organelle membranes. This role has suggested to some workers that ATP-utilizing enzymes and ion pumps, which are assayed as ion-dependent ATP-hydrolyzing enzymes (ATPases), may be involved in the neurotoxic actions of pyrethroids. Early studies in this area (reviewed in 29) focused on the $Na^+,K^+$-ATPase, the pump that maintains the potassium and sodium ion gradients across the cell membrane of neurons and other cells, and the mitochondrial oligomycin-sensitive $Mg^{2+}$-ATPase, which participates in ATP biosynthesis, as possible targets for the action of DDT. These studies concluded that the latter enzyme was the more sensitive to inhibition by DDT. Recent studies have shown that Type I and Type II pyrethroids also inhibit oligomycin-sensitive $Mg^{2+}$-ATPases (88).

A second group of ATPases implicated in the action of pyrethroids comprises the various $Ca^{2+}$-ATPases, which are thought to be involved in the tight homeostatic regulation of intracellular calcium levels. The rationale for considering calcium regulation as a target of insecticide action and the early studies demonstrating effects of DDT and pyrethroids on $Ca^{2+}$-dependent ATPases have been reviewed elsewhere (6). Clark & Matsumura (24, 25) identified two $Ca^{2+}$-ATPase activities in squid and cockroach nerve preparations that were inhibited by pyrethroids: a $Na^+$-$Ca^{2+}$-ATPase thought to represent an ATP-modulated sodium-calcium exchange transporter, which was more sensitive to inhibition by Type I pyrethroids such as allethrin; and a $Ca^{2+}$ + $Mg^{2+}$-ATPase thought to represent an energy-dependent calcium extrusion system, which was more sensitive to inhibition by Type II pyrethroids such as cypermethrin. Recently, pyrethroids have been shown to inhibit a $Ca^{2+}$-ATPase in rat brain synaptosomes that is stimulated by calmodulin, a calcium-binding protein, without affecting the basal unstimulated activity (93). Comparison of a variety of pyrethroids as inhibitors showed that Type II compounds were more potent than Type I compounds in this assay. These studies also showed that pyrethroids inhibited calmodulin-stimulated adenylate cyclase activity, thereby suggesting that all of the effects in this system were mediated through an interaction with calmodulin rather than with either the ATPase or adenylate cyclase. DDT and pyrethroids have also been reported to inhibit calmodulin stimulation of phosphodiesterase activity (89).

## SUMMARY AND CONCLUSIONS

Pyrethroid insecticides interact with a variety of neurochemical processes, but not all of these actions are likely to be involved in the disruption of nerve function. Several lines of evidence suggest that the voltage-sensitive sodium channel is the single principal molecular target site for all pyrethroids and DDT analogs in both insects and mammals. The alterations of sodium channel functions identified in both biophysical and biochemical studies are directly related to the effects of these compounds on intact nerves. The pyrethroid recognition site of the sodium channel exhibits the stringent stereospecificity predicted by in vivo estimates of intrinsic neurotoxicity in both insects and mammals. Type I and Type II compounds produce qualitatively different effects on sodium channel tail currents, divergent actions on intact nerves, and different effects on the excitability of vertebrate skeletal muscle. Moreover, compounds that are defined as intermediate in the Type I/Type II classification scheme are also intermediate in their effects on sodium channel kinetics. The range of different actions on sensory and motor nerve pathways arising from these qualitatively different effects at the level of the sodium channel appear to be sufficient to explain the distinct poisoning syndromes that have been identified in both insects and mammals. Thus, it does not appear necessary to invoke different primary target sites for Type I and Type II compounds to explain their actions in whole animals.

Although the voltage-sensitive sodium channel is likely to be the principal site of pyrethroid action, it is probably not the only site involved in intoxication. Insect neurosecretory neurons are sensitive to very low concentrations of pyrethroids, and disruption of the neuroendocrine system has been implicated as a factor contributing to the irreversible effects of pyrethroid intoxication in insects. Since action potentials in these nerves are carried by calcium ions through TTX-insensitive voltage-gated cation channels, these findings provide evidence that pyrethroids can alter neuronal excitability through an action on voltage-sensitive channels other than the sodium channel. Actions on voltage-sensitive calcium channels may also be involved in the effects of pyrethroids on neurotransmitter release in mammals. The proconvulsant actions of pyrethroids mediated through the peripheral-type benzodiazepine receptor may also contribute to pyrethroid intoxication. Both Type I and Type II compounds are potent proconvulsants in vivo at doses well below those required to produce pyrethroid-dependent intoxication. This finding suggests that effects of pyrethroids at this site do not constitute a primary neurotoxic lesion or identify a secondary site that is unique to Type II compounds, but they may contribute to the overall intoxication process in mammals. The substantial differences in benzodiazepine pharmacology between insects and mammals make an evaluation of the role of this site in insect intoxication difficult.

The actions of pyrethroids on other neurochemical targets in in vitro assays do not appear to be relevant to the neurotoxic effects of these compounds in vivo. Although Type II pyrethroids bind to the chloride ionophore component of the GABA receptor complex and inhibit GABA-dependent chloride flux, the concentrations required to affect this system are several orders of magnitude higher than those capable of disrupting sodium channel function. Moreover, actions of pyrethroids at the GABA receptor fail to exhibit the stringent stereospecificity predicted from acute neurotoxicity determinations in vivo. In contrast to the actions of pyrethroids on the GABA receptor complex, the binding of pyrethroids to the nicotinic acetylcholine receptor appears to have no effect on the functional properties of the latter. This conclusion is consistent with the failure of physiological studies to identify receptor-mediated postsynaptic effects of pyrethroids (see above) and implies that the binding interactions of pyrethroids at the nicotinic acetylcholine receptor are unrelated to the neurotoxicity of these compounds. Finally, the relationship between the inhibitory effects of DDT and pyrethroids on ATPases and calmodulin and the neurotoxicity of these compounds is obscure. The measured effects of DDT and pyrethroids on $Na^+,K^+$-ATPase and mitochondrial $Mg^{2+}$-ATPase activities do not explain the effects of these compounds on whole nerves (14). Regulation of intraneuronal calcium concentration is more directly related to nerve function, but the modification of calcium homeostasis by pyrethroids cannot directly account for the effects of these compounds on neuronal excitability.

ACKNOWLEDGMENTS

Studies reviewed here originating from the senior author's laboratory were supported in part by grant number ES02160 from the National Institutes of Health.

*Literature Cited*

1. Abalis, I. M., Eldefrawi, M. E., Eldefrawi, A. T. 1983. Biochemical identification of putative GABA/benzodiazepine receptors in house fly thorax muscles. *Pestic. Biochem. Physiol.* 20:39–48
2. Abalis, I. M., Eldefrawi, M. E., Eldefrawi, A. T. 1986. Effects of insecticides on GABA-induced chloride influx into rat brain microsacs. *J. Toxicol. Environ. Health* 18:13–23
3. Abbassy, M. A., Eldefrawi, M. E., Eldefrawi, A. T. 1982. Allethrin interactions with the nicotinic acetylcholine receptor channel. *Life Sci.* 31:1547–52
4. Abbassy, M. A., Eldefrawi, M. E., Eldefrawi, A. T. 1983. Pyrethroid action on the nicotinic acetylcholine receptor/channel. *Pestic. Biochem. Physiol.* 19:299–308
5. Abbassy, M. A., Eldefrawi, M. E., Eldefrawi, A. T. 1983. Influence of the alcohol moiety of pyrethroids on their interactions with the nicotinic acetylcholine receptor. *J. Toxicol. Environ. Health* 12:575–90
6. Beeman, R. W. 1982. Recent advances in mode of action of insecticides. *Ann. Rev. Entomol.* 27:253–81
7. Berlin, J. R., Akera, T., Brody, T. M., Matsumura, F. 1984. The inotropic effects of a synthetic pyrethroid decamethrin on isolated guinea pig atrial muscle. *Eur. J. Pharmacol.* 98:313–22
8. Bloomquist, J. R., Adams, P. M., Soderlund, D. M. 1986. Inhibition of $\gamma$-aminobutyric acid–stimulated chloride flux in mouse brain vesicle by polychlorocycloalkane and pyrethroid insecticides. *NeuroToxicology* 7:11–20
9. Bloomquist, J. R., Miller, T. A. 1985. Carbofuran triggers flight motor output in pyrethroid-blocked reflex pathways of the house fly. *Pestic. Biochem. Physiol.* 23:247–55
10. Bloomquist, J. R., Soderlund, D. M. 1985. Neurotoxic insecticides inhibit GABA-dependent chloride uptake by mouse brain vesicles. *Biochem. Biophys. Res. Commun.* 133:37–43
11. Bloomquist, J. R., Soderlund, D. M. 1988. Pyrethroid insecticides and DDT modify alkaloid-dependent sodium channel activation and its enhancement

by sea anemone toxin. *Mol. Pharmacol.* 33:543–50

12. Breer, H., Jeserich, G. 1980. A microscale flotation technique for the isolation of synaptosomes from nervous tissue of *Locusta migratoria*. *Insect Biochem.* 10:457–63

13. Breer, H., Knipper, M. 1984. Characterization of acetylcholine release from insect synaptosomes. *Insect Biochem.* 14:337–44

14. Brooks, G. T. 1980. Biochemical targets and insecticide action. In *Insect Neurobiology and Pesticide Action (Neurotox 79)*, pp. 41–55. London: Soc. Chem. Ind.

15. Brooks, M. W., Clark, J. M. 1987. Enhancement of norepinephrine release from rat brain synaptosomes by alpha cyano pyrethroids. *Pestic. Biochem. Physiol.* 28:127–39

16. Brown, G. B., Olsen, R. W. 1984. Batrachotoxin-benzoate binding as an index of pyrethroid interaction at $Na^+$ channels. *Soc. Neurosci. Abstr.* 10:865

17. Casida, J. E. 1980. Pyrethrum flowers and pyrethroid insecticides. *Environ. Health Perspect.* 34:189–202

18. Casida, J. E., Gammon, D. W., Glickman, A. H., Lawrence, L. J. 1983. Mechanisms of selective action of pyrethroid insecticides. *Ann. Rev. Pharmacol. Toxicol.* 23:413–38

19. Casida, J. E., Maddrell, S. H. P. 1971. Diuretic hormone release on poisoning *Rhodnius* with insecticide chemicals. *Pestic. Biochem. Physiol.* 1:71–83

20. Chalmers, A. E., Miller, T. A., Olsen, R. W. 1987. Deltamethrin: a neurophysiological study of the sites of action. *Pestic. Biochem. Physiol.* 27:36–41

21. Chalmers, A. E., Osborne, M. P. 1986. The crayfish stretch receptor organ: a useful model system for investigating the effects of neuroactive substances. *Pestic. Biochem. Physiol.* 26:128–38

22. Chang, C. P., Plapp, F. W. Jr. 1983. DDT and pyrethroids: receptor binding and mode of action in the house fly. *Pestic. Biochem. Physiol.* 20:76–85

23. Chinn, K., Narahashi, T. 1986. Stabilization of sodium channel states by deltamethrin in mouse neuroblastoma cells. *J. Physiol.* 380:191–207

24. Clark, J. M., Matsumura, F. 1982. Two different types of inhibitory effects of pyrethroids on nerve Ca- and Ca + Mg-ATPase activity in the squid, *Loligo pealei*. *Pestic. Biochem. Physiol.* 18:180–90

25. Clark, J. M., Matsumura, F. 1987. The action of two classes of pyrethroids on the inhibition of brain Na-Ca and Ca +

Mg ATP hydrolyzing activities of the American cockroach. *Comp. Biochem. Physiol.* C 86:135–45

26. Clements, A. N., May, T. E. 1977. The actions of pyrethroids upon the peripheral nervous system and associated organs in the locust. *Pestic. Sci.* 8:661–80

27. Cohen, E., Casida, J. E. 1986. Effects of insecticides and GABAergic agents on a house fly [$^{35}S$]$t$-butylbicyclophosphorothionate binding site. *Pestic. Biochem. Physiol.* 25:63–72

28. Crofton, K. M., Reiter, L. W., Mailman, R. B. 1987. Pyrethroid insecticides and radioligand displacement from the GABA receptor chloride ionophore complex. *Toxicol. Lett.* 35:183–90

29. Cutkomp, L. K., Koch, R. B., Desaiah, D. 1982. Inhibition of ATPase by chlorinated hydrocarbons. In *Insecticide Mode of Action*, ed. J. R. Coats, pp. 45–69. New York: Academic

30. Devaud, L., Murray, T. F. 1987. Interactions of pyrethroid insecticides with the peripheral-type benzodiazepine receptor. *Soc. Neurosci. Abstr.* 13:1230

31. Devaud, L. L., Szot, P., Murray, T. F. 1986. PK 11195 antagonism of pyrethroid-induced proconvulsant activity. *Eur. J. Pharmacol.* 120:269–73

32. Doherty, J. D., Lauter, C. J., Salem, N. Jr. 1986. Synaptic effects of the synthetic pyrethroid resmethrin in rat brain *in vitro*. *Comp. Biochem. Physiol.* C 84:373–79

33. Doherty, J. D., Nishimura, K., Kurihara, N., Fujita, T. 1987. Promotion of norepinephrine release and inhibition of calcium uptake by pyrethroids in rat brain synaptosomes. *Pestic. Biochem. Physiol.* 29:187–96

34. Donnellan, J. F., Alexander, K., Chendlik, R. 1976. The isolation of cholinergic nerve terminals from fleshfly heads. *Insect Biochem.* 6:419–23

35. Dwivedy, A. K. 1985. Cholinergic properties of pinched-off synaptic nerve endings from the central nervous system of the cockroach *Periplaneta americana*. *Pestic. Sci.* 16:615–26

36. Eells, J. T., Dubocovich, M. L. 1988. Pyrethroid insecticides evoke neurotransmitter release from rabbit striatal slices. *J. Pharmacol. Exp. Ther.* In press

37. Eldefrawi, M. E., Sherby, S. M., Abalis, I. M., Eldefrawi, A. T. 1985. Interactions of pyrethroid and cyclodiene insecticides with nicotinic acetylcholine and GABA receptors. *NeuroToxicology* 6:47–62

38. Elliott, M. 1980. Established pyrethroid insecticides. *Pestic. Sci.* 11:119–28
39. Elliott, M., Farnham, A. W., Janes, N. F., Needham, P. H., Pulman, D. A. 1974. Insecticidally active conformations of pyrethroids. In *Mechanism of Pesticide Action*, ed. G. K. Kohn, pp. 80–91. Washington, DC: Am. Chem. Soc.
40. Elliott, M., Farnham, A. W., Janes, N. F., Needham, P. H., Pulman, D. A., et al. 1973. A photostable pyrethroid. *Nature* 246:169–70
41. Elliott, M., Janes, N. F. 1978. Synthetic pyrethroids—a new class of insecticide. *Chem. Soc. Rev.* 7:473–505
42. Elliott, M., Janes, N. F., Potter, C. 1978. The future of pyrethroids in insect control. *Ann. Rev. Entomol.* 23:443–69
43. Evans, M. H. 1976. End-plate potentials in frog muscle exposed to a synthetic pyrethroid. *Pestic. Biochem. Physiol.* 6:547–50
44. Forshaw, P. J., Bradbury, J. E. 1983. Pharmacological effects of pyrethroids on the cardiovascular system of the rat. *Eur. J. Pharmacol.* 91:207–13
45. Forshaw, P. J., Lister, T., Ray, D. E. 1987. The effect of two types of pyrethroids on rat skeletal muscle. *Eur. J. Pharmacol.* 134:89–96
46. Forshaw, P. J., Ray, D. E. 1986. The effects of two pyrethroids, cismethrin and deltamethrin, on skeletal muscle and the trigeminal reflex system in the rat. *Pestic. Biochem. Physiol.* 25:143–51
47. Gammon, D. W., Brown, M. A., Casida, J. E. 1981. Two classes of pyrethroid action in the cockroach. *Pestic. Biochem. Physiol.* 15:181–91
48. Gammon, D., Casida, J. E. 1983. Pyrethroids of the most potent class antagonize GABA action at the crayfish neuromuscular junction. *Neurosci. Lett.* 40:163–68
49. Gammon, D. W., Lawrence, L. J., Casida, J. E. 1982. Pyrethroid toxicology: protective effects of diazepam and phenobarbital in the mouse and the cockroach. *Toxicol. Appl. Pharmacol.* 66:290–96
50. Gammon, D. W., Sander, G. 1985. Two mechanisms of pyrethroid action: electrophysiological and pharmacological evidence. *NeuroToxicology* 6:63–86
51. Ghiasuddin, S. M., Soderlund, D. M. 1985. Pyrethroid insecticides: potent, stereospecific enhancers of mouse brain sodium channel activation. *Pestic. Biochem. Physiol.* 24:200–6
52. Gray, A. J., Rickard, J. 1982. Toxicity of pyrethroids to rats after direct injection into the central nervous system. *NeuroToxicology* 3:25–35
53. Gray, A. J., Soderlund, D. M. 1985. Mammalian toxicology of pyrethroids. In *Progress in Pesticide Biochemistry and Toxicology*, ed. D. H. Hutson, T. R. Roberts, 5:193–248. New York: Wiley
54. Hamill, O. P., Marty, A., Neher, E., Sakmann, B., Sigworth, F. J. 1981. Improved patch-clamp techniques for high-resolution current recording from cells and cell-free membrane patches. *Pflügers Arch.* 391:85–100
55. Harris, R. A., Allan, A. M. 1985. Functional coupling of $\gamma$-aminobutyric acid receptors to chloride channels in brain membranes. *Science* 228:1108–10
56. Holan, G., Frelin, C., Lazdunski, M. 1985. Selectivity of action between pyrethroids and combined DDT-pyrethroid insecticides on $Na^+$ influx into mammalian neuroblastoma. *Experientia* 41:520–22
57. Holan, G., O'Keefe, D. F., Virgona, C., Walser, R. 1978. Structural and biological link between pyrethroids and DDT in new insecticides. *Nature* 272:734–36
58. Holloway, S. F., Salgado, V. L., Wu, C. H., Narahashi, T. 1984. Maintained opening of single Na channels by fenvalerate. *Soc. Neurosci. Abstr.* 10:864
59. Hue, B., Mony, L. 1987. Actions of deltamethrin and tralomethrin on cholinergic synaptic transmission in the central nervous system of the cockroach *(Periplaneta americana)*. *Comp. Biochem. Physiol. C* 86:349–52
60. Jacques, Y., Romey, G., Cavey, M. T., Kartalovski, B, Lazdunski, M. 1980. Interaction of pyrethroids with the $Na^+$ channel in mammalian neuronal cells in culture. *Biochim. Biophys. Acta* 600:882–97
61. Lawrence, L. J., Casida, J. E. 1982. Pyrethroid toxicology: mouse intracerebral structure-toxicity relationships. *Pestic. Biochem. Physiol.* 18:9–14
62. Lawrence, L. J., Casida, J. E. 1983. Stereospecific action of pyrethroid insecticides on the $\gamma$-aminobutyric acid receptor–ionophore complex. *Science* 221:1399–401
63. Lawrence, L. J., Gee, K. W., Yamamura, H. I. 1985. Interactions of pyrethroid insecticides with chloride ionophore–associated binding sites. *NeuroToxicology* 6:87–98
64. Leeb-Lundberg, F., Olsen, R. W. 1980. Picrotoxinin binding as a probe of the GABA postsynaptic membrane recep-

tor–ionophore complex. In *Psychopharmacology and Biochemistry of Neurotransmitter Receptors*, ed. H. I. Yamamura, R. W. Olsen, E. Usdin, pp. 593–606. New York: Elsevier

65. Leibowitz, M. D., Schwarz, J. R., Holan, G., Hille, B. 1987. Electrophysiological comparison of insecticide and alkaloid agonists of Na channels. *J. Gen. Physiol.* 90:75–93

66. Leibowitz, M. D., Sutro, J. B., Hille, B. 1986. Voltage-dependent gating of veratridine-modified Na channels. *J. Gen. Physiol.* 87:25–46

67. Lummis, S. C. R., Chow, S. C., Holan, G., Johnston, G. A. R. 1987. γ-Aminobutyric acid receptor ionophore complexes: differential effects of deltamethrin, dichlorodiphenyltrichloroethane, and some novel insecticides in a rat brain membrane preparation. *J. Neurochem.* 48:689–94

68. Lummis, S. C. R., Sattelle, D. B. 1986. Binding sites for [³H]GABA, [³H]flunitrazepam and [³⁵S]TBPS in insect CNS. *Neurochem. Int.* 9:287–93

69. Lund, A. E., Narahashi, T. 1982. Dose-dependent interaction of the pyrethroid isomers with sodium channels of squid axon membranes. *NeuroToxicology* 3:11–24

70. Lund, A. E., Narahashi, T. 1983. Kinetics of sodium channel modification as the basis for the variation in the nerve membrane effects of pyrethroids and DDT analogs. *Pestic. Biochem. Physiol.* 20:203–16

71. McLaughlin, G. A. 1973. History of pyrethrum. In *Pyrethrum—The Natural Insecticide*, ed. J. E. Casida, pp. 3–15. New York: Academic

72. Narahashi, T. 1962. Effect of the insecticide allethrin on membrane potentials of cockroach giant axons. *J. Cell. Comp. Physiol.* 59:61–65

73. Narahashi, T. 1962. Nature of the negative after-potential increased by the insecticide allethrin in cockroach giant axons. *J. Cell. Comp. Physiol.* 59:67–76

74. Narahashi, T. 1969. Mode of action of DDT and allethrin on nerve: cellular and molecular mechanisms. *Residue Rev.* 25:275–88

75. Narahashi, T. 1984. Nerve membrane sodium channels as the target of pyrethroids. In *Cellular and Molecular Neurotoxicology*, ed. T. Narahashi, pp. 85–108. New York: Raven

76. Narahashi, T. 1985. Nerve membrane ionic channels as the primary target of pyrethroids. *NeuroToxicology* 6:3–22

77. Narahashi, T. 1986. Toxins that modulate the sodium channel gating mechanism. In *Tetrodotoxin, Saxitoxin, and the Molecular Biology of the Sodium Channel*, ed. C. Y. Kao, S. R. Levinson, pp. 133–51. New York: NY Acad. Sci.

78. Narahashi, T. 1986. Mechanisms of action of pyrethroids on sodium and calcium channel gating. In *Neuropharmacology of Pesticide Action*, ed. M. G. Ford, G. G. Lunt, R. C. Reay, P. N. R. Usherwood, pp 36–60. Chichester, UK: Horwood

79. Nicholson, R. A., Baines, P., Robinson, P. S. 1987. Insect synaptosomes in superfusion. A technique to investigate the actions of ion channel directed neurotoxicants by monitoring their effects on transmitter release. In *Sites of Action for Neurotoxic Pesticides*, ed. R. M. Hollingworth, M. B. Green, pp. 262–72. Washington, DC: Am. Chem. Soc.

80. Nicholson, R. A., Wilson, R. G., Potter, C., Black, M. H. 1983. Pyrethroid- and DDT-evoked release of GABA from the nervous system *in vitro*. In *Pesticide Chemistry: Human Welfare and Environment*, ed. J. Miyamoto, P. C. Kearney, 3:75–78. Oxford, UK: Pergamon

81. Orchard, I. 1980. Electrical activity of neurosecretory cells and its modulation by insecticides. See Ref. 14, pp. 321–28

82. Orchard, I. 1980. The effects of pyrethroids on the electrical activity of neurosecretory cells from the brain of *Rhodnius prolixus*. *Pestic. Biochem. Physiol.* 13:220–26

83. Orchard, I., Osborne, M. P. 1979. The action of insecticides on neurosecretory neurons in the stick insect, *Carausius morosus*. *Pestic. Biochem. Physiol.* 10:197–202

84. Osborne, M. P. 1986. Insect neurosecretory cells—structural and physiological effects induced by insecticides and related compounds. See Ref. 78, pp. 203–43

85. Osborne, M. P., Hart, R. J. 1979. Neurophysiological studies of the effects of permethrin upon pyrethroid resistant (*kdr*) and susceptible strains of dipteran larvae. *Pestic. Sci.* 10:407–13

86. Ozoe, Y., Eto, M., Mochida, K., Nakamura, T. 1986. Characterization of high affinity binding of [³H]propyl bicyclic phosphate to house fly head extracts. *Pestic. Biochem. Physiol.* 26:263–74

87. Payne, G. T., Soderlund, D. M. 1988. *Allosteric interactions of insecticides with the [³H]BTX-B binding site of the voltage-sensitive sodium channel.* Pre-

sented at Neurotox '88, Int. Symp. Mol. Basis Drug Pestic. Action, Nottingham, UK

88. Rao, K. S. P., Chetty, C. S., Desaiah, D. 1984. *In vitro* effects of pyrethroids on rat brain and liver ATPase activities. *J. Toxicol. Environ. Health* 14:257–65

89. Rashatwar, S. S., Matsumura, F. 1985. Interaction of DDT and pyrethroids with calmodulin and its significance in the expression of enzyme activities of phosphodiesterase. *Biochem. Pharmacol.* 34:1689–94

90. Ruigt, G. S. F. 1985. Pyrethroids. In *Comprehensive Insect Physiology, Biochemistry and Pharmacology,* ed. G. A. Kerkut, L. I. Gilbert, 12:183–262. Oxford, UK: Pergamon

91. Ruigt, G. S. F., Klis, J. F. L., van den Bercken, J. 1986. Pronounced repetitive activity induced by the pyrethroid insecticide, fenfluthrin, in the slowly adapting stretch receptor neuron of the crayfish. *J. Comp. Physiol.* 159:43–53

92. Ruigt, G. S. F., van den Bercken, J. 1986. Action of pyrethroids on a nerve-muscle preparation of the clawed frog, *Xenopus laevis. Pestic. Biochem. Physiol.* 25:176–87

93. Sahib, I. K. A., Rao, K. S. P., Desaiah, D. 1987. Pyrethroid inhibition of basal and calmodulin stimulated $Ca^{2+}$ ATPase and adenylate cyclase in rat brain. *J. Appl. Toxicol.* 7:75–80

94. Salgado, V. L., Irving, S. N., Miller, T. A. 1983. Depolarization of motor nerve terminals by pyrethroids in susceptible and *kdr*-resistant house flies. *Pestic. Biochem. Physiol.* 20:100–14

95. Salgado, V. L., Irving, S. N., Miller, T. A. 1983. The importance of nerve terminal depolarization in pyrethroid poisoning of insects. *Pestic. Biochem. Physiol.* 20:169–82

96. Scott, J. G., Matsumura, F. 1983. Evidence for two types of toxic actions of pyrethroids on susceptible and DDT-resistant German cockroaches. *Pestic. Biochem. Physiol.* 19:141–50

97. Seifert, J., Casida, J. E. 1985. Solubilization and detergent effects on interactions of some drugs and insecticides with the *t*-butylbicyclophosphorothionate binding site within the γ-aminobutyric acid receptor–ionophore complex. *J. Neurochem.* 44:110–16

98. Shankland, D. L. 1976. The nervous system: comparative physiology and pharmacology. In *Insecticide Biochemistry and Physiology,* ed. C. F. Wilkinson, pp. 229–70. New York/London: Plenum

99. Singh, G. J. P., Barker, J. F., Kundu, S. C. 1982. Bioresmethrin-induced alterations in the ultrastructure of neurosecretory cells of insect corpora cardiaca. *Pestic. Biochem. Physiol.* 18:158–68

100. Singh, G. J. P., Orchard, I. 1982. Is insecticide-induced release of insect neurohormones a secondary effect of hyperactivity of the central nervous system? *Pestic. Biochem. Physiol.* 17:232–42

101. Singh, G. J. P., Orchard, I. 1983. Action of bioresmethrin on the corpus cardiacum of *Locusta migratoria. Pestic. Sci.* 14:229–34

102. Singh, G. J. P., Orchard, I., Loughton, B. G. 1982. Pharmacology of insecticide-induced release of hyperlipemic hormone in the locust, *Locusta migratoria. Gen. Pharmacol.* 13:471–75

103. Soderlund, D. M. 1979. Pharmacokinetic behavior of enantiomeric pyrethroid esters in the cockroach, *Periplaneta americana* L. *Pestic. Biochem. Physiol.* 12:38–48

104. Soderlund, D. M., Bloomquist, J. R., Ghiasuddin, S. M., Stuart, A. M. 1987. Enhancement of veratridine-dependent sodium channel activation by pyrethroids and DDT analogs. See Ref. 79, pp. 251–61

105. Soderlund, D. M., Ghiasuddin, S. M., Helmuth, D. W. 1983. Receptor-like stereospecific binding of a pyrethroid insecticide to mouse brain membranes. *Life Sci.* 33:261–67

106. Soderlund, D. M., Hessney, C. W., Helmuth, D. W. 1983. Pharmacokinetics of *cis*- and *trans*-substituted pyrethroids in the American cockroach. *Pestic. Biochem. Physiol.* 20:161–68

107. Squires, R. F., Casida, J. E., Richardson, M., Saederup, E. 1983. [$^{35}$S]*t*-Butylbicyclophosphorothionate binds with high affinity to brain-specific sites coupled to γ-aminobutyric acid–A and ion recognition sites. *Mol. Pharmacol.* 23:326–36

108. van den Bercken, J., Kroese, A. B. A., Akkermans, L. M. A. 1979. Effects of insecticides on the sensory nervous system. In *Neurotoxicology of Insecticides and Phermones,* ed. T. Narahashi, pp. 183–210. New York: Plenum

109. Verschoyle, R. D., Aldridge, W. N. 1980. Structure-activity relationships of some pyrethroids in rats. *Arch. Toxicol.* 45:325–29

110. Vijverberg, H. P. M., de Weille, J. R. 1985. The interaction of pyrethroids with voltage-dependent Na channels. *NeuroToxicology* 6:23–34

111. Vijverberg, H. P. M., de Weille, J. R., Ruigt, G. S. F., van den Bercken, J. 1986. The effect of pyrethroid structure on the interaction with the sodium channel in the nerve membrane. See Ref. 78, pp. 267–85

112. Vijverberg, H. P. M., Ruigt, G. S. F., van den Bercken, J. 1982. Structure-related effects of pyrethroid insecticides on the lateral-line sense organ and on peripheral nerves of the clawed frog, *Xenopus laevis*. *Pestic. Biochem. Physiol.* 18:315–24

113. Vijverberg, H. P. M., van den Bercken, J. 1982. Action of pyrethroid insecticides on the vertebrate nervous system. *Neuropathol. Appl. Neurobiol.* 8:421–40

114. Vijverberg, H. P. M., van der Zalm, J. M., van den Bercken, J. 1982. Similar mode of action of pyrethroids and DDT on sodium channel gating in myelinated nerves. *Nature* 295:601–3

115. Vijverberg, H. P. M., van der Zalm, J. M., van Kleef, R. G. D. M., van den Bercken, J. 1983. Temperature- and structure-dependent interaction of pyrethroids with the sodium channels in frog node of ranvier. *Biochim. Biophys. Acta* 728:73–82

116. Wouters, W., van den Bercken, J., van Ginneken, A. 1977. Presynaptic action of the pyrethroid insecticide allethrin in the frog motor end plate. *Eur. J. Pharmacol.* 43:163–71

117. Wright, C. D. P., Forshaw, P. J., Ray, D. E. 1988. Classification of the actions of ten pyrethroid insecticides in the rat, using the trigeminal reflex and skeletal muscle as test systems. *Pestic. Biochem. Physiol.* 30:79–86

118. Yamamoto, D., Quandt, F. N., Narahashi, T. 1983. Modification of single sodium channels by the insecticide tetramethrin. *Brain Res.* 274: 344–49

*Ann. Rev. Entomol. 1989. 34:97–116*

# IMMEDIATE AND LATENT EFFECTS OF CARBON DIOXIDE ON INSECTS

## Gérard Nicolas

Laboratoire de Biologie de l'Insecte, Université Paris-Sud, Bâtiment 440-443, 91405 Orsay Cedex, France

## Daniel Sillans

Laboratoire de Biométrie, Université Claude-Bernard, Lyon I, 43 boulevard du 11 Novembre 1918, 69622 Villeurbanne, France

## PERSPECTIVES AND OVERVIEW

A considerable amount of information has been published concerning the effects of $CO_2$ on various ethological, biological, physiological, and metabolic characteristics of insects. Descriptions of the effects of $CO_2$ are scattered in a large number of papers, and the data are gathered in different ways according to the authors' scope of research. Up to now, no general review covering all the aspects of immediate and latent $CO_2$ effects has been available. We therefore compile the main results of various fields of activities in this area. We emphasize the often ignored wide range of effects of small $CO_2$ variations in the insects' environment as well as the routine use of this gas at high concentrations for practical or experimental purposes. A knowledge of the effects of $CO_2$ has led to important applications, mainly in medical and veterinary fields and in stored-product entomology.

Some species of insects, such as bees inside hives, live in environments with a $CO_2$ content much higher than that of the atmosphere, which contains about 0.035% $CO_2$ in volume or 350 parts per million (ppm). These insects react to $CO_2$ increases and regulate their abiotic environment.

$CO_2$ has a significant role in the complex interactions between insects and plants and between insects and mammals in their natural environment. $CO_2$ is a known attractant for many invertebrates, which perform oriented responses

97

to a $CO_2$ gradient. $CO_2$ is an important cue for host location in phytophagous, saprophagous, and hematophagous insects; this role is particularly important in biting flies, and increasing attention has been paid to the development of artificial olfactory baits, which provide a basis for controlling insect vectors of diseases. Another expanding field is related to the artificial raising of $CO_2$ levels to induce sublethal or lethal effects and to prevent and control infestations in stored food. This method is a viable alternative to chemical fumigation and has the advantage of leaving no pesticide residues. In agricultural entomology, studies have been undertaken in the last few years to predict how the gradually increasing $CO_2$ content in the atmosphere will affect plant productivity and the impact of herbivores in the next century. At present few data are available, and further research is needed.

Immobilization of insects is necessary for various procedures such as inseminating queen bees or performing surgery. $CO_2$ remains probably the most popular anesthetic in entomological research. Many workers have stated that narcosis has no repercussion or only minor and transitory ones. However, many quantitative studies on numerous insect species have demonstrated various deferred $CO_2$ effects that may continue for hours or even days. An ideal anesthetic should provide no harmful side effects, but all procedures used for immobilization are known to involve risks.

The means by which the $CO_2$ operates as an anesthetic is a frequent object of speculation. Compared to the amount of work describing the effects of $CO_2$, little effort has been made to determine the underlying mechanisms of the induced changes. There is controversy as to whether the actions of $CO_2$ result from a specific effect of this agent, of anoxia, and/or of pH variations. This aspect of $CO_2$ research cannot be incorporated here (but see 110, 130, 167). Since 1980, the responses of some blood-sucking insects to a $CO_2$ source (153), the effects of controlled atmospheres on stored-product pests (7), and the effects of $CO_2$ on subsocial insects (47) have been reviewed.

## NATURAL VARIATIONS OF ENVIRONMENTAL $CO_2$ CONCENTRATIONS

### Effects of Naturally High $CO_2$ Content

Some species of insects, at least at some stage of their life, live in environments with higher $CO_2$ and lower $O_2$ contents than those of the normal atmosphere. Excess $CO_2$ is likely to occur in overwintering chambers located under the bark of trees or stumps (113), in the soil when it is covered by ice and snow (27), or inside decomposing dung used as a food resource (3). In caves, the $CO_2$ concentration is more than twice that found in the surface atmosphere (61). Intragranular $CO_2$ concentrations of 2% or higher have been recorded in nonairtight steel farm bins containing insect-infested seeds (137). A high $CO_2$ concentration is also characteristic of the abiotic environment of social insects. Within the nests of termites of the subfamily Macrotermitinae, CO$_2$ concentrations ranging from 0.8 to 2.9% (86) and from 1.2 to 5.2% (89) have been recorded. Inside beehives the CO$_2$ content varies from 0.2 to 6% (reviewed in 18, 135). Large seasonal variations occur in the CO$_2$ level in the environment of the social insects, and daily fluctuations have also been described (29, 125, 135). In hives, there appears to be an inverse relationship between CO$_2$ concentration and daily air temperature, and the highest flight traffic out of the hive occurs during the period of lowest CO$_2$ concentration (29). Ventilation of honey bee nests through fanning activity has long been recognized as a social thermoregulatory measure; however, in hives in which the temperature was stable an increase in CO$_2$ has been shown to induce fanning behavior (127). Some termites rebuild porous parts of their nests if the internal CO$_2$ concentration increases from 1 to 2% (125). In addition, an increase in the CO$_2$ level may induce digging behavior in ants (55).

## Effects on Host-Seeking Behavior

CO$_2$ AS A STIMULUS FOR HOST-PLANT FINDING    Since the discovery in 1957 that the larvae of three species of soil-dwelling insects could locate plant roots or a CO$_2$ source (71), CO$_2$ has been shown to stimulate an oriented response in several phytophagous larvae feeding below ground and in saprophagous larvae (67). Rather than locating their food simply by random wandering, such insects respond to chemical factors, particularly a CO$_2$ gradient emanating from the plant or from decaying plant or animal matter. Removal of CO$_2$ from an airstream emanating from seeds eliminates the attractancy (32, 66, 140, 144). Wireworm larvae can locate a food source from distances of up to 20 cm and respond to a CO$_2$ concentration increase as small as 0.002% (32) or 0.003% (72). Depending on the insect species, CO$_2$ concentrations ranging from 1.5% (32) to 6.5% (72) have a repellent effect.

For polyphagous soil-inhabiting insects a response to CO$_2$ emanating from plant roots is an adequate means of finding different potential hosts; the CO$_2$ acts as an unspecific attractant (72). Mono- and oligophagous larvae, which need to find their hosts among a complex of nonhost plants, have a more specific search mechanism involving other plant metabolites (67). CO$_2$ might also incite biting, which would lead to the release of plant juices and the acceptance or rejection of the plants (32). CO$_2$ probably also plays a part in the oviposition-related behavior of the Japanese pine sawyer beetle, since the adult beetles are attracted at the same time that CO$_2$ is released from trunks (73).

CO$_2$ AS A STIMULUS FOR HOST LOCATION BY BLOOD-SUCKING IN-SECTS    Many factors are involved in the host-seeking behavior of blood-sucking insects; olfaction plays an important part and is often integrated with other senses, especially vision (2). Laboratory and field experiments have shown that among the chemicals emanating from the host, $CO_2$ is an important cue for location of hosts by biting flies. $CO_2$ is a natural component of human and animal breath. In humans, expired breath contains about 4.5% $CO_2$; therefore the exhalations have to be diluted nearly 130 times before the $CO_2$ concentration falls to the atmospheric background level. The effectiveness of $CO_2$ in drawing hematophagous insects to the immediate vicinity of the source is well documented for Culicidae (52), Glossinidae (153), and Muscidae (159, 160), but is also known for other groups such as Tabanidae (148) and Simulidae (145). $CO_2$ is also attractive for nonbiting flies (153). The attractiveness of animal baits or a $CO_2$ source varies widely according to the insect species (53, 164). Flies respond to an increase in $CO_2$ rather than to its concentration; they rapidly get habituated to new levels of $CO_2$, and regardless of the background concentration a very small change may influence their behavior (52, 159).

In some species of Culicidae and Simulidae, hosts attract the insects over a greater distance than an equivalent artificial source of $CO_2$ (15, 53). In these species, the upwind flight behavior of host-seeking involves separate mechanisms for long-, middle-, and close-range orientation. Long-range attraction involves host-specific odors, while a combination of $CO_2$ and odors is probably the main guide in the insect's host-directed orientation at middle range (20). In close-range orientation, where most directed visual responses occur, behavior is affected by the physical makeup of the host: color, size, shape, and movement (15, 54); $CO_2$ could also have a combined action with warm, moist convection currents (52). In Muscidae, $CO_2$ is initially involved as an activator before the insect becomes oriented to a source (160). The physiological state of a specimen may influence its responses to $CO_2$ (76, 159).

Laboratory experiments have been carried out to define the role of $CO_2$ and breath odors on the activity of hematophagous insects. A quantitative relationship has been demonstrated between the number of flights and $CO_2$ concentrations up to 1% for Muscidae (159) or 5% for Glossinidae (152). Expired breath elicits a greater increase in flight activity than could be predicted from its $CO_2$ content; the magnitude of the response results either from an additivity of the responses to $CO_2$ and to host odors (152) or from a synergistic effect (156, 159).

The correspondence between field and laboratory results is not perfect. For example, the kinetic effect of low $CO_2$ concentration that characterizes the laboratory behavior of some flies has not been convincingly demonstrated in the field (discussed in 20).

The combination of $CO_2$ with active trapping devices has been used to catch blood-sucking Diptera and to assess their density (51). Vale & Hall (154) have identified the attractive host-odor components and have evaluated the efficiency of mixtures of specific components with $CO_2$.

## Interrelationship Between Plants and Pollinators

The behavior of wasp pollinators of different *Ficus* spp. depends on variations in the internal gaseous composition of figs at various stages of ripening (49). The CO$_2$ content in the fig cavity may rise to as much as 10%. Laboratory experiments have shown that males may require such a high CO$_2$ content to carry out their normal activities and to fertilize the females, which remain inactive in the syconial cavity as long as the CO$_2$ concentration is higher than 3–4%. After having fertilized the females, the males perforate the syconial wall, and the atmosphere within the fig becomes balanced with the external atmosphere. As a result, the females become active, fill their pockets with pollen, and escape from the fig, directed toward the exit holes by a chemotactic response to a decreasing CO$_2$ concentration gradient (49).

## Projected Consequences of CO$_2$ Accumulation in the Earth's Atmosphere

The CO$_2$ concentration in the atmosphere is gradually increasing, mainly as a result of fossil fuel combustion and deforestation. While it was about 260–270 ppm during the preindustrial period, it has now reached 350 ppm. It is estimated that if the accumulation continues, the atmospheric CO$_2$ concentration could reach 600 ppm in the middle of the next century (122). It has been projected that the continuing increase in atmospheric CO$_2$ concentration will enhance plant productivity. Plants using the Calvin photosynthetic pathway have been grown in chambers in which CO$_2$ was enriched to anticipated levels. The leaves of these plants contained more carbohydrates but had a lower nitrogen content, lower protein concentration, and reduced nutritive value for herbivores (80). This modification of plant metabolism may alter the feeding behavior of leaf-eating insects and consequently influence plant productivity. Recent studies with Lepidoptera have shown that higher leaf consumption on plants grown in a CO$_2$-enriched environment is offset by an increased amount of leaf material produced (79, 80, 112); moreover, change in plant growth may not affect homopteran dynamics (22). These results suggest that the impact of herbivores on their host plant will not increase as the level of atmospheric CO$_2$ rises, but this hypothesis has to be confirmed for other insects for which no data are available.

## CO$_2$-Sensitive External Receptors

CO$_2$-sensitive external receptors have been reported only on the antennae in Hymenoptera (11, 75, 142) and only on the labial palps in Lepidoptera (14). In Diptera, such receptors have been located on the maxillary palps of the Culicidae (70, 111) and on the antennae of the Muscidae (161). In Coleoptera, they are found on both maxillary and labial palps, although the maxillary palps appear to be more important than the labial palps in CO$_2$ reception (31, 92). The receptors are also found on the antennae of a coleopteran cave species (28). $CO_2$-sensitive external receptors have not been reported for other groups of insects, perhaps because the gas has not been tested as a stimulus.

While the biological significance of the $CO_2$ receptors is well understood for hematophagous, saprophagous, and root-feeding insects (host-food location) and social insects (hive or mound ventilation), it remains obscure for Lepidoptera. The receptors of Lepidoptera are broadly similar to those of other insects in their responses to $CO_2$, but they differ by their sensitivity to several other odorants. Moreover, in an ambient atmosphere the lepidopteran receptor cells are already in an excited condition (14). In Hymenoptera, some of the $CO_2$ receptors respond to a gas mixture containing the normal $CO_2$ concentration of air. In mosquitoes, the $CO_2$ receptors show different responses to a 0.01% $CO_2$ increase (70). This seems to be the lowest range of variation in the $CO_2$ concentration tested at the peripheral sensory receptor level. The change in the responsiveness of sensory cells could probably be induced by a much smaller increase in the $CO_2$ concentration, since behavioral studies have shown that some insects react to a $CO_2$ increase of as little as 0.002–0.003% (32, 72).

Gases with an anesthetic effect such as $N_2O$ and xenon interact with honeybee $CO_2$ receptors in a way that is antagonistic to the effect of $CO_2$ (142). The mechanism of $CO_2$ reception needs to be studied (141).

## Physiological Effects of $CO_2$

Few indications are available concerning the physiological repercussions of natural increases in $CO_2$ concentration in the environment of insects. It has been suggested that the low metabolic rate of troglobites may be partly due to a physiological response to the high $CO_2$ content of their environment (61). In beetles overwintering in chambers under the bark of dead trees, which has a high $CO_2$ concentration, the hemolymph shows a considerable increase of lactic acid (113). At present, the only evidence of an influence of microclimatic $CO_2$ variation on hormonal levels is found in bees; along with the temperature, the $CO_2$ concentration controls juvenile hormone titers of the hemolymph and whole-body extracts (18). The ways in which $CO_2$ influences juvenile hormone metabolism have not been investigated.

# EXPERIMENTALLY INDUCED HIGH $CO_2$ CONCENTRATIONS

## Anesthesia Mechanisms

The mechanisms of anesthesia appear to be similar whatever the nature of the anesthetic and to involve a physical phenomenon leading to changes in the permeability of the neuron cell membranes (98). A theoretical mechanism has been proposed for the deformation of these membranes (150). In vitro studies have in fact shown that general anesthetics modify the ordered array of phospholipids and deform the proteins of these biological membranes (40, 98, 128, 136).

It is not surprising that the same anesthetics are efficient both in vertebrates and in insects, since the neural mechanisms are similar in all animal groups. Poikilotherms, contrary to warm-blooded animals, can be used to study how the anesthetic potency varies with temperature (129). In *Drosophila* species, the anesthetic potency of CO$_2$ decreases as the temperature increases. This agrees with certain theoretical mechanisms such as those that involve hydrate crystal formation around the neuronal proteins (95, 114). But since both CO$_2$ solubility in lipids (134) and CO$_2$ hydrate crystal stability decrease as the temperature increases, temperature-related studies cannot resolve the site of the anesthetic action.

Since the CO$_2$ hydrate is carbonic acid, it must produce acidification and act indirectly on the membranes by modifying their permeability. However, CO$_2$ anesthesia in insects does not seem to result from pH variations, as it does not result from anoxia (see 130). The mechanisms of CO$_2$ anesthesia, as of other anesthetics, are not yet fully elucidated.

## Nervous System

CO$_2$ seems to exert its anesthetic effect directly on the nervous system via the tracheae and not via the blood (167). In experiments with locusts and $^{14}$C-labeled CO$_2$, $^{14}$C was detected in all insect body parts, but its relative persistence was highest (up to 24 hr) in the central nervous system or its immediate vicinity (105, 106). Pure CO$_2$ has an inhibitory effect on the bioelectrical responses of the nervous system, while a smaller concentration (15%) has a stimulatory effect (11). CO$_2$ has also been reported to induce depolarization of the neurons (24).

Short-term exposure of insects to pure CO$_2$ or high concentrations greatly reduces the amplitude of the electroretinographic responses induced by light stimulation (110, 143, 165) and leads to changes in the critical fusion frequency of the eye (82). CO$_2$ may also inhibit the pigment dispersal caused by light in superposition eyes (4).

γ-Aminobutyric acid, which is a synaptic transmission inhibitor, has been implicated in a possible mechanism by which CO$_2$ might stimulate the nervous system (discussed in 69). CO$_2$ could also have an effect on acetylcholine, which is another synaptic transmission modulator (26). It also influences the quantity of neurotransmitters such as dopamine and octopamine (46).

## Muscles, Respiration, and Circulation

Gas exchanges take place almost exclusively at the spiracular level, and the local action of CO$_2$ on spiracles is well known (94). Spiracles generally possess a closer muscle innervated by a pair of motor neurons. Low $CO_2$ concentrations cause the closer muscle to relax in the face of continued motor neuron excitations. Thus $CO_2$ acts directly on events in the neuromuscular transmission sequence (62). However, in some insect species, if $CO_2$ initially causes the valves to open by local action on the muscle, when it reaches the central nervous system it causes a reduction in the tonic discharge to the closer muscle, which may allow the valve to open further. Finally, although the peripheral action of $CO_2$ probably plays an important part in control of the valve, it also acts on the ganglia to adjust the motor discharge to the spiracular muscle (94). Increasing the external $CO_2$ concentration to 10% seems to involve permanent valve opening (94). An increase in external $CO_2$ also affects the ventilatory rate, which fluctuates in frequency in relation to the $CO_2$ level and may become steady when external $CO_2$ reaches 5–10% (93). Insects may be anesthetized in the presence of $O_2$ (37), but if this gas is maintained at normal atmospheric tension, $CO_2$ has no appreciable effect on oxygen consumption (36). An important consequence of increased $CO_2$ concentration is the permanent opening of the spiracles, which induces water loss and may cause mortality (19). The water loss is higher at high $CO_2$ concentrations and more pronounced at low relative humidities (99). $CO_2$ has not been found to have any major effect on water movement from the rectal lumen (60).

The insect circulatory system contributes very slightly or not at all to respiratory exchanges and gas transport (97); thus the heart can stop beating without damage to the insect. A high $CO_2$ concentration will arrest the heart activity of many insects (65, 133), while a low $CO_2$ level (5%) stimulates it. As soon as the insect is in contact with pure $CO_2$ the heartbeat stops. During recovery, the rhythm accelerates rapidly to a maximum value above that measured before anesthesia and then returns to normal. Both the maximum amplitude and the recovery time are dependent on the duration of anesthesia (133). The same result is obtained with a 50% $CO_2$–50% $O_2$ mixture (131). When the $CO_2$ concentration is less than 50% the heart rate is slowed in proportion to the concentration. When the $CO_2$ rate is varied according to a sinusoidal rhythm (minimum 0%, maximum 50%), the silkworm, *Bombyx mori,* immediately regulates its cardiac rhythm in relation to the composition of gases. The rapidity of the responses suggests an intervention at the nervous-system level more than at the metabolic level (130, 132). Other authors have also observed a marked decrease in the heartbeat rate as the $CO_2$ level increases, with the heartbeat stopping in 100% $CO_2$ (37, 158). Often the heartbeat continues after the insects are immobilized and resumes before they recover activity, which is probably evidence of the myogenicity of the insect heart (90).

A $CO_2$ effect has also been reported on other muscles besides those of the spiracles and heart. A comparison of the electrical activity in single fibrillar muscle fibers of flight muscles from several orders of insects revealed differences in neuromuscular mechanisms and in the excitability of the muscle membrane in response to CO$_2$ (91). In wasps and flies, CO$_2$ induces a rapid depolarization of the resting fiber membrane, while for many fibers of beetles no such effect was detected (91). A brief exposure to pure CO$_2$ did not abolish the electrical coupling between certain distinct skeletal muscles (126).

## Behavior

A single brief anesthesia by CO$_2$ may have strong ethological repercussions. One exposure to CO$_2$ causes worker honey bees, *Apis mellifera,* to change from hive to field activities at an earlier age, leads to a marked reduction of pollen gathering (123), modifies hoarding behavior (88), and affects the age-dependent temperature preference (56); some of these effects are observed after only a short anesthesia (30 sec) with pure CO$_2$ (33). CO$_2$ may destroy the activity rhythm (121); one CO$_2$ anesthesia results in an increase in the activity of adult *Drosophila melanogaster,* which can last many hours (155), while it reduces locomotion in *Locusta migratoria* larvae (106). A single CO$_2$ anesthesia is also known to reduce social aggregation in *L. migratoria* (106), to delay mating activity in *Empoasca devastans* (74), and to cause a 12-hr inhibition of drinking and a 24-hr inhibition of feeding in *Acheta domesticus* (167).

Few data are available concerning the effects of repeated anesthesia on behavior. In gregarious *L. migratoria* larvae, daily 1-min doses of 100% CO$_2$ reduced the number of displacements without any effect on the circadian distribution of marching (103) and also reduced the tendency to aggregate (50).

Behavioral aberrations resulting from anesthesia have also been reported. Anesthesia increases brood cannibalism in ants (138) and leads to the ejection of living larvae from the nests of several *Bombus* species (115).

In *Plodia interpunctella,* a 15-min exposure of emerging males to CO$_2$ prevented wing inflation, thus preventing the insects from flying and indirectly impeding mating. This CO$_2$ effect has been suggested as a way to inhibit reproduction to control this insect, which infests stored products (84). In another stored-product insect, *Ephestia cautella,* CO$_2$ may play a part in allowing the insect to initiate new infestations, since the addition of 30% CO$_2$ to a grain odor stream increased the attractiveness of the odor and the oviposition responses near the source. Since storage tanks are not usually absolutely gastight, the atmosphere emanating from a grain store containing high CO$_2$ levels will provide olfactory cues that will cause females to aggregate at the tank and oviposit in the vicinity of the leaks (5).

## Learning and Memory

$CO_2$ can interfere with learning or disrupt memory. In bees trained to feed at specific times, a prolonged narcosis of 2 hr in 100% $CO_2$ induced a modification in the timing of visits to a feeding dish. The first visit occurred at the usual time, but the second occurred with a delay equal to the duration of the anesthesia. Thus this $CO_2$ treatment produced an important alteration of the endogenous timing mechanisms (81).

$CO_2$ has not been widely used in memory disruption experiments, but it has been shown to produce retrograde amnesia in bees (8) and cockroaches (41, 83). In cockroaches, a mere 45-sec administration of 100% $CO_2$ immediately after training definitely interfered with the retention of learning (83). If a longer-term memory is established by overtraining, the sensitivity to disruption disappears (38).

## Development, Growth, and Morphometrics

$CO_2$ may affect developmental processes, but the results vary according to the species and treatment. Larval duration may be retarded by repeated short exposures (16, 37, 146, 147). In *Blatella germanica* a 3-min exposure to a high $CO_2$ concentration only once a week consistently prolonged larval life by 14–53% (16), while in *L. migratoria* a daily 3-min anesthesia with pure $CO_2$ had no serious consequences (106). Repeated $CO_2$ anesthesia may induce supernumerary instars (106, 146), reduce weight (37, 167), and affect size and morphometrics (104, 106, 147). The $CO_2$ effects on morphometrics are modulated by genetic factors (77). Brief exposures to a mixture containing high $CO_2$ concentrations and air had less adverse effects on growth and weight loss than exposures to 100% $CO_2$ (37, 167). Repeated short periods of anesthesia have also been shown to decrease the intensity of the larval diapause induced by climatic factors in *Gryllus campestris* (64).

## Reproduction

$CO_2$ may influence reproduction in several ways. A short anesthesia may delay or impede mating activity (74, 84, 162). In some *Glossina* species a 15-sec $CO_2$ anesthesia suppresses the subsequent insemination frequency of the females for at least 72 hr and reduces the insemination capability of the males (96). In certain insects a 30-min to 1-hr exposure to a high $CO_2$ concentration reduces egg production and hatchability (1, 5, 59, 85); these adverse effects may be significantly enhanced by repeating and/or extending the anesthesia (116). Daily, repeated 2-hr exposures of adult *Tribolium castaneum* to $CO_2$ before maturation suppressed oocyte development in the ovarioles (117).

In *L. migratoria* a daily short anesthesia throughout larval development had no detrimental effect on reproduction (102, 106); the $CO_2$ treatment reduced the delay of sexual maturation without affecting the egg number. Moreover, a daily CO$_2$ treatment stopped the inhibition of ovarian maturation established in some *L. migratoria* strains by the photoperiodic conditions (107). In *Pyrrhocoris apterus,* however, daily treatment considerably reduced the egg number in reproducing females (58).

Honey bee queens given two CO$_2$ anesthetic treatments (10 and 3 min) on consecutive days began oviposition earlier than untreated queens (87). CO$_2$ treatment of unmated queens resulted in an almost immediate stimulation of vitellogenin synthesis (39)

## Mortality

Much information is available on the time required to kill insects in various modified atmospheres (reviewed in 6, 7). The time required varies with the species and is affected by the life stage, the climatic conditions (temperature, humidity), and the composition of the atmosphere. Lowering the temperature or raising the relative humidity generally increases the exposure time needed to obtain lethal effects. Moreover, mixtures of CO$_2$ and air may be more rapidly lethal to some species than pure CO$_2$ (6, 7). A complete insect kill seems necessary for control, since the offspring of the survivors may show increased resistance, which would necessitate longer exposure periods to the same CO$_2$ concentration to obtain similar mortality (13, 100). Moreover, in certain insects exposed to 80% CO$_2$, which resulted in 96% mortality, the survivors multiplied normally (120).

Adding CO$_2$ to certain fumigants increases their toxicity. When 10% CO$_2$ was added to methyl iodide there was an 80% increase in mortality of *Sitophilus orizae* (119). The toxicity of acrylonitrite for *Tribolium castaneum* increased sevenfold in the presence of 30% CO$_2$ (12). The effectiveness of fumigants combined with CO$_2$ declines with a drop in temperature (12). Adding CO$_2$ to fumigants may permit reduction of either concentration or exposure times (12).

Under laboratory conditions, immobilization of insects with CO$_2$ before applying or injecting pesticides may enhance the action of the chemicals (139, 157, 163, 168). In *Ceratitis capitata* (157) and *T. castaneum* (168) anesthesia as brief as 3 or 5 min potentiates the toxicity of malathion, while for *Tribolium confusum* an anesthesia of up to 60 min is needed (168). Prior immobilization of 9 and 5 min in grasshoppers and cockroaches, respectively, had no significant effect on the insects' susceptibility to insecticides (10, 34). In locusts, a daily CO$_2$ anesthesia for 1 min led to a lower resistance to lindane ingested with foods (30).

Since modifications in the radiosensitivity of various species have been demonstrated when insects have been exposed to various gases, the effect of CO$_2$ on the mortality of irradiated insects was investigated as a means of control. However, insects irradiated in $CO_2$ atmospheres appeared to be more resistant to radiation damage (23). A prior $CO_2$ anesthesia for 30 min had no significant effect on mortality (149).

## Hormonal Levels

Few investigations have been undertaken concerning the consequences of $CO_2$ treatments on the complex made up of the neurosecretory cells in the brain, the corpora allata (CA), and the ventral gland, which regulates many functions in insects. Woodring et al (167) suggested that a brief anesthesia causes a long term disruption of certain neuroendocrine functions in *A. domesticus*. In *L. migratoria*, $CO_2$ treatments and either implantations of the pars intercerebralis or additions of CA or synthetic juvenile hormone may have similar effects on pigmentation, the weight of newborn larvae, the inhibition of ovarian maturation that appears in some strains, and the persistence of the ventral gland in gregarious adults (106, 108, 109). These similarities give indirect evidence concerning the effect of $CO_2$ on brain neurosecretion and juvenile hormone production. Histological studies have provided direct evidence of a $CO_2$ effect on the neurosecretory cells of the brain and on the CA, which are both strongly affected by anesthesia (9, 57). The best proof of hormonal changes following $CO_2$ treatments is an increase in juvenile hormone titers (18, 48).

## Biochemical Processes

$CO_2$ actions on some biochemical processes are known. $CO_2$ increases the concentration of certain blood proteins and affects the blood lipid and carbohydrate contents (42, 43, 101, 167). High levels of $CO_2$ in the atmosphere bring about many modifications in metabolic pathways whose interactions are not well known. After $CO_2$ treatment the consumption of glycogen tends to be higher, although the glucose level is not modified (42). A single 1–2-min anesthesia results in a rapid and substantial increase in neuronal glycogen which may persist for a least six days (166). $CO_2$ inhibits succinic dehydrogenase (35) and modulates NADPH production (45). $CO_2$ exposure also inhibits the biosynthesis of glutathione (44). High $CO_2$ concentrations (40–96%) caused *T. castaneum* to secrete considerable quantities of parabenzoquinones (63).

## Virus-Induced $CO_2$ Sensitivity

Although $CO_2$ exerts its normal anesthetic action in certain individuals of *Drosophila melanogaster*, in others it induces a fatal toxic effect (78). The sensitive flies do not wake or have an abnormal awakening and show characteristics of paralysis. These symptoms are induced by the sigma virus, which is naturally and hereditarily present in some strains (78), and also by serotypes of the vesicular stomatitis virus (VSV), which can be inoculated artificially (21, 118). If *D. melanogaster* is infected by the sigma virus after inoculation, the interval between the inoculation and the appearance of symptoms of CO$_2$ sensitivity depends on the site of inoculation; the most precocious sensitivity appears when the virus is inoculated near or in the cephalic or thoracic nerve center (21). The lethal effect of CO$_2$ appears when the nervous system, and most probably the thoracic ganglion, has become the site of viral multiplication (21); paralysis is attributed to the fixation of this gas on the cell membranes modified by viral maturation (17).

CO$_2$ sensitivity can also be induced in *D. melanogaster* by injecting the iota virus extracted from *Drosophila immigrans* (68). In this case, however, the CO$_2$ sensitivity is brief and restricted to the last two or three days of adult life.

CO$_2$ seems to have a specific action in the induction of sensitivity symptoms in *D. melanogaster* infected with the sigma virus, since the other gases tested do not have any lethal consequence (78). In infection with the iota virus, propane produces the same effect as CO$_2$ (68). A natural CO$_2$ sensitivity caused by a hereditary virus has also been found in mosquitoes (124), as has an induced sensitivity after an infection with the encephalitis virus (151).

A chromosomally inherited CO$_2$ sensitivity has also been described in *D. melanogaster* (25). The flies show only a very slow recovery after a standard exposure to CO$_2$. CO$_2$ sensitivity induced by the presence of a virus is known only in Diptera.

## CONCLUDING REMARKS

Perhaps the most striking impression left by the collective papers on CO$_2$ actions is that of the diversity of these actions and the wide variety of levels at which this gas may act. It is clear from this review that there are several inconsistencies and gaps in our knowledge on CO$_2$ actions and on their underlying mechanisms. Medical and economic interests in CO$_2$ effects under natural atmospheric variations have led to a bias toward methods for insect control; little is known about the mechanisms of CO$_2$ perception and transmission that initiate the responses or about the physiological basis of these responses. No unifying explanation has emerged concerning behavioral and biological consequences of exposure of insects to high CO$_2$ concentrations, even though major and probably interactive effects of the gas have been demonstrated on the brain, the endocrine system, metabolic processes, and the respiratory and circulatory systems. There are many difficulties in comparing data owing to differences in the insect populations tested, the rearing methods, and the experimental techniques (e.g. pure CO$_2$ or CO$_2$ in various combinations with air or other gases; single or repeated exposures of short or long duration). An important point is that one short episode of anesthesia has long-term consequences. Therefore the use of $CO_2$ as an anesthetic should be minimized in the absence of careful analysis of its effects. $CO_2$ can also be used to interfere with some characteristics for theoretical studies on learning and memory (83), endogenous clocks (81), development (146), and social interactions (47, 106).

## Literature Cited

1. Aliniazee, M. T., Lindgren, D. L. 1970. Egg hatch of *Tribolium confusum* and *Tribolium castaneum* in different carbon dioxide and nitrogen atmospheres. *Ann. Entomol. Soc. Am.* 63:1010–12
2. Allan, S. A., Day, J. F., Edman, J. D. 1987. Visual ecology of biting flies. *Ann. Rev. Entomol.* 32:297–316
3. Anderson, J. M., Coe, M. J. 1974. Decomposition of elephant dung in an arid tropical environment. *Oecologia Berlin* 14:111–25
4. Banister, M. J., White, R. H. 1987. Pigment migration in the compound eye of *Manduca sexta*: effects of light, nitrogen and carbon dioxide. *J. Insect Physiol.* 33:733–43
5. Barrer, P. M., Jay, E. G. 1980. Laboratory observations on the ability of *Ephestia cautella* (Walker) (Lepidoptera: Phycitidae) to locate, and to oviposit in response to a source of grain odour. *J. Stored Prod. Res.* 16:1–7
6. Bailey, S. W., Banks, H. J. 1975. The use of controlled atmospheres for the storage of grain. *Proc. 1st Int. Working Conf. Stored-Prod. Entomol., Savannah, Ga., 1974* pp. 362–74
7. Bailey, S. W., Banks, H. J. 1980. A review of recent studies of the effects of controlled atmospheres on stored product pests. In *Controlled Atmosphere Storage of Grains,* ed. J. Shybal, pp. 101–18. New York: Elsevier
8. Beckmann, H. E. 1974. The damaging effect of supercooling, narcosis, and stress on the memory of the honey-bee. *J. Comp. Physiol.* 94:249–66
9. Biedermann, M. 1964. Neurosekretion bei Arbeiterinnen und Königinnen von *Apis mellifica* L. unter natürlichen und experimentallen Bedingungen. *Z. Wiss. Zool.* 170:256–308
10. Blickenstaff, C. C. 1973. Grasshoppers: effect of $CO_2$ anaesthesia. *J. Econ. Entomol.* 66:538–39
11. Boistel, J. 1960. *Caractéristiques fonctionnelles des fibres nerveuses et des récepteurs tactiles et olfactifs des insectes.* Paris: Arnette. 147 pp.
12. Bond, E. J., Buckland, C. T. 1978. Control of insects with fumigants at low temperatures: toxicity of fumigants in atmospheres of carbon dioxide. *J. Econ. Entomol.* 71:307–9
13. Bond, E. J., Buckland, C. T. 1979. Development of resistance to carbon dioxide in the granary weevil. *J. Econ. Entomol.* 72:770–71
14. Borgner, F., Boppre, M. Ernst, K. D., Boeckh, J. 1986. $CO_2$ sensitive receptors on labial palps of *Rhodogastria* moths (Lepidoptera: Arctiidae): physiology, fine structure and central projection. *J. Comp. Physiol. A* 158:741–49
15. Bradbury, W. C., Bennett, G. F. 1974. Behavior of adult Simuliidae (Diptera). II. Vision and olfaction in near-orientation and landing. *Can. J. Zool.* 52:1355–64
16. Brooks, M. A. 1957. Growth-retarding effect of carbon dioxide anaesthesia on the German cockroach. *J. Insect Physiol.* 1:76–84
17. Brun, G., Plus, N. 1980. The viruses of *Drosophila.* In *The Genetics and Biology of* Drosophila, ed. M. Ashburner, T. R. F. Wright, 2:625–702. New York/London: Academic. 702 pp.
18. Buhler, A., Lanzrein, B., Wille, H. 1983. Influence of temperature and carbon dioxide concentration on juvenile hormone titre and dependent parameters of adult worker honey-bees (*Apis mellifera* L.). *J. Insect Physiol.* 29:885–93
19. Bursell, E. 1974. Environmental aspects—humidity. In *The Physiology of Insecta,* ed. M. Rockstein, 2:43–84. New York/London: Academic. 568 pp.
20. Bursell, E. 1987. The effect of wind borne odours on the direction of flight in tsetse flies, *Glossina* spp. *Physiol. Entomol.* 12:149–56
21. Bussereau, F. 1973. *Etude du symptôme de la sensibilité au $CO_2$ provoquée par des rhabdovirus chez* Drosophila melanogaster *Meigen.* Thesis. Univ. Paris XI, Orsay. 106 pp.
22. Butler, G. D., Kimball, B. A., Mauney, J. R. 1986. Populations of *Bemisia tabaci* (Homoptera: Aleyrodidae) on cotton grown in open-top field chambers en-

riched with $CO_2$. *Environ. Entomol.* 15:61–63

23. Clark, A. M., Herr, A. B. Jr. 1955. The effects of certain gases on the radiosensitivity of *Habrobracon* during development. *Radiat. Res.* 2:538–43

24. Clark, M. A., Eaton, D. C. 1983. Effect of $CO_2$ on neurons of the house cricket, *Acheta domestica. J. Neurobiol.* 14: 237–50

25. Clark, M. A., McCrady, W. B., Fielding, C. L. 1979. Independent production of $CO_2$ sensitivity by nuclear gene *Dly* and a transmissible agent from delayed recovery *Drosophila melanogaster. Genetics* 92:503–10

26. Colhoun, E. H. 1963. The physiological significance of acetylcholine in insects and observations upon other pharmacologically active substances. *Adv. Insect Physiol.* 1:1–46

27. Conradi-Larsen, E. M., Somme, L. 1973. The overwintering of *Pelophila borealis* Payk. II. Aerobic and anaerobic metabolism. *Nor. Entomol. Tidsskr.* 20: 325–32

28. Corbière-Tichané, G., Bermond, N. 1971. Ultrastructure et electrophysiologie des styles antennaires de la larve de *Speophyes lucidulus* Delar. (Coléoptère: Bathysciinae). *Ann. Sci. Nat. Zool.* 13: 505–42

29. Dietlein, D. G. 1985. Measurement of carbon dioxide levels in honey bee colonies using a thermistor bridge. *Am. Bee J.* 125:773–74

30. Djob Bikoi, J., Fuzeau-Braesch, S. 1979. Sensibilité phasaire de *Locusta migratoria* aux résidus d'un insecticide, le lindane, issu de semences de blé traitées. *Phytiatr.-Phytopharm.* 28:185–92

31. Doane, J. F., Klinger, K. 1978. Location of $CO_2$-receptive sensilla on larvae of wireworms *Agriotes lineatus-obscurus* and *Limonius californicus. Ann. Entomol. Soc. Am.* 71:357–60

32. Doane, J. F., Lee, Y., Klinger, J., Wescott, N. D. 1975. The orientation response of *Ctenicera destructor* and other wireworms (Coleoptera: Elateridae) to germinating grain and to carbon dioxide. *Can. Entomol.* 107:1233–52

33. Ebadi, R., Gary, N. E., Lorenzen, K. 1980. Effects of carbon dioxide and low temperature narcosis on honey bees, *Apis mellifera. Environ. Entomol.* 9: 144–47

34. Edward, A. J. 1981. Effects of carbon dioxide anaesthesia and crowding on the susceptibility of cockroaches to insecticides. *Entomol. Exp. Appl.* 29:339–44

35. Edwards, L. J. 1968. Carbon dioxide anaesthesia and succinic dehydrogenase in the corn earworm, *Heliothis zea. J. Insect Physiol.* 14:1045–48

36. Edwards, L. J., Batten, R. W. 1973. Oxygen consumption in carbon dioxide anesthetized house flies, *Musca domestica* Linn. (Diptera: Muscidae). *Comp. Biochem. Physiol. A* 44:1163–67

37. Edwards, L. J., Patton, R. L. 1965. Effects of carbon dioxide anaesthesia on the house cricket, *Acheta domesticus* (Orthoptera: Gryllidae). *Ann. Entomol. Soc. Am.* 58:828–32

38. Eisenstein, E. M., Reep, R. L. 1985. Behavioral and cellular studies of learning and memory in insects. In *Comprehensive Insect Physiology, Biochemistry and Pharmacology,* ed. G. A. Kerkut, L. I. Gilbert, 9:513–48. New York: Pergamon. 735 pp.

39. Engels, W., Goncalves, L. S., Engels, E. 1976. Effects of carbon dioxide on vitellogenin metabolism in unmated queen honeybees. *J. Apic. Res.* 15:3–10

40. Franks, N. P., Lieb, W. R. 1982. Molecular mechanisms of general anesthesia. *Nature* 300:487–93

41. Freckleton, W. C. Jr., Wahlsten, D. 1968. Carbon dioxide–induced amnesia in the cockroach *Periplaneta americana. Psychonomic Sci.* 12:179–80

42. Friedlander, A., Navarro, S. 1979. The effect of controlled atmospheres on carbohydrate metabolism in the tissue of *Ephestia cautella* (Walker) pupae. *Insect Biochem.* 9:79–83

43. Friedlander, A., Navarro, S. 1979. Triglyceride metabolism in *Ephestia cautella* pupae exposed to carbon dioxide. *Experientia* 35:1424–25

44. Friedlander, A., Navarro, S. 1984. The glutathione status of *Ephestia cautella* (Walker) pupae exposed to carbon dioxide. *Comp. Biochem. Physiol. C* 79: 217–18

45. Friedlander, A., Navarro, S., Silhacek, D. L., 1984. The effect of carbon dioxide on NADPH production in *Ephestia cautella* (Wlk.) pupae. *Comp. Biochem. Physiol. B* 77:839–42

46. Fuzeau-Braesch, S. 1977. Comportement et taux de catécholamines: Etude comparative des insectes grégaires, solitaires et traités au gaz carbonique chez *Locusta migratoria. C. R. Acad. Sci.* 284:1361–64

47. Fuzeau-Braesch, S., Nicolas, G. 1981. Effect of carbon dioxide on subsocial insects. *Comp. Biochem. Physiol. A* 68: 289–97

48. Fuzeau-Braesch, S., Nicolas, G., Baehr, J. C., Porcheron, P. 1982. A

study of hormonal levels of the locust *Locusta migratoria cinerascens* artificially changed to the solitary state by a chronic $CO_2$ treatment of one minute per day. *Comp. Biochem. Physiol. A* 71:53–58

49. Galil, J., Zeroni, M., Bar Shalom (Bogoslavsky), D. 1973. Carbon dioxide and ethylene effects in the coordination between the pollinator *Blastophaga quadraticeps* and the syconium in *Ficus religiosa. New Phytol.* 72:1113–27

50. Gillett, S., Nicolas, G., Fuzeau-Braesch, S. 1972. Comportement de *Locusta migratoria cinerascens* (Fab.). Evolution vers le type solitaire sous l'action périodique du gaz carbonique. II. Agrégation. *Rev. Comp. Anim.* 6:263–72

51. Gillies, M. T. 1974. Methods for assessing the density and survival of bloodsucking Diptera. *Ann. Rev. Entomol.* 19:345–62

52. Gillies, M. T. 1980. The role of carbon dioxide in host-finding by mosquitoes (Diptera: Culicidae): a review. *Bull. Entomol. Res.* 70:525–32

53. Gillies, M. T., Wilkes, T. J. 1972. The range of attraction of animal baits and carbon dioxide for mosquitoes. Studies in a freshwater area of West Africa. *Bull. Entomol. Res.* 61:389–404

54. Gillies, M. T., Wilkes, T. J. 1982. Responses of host-seeking *Mansonia* and *Anopheles* mosquitoes (Diptera: Culicidae) in West Africa to visual features of a target. *J. Med. Entomol.* 19:68–72

55. Hangartner, W. 1969. Carbon dioxide, releaser for digging behaviour in *Solenopsis geminata* (Hymenoptera: Formicidae). *Psyche* 76:58–67

56. Heran, H. 1952. Untersuchungen über den Temperatursinn der Honigbiene (*Apis mellifera*) unter besonderer Berücksichtigung der Wahrnehmung strahlender Wärme. *Z. Vgl. Physiol.* 34:179–206

57. Herrman, H. 1969. Die neurohormonale Kontrolle der Paarungsflüge und der Eilegetätigkeit bei der Bienenkönigin. *Z. Bienenforsch.* 9:509–44

58. Hodkova, M., Fuzeau-Braesch, S. 1988. Effet du gaz carbonique sur la reproduction de *Pyrrhocoris apterus* sous deux régimes lumineux: inhibition de la reproduction en jours longs. *C. R. Acad. Sci.* 306:121–24

59. Hooper, G.H.S. 1970. Use of carbon dioxide, nitrogen, and cold to immobilize adults of the Mediterranean fruit fly. *J. Econ. Entomol.* 63:1962–63

60. Hopkins, T. L., Srivastava, B. B. L.

61. Howarth, F. G. 1983. Ecology of cave arthropods. *Ann. Rev. Entomol.* 28:365–89

62. Hoyle, G. 1960. The action of carbon dioxide gas on an insect spiracular muscle. *J. Insect Physiol.* 4:63–79

63. Irwin, D. G., Smith, L. W. Jr., Pratt, J. J. Jr. 1972. Effects of carbon dioxide and nitrogen on the secretion of parabenzoquinones by *Tribolium castaneum* (Herbst). *J. Stored Prod. Res.* 8:213–19

64. Ismail, S., Fuzeau-Braesch, S. 1983. Action de l'hormone juvénile, de l'octopamine et du gaz carbonique sur la diapause de *Gryllus campestris* L. (Orthoptère). *Rev. Can. Biol. Exp.* 42: 229–33

65. Jones, J. C. 1974. Factors affecting heart rates in insects. See Ref. 19, 5:119–67. 648 pp.

66. Jones, O. T., Coaker, T. H. 1977. Oriented responses of carrot fly larvae, *Psila rosae,* to plant odours, carbon dioxide and carrot root volatiles. *Physiol. Entomol.* 2:189–97

67. Jones, O. T., Coaker, T. H. 1978. A basis for host plant finding in phytophagous larvae. *Entomol. Exp. Appl.* 24:272–84

68. Jousset, F. X. 1972. Le virus iota de "*Drosophila immigrans*" étudié chez "*Drosophila melanogaster*": symptôme de la sensibilité au $CO_2$, descriptions des anomalies provoquées chez l'hôte. *Ann. Inst. Pasteur Paris* 123:275–88

69. Kashin, P. 1973. Reversal of *gamma*-aminobutyric acid inhibition by carbon dioxide. *Comp. Biochem. Physiol. A* 44:829–50

70. Kellogg, F. E. 1970. Water vapour and carbon dioxide receptors in *Aedes aegypti. J. Insect Physiol.* 16:99–108

71. Klinger, J. 1957. Über die Bedeutung des Kholendioxyds für die Orientierung der Larven von *Otiorrhynchus sulcatus* F., *Melolontha* und *Agriotes* (Col.) im Boden (Vorläufige Mitteilung). *Mitt. Schweiz. Entomol. Ges.* 30:317–22

72. Klinger, J. 1958. Die Bedeutung der Kohlendioxyd-Ausscheidung der Wurzeln für die Orientierung der Larven von *Otiorrhynchus sulcatus* F. und anderer bodenbewohnender phytophager Insektenarten. *Mitt. Schweiz. Entomol. Ges.* 31:205–26

73. Kobayashi, F. Yamane, K., Ikeda, T. 1984. The Japanese pine sawyer beetle as the vector of pine wilt disease. *Ann. Rev. Entomol.* 29:115–35

74. Kumar, K., Saxena, K. N. 1978. Mating behavior of the cotton leafhopper, *Empoasca devastans*, in relation to its age, ovarian development, diurnal cycle, and CO$_2$ treatment. *Ann. Entomol. Soc. Am.* 71:108–10

75. Lacher, V. 1967. Verhaltenreaktionen der Bienenarbeiterin bei Dressur auf Kohlendioxid. *Z. Vgl. Physiol.* 54:75–84

76. Leprince, D. J., Jolicoeur, P. 1986. Response to carbon dioxide of *Tabanus quinquevittatus* Wiedemann females (Diptera: Tabanidae) in relation to relative abundance, parity, follicle development, and sperm and fructose presence. *Can. Entomol.* 118:1273–77

77. Lespinasse, R., Nicolas G. 1981. Correlation between B chromosome frequency and solitary phenotype production in crowded populations treated with CO$_2$ in different strains of *Locusta migratoria* L. In *The Biosystematics of Social Insects*, ed. P. E. Howse, J. L. Clément, pp. 95–106. London/New York: Academic. 346 pp.

78. L'Heritier, Ph., Teissier, G. 1937. Une anomalie physiologique héréditaire chez la drosophile. *C. R. Acad. Sci.* 205: 1099–101

79. Lincoln, D. E., Couvet, D., Sionit, N. 1986. Response of an insect herbivore to host plants grown in carbon dioxide enriched atmospheres. *Oecologia Berlin* 69:556–60

80. Lincoln, D. E., Sionit, N., Strain, B. R. 1984. Growth and feeding response of *Pseudoplusia includens* (Lepidoptera: Noctuidae) to host plants grown in controlled carbon dioxide atmospheres. *Environ. Entomol.* 13:1527–30

81. Lindauer, M., Medugorac, I. 1967. Ein sozialer Zeitgeber im Bienenstaat. In *L'effet de groupe chez les animaux*, pp. 15–25. Paris: CNRS. 390 pp.

82. Loew, E. R. 1974. Component analysis of the mass electrical responses (ERG) of the fly *Sarcophaga bullata*. *J. Insect Physiol.* 20:1737–48

83. Lovell, K. L., Eisenstein, E. M. 1973. Dark avoidance learning and memory disruption by carbon dioxide in cockroaches. *Physiol. Behav.* 10:835–40

84. Lum, P. T. M. 1974. Effect of carbon dioxide anaesthesia at eclosion upon mating efficiency of male *Plodia interpunctella* (Lepidoptera: Pyralidae). *J. Stored Prod. Res.* 10:69–71

85. Lum, P. T. M., Flaherty, B. R. 1972. Effect of carbon dioxide on production and hatchability of eggs of *Plodia interpunctella* (Lepidoptera: Phycitidae). *Ann. Entomol. Soc. Am.* 65:976–77

86. Luscher, M. 1961. Air conditioned termite nests. *Sci. Am.* 205:138–45

87. Mackensen, O. 1947. Effect of carbon dioxide on initial oviposition of artificially inseminated and virgin queen bees. *J. Econ. Entomol.* 40:344–49

88. Mardan, M., Rinderer, T. E. 1980. Effects of carbon dioxide and cold anaesthesia on the hoarding behaviour of the honeybee. *J. Apic. Res.* 19(3):149–53

89. Matsumoto, T. 1977. Respiration of fungus comb and CO$_2$ concentration in the center of mounds of some termites. *Proc. 8th IUSSI (Int. Union Study Soc. Insects) Congr.*, pp. 104–5. Wageningen, the Netherlands: Cent. Agric. Publ. Doc. 325 pp.

90. McCann, F. V. 1970. Physiology of insect hearts. *Ann. Rev. Entomol.* 15:173–200

91. McCann, F. V., Boettiger, E. G. 1961. Studies on the flight mechanism of insects. I. The electrophysiology of fibrillar flight muscle. *J. Gen. Physiol.* 45:125–42

92. Meeking, J. M., Seabrook, W. D., Paim, U. 1974. Perception of carbon dioxide by larvae of *Orthosoma brunneum* (Coleoptera: Cerambycidae) as indicated by recordings from the ventral nerve cord. *Can. Entomol.* 106:257–62

93. Mill, P. J. 1985. Structure and physiology of the respiratory system. See Ref. 38, 3:517–93. 625 pp.

94. Miller, P. L. 1974. Respiration. Aerial gas transport. See Ref 19, 6:119–67. 548 pp.

95. Miller, S. L. 1961. A theory of gaseous anesthetics. *Proc. Natl. Acad. Sci.* 47: 1515–24

96. Moloo, S. K., Kutuza, S. B. 1975. Effects of carbon dioxide anesthetic on *Glossina*. *Acta Trop.* 32:159–65

97. Mullins, D. E. 1985. Chemistry and physiology of the hemolymph. See Ref. 38, 3:355–400. 625 pp.

98. Mullins, L. J. 1975. Anesthesia: an overview. In *Progress in Anesthesiology*, Vol. 1, *Molecular Mechanisms of Anesthesia*, ed. B. R. Fink, pp. 237–42. New York: Raven. 652 pp.

99. Navarro, S., Calderon, M. 1974. Exposure of *Ephestia cautella* (Wlk.) pupae to carbon dioxide concentrations at different relative humidities: the effect on adult emergence and loss in weight. *J. Stored Prod. Res.* 10:237–41

100. Navarro, S., Dias, R., Donahaye, E. 1985. Induced tolerance of *Sitophilus orizae* adults to carbon dioxide. *J. Stored Prod. Res.* 21:207–13

101. Navarro, S., Friedlander, A. 1975. The

effect of carbon dioxide anesthesia on the lactate and pyruvate levels in the hemolymph of *Ephestia cautella* (Wlk.) pupae. *Comp. Biochem. Physiol. B* 50:187–89

102. Nicolas, G. 1972. Evolution vers le type solitaire chez le criquet grégaire, *Locusta migratoria cinerascens* (Fab.) soumis à l'action périodique du gaz carbonique. *Acrida* 1:97–110

103. Nicolas, G. 1972. Comportement de *Locusta migratoria cinerascens* (Fab.): evolution vers le type solitaire sous l'action périodique d'une minute de gaz carbonique. I. Activité locomotrice. *Rev. Comp. Anim.* 6:255–61

104. Nicolas, G. 1973. Etude morphométrique du criquet migrateur, *Locusta migratoria cinerascens* (Fab.). Evolution vers le type solitaire sous l'action périodique du gaz carbonique. *C. R. Acad. Sci.* 276:1475–78

105. Nicolas, G. 1977. La narcose par le gaz carbonique. Durée de séjour de l'anesthésique dans le corps du criquet migrateur et recherche des sites de rétention par l'utilisation du $^{14}$C. *Bull. Soc. Zool. Fr.* 102:329

106. Nicolas, G. 1978. *Etude descriptive de la réalisation expérimentale du phénotype solitaire chez le criquet migrateur,* Locusta migratoria L., *élevé en groupement: influence du gaz carbonique, recherche des mécanismes d'action.* Thesis. Univ. Paris XI, Orsay. 417 pp.

107. Nicolas, G. 1979. Influence du $CO_2$ sur la fécondité de deux souches de *Locusta migratoria* L. sensibles à la photopériode; levée de l'inhibition de la maturation ovarienne sous longue photophase. *C. R. Acad. Sci.* 288:251–54

108. Nicolas, G., Cassier, P., Fain-Maurel, A. M. 1969. Evolution vers le phenotype solitaire et persistance des glandes de mue chez *Locusta migratoria cinerascens*, phase grégaire, sous l'influence du gaz carbonique. Etude expérimentale et ultrastructurale. *C. R. Acad. Sci.* 268:1532–34

109. Nicolas, G., Ismail, S. 1978. Réalisation expérimentale de l'assombrissement tégumentaire chez le criquet migrateur *Locusta migratoria* L. Influence des neurosécrétions cérébrales: comparaison avec les effets du $CO_2$. *C. R. Seances Soc. Biol. Paris* 172:1075–78

110. Nicolas, G., Queinnec, Y. 1984. Influence du gaz carbonique sur les réponses électrorétinographiques de *Locusta migratoria cinerascens* au cours de l'ontogenèse: influence du mode de vie de l'animal. *J. Physiol. Paris* 79:149–61

111. Omer, S. M., Gillies, M. T. 1971. Loss of response to carbon dioxide in palpectomized female mosquitoes. *Entomol. Exp. Appl.* 14:251–52

112. Osbrink, W. L. A., Trumble, J. T., Wagner, R. E. 1987. Host suitability of *Phaseolus lunata* for *Trichoplusia ni* (Lepidoptera: Noctuidae) in controlled carbon dioxide atmospheres. *Environ. Entomol.* 16:639–44

113. Pasche, A., Zachariassen, K. E. 1973. Tolerance to hypoxia and hypercapnia in adult *Rhagium inquisitor* L. (Coleoptera: Cerambycidae). *Nor. Entomol. Tidsskr.* 20:323–24

114. Pauling, L. 1961. A molecular theory of general anaesthesia. *Science* 134:15–21

115. Pomeroy, N., Plowright, R. C. 1979. Larval ejection following $CO_2$ narcosis of bumble bees (Hymenoptera: Apidae). *J. Kans. Entomol. Soc.* 52:215–17

116. Press, J. W., Flaherty, B. R. 1973. Hatchability of *Plodia interpunctella* eggs exposed to a carbon dioxide atmosphere: relationship of egg age to exposure time. *J. Ga. Entomol. Soc.* 8:210–13

117. Press, J. W., Flaherty, B. R., Arbogast, R. T. 1976. Oocyte maturation in *Tribolium castaneum* after repetitive sublethal carbon dioxide exposures. *Ann. Entomol. Soc. Am.* 66:480–81

118. Printz, P. 1967. Mise en évidence d'un variant du virus de la stomatite vésiculaire (souche Indiana) conférant une sensibilité retardée au gaz carbonique chez *Drosophila melanogaster*. *C. R. Acad. Sci.* 264:169–72

119. Rajendran, S., Kashi, K. P., Muthu, M. 1977. The enhanced toxicity of methyl iodide to the adults of *Sitophilus orizae* (L.) and *Tribolium castaneum* (Herbst) in the presence of carbon dioxide. *J. Food Sci. Technol.* 14:217–19

120. Rajendran, S., Shivaramaiah, H. 1985. The differential effects of selected fumigants on the multiplicative potential of *Rhysopertha dominica* F. (Coleoptera: Bostrichidae). *Entomon* 10:7–12

121. Ralph, C. L. 1959. Modification of activity rhythm of *Periplaneta americana* (L.) induced by carbon dioxide and nitrogen. *Physiol. Zool.* 32:57–62

122. Revelle, R. 1982. Carbon dioxide and world climate. *Sci. Am.* 247:33–41

123. Ribbands, C. R. 1950. Changes in the behavior of honey-bees following their recovery from anaesthesia. *J. Exp. Biol.* 27:302–10

124. Rosen, L. L., Shrogr, D. A. 1981. Natural $CO_2$ sensitivity in mosquitoes *Culex quinquefasciatus* caused by a hereditary virus. *Ann. Virol.* 132:543–48

125. Ruelle, J. E. 1964. L'architecture du nid de *Macrotermes natalensis* et son sens fonctionnel. In *Etudes sur les termites africains*, ed. A. Bouillon, pp. 327–62. Léopoldville, the Congo: Léopoldville Univ. Press

126. Schwartz, L. M. 1981. Transegmental electrical coupling between adjacent intersegmental muscles in the tobacco hawkmoth (*Manduca sexta*). *J. Insect Physiol.* 27:727–34

127. Seeley, T. D. 1974. Atmospheric carbon dioxide regulation in honey-bee (*Apis mellifera*) colonies. *J. Insect Physiol.* 20:2301–5

128. Seeman, P. 1975. The membrane expansion theory of anesthesia. See Ref. 98, pp. 243–52

129. Sillans, D. 1973. Influence de la température subie au cours du réveil sur la réponse à l'anesthésie de *Drosophila melanogaster. Anesth. Analg. Réanim.* 30:261–70

130. Sillans, D. 1978. *Contribution à l'étude des mécanismes et des effets de l'anesthésie chez les insectes.* Thesis. Univ. C. Bernard, Lyon, France. 148 pp.

131. Sillans, D. 1979. Etude des mécanismes de l'anesthésie au CO$_2$ chez *Bombyx mori* à l'aide des variations du rythme circulatoire. *Nat. Can.* 106:349–409

132. Sillans, D., Biston, J. 1979. Studies on the anesthetic mechanism of carbon dioxide by using *Bombyx mori* larvae. *Biochimie* 61:153–56

133. Sillans, D., Esteve, J., Legay, J. M. 1969. Influence de l'anesthésie à l'anhydride carbonique sur le rythme circulatoire des chenilles de *Bombyx mori. C. R. Acad. Sci.* 269:1209–12

134. Simon, S. A., Gutknecht, J. 1980. Solubility of carbon dioxide in lipid bilayer membranes and organic solvents. *Biochim. Biophys. Acta* 596:352–58

135. Simpson, J. 1968. Le microclimat à l'intérieur de la grappe d'abeilles. In *Traité de Biologie de l'Abeille,* 1:224–34. Paris: Masson & Cie. 547 pp.

136. Singer, S. J., Nicolson, G. L. 1972. The fluid mosaic model of the structure of cell membranes. *Science* 175:720–31

137. Sinha, R. N., Waterer, D., Muir, W. E. 1986. Carbon dioxide concentrations associated with insect infestations of stored grains. 1. Natural infestation of corn, barley and wheat in farm granaries. *Sci. Aliment.* 6:91–98

138. Sorensen, A. A., Busch, T. M., Vinson, S. B. 1983. Factors affecting brood cannibalism in laboratory colonies of the imported fire ant, *Solenopsis invicta* Buren (Hymenoptera: Formicidae). *J. Kans. Entomol. Soc.* 56:140–50

139. Speirs, R. D., Zettler, J. L. 1972. Cold vs CO$_2$ immobilization for handling red flour beetles *Tribolium castaneum* in topical application toxicity studies (Coleoptera: Tenebrionidae). *J. Ga. Entomol. Soc.* 7:115–18

140. Stadler, E. 1971. Über die Orientierung und das Wirtswahlverhalten der Mohrenfliege, *Psila rosae* F. (Diptera: Psilidae). I. Larven. *Z. Angew. Entomol.* 69:425–38

141. Stange, G. 1974. The influence of a carbonic anhydrase inhibitor on the function of the honeybee antennal CO$_2$-receptors. *J. Comp. Physiol.* 91:147–59

142. Stange, G., Diesendorf, M. 1973. The response of the honeybee antennal CO$_2$-receptors to N$_2$O and Xe. *J. Comp. Physiol.* 86:139–58

143. Stark, W. S. 1972. The effect of ether and carbon dioxide on the components of the ERG of *Drosophila. Drosophila Inf. Serv.* 48:82

144. Strnad, S. P., Bergman, M. K., Fulton, W. C. 1986. First-instar western rootworm (Coleoptera: Chrysomelidae) response to carbon dioxide. *Environ. Entomol.* 15:839–42

145. Sutcliffe, J. F. 1986. Black fly host location: a review. *Can. J. Zool.* 64:1041–53

146. Tanaka, A. 1982. Effects of carbon dioxide anaesthesia on the number of instars, larval duration and adult body size of the German cockroach, *Blatella germanica. J. Insect Physiol.* 28:813–21

147. Tanaka, A. 1985. Further studies on the multiple effects of carbon dioxide anesthesia in the German cockroach. *Blatella germanica. Growth* 49:293–305

148. Thornhill, A. L., Hays, K. L. 1972. Dispersal and flight activities of some species of *Tabanus* (Diptera: Tabanidae). *Environ. Entomol.* 1:602–6

149. Tilton, E. W., Burkholder, W. E., Gogburn, R. R. 1965. Notes on the effect of preconditioning confused flour beetles with temperature variations or carbon dioxide prior to gamma irradiation. *J. Econ. Entomol.* 58:179–80

150. Trudell, J. R. 1977. A unitary theory of anesthesia based on lateral phase separations in nerve membranes. *Anesthesiology* 46:5–10

151. Turell, M. J., Hardy, J. L. 1980. Carbon dioxide sensitivity of mosquitoes infected with California encephalitis virus. *Science* 209:1029–30

152. Turner, D. A. 1971. Olfactory perception of live hosts and carbon dioxide by tsetse fly *Glossina morsitans orientalis* Vanderplank. *Bull. Entomol. Res.* 61:75–96

153. Vale, G. A. 1980. Field studies on the responses of tsetse flies (Glossinidae) and other Diptera to carbon dioxide, acetone and other chemicals. *Bull Entomol. Res.* 70:563–70

154. Vale, G. A., Hall, D. R. 1985. The use of 1-octen-3-ol, acetone and carbon dioxide to improve baits for tsetse flies, *Glossina* spp. (Diptera: Glossinidae). *Bull. Entomol. Res.* 75:219–31

155. van-Dijken, F. R., van Sambeek, J. P. W., Scharloo, W. 1977. Influence of anaesthesia by carbon dioxide and ether on locomotor activity in *Drosophila melanogaster*. *Experientia* 33:1360–61

156. Vickery, C. A. Jr., Meadows, K. E., Baughman, I. E. 1966. Synergism of carbon dioxide and chick as bait for *Culex nigripalpus*. *Mosq. News* 26:507–8

157. Vinuela, E. 1982. Influence of cold and carbon dioxide anaesthesia on the susceptibility of adults of *Ceratitis capitata* to malathion. *Entomol. Exp. Appl.* 32:296–98

158. Ward, S. C. 1971. Carbon dioxide anesthesia of the heart of the horned passalus beetle, *Popilius disjonctus*. *Ann. Entomol. Soc. Am.* 64:430–37

159. Warnes, M. L., Finlayson, L. H. 1985. Responses of the stable fly, *Stomoxys calcitrans* (L.) (Diptera: Muscidae), to carbon dioxide and host odours. I. Activation. *Bull. Entomol. Res.* 75:519–27

160. Warnes, M. L., Finlayson, L. H. 1985. Responses of the stable fly, *Stomoxys calcitrans* (L.) (Diptera: Muscidae), to carbon dioxide and host odours. II. Orientation. *Bull. Entomol. Res.* 75:717–27

161. Warnes, M. L., Finlayson, L. H. 1986. Electroantennogram responses of the stable fly, *Stomoxys calcitrans,* to carbon dioxide and other odours. *Physiol. Entomol.* 11:469–73

162. Whisenant, B. R., Brady, V. E. 1967. Effects of anesthesia on the subsequent mating behavior of *Plodia interpunctella* males. *J. Ga. Entomol. Soc.* 2:27–30

163. Williams, P. 1985. Toxicity of methyl bromide in carbon dioxide–enriched atmospheres to beetles attacking stored grain. *Gen. Appl. Entomol.* 17:17–24

164. Wilson, B. H., Richardson, C. G. 1970. Attraction of deer flies (*Chrysops*) (Diptera: Tabanidae) to traps baited with dry ice under field conditions in Louisiana. *J. Med. Entomol.* 7:625

165. Wong, P. T., Kaplan, W. D., Trout, W. E., Hanstein, B. 1972. Carbon dioxide induced changes in the electroretinogram of *Drosophila* unaccompanied by altered phototactic behaviour. *J. Insect Physiol.* 18:1493–500

166. Wood, M. R., Argiro, V., Pelikan, P., Cohen, M. J. 1980. Glycogen in the central neurons of insects: massive aggregations induced by anoxia or axotomy. *J. Insect Physiol.* 26:791–99

167. Woodring, J. P., Clifford, C. W., Roe, R. M., Beckman, B. R. 1978. Effects of $CO_2$ and anoxia on feeding, growth, metabolism, water balance and blood composition in larval female house crickets. *Acheta domesticus. J. Insect Physiol.* 24:499–509

168. Young, S. Y., McDonald, L. L. 1970. Effect of $CO_2$ anaesthesia on malathion toxicity to four species of stored product insects. *Ann. Entomol. Soc. Am.* 63:381–82

*Ann. Rev. Entomol. 1989. 34:117–37*

# ECOLOGICAL CONSIDERATIONS IN THE MANAGEMENT OF *DELIA* PEST SPECIES IN VEGETABLE CROPS

## Stan Finch

AFRC Institute of Horticultural Research, Wellesbourne, Warwick CV35 9EF, United Kingdom

## INTRODUCTION

Flies belonging to the genus *Delia* are members of the Anthyomyiidae, a taxonomically difficult subfamily of the Muscidae (Diptera) as many of the characters used to differentiate the genera and species are not always constant (13). The various keys (76, 81) for separating the species can be difficult to use, even when large numbers of different species are available. To the field ecologist, keys to indicate how pest species differ from other insects collected in samples (5) may be more useful than keys restricted solely to anthomyiid species.

Certain common *Delia* species are pests of economic importance. The larvae of *D. coarctata* and *D. platura* attack cereals, those of *D. radicum* and *D. floralis* the roots of brassica crops, and those of *D. antiqua, D. platura,* and *D. florilega* the roots of onions (13). This review concentrates on *Delia* species that attack vegetable crops and in particular on *D. radicum* (the cabbage maggot) and *D. antiqua* (the onion maggot), the two pests that have been studied in greatest detail. As parallel work has often been carried out on both species, certain points will be illustrated by reference to only one species to save space.

Like most pest species, these anthomyiid flies have experienced frequent changes of name. Within the last 20 years, the genus of the cabbage maggot has been called *Hylemya* in the United States and Canada; *Hylemyia* in Norway; *Phorbia* or *Chortophila* in France, Germany, and Russia; and *Delia* or *Erioischia* in Belgium and England (10). Until 1981, *brassicae* was

117

accepted universally as the specific name for the cabbage maggot, but with the finding of a specimen named *Musca radicum* (the root-eating fly) in the Linnean collection, *radicum* superseded *brassicae* (71). Synonyms for *D. antiqua* have been given by Loosjes (63). Common names are equally confusing. *D. platura*, the seedcorn maggot in the United States, is referred to as the bean seed fly in Europe. Care is required when reviewing literature on *Delia* damage to onion crops, as some authors refer to the "onion fly/onion maggot," specifically *D. antiqua*, whereas others refer to "onion flies/onion maggots," which can be *D. antiqua*, *D. platura*, *D. florilega*, a mixture of any two, or all three.

This review aims to assess the merits of methods that might reduce the amounts of insecticide applied to protect crops against *Delia* pests. It includes insecticidal studies, as insecticides will undoubtedly remain the major method of controlling these species for the foreseeable future. With numbers of available insecticides decreasing and the likelihood that new insecticides will be limited, certain authors (e.g. 45) have stressed the need for a more integrated approach to crop protection. The main emphasis of this review is on the progress and the difficulties of successfully integrating noninsecticidal methods of crop protection into everyday control practices.

## PEST DISTRIBUTION, CROP DAMAGE, AND LOSS ASSESSMENT

The truly phytophagous *D. radicum*, *D. antiqua, and D. floralis* (the turnip maggot) and the saprophagous *D. platura* are restricted largely to the temperate zone of the Holarctic region (35–60° N). The life histories and the numbers of generations each year are shown in agricultural textbooks (1, 55). Female flies of the phytophagous species generally lay their eggs in the soil alongside their selected host plants (22, 29), but the saprophagous *D. platura* oviposits alongside decaying plant material, including onions damaged by *D. antiqua* and seeds that are slow to germinate. *Delia platura* is not a primary pest; it invades large seeds sown in cold soils, but only when the seed coats have become infested with pathogens before the plants germinate. The odors associated with microbial decomposition of the seed coat stimulate *D. platura* to oviposit (2), and once the larvae start feeding they invariably attack the living seedlings.

Seedlings not protected by insecticide are often damaged severely by *Delia* pests. In some seasons, *D. radicum* infestations kill 90% of the plants in some untreated brassica crops in North America (59) and mainland Europe (10). In the United Kingdom, plant losses may reach 60% in untreated crops, but estimates made twenty years ago (the only accurate ones available) suggest that, on average, losses are nearer 24% (85). Depending upon crop type, losses of untreated onion plants to *D. antiqua* average 24–40% in Canada (45). When the onion fly developed resistance to organochlorine insecticides, plant losses in the Netherlands were 50–100% depending on soil type (63). Vigorously growing crops can support large populations of larvae without showing signs of attack, but crops that sustain larval damage to the part of the plant used for human consumption, e.g. brussels sprouts, swede (rutabaga), and onion, are reduced in quality (102). This type of damage is the main reason why it is often necessary to control small populations of *Delia*.

## SAMPLING METHODS

Methods for assessing the numbers of each of the stages in the life cycle of *Delia* species have been reviewed (26). Although accurate sampling methods have been developed to answer specific research questions (26), few have been developed specifically to obtain the best estimate of the pest population for a given cost (90). Consequently, most sampling methods are inappropriate for use in commercial management systems (107), as the number of samples required is often prohibitively large (26). However, as advisors in the field and farmers generally have little time in which to assess pest numbers, only simple techniques of sampling are acceptable for use on a field scale. Frequently, crop protection decisions must be made on the basis of a few samples taken from a large number of fields, with comparisons among fields used to indicate those most likely to benefit from pest control measures.

## ADVANCES IN CROP PROTECTION TACTICS

### Cultural Control

CROP HYGIENE   Although crop hygiene is considered important in pest control systems, much of the advice given has remained unchanged for many years, even though farming practices may have changed radically during the same period. Therefore, methods of cultural control may need to be challenged before being accepted. For example, covering cull piles of onions with soil and removing volunteer plants from onion fields when the spring population of flies had laid eggs on them were considered important in reducing populations of *D. antiqua* (64). However, recent studies indicated that neither source was important in supporting *D. antiqua* populations (31). Bulbs damaged or crushed during harvesting and left in the field at the end of the season provided the major source of this pest (31, 64). Hence, to avoid the buildup of overwintering populations of *D. antiqua,* crops should be harvested using methods that minimize bulb damage (20, 31). During winter, few *Delia* pupae are killed by low temperatures (106) or predators (37). However, survival of *D. radicum* pupae was reduced by 75% when infested fields were plowed in the autumn (37). Spring plowing reduced pupal survival by only 40%. Thus, whenever possible, crop fields should be plowed in early winter to reduce *D. radicum* populations. Whether the same is true for *D. antiqua* populations needs to be tested.

CROP ROTATION    Crop rotation is used primarily to avoid depleting soils of nutrients and to avoid the buildup of soil pathogens. There have been few attempts to rotate crops specifically to avoid insects. This is understandable, as specialist feeders such as *Delia* spp. are capable of dispersing thousands of meters in flight (36, 63). The use of rotation to isolate crops is often impractical on an individual-farm basis because of the large areas involved in infestations, but it would be effective if it could be implemented on a regional basis.

It has often been proposed that wild host plants serve as bridges that prevent effective crop isolation. However, *D. antiqua* has few wild hosts (63), and although *D. radicum* can develop on a wide range of wild plants, root flies are rarely found on these species in the field (29). Therefore, the influence of wild host plants in maintaining populations of *Delia* spp. may have been over-estimated.

INTRODUCTION OF NEW CROPS    It was feared that the introduction of oilseed rape ("Canola" in Canada) as a break crop in the United Kingdom would considerably increase pest problems in cruciferous vegetable crops. Fortunately, a series of agronomic changes inadvertently prevented this. When first introduced 15 years ago, the oilseed crop was sown largely in the spring, and its seedling stage coincided with the first-generation (spring) attack by *D. radicum*. Rapidly growing oilseed rape seedlings produce numerous visual (74, 75) and chemical (25, 49, 66, 84, 97) stimuli that make the seedlings highly attractive to *D. radicum* at a stage in their growth when they are most susceptible to damage. Hence initially *D. radicum* was regarded as a considerable threat to spring-sown oilseed rape crops (101). Despite this drawback, oilseed rape was welcomed by cereal growers because, unlike earlier break crops such as potatoes, it could be harvested with the machinery used for cereals. Unfortunately, the harvest of spring-sown crops coincided with the cereal harvests; therefore overwintering crops were favored, as they permitted the oilseed to be harvested before the cereals. For this reason, practically all of the British crop, which now covers an acreage approximately five times larger than that planted with cruciferous vegetables (101), is now sown in autumn. Unless *D. radicum* damages the roots of the oilseed crop in the autumn, it cannot become established in the following year because healthy roots become extremely tough during winter (33). The steady expansion of the oilseed rape crop, its high plant density (100–140 plants/m$^2$), and the crop's ability to compensate for dead plants have helped reduce the impact of *D. radicum* infestations. Only the few crops sown in early August are now established sufficiently to coincide with *D. radicum* attacks in those years when a third generation of flies occurs (27). Hence, a new crop that threatened to increase *D. radicum* infestations appreciably in vegetable-growing areas in the United Kingdom has not done so.

Although only spring-sown oilseed rape crops can be grown in northern Canada, plant mortality from the combined attacks of *D. radicum* and *D. floralis* have ranged from only 1 to 14% (44). Provided oilseed crops continue to be rotated sufficiently, plant mortality could remain low, as both pests are restricted to one midsummer generation in northern Canada.

SOIL AND CROP COVERS    It is possible to prevent *Delia* species from laying eggs on brassica plants by covering seed beds with cheesecloth or by covering the soil around the base of plants with tarred felt discs (83). Although the disc type of protection has been improved (82), the complete crop cover approach has been ignored, largely because the original covers were detrimental to crop growth, expensive, and difficult to apply. However, changes in agronomic practices have overtaken pest control systems. The cheap lightweight covers developed to extend the cropping season of vegetables in northern temperate climates also keep foliage- and root-feeding pests off crops (53). Crop covers could also be used effectively to enhance biological control by released parasites and particularly pathogens. The covers produce high-humidity microclimates that favor the spread of certain pathogens.

SOWING, PLANTING, AND HARVESTING TIMES    A common method of protecting crops against *Delia* spp. is to vary the times of sowing or planting crops to avoid invasion by migrants or vulnerability during the main egg-laying periods (9). Sowing or planting early in the season allows plants to become sufficiently established to tolerate moderate levels of root fly attack. Sowing or planting late to avoid attack is also an effective strategy against *Delia* species (9). These strategies may not be possible against certain species. For example, *D. platura* populations can be extremely high at favored sites even when the overall population is in decline (52). Hence, planting at "low risk" dates does not seem to be a feasible technique for reducing damage caused by *D. platura*. Attempts to establish crops between the pest generations requires more detailed information about the biology of the pest. With pests that have an innate tendency to emerge late in certain localities (27), delayed planting may increase, rather than decrease, crop damage.

INTERCROPPING    Growing two crops at the same time on one piece of land, usually called "intercropping," is a cultural system practiced commonly in the tropics. It can provide higher crop yields through reduced weed competition, soil conservation, and better use of incident radiation, water, and soil nutrients (9). Also, insect attack is often less than on monocultures (9). Root (77) suggested that intercropping reduces the numbers of phytophagous pests because mixed plantings attract more predators and reduce the ability of the pest insect to find its host plants. Coaker (9) supported these proposals, citing examples (16, 78, 89, 98) in which intercropping brassicas with taxonomically unrelated crops reduced infestations of *D. radicum*. An alternative hypothesis is that the cruciferous plants in these intercropped systems may have attracted fewer pest insects than the plants in the monoculture because they were smaller and thus less attractive to insects during the host-plant selection process (28). Potential advantages of intercropping require substantiation using strict experimental procedures. In the experiments carried out to date, spatial distribution and plant growth have been varied at the same time. Until these factors are investigated independently, conclusions concerning benefits to be derived from intercropping must remain tenuous.

## Development of Insect Traps

COMPARISON OF TRAP TYPES    Traps for capturing *Delia* adults include cones (15, 19, 100), flight-interception traps (63), and a wide range of different sizes and shapes of sticky (64, 79) and water traps (26). Frequently traps have been used to determine how *Delia* species locate host crops (26, 38, 48, 70). Some authors consider such information useful for pest control, but no firm suggestions have been made as to how the information could be used.

TRAPS FOR MONITORING FLUCTUATIONS IN PEST NUMBERS    These traps are used specifically to indicate the most appropriate times to apply insecticide treatments. Although relatively ineffective traps will provide this information, the trend has been to produce highly effective traps whenever possible. Only a few traps can be used in each field for economic reasons, and failure to monitor the start of a new generation can prove costly in terms of crop damage. Traps should monitor changes in the pest population without catching too many other insects. Attempts have been made to identify the most attractive color for traps for monitoring *Delia* spp. (47, 61, 99). Most traps are some shade of yellow, a color that also attracts beneficial insects. Vernon (99) concluded that blue and purple traps increase trap selectivity, but since these colors were tested only in late May, season-long comparison with yellow traps is still needed. Usually, the capture of too few pest individuals can be rectified by increasing the size of the trap. It is not helpful to catch more flies than are needed to obtain a reasonable estimate of fluctuations in the size of the population, as identifying and counting large numbers of flies is time-consuming. Attractants should not be added to make monitoring traps more effective; attractants derived from food-based baits, which deteriorate rapidly, give variable results (18, 66). Similarly, traps baited with pure synthetic host-plant chemicals are rarely consistently attractive to *D. antiqua* (65) or *D. radicum* (39). It seems advisable that traps used for monitoring populations of *Delia* spp. should be based solely on physical attractants.

TRAPS FOR DETERMINING WHETHER TO APPLY INSECTICIDES    In Canada, the numbers of *D. radicum* flies caught on sticky traps are used to decide whether or not to apply insecticide sprays. In rutabaga crops, where little or no damage can be tolerated, the spray thresholds are very low, e.g. one fly per trap per day (79). Such a threshold indicates little more than the presence or absence of flies, and it would be exceeded practically every day in most brassica crops in Europe. Unless spray thresholds are higher, the value of using traps to determine whether they have been surpassed must be questioned, as the information obtained is no better than that from monitoring traps. The major difficulty in using trap data to decide whether or not to apply insecticide is in deciding where to place the traps so that the number of flies caught represents the infestation in the field as a whole. Often, the most easily accessible site is chosen. Hence, traps are frequently placed alongside hedgerows, where they catch high numbers of males; these catches may or may not correctly indicate the numbers of females in the vicinity. Attempts to relate numbers of root flies to subsequent crop damage have not been successful (44).

TRAPS FOR DIRECT CONTROL    Water traps have provided successful direct control of *D. radicum* in only one instance, and even then the number of traps required was prohibitively high (180 traps/ha) (34). During September in England, when *D. radicum* populations are large, a water trap releasing the attractant allylisothiocyanate (ANCS) can catch up to 20 times as many *D. radicum* flies as a standard water trap. However, attempts to capitalize on this trap have been unsuccessful. Although three ANCS traps caught 12,500 *D. radicum* females from a plot of 5000 cauliflower plants during the third generation of flies, similar numbers of eggs (5 per plant per week) were recovered both from this plot and from the control plot, where fly activity was only monitored. Trapping high numbers of females in the autumn does not necessarily improve crop protection, as it appears that ANCS traps preferentially catch females that are not going to lay eggs. High numbers of *Delia* adults associated with low numbers of eggs is a regular phenomenon at the end of the season (39, 44, 80).

## Chemical Control

REDUCING AMOUNTS OF INSECTICIDE PER APPLICATION    Considerable work has been carried out over the past 30 years to increase the proportion of the insecticide dose that reaches the target pest and to decrease the amounts of insecticide entering the environment at large. The trend has been away from broadcast treatments and toward more localized treatments in which insecticide is confined around the roots of crop plants. Less than 50% of the insecticide needed for continuous band treatments of crop rows may be needed for spaced crops when products are applied as spot treatments around the base of the plant (88). More effective use of insecticides can also be achieved by improving their placement in the soil. For example, deep distribution of a granular insecticide, using a Matco Verba® vertical band applicator during drilling, protected long white radish *(Raphanus sativus)* against *D. radicum* more effectively than a shallowly incorporated, surface band application (92). The precise placement of low-volume liquid insecticide formulations under radish seeds at drilling halved the amount of insecticide required to reduce the numbers of *D. radicum* larvae by 90% (93). It is likely that the amount of insecticide required to protect some vegetable crops against root-feeding *Delia* pests can be reduced further by film-coating the seeds (88).

REDUCING THE NUMBER OF PESTICIDE APPLICATIONS    Practical systems for forecasting the times of attack by *Delia* species have been devised so that insecticide sprays may be applied to coincide with pest attacks. This approach usually reduces the number of insecticide sprays necessary in any one season. Models for forecasting the times of insect attack are generally based on some form of accumulated temperature (19, 69, 86, 109), often referred to as accumulated day-degrees ($D°$), as the rate of insect development depends directly upon temperature. The models are devised by determining the pests' rate of development under a range of laboratory conditions, monitoring the pests' activity under field conditions, and finally combining the two approaches. The advantage of using a physiological time scale (here accumulated $D°$) rather than a fixed calendar date for insecticide applications is that 10% oviposition by second and third generation *D. radicum* can vary by as much as 4 wk from a warm to a cool year. Thus insecticide applied on the mean date could be applied either two weeks early or two weeks late. Adopting this modeling approach, Collier & Finch (12) showed that after the overwintering population had emerged *D. radicum* required 580 $D°$ accumulated above a base temperature of 6°C at a depth of 6 cm in the soil to complete a generation. Similar models have been developed for forecasting the times of attack by populations of *D. platura* (43), *D. florilega* (94), and *D. antiqua* (3, 62, 105).

The accuracy of the *D. radicum* model, initially restricted to accumulated temperatures, has now been improved considerably by the inclusion of parameters for the varied times of spring emergence, the effect of the summer resting phase (estivation), and the effect of shortening daylength (27).

## Biological Control

INTERSPECIFIC COMPETITION    Biologically derived chemicals that influence insect behavior are of great interest for broadening the basis of pest control in an acceptable manner (58). The presence of one pest species on a plant may influence whether another will select and successfully colonize that plant. Often the insect already present secretes or excretes repellent, alarm, or warning chemicals that signal others of the presence of the insect. For example, sinapic acid (57) in the frass of larvae of the garden-pebble moth, *Evergestis forficalis,* deterred *D. radicum* from laying eggs on otherwise acceptable brassica plants (58). Few studies have been made on how one insect species affects another during host-plant selection (42). The numbers of biologically derived deterrent chemicals involved in such systems could be prodigious; identification of such chemicals is likely to prove rewarding.

FUNGI    In the field, the fungus *Entomophthora muscae* regularly kills adults of *Delia* species. It becomes particularly common following warm moist conditions, when dead, diseased flies can be seen clinging in a characteristic manner to the foliage of many hedgerow plants. A second fungus, *Strongwellsea castrans,* which sterilizes the flies, is reported more frequently from the United Kingdom than from North America (68), probably because the fungus is easier to detect in insects caught in water traps than in those caught in sticky traps. At certain times the sunken white body between the abdominal sclerites caused by *S. castrans* is found in as many as 90% of the flies (41).

Both fungi have been suggested as possible biological control agents for reducing infestations of *D. antiqua* (7), *D. radicum*, and *D. floralis* (60), though the authors clearly recognize the difficulties inherent in this approach. Fungal pathogens are usually costly to produce, and individual hosts within an insect population show a great variation in susceptibility to infection (6). The major constraint with these pathogens is their susceptibility to environmental factors, which in broad terms also include pesticides (see below). Conidia of both fungi appear to be most effective along crop boundaries, where the moisture that typifies grassy hedgerows protects conidia from desiccation and harmful ultraviolet radiation (7). The onion canopy itself may not be important in the secondary infection of *D. antiqua* adults by *E. muscae* because spore germination is completed in the crop between 6:00 and 10:00 AM (7). However, as this is one of the periods in the day when flies are most active in the crop (32), infection in the crop needs to be studied further. In comparison to the area of the crop, the boundary area is relatively small; although it may contain more insects per unit area than the crop, most of the insects, and particularly the females, are in the crop. Hence, a relatively low level of infection in the crop may have a greater effect than reduction of boundary sources in regulating the overall fly population. A further difficulty limiting fungi as potential control agents is that insects are often not infected until after they have laid their eggs. Killing a large proportion of the male flies that aggregate along crop boundaries may have little value in population control, as *Delia* females mate only once and individual males can mate successfully with as many as 40 females (73).

Other studies on fungal control of *Delia* spp. have been restricted to laboratory experiments. In one, isolates of *Paecilomyces farinosus* and *Paecilomyces fumosoroseus* consistently killed more pupae and emerging adults of *D. antiqua* than nine other entomopathogenic hyphomycetes (72). The authors suggested that these fungi might be evaluated against *D. antiqua* as soil-applied biological control agents, but they failed to suggest how such treatments could be applied effectively and/or economically. Although low concentrations of spores (2.5 and 8.4 $\times$ $10^5$ spores/ml) killed 50% of the pupal population, large inocula would be required against pupae, which are often distributed 20 cm deep in organic field soils. The fungi might be incorporated into the soil at drilling, but as they have little effect against *D. radicum* larvae (72), the infectivity of the initial inoculum may decrease considerably in the 5–10 wk period before the pest insects pupate. It is likely that most conidia sprayed onto germinated crops would become concentrated in the surface layers of the soil, particularly in highly organic soils. The incorporation of fungi into the soil to reduce overwintering populations of *D. antiqua* after onion harvesting would also require a large fungal inoculum. Also, the fungi may not be as effective during winter, when temperatures are low, as in the summer. Diapausing pupae may be less susceptible to the fungi than nondiapausing pupae. Finally, the original results were obtained by keeping the test insects in closed containers to maintain the relative humidity close to saturation. Even with intensive irrigation, such conditions would be difficult to produce in the field. Species of fungi that are effective in soil much drier than field capacity may eventually prove more valuable than those selected under high-humidity conditions in the laboratory.

NEMATODES    The results of experiments testing the effectiveness of nematodes against *Delia* spp. have been inconsistent, probably because most experiments were done under conditions that were optimal for the nematodes but suboptimal for the pests. For example, Morris (67) described the susceptibility of third-instar larvae of *D. radicum* and *D. antiqua* to the nematodes *Steinernema feltiae* and *Heterorhabditis bacteriophora* at 25°C in sealed petri dishes containing free water. Food was not provided for the insects. The success of the nematodes was judged by dissecting the dead insects and looking for the nematodes after 8 days of exposure. As few as 125 nematodes per insect killed 80% of the *D. radicum* larvae tested, whereas only 83 nematodes per insect were needed to kill all of the *D. antiqua* larvae. However, maintaining *D. antiqua* and *D. radicum* under such conditions is far from ideal. Morris (67) suggested that these *Delia* spp. should be good candidates for control by nematodes, because their eggs are laid and hatch at or near the soil surface, where the insect larvae would be exposed to nematodes applied as a spray. To protect crops effectively, the nematodes must be as effective against first- and second-instar larvae, which usually cause most of the crop damage, as against large third-instar larvae. In evaluation experiments, nematodes should be tested against the appropriate instar of the pest insect at temperatures optimal to the insect. Most of all, the nematodes should be tested under conditions as similar as possible to those of the insects' natural environment because environmental factors may influence the efficacy of control. For example, breakdown products from the host plants may confer some protection for pest insects by deterring nonadapted organisms from entering the environment.

PARASITOIDS    In the field, many *Delia* pupae are parasitized by the cynipid wasp *Trybliographa rapae* and the staphylinid beetles *Aleochara bilineata* and *Aleochara bipustulata*. European attempts to use parasitoids in biological control systems have concentrated on the beetles. Not only do *Aleochara* larvae parasitize the fly pupae, but moreover, adults of this genus have been shown to consume about 10 fly eggs or first-instar larvae each day in laboratory experiments (51). Whether they eat these numbers of eggs in the food choice situations available to them in the field remains to be answered. In the field the beetles emerge after most of the fly eggs have been laid, so growers would have to release them early to capitalize on their predatory behavior. They would then have to remain near the crop plants for some weeks to be effective as parasitoids. Despite these reservations, attempts are being made to rear *A. bilineata* in numbers sufficient for inundative release (4, 53). The numbers of beetles required are speculative and have been estimated at between 20,000 (4) and 650,000 (51) beetles per hectare. Using the rearing technique described by Whistlecraft et al (103), 10 hr of labor would be needed to produce 20,000 *Aleochara*. The work, and thus the costs, involved in rearing enough parasitoids to treat large areas would be daunting.

The use of hymenopterous parasitoids against *D. antiqua* seems more promising. The parasitoid *Aphaereta pallipes* can be produced in large numbers (average 15 per pupa) from mass-reared *D. antiqua* (45, 104). Unfortunately, the parasitoid females do not leave marking pheromones on the host larvae in which they have laid eggs. Hence many eggs may be laid in easily accessible pest larvae and none in others. In addition, *A. pallipes* cannot reduce damage caused by the existing *Delia* infestation because, irrespective of when eggs are laid, they do not hatch until the *Delia* larvae start to pupate. In a small-scale field experiment, parasitism of *D. antiqua* in a release area was 10%, compared to < 1% in the control area (96). However, under field conditions and in miniature mass-rearing beds designed to attract high numbers of parasitoids, overall parasitism rarely exceeded 20%. The contribution from this parasitoid appears to be restricted largely to stabilizing the population from one generation to the next.

PREDATORS    In the late 1950s and early 1960s, detailed information concerning the detrimental effects of organochlorine insecticides on the soil fauna in vegetable crops was obtained at Wellesbourne, United Kingdom, by three different groups of research workers (11, 54, 108). All three groups estimated that predatory carabid and staphylinid ground beetles ate about 95% of the eggs and early larval instars of *D. radicum*. They therefore stressed the importance of these beetles in helping to reduce infestations of this fly. However, their estimates of predation were obtained in soil treated regularly with DDT, in which the numbers of many soil-inhabiting organisms such as Collembola and earthworms were depleted severely. Hence surviving ground beetles relied on immigrating plant feeders, such as *Delia* spp., for food. It is not surprising that under such conditions estimates of predation of *Delia* eggs were extremely high.

Following the replacement of organochlorine insecticides by less persistent, often ecologically less detrimental "softer" organophosphorus and carbamate insecticides, the beneficial soil fauna appears to have reverted to its status from prior to the organochlorine era. Recent results have indicated that predation of *D. radicum* eggs by predatory ground beetles rarely eliminates more than 30% of the eggs laid around plants (40). This is probably because the ground beetles now have many other types of animal material available as food and often feed on these in preference to *D. radicum* eggs.

Many (60–100) species of carabid and staphylinid ground beetles are considered important predators of eggs and larvae of *Delia* spp. Numerous authors have suggested that some of these beetle species could be used to control *Delia* spp. directly. However, a factor often overlooked is that not one of the many beetle species preys specifically on *Delia* eggs and/or larvae. The beetles are opportunistic feeders with limited powers of dispersal. Their beneficial effect is in maintaining pest populations at more or less constant levels from year to year. For insecticides to be consistently effective at reduced application rates (88, 92, 93) it is essential that a healthy ground beetle fauna is maintained. Wide fluctuations in the fauna place great stress on insecticides, which all act in a density-independent manner.

STERILE INSECTS    To control field populations of *Delia* spp., flies of the same species may be sterilized in the laboratory and then released into the natural population, or the field population may be sterilized in the field. The latter method was tested against *D. radicum* in England to avoid problems associated with the mass-rearing and release of sterile flies (35). Although baits containing the chemosterilant tepa [tris-(1-aziridinyl)phosphine oxide] sterilized up to 94% of the flies in large field cages, sterility in the open field rarely exceeded 30% of the egg population, even when bait stations were spaced only 2.5 m apart in experimental crops (35). Releasing laboratory-reared sterile flies was no more effective than using baits against *D. radicum,* even though the number of sterile insects released was 30 times the number of flies in the field populations (50).

Most of the early success with sterile insects released against *Delia* species was achieved by a group of entomologists in the Netherlands that was funded specifically to determine the efficacy of this technique for control of *D. antiqua* populations (95). The Netherlands project involved determining the dispersal range of the flies to measure "isolated areas" (63), sterilizing the insects without lowering their competitiveness (95), and determining the numbers of sterile flies required for release into the natural population (91). In addition, efficient methods were developed for mass-rearing the flies, and more importantly, flies were produced that could compete with or even outcompete their wild counterparts (95). Although the sterile-male technique is now used to control *D. antiqua* in the Netherlands, it has not received unanimous support there, partly because sterile flies have to be released for at least five years before the method becomes economically attractive (95). In addition, some growers prefer insecticides that, on heavier soils, also reduce damage to onions caused by stem nematodes.

The sterile-male release technique has also been tested against *D. antiqua* in Canada (45). The Canadian workers were dissatisfied with the quality of the flies produced in the laboratory, and instead produced large numbers of flies in cold-frames, using less than half the man-hours needed for laboratory rearing (46). In addition, the flies produced in this way were larger and more vigorous than those reared in the laboratory. Although the ratio of sterile to wild insects trapped in Canada was only 2.5:1 in the year following release, the sterile insects produced a marked effect, albeit in an area where insecticides were also being used.

Sterile-male release may be employed most effectively in an integrated approach in which the intrinsic rate of increase from one generation to the next, calculated by the Dutch researchers to be fivefold (63), is prevented by the judicious use of insecticides. It may be most advantageous to rely on soil-applied insecticides to establish the crops and then to use the sterile-male approach as a substitute for routine midseason sprays (45), particularly as many of these sprays appear to do more harm than good. (30). The number of insects required would then be much lower than the 20:1 ratio considered by the Dutch workers to be necessary during the first year in a system relying totally on the sterile-male approach. Onion crops grown in small isolated pockets of muck soil, as in Canada and the United States, are more amenable to this type of area management than large areas of contiguous crops.

## Resistant Cultivars

Breeding crops resistant to pests is the most obvious method for protecting the crops without using insecticides. It is now generally accepted that there is little chance of producing a cruciferous cultivar that is resistant to all insect pests. For example, although three red cabbage cultivars had low attractancy for egg-laying by *D. radicum,* none showed any resistance to attack by flea beetles (*Phyllotreta* spp.) or cabbage stem weevil (*Ceutorhynchus quad-ridens*) (23). Consequently, the main emphasis in northern temperate regions has been on producing cultivars resistant to the major pests, namely the *Delia* species (27). Despite concerted efforts in the last 15 years, little resistance to attack by *Delia* species has been found in cultivated cruciferous crops, and the resistance identified has not always been durable. Attempts to select radish resistant to *D. radicum* were unsuccessful, largely because the relative attractiveness of the different radish families changed as the plants aged (14). Similarly, attempts to produce cauliflower breeding lines resistant to this pest were unsuccessful, for reasons that have not yet been explained (24).

Bulb onion (*Allium cepa*) cultivars have been screened extensively for resistance to *D. antiqua.* To distinguish plants with true resistance from those that escaped damage because they were not attacked, fly infestations in screening trials must be severe. Usually, appropriate levels of discrimination are achieved when the pest population damages 80–90% of the plants of each susceptible cultivar tested (22). Under these conditions, cultivars of onions with some degree of resistance have been identified. Although certain accessions of *Allium fistulosum* (Welsh onion) were more resistant to damage (22) than *A. cepa* cultivars, others, particularly those of West German origin, were not (17). It is difficult to predict the levels of resistance that may eventually be attained against soil-inhabiting *Delia* pests.

Unfortunately, the flies lay eggs alongside brassica and onion seedlings, which because of their small size, are extremely susceptible even to low pest infestations. Older brassica plants can generally tolerate damage, though the damage is only acceptable if it is indirect. With onions, protection of crops is more exacting, since all damage is direct. Even so, undamaged bulbing plants seem able to resist penetration by maggots (30, 32). Current crop protection strategies could be improved by the selection of bulb onion cultivars with mid-season resistance to maggot invasion (30). Benefits from physical resistance of this type are vitiated immediately if bulbs are damaged during cultural operations (20, 30, 32). It is ironic that much of the crop damage that assists *D. antiqua* populations to remain at high levels in onion crops is produced by the tractors applying routine midseason sprays against the flies (32).

## Interactions Among Control Tactics

Regrettably, entomologists have little influence over chemicals other than insecticides that are applied to ensure that crop yields are acceptable. It is now unusual for any crop to be grown without herbicides, yet only a few workers (8, 21) have considered how these chemicals affect the beneficial fauna in cultivated fields. More is known about the effects of insecticides, though little effort has been made to incorporate the findings into practice. Hence, even though insecticides such as chlorfenvinphos are about 1000 times less toxic than other insecticides to certain beneficial parasitoids (45), toxicity to beneficial organisms is only rarely used as the basis for selecting which chemical to apply in the field. Apart from *Aphaereta pallipes*, most of the other parasitoids of *Delia* spp. attack late in the life cycle of the pest and consequently are subjected to relatively low concentrations of the pesticides. In addition, the parasitoids do not search for host plants at random, and they do not go to those without *Delia* larvae, where the pesticide concentrations may be high. Repellent effects of pesticides and the odors, sounds, and vibrations produced by feeding larvae of the pest species probably help to direct parasitoids away from the higher concentrations of pesticides and toward infested hosts (56).

The direct and residual effects of three herbicides, three fungicides, and one insecticide indicated that the parasitoid *A. pallipes* was more susceptible than *D. antiqua* to each of the three different types of pesticide (8). Applications of certain fungicides were also detrimental to the survival of conidia of *Entomophthora muscae;* conidial mortality approached 100% immediately after standard fungicide sprays were applied to onion fields (7). An unexpected interaction was that, although sprays of the insecticide malathion had no differential effect on the mortality of healthy flies versus flies infected with *E. muscae,* the behavior of the infected flies changed after the sprays had been applied. Instead of typically climbing to the top of the plant foliage, a good point for the subsequent wind dispersal of the spores, the flies moved downward, and the fungi thus sporulated on the soil. This alteration in the flies' behavior was considered highly significant in limiting the spread of this pathogen in the field (8).

## CONCLUSIONS

The successful integration into crop protection systems of methods of controlling *Delia* spp. that are less environmentally hazardous than insecticides will depend largely on the other pesticides that are applied to protect crops against other pest insects, pathogens, and weeds. Unless systems involving biological control are carefully integrated with other control measures, benefits that may be derived from their use could be lost readily by the inadvertent use of inappropriate pesticides. Biological control is more suited to crops like onions in which the pest complex is relatively restricted, and the chance application of an inappropriate pesticide less likely, than to crops such as crucifers that have a wide range of pests and pathogens, several of which may need to be controlled with pesticides in any particular year. The most reliable crop protection strategy in the latter situation is to use selective rather than broad-spectrum pesticides. However, it is not likely that highly selective chemicals will be developed by agrochemical companies, because owing to the small markets they would not be likely to recoup the costs of product development, let alone make a profit. It is also possible that broad-spectrum chemicals may be advantageous in controlling more than one pest and that an emphasis on greater selectivity may create as many problems as it solves.

Apart from the use of plant resistance, the simplest way to minimize pest damage without using insecticides would be to cover crops at the times of peak immigration of the major pest. Although this puts no outside constraints on the subsequent use of fungicides or herbicides, it would raise the cost of crop production, as would most other methods involving integrated pest management approaches (45). An alternative strategy is to move toward systems of organic crop production, growing the crops out of season to avoid attacks by particularly damaging pests and accepting either lower-quality produce or lower yields. Both consequences would raise food prices. Whether or not this would prove acceptable remains to be seen. Whatever happens, the switch from insecticides to alternative methods of pest control is likely to occur slowly unless soil-applied insecticides are no longer effective (87) or no longer available. The biological information required to produce damage-free crops using insecticides is infinitesimal compared to the biological information required to produce a comparable crop with the methods of crop protection currently suggested as alternatives to insecticides. If the era in which we can rely solely on insecticides is drawing to a close (45), research on alternative methods of pest control needs to be expanded.

ACKNOWLEDGMENT

I sincerely thank A. R. Thompson for his valuable comments and critical review of an earlier draft of this manuscript.

## Literature Cited

1. Balachowsky, A., Mesnil, L. 1936. *Les insectes nuisibles aux plantes cultivées. Leur moers, leur destruction,* Vol. 2. Paris: Busson. 780 pp.
2. Barlow, C. A. 1965. Stimulation of oviposition in the seed-corn maggot, *Hylemya cilicrura* (Rond.) (Dipt., Anthomyiidae). *Entomol. Exp. Appl.* 8:83–95
3. Boivin, G., Benoit, D. L. 1987. Predicting onion maggot (Diptera: Anthomyiidae) flights in southwestern Quebec, Canada using degree-days and common weeds. *Phytoprotection* 68:65–70
4. Bromand, B. 1980. Investigations on the biological control of the cabbage root fly *(Hylemya brassicae)* with *Aleochara bilineata. IOBC-WPRS Bull.* III/1:49–62
5. Brooks, A. R. 1951. Identification of the root maggots (Diptera: Anthomyiidae) attacking cruciferous garden crops in Canada, with notes on biology and control. *Can. Entomol.* 83:109–20
6. Burges, H. D. 1981. *Microbial Control of Pests and Plant Diseases 1970–1980.* London: Academic. 949 pp.
7. Carruthers, R. I., Haynes, D. L. 1986. The effect of temperature, moisture and habitat on *Entomophthora muscae* conidial germination and survival in the onion agroecosystem. *Environ. Entomol.* 15:1154–60
8. Carruthers, R. I., Whitfield, G. H., Haynes, D. L. 1985. Pesticide-induced mortality of natural enemies of the onion maggot, *Delia antiqua* (Dip.: Anthomyiidae). *Entomophaga* 30:151–61
9. Coaker, T. H. 1987. Cultural methods: the crop. In *Integrated Pest Management,* ed. A. J. Burn, T. H. Coaker, P. C. Jepson, pp. 69–99. London: Academic. 474 pp.
10. Coaker, T. H., Finch, S. 1971. The cabbage root fly *Erioischia brassicae* (Bouché). In *Report of the National Vegetable Research Station 1970,* pp. 23–42. Wellesbourne, UK: Natl. Veg. Res. Stn. 139 pp.
11. Coaker, T. H., Williams, D. A. 1963. The importance of some Carabidae and Staphylinidae as predators of the cabbage root fly, *Erioischia brassicae* (Bouché). *Entomol. Exp. Appl.* 6:156–64
12. Collier, R. H., Finch, S. 1985. Accumulated temperatures for predicting the time of emergence in the spring of the cabbage root fly, *Delia radicum*

13. (L.) (Diptera: Anthomyiidae). *Bull. Entomol. Res.* 75:395–404
13. Colyer, C. N., Hammond, C. O. 1968. *Flies of the British Isles.* London/New York: Warne. 383 pp.
14. Crisp, P., Johnson, A. G., Ellis, P. R., Hardman, J. A. 1977. Genetical and environmental interactions affecting resistance in radish to cabbage root fly. *Heredity* 38:209–18
15. Dapsis, L. J., Ferro, D. N. 1983. Effectiveness of baited cone traps and colored sticky traps for monitoring adult cabbage maggots. *Entomol. Exp. Appl.* 33:35–42
16. Dempster, J. P., Coaker, T. H. 1974. Diversification of crop ecosystems as a means of controlling pests. In *Biology in Pest and Disease Control,* ed. D. Price Jones, M. E. Solomon, pp. 106–14. Oxford, UK: Blackwell. 398 pp.
17. de Ponti, O. M. B. 1980. Breeding onion *(Allium cepa)* for resistance to onion fly *(Delia antiqua).* In *Integrated Control of Insect Pests in the Netherlands,* ed. A. K. Minks, P. Gruys, pp. 173–75. Wageningen, the Netherlands: Pudoc. 304 pp.
18. Dindonis, L. L., Miller, J. R. 1980. Host-finding responses of onion and seed corn flies to healthy and decomposing onions and several synthetic constituents of onion. *Environ. Entomol.* 9:467–72
19. Eckenrode, C. J., Chapman, R. K. 1972. Seasonal adult cabbage maggot populations in the field in relation to thermal-unit accumulations. *Ann. Entomol. Soc. Am.* 65:151–56
20. Eckenrode, C. J., Nyrop, J. P. 1986. Impact of physical injury and commercial lifting on damage to onion bulbs by larvae of onion maggot (Diptera: Anthomyiidae). *J. Econ. Entomol.* 79: 1606–8
21. Edwards, C. A., Thompson, A. R. 1973. Pesticides and the soil fauna. *Residue Rev.* 45:1–79
22. Ellis, P. R., Eckenrode, C. J. 1979. Factors influencing resistance in *Allium* sp. to onion maggot. *Bull. Entomol. Soc. Am.* 25:151–53
23. Ellis, P. R., Hardman, J. A., Crisp, P. 1988. Investigations of the resistance of cabbage cultivars and breeders lines to insect pests at Wellesbourne. In *Progress on Pest Management in Field Vegetables,* ed. R. Cavalloro, C. Pelerents, P. P. Rotondo. Luxembourg: Off. Off. Publ. Eur. Communities. In press

24. Ellis, P. R., Hardman, J. A., Crisp, P., Gray, A. R. 1985. Investigations of resistance in cauliflower to cabbage root fly attack. *Proc. Better Brassica '84 Conf., St. Andrews, Scotland,* pp. 202–6. Invergowrie, UK: Scott. Crops Res. Inst. 252 pp.

25. Finch, S. 1980. Chemical attraction of plant-feeding insects to plants. *Appl. Biol.* 5:67–143

26. Finch, S. 1980. Pest assessment—as it relates to *Hylemya brassicae* populations. *IOBC-WPRS Bull.* III/1:1–9

27. Finch, S. 1987. Horticultural crops. See Ref. 9, pp. 257–93

28. Finch, S. 1988. Entomology of crucifers and agriculture—diversification of the agroecosystem in relation to pest damage in cruciferous crops. In *The Entomology of Indigenous and Naturalized Systems in Agriculture,* ed. M. K. Harris, C. E. Rogers. Boulder, Colo: Westview. In press

29. Finch, S., Ackley, C. M. 1977. Cultivated and wild host plants supporting populations of the cabbage root fly. *Ann. Appl. Biol.* 85:13–22

30. Finch, S., Cadoux, M. E., Eckenrode, C. J., Spittler, T. D. 1986. Appraisal of current strategies for controlling onion maggot (Diptera: Anthomyiidae) in New York State. *J. Econ. Entomol.* 79:736–40

31. Finch, S., Eckenrode, C. J. 1985. Influence of unharvested, cull-pile, and volunteer onions on populations of onion maggot (Diptera: Anthomyiidae). *J. Econ. Entomol.* 78:542–46

32. Finch, S., Eckenrode, C. J., Cadoux, M. E. 1986. Behaviour of onion maggot (Diptera: Anthomyiidae) in commercial onion fields treated regularly with parathion sprays. *J. Econ. Entomol.* 79:107–13

33. Finch, S., Jones, T. H. 1987. Factors influencing the severity of cabbage root fly infestations in crops of oilseed rape. *IOBC-WPRS Bull.* X/4:85–92

34. Finch, S., Skinner, G. 1972. Studies on the adult cabbage root fly. Trapping. In *Report of the National Vegetable Research Station 1971,* pp. 70–71. Wellesbourne, UK: Natl. Veg. Res. Stn. 101 pp.

35. Finch, S., Skinner, G. 1973. Chemosterilization of the cabbage root fly under field conditions. *Ann. Appl. Biol.* 73:243–58

36. Finch, S., Skinner, G. 1975. Dispersal of the cabbage root fly. *Ann. Appl. Biol.* 81:1–19

37. Finch, S., Skinner, G. 1980. Mortality of overwintering pupae of the cabbage root fly *(Delia brassicae). J. Appl. Ecol.* 17:657–66

38. Finch, S., Skinner, G. 1982. Upwind flight by the cabbage root fly, *Delia radicum. Physiol. Entomol.* 7:387–99

39. Finch, S., Skinner, G. 1983. Cabbage root fly. Monitoring fly numbers. In *Report of the National Vegetable Research Station 1982,* p. 33. Wellesbourne, UK: Natl. Veg. Res. Stn. 176 pp.

40. Finch, S., Skinner, G. 1988. Mortality of the immature stages of the cabbage root fly. See Ref. 24, In press

41. Finch, S., Skinner, G., Ackley, C. M. 1981. Cabbage root fly. Monitoring fly populations. In *Report of the National Vegetable Research Station 1980,* pp. 37–38. Wellesbourne, UK: Natl. Veg. Res. Stn. 196 pp.

42. Fletcher, B. S. 1987. The biology of dacine fruit flies. *Ann. Rev. Entomol.* 32:115–44

43. Funderburk, J. E., Higley, L. G., Pedigo, L. P. 1984. Seedcorn maggot *Delia platura* (Diptera: Anthomyiidae) phenology in central Iowa, USA and examination of a thermal unit system to predict development under field conditions. *Environ. Entomol.* 13:105–9

44. Griffiths, G. C. D. 1986. Phenology and dispersion of *Delia radicum* (Diptera: Anthomyiidae) in canola fields at Morinville, Alberta, Canada. *Quaest. Entomol.* 22:29–50

45. Harris, C. R., Svec, H. J., Tolman, J. H., Tomlin, A. D., McEwen, F. L. 1981. A rational integration of methods to control onion maggot in south western Ontario. *Proc. Br. Crop Prot. Conf.* 3:789–99

46. Harris, C. R., Whistlecraft, J. W., Svec, H. J., Tolman, J. H., Tomlin, A. D. 1984. Outdoor rearing technique for mass production of onion maggots, *Delia antiqua* (Diptera: Anthomyiidae). *J. Econ. Entomol.* 77:824–27

47. Harris, M. O., Miller, J. R. 1983. Color stimuli and oviposition behavior of the onion fly *Delia antiqua* (Diptera: Anthomyiidae). *Ann. Entomol. Soc. Am.* 76:766–71

48. Havukkala, I. 1987. Odor source finding behaviour of the turnip root fly *Delia floralis* Fall. (Diptera: Anthomyiidae) in the field. *J. Appl. Entomol.* 104:105–10

49. Hawkes, C., Coaker, T. H. 1979. Factors affecting the behaviour responses of the adult cabbage root fly, *Delia brassicae,* to host plant odour. *Entomol. Exp. Appl.* 25:45–58

50. Hertveldt, L., Van Keymeulen, M., Pelerents, C. 1980. Development of the sterile insect release method against the

cabbage root fly, *Delia brassicae* (B.), in north Belgium. *IOBC-WPRS Bull.* III/1:63–87

51. Hertveldt, L., Van Keymeulen, M., Pelerents, C. 1984. Large scale rearing of the entomophagous rove beetle *Aleochara bilineata* (Coleoptera: Staphylinidae). *Mitt. Biol. Bundesanst. Land-Forstwirtsch.* 218:70–75

52. Higley, L. G., Pedigo, L. P. 1984. Seedcorn maggot (Diptera: Anthomyiidae) population biology and aestivation in central Iowa. *Environ. Entomol.* 13:1436–42

53. Hough-Goldstein, J. A. 1987. Tests of a spun polyester row cover as a barrier against seedcorn maggot (Diptera: Anthomyiidae) and cabbage pest infestations *J. Econ. Entomol.* 80:768–72

54. Hughes, R. D. 1959. The natural mortality of *Erioischia brassicae* (Bouché) (Dipt., Anthomyiidae), during the egg stage of the first generation. *J. Anim. Ecol.* 28:343–57

55. Jones, F. G. W., Jones, M. G. 1984. *Pests of Field Crops.* London: Arnold. 392 pp. 3rd ed.

56. Jones, T. H. 1986. *Patterns of parasitism by Trybliographa rapae (Westw.), a cynipid parasitoid of the cabbage root fly.* PhD thesis. Univ. London. 317 pp.

57. Jones, T. H., Cole, A., Finch, S. 1988. A cabbage root fly oviposition deterrent in the frass of garden pebble moth caterpillars. *Entomol. Exp. Appl.* 47:In press

58. Jones, T. H., Finch, S. 1987. The effect of a chemical deterrent, released from the frass of caterpillars of the garden pebble moth, on cabbage root fly oviposition. *Entomol. Exp. Appl.* 45:283–88

59. King, K. M., Forbes, A. R. 1954. Control of root maggots in rutabagas. *J. Econ. Entomol.* 47:607–15

60. Lamb, D. J., Foster, G. N. 1986. Some observations on *Strongwellsea castrans* (Zygomycetes: Entomophthorales), a parasite of root flies (*Delia* spp.), in the south of Scotland. *Entomophaga* 31:91–97

61. Laska, P., Zelenkova, I., Bicik, V. 1986. Color attraction in species of genera *Delia* (Diptera: Anthomyiidae), *Ceutorhynchus*, *Meligethes* and *Phyllotreta* (Coleoptera: Curculionidae, Nitidulidae, Chrysomelidae). *Acta. Entomol. Bohemoslov.* 83:418–24

62. Liu, H. J., McEwen, F. L., Ritcey, G. 1982. Forecasting events in the life cycle of the onion maggot *Hylemyia antiqua* (Diptera: Anthomyiidae): application to control schemes. *Environ. Entomol.* 11:751–55

63. Loosjes, M. 1976. *Ecology and Genetic Control of the Onion Fly, Delia antiqua (Meigen).* Wageningen, The Netherlands: Pudoc. 179 pp.

64. Madder, D. J., McEwen, F. L. 1982. Integrated pest management in onions and carrots. *Rep. Dep. Environ. Biol. Univ. Guelph, Canada.* 94 pp.

65. Miller, J. R., Haarer, B. K. 1981. Yeast and corn hydrolysates and other nutritious materials as attractants for onion and seed flies. *J. Chem. Ecol.* 7:555–62

66. Miller, J. R., Strickler, K. L. 1984. Finding and accepting host plants. In *Chemical Ecology of Insects*, ed. W. J. Bell, R. T. Cardé, 1:127–57. London/New York: Chapman & Hall. 524 pp.

67. Morris, O. N. 1985. Susceptibility of 31 species of agricultural insect pests to the entomogenous nematodes *Sternernema feltiae* and *Heterorhabditis bacteriophora*. *Can. Entomol.* 117:401–08

68. Nair, K. S. S., McEwen, F. L. 1973. *Strongwellsea castrans* (Phycomycetes: Entomophthoraceae), a fungal parasite of adult cabbage maggot, *Hylemya brassicae* (Diptera: Anthomyiidae). *J. Invertebr. Pathol.* 22:442–49

69. Nair, K. S. S., McEwen, F. L. 1975. Ecology of the cabbage maggot, *Hylemya brassicae* (Diptera: Anthomyiidae), in rutabaga in southwestern Ontario, with some observations on other root maggots. *Can. Entomol.* 107:343–54

70. Nottingham, S. F., Coaker, T. H. 1987. Changes in flight track angles of cabbage root fly, *Delia radicum*, in diffuse clouds and discrete plumes of the host-plant volatile allylisothiocyanate. *Entomol. Exp. Appl.* 43:275–78

71. Pont, A. C. 1981. The Linnean species of the families Fanniidae, Anthomyiidae and Muscidae (Insecta: Diptera). *Biol. J. Linn. Soc.* 15:165–75

72. Poprawski, T. J., Robert, P. H., Majchrowicz, I., Boivin, G. 1985. Susceptibility of *Delia antiqua* (Diptera: Anthomyiidae) to eleven isolates of entomopathogenic hyphomycetes. *Environ. Entomol.* 14:557–61

73. Price, J. L. 1974. *Studies on the mating behaviour of the cabbage root fly Erioischia brassicae (Bouché).* PhD thesis. Univ. Birmingham, UK. 208 pp.

74. Prokopy, R. J., Collier, R. H., Finch, S. 1983. Visual detection of host plants by cabbage root flies. *Entomol. Exp. Appl.* 34:85–89

75. Prokopy, R. J., Owens, E. D. 1983.

Visual detection of plants by herbivorous insects. *Ann. Rev. Entomol.* 28:337–64

76. Ringdahl, O. 1959. Provisional key to female Anthomyiidae. *Insektfauna Sven.* 11 (3):294–300

77. Root, R. B. 1973. Organization of a plant-arthropod association in simple and diverse habitats: the fauna of collards *(Brassicae oleracea). Ecol. Monogr.* 43:95–120

78. Ryan, J., Ryan, M. F., McNaeidhe, F. 1980. The effect of interrow cover on populations of the cabbage root fly *Delia brassicae* (Wiedemann). *J. Appl. Ecol.* 17:31–40

79. Sears, M. K. 1983. Feasibility of monitoring cabbage maggot flies in rutabagas. *Rep. Prov. Lottery Res. Programs Univ. Guelph, Canada.* 42. pp.

80. Sears, M. K., Dufault, C. P. 1986. Flight activity and oviposition of the cabbage maggot, *Delia radicum* (Diptera: Anthomyiidae), in relation to damage to rutabagas. *J. Econ. Entomol.* 79:54–58

81. Séguy, E. 1923. *Faune de France,* Vol. 6, *Diptères Anthomyides.* Paris: Lechevalier. 393 pp.

82. Skinner, G., Finch, S. 1986. Reduction of cabbage root fly *(Delia radicum)* damage by protective discs. *Ann. Appl. Biol.* 108:1–10

83. Slingerland, M. V. 1894. The cabbage maggot, with notes on the maggot and allied insects. *Bull. NY State Agric. Exp. Stn.* 78:479–77

84. Stadler, E. 1978. Chemoreception of host plant chemicals by ovipositing females of *Delia brassicae. Entomol. Exp. Appl.* 24:511–20

85. Strickland, A. H. 1965. Pest control and productivity in British Agriculture. *J. R. Soc. Arts* 113:62–81

86. Strong, F. E., Apple, J. W. 1958. Studies on the thermal constants and seasonal occurrence of the seedcorn maggot under controlled conditions. *J. Econ. Entomol.* 51:704–7

87. Suett, D. L. 1987. Influence of treatment of soil with carbofuran on the subsequent performance of insecticides against cabbage root fly *(Delia radicum)* and carrot fly *(Psila rosae). Crop Prot.* 6:371–78

88. Suett, D. L., Thompson, A. R. 1985. The development of localised insecticide placement methods in soil. *Br. Crop Prot. Counc. Monogr.* 28:65–74

89. Theunissen, J., den Ouden, H. 1980. Effects of intercropping with *Spergula arvensis* on pests of Brussels sprouts. *Entomol. Exp. Appl.* 27:260–68

90. Theunissen, J., den Ouden, H. 1985. Tolerance levels for supervised control of insect pests in brussels sprouts and white cabbage. *Z. Angew. Entomol.* 100:84–87

91. Theunissen, J., Loosjes, M., Noordink, J. P. W., Noorlander, J., Ticheler, J. 1975. Small-scale field experiments on sterile-insect control of the onion fly, *Hylemya antiqua* (Meigen). In *Controlling Fruit Flies by the Sterile-Insect Technique,* pp. 83–91. Vienna: Int. At. Energy Agency. 175 pp.

92. Thompson, A. R., Percivall, A. L., Edmonds, G. H. 1986. The performance of carbofuran soil treatments against cabbage root fly *(Delia radicum)* on long white radish. *Meded. Fac. Landbouwwet. Rijks. Univ. Gent* 51:1157–65

93. Thompson, A. R., Rowse, H. R., Himsworth, A. D., Edmonds, G. H. 1988. Improving the performance of carbosulfan against cabbage root fly with low volume liquid treatments applied under field-sown seed. *Br. Crop Prot. Counc. Monogr.* 39:387–94

94. Throne, J. E., Eckenrode, C. J. 1986. Development rates for the seed maggots *Delia platura* and *Delia florilega* (Diptera: Anthomyiidae). *Environ. Entomol.* 15:1022–27

95. Ticheler, J., Loosjes, M., Noorlander, J. 1980. Sterile-male technique for control of the onion maggot, *Delia antiqua.* See Ref. 17, pp. 93–97

96. Tomlin, A. D., Miller, J. J., Harris, C. R., Tolman, J. H. 1985. Arthropod parasitoids and predators of the onion maggot *Delia antiqua* (Diptera: Anthomyiidae) in southwestern Ontario Canada. *J. Econ. Entomol.* 78:975–81

97. Traynier, R. M. M. 1967. Effect of host plant odour on the behaviour of the adult cabbage root fly, *Erioischia brassicae. Entomol. Exp. Appl.* 10:321–28

98. Tukahirwa, E. M., Coaker, T. H. 1982. Effect of mixed cropping on some insect pests of brassicas; reduced *Brevicoryne brassicae* infestations and influences on epigeal predators and the disturbance of oviposition behaviour in *Delia brassicae. Entomol. Exp. Appl.* 32:129–40

99. Vernon, R. S. 1986. A spectral zone of color preference for the onion fly, *Delia antiqua* (Diptera: Anthomyiidae), with reference to the reflective intensity of traps. *Can. Entomol.* 118:849–56

100. Vincent, C., Stewart, R. K. 1981. Evaluation of two types of traps used in monitoring adults of the cabbage maggot, *Hylemya brassicae* (Diptera: Anthomyiidae). *Ann. Soc. Entomol. Quebec* 26:41–50

101. Wheatley, G. A., Finch, S. 1984. Effects of oilseed rape on the status of insect pests of vegetable brassicas. *Proc. Br. Crop Prot. Conf.* 2:807–14
102. Wheatley, G. A., Thompson, A. R. 1981. Pest damage and insecticide performance in vegetable crops for processing. In *Quality In Stored And Processed Vegetables and Fruit,* ed. P. W. Goodenough, R. K. Atkin, pp. 165–75. London: Academic. 398 pp.
103. Whistlecraft, J. W., Harris, C. R., Tolman, J. H., Tomlin, A. D. 1985. Mass-rearing technique for *Aleochara bilineata* (Coleoptera: Staphylinidae). *J. Econ. Entomol.* 78:995–97
104. Whistlecraft, J. W., Harris, C. R., Tomlin, A. D., Tolman, J. H. 1984. Mass rearing technique for a braconid parasite *Aphaereta pallipes* (Hymenoptera: Braconidae). *J. Econ. Entomol.* 77:814–16
105. Whitfield, G. H., Carruthers, R. I., Haynes, D. L. 1985. Phenology and control of the onion maggot (Diptera: Anthomyiidae) in Michigan onion production. *Agric. Ecosyst. Environ.* 12:189–200
106. Whitfield, G. H., Drummond, F. A., Haynes, D. L. 1986. Overwintering survival of the onion maggot *Delia antiqua* (Diptera: Anthomyiidae) in Michigan, USA. *J. Kans. Entomol. Soc.* 59:197–99
107. Wilson, L. T. 1985. Estimating the abundance and impact of arthropod natural enemies in IPM systems. In *Biological Control in Agricultural IPM Systems,* ed. M. A. Hoy, D. C. Herzog, pp. 303–22. Orlando, Fla: Academic. 589 pp.
108. Wright, D. W., Hughes, R. D., Worrall, J. 1960. The effect of certain predators on the numbers of cabbage root fly (*Erioischia brassicae* (Bouché)) and the subsequent damage caused by the pest. *Ann. Appl. Biol.* 48:756–63
109. Wyman, J. A., Libby, J. L., Chapman, R. K. 1977. Cabbage maggot management aided by predictions of adult emergence. *J. Econ. Entomol.* 70:327–31

Ann. Rev. Entomol. 1989. 34:139–61

# BIOLOGY, HOST RELATIONS, AND EPIDEMIOLOGY OF *SARCOPTES SCABIEI*

## Larry G. Arlian

Department of Biological Sciences, Wright State University, Dayton, Ohio 45435

## HISTORY OF SCABIES

Scabies is a contagious disease of humans and other mammals. It is caused by the mite *Sarcoptes scabiei*, which burrows in the lower stratum corneum of the skin. Scabies was one of the first diseases in humans with a known cause (41, 85). The Italians Giovanni Cosimo Bonomo and Diancinto Cestoni first described and illustrated the mite in 1689 in a now-famous letter to Francesco Redi (2, 80, 89). However, it was not until 200 years later that scabies was generally accepted as a parasitic disease (85). Endemic and enzootic levels of human and animal scabies, respectively, continue to occur despite the availability of various therapies. Currently, sporadic outbreaks and epidemics in communities (17, 22, 23, 44, 59, 69, 82–85, 91, 94, 102, 119), nursing homes (74, 90), schools (54), hospitals (24, 48, 62, 97), and other institutions (73) and epizootics in wild and domestic animal populations (18, 45, 47, 56, 57, 68, 80, 93, 106, 111) are frequently reported.

Historically, epidemics of human scabies have occurred on a worldwide basis in 30-year cycles with 15-year gaps between them (83, 85). Although many opinions have been given, there is no satisfactory explanation for the significant fluctuations in scabies prevalence. Two notable increases occurred in the last 50 years; one peaked in the mid 1940s, and a second began in the mid to late 1960s on a worldwide basis and in the early 1970s in the United States (25, 69, 70, 83, 85, 87). Scabies sufferers constituted about 2–4% of patients seen by dermatologists in the United States in 1975 (102). The

National Disease and Therapeutic Index for the 12-mo period ending in December 1978 reported approximately one million doctor visits for scabies, a 17-fold increase over 1970 and a threefold increase over the 1974 estimate. The Information Medical Service reported 859,000 visits for scabies in 1983. The increase in scabies in the United States seems to have leveled off, but it has continued at a steady level in the 1980s. Reports from Czechoslovakia (88), Denmark (22), Great Britain (23, 107), India (77, 101), Australia (61), Greece (98), and Africa (110) all called attention to the rise in scabies over the past 20 years. In the United States and most other countries human scabies is not a reportable disease, so it is difficult to know its actual prevalence. However, the above data, coupled with the continued case reports, indicate that scabies is still common. In spite of the latest increase in incidence, scabies has always been a common disease and more common than was generally supposed. The disease in humans was previously thought to be associated with overcrowding, poverty, and poor hygiene. However, the socioeconomic characteristics of patients with scabies during the recent re-surgence in the United States seem to be representative of the overall popula-tion (8, 54, 82–85). Studies employing house-to-house survey techniques demonstrated significant scabies prevalence among the middle to high socio-economic classes and among persons with good hygiene (8, 44, 78). A recent study (8) found that among scabietic patients seen by a dermatologist in southwestern Ohio, 54% of the infested families or individuals were of average socioeconomic level and 33.3% were of above average or high socioeconomic level. Hygienic standards were at least average for both these classes of patients.

A number of studies have reported epizootics or cases of scabies in animals in the last 20 years. Most of these have involved wild canines. Notable epizootics or individual cases of scabies were reported in coyotes in Texas and Kansas (42, 93), coyote–red wolf hybrids in Texas and Louisiana (92), coyotes and wolves in Alberta, Canada (114), red foxes and coyotes in Wisconsin (115), red foxes in Pennsylvania (95), dogs (109), wild canines in New York (111, 112), cattle in New York, Germany, and Denmark (45, 64, 80), mice and peccaries in the New York Zoological Society (68), cats (65, 47), and horses, tapirs, and chamois (55, 57, 59). Hourrigan (53) has outlined the history of scabies in cattle in the United States. Surveys in the United States, New Zealand, Ireland, Australia, and England have shown that 30–35% of the domestic pigs surveyed in some populations were infested with *S. scabiei* (67, 103–105).

Most published information on scabies has focused on the clinical and epidemiological aspects of the disease in humans (reviewed in 2, 86, 87). Several reviews have given an interesting historical perspective and early biological information (41, 49, 76). Until recently, few modern studies had directly investigated the basic biology of the parasite, the host-parasite interactions, and the host immune and physiological responses to the mite and its products. However, studies along these lines are important to develop a better understanding of the epidemiology of this disease. Therefore, although this chapter reviews some basic biology and clinical aspects of the parasite and the disease, it focuses primarily on experimental studies of the parasite and the host response conducted in the last 10 years.

## MORPHOLOGY

*Sarcoptes scabiei* belongs to the cohort Astigmata (order Acariformes, suborder Sarcoptiformes) (81). *Sarcoptes scabiei* is recognized by the characteristic oval, ventrally flattened, and dorsally convex tortoise-like body, stout dorsal setae, numerous cuticular spines, and transversely ridged cuticular striations. The male (213–285 $\mu$m long by 162–210 $\mu$m wide) is about two thirds the size of the female (300–504 $\mu$m long by 230–420 $\mu$m wide) (30). The female exhibits an external copulatory papilla of the bursa copulatrix on the posterior idiosoma anterior to the posterior-dorsal anal opening. The tarsus of legs I and II on females and males and the tibio-tarsus of leg IV of males bear a stalked empodium that terminates in a broad pad. Legs III and IV of females and leg III of males terminate in a long seta. In addition, the tarsus of legs I and II and tibio-tarsus III bear two spur-like claws in both sexes. Tibio-tarsus IV bears one spur-like claw in males and two in females. The anterior stubby legs extend beyond the anterior-lateral margin of the propodosoma, while the posterior legs do not extend beyond the body margin. Five pairs of dorsal setae, five pairs of lateral setae, and anterior pairs of internal and external scapular setae are present on the dorsal surface. The internal scapular setae and the first dorsal and lateral setae are lamellate. Anterior, medial, and posterior genital setae are present in the male; females lack the medial genital setae. Additional descriptive details, in particular of leg chaetotaxy and larval and nymphal stages, can be found in Reference 30.

## LIFE CYCLE

Until recently, little was directly known about the life cycle of *S. scabiei;* knowledge was primarily based upon pre–World War II anecdotal observations, principally of *S. scabiei* var. *hominis* (41, 49, 76, 117). A recent in vivo study of the life cycle of *S. scabiei* var. *canis* on a rabbit host, coupled with a detailed morphological study of active and quiescent life stages, for the first time directly revealed that development of both males and females consists of egg, larval, protonymphal, and tritonymphal instars (13, 30). These data are in contrast to several older reports that development consists of only one nymphal stage in males but of two in females (41, 43, 49, 76, 117). Because the male tritonymph is smaller than the female tritonymph and only slightly larger than the protonymph, this life stage may be confused with the protonymph upon examination with a stereoscope. Protonymphs that eventually become males pass through a small tritonymphal stage. However, protonymphs that give rise to females pass through a larger tritonymphal stage than that of the males. In vivo studies revealed that females are oviparous (13), in spite of a report to the contrary (79).

S. scabiei var. canis eggs hatch after a 50–53-hr incubation. Durations of the larval and protonymphal stages are 3.22 ± 1.52 to 4.20 ± 1.52 and 2.33 ± 0.66 to 3.40 ± 0.84 days, respectively. Tritonymphal development requires 2.42 ± 0.51 to 3.42 ± 0.51 days for males and 2.22 ± 1.01 to 3.22 ± 0.97 days for females. Development from egg to adult requires 10–13 days.

It is presumed that the life cycle of S. scabiei from other hosts is similar, but this remains to be directly determined. Other aspects of the life history such as fecundity, copulation, the molting process, quiescence, longevity of adults, feeding behavior, and pheromonal activity are still unknown for all scabies strains.

## HOST SPECIFICITY AND CROSS-INFESTIVITY

Sarcoptes scabiei infests many different mammalian hosts in 17 families and 7 orders (30). The mites from different hosts exhibit little or no morphological differences; therefore, based on morphology alone, it is unclear if Sarcoptes mites associated with different hosts represent different species or simply different varieties of one species. Fain (30, 31) does not consider the variation between strains from different hosts taxonomically significant and proposed that the genus Sarcoptes contains only one valid but variable species. Most of the morphological variations among strains from different hosts are in size, the dorsal field of spines, and the ventrolateral spines. For example, most or all specimens from humans exhibit a bare area (lack 1–5 spines) in the dorsal field of spines and lack ventrolateral spines, while those from pigs exhibit a similar dorsal bare area but have ventrolateral spines. Most S. scabiei mites that parasitize dogs have no bare area and have ventrolateral spines, while those that parasitize cattle lack ventrolateral spines. Despite similar morphology, Sarcoptes mites that parasitize some hosts are recognized as distinct species (56, 57, 59).

Biological evidence indicates that there are physiological differences among scabies mites from different hosts and that scabies mites from different hosts are largely host specific (10, 14, 55, 57, 58). Experimental attempts to transfer scabies from dogs to mice, nude (thymus-deficient) mice, hairless mice, rats, guinea pigs, pigs, cattle, cats, goats, and sheep were unsuccessful

(10, 14). Most of these hosts are known to be parasitized in nature by *S. scabiei*. Scabies from pigs could not be transferred to dogs. However, scabies from dogs could be transferred to New Zealand white laboratory rabbits, although scabies from humans and pigs could not be transferred to the same rabbits (10, 14). These experiments suggest that the transfer of *S. scabiei* from one host species to another does not usually occur. The fact that *S. scabiei* var. *canis* infested the laboratory rabbit while the mites from humans and pigs could not is clear evidence of strain differences. Since the same host rabbits were used in all three cases, the host factors limiting exploitation on the rabbit were identical for the three strains.

Host specificity may be attributed to a number of parasite and host factors and their interactions. The mechanisms for the host specificity of scabies are largely unknown, but specific factors generally do not allow some strains to survive and proliferate for extended periods on strange hosts. Host-recognition and host-seeking stimuli such as odor or body temperature may be involved in host specificity for some strains, but this possibility remains to be investigated for a range of strains. Host odor was not a factor in host specificity for *S. scabiei* var. *canis* in the previously mentioned cross-host transfer experiments (11, 14). When dislodged from a host, this mite was attracted to the odor of cat, rabbit, goat, calf, guinea pig, mouse, and rat (Figure 1), although none of these hosts could be permanently infested (11, 14). Since these mites reside in the host's epidermis, it is probable that some limiting factors and processes are located there. These factors may include physiological differences in the requirements of mite strains; differences in dietary and nondietary properties of the host skin environment; ability of the host to mount an immune response; antigenicity of the parasite, which provokes the immune response in the host; and resistance of the mite to the host immune response.

Humans occasionally become infested with animal scabies strains when they handle or live with infested animals (31, 45, 56, 58, 91, 109, 113). Just how often this occurs is unknown, since the various forms are morphologically similar and not easily distinguished with the usual oil scrape preparations used for clinical diagnosis. The clinical signs of natural and unnatural scabies infestations are identical. Likewise, the extent of cross-infestation between other animals is unknown.

Based on host transfer experiments, most human and animal cross-infestations are probably self-limiting. However, cross-infestations involving animal strains can result in temporary infestations that last several months, during which mites reproduce (Table 1). Transfer experiments indicated that canine mites are capable of burrowing, feeding, and producing eggs in human skin (29). Within 24 hr these infestations produced intense itching and vesicular and pustular lesions around each burrowed mite. These infestations

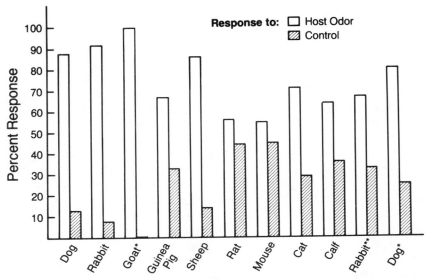

*Figure 1*  Response of *S. scabiei* var. *canis* after female mites were placed midway between a host odor stimulus and scrubbed air (control). The host odor and control were 5 cm apart except for * = 2 cm and ** = 7 cm.

were temporary and self-limiting. After 1–3 days only dead mites were found in the center of each of these lesions (5).

The ability of strains to develop temporary infestations and reproduce on strange hosts has raised the possibility that temporary hosts may serve as reservoirs for transmission of mites back to natural hosts (14). Self-limiting cross-infestations on some experimentally infested strange hosts lasted up to 13 wk (Table 1). Most of these infestations spread from the point of inoculation as mites reproduced and increasing parasitemia developed. Hosts infested with strange scabies strains exhibited hyperkeratosis similar to that seen in advanced infestations on natural hosts before the infestation gradually waned (14). Transfer of scabies from unnatural hosts back to natural ones may explain apparent treatment failures in domestic animals.

## SENSORY PHYSIOLOGY

*Sarcoptes scabiei* var. *canis* perceives both host odors and thermal stimuli. When dislodged from the host, mites are attracted to the source of these stimuli (11, 14). Likewise, *S. scabiei* var. *hominis* exhibits a strong thermotaxis and moves to the warmest part of a temperature gradient (72). Females of *S. scabiei* var. *canis* placed at varying distances from a rabbit host on a rod

**Table 1**  Duration of temporary infestations on strange hosts experimentally infested with different strains of *S. scabiei*[a]

| Host | *S. scabiei* strain | Duration of infestation |
|------|------------------|----------------------|
| Guinea pig | var. *canis* | 51 days |
| Mouse | var. *canis* | 1–2 days |
| Rat | var. *canis* | 6–8 wk |
| Dog | var. *suis* | 2–4 wk |
| Rabbit | var. *suis* | 2–4 wk |
| Sheep | var. *canis* | 1–3 wk |
| Goat | var. *canis* | 10–13 wk |
| Calf | var. *canis* | 5–7 wk |
| Cat | var. *canis* | 8–10 wk |
| Pig | var. *canis* | 4–10 wk |

[a]Data from References 10 and 14.

leading to and touching the rabbit migrated toward the host from up to 15.4 cm away (11). More than 63% of the specimens moved toward the host when they were placed at 5.6 cm or less from the rabbit. In these experiments the percentage of test mites moving toward the host increased as the distance was decreased. Within this experimental design, host body temperature, odor, and $CO_2$ were possible stimuli. Similar experiments in which the host was replaced by a 32°C abiotic thermal stimulus, which eliminated host odor and $CO_2$ stimuli, gave similar results; 83% of mites placed 5.6 cm from the heat source migrated toward the thermal stimulus. Female mites placed midway between a host and a thermal stimulus at close proximity responded to both with no preference. As the distance between the two stimuli was increased, most mites that responded moved toward the host. Therefore, a combination of host odor and body temperature was a stronger stimulus, or mites perceived odor from a greater distance than temperature alone. Additional experiments showed that all life stages of *S. scabiei* var. *canis* perceived and responded to host odor in the absence of $CO_2$ (14).

## SURVIVAL AND INFESTIVITY

Although it was not directly investigated, the role of fomites in the transmission of human scabies was formerly thought to be minimal (82, 84, 85, 91, 102, 119). However, recent direct studies of the life cycle, behavior, survival, infestivity, and prevalence of mites in the host environment indicates that fomites can be important sources of mites for infestation or reinfestation (8, 9, 11, 13, 14, 16).

Life cycle studies have shown that all life stages frequently leave the burrow and wander on the skin, and they may fall or become dislodged from the host (13). Live mites can be recovered from the environment of scabies patients (8, 9) and presumably from that of other hosts. Survival off the host, coupled with retained infestivity and host-seeking behavior, make it probable that dislodged mites can be important in scabies transmission in homes, barns, outdoor livestock enclosures, and kennels.

Survival off the host is greatly dependent on ambient temperature and relative humidity (RH) (9, 16). Lower temperatures and higher RH, which reduce desiccation, generally favor longer survival, while higher temperatures and lower RH result in significantly reduced survival. Female mites survived only a few hours at 45°C and 25 and 45% RH, but 8 and 19 days at 10°C and 25 and 97% RH, respectively. Females died after 10 min at 50°C; however, those that survived 5 min at this temperature immediately burrowed into the skin when placed back on the host. At the other extreme, some females survived 1 hr at −25°C, but survivors were not infestive and would not penetrate host skin, although they were active and walked about. Therefore, natural or manipulated high or low temperatures could be used to control these mites in the environment.

Generally, all life stages of *S. scabiei* var. *canis* survived 1–9 days at 15–25°C and 25–97% RH. Under comparable conditions males and larvae survived one third to one half as long as other life stages. Tritonymphs of females generally exhibited the longest survival (21 days) of any life stage at 10°C and 97% RH, but they survived only 5–10 hr when survival was determined at 25, 30, 35, and 45°C and 97% RH. Significantly, these mites generally survived 2–5 days at temperature and RH comparable to outdoor summer conditions or those in heated homes and barns. Lower temperatures comparable to winter conditions extended survival.

How long mites retain the ability to penetrate a host (remain infestive) is an important aspect of their survival time off the host. When held off the host at room conditions, *S. scabiei* var. *suis* from pigs burrowed into the skin 24 hr later (9). Likewise, *S. scabiei* var. *canis* that had burrowed in the skin remained infestive 36 hr after the death of the host.

When *S. scabiei* var. *canis* mites were removed from the host, held at 75% RH and 22–24°C for 24 hr, and then placed back on the skin they began burrowing into the skin within 6 min; within 1 hr these specimens were half to fully submerged in the stratum corneum (9, 16). The mean time for complete penetration of these mites was 35.4 ± 17 min. Under these conditions no mites survived 40 hr off the host. Those surviving 36 hr initiated penetration within 3 ± 3 min when placed back on the host, but required 97 ± 36 min to penetrate completely. Nutritional and water requirements stimulated specimens that had been off the host longer to initiate penetration sooner than fresh specimens. However, because of their weakened condition the time required for complete penetration increased as a function of the time off the host. In addition, live specimens of *S. scabiei* var. *hominis* recovered from the bed linen of scabietic patients penetrated and burrowed when placed on a rabbit. Specimens removed from patients with crusted scabies and held alternately under room conditions (21°C and 45% RH) and refrigeration (4 or 10°C and 95% RH) for 12-hr periods remained infestive and burrowed into rabbit skin after 4 days. These observations and experiments indicate that mites remain infestive for at least one half to two thirds of their survival time when dislodged from the host. By comparison, the average time needed for fresh *S. scabiei* var. *hominis* specimens to initiate penetration was 9.6 ± 6.2 min, and these specimens were completely submerged in 31.0 ± 15.0 min. Fresh females of *S. scabiei* var. *canis* initiated penetration within 20.5 ± 16.5 min and required 23 ± 5.9 min after initial penetration to become completely submerged in the stratum corneum. Males, nymphs, and larvae of *S. scabiei* var. *canis* penetrate and become completely submerged in the skin in less time than females (9). Larvae, compared with other life stages, complete penetration most rapidly (Table 2).

## PREVALENCE OF SCABIES IN THE HOST ENVIRONMENT

Only a few studies have directly investigated the occurrence of scabies mites in the environment of scabietic patients (8, 9, 21, 51). In detailed studies of the home environment of 37 scabietic patients, 44% of the homes contained scabies mites in the dust vacuumed from floors, overstuffed chairs or couches, and mattresses (8). Dust samples from 64% of the positive homes contained live mites. Live mites were most often recovered from bedroom floors and chairs or couches. Some dust samples were not analyzed until 72 hr after collection; the prevalence of live mites would likely have been higher had the dust samples been analyzed immediately upon collection. The recov-

**Table 2**  Time required for *S. scabiei* var. *canis* to penetrate the skin of a rabbit host[a]

| Life stage | Time required to penetrate (min) | Number tested |
|---|---|---|
| Female | 43.5 | 17 |
| Male | 21.2 | 19 |
| Nymph | 16.8 | 18 |
| Larva | 11.1 | 18 |

[a]Data from Reference 9.

ery of live mites from the homes of scabietic patients, coupled with mite survival of 1–5 days at room conditions (9, 16), the host-seeking behavior of all life stages (11, 14), and the mite's rapid penetration of the host's skin once on the host (9), indicates that fomites in homes can be a source for transmission.

Interestingly, detailed study of the environment of nursing homes housing scabietic patients gave different results (8). Live mites were less frequently recovered from bedding, floors, and furniture than in patient homes. Regular housekeeping practices, frequent bed linen changes, and stringent hygienic standards in the nursing homes studied minimized fomite contamination, and higher room temperatures resulted in higher mite mortality due to desiccation. Personal contact was probably the primary means of transmission among patients, staff, and family members in these nursing homes.

## HOST/PARASITE ENERGETICS

Few studies have quantitated the effects of scabies infestations on the health, weight, and physiology of parasitized animals. Lightly infested hosts usually do not exhibit obvious disease manifestations. Chronic and severe scabies infestations cause visible deterioration of the host's physical condition, progressive emaciation or reduced weight gain in growing animals, reduced milk production in cattle, and eventually even death (3, 6, 7, 20, 80, 93, 103). These signs may be accompanied by changes in specific serology and blood biochemistry in some hosts, depending upon the host species and the duration and severity of the infestation (6, 93). The reasons for weight changes, altered physiology, and death are unknown.

These mites directly drain energy from the host by consuming the host's lymph and lysed tissue. Indirectly, host energy is drained by the mite-stimulated hyperplasia and the subsequent loss of stratum corneum and by increased physical activity caused by response to the parasitic infestation. Study of the energetics of rabbits parasitized by *S. scabiei* var. *canis* has shown that even under heavy parasitemia mites do not deplete sufficient energy from the host to account for the weight loss or reduced weight gain of heavily parasitized hosts (7). Mite density at the crusted stratum corneum/stratum lucidum interface of heavily parasitized rabbits was about 1483 mites per $cm^2$ (7). A typical mite population on this host comprised 18% females, 18% males, 13% female tritonymphs, 24% protonymphs and male tritonymphs combined, and 27% larvae. Measured $O_2$ consumption at 34°C and 75% RH for female and male *S. scabiei* was 2 and 0.76 nl $O_2$ per hr per mite, respectively. Based on proportional size and activity, nymphs and larvae utilized an estimated 0.76 and 0.38 nl $O_2$ per hr per mite, respectively. With this mite density and these $O_2$ requirements, it has been estimated that the mite population would utilize 150.48 $\mu$l $O_2$ per day per $cm^2$ of infested surface. Since 1 $\mu$l $O_2$ consumed corresponds to approximately 0.0048 cal derived through metabolism of food, 0.722 cal per day per $cm^2$ was obtained from the host. This amount was only 0.25–0.40% of the host's daily metabolism. Based on weight gain by matched uninfested control rabbits, mite consumption reduced the weekly weight gain of growing rabbits by only 1.9%, or less than 0.01 g of host biomass (carbohydrate, fat, or protein) per wk. Clearly, direct energy drain does not cause reduced weight gain or weight loss in heavily parasitized animals (7). Indirect energy drains, i.e. the hyperplasia and loss of stratum corneum that may accompany heavy infestations, are probably much more significant.

Controlled laboratory experiments demonstrated that during development of a scabies infestation, young rabbits grew at a rate comparable to that of noninfested control rabbits during the first 17 wk of infestation. However, infested rabbits consumed more food than controls, which indicates a lower food-converting efficiency. As the infestation spread and parasitemia became heavy, growth of the rabbits stopped abruptly. A steady weight loss occurred over the next 25 wk until the rabbits were treated. Control rabbits continued to grow. Interestingly, food consumption by infested rabbits during the period of weight loss was similar to that of controls. Therefore, for unknown reasons, food-converting efficiency was much reduced. Two similar studies reported that weight gains in young pigs parasitized by *S. scabiei* were reduced by 3.3 and 5.5% when compared to gains of noninfested controls (3,20). However, the infested pigs of the Cargill & Dobson study (20) showed depressed food efficiency, whereas those in the study of Alva-Valdes et al (3) showed no significant difference compared with the controls. By contrast, another study reported that pigs lightly infested with *S. scabiei* responded similarly to parasitized rabbits in that the infestation had no significant effect on growth (103, 106). Unlike the rabbits, the pigs showed no change in food-converting efficiency.

Although data are limited, it generally appears that light infestations have little effect on the growth rate of young hosts, but the food-converting efficiency may or may not be influenced. In chronic heavy scabies infestations weight is affected. It is unknown what causes the reduced weight. Reports of pathophysiology in infested hosts suggest that toxic secretions from the mite, coupled with the host's immune responses and possibly with secondary bacterial infections, contribute to disease manifestations (5, 6, 93, 104, 105).

## PATHOPHYSIOLOGY

Animals with chronic heavy scabies infestations exhibit varied changes in serology and blood biochemistry. Parasitized rabbits and pigs develop anemia as evidenced by significantly reduced hemoglobin, hematocrit, and mean corpuscular hemoglobulin levels (6, 103). Sheahan (103) suggested that in pigs the anemia was related to an iron deficiency caused by loss of iron in shed skin scales and reduced ability to absorb iron from the intestine. Reduced serum iron concentrations in human patients with scaly dermatoses have been reported (66, 108). Rabbits exhibited highly variable serum iron levels, but although they were lower than those of controls, they were not statistically different. Whether or not there is a relationship between anemia caused by scabies infestation and dietary iron deficiency, altered iron absorption, or iron metabolism is still unknown. Dogs and rabbits provided ad lib complete diets in the laboratory and both iron-treated and iron-deprived pigs developed severe scabietic lesions, which suggests that dietary iron deficiency does not predispose to scabies. However, the iron-deprived pigs of the Sheahan study (103) showed significantly reduced weight gains and lower hematological values, which had no apparent relationship to the severity of the scabietic lesions. These data indicate that iron deprivation may not affect susceptibility but that it does affect severity of anemia and expression of pathological manifestations (possibly by increasing the iron requirements). In contrast, Kutzer (55) considers dietary vitamin A and mineral supplies to be important factors in predisposition to scabies. One study reported that blood hematocrit, hemoglobin, and total serum protein levels in cattle moderately or severely infested with *S. scabiei*, in contrast to those in rabbits and pigs, were normal (80).

Two studies have examined in detail selected hematological and blood biochemical parameters usually used as prognostic indicators of disease in infested and uninfested hosts. Abnormal values in some of these parameters indicated that scabies causes pathological conditions in severely infested hosts (6, 93). Similar neutrophil, lymphocyte, monocyte, eosinophil, and basophil levels in scabies-infested and uninfested wild coyotes have been reported (93). Interestingly, infested humans show blood eosinophilia (37). By contrast, scabies-infested laboratory rabbits exhibited elevated neutrophil levels and reduced lymphocyte, eosinophil, and basophil levels (6). The reduced levels in rabbits may reflect cell migration to the infested dermal/epidermal area and sequestering of B cells in lymphoid tissue. Generally, deviant serum biochemical parameters that accompanied the infestation in both rabbit and coyote hosts were similar except that total serum protein was elevated in rabbits and reduced in coyotes; glucose and alkaline phosphatase were normal in rabbits but reduced in coyotes; blood urea nitrogen (BUN) was reduced in rabbits and elevated in coyotes; and bilirubin was normal in rabbits and elevated in coyotes (Table 3). Slightly and moderately infested rabbits and coyotes showed fewer or no deviations in blood chemistry levels. These indexes return to normal in rabbits following treatment and recovery, which indicates that the pathophysiological conditions that develop as a result of

**Table 3**  Serum biochemical parameters for *S. scabiei* var. *canis*–infested rabbits (6) and coyotes (93)

| Serum biochemistry | Mean value infested host relative to control | |
|---|---|---|
| | Rabbit | Coyote |
| Total protein | Elevated | Reduced[a] |
| Total globulins | Elevated | — |
| Alpha globulins | — | Reduced |
| Gamma globulins | — | Elevated |
| Albumin | Reduced | Reduced |
| Albumin/globulin ratio | Reduced | Reduced[a] |
| Triglycerides | Elevated | — |
| Cholesterol | Normal[b] | — |
| Glucose | Normal[b] | Reduced[a] |
| Blood urea nitrogen (BUN) | Reduced | Elevated |
| Creatinine | Reduced | Reduced[a] |
| BUN/creatinine ratio | Normal[b] | — |
| Uric acid | Normal | — |
| Lactic acid dehydrogenase | Normal[b] | Elevated |
| Glutamic oxaloacetic transaminase | Normal[b] | — |
| Total bilirubin | Normal | Elevated[a] |
| Gamma glutamyltransferase | Reduced[b] | — |
| Glutamic pyruvic transaminase | Reduced | — |
| Alkaline phosphatase | Normal[b] | Reduced |
| Iron | Reduced[a] | — |
| Calcium | Reduced | Reduced |
| Phosphorus | Normal | Elevated[a] |
| Sodium | Normal | — |
| Potassium | Normal | — |
| Chloride | Elevated | — |
| Creatinine phosphokinase | — | Reduced |

[a]Not reported as significant.
[b]Some individuals significantly different from controls.

scabies infestation are reversible. Sheahan (104, 105) has also reported significant increases in mean total serum protein and beta and gamma globulins in infested pigs. These pigs also exhibited large deposits of glycogen and acid mucopolysaccharide in the scabietic lesions.

## CLINICAL SCABIES

The usual signs and symptoms of human scabies infestations are pruritic, erythematous, papular, and vesicular lesions that are associated with the burrowed mite. Some children may develop nodular scabies, which is characterized by reddish-brown infiltrated nodules that may persist for months despite therapy (85). In some patients an additional rash with numerous small mite-free urticarial papules develops on other mite-free areas of the body (37, 46, 63, 99). The cause of these secondary lesions is unknown, but it has been suggested that they result from an allergic sensitivity or autosensitization reaction that is due to either cell-mediated immune responses or circulating immune complexes (46, 63). Itching associated with the mite may be mild to severe, and it is reported by many to be most noticeable at night (2). It is not clear if the nighttime intense itching is due to increased body sensitization during that time, increased mite activity (burrowing and associated secretions and defecation), or simply an increased awareness of discomfort in a resting individual. Metabolic studies of *S. scabiei* off the host have shown that there are no intrinsic diurnal changes in mite $O_2$ consumption that would reflect changes in activity (7). However, on the host, diurnal changes in the host's physiology could act as extrinsic stimuli to prompt parallel changes in mite activity.

In humans, mites and lesions most often tend to localize in the webs of the fingers, the volar aspect of the wrists and arms, and the extensor aspect of the elbows (2, 25, 69–71). Other areas commonly involved, in order of declining frequency, are the soles and insteps of the feet, genitals, buttocks, axillae, waistline, and the area around the nipples (2, 25, 69). It is not clear what characteristics of the skin direct infestation to these body areas. In humans, the clinical signs and symptoms of early to moderate stages of scabies infestation mimic other skin diseases such as dermatitis, herpetiformis, eczema, syphilis, impetigo, furunculosis, papular urticaria, allergic reactions, and response to other ectoparasitic mites and insects (e.g. chiggers, *Pyemotes* spp., lice, fleas, and bedbugs) (2, 19, 26).

The severe itching and associated lesions of a primary infestation are generally reported to take 4–8 wk to develop (70). Therefore, the disease is usually not diagnosed until long after the patient's contact with scabies. This latent period of primary infestation is thought to be associated with the development of sensitization. However, with subsequent reinfestations (secondary infestations) sensitization and similar signs and symptoms are evident within 24–48 hr and develop rapidly, presumably because the host has been sensitized by the previous infestation (70). A few recent reports have indicated that lesions and the associated pruritus of primary infestations have developed within a few days to 2 wk in some cases (62, 90, 97). The reason for this accelerated development is unknown. It may be associated with these scabies patients' previously developed sensitivity to the house dust mite (15, 75). It has recently been discovered that some antigens of scabies and house dust mites are cross-reactive, and atopic patients with dust mite sensitivity and no history of scabies exhibit circulating IgE directed at determinants of *S. scabiei* antigens (15). On the other hand, it is interesting that primary (temporary) canine scabies in humans exhibits pruritic, erythematous vesicular lesions within 24–36 hr. The rapid skin reaction of humans to *S. scabiei* var. *canis* and the generally more delayed skin reaction to *S. scabiei* var. *hominis* indicate that there are antigenic differences between the two strains.

Mites produce a serpiginous burrow in the stratum corneum as they digest and consume this tissue. This burrow is frequently described as a clinical sign for the diagnosis of scabies. Unfortunately, in most hosts these burrows are rarely visible with the unaided eye or without enhancement. Generally they are only visible when there is sufficient inflammation associated with the burrow and the accumulation of ingrained dirty material. They are always present, however, and they may be made visible by placing a drop of ink on or rubbing an ink pen over the suspected skin (120). After a few minutes ink will seep into the burrow, and when the excess is wiped from the skin surface the ink-filled burrow will be clearly visible. The burrow is also enhanced by applying liquid tetracycline (27). The burrows then fluoresce bright yellow-green under a Woods light.

The clinical skin signs of scabies infestation are similar for most mammals except that in some hosts mite proliferation and thus progression of the disease is more rapid than in others. For example, experimental inoculation of a small area on the back of a dog led to total body involvement and localized hyperkeratosis in 3–4 wk (10). By comparison, similar inoculation of rabbits with *S. scabiei* var. *canis* resulted in full body involvement and localized hyperkeratosis with crusting only after many months. Likewise, humans develop localized regions of hyperkeratosis and full body involvement only after many months with infestations of *S. scabiei* var. *hominis*.

Untreated scabies in all mammals advances to scaly skin and eventually heavy crusting from hyperkeratosis. The crusted state in humans is referred to as Norwegian scabies. Hyperkeratotic crusts in humans and animals vary in thickness from a few to 20 mm. Crusts are dry and porous with numerous vacant mite burrows (10, 116). Crusts can be readily pried off the host. Mites are found on the undersurface of the crusts at the moist interface of the stratum corneum/stratum lucidum and the stratum granulosum of the epidermis. Mite density at this interface may exceed 1400 mites per $cm^2$ of skin surface (7).

Scabies is indisputably diagnosed in all hosts by positive skin scrapings. Suspected areas are scraped with a scalpel or razor blade, and the removed epidermis is placed on a microscope slide in a drop of mineral oil and then overlaid with a coverslip. Eggs, all active life stages, fecal pellets, and in some cases partial or complete burrows are visible under both the dissecting and compound microscopes.

## SCABIES ANTIGENS

Scabies mites reside in the lower stratum corneum of the epidermis near the stratum corneum/stratum lucidum interface (28, 29). Histological, nutritional, and water-procurement studies suggest that intercellular fluid from lower skin zones may seep into the burrows or at least into close proximity of the burrowed mites and their mouthparts. Apparently this aqueous material provides a medium for soluble antigens (presumably from the mite's body), saliva and other body secretions, and feces to diffuse into the dermis and stimulate an inflammatory and immune response. As yet little is known about the specific source and properties of these antigens for most scabies strains. Until recently the inadequate supplies of scabies mites and antisera against them limited investigations both to characterize scabies antigens and to determine the host immune response to them. However, the recently reported successful transfer of *S. scabiei* var. *canis* to laboratory rabbits has provided an adequate supply of mites for preparation of antigenic extracts and antiscabies sera for immunologic study (10, 12). Using this rabbit model, antigens of *S. scabiei* var. *canis* have been partially characterized by crossed immunoelectrophoresis (CIE), crossed radioimmunoelectrophoresis (CRIE), and SDS-PAGE coupled with Western blotting and autoradiography (12, 15).

CIE of *S. scabiei* var. *canis* extract reacted with antisera from rabbits heavily parasitized with this strain demonstrated the presence of nine antigens; seven antigens were anodically moving (electronegative) and two were cathodically moving (electropositive) (12, 15). Fractionation of *S. scabiei* var. *canis* extract by SDS-PAGE resolved more than 30 protein/peptide bands (15). At least 20 bands were in the 14–66 kd range. Following blot transfer of the protein bands onto nitrocellulose membranes and incubation of individual tracks in human allergic reference sera and radiolabeled anti-human IgE, autoradiograms showed that 15 different protein/peptide bands bound IgE antibodies. Since nine antigens were identified by CIE but 15 SDS-PAGE–resolved bands bound human IgE, some of the antigens must exhibit multiple antigenic determinants. It is not known if multiple determinants on a single antigen are similar or different. The finding that determinants of *S. scabiei* antigens bind human IgE is significant, since IgE is the antibody type associated with allergic disease.

Preliminary indirect immunologic evidence indicated at least some cross-antigenicity between antigens of *S. scabiei* strains from different hosts (15). At least two antigen-antibody precipitates formed by CIE reaction of antigens of *S. scabiei* var. *canis* extract with anti–*S. scabiei* var. *canis* rabbit sera following incubation of CIE gels in scabietic (*S. scabiei* var. *hominis*) patient sera exhibited IgE binding. The fact that patients infested with *S. scabiei* var. *hominis* exhibited circulating IgE antibodies that bound to antigens of *S. scabiei* var. *canis* is direct evidence (provided that this binding was not cross-reactivity involving house dust mites) that there is at least partial, if not complete, identity between antigens of these strains. Falk et al (36) reported that rabbit antiserum raised against the house dust mite *Dermatophagoides farinae* recognized four *S. scabiei* var. *hominis* antigens. By comparison, anti–*D. farinae* rabbit serum recognized six *S. scabiei* var. *canis* antigens (15). These results are indirect evidence that the *S. scabiei* strains from dogs and humans may share at least four antigens or epitopes that cross-react with house dust mite antigens (15). Direct investigations of the similarities and differences in antigenicity of scabies from different hosts are still necessary to establish clearly the immunologic relationships among strains and their implications.

## CELL-MEDIATED IMMUNE RESPONSE

The cell-mediated host immune response to scabies antigens is characterized by cell infiltrates in or near the skin lesions. These infiltrates may be in the superficial and deep dermis, in tissue surrounding small blood vessels, around sweat glands in the lower dermis, and in subcutaneous fat (1, 37–39, 50). They consist of mainly lymphocytes and fewer histocytes, neutrophils, and eosinophils (37). The density of the infiltrates varies according to the level of infestation and varies among lesions, but the predominant lymphocytes in the infiltrates are T lymphocytes (38). The fact that the ratio of T lymphocytes to B lymphocytes is greater in the infiltrates than in peripheral blood reflects selective movement of T cells into the dermis (38). T lymphocytes have a major role in delayed hypersensitivity skin reactions. Their accumulation in the dermis suggests that this type of reaction occurs in scabies, and their accumulation in lesions shows the importance of a cell-mediated immune response in the pathogenesis of scabies (38). Prick and intracutaneous skin testing of patients within a year after scabies infestation indicated that immediate hypersensitivity reactions may occur (35).

Scabies-infested pigs developed immediate and delayed hypersensitivity reactions following local inoculation with a crude extract of *S. scabiei* (105). The immediate reaction resembled a local anaphylactic response. Interestingly, iron-deprived pigs reacted less to the intradermal inoculation than iron-treated littermates. These reactions appeared similar to reactions in human skin.

## HUMORAL IMMUNE RESPONSE

It is clear that scabies antigens provoke a humoral immune response. However, few controlled studies have investigated and thoroughly characterized this response for any host. Both immunized and naturally infested rabbits, pigs, and guinea pigs exhibit serum antibodies directed at *S. scabiei* antigens (12, 15, 104, 118, 121, 122). Direct reaction of *S. scabiei* var. *canis* extract by CIE with sera from scabies-infested rabbits demonstrated that the rabbits built antibodies to nine different antigens (12). Varied alterations in serum isotypic immunoglobulin levels in scabietic patients relative to posttreatment levels or levels in serum of uninfested patients have provided indirect evidence of a humoral response in humans. Hancock et al (46) reported that circulating IgA levels may be significantly lower in scabies patients than in normal persons. Likewise, Falk (32, 33) found that IgA levels of scabietic patients were significantly lower than those of the same patients 6 and 9 mo later following successful treatment. Elevated serum IgG and IgM levels in scabietic patients, which returned to normal following treatment, have also been reported (32).

Investigations of serum IgE levels of scabietic patients give conflicting evidence of whether or not scabies antigens provoke an IgE-mediated reaction to scabies. Hancock et al (46) found that 99 of 100 patients examined in a study had total IgE levels within normal limits. By contrast, Araujo-Fontaine et al (4), Falk (32), Falk & Bolle (34, 35), and Larregue et al (60) found elevated IgE values in 15, 20, 42, 45, and 65% of their patients, respectively. Since specific scabies antigens were not used as probes in these studies, it was not known if IgE was induced by scabies antigens or was the result of atopy. Most of the patients in Falk's study (32) had no obvious history of atopy, but their histories were not confirmed. Two more recent studies using scabies extracts have directly demonstrated *S. scabiei*–specific IgE in scabietic sera, but the number of patients examined was small (5, 96).

Since these initial immunologic investigations of scabietic patient sera, understanding of the antibody response to scabies antigens has become even more uncertain with the recent finding that scabies and house dust mites are highly cross-antigenic; the IgE built against *D. farinae* recognizes antigenic determinants on *S. scabiei*. Because of this cross-reactivity and the lack of human studies with patients skin-tested to house dust mites, the IgE and other isotypic responses to scabies have not yet been confirmed.

Apparently, mite antigens promote selected diffusion of circulating antibodies and C3 complement out of the vasculature and their deposition in dermal vessels, epidermis, and the dermoepidermal junction in the vicinity of the mites. Immunofluorescence studies of biopsied specimens from involved skin of 11 patients revealed IgE in vessel walls from four patients, IgM and C3 deposits at the dermoepidermal junction in one patient, and only C3 deposits in the latter area in one patient (40). Deposits of IgG, IgM, and C3 in dermal vessels and IgG, IgA, and C3 at the dermoepidermal junction have also been noted in similar studies (52, 100). The biological importance of these responses in determining host susceptibility and resistance to scabies infestations and host specificity remain to be determined. Variations in resistance and susceptibility in natural hosts as well as host specificity may reflect both differences in the hosts' ability to mount these responses and the parasites' resistance to them.

# SUMMARY

Scabies continues to be an important parasitic disease of humans and other mammals. Surprisingly for a disease that has afflicted humans since antiquity, little is directly known about the basic biology of the parasite, the host-parasite interactions, the host immune response, and host susceptibility. Much more research in these areas is needed if we are to understand fully the occurrence, transmission, and epidemiology of both human and animal scabies.

ACKNOWLEDGMENTS

Much of the research in this review was supported by grant No. AI 17252 from the National Institute of Allergy and Infectious Diseases.

I express my thanks to D. L. Vyszenski-Moher for editorial assistance during preparation of the manuscript.

*Literature Cited*

1. Ackerman, A. B. 1977. Histopathology of human scabies. See Ref. 87, pp. 88–95
2. Alexander, J. O. 1984. Scabies. *Arthropods and Human Skin*, pp. 227–92. Berlin: Springer-Verlag. 422 pp.
3. Alva-Valdes, R., Wallace, D. H., Foster, A. G., Ericsson, G. F., Wooden, J. W. 1986. The effects of sarcoptic mange on the productivity of confined pigs. *Vet. Med.* 81(3):258–62
4. Araujo-Fontaine, A., Thiery, R., Heid, E. 1977. Serum IgE levels in scabies; studies of about 100 cases. *Ann. Dermatol. Venereol.* 104:203–5
5. Arlian, L. G. 1988. Host-parasite interaction of *Sarcoptes scabiei*. In *Progress in Acarology*, ed. G. P. Channa-Basavanna, B. K. Negeshchandra. New Delhi: Oxford & IBH. In press
6. Arlian, L. G., Ahmed, M., Vyszenski-Moher, D. L. 1988. Effects of *S. scabiei* on blood indexes of parasitized rabbits. *J. Med. Entomol.* 25:In press
7. Arlian, L. G., Ahmed, M., Vyszenski-Moher, D. L., Estes, S. A., Achar, S. 1988. Energetic relationships of *Sarcoptes scabiei* var. *canis* with the laboratory rabbit. *J. Med. Entomol.* 25:57–63
8. Arlian, L. G., Estes, S. A., Vyszenski-Moher, D. L. 1988. Prevalence of *Sarcoptes scabiei* in the environment of scabietic patients. *J. Am. Acad. Dermatol.* 19(5):In press
9. Arlian, L. G., Runyan, R. A., Achar, S., Estes, S. A. 1984. Survival and infestivity of *Sarcoptes scabiei* var. *canis* and var. *hominis*. *J. Am. Acad. Dermatol.* 11:210–15
10. Arlian, L. G., Runyan, R. A., Estes, S. A. 1984. Cross-infestivity of *Sarcoptes scabiei*. *J. Am. Acad. Dermatol.* 10: 979–86
11. Arlian, L. G., Runyan, R. A., Sorlie, L. B., Estes, S. A. 1984. Host-seeking behavior of *Sarcoptes scabiei*. *J. Am. Acad. Dermatol.* 11:594–98
12. Arlian, L. G., Runyan, R. A., Sorlie, L. B., Vyszenski-Moher, D. L., Estes, S. A. 1985. Characterization of *Sarcoptes scabiei* var. *canis* (Acari: Sarcoptidae) antigens and induced antibodies in rabbits. *J. Med. Entomol.* 22:321–23
13. Arlian, L. G., Vyszenski-Moher, D. L. 1988. Life cycle of *Sarcoptes scabiei* var. *canis*. *J. Parasitol.* 74:427–30
14. Arlian, L. G., Vyszenski-Moher, D. L., Cordova, D. 1988. Host specificity of *S. scabiei* var. *canis* and the role of host odor. *J. Med. Entomol.* 25:52–56
15. Arlian, L. G., Vyszenski-Moher, D. L., Gilmore, A. M. 1988. Cross-reactivity of *Sarcoptes scabiei* and the house dust mite, *Dermatophagoides farinae*. *J. Med. Entomol.* 25:240–47
16. Arlian, L. G., Vyszenski-Moher, D. L., Pole, M. J. 1988. Survival of adults and developmental stages of *Sarcoptes scabiei* var. *canis* when off the host. *Exp. Appl. Acarol.* In press
17. Blumenthal, D. S., Taplin, D., Schultz,

M. G. 1976. A community outbreak of scabies. *Am. J. Epidemiol.* 104:667–72

18. Brownlie, W. M., Harrison, I. R. 1960. Sarcoptic mange in pigs. *Vet. Rec.* 72:1022–23

19. Busvine, J. R. 1977. Dermatoses due to arthropods other than the scabies mite. See Ref. 87, pp. 132–38

20. Cargill, C. F., Dobson, K. J. 1979. Experimental *Sarcoptes* infestations in pigs: (2) Effects on production. *Vet. Rec.* 104:33–36

21. Carslaw, R. W., Dobson, R. M., Hood, A. J. K., Taylor, R. N. 1975. Mites in the environment of cases of Norwegian scabies. *Br. J. Dermatol.* 92:333–37

22. Christophersen, J. 1978. The epidemiology of scabies in Denmark, 1900–1975. *Arch. Dermatol.* 114:747–50

23. Church, R. E., Knowelden, J. 1978. Scabies in Sheffield: a family infestation. *Br. Med. J.* 1:761–63

24. Cooper, C. L., Jackson, M. M. 1986. Outbreak of scabies in a small community hospital. *Am. J. Infect. Control* 14: 173–79

25. Epstein, E., Orkin, M. 1985. Scabies: clinical aspects. See Ref. 86, pp. 19–22

26. Estes, S. A. 1978. Scabies: the diagnostic dilemma. *Ariz. Med.* 35:477–79

27. Estes, S. A. 1982. Diagnosis and management of scabies. *Med. Clin. North Am.* 66:955–63

28. Estes, S. A., Arlian, L. G. 1984. Reply. *J. Am. Acad. Dermatol.* 10:676–77

29. Estes, S. A., Kummel, B., Arlian, L. G. 1983. Experimental canine scabies in humans. *J. Am. Acad. Dermatol.* 9:397–401

30. Fain, A. 1968. Etude de la variabilité de *Sarcoptes scabiei* avec une revision des Sarcoptidae. *Acta Zool. Pathol. Antverp.* 47:1–196

31. Fain, A. 1978. Epidemiological problems of scabies. *Int. J. Dermatol.* 17: 20–30

32. Falk, E. S. 1980. Serum immunoglobulin values in patients with scabies. *Br. J. Dermatol.* 102:57–61

33. Falk, E. S. 1981. Serum IgE before and after treatment for scabies. *Allergy* 36: 167–74

34. Falk, E. S., Bolle, R. 1980. IgE antibodies to house dust mite in patients with scabies. *Br. J. Dermatol.* 102:283–88

35. Falk, E. S., Bolle, R. 1980. In vivo demonstration of specific immunological hypersensitivity to scabies mite. *Br. J. Dermatol.* 103:367–73

36. Falk, E. S., Dale, S., Bolle, R. Haneberg, B. 1981. Antigens common to scabies and house dust mites. *Allergy* 36:233–38

37. Falk, E. S., Eide, T. J. 1981. Histologic and clinical findings in human scabies. *Int. J. Dermatol.* 20:600–5

38. Falk, E. S., Matre, R. 1982. In situ characterization of cell infiltrates in the dermis of human scabies. *Am. J. Dermatopathol.* 4:9–15

39. Fernandez, N., Torres, A., Ackerman, B. 1977. Pathologic findings in human scabies. *Arch. Dermatol.* 113:320–24

40. Frentz, G., Veien, N. K., Eriksen, K. 1977. Immunofluorescence studies in scabies. *J. Cutaneous Pathol.* 4:191–93

41. Friedman, R. 1947. *The Story of Scabies,* Vol. 1. New York: Froben. 468 pp.

42. Gier, H. T., Kruckenberg, S. M., Marler, R. J. 1978. Parasites and diseases of coyotes. In *Coyotes, Biology, Behavior, and Management,* ed. M. Bekoff, pp. 37–69. New York: Academic. 384 pp.

43. Gordon, R. M., Lavoipierre, M. M. J. 1962. *Entomology for Students of Medicine.* Oxford: Blackwell. 258 pp.

44. Gulati, P. V., Braganza, C., Singh, K. P., Borker, V. 1977. Scabies in a semiurban area of India: an epidemiologic study. *Int. J. Dermatol.* 16:594–98

45. Haarløv, N., Møller-Madsen, M. 1982. Sarcoptid mange in a Danish cattle herd. In *Acarology VI,* ed. D. A. Griffith, C. E. Bowman, 2:1138–42. West Sussex, England: Horwood. 1296 pp.

46. Hancock, B. W., Ward, A. M., Path, M. R. C. 1974. Serum immunoglobulin in scabies. *J. Invest. Dermatol.* 63:482–84

47. Hawkins, J. A., McDonald, R. K., Woody, B. J. 1987. *Sarcoptes scabiei* infestation in a cat. *J. Am. Vet. Med. Assoc.* 190:1572–73

48. Haydon, J. R., Caplan, R. M. 1971. Epidemic scabies. *Arch. Dermatol.* 103: 168–73

49. Heilesen, B. 1946. Studies on *Acarus scabiei* and scabies. *Acta Derm.-Venereol. Suppl.* 26:1–370

50. Hejazi, N., Mehregan, A. H. 1975. Scabies: histological study of inflammatory lesions. *Arch. Dermatol.* 111:37–39

51. Hewitt, M., Barrow, G. I., Miller, D. C., Turk, F., Turk, S. 1973. Mites in the personal environment and their role in skin disorders. *Br. J. Dermatol.* 89: 401–9

52. Hoefling, K. K., Schroeter, A. L. 1980. Dermatoimmunopathology of scabies. *J. Am. Acad. Dermatol.* 3:237–40

53. Hourrigan, J. L. 1979. Spread and de-

tection of psoroptic scabies of cattle in the United States. *J. Am. Vet. Med. Assoc.* 175:1278–80

54. Juranek, D. D., Schultz, M. G. 1977. Epidemiologic investigations of scabies in the United States. See Ref. 87, pp. 64–72

55. Kutzer, E. 1966. Zur Epidemiologie der *Sarcoptes*räude. *Angew. Parasitol.* 7: 241–48

56. Kutzer, E. 1970. Merkblätter über angewandte Parasitenkunde und Schädlingsbekämpfung, Merkblatt Nr 17. *Sarcoptes*-Milben und *Sarcoptes*räude der Haustiere. Beilage zu. *Angew. Parasitol.* 11:1–22

57. Kutzer, E., Grünberg, W. 1967. Sarcoptesräude (*Sarcoptes tapiri* nov. spec.) bei Tapiren (*Tapirus terrestris* L.). *Z. Parasitenkd.* 29:46–60

58. Kutzer, E., Grünberg, W. 1969. Zur frage der übertragung tierischer Sarcoptesräuden auf den Menschen. *Berl. Münch. Tierärztl. Wochenschr.* 82:311–14

59. Kutzer, E., Onderscheka, W. 1966. Die Räude der Gemse und ihre Bekämpfung. *Z. Jagdwiss.* 12:63–84

60. Larregue, M., Gombert, P., Levy, C., Gallet, P., Rat, J. P. 1976. Elevation des IgE dan la gale chez l'enfant. *Bull. Soc. Fr. Dermatol. Syphilig.* 83:54

61. Lee, D. J. 1971. High incidence of scabies. *Med. J. Aust.* 2(9):500–1

62. Lerche, N. W., Currier, R. W., Juranek, D. O., Baer, W., Dubay, N. 1983. Atypical crusted "Norwegian" scabies: report of nosocomial transmission in a community hospital and an approach to control. *Cutis* 31:637–84

63. Levene, G. M., Turk, J. L. 1970. Immunological aspects of skin disease. *Br. J. Hosp. Med.* 3:811–18

64. Liebisch, A., Petrich, J. 1977. Zur gegenwärtigen Verbreitung und Bekämpfung der Rinderräude in Norddeutschland. *Dtsch. Tierärztl. Wochenschr.* 84:424–27

65. Lindquist, W. D. 1973. Sarcoptic mange in a cat. *J. Am. Vet. Med. Assoc.* 162:639–40

66. Marks, J., Shuster, S. 1968. Iron metabolism in skin disease. *Arch. Dermatol.* 98:469–75

67. McPherson, E. A. 1960. Sarcoptic mange in pigs. *Vet. Rec.* 72:869–70

68. Meierhenry, E. F., Clausen, L. W. 1977. Sarcoptic mange in collared peccaries. *J. Am. Vet. Med. Assoc.* 71: 983–84

69. Mellanby, K. 1977. Epidemiology of scabies. See Ref. 87, pp. 60–63

70. Mellanby, K. 1977. Scabies in 1976. *R. Soc. Health J.* 97:32–36

71. Mellanby, K. 1985. Biology of the parasite. See Ref. 86, pp. 9–18

72. Mellanby, K., Johnson, C. G., Bartley, W. C., Brown, P. 1942. Experiments on the survival and behavior of the itch mite, *Sarcoptes scabiei* DeG. var. *hominis*. *Bull. Entomol. Res.* 33:267–71

73. Melton, L. J., Brazin, S. A., Damm, S. R. 1978. Scabies in the United States Navy. *Am. J. Public Health* 68:776–78

74. Moberg, S. A. W., Lowhagen, G. E., Hersle, K. S. 1984. An epidemic of scabies with unusual features and treatment resistance in a nursing home. *J. Am. Acad. Dermatol.* 11:242–44

75. Morrison-Smith, J., Disney, M. E., Williams, J. P., Goels, Z. A. 1969. Clinical significance of skin reactions to mite extracts in children with asthma. *Br. Med. J.* 2:723

76. Munro, J. W. 1919. Report of scabies investigation. *J. R. Army Med. Corps* 33:1–41

77. Nair, B. K. H., Joseph, A., Kandamuthan, M. 1977. Epidemic scabies. *Indian J. Med. Res.* 65:513–18

78. Nair, B. K. H., Joseph, A., Narayanan, P. L., Chacko, K. V. 1973. Epidemiology of scabies. *Indian J. Dermatol. Venereol.* 39:101

79. Nitzulescu, V. 1973. Sur la viviparité possible chez la sarcopte de la gale humaine. *Ann. Parasitol.* 48:355–58

80. Nusbaum, S. R., Drazek, F. J., Holden, H., Love, T. J., Marvin, J., et al. 1975. *Sarcoptes scabiei bovis*—a potential danger. *J. Am Vet. Med. Assoc.* 166: 252–56

81. OConnor, B. 1982. Evolutionary ecology of astigmatid mites. *Ann. Rev. Entomol.* 27:385–409

82. Orkin, M. 1971. Resurgence of scabies. *J. Am. Med. Assoc.* 217:593–97

83. Orkin, M. 1975. Today's scabies. *J. Am. Med. Assoc.* 233:882–85

84. Orkin, M., Maibach, H. I. 1978. This scabies pandemic. *N. Engl. J. Med.* 298:496–98

85. Orkin, M., Maibach, H. I. 1978. Scabies in children. Symposium on Pediatric Dermatology. *Pediatr. Clin. North Am.* 25:371–86

86. Orkin, M., Maibach, H. I., eds. 1985. *Cutaneous Infestations and Insect Bites.* New York: Dekker. 321 pp.

87. Orkin, M., Maibach, H. I., Parish, L. C., Schwartzman, R. M., eds. 1977. *Scabies and Pediculosis.* Philadelphia: Lippincott. 203 pp.

88. Palicka, P., Merka, V. 1971. Contemporary epidemiological problems of scabies. *J. Hyg. Epidemiol. Microbiol. Immunol.* 15:457–61
89. Parish, L. C. 1977. History of scabies. See Ref. 87, pp. 1–7
90. Parish, L. C., Millikan, L. E., Witkowski, J. A., Schwartzman, R. 1983. Scabies in the extended care facility. *Int. J. Dermatol.* 22:380–82
91. Parlette, H. L. 1975. Scabietic infestations of man. *Cutis* 16:47–52
92. Pence, D. B., Casto, S. D., Carley, C. J. 1981. Ectoparasites of wild canids from the Gulf Coastal prairies of Texas and Louisiana. *J. Med. Entomol.* 18:409–12
93. Pence, D. B., Windberg, L. A., Pence, B. C., Sprowls, R. 1983. The epizootiology and pathology of sarcoptic mange in coyotes, *Canis latrans*, from South Texas. *J. Parasitol.* 69:1100–15
94. Poindexter, H. A. 1978. Scabies. *J. Natl. Med. Assoc.* 70:525–26
95. Pryor, L. B. 1956. Sarcoptic mange in wild foxes in Pennsylvania. *J. Mammal.* 37:90–93
96. Rantanen, T., Bjorksten, F., Reunala, T., Salo, O. P. 1981. Serum IgE antibodies to scabies mite. *Acta Derm.-Venereol.* 61:358–60
97. Reilly, S., Cullen, D., Davies, M. G. 1984. An outbreak of scabies in a hospital and community. *Br. Med. J.* 291:1031–32
98. Rigatos, G. A., Kappos-Rigatos, I. 1976. Scabies in Greece. *Arch. Dermatol.* 112:1466
99. Rook, A. 1972. Skin diseases caused by arthropods and other venomous or noxious animals. In *Textbook of Dermatology*, ed. A. Rook, D. S. Wilkinson, F. J. Ebling, 1:845–84. Oxford: Blackwell. 1060 pp. 2nd ed.
100. Salo, O. P., Reunala, T., Kalimo, K., Rantanen, T. 1982. Immunoglobulin and complement deposits in the skin and circulating immune complexes in scabies. *Acta Derm.-Venereol.* 62:73–76
101. Sehgal, V. N., Rao, R. L., Rege, V. L., Vadira, S. N. 1972. Scabies: a study of incidence and a treatment method. *Int. J. Dermatol.* 11:106–11
102. Shaw, P. K., Juranek, D. D. 1976. Recent trends in scabies in the United States. *J. Infect. Dis.* 134:414–16
103. Sheahan, B. J. 1974. Experimental *Sarcoptes scabiei* infection in pigs: clinical signs and significance of infection. *Vet. Rec.* 94:202–9
104. Sheahan, B. J. 1975. Pathology of *Sarcoptes scabiei* infection in pigs. I. Naturally occurring and experimentally induced lesions. *J. Comp. Pathol.* 85:87–95
105. Sheahan, B. J. 1975. Pathology of *Sarcoptes scabiei* infection in pigs. II. Histological, histochemical and ultrastructural changes at skin test sites. *J. Comp. Pathol.* 85:97–110
106. Sheahan, B. J., O'Connor, P. J., Kelly, E. P. 1974. Improved weight gains in pigs following treatment for sarcoptic mange. *Vet. Rec.* 94:169–70
107. Shrank, A. B., Alexander, S. L. 1967. Scabies: another epidemic? *Br. Med. J.* 1:669–71
108. Shuster, S., Marks, J. 1967. Dermatopathic anaemia. *Br. J. Dermatol.* 79:393–97
109. Smith, E. B., Claypoole, T. F. 1967. Canine scabies in dogs and in humans. *J. Am. Med. Assoc.* 199:95–100
110. Stamps, T. J. 1969. Scabies in Negroes. *Br. Med. J.* 1:513
111. Stone, W. B., Parks, E., Weber, B. L., Parks, F. J. 1972. Experimental transfer of sarcoptic mange from red foxes and wild canids to captive wildlife and domestic animals. *NY Fish Game J.* 19:1–11
112. Stone, W. B., Tullar, B. F., Zeh, J. B., Weber, B. L. 1974. Incidence and distribution of mange mites in foxes in New York. *NY Fish Game J.* 21:163–66
113. Thomsett, L. R. 1968. Mite infestations of man contracted from dogs and cats. *Br. Med. J.* 3:93–95
114. Todd, A. W., Gunson, J. R., Samuel, W. M. 1981. Sarcoptic mange, an important disease of coyotes and wolves of Alberta, Canada. In *World-wide Furbearer Conf. Proc.*, ed. J. A. Chapman, D. Pursley, pp. 706–29
115. Trainer, D. O., Hale, J. B. 1969. Sarcoptic mange in red foxes and coyotes of Wisconsin. *Bull. Wildl. Dis. Assoc.* 5:387–91
116. Van Neste, D., Lachapelle, J. M. 1981. Host-parasite relationship in hyperkeratotic (Norwegian) scabies: pathological and immunological findings. *Br. J. Dermatol.* 105:667–78
117. Van Neste, D., Mrena, E., Marchal, G. 1981. Le cycle evolutif du *Sarcoptes scabiei* (var. *hominis*): une etude en microscopie electronique a balayage. *Ann. Dermatol. Venereol.* 108:355–61
118. Van Neste, D., Salmon, J. 1978. Circulating antigen antibody complexes in scabies. *Dermatologica* 157:221–24

119. Witkowski, J. A., Parish, L. C. 1978. Scabies update 1978. *Int. J. Dermatol.* 17:401–2
120. Woodley, D., Saurat, I. 1981. The burrow ink test and the scabies mite. *J. Am. Acad. Dermatol.* 4:715–22
121. Wooten, E. L., Gaafar, S. M. 1984. Detection of serum antibodies to sarcoptic mange mite antigens by the passive hemagglutination assay in pigs infested with *Sarcoptes scabiei* var *suis*. *Vet. Parasitol.* 15:309–16
122. Wooten, E. L., Gaafar, S. M. 1984. Hemagglutinating factor in an extract of *Sarcoptes scabiei* var. *suis* (De Geer). *Vet. Parasitol.* 15:317–23

*Ann. Rev. Entomol. 1989. 34:163–90*
*Copyright © 1989 by Annual Reviews Inc. All rights reserved*

# BIONOMICS OF THE LARGE CARPENTER BEES OF THE GENUS *XYLOCOPA*

## D. Gerling

Department of Zoology, George S. Wise Faculty of Life Sciences, Tel Aviv University, Ramat Aviv 69978, Israel

## H. H. W. Velthuis

Department of Comparative Physiology, University of Utrecht, Jan van Galenstraat 40, 3572 LA Utrecht, the Netherlands

## A. Hefetz

Department of Zoology, George S. Wise Faculty of Life Sciences, Tel Aviv University, Ramat Aviv 69978, Israel

## INTRODUCTION

The large carpenter bees of the genus *Xylocopa* include more than 730 species, which are grouped into 48 subgenera (52). As their name implies, they include some of the largest bees, often exceeding 3 cm in length. Females of most species are conspicuous in their black or blue coloration, which may be variegated with lighter-colored pubescence. Males either resemble females or may be completely covered with a light brown, light green, or yellowish green pubescence. Most species occur in the tropics or subtropics. Less than 10% extend into the temperate zones, reaching as far north as the steppes of Russia (72). Nests are usually constructed in dead plant material; an almost infinite variety of wood, cane, and flowering stalks is used (49, 72).

Recorded studies of carpenter bees date back almost 300 years (52) and now include observations on about 60 different species. However, apart from

163

0066-4170/89/0101-0163$02.00

several early landmark studies (72, 73, 78, 83), most investigations of carpenter bee bionomics have been conducted during the last 30 years. Hurd & Moure's 1963 monograph (52) apparently stimulated many of the studies on behavior and chemical communication (5, 13–15, 35, 38, 40, 87, 97, 101, 105).

Although we have tried to cover most of the published material on carpenter bee bionomics, space limitation prevents us from discussing all aspects or even citing all works. The first part of this review is devoted to nesting activity, flower associations, and natural enemies. In the second part, interactions between conspecific bees are discussed in relation to mating and sociality. This latter aspect is particularly timely because during the past 15 years the social structure of several species has been unraveled. The bees' readiness to nest in the various materials presented to them and their relatively large size allow for easy viewing of bee interactions outside or within the nest, by direct observation or by X-ray technology. These studies have shown carpenter bees to be K-selected species that may employ diverse intraspecific strategies for mate location. The bees display a gradation of levels of social interaction, ranging from solitary to eusocial, both on temporal and species-specific scales. Furthermore, being often of a facultative nature, these social interactions are strongly influenced by ecological conditions. In this review we hope to demonstrate that carpenter bees are an excellent group for the study of social behavior and that ecological constraints had an important role in the shaping of the group's evolution.

## SEASONAL ACTIVITY AND LIFE CYCLES

Voltinism in carpenter bees is limited by temperature and the availability of food. The number of generations per year depends upon the climate and ranges from one to four (15). Successive broods may often be produced by the same female.

All nesting cycles show certain common features (a) The bees have a period of reproductive quiescence in which both sexes cohabit in groups. This may occur during the dry season; for example, in Mbita, eastern Kenya, X. flavocincta, X. inconstans, and X. senior were reproductively inactive during March 1984 (D. Gerling, personal observations). Alternatively, bees may enter a nonreproductive phase during the cooler months. In the Northern Hemisphere, quiescence usually lasts from October through February or March (11, 38, 72, 83). Some bivoltine species are quiescent following each adult emergence (22). (b) Most quiescent females are unmated (38, 40). (c) Mating occurs, often following territorial flights by the males, before nest establishment (5, 38, 101). Although males of X. virginica and X. pubescens mate mainly during the spring (11, 38), they may also defend territories and occasionally mate in the fall (40). X. virginica is the only known species in which one-year-old females cohabit with two-year-old females while only the older females perform nesting activities. (*d*) First-laid eggs produce the first bees to emerge as adults. These adults break all of the partitions in their tunnel and may push the remaining pupae to the distal end of the nest (40). In some species males develop faster than females, even though the eggs that produce them are laid after those that produce females (51, 72, 95a). (*e*) Emerging bees remain teneral for one to several weeks (5, 78, 94). During this period they feed on pollen brought by their mother (100) and/or sisters (22).

In multivoltine species, the founding females may live for several months and deposit more than one round of progeny. Following their teneral period, the emerging bees may join the mother's nest, dig adjacent nests, or disperse (11, 21, 22, 40, 105). Nests may be founded by one or more females (see section on origins of sociality).

# NESTING

On the whole, the nesting habits of carpenter bees are similar, since all build their nests in plant material. However, details of the exact habits and the nature of the nesting materials are still lacking for species in 21 of the 48 subgenera listed by Hurd & Moure (52).

## Nest Types

Two major nest-construction patterns can be discerned: those with unbranched and those with branched tunnels. The former are constructed in culms that are hollow or have a soft interior (21, 40, 49, 73); the latter are constructed in tree trunks and twigs or structural timber.

UNBRANCHED NESTS    Bees of eight subgenera build only unbranched nests. Four of these subgenera are quite flexible in their choice of nesting substrates; four are more rigid (16, 40, 46, 49, 50).

The unbranched tunnel may extend to one or both sides of the nest entrance, depending upon its position. Each tunnel typically contains 4–7 closed cells (51, 52, 73), but as many as 15 per tunnel and 19 per nest have been found in *X. sulcatipes* (D. Gerling, personal observations); the latter were probably produced by more than one female.

BRANCHED NESTS    Branched nests may be found in a range of wooden substrates including both dead tree parts and structural timber. Bees from five subgenera may construct either branched nests in solid wood or straight nests in reeds or culms (52). Nests may also be constructed in man-made materials such as lead (92), styrofoam, fiberboard (11), and fiberglass (D. Gerling, personal observations).

Some multivoltine species, such as *X. pubescens* (11, 40), make relatively short tunnels with only 1–3 cells. Having completed the first tunnel, they may dig additional tunnels to accommodate brood cells. Others, such as *X. sonorina* (36), dig out a tunnel that is long enough to lay one generation's complement of eggs (usually 4–6 eggs per tunnel). No additional eggs are then laid until the first group has reached maturity. The univoltine species, e.g. *X. violacea, X. virginica,* and *X. valga,* build and provision 1–4 tunnels in succession (38, 55, 72).

## Nesting Activities

Nesting activities of the female include site selection and probing, initial digging, supplementary activities, nest care, cell preparation, provisioning and ovipositing, cell-sealing, subsequent digging, and nest defense against conspecifics and other organisms. The subjects of provisioning, ovipositing, and cell-sealing, which have been relatively well covered in the recent literature (5, 40, 53, 93, 100, 104, 107, 108), are not discussed owing to space limitations.

SITE AND MATERIAL SELECTION    The little that is known about nesting-site selection is anecdotal rather than experimental. The culm-nesting *X. iris* occupies *Peucedanum alsaticum* in open fields (73), whereas many of the solid-wood nesters, e.g. *X. gualanensis, X. pubescens,* and *X. virginica,* are most often found in shaded places or within structures (38, 83, 87; D. Gerling, A. Hefetz & H. H. W. Velthuis, personal observations). Site selection may be affected by the presence of other active carpenter bee nests (54, 57). In Sumatra, *X. traquebarica* nests were never found near each other, and nesting *X. confusa* and *X. latipes* females were seen preventing newcomers from digging nests in the immediate vicinity (54). Frequently observed nest aggregations may be due to the predilection of the bees to nest where others do, but they may also be caused by the clumped distribution of nesting material.

Selection of nesting material may be elaborate. The same species may choose either soft, easily penetrable material, which makes a less durable nesting site, or a harder material that offers better protection (105). Painting or staining structural timber did not hamper nesting activities by *X. sonorina* (36) or *X. virginica* (9), but Balduf (7) found that *X. virginica* avoided white-stained entrances.

Shallow excavations at the nesting sites (20, 36, 48) can be regarded either as examination of the substrate prior to digging or as unsuccessful attempts to start nests. Selection pertaining to substrate thickness has been demonstrated for *X. californica arizonensis;* these bees used inflorescence stalks of *Yucca elata* that were thick enough to support the nest but thin enough to allow breakage and to facilitate the seed dissemination of the plant (33).

NEST CONSTRUCTION AND CELL PREPARATION    The entrance hole is usually somewhat elliptical, about 3–5 mm narrower than the diameter of the tunnel, and only slightly wider than the diameter of the bee (71). Digging is a more or less continuous process that may last through the night (5, 72; M. J. Orlove, personal communication), take up to a few days, or be carried out in sessions over several days (36, 38). Tunneling may start on the horizontal or vertical surface of the substrate or on its broken or cut surface (e.g. 11, 38, 40, 51, 62, 78, 83). With the exception of *X. violacea* (55), bees that dig in culms such as bamboo or *Arundo donax* do not usually extend their nest tunnels through the internodal partitions.

The interior of the nest consists of an antechamber or vestibule from which the nesting tunnels start. In all natural substrates main tunnels are oriented parallel to the grain of the wood, sometimes even through wood knots (78). However, this orientation is lost when no wood grain is available, as in styrofoam (11). Ascending tunnels are dug first (5). Sawdust is pushed out of the nest with the head or abdomen and legs, and may pile up at the nest entrance (5, 12, 83, 105). Supplementary activities associated with nest preparation have been reported for *X. iris* and for *X. fimbriata*. The former species makes the nest less susceptible to weather hazards by cutting the culm above the nesting tunnel and breaking the top off (73). *Xylocopa fimbriata* constructs special vertical shafts used only for ventilation (104). It has not been noted how these 1-m long shafts are constructed.

Nest care consists mainly of cleaning and grooming the nest burrows. Refuse, e.g. sawdust, broken cell partitions, dead bees and their parts, exuviae, and feces, is either pushed out of the entrance or stored in old cells or at the distal ends of a burrow in a culm (40). Bees groom their nests throughout occupancy by coating the walls with glandular secretions. This is most commonly done by licking, but bees may also coat the walls with intersegmental gland secretions by rubbing the abdomen on the walls (41). Licking may follow bending of the head toward the open sting chamber, from which secretory materials are apparently gathered (39). It appears most intense during cell preparations, when the bees lick the entire interior of the cell prior to making bee bread (40), which is a mixture of pollen, nectar, and glandular secretions.

We interpret these grooming activities as the formation of brood-cell linings. In this respect we differ with Sakagami & Michener (89), who claimed that carpenter bees lack such structures. Hydrocarbons of bee origin have been isolated from tunnels of *X. sulcatipes* (see section on chemical communication). The hydrophobic nature of this lining protects the cell from desiccation or excessive humidity. The lining may also contain antibiotics to prevent the growth of microorganisms. The cell lining of carpenter bee nests can be seen as a thin layer in X-ray images (40, 101) and is sometimes apparent as a yellowish or brownish shining coat on the cell wall or on the outer side of the cell partition (H. H. W. Velthuis, personal observation).

Bee bread preparation follows the deposition of pollen loads in the future cell. The duration of foraging trips and the amount of pollen brought vary inter- and intraspecifically with the availability, distance, and quality of suitable flowers on the one hand and with certain nest characteristics on the other. These include the presence of guards and of progeny that feed on the pollen and nectar stores (see also sections on associations with flowers and sociality) (7, 11, 78, 100) or directly on the collected pollen (105). Consequently, reports on the duration of cell provisioning vary with the species studied and with the specific conditions of the observation (e.g. 11, 55, 59).

The final size of the bee bread depends on the species (5, 13, 72). However, intraspecific size differences in bee breads, and corresponding differences in the size of emerging adults, have been recorded for *X. sulcatipes* (107). In this case, the large variability might be related to the harsh desert environment.

Once the eggs have been laid, the mother primarily guards the nest. Wagner (104) saw females of *Xylocopa* sp. from the Yucatan peninsula in Mexico stuff the space between the last sealed cell and the entrance with long animal hairs.

Subsequent digging is associated with nest extension or reuse. During nest extension bees usually avoid breaking the partitions between tunnels and the outside world (5), although *X. sonorina* nests sometimes have two entrances, of which one originated from such a breakage (36). Reuse usually involves tunnel extension (36, 38, 40, 87); the next generation is raised in the freshly dug section.

## ASSOCIATIONS WITH FLOWERS

Flowers provide the sole source of food and most of the water for carpenter bees. Since the bees are long lived and have a wide geographical range, they need to be polytropic (32) and may visit a large number of different plants. *Xylocopa darwini,* for example, visited 160 species belonging to 28 families in the Galapagos Islands (67). The bees collect readily from different flower species according to the availability of the resources and their needs, but they may show flower constancy (32) associated with the abundance of a certain flower species at a fixed time of the day (67; D. Gerling & H. H. W. Velthuis, personal observations).

Whereas most species are diurnal, some are crepuscular and matinal, flying under very low light intensity. The best studied species in the latter category is *X. tabaniformis* (48, 56, 80), whose 10 subspecies visit flowers after sunset and before dawn.

Diurnal *Xylocopa* spp. may visit different flower species at different hours. *Xylocopa gualanensis* collected nectar from *Delonix regia* all day, with peaks shortly after sunrise; from *Crotalaria* sp. starting at 10:00 AM; and from *Callinandra* sp. between 3:00 and 4:00 PM (86). Visits to passion fruit flowers coincided with their open hours. *Xylocopa mordax* on St. Vincent Island and *X. sonorina* in Hawaii paid most of their visits between 11:00 AM and 3:00 PM (25, 79).

*Xylocopa* species avoid revisiting flowers shortly after previous visits (30, 34, 82). This feature, which probably enhances the efficiency of both food collection and flower pollination, was associated with the deposition of marking pheromones by *X. virginica* on *Passiflora edulis* (34) and by *X. pubescens* and *X. sulcatipes* on *Calotropis procera* (30). In the latter case, each bee species was repelled both by her own and by the second species' markings.

## Nectar

The bee "legitimately" obtains nectar by reaching the source from within or upon the flower, thus usually effecting its pollination. However, some flowers possess a long tubular corolla in which nectar is inaccessible to the bees; they must therefore obtain it through "nectar theft" or "robbery" (32, 53), imbibing the nectar through a slit that they make with their maxillary galeae in the wall of the calyx or corolla (90)

The range of robbed flowers differs among the bee species. Smaller species may be able to reach the nectar sources in the legitimate way, but may have difficulty in piercing the walls of the petals or calyxes (95b). Young bees of different species apparently vary in their readiness to learn how to rob new flowers (36). Finally, each species has a repertoire of plants with which it evolved. This repertoire depends upon the local flora in addition to the characteristics of the bee.

In the tropics, plants protect themselves from robbery with morphological structures, such as a strong wall separating the "anteroom" and "nectar room," or by producing extra floral nectar, which is visited by ants, which in turn inhibit the bees from robbing nectar (95b). However, such ant guards failed to prevent extensive nectar robbery from *Hibiscus* flowers in Hawaii (36). Ant guards appear to function only in the tropics, where they evolved (82). Euphilic, or specialized, *Xylocopa* host flowers have usually developed barriers that prevent the collection of nectar unless considerable force is used (90). For example, in *Thunbergia grandiflora*, separation of the dorsal stamen filaments, which block the way to the nectar, requires considerable pressure (82). In *Calotropis gigantea* and *C. procera,* force is needed to penetrate the nectar reservoir (cuculus) (29).

Sugar concentration in the nectar of many flower species visited by *Xylocopa* spp. varies between 25 and 27% (32), whereas a concentration of 60–62% is used for bee bread preparation (25). To obtain the required concentration, the bee exposes the collected nectar on its proboscis drop by drop until the excess water has evaporated. This nectar ripening or dehydration has been observed principally under tropical and subtropical conditions. It may occur when the bee is on the plant (D. Gerling, personal observations) or at the entrance of the nest (20, 22, 25, 36, 78, 105). Nectar dehydration by males increases the energy/volume ratio of the nectar in their crops, thus enabling them to stay longer in their territory (110).

## Pollen

Schremmer (90) has discussed the collection and carrying of pollen of carpenter bees in detail. Bees may obtain pollen by actively brushing it off anthers with exposed pollen such as those of *Lonicera etrusca*, *Leucaena glauca*, and *Luffa cylindrica* (40, 59). Alternatively, they may obtain pollen passively when it is deposited on their frons, venter, or dorsum according to the position of the anthers relative to the nectar-gathering bee, as with *Gliricidia sepium*, *T. grandiflora*, and *Justicia galapagana* (25, 54, 82). Pollen is also obtained from anthers that are poricidally dehiscent (17). Carpenter bees, with the exception of *X. imitator* (5), use the "buzz pollination" technique (17), whereby the pollen is released from the anther following vibration of the indirect flight muscles of the bee, which is suspended by its legs from the anthers or corolla.

Pollen may be carried in the crop or on the hind legs. When it is deposited on the notum or fore- and middle legs, it is moved to the foretarsi, scraped off by a special comb on the maxillary stipes, and swallowed. The crop then serves as a pollen-carrying device, as in *Hylaeus* bees (90). Pollen that is deposited on the abdomen, middle legs (5), or hind legs (90) is carried on the hind legs to the nest.

## Physiological Considerations

Flight activity and hours of flower visitation are limited by physiological requirements. *Xylocopa pubescens* does not usually forage at temperatures that are lower than 18°C, whereas the more thermophilic *X. sulcatipes* (37) and *X. capitata* (77) start foraging at ambient temperatures of 22–32°C. In all of the studied carpenter bees, thoracic temperatures during flight reached 39–40°C, indicating that the bees are endothermic. Their ability to maintain high thoracic temperatures seems to be independent of the ambient temperatures. *Xylocopa varipuncta* females maintained a thoracic temperature of 33–46°C during flight at ambient temperatures ranging from 12–40°C, and the thoracic temperature excess did not remain constant over this range (47). Therefore, it was deduced that the bees are capable of thermoregulation. Despite their ability to raise their thoracic temperatures to flight thresholds even at an ambient temperature as low as 12°C, carpenter bees do not usually forage at such low temperatures; they may be constrained by the high energy costs of maintaining the required thoracic temperature and by their high thermal conductance. Glabrous and furry species differ in the latter characteristic; cooling rates from the furry thorax of *X. pubescens* are slower than those from the glabrous thorax of *X. sulcatipes*. Consequently, *X. pubescens* females may forage at lower temperatures than *X. sulcatipes* because of lower heat loss when the bee alights to probe for nectar or to collect pollen.

The thermoregulatory abilities of carpenter bees are also expressed at high ambient temperatures. In *X. varipuncta* the thoracic temperature excess at an ambient temperature of 40°C is only 6°C as compared to 24°C at low ambient temperatures (47). This cooling of the thorax is apparently not achieved by evaporative water loss, as inferred from findings for *X. capitata* in which water loss was proportional to oxygen consumption (77). Rather, it may be due to conductive heat loss, especially in glabrous species, or by shunting heat to the abdomen. The smooth, flat, and relatively large abdomen provides very good heat dissipation. A convective cooling mechanism was also suggested (24, 47) when an increase in flight speed at high ambient temperatures was noted. Convection may be the main way of cooling the head, since evaporative heat loss does not seem to occur in these species. From his conductance values, Chappell (24) calculated that at ambient temperatures of 35–37°C, *X. californica* bees could not hover because they might overheat. However, males of *X. pubescens* engaged in territorial defense in Hatzevah, Israel spend a considerable time hovering at ambient temperatures of 35–40°C (A. Hefetz, personal observations). Unfortunately, nothing is known about their thermoregulation or thermal conductance.

## NATURAL ENEMIES AND MUTUALISTS

Their large size and the habit of storing food that is high in energy and protein content make carpenter bees an attractive target for natural enemies. Recorded natural enemies (Table 1) include 19 families of predators and 16 of parasites (including possible mutualistic symbionts). Numbers and ratios of parasites and predators differ significantly among global regions. These differences probably reflect true faunistic differences (50, 72, 105) except in Asia, where natural enemies have hardly been studied.

The predators include omnivores (e.g. ants), wood destroyers (termites), and insectivores (e.g. various beetles, orioles, shrikes, and woodpeckers). Humans have a special place in this complex, since they may act as specific predators, seeking the bees for use of their products (50, 105) or destroying them because of their pestiferous nature (9). However, the most detrimental

**Table 1** Predators, parasites, and cleptoparasites of *Xylocopa* species[a]

| Vertebrate predators | Invertebrate predators | Parasites | Cleptoparasites |
|---|---|---|---|
| Humans [c (106)] | Termites [M (40)] | Asilidae [IIa, A (95)] | Gasteruptiidae [IIa, A (105)] |
| Baboons [A (106)] | Dermestidae [IIc, U (50)] | *Hyperechia* spp. | *Gasteruption* sp. |
| Orioles [U (50)] | *Attagenus* sp. | Bombyliidae | Ichneumonidae [Ic, E (13)] |
| Wrens [U (50)] | *Trogoderma* sp. | *Anthrax* spp. [IIa, NM (38)] | *Hoplocryptus femoralis* |
| Crows [M ([b])] | Mantidae [MP (40, 50)] | *Villa* sp. [? (50)] | Meloidae [IIb (106)] |
| Woodpeckers [UAM (40, 50, 106)] | Ants [AMP (5, 40)] | Conopidae [IIa, AS (50, 105)] | *Cissites* spp. [NI (42, 49)] |
| Shrikes [A (105)] | Asilidae [A (95)] | *Physocephala* spp. | *Horia* spp. [A (26, 72)] |
| Bee-eaters [M (P)] | *Hyperechia* | Encyrtidae | *Huarpea* sp. [A (11, 81)] |
| Squirrels [E (72)] | Phycitidae [IIc, U (78)] | *Coelopencyrtus* spp. [Ic, AM (40, 105)] | *Sitaris* spp. [A (16)] |
| Tomtits[c] [E (73)] | Pyralidae [IIc, U (78)] | *Giraultella lopezi* [?, S (50)] | *Synhoria* spp. [A (106)] |
| | *Paralipsa exacta* (105) | Evaniidae | Sapygidae [Ia, EMS (72)] |
| | *Vitula serentiella* (50) | *Trigonophenus xylocopae* [?, U (72)] | *Polochrum* spp. [50, 72, [b])] |
| | Tenebrionidae [IIc, U (78)] | Leucospidae [Ib, SA (50, 51)] | Acaridae [IIb, C (62, 78)] |
| | *Aphanotus brevicornis* | *Leucospis* spp. | *Horstia* spp. |
| | Cheyletidae [Ib, IAS (61)] | Perilampidae sp. [?, U (8)] | Chaetodactylidae [IIb, c (8, 94, [b])] |
| | *Chaletophyes* spp.[d] | Torymidae [?, U (72)] | *Sennertia* spp. |
| | | *Monodontomerus montivagus* | Suidasiidae [IIb, c (62)] |
| | | Laelapidae [Ib, AMI (68, 94)] | *Tortonia* spp. |
| | | *Dinogamasus* spp.[d] | |

[a]Modes of nest invasion by parasites: I, by adult; II, by immature; ?, unknown; a, prior to cell closure, self-propelled; b, prior to cell closure, phoretic on the bee; c, following closure. Distribution: A, Africa; C, cosmopolitan; E, Europe; I, Asia; M, Middle East (mainly Israel); N, New World; P, Pacific (Hawaii); S, South America; U, North America.
[b]D. Gerling, personal observations.
[c]Identity uncertain.
[d]Possibly mutualists.

interference by humans is nest and wood destruction and habitat modification during land clearing and construction projects.

Numerous parasites attack the larvae. They include four well studied genera (*Hyperechia, Anthrax, Coelopencyrtus, Leucospis*), and three less well studied ones. In addition, five groups of cleptoparasites, which devour or kill the egg and develop on the bee bread, are recognized. These include meloids, several well studied cases of sapygids and gasteruptiids, one ichneumonid, and various mites. One parasite genus, the conopid *Physocephala*, attacks adult bees. Several omnivorous beetle and moth larvae have been found penetrating the nests and feeding on their contents either as predators or parasites (48; Table 1).

At least five lineages of mites have produced specialized associates of large carpenter bees. All have special phoretic instars that disperse via nest-founding female bees; the mites then complete their life cycles in the brood cells. Deutonymphs of three genera of astigmatid mites (*Sennertia, Horstia,* and *Torotnia*) are commonly phoretic on carpenter bees throughout the world. The other instars of these mites feed on bee bread in the sealed cells, possibly after killing the eggs or young larvae (8, 63, 68). Predatory mites of the genus *Cheletophyes* occur in nests of tropical and subtropical *Xylocopa* spp. (61). They presumably feed on the astigmatid mites there and thus might be mutualistic with the bees.

*Dinogamasus* mites are phoretic in special abdominal pouches in the Old World subgenera *Koptortosoma, Afroxylocopa,* and *Mesotrichia* (105). In the brood cells the mites appear to feed on larval and pupal exudates and are possibly parasitic on the immature bees themselves (69, 94). The special pouches on the adult bees suggest a mutualistic relationship, but the immature bees develop normally when the mites are experimentally removed from the nests (94). Perhaps the mites enhance development of the immature bees through direct or indirect inhibition of microbial contamination.

As summarized in Table 1, enemies may gain entrance as adults, which may walk in (Ia) or be carried in by the bees (Ib). Others enter as larvae, by crawling in before cell closure (IIa), by being carried in as phoretic instars or triungulins (IIb), or by breaking in (IIc). The female of *Hoplocryptus femoralis* oviposits through the wall of the nest (13), and that of *Coelopencyrtus* spp. cuts through the cell partitions to oviposit (Ic) (105).

Other bees, including the magachilid *Heriades* (106), congenerics, and conspecifics, constitute some of the most important natural enemies of carpenter bees through robbery of food stores, eviction of the nesting bee, destruction of progeny in the nest, and usurpation (40, 101, 106). In *X. sulcatipes* these phenomena appear to correlate directly with the density of the bees and inversely with the availability of nest sites (84).

A study in South Africa (106) showed that 17% of carpenter bee mortality was due to parasites; 41% of these deaths were caused by *Coelopencyrtus* spp. Predation caused only 6.5% of the mortality, with woodpeckers accounting for 34% of total predation. Noting that the bee population did not increase in spite of these relatively low rates of mortality, Watmough (106) concluded that additional mortality factors must have been present.

Although bees may take aggressive action when defending their nests against congeners (54, 57, 84), their defense against other predators or parasites is usually nonaggressive and may amount to building well constructed cell partitions (78), blocking the nest entrance in various ways (5, 58), squirting material from the anal opening, or expelling all the progeny of a nest that has been disturbed (15). The difference between reactions to bees and reactions to other enemies reflects the possible effectiveness of the kinds of defense measures and may point to the relative importance of depredation by congenerics versus that by other organisms.

## MATING BEHAVIOR

Since male carpenter bees do not exhibit parental care for their progeny and occur roughly in the same proportion as the females, they tend to be polygynous and often compete for the opportunity to mate. In theoretical treatments of polygynous mating systems, three patterns have been distinguished (1, 3): female defense, resource defense, and male dominance polygyny. All are represented in the carpenter bees.

The variety of cues used in these systems leads to diverse mating behaviors. This diversity allows for an analysis of the ecological factors shaping reproductive behavior and an attempt to trace the evolutionary history of these reproductive systems. Partly based on other reviews (3, 18, 28, 31, 81) we have distinguished five hypothetical evolutionary steps:

1. The species is polygamous and both sexes search randomly for a partner. This condition is, however, rather unstable because polygamy has different meanings for the two sexes. Male reproductive success depends primarily on the number of matings, as opposed to female reproductive success, which depends primarily on the number of eggs produced. Therefore, multiple matings of females are directly related to the frequency of encounters with males, especially when mating might be easier than averting males.

2. Males could increase their frequency of matings by waiting at places frequented by females, such as nest sites (female defense) and feeding sites (resource defense). In dense nest aggregations the male strategy mainly involves patrolling at the nest site and competing directly for each emerging female, but some territoriality also occurs. In aggregations with widely spaced nests competition may lead to territoriality either at the nest sites or near flowers, depending upon their relative availability and dispersal (7, 27,

38, 99). The same species may practice both types of territoriality (5, 10, 27, 99; M. J. Orlove, personal communication).

An example of facultative territorial site choice is found in *X. sulcatipes*. In the spring, when favored food sources are concentrated, the males clump around single *Moringa peregrina* trees and exhibit almost no mutual aggression. At times a male may guard a bunch of flowers, but only temporarily. As the season progresses and flower sources become sparser, the same male population shifts to territorial behavior either at nesting sites or at food resources (40; D. Gerling, A. Hefetz & H. H. W. Velthuis, personal observations).

Cruising around *Cassia* bushes (5; H. H. W. Velthuis, personal observations) by males of *X. sulcatipes* probably constitutes a special case of food resource territoriality, since these bushes provide only pollen, which virgin females do not seek. However, virgin females may make exploratory foraging flights and be attracted from a distance to potential food sources by visual and olfactory cues. Males may rely on similar cues rather than on actual food collection by females when choosing their territories.

3. Non-resource-based territoriality (1) may evolve from food source territoriality as a response to a relatively low density of bees in relation to plant density. In this situation the probability of a single male encountering females may be very low, and the production of attractants by the male would be advantageous to both sexes. The attractants may be signals such as loud buzzing by the territorial male and/or the addition of an odor to the flowers; the males may use these signals either when patrolling along extended paths (cf 43) or when hovering at a specific site (5, 56, 66).

Female mate-searching behavior might involve waiting for the male at the flower on which she discovers his markings or a directed flight toward a marked spot from a distance. At this point real male dominance polygyny begins, for males may produce quantitatively or qualitatively individual pheromonal signals that may reflect their fitness, and the female could react correspondingly by selecting among the males.

4. The use of pheromonal cues enables males to relocate their territories from resource sites to prominent sites such as hilltops, various protrusions, or trees (2, 40, 97, 98). At this point, long- and/or short-range pheromones may direct the female into the territory and possibly into a focal point within the territory. Short-range marking of a focal point on which an attracted female is most likely to land before copulation has been recorded (56, 74, 97).

A cooperative display, whereby the males ceased to be mutually aggressive, was observed in *X. frontalis* (19) and in *X. hirsutissima* (97). It is possible that such behavior enhances the attractiveness of the territory by creating stronger signals. However, until a comparative study of the reproductive success in single-male versus cooperative-display territories is conducted, this point remains speculative.

5. Competition among males for non resource sites may lead to various subsequent adaptations: (*a*) Odor deposited at a favorable site makes it even more attractive to other males. By replacing the resident male, an intruder may benefit from the remaining odors of his predecessors because his quality would be judged by the overall strength of the odor signal. (*b*) Instead of placing the pheromone on an object in the territory, the male would do better if he placed at least part of it on his own body and carried it along in case of eviction from his territory. The pheromone may originate in the mandibular glands (27, 45, 97) or in the mesosomal gland (4).

Female behavior has evolved concomitantly with male behavior. While they may have been passive in the earlier stages of the proposed evolutionary sequence, from Step 3 onward females have actively searched for males in order to become inseminated. This searching has opened the way for the development of more specific signals such as female pheromones or landing on focal points that have been previously marked by the males. The specific search for a mate by females has additional consequences. As far as is known, carpenter bees only start egg-laying after having been inseminated. Mating, therefore, may be the physiological trigger for oogenesis and for nesting behavior. If males are met only away from resources, the way is open for a shift from multiple to single insemination. Based on the haplo-diploid mechanism of sex determination in the Hymenoptera, single insemination leads to a high degree of relatedness of sisters (44) and thus fulfills one of the prerequisites for the start of female social evolution.

# CHEMICAL COMMUNICATION AND EXOCRINOLOGY

Carpenter bees possess a series of exocrine glands including mandibular, intersegmental abdominal, and Dufour's glands in females, and mandibular and mesosomal glands in males. While the chemistry of the secretions of a few species has been established, only fragmentary information exists about their functions.

## Mandibular Glands

The activity of the mandibular glands of females has been investigated only in *X. sulcatipes* and was found to be involved in mating behavior. The females apparently use the secretion as a sex pheromone. Males that were experimentally smeared with a female's mandibular gland and then introduced into another male's territory elicited exploratory behavior instead of the aggressive behavior that an intruding male usually evokes (99). The pheromone is prominent in virgin or recently mated females, but probably disappears thereafter. Nesting females that enter a male's territory are immediately inspected but are ignored after a few seconds. The chemical nature of the pheromone is unknown.

Male mandibular gland secretions have been investigated in *X. sulcatipes* and *X. hirsutissima*. Although there are common components in the glandular exudates of the two species, the secretion can be considered species specific. The secretion of *X. hirsutissima* contains *cis*-2-methyl-5-hydroxyhexanoic acid lactone and its *bis* homolog, benzaldehyde, benzoic acid, *p*-cresol, vanillin, and aliphatic hydrocarbons (107). While cresol and vanillin are also components in the secretion of *X. sulcatipes,* guaiacol is unique to this species. These aromatic compounds are accompanied by a series of paraffins, whose source may be the mandibular cuticle (45).

Males of both species use the secretion from mandibular glands for marking their territories (45, 97, 98). Marking is most intense at the onset of territorial establishment and at the borders with neighboring males. Females in both species approached the territory downwind, which suggests that the secretion has an attractant component. The scent of the secretion also aids in the detection of intruding males. In *X. sulcatipes,* immobilized males or mated females that were smeared with male mandibular gland secretion and then hidden in a male's territory were easily detected and attacked. Guaiacol seems to be the aggression elicitor, whereas vanillin is attractive to females that visit the territory. Thus the secretion has a dual function: betrayal of males and attraction of females (45). Interestingly, males that visit flowers within a territory are not attacked, while potential competitors for females are attacked. It is not known how the defending male distinguishes between the two activities or whether any chemicals are involved. In *X. hirsutissima,* excised mandibular glands applied on filter paper elicited aggressive behavior of territorial males (97, 98). However, synthetic *cis*-2-methyl-5-hydroxyhexanoic acid lactone alone or in combination with vanillin could not elicit the same behavior. Unfortunately, owing to a shortage of synthetic products the number of experiments was insufficient for unequivocal conclusions.

## Mesosomal Glands

Mesosomal glands are male-specific exocrine glands situated in the thorax (103). The secretions in four species have been characterized and exhibit species specificity. The secretions of *X. varipuncta* and *X. gualanensis* both include geranylgraniol and farnesal as major products but differ in the identity of minor components; the former contains an isomer of 3,7,10-trimethyl-2, 7,10-dodecatrienal (4), whereas the latter contains small amounts of linalool, nerol, and farnesol (109). The secretions of *X. fimbriata* and *X. micheneri* have unique compositions. The former possesses 2-heptadecanone, farnesol, and farnesyl acetate as major components, which are accompanied by various terpenes in minor amounts. The secretion of *X. micheneri* lacks terpenes but includes isopropyl oleate, (z)-11-eicosenol, oleyl alcohol, and traces of methyl palmitate (4).

The gland opens into the notum, and the secretion apparently oozes out. Males possessing mesosomal glands are often observed smearing the secretion onto their bodies using their legs (103). All species in which a well-developed mesosomal gland has been found to date occupy nonresource territories; the secretion apparently acts as a territorial pheromone. This mode of secretion is especially adaptive for territorial behavior (see section on mating behavior).

## Intersegmental Glands

As in other bees, females of *Xylocopa* spp. have several pairs of glands situated in the abdomen and opening into the intersegmental membranes [also called yellow glands (41)]. In *X. pubescens* each gland is composed of numerous unicellular secretory elements connected to a ductule that empties into the intersegmental membrane (41). The glands develop cyclically. In nonnesting bees they are compact and white, but increase in size and turn yellow as the breeding season progresses. The activity of the different cells in each gland is asynchronous, which prolongs the period of glandular activity to coincide with the long nesting period of this species. Ultrastructural and histochemical studies revealed rich, rough endoplasmic reticulum indicative of protein synthesis, as well as mucopolysaccharides containing carboxylic groups of the hyaluronic acid type. Gerling et al (41) suggested that the secretion of the glands serves to coat the inside of the cells. In *X. sulcatipes,* at least part of the secretion is pentane soluble (A. Hefetz, unpublished) and contains high–boiling point hydrocarbons ($C_{23}$–$C_{31}$) similar to those found in the cell lining (64).

## Dufour's Glands

From the chemical point of view, the Dufour's gland is the most studied exocrine gland in *Xylocopa* species. It is a long tubular gland occupying much of the abdomen, composed of a single epithelial cell layer that surrounds the reservoir (23, 27a, 64), which opens into the sting chamber.

Development of the gland in teneral bees begins with feeding on bee bread, and its secretory activity is most pronounced during active nesting. In laboratory-reared teneral bees the gland contained ~60 $\mu$g of secretion, whereas larger quantities were found in bees that were fed on bee bread. Field-collected, actively nesting bees contained 0.7–1 mg of secretion (A. Hefetz & S. Kronenberg, personal observation).

The gland chemistry of six species has been investigated and reveals an interesting pattern (Table 2). Two of the species, *X. pubescens* (A. Hefetz, H. J. Willimas & S. B. Vinson, personal observations) and *X. micans* (109) are exclusively hydrocarbon producers, while in *X. sulcatipes* the secretion is dominated by ethyl esters with only minor amounts of two hydrocarbons (64). The secretion of *X. iris* has both esters and hydrocarbons in appreciable amounts, while in *X. virginica texana* (102, 108) the esters are in the minority. The secretion of *Proxylocopa olivieri* is composed exclusively of hydrocarbons (64). Considering the suggestion that *Proxylocopa* is a progenitor of *Xylocopa* (49), two evolutionary pathways might have occurred; one retained hydrocarbon synthesis, while the other slowly shifted to ester synthesis.

In *X. virginica texana* (34) the function of the Dufour's gland is the production of a scent that the bee deposits on the flower following the collection of nectar. Flowers marked in this way were avoided by conspecifics for 10 min (34). Experiments with synthetic components showed that only part of the complex blend occurring in the gland was sufficient for deterring conspecifics. Flowers treated with methyl myristate or methyl palmitate were probed for nectar significantly less than untreated ones. Pentadecane or hexadecane by themselves were ineffective repellents, but in combination with the two above-mentioned methyl esters they were as effective as Dufour's gland extract (102). *Xylocopa pubescens* and *X. sulcatipes* exhibited similar avoidance behavior when visiting marked *Calotropis* flowers (30; A. Hefetz, personal observations).

The presumed original role of the Dufour's gland, i.e. providing the brood cell lining, which still exists in *Proxylocopa,* was apparently lost in *Xylocopa* (46). In fact, the lining of brood cells of *X. sulcatipes* that were constructed in *Ferula* sp. contained only aliphatic hydrocarbons and lacked the esters characteristic of Dufour's gland secretions. Since these hydrocarbons did not originate in the *Ferula* sp. they must have originated in an exocrine gland, presumably the intersegmental glands. The large existing Dufour's gland apparently became adapted for other functions such as flower marking. One cannot exclude the possibility that despite the loss of its function in the production of the cell lining, the Dufour's gland retained its communicative role as a nest marker (46). Consequently, flower marking can be regarded as a secondary adaptation in some species. *Xylocopa (Mesotrichia) caffra* uses scent for nest location. These bees were observed persistently rubbing the tips of their abdomens around nest entrances, which suggests that abdominal glands are the source of this odor (94). Similar smearing of secretions was observed for *X. fimbriata,* but was interpreted as defensive (58). The use of scent for locating the nest entrance was also observed in *X. flavorufa, X. imitator,* and *X. torrida* (6) and in *X. sulcatipes* (A. Hefetz, personal observations). Furthermore, single-gland analysis in *X. sulcatipes* revealed individual

**Table 2**  Dufour's gland composition in six *Xylocopa* species

| Compound | X. pubescens[a] | X. micans[b] | X. virginica[b] | X. sulcatipes[c] | X. iris[a] | X. valga[a,d] |
|---|---|---|---|---|---|---|
| Methyl myristate | | | t | | | |
| Ethyl myristate | | | t | | | |
| Methyl palmitate | | | ++++ | | | |
| Ethyl palmitate | | | + | | | |
| Heineicosene | | t[e] | +++ | | | |
| Heineicosane | | t | + | | | |
| Ethyl stearate | | | t | + | | |
| Tricosene | ++ | ++ | + | | | ++++ |
| Tricosane | ++ | ++ | +++ | | | ++ |
| Tetracosene | | | ++++ | | | |
| Tetracosane | | | ++++ | | | |
| Pentacosene | ++ | ++++ | | | | ++ |
| Pentacosane | ++ | ++++ | | | | + |
| Methyl eicosanoate | | | | + | t | |
| Ethyl eicosanoate | | | | ++++ | ++ | t |
| Methyl docosanoate | | | ++ | + | + | |
| Ethyl docosanoate | | | + | +++ | ++ | |
| Hexacosane | | | ++ | | | |
| Heptacosene | ++++ | ++ | +++ | | ++++ | |
| Heptacosane | +++ | ++ | | | + | ++ |
| Methyl tetracosanoate | | | | | | |
| Ethyl tetracosanoate | | | | | t | |
| Nonacosene | + | + | + | ++ | ++++ | + |
| Nonacosane | + | | + | | +++ | |
| Hentriacontene | + | ++ | | | | |
| Hentriacontane | ++ | | + | + | ++++ | |

[a] From A. Hefetz, H. J. Williams, & S. B. Vinson, personal observations.
[b] From Reference 107.
[c] From Reference 62.
[d] Based on incomplete analysis.
[e] t = traces.

compositional differences in Dufour's gland secretions as required for a nest recognition pheromone (A. Hefetz & S. Kronenberg, unpublished).

## ORIGINS AND FORMS OF SOCIALITY

The main selective pressures leading to sociality in the large carpenter bees are intraspecific and intrageneric competition for the nest substrate, existing nests, and nest provisions. Competition for the existing nest can be related to the rarity of the nest substrate or to the value of the nest once it has been constructed; this competition necessitates efficient defense (40, 58, 78). Another factor is temporary food shortage, which may be the origin of the frequently developed habit of robbing congeneric or conspecific nests. The need for nest defense may interfere with the efficiency of cell provisioning by forcing females to make short trips that result in the collection of small pollen loads and the construction of only a few brood cells or even none at all (8, 11, 40, 95a, 101).

Competitive conditions promote a prolonged life of the mother and the potential for matrifilial societies. In habitats allowing more than one brood cycle per year, an active colony, distinct from a hibernation aggregation or a prereproductive assemblage (76), could evolve, which could easily become semisocial or communal. In either case, the presence of additional bees at the nest that assume the role of guards makes long foraging trips possible without risk to the existing nest and provisions. Semisocial colonies, not necessarily involving related females, may also be formed by the joining of nests before brood production starts (21, 84). A social level above the communal probably exists in more than the few heretofore-studied species. The best candidates are the species that have branched nests, some of which are now listed as communal (15, 87). In a communal system dominance interactions could easily turn the nest into a semisocial one.

In every case of social nesting the division of labor is based on differences in activity. Only the foraging female constructs galleries and cells and lays the eggs, while the other bees mostly guard the nest. In return they receive nectar trophallactically (15, 40, 83, 87) and consume pollen from the cell that is being provisioned (95a, 105). Interactions are predominantly of an overtly aggressive nature (100), but allogrooming also occurs, and food donation may be voluntary (95a). This form of sociality is exclusively found in the large carpenter bees. In the other social Hymenoptera the subordinate members of the colony do the foraging and often several other tasks.

Intraspecific competition led to various adaptations that reduce the risk for the immature. These are related to a shortening of the preimaginal stage. The adaptations occur in various degrees in the different species. The large egg of carpenter bees is a means of starting with a relatively large larva, which may shorten the duration of larval development. The large egg also facilitates the production of more offspring, while requiring only limited flight activity of the mother during the initial, most vulnerable stage of the nest (96). The shortening of the developmental period results in the production of teneral bees that need supplementary nutrition before they can assume regular adult functions. This nutrition is obtained through feeding on pollen and bee bread. (95a, 96).

At present sociality has been demonstrated in eight species. Bonelli's (15) pioneering studies indicated the occurrence of eusocial colonies in the African *X. combusta*. Here a solitary female initiated a nest and made a brood consisting of only three cells. The oldest daughter guarded the nest while the mother prepared a second brood of two to four bees. Then the mother died and the guarding daughter became reproductive. The use of marked bees and X-ray technology (38) revealed that in *X. virginica* many females lived for two years and that often two or three bees lived together in the same branched nest, but only one female was active. The guarding female became the egg layer in the following year in this univoltine species. In the Hawaiian *X. sonorina* often several females were present in the nest. The group was either matrifilial or semisocial as a result of other bees joining the nest. Occasionally unmated females with worn wings were observed (36), which is suggestive of the presence of a nonreproductive caste.

In *X. frontalis*, *X. grisescens* (20), and *X. suspecta* (22) a period of inactivity was observed twice a year, after which activity was resumed. If the colony consisted of an active mother and her offspring, 4–6 wk of cohabitation allowed the mother to make a few brood cells while the daughters were still unmated. In a two-female nest, which could involve sisters, cohabitation lasted from 5–14 wk, and brood production was more than twice that in matrifilial nests. It was often difficult to tell which female was reproductive, since both had activated ovaries (21).

The best-documented cases of sociality concern two Israeli species, *X. sulcatipes* and *X. pubescens* (11, 40, 100, 101). *X. sulcatipes* culm nesters have long-lasting associations of the mother with her dependent adult offspring. This species may also form single-generation matrifilial colonies, in which both females may forage and provision cells (40). Since the conditions of the ovaries are sometimes similar in the two females it is probable that both have oviposited (96). In most nests having two females, however, only one is active. In such cases, the nonreproductive female provides protection for the nest and its contents while the active bee is on a foraging trip (96). Robbery, especially under high bee density, promotes effective nest guarding and frequently encourages joining of solitary females (84). Even for the nonreproducing bee this situation might be more advantageous than being on her own, for she may have a chance to take over the nest and become the dominant female.

*Xylocopa pubescens* generally nests in logs (11), where it may form extended aggregations, or in cane (40, 60). One entrance may lead to several nests, probably made by the daughters of a solitary foundress. Among the teneral progeny a hierarchy leads to differential feeding by trophallaxis (40). In addition to receiving nectar, teneral adults consume pollen in amounts that equal those that they receive as larvae (95a, 100). They obtain this pollen either directly from the entering mother or, more commonly, from the brood cells that she is preparing. Teneral adults compete for positioning in the nest opening, because the first bees to contact the returning mother receive the most, especially when foraging is limited. Guarding, therefore, can be considered a side effect of competition (100). The dominance relations observed in a matrifilial nest of *X. pubescens* were often unstable (95a). In about half of the nests the mother succeeded in raising at least one additional brood, but in the other half the daughter became dominant. In the latter case the mother either could be evicted or could remain as a guard. One mother remained a guard for three consecutive broods of her daughter. Evicted mothers may establish successful nests elsewhere.

It is not very illuminating to distinguish among communal, semisocial, and eusocial levels in the carpenter bees, as such distinctions concern phases rather than the climax of any development. For this reason the terms metasocial (40) and eosocial (88) have been proposed. The various forms of sociality are similarly transient in several small carpenter bees (91). The slow adult development of the teneral young leads to a division of labor that may last for several weeks, allowing the mother to produce another brood. In semisocial nests involving sisters or unrelated females, dominance relationships may cause the subordinate to develop more slowly, even though it obtains sufficient nourishment, while the dominant female may be physiologically activated. If such mechanisms operate, sociality of the carpenter bees is more than a protracted dispersal and development.

## ECONOMIC IMPORTANCE

Carpenter bees are notorious for damaging man-made wooden structures (9), lead cables, fiberglass insulation, and styrofoam. Despite their large size, widespread distribution, and use of wood as substrate, they do not usually cause severe economic damage because of their slow development, their reuse of nests, and the preference of some species to develop in trees rather than in structural timber.

The greatest utility of carpenter bees lies in pollination. Although only a few studies have been published on the value of the large carpenter bees as crop pollinators, their large body size makes them the natural pollinators of large flowers such as *Passiflora* and *Luffa* species and of other cultivated fruits and vegetables (59, 67). Carpenter bees contribute measurably to pollination and fruit setting of *Passiflora edulis* (25, 79, 85). In a Brazilian study (85) a single carpenter bee pollinated as many *Passiflora* flowers as an entire colony of honey bees. A major reason for the failure of attempted *Xylocopa* utilization has probably been that no food sources were available to the bees once the crop to be pollinated was no longer in bloom (79).

Before agricultural benefits can be gained from carpenter bees the management of semiwild populations must first be achieved. Our fundamental knowledge of this area is still so rudimentary that quick gains are hardly to be expected. Instead, a step-by-step approach should be made. However, we believe that the large carpenter bees have the potential to become an important aid in agricultural systems, especially where other pollinators, such as the honey bee, are not effective.

## DISCUSSION

Carpenter bees are well suited for the study of evolution of social behavior. Comparisons among species exhibiting the different forms of sociality provide us with an understanding of the selective pressures leading to these incipient yet varied forms. Some of the prerequisites of semisociality have been exhibited by the studied species. These characteristics include matrifilial nests with temporary or permanent division of labor and reproduction between two females of the same or of successive generations. Cohabitation brings about the need to recognize nestmates, and arguments derived from kin selection models imply that it is essential to recognize kin (41a). The limited information available on nest recognition and exocrine secretions indicates that individual recognition can readily be realized.

Cohabitation not only leads to a better defense against competitors from the outside, but also involves competition within the nest. Successful cohabitation implies marked physiological differences between the partners, which are expressed as temporal differences in function and behavior. Further studies, especially on the dynamics of reproductive physiology, are necessary for a better understanding of how this temporal association could have evolved. A second element making the sociality of carpenter bees especially interesting is their taxonomic position as the sister group of the Apidae (88). A comparative treatment of sociality in these bee groups (70, 71, 75, 88, 91, 96) could indicate the relations between the various mechanisms operating in colonies of increasing complexity and the shared evolutionary constraints and benefits of sociality.

The males of many species evolved ingenious mating behaviors. Often males from different populations of a given species or males from a single population apply different strategies to locate their mates. Are such strategies entirely genetically fixed, so that the variation within a population reflects fluctuating environmental conditions, or do individual experiences lead to alternative behavioral patterns?

Sensory physiologists might find it rewarding to investigate whether carpenter bees possess hearing. Such a capacity is not only suggested by the audible males of several species, but also by the behavior of the teneral bees upon the approach of their mother to the nest. We were able to recognize individual mothers by the sounds they produced in flight, and we observed teneral young of *X. pubsecens* rushing toward the nest entrance when their approaching mother was still airborne several meters away (D. Gerling, H. H. W. Velthuis & A. Hefetz, personal observations).

Finally, we would like to draw attention to the relation between carpenter bees and the human population explosion. Undoubtedly many bee population changes will occur in the future owing to the human hunger for land and its products. On the other hand, carpenter bees could be of great help in the production of agricultural crops. Considering that we use only a few of the 20,000 existing species of Apoidea in pollination, and considering the special niche that large bees occupy in pollination biology, it is probable that carpenter bees could increase crop production. For successful introduction of these insects, as for any other new enterprise, it is first necessary to expand the accumulation of basic knowledge.

## Acknowledgments

We are indebted to numerous people who rendered help, advice, and information. Special thanks are due to J. Alcock, G. Eickwort, D. Eisikowitch, J. Kugler, C. D. Michener, and E. Rosenberg for critical reading of all or parts of the manuscript and for rendering many helpful suggestions; to J. F. Anderson, S. L. Buchmann, E. Camillo, G. W. Frankie, C. A. Garofalo, R. L. McKinley, C. D. Michener, M. J. Orlove, R. D. Plattner, J. van der Blom, S. B. Vinson, D. Weisleder, and H. J. Williams for permission to use their unpublished data; and to Fini Kaplan, Tova Feler, and the typists N. Paz and C. Shapiro for invaluable technical assistance.

## Literature Cited

1. Alcock, J., Barrows, E. M., Gordh, G., Hubbard, L. J., Kirkendall, L., et al. 1978. The ecology and evolution of male reproductive behaviour in the bees and wasps. *Zool. J. Linn. Soc.* 64:293–326
2. Alcock, J., Smith, A. P. 1987. Hilltopping, leks and female choice in the carpenter bee *Xylocopa (Neoxylocopa) varipuncta. J. Zool.* 211:1–10
3. Alexander, R. D. 1975. Natural selection and specialized chorusing behavior in acoustical insects. In *Insects, Science and Sociality,* ed. D. Pimentel, pp. 35–77. New York: Academic. 284 pp.
4. Anderson, J. F., Buchmann, S. L., Weisleder, D., Plattner, R. D., Minckley, R. L. 1988. Identification of thoracic gland constituents from male *Xylocopa* spp. Latreille (Hymenoptera: Anthophoridae) from Arizona. *J. Chem. Ecol.* 14(4):1153–62
5. Anzenberger, G. 1977. Ethological study of African carpenter bees of the

genus *Xylocopa* (Hymenoptera, Anthophoridae). *Z. Tierpsychol.* 44:337–74

6. Anzenberger, G. 1986. How do carpenter bees recognize the entrance of their nests? An experimental investigation in a natural habitat. *Ethology* 71:54–62

7. Balduf, W. V. 1962. Life of the carpenter bee, *Xylocopa virginica* (Linn.) (Xylocopidae, Hymenoptera). *Ann. Entomol. Soc. Am.* 55:263–71

8. Barajas, R. A. 1984. *Observaciones de la biologia de* Xylocopa tabaniformis azteca *Cresson (Hymenptera: Anthophoridae)*. PhD thesis. Univ. Mexico, Mexico City. 68 pp.

9. Barrows, E. M. 1980. Results of a survey of damage caused by the carpenter bee *Xylocopa virginica* (Hymenoptera: Anthophoridae). *Proc. Entomol. Soc. Wash.* 82(1):44–47

10. Barrows, E. M. 1983. Male territoriality in the carpenter bee *Xylocopa virginica virginica. Anim. Behav.* 31:806–13

11. Ben Mordechai, Y., Cohen, R., Gerling, D., Moscovitz, E. 1978. The biology of *Xylocopa pubescens* (Spinola) (Hymenoptera: Anthophoridae) in Israel. *Isr. J. Entomol.* 12:107–21

12. Bodkin, G. E. 1918. Notes on some British Guiana Hymenoptera (exclusive of Formicidae). *Trans. R. Entomol. Soc. London* 1919:297–321

13. Bonelli, B. 1967. Osservazioni biologiche sugli Imenotteri melliferi e predatori della Val di Fiemme. 24. *Xylocopa cynaescens* Brulle (*iris* Christ). *Boll. Ist. Dei Entomol. Univ. Studi Bologna* 28:253–63

14. Bonelli, B. 1974. Osservazioni etoecologiche sugli Imenotteri aculeati dell'Etiopia. VI. *Boll. Ist. Dei Entomol. Univ. Studi Bologna* 32:105–32

15. Bonelli, B. 1976. Osservazioni etoecologiche sugli Imenotteri aculeati dell'Etiopia. VII. *Xylocopa (Mesotrichia) combusta* Smith (Hymenoptera—Anthophoridae). *Boll. Ist. Dei Entomol. Univ. Studi Bologna* 33:1–31

16. Brauns, H. 1913. Biologie Südafrikanischer Apiden. *Z. Wiss. Insektenbiol.* 9:116–20

17. Buchmann, S. L. 1983. Buzz pollination in angiosperms. In *Handbook of Experimental Pollination Biology,* ed. C. E. Jones, R. J. Little, pp. 73–117. New York: Van Nostrand Reinhold. 558 pp.

18. Cade, W. 1979. The evolution of alternative reproductive strategies in field crickets. In *Sexual Selection and Reproductive Competition in Insects,* ed. M. S. Blum, N. A. Blum, pp. 343–79. New York: Academic. 463 pp.

19. Camargo, J. M. F., Velthuis, H. H. W.

1979. Sobre o comportamento de *Xylocopa (Megaxylocopa) frontalis* (Olivier) (Hymenoptera—Anthophoridae). *Dusenia* 11:35–39

20. Camillo, E., Garofalo, C. A. 1982. On the bionomics of *Xylocopa frontalis* (Oliver) and *Xylocopa grisescens* (Lepeletier) in southern Brazil. I. Nest construction and biological cycle. *Rev. Bras. Biol.* 42(3):571–82

21. Camillo, E., Garofalo, C. A. 1988. Social organization in reactivated nests of three species of *Xylocopa* (Hymenoptera, Anthophoridae) in southern Brazil. *Insectes Soc.* In press

22. Camillo, E., Garofalo, C. A., Muccillo, G. 1986. On the bionomics of *Xylocopa suspecta* (Moure) in southern Brazil: nest construction and biological cycle (Hymenoptera, Anthophoridae). *Rev. Bras. Biol.* 46(2):383–93

23. Champ, G. B., Barrows, E. M. 1986. Ultrastructural features of the cells of Dufour's gland and associated structure in the carpenter bee *Xylocopa virginica* (L.) (Hymenoptera: Anthophoridae). *Ann. Entomol. Soc. Am.* 79:1008–18

24. Chappell, M. A. 1982. Temperature regulation and control of carpenter bees (*Xylocopa californica*) foraging in the Colorado Desert of Southern California. *Physiol. Zool.* 55:267–80

25. Corbet, S. A., Willmer, P. G. 1980. Pollination of the yellow passionfruit: nectar, pollen and carpenter bees. *J. Agric. Sci.* 95:655–66

26. Croz, A. 1929. *Horia testacea* Fab. (= *H. africana* Auriv.) sa larva primaire. *Bull. Soc. R. Entomol. Egypte* 13(1/3):1–7

27. Cruden, R. W. 1966. Observations on the behavior of *Xylocopa c. californica* and *X. tabaniformis orpifex* (Hymenoptera: Apoidea). *Pan-Pac. Entomol.* 42:111–19

27a. de Lello, E. 1969. Glandulas anexas ao aparelho de ferro das abelhas; anatomia e histologia. IV. Hymenoptera: Anthophoridae. *Cienc. Cult. São Paulo* 23(6):765–72

28. Eickwort, G. C., Ginsberg, H. S. 1980. Foraging and mating behavior in Apoidea. *Ann. Rev. Entomol.* 25:421–46

29. Eisikowitch, D. 1986. Morphoecological aspects on the pollination of *Calotropis procera* (Asclepiadaceae) in Israel. *Plant Syst. Evol.* 152:185–94

30. Eisikowitch, D. 1987. *Calotropis procera* (Ait.) Ait. F. (Asclepiadaceae) and *Xylocopa* spp.: a study of interrelationships. In *Insects–Plants, Proc.*

*6th Int. Symp. Insect-Plant Relat., PAU 1986*, ed. V. Labeyrie, G. Fabres, D. Lachaise, pp. 341–45. Dordrecht, the Netherlands: Junk. 459 pp.

31. Emlen, S. T., Ohring, L. W. 1977. Ecology, sexual selection, and the evolution of mating systems. *Science* 197:215–23

32. Faegri, K., van der Pijl, L. 1979. *The Principles of Pollination Ecology*. Oxford, UK: Pergamon. 244 pp. 3rd ed.

33. Fowler, H. G. 1983. La abeja *Xylocopa californica arizonensis* (Hymenoptera: Anthophoridae) e *Yucca elata* (Agavaceae); nidos, poblaciones, comportamiento e importancia al ciclaje de nutrientes en el desierto de Chihuahua. *Folia Entomol. Mex.* 56:75–83

34. Frankie, G. W., Vinson, S. B., 1977. Scent marking of passion flowers in Texas by females of *Xylocopa virginica texana* (Hymenoptera: Anthophoridae). *J. Kans. Entomol. Soc.* 50:613–25

35. Frankie, G. W., Vinson, S. B., Lewis, A. 1979. Territorial behavior in male *Xylocopa micans* (Hymenoptera: Anthophoridae). *J. Kans. Entomol. Soc.* 52:313–23

36. Gerling, D. 1983. Nesting biology and flower relationships of *Xylocopa sonorina* Smith in Hawaii (Hymenoptera: Anthophoridae). *Pan-Pac. Entomol.* 58(4):336–51

37. Gerling, D., Hefetz, A. 1981. The ecology of the carpenter bee, *Xylocopa sulcatipes* Maa in Israel. In *Development in Arid Zone Ecology and Environmental Quality*, ed. H. Shuval, pp. 71–74. Philadelphia: Balshan. 418 pp.

38. Gerling, D., Hermann, H. R. 1978. Biology and mating behaviour of *Xylocopa virginica* L. (Hymenoptera, Anthophoridae). *Behav. Ecol. Sociobiol.* 3:99–111

39. Gerling, D., Hurd, P. D. Jr., Hefetz, A. 1981. In-nest behavior of the carpenter bee, *Xyloxopa pubescens* Spinola (Hymenoptera: Anthophoridae). *J. Kans. Entomol. Soc.* 54(2):209–18

40. Gerling, D., Hurd, P. D. Jr., Hefetz, A. 1983. Comparative behavioral biology of two Middle East species of carpenter bees (*Xylocopa* Latreille) (Hymenoptera: Apoidea). *Smithson. Contrib. Zool.* 369:1–33

41. Gerling, D., Orion, T., Ovadia, M. 1979. Morphology, histochemistry, and ultrastructure of the yellow glands of *Xylocopa pubescens* Spinola (Hymenoptera, Anthophoridae). *Int. J. Insect Morphol. Embryol.* 8(2):123–34

41a. Grafen, A. 1984. Natural selection, kin selection and group selection. In *Behavioral Ecology*, ed. J. R. Krebs, N. B. Davies, pp. 62–84. Sunderland, Mass: Sinauer. 439 pp. 2nd ed.

42. Green, E. E. 1902. On some parasites of *Xylocopa tenuiscapa* Westw. *Entomol. Mon. Mag.* 38:232–33

43. Haas, A. 1960. Vergleichende Verhaltensstudien zum Paarungsschwarm solitarer Apiden. *Z. Tierpsychol.* 17:402–16

44. Hamilton, W. D. 1964. The genetical evolution of social behavior, I and II. *J. Theor. Biol.* 7:1–52

45. Hefetz, A. 1983. Function of secretion of mandibular gland of male in territorial behavior of *Xylocopa sulcatipes* (Hymonoptera: Anthophoridae). *J. Chem. Ecol.* 9(7):923–31

46. Hefetz, A. 1987. The role of Dufour's gland secretions in bees. *Physiol. Entomol.* 12:243–53

47. Heinrich, B., Buchmann, S. L. 1986. Thermoregulatory physiology of the carpenter bee, *Xylocopa varipuncta*. *J. Comp. Physiol.* 156B:557–62

48. Hurd, P. D. Jr. 1955. The carpenter bees of California (Hymenoptera: Apoidea). *Bull. Calif. Insect Surv.* 4(2):35–72

49. Hurd, P. D. Jr. 1958. Observations on the nesting habits of some New World carpenter bees with remarks on their importance in the problem of species formation (Hymenoptera, Apoidea). *Ann. Entomol. Soc. Am.* 51:365–75

50. Hurd, P. D. Jr. 1978. *An Annotated Catalog of the Carpenter Bees (Genus Xylocopa Latreille) of the Western Hemisphere (Hymenoptera: Anthophoridae)*. Washington, DC: Smithson. Inst. 106 pp.

51. Hurd, P. D. Jr., Moure, J. S. 1960. A new world subgenus of bamboo-nesting carpenter bees belonging to the genus *Xylocopa* Latreille (Hymenoptera: Apoidea). *Ann. Entomol. Soc. Am.* 53:809–21

52. Hurd, P. D. Jr., Moure, J. S. 1963. *A Classification of the Large Carpenter Bees (Xylocopini) (Hymenoptera, Apoidea)*. *Univ. Calif. Berkeley Publ. Entomol.* Vol. 29. 365 pp.

53. Iwata, K. 1976. *Evolution of Instinct. Comparative Ethology of Hymenoptera*. New Delhi: Amerind. 535 pp.

54. Jacobson, E. 1927. Fauna sumatrensis: Xylocopinae, biologie. *Suppl. Entomol.* 16:93–103

55. Janvier, H. 1977. Comportamiento de *Xylocopa violacea* Linneo, 1758. (Hymenoptera). *Graellsia* 32:193–213

56. Janzen, D. H. 1964. Notes on the behavior of four subspecies of the carpenter bee, *Xylocopa (Notoxylocopa) taba-*

*niformis*, in Mexico. *Ann. Entomol. Soc. Am.* 57:296–301

57. Janzen, D. H. 1966. Notes on the behavior of the carpenter bee, *Xylocopa fimbriata* in Mexico (Hymenoptera: Apoidea). *J. Kans. Entomol. Soc.* 39:633–41

58. Kapil, R. P., Dhaliwal, J. S. 1968. Defence of nest by the female of *Xylocopa fenestrata* Fab. (Xylocopinae, Hymenoptera). *Insectes Soc.* 15(4):419–22

59. Kapil, R. P., Dhaliwal, J. S. 1969. Biology of *Xylocopa* species II. Field activities, flight range and trials on transportation of nests. *J. Res. Punjab Agric. Univ.* 6(1):262–71 (Suppl.)

60. Kapil, R. P., Dhaliwal, J. S. 1970. Biology and management of pollinating bees *Xylocopa pubescens* (Spinola) and *Xylocopa fenestrata* (Fabricius) (Xylocopinae; Hymenoptera). In *Biology, Ecology and Utilization of Insects Other Than Honeybees in the Pollination of Crops. Final Res. Rep. PL-480 Proj., A7-ENT-19*, Punjab Agric. Univ., Ludhiana, India

61. Klompen, J. S. H., Mendez, E., Lukoschus, F. S. 1984. A new species of the genus *Chelotophyos* Oudemana 1914 (Prostigmata: Cheyletidae) from the nest of a carpenter bee in Panama. *Acarologia* 25:249–51

62. Kojima, J.-I. 1979. Observations on nest structure of the Japanese large carpenter bee, *Xylocopa appendiculata circumvolans* Smith (Hymenoptera, Anthophoridae). *New Entomol.* 28(2):33–38

63. Krombein, K. V. 1962. Biological notes on acarid mites associated with solitary wood-nesting wasps and bees. *Proc. Entomol. Soc. Wash.* 64:11–19

64. Kronenberg, S., Hefetz, A. 1984. Comparative analysis of Dufour's gland secretion of two carpenter bees (Xylocopinae: Anthophoridae) with different nesting habits. *Comp. Biochem. Physiol.* 79B:421–25

65. Deleted in proof

66. Lieftinck, M. A. 1955. The carpenter-bees (*Xylocopa* Latr.) of the Lesser Sunda Islands and Tanimbar (Hymenoptera, Apoidea). *Verh. Naturforsch. Ges. Basel* 66:5–32

67. Linsley, E. G., Rick, C. M., Stephens, S. G. 1966. Observations on the floral relationships of the Galapagos carpenter bee (Hymenoptera, Apoidea). *Pan-Pac. Entomol.* 42(1):1–18

68. Lombert, H. A. P. M., O'Conner, B. M., Lukoschus, F. S., Whitaker, J. O., Jr. 1987. Ontogeny, systematics and ecology of *Sennertia (Amsennertia)*

*americana* Delfinado & Baker, 1976, (Acari: Chaetodactylidae) from the nest of the carpenter bee, *Xylocopa virginica* (Hymenoptera: Anthophoridae). *Int. J. Acarol.* 13:113–29

69. Madel, G. 1975. Vergesellschaftung der Milbenart *Dinogamasus vollosior* mit der Ostafrikanischer Holzbiene *Xylocopa flavorufa* (Acarina: Laelaptidae/Hymenoptera: Xylocopidae). *Entomol. Ger.* 1(2):144–50

70. Maeta, Y., Sakagami, S. F., Michener, C. D. 1985. Laboratory studies on the life cycle and nesting biology of *Braunsapis sauteriella*, a social xylocopine bee (Hymenoptera: Apidae). *Sociobiology* 10:17–41

71. Maeta, Y., Sakagami, S. F., Shiokawa, M. 1985. Observations on a nest aggregation of the Taiwanese bamboo carpenter bee *Xylocopa (Biluna) tranquebarorum* (Hymenoptera, Anthophoridae). *J. Kans. Entomol. Soc.* 58(1):36–41

72. Malyshev, S. J. 1931. Lebensgeschichte der holzbienen *Xylocopa* Latr. (Apoidea). *Z. Morphol. Oekol. Tiere* 23(3/4):754–809

73. Malyshev, S. J. 1947. The life and instincts of the dwarf carpenter bee *Xylocopa iris* Christ. *Izv. Akad. Nauk. SSSR Ser. Biol.* 1:53–77

74. Marshall, L. D., Alcock, J. 1981. The evolution of the mating system of the carpenter bee *Xylocopa varipuncta* (Hymenoptera: Anthophoridae). *J. Zool.* 193:315–24

75. Michener, C. D. 1974. *The Social Behavior of the Bees.* Cambridge, Mass: Belknap. 404 pp.

76. Michener, C. D. 1988. Caste in Xylocopine bees. In *Caste and Reproduction in Insect Evolution*, ed. W. Engel. New York: Springer. In press

77. Nicholson, S., Louw, G. 1982. Simultaneous measurements of evaporative water loss, oxygen consumption, and thoracic temperature during flight in a carpenter bee. *J. Exp. Zool.* 222:287–96

78. Nininger, H. H. 1916. Studies on the life histories of two carpenter bees of California, with notes on certain parasites. *J. Entomol. Zool.* 8:158–64

79. Nishida, T. 1963. Ecology of the pollination of the passion fruit. *Hawaii Agric. Exp. Stn. Tech. Bull.* 55:21–31

80. O'Brian, L. B., Hurd, P. D. Jr. 1964. Carpenter bees of the subgenus *Notoxylocopa* (Hymenoptera, Apoidea). *Ann. Entomol. Soc. Am.* 58(2):175–96

81. Parker, G. A. 1978. Evolution of competitive mate searching. *Ann. Rev. Entomol.* 23:173–96

82. Deleted in proof
83. Rau, P. 1933. *The Jungle Bees and Wasps of Barro Colorado Island (With Notes on Other Insects)*. Kirkwood, Mo: Rau. 324 pp.
84. Reiss, A. 1986. *Ecological and behavioral aspects of the relationships between the individuals of the carpenter bee* Xylocopa sulcatipes *Maa*. MSc thesis. Tel Aviv Univ., Israel. 71 pp. (In Hebrew with English summary)
85. Ruggiero, C., Lam-Sanchez, A., Banzatto, D. A. 1974. Estudos de polinizacao do Maracuja amarelo. *Resumos Encontro. Reg. Soc. Bras. Prog. Cienc., Jaboticabal, Brazil*, pp. 152–53 (Abstr.)
86. Sage, R. D. 1968. Observations of feeding, nesting and territorial behavior of carpenter bees genus *Xylocopa* in Costa Rica. *Ann. Entomol. Soc. Am.* 61(4): 884–89
87. Sakagami, S. F., Laroca, S. 1971. Observations on the bionomics of some Neotropical xylocopine bees, with comparative and biofaunistic notes (Hymenoptera, Anthophoridae). *J. Fac. Sci. Hokkaido Univ. Ser. 6* 18:57–127
88. Sakagami, S. F., Maeta, Y. 1987. Multifemale nests and rudimentary castes of an "almost" solitary bee, *Ceratina flavipes,* with additional observations on multifemale nests of *Ceratina japonica* (Hymenoptera, Apoidea). *Kontyu* 55: 391–409
89. Sakagami, S. F., Michener, C. D. 1987. Tribes of Xylocopinae and origin of the Apidae (Hymenoptera: Apoidea). *Ann. Entomol. Soc. Am.* 80:439–50
90. Schremmer, F. 1972. Der stechsaugruessel, der nektarraub, das pollensammeln und der blutenbesuch der holzbienen (*Xylocopa*) (Hymenoptera, Apidae). *Z. Morphol. Tiere* 72:263–94
91. Schwarz, M. P. 1986. Persistent multifemale nests in an Australian allodapine bee, *Exoneura bicolor* (Hymenoptera, Anthophoridae). *Insectes Soc.* 33:258–77
92. Scott, H. 1932. Carpenter bees eating lead cable-covers. *Entomologist London* 68:8
93. Shields, O. 1967. Hilltopping. *J. Res. Lepid.* 6:69–178
94. Skaife, S. H. 1952. The yellow-banded carpenter bee, *Mesotrichia caffra* Linn., and its symbiotic mite, *Dinogamasus braunsi* Vitzthun. *J. Entomol. Soc. South. Afr.* 15:63–76
95. Tsacas, L., Desmier de Chenon, R., Couten, R. 1970. Observations sur le parasitisme larvaire d'Hyperechia bomboides (Dipt. Asilidae). *Ann. Soc. Entomol. Fr.* (NS) 6(2):493–512

95a. van der Blom, J., Velthuis, H. H. W. 1988. Social behavior of the carpenter bee *Xylocopa pubescens* Spinola. *Ethology* In press
95b. van der Pijl, L. 1954. *Xylocopa* and flowers in the tropics I, II, III. *Proc. K. Ned. Akad. Wet. Ser. C* 57:413–23, 541–51, 552–62
96. Velthuis, H. H. W. 1987. The evolution of sociality: ultimate and proximate factors leading to primitive social behavior in carpenter bees. *Experientia Suppl.* 54:405–30
97. Velthuis, H. H. W., Camargo, J. M. F. 1975. Observations on male territories in a carpenter bee, *Xylocopa (Neoxylocopa) hirsutissima* Maidl (Hymenoptera, Anthophoridae). *Z. Tierpsychol.* 38: 409–18
98. Velthuis, H. H. W., Camargo, J. M. F. 1975. Further observations on the function of male territories in the carpenter bee *Xylocopa (Neoxylocopa) hirsutissima* Maidl (Hymenoptera, Anthophoridae). *Neth. J. Zool.* 25(4):516–28
99. Velthuis, H. H. W., Gerling, D. 1980. Observations on territoriality and mating behaviour of the carpenter bee *Xylocopa sulcatipes*. *Entomol. Exp. Appl.* 28:82–91
100. Velthuis, H. H. W., Gerling, D. 1983. At brink of sociality: interactions between adults of the carpenter bee *Xylocopa pubescens* Spinola. *Behav. Ecol. Sociobiol.* 12:209–14
101. Velthuis, H. H. W., Wolf, Y. M., Gerling, D. 1984. Provisioning and preparation of the brood cell in two carpenter bees, *Xylocopa sulcatipes* Maa and *Xylocopa pubescens* Spinola (Hymenoptera, Anthophoridae). *Isr. J. Entomol.* 18:39–51
102. Vinson, S. B., Frankie, G. W., Blum, M. S., Wheeler, J. W. 1978. Isolation, identification and function of Dufour's gland secretion of *Xylocopa virginica texana* (Hymenoptera: Anthophoridae). *J. Chem. Ecol.* 4:315–23
103. Vinson, S. B., Frankie, G. W., Williams, H. J. 1986. Description of a new dorsal mesosomal gland in two *Xylocopa* species (Hymenoptera: Anthophoridae) from Costa Rica. *J. Kans. Entomol. Soc.* 59:185–89
104. Wagner, H. W. 1958. Beitrag zum Verhalten einiger mexikanischer holzbienen (*Xylocopa*). *Z. Tierpsychol.* 14(3):303–8
105. Watmough, R. H. 1974. Biology and behaviour of carpenter bees in southern Africa. *J. Entomol. Soc. South. Afr.* 37(2):261–81
106. Watmough, R. H. 1983. Mortality, sex

ratio and fecundity in natural populations of large carpenter bees (*Xylocopa* spp.). *J. Anim. Ecol.* 52:111–25

107. Wheeler, J. W., Evans, S. L., Blum, M. S., Velthuis, H. H. W., Camargo, J. M. F. 1976. *cis*-2-Methyl-5-hydroxyhexanoic acid lactone in the mandibular gland secretion of a carpenter bee. *Tetrahedron Lett.* 45:4029–32

108. Williams, H. J., Elzen, G. W., Strand, M. R., Vinson, S. B. 1983. Chemistry of Dufour's gland secretions of *Xylocopa virginica texana* and *Xylocopa micans* (Hymenoptera: Anthophoridae)—A comparison and reevaluation of previous work. *Comp. Biochem. Physiol.* 74B:759–61

109. Williams, H. J., Vinson, S. B., Frankie, G. W. 1987. Chemical content of the dorsal mesosomal gland of two *Xylocopa* species (Hymenoptera: Anthophoridae) from Costa Rica. *Comp. Biochem. Physiol.* 86B:311

110. Wittmann, D., Scholz, E. 1987. Nectar dehydration and body weight reduction: the cues of *Xylocopa nigrocincta* to prolong duration of territorial flights. In *Chemistry and Biology of Social Insects, Int. Congr. Social Insects, Munich*, pp. 724–25. Munich: Peperny

*Ann. Rev. Entomol. 1989. 34:191–210*

# FORAGING STRATEGIES OF ANTS

*James F. A. Traniello*

Department of Biology, Boston University, Boston, Massachusetts 02215

## OVERVIEW: ANT FORAGING STRATEGY AS AN INDIVIDUAL AND SOCIAL PROCESS

Ant foraging is a collective process composed of the activities of individuals as well as behaviorally integrated groups. Therefore, a great challenge in the socioecology of foraging is to explain how the behavior of such a potentially large and complex system as an ant colony emerges as a function of the properties of its individual components. The task of studying foraging in ants in simplified by the fact that foragers, owing to their sterility, may do little more than forage during their tenure as food harvesters. However, eusociality penetrates virtually every facet of foraging strategy and adds additional complicating social dimensions to its analysis.

Individual and social processes of foraging can be categorized into components to facilitate the analysis of foraging at the level of the single worker, the colony, or the integration of both (Table 1). A generalized ethogram of forager behavior can be described to illustrate these processes. A forager leaves the nest entrance in either a random or a consistent direction. A travel phase ensues, during which the worker maintains a constant compass bearing and moves directly away from the nest. At some point during the travel phase the forager shows a high frequency of turning, marking the beginning of search. During search, food resources are encountered and selected based upon a forager's physical caste, age, and prior experience, the trip distance, thermal stress, resource quality, and the colony's current nutritional status. Depending on the ant species and the size, density, or quality of the food, the forager may communicate information to nestmates about its location and recruit additional foragers.

Summing the above-described individual activities across all foragers yields the most basic aspect of cooperative foraging. Changing nutritional demands associated with the maturation of reproductives may shift a colony's

191

0066-4170/89/0101-0191$02.00

**Table 1**  Individual and social components of ant foraging systems, with representative references

| Individual | Social |
|---|---|
| 1. Trip direction, distance, and duration (92, 112, 118, 119) | 1. Summation of individual efforts (41, 119) |
| 2. Search pattern and duration (91, 112, 118, 119) | 2. Temporal and physical caste; colony demographic distribution (12, 24, 33, 59, 75, 78) |
| 3. Loading capacity and retrieval (3, 34, 40, 63, 73, 89) | 3. Tempo (60, 75) |
| 4. Resource assessment and selection (6, 23, 55, 64, 70, 79, 86, 89, 92, 99, 100, 116, 117, 119) | 4. Communication and cooperative retrieval (1, 34, 50, 53, 109, 110) |
| 5. Induction of cooperative foraging (1, 19, 109) | 5. Resource assessment and colony regulation of foraging (96, 100, 101, 104–106) (see also first column, component 4) |

foraging activity among different food types. Communication mediates cooperation among foragers during search and resource retrieval and serves as a control mechanism of colony-level foraging responses. As a result it provides regulatory feedback during the daily and seasonal organization of foraging.

In this review I treat foraging strategy as a system divisible into elements composed of individual and collective action, recognizing that the distinction between the two may often be blurred because of the high degree of social integration of colony foraging activity. My goal is to provide a synthesis of selected current literature to supplement available reviews (9–11, 14, 43, 57, 75, 76, 98, 120).

## FORAGING SYSTEM EVOLUTION: A COMPONENT ANALYSIS APPROACH

The traits that are used to categorize ant foraging are highly variable in their distribution among the more primitive (e.g. ponerine) or advanced (e.g. formicine) subfamilies of ants. For example, "diffuse foraging" (75), in which foragers leave the colony singly and retrieve food solitarily, occurs in the Ponerinae and Myrmeciinae. The formicines *Cataglyphis* spp. and *Ocymyrmex* spp. also collect randomly dispersed food largely through individual effort (118, 119), but the ponerines *Pachycondyla laevigata* and *Leptogenys* spp. have chemically mediated cooperative foraging (54, 68). Ecological influences on foraging systems, therefore, seem to override phylogenetic tendencies. Also, any one foraging mode must not be mistaken for a species-typical characteristic when in fact it may be a behaviorally flexible component of total foraging strategy. The use of such a foraging categorization to indicate the outcome of higher-level ecological processes may thus yield conclusions that are not entirely valid. For example, the gross categorization of ant species as being "individual," "recruit," and/or "group" foragers (4, 5) may be "too weak and artificial to be useful" (52) when used to define foraging systems in studies of community organization. Given the wide variety of feeding habits and levels of organization and complexity, I suggest that foraging systems be analyzed as a series of components that may each have individual and social influences and that have all evolved in response to resource distribution patterns, competition, and predation. The designation of systems or species as "individual foraging" should be dropped because the behavior of a forager is probably never completely independent of the activities of other foragers and the state of the colony as a whole. Therefore, the idea of individual foraging is erroneous or at least misleading, even in species lacking recruitment communication. Some components of foraging behavior, such as search or retrieval, may largely be individual activities. Search and retrieval tactics will depend on the size of resources, their temporal and spatial distribution patterns, and their resistance to retrieval, as well as on the loading capacities of foragers. Therefore, retrieval may be an individual or group process when viewed as a component of foraging, but this one feature should not be used to categorize the entire system. In summary, any single categorization of ant foraging systems is inadequate in that it does not consider the components of foraging behavior and their ecological influences, which are necessary to fully understand the total expression of a colony's foraging strategy.

## Caste Evolution and Foraging Strategy

Because foragers are sterile they do not have conflicting demands on their time and energy budgets that involve trade-offs between searching for food and searching for mates. As a result, they are often viewed as the products of selection for traits that maximize energy return to the colony. Several social traits have consequences that affect a colony's economy; among these are temporal polyethism and senescence, caste polymorphism, and foraging tempo (60, 75, 78). Selection at the colony level can not only affect the behavioral repertories of foragers to enhance their food harvesting ability, but can also shape patterns of age demography and mortality. In *Pogonomyrmex owyheei,* the life expectancy of a forager averages 14 days under natural conditions; survivorship is similar in the laboratory under conditions of starvation (78). However, in *Cataglyphis bicolor,* foragers have a life expectancy of 6.1 days in the field but live for months or longer in the laboratory (93). Colony fitness may be correlated with the morphology and size frequency distribution of foragers. Their structural morphology per se may reflect feeding habit. Although it is clear in the case of some specialist predators such as *Amblyopone, Odontomachus,* and *Strumigenys* species that mandibular shape and head structure reflect feeding habits, the relationship between morphology and foraging ecology is often unclear. Recent studies using digitization analysis to detect the ecological correlates of variation in head morphology have shown that head shape, as differentiated from size, is related to diet (B. Cole, personal communication). Other studies have ascribed a role to vertebrate predation as a selective agent of form (56).

## Caste, Competition, and Community Organization

Among monomorphic species, body size may shift owing to interspecific competition and may facilitate coexistence. It is assumed that there is a correlation between forager size and food-item size, so that morphological divergence produces resource partitioning. Body size of *C. bicolor* foragers is considerably smaller in populations in Greece than in Tunisia, apparently owing to the sympatry of the smaller-bodied *Cataglyphis albicans* in the latter environment (119). In Greece, where *C. albicans* is absent, the body size of *C. bicolor* foragers is intermediate in size between that of *C. albicans* and *C. bicolor* in Tunisia, which suggests that a morphological shift has occurred because of local competition. Several studies have indicated a correlation between body size and food-item size for desert seed-eating ant communities and communities of generalist ant species (16, 17, 21, 38, 75, 110, 118). However, given the limitations of the use of body size as an indicator of food choice and niche (46, 47, 122) and the fact that the food recruitment systems of ants make them less constrained by their morphology than other animals (14, 20, 110), size-based partitioning of resources among sympatric species may be the exception rather than the rule in ants (21).

The species diversity of granivorous ants of the southwest desert in North America is related to productivity, and coexistence is based upon body size differences or foraging strategies that are correlated with seed size selection or seed density distributions (21, 22). In the Australian arid zone, species density is similar to that of North American deserts, but there is less separation by size among sympatric species (72). Also, differences in productivity among sites do not appear to have a strong influence on associated harvester ant communities in Australia as in North America, seemingly because population density is not limited by food availability (7, 8). Vegetation structure, and perhaps productivity, appears to be related to community organization (2, 7, 8, 25). Worker size/seed size correlates can be found among some species, but other factors such as seed morphology and biochemistry affect seed choice as well (25). In the Negev Desert one of three sympatric *Messor* species shows a small but significant correlation between forager size and seed size harvested (S. Rissing, personal communication). In North-Temperate open field ants and other communities of ants with similar generalist feeding habits, prey size is correlated with body size but mechanisms of cooperative foraging and interference also seem to be important determinants of prey selection and community-level interactions (1, 22, 29, 31, 49, 53, 65, 109, 110).

## Caste Polymorphism and Foraging Ecology

Although it occurs in a small fraction of ant species, the role of caste polymorphism in foraging has received much attention. Has natural selection adaptively shaped colony demography? Do the proportions of workers of different size castes match the characteristics of food resources, and if so, how closely (75)? Size matching, the pairing of forager size and food-item size, has been suggested for a number of species (21, 24, 75, 80), but the correlation explains little of the variance in food choice (83, 84). Competition, diet expansion, and polymorphism might be linked in *Pogonomyrmex* spp. (75), but in the polymorphic *Pogonomyrmex badius* there is only a low, insignificant correlation between polymorphism and diet (J. F. A. Traniello & S. N. Beshers, unpublished). Colonies of *Veromessor pergandei* show increased size variation of workers in areas of relaxed interspecific competition (24). However, less than 4% of the variance in the size of seeds collected by *V. pergandei* is explained by forager body size, and worker body size changes during seasonal cycles, showing little correlation with seed size or the performance of any task (83, 84). In this species, worker size may change according to levels of resource availability, usable foraging time, and alate production, rather than according to selection for size pairing (83).

Polymorphism in *Eciton burchelli* is thought to enhance foraging efficiency by lowering prey retrieval costs. Submajors, which comprise 3% of the colony, are disproportionately active in prey transportation, representing 26% of all prey carriers (33). Leg length allometry may increase foraging velocity. Across physical castes, prey dry weight is significantly correlated with worker dry weight; submajors carry the largest prey loads, although loaded submajors do not transport prey more rapidly than other castes (33). The inverse relationship between ant body size and respiratory rate (3, 63) and computer simulations (80) support the notion of enhanced foraging efficiency in a polymorphic series, but critical demonstrations are lacking.

FORAGER LOAD CAPACITY AND CASTE EFFICIENCY     One of the most striking features of ant biology that bears on foraging efficiency is the ability of an individual forager to carry a resource load that is considerably heavier than its own body weight. The loading ability of a forager, measured as burden ($M^{ant} + M^{load}/M^{ant}$, where $M$ is mass (3, 82), is 1.5–7.5 in *Atta cephalotes* (89), 1.2–1.5 in *E. burchelli* (34), on average 2.2 in *Pogono-*

*myrmex occidentalis* (32), and 1.1–15.3 in *Formica schaufussi* (J. F. A. Traniello, unpublished.) *Eciton burchelli* submajors have disproportionately long legs and mandibles, and can easily straddle and transport prey. While this mode of rapid prey transport appears efficient, the burden capacity of lone foragers is greater in species such as *F. schaufussi* that drag rather than lift their prey. In this latter case prey transport speed decreases sharply with increasing load weight, most likely because of frictional resistance (J. F. A. Traniello, unpublished). In polymorphic species, large workers may carry larger loads, but the cost of producing these workers and the energetic gain from their input to a colony's nutritional economy are unclear. For example, in *Eciton* spp. increasing forager weight increases load capacity and possibly efficiency (12, 34). However, a standard calculation of burden reveals load ratios are 1.2, 1.3, and 1.5, respectively, for 2.5-, 5-, and 10-mg workers; therefore, larger workers have only slightly greater load capacities. Transport costs decrease with increasing body size in ants in general, and in *Eciton hamatum* the gross and net costs of transport decrease with body mass and running speed (3). However, the cost for polymorphic foragers of transporting average-size burdens must be estimated before transport efficiency can be fully evaluated.

RESOURCE DISTRIBUTION, POLYMORPHISM, AND COOPERATIVE RETRIEVAL   A significant body size/prey size correlation does not necessarily imply that competition has led to divergence in worker morphology. First, morphology permits a worker to collect not only prey matched to its body size, but also any smaller load as well. This conclusion is supported by the fact that for some species variance in prey size increases as body size increases (110). Second, the range of prey sizes available or actually taken may be extremely large in comparison to the range of worker body sizes in some sympatric ant species, so body size variation can at best adapt a colony to collecting a limited range of available prey. Third, recruitment communication, which is common in many species, provides a flexible mechanism for simulating a larger-bodied forager by organizing a cooperative retrieval group (34, 53, 109). Given the unpredictable distribution patterns of the food sources of many ant species, the flexibility of caste simulation through cooperative retrieval appears to best adapt a colony to temporal, spatial, and size variability; resource distribution patterns may vary on a relatively brief ecological time scale, but worker size demography cannot be shifted within the same time frame.

Cooperative prey retrieval occurs in many species of ants, but its evolution, mechanics, and energetics are poorly understood. In *Lasius neoniger, F. schaufussi*, and *E. burchelli*, transport group size is correlated with prey weight, seemingly to maximize carriage efficiency (34, 109; J. F. A. Traniello, unpublished). In *E. burchelli*, groups are structured teams composed of major workers plus one submajor; this group composition seems to minimize rotational forces during prey movement (34). In other ants, however, groups often seem to lack structure beyond a forager number/prey weight relationship and to lack coordination. In *F. schaufussi* only some of the ants in a recruited group are involved in load carriage; some workers ride atop the prey while others pull in opposing directions. The loading capacity of cooperative retrieval groups ranges from 1.7 to 8.9, which is lower than the loading capacity of single workers. In this species, and in *Novomessor* spp., cooperative retrieval may be organized to decrease interference competition through resource defense rather than to move prey with greater energetic efficiency (J. F. A. Traniello, unpublished).

# FORAGING THEORY AND DIET SELECTION

It has been argued that ants are ideally suited to test predictions of theories of feeding strategies because of their eusociality; natural selection has made foragers efficient food harvesters that maximize a colony's energy intake (75). Equally convincing is the argument that difficulties arise in the application of optimization models to ants because the forager population of a colony is often large and sometimes polymorphic, and worker foraging activity is severely constrained by temperature (80). Although there have been many demonstrations of food selection in ants (6, 23, 30, 55, 64, 70, 79, 81, 86, 89, 92, 99, 100, 116, 117, 119), tests of theory have shown varying degrees of correspondence between ant foraging behavior and the predictions of models (23, 48, 94, 100). The seed-harvester ants *Pogonomyrmex rugosus* and *Pogonomyrmex barbatus* increased diet specialization with increasing distance of seed patches from the nest, but showed partial preferences (23). Similar tests conducted with *P. occidentalis* and *Pogonomyrmex californicus* showed no significant change or a nonsignificant trend in seed size choice as a function of distance (48,100). However, distances at which seed patches were offered were variable for the different species (3–12 m vs 3–7 m), and for *P. californicus* the maximum distance was 85 cm from the nest. The basic premise of these studies is that the net energy value of seeds of all sizes decreases with increasing distance from the nest owing to the added cost of travel and transport, so that energy maximization is achieved at the more distant patches through the selection of large seeds. The assumption is that energetic costs of search or travel and transport are significant relative to energy gain. However, foraging costs for *P. occidentalis* workers are less than 0.1% of the caloric content of an average collected seed (32). This suggests that feeding selectivity may not change substantially with increasing travel costs and that time cost and net energetic gain rate may be important criteria in foraging strategy (32).

Ants are poikilothermic, and their foraging activity is therefore constrained by temperature, water stress, and other physical factors that might affect the energetic costs of foraging or the use of time. Several studies have demonstrated a significant effect of ambient temperature on metabolic rate as measured by oxygen consumption (26, 77). In *F. schaufussi* the elevated rate of oxygen consumption appears to be an important component of foraging cost that influences prey selection (114). The role of travel distance and its importance relative to temperature as a foraging cost has also been examined in *F. schaufussi*. Prey of two different sizes were offered at distances of 1 and 5 m from the nest at low and high temperatures. At both 1 and 5 m, increase in temperature produced a decrease in selectivity; increasing travel costs at either low or high temperature, however, did not affect selectivity (R. V. Bowen & J. F. A. Traniello, unpublished). The magnitudes of temperature-induced costs and travel costs seem therefore to be very different; temperature produces the more important constraint, if in fact temperature represents a foraging cost rather than a thermal risk factor.

Laboratory studies of patch use by *Solenopsis geminata* have shown that colonies recruit more foragers to sucrose patches that are closer to the nest, larger, or of higher concentration than to more distant, lower-quality, or smaller patches (100). Patterns of recruitment and patch use in *S. geminata* support models of energy maximization rather than time minimization. Similar support for the hypothesis of energy maximization is provided by *P. rugosus* and *P. barbatus*, which recruit more foragers to seed patches close to the colony (100).

Leafcutting ants select tree species based on the presence or absence of secondary plant compounds, water content, and leaf density (6, 55, 86, 116, 117). Colonies are more selective of species the more distant they are from the nest, although at these foraging loci they sometimes do not harvest the leaves of neighboring palatable species (86). Foraging by leafcutting ants follows the prediction of optimal diet theory, which states that colonies should show greater diet diversity and lower resource constancy when average resource quality is low; colonies indeed shift their leafcutting among a greater number of tree species when the average palatability of resource trees is low, and specialize in richer-resource environments (86, 95). Also, 70% of observed colonies cut relatively close to the nest, and foragers cut leaves from higher-quality tree species with increasing distance from the nest (86).

Risk as a factor influencing foraging decisions has been examined through studies of both simulated and natural predation (37, 66, 121) and competition (49, 113), which result in a decrease or cessation of foraging or a change in foraging direction. The influence of competitors on the spatial patterning of foragers and resource use has been described in *Pogonomyrmex* spp. (49) and in *L. neoniger* (113). In these species confrontation between foragers from adjacent colonies leads to altered trail system directionality and search pattern, respectively, without a reduction in overall foraging activity. In *Lasius pallitarsus*, resource use (patch choice) is affected by the presence of *Formica subnuda* workers; *L. pallitarsus* workers shift foraging activity to a safe, competitor-free patch, even if it is of lower quality (74).

Theories of foraging strategy provide an important organizational framework with an emphasis on energetics that is valuable for the study of the foraging economy of ants. But optimization models should be cautiously applied because the economics of foraging cannot be studied in isolation from other aspects of ecology, energetics, social organization, and the intrinsic behavioral limitations of ants. For example, some models (97) assume forager knowledge of the resource base as a requirement for optimal feeding strategy, yet the relatively short life span of a forager may limit an individual ant's experience. Moreover, if resources are unpredictably distributed, it is difficult to see how such information would modify future individual or colony decisions. The spatial pattern of foraging may also limit the acquisition of information about resource distribution. Low encounter rates and foraging success may be additional complications. Little is known of the memory capability of ants in regard to food choice; a few studies show they have the ability to store information and use it in foraging decisions (81, 87, 114; R. Johnson, personal communication). Should foragers have knowledge of resource availability, how are the preferences of individuals summed to yield a colony-wide pattern of choice? The role of these factors in the implementation of foraging strategy is poorly understood.

The social organization of foraging may influence spatial patterns of search and therefore resource assessment. The trunk trail systems of *Atta* spp., for example, originate from pheromone deposition and guide foragers to a tree to be attacked; ultimately, a path between the nest and the tree is cleared of obstruction. While this path enhances leaf transport efficiency, it limits ants to searching for new host trees only near the trunk route's end and sides (86, 94). Similarly, the use of paths cleared of vegetation by *P. occidentalis* foragers enhances gain rate but probably canalizes search (32).

## SEARCH PATTERN

Search pattern can be analyzed at two levels. First, search concerns the use of foraging space by colonies within populations, and intra- and interspecific territorial interactions may contribute to the partitioning of space between colonies (42, 49, 51, 62, 90, 113, 115). Therefore, search pattern involves colony-level regulation of the spatial allocation of foragers. Second, each trip made by a forager has its own ecologically influenced spatial and temporal organization (41, 91, 112, 118, 119). There is also a relationship between individual and colony-wide search patterns, since a forager may influence the search patterns of nestmates through communication. At both levels, resource distribution in space and time and competition should be major influences on the organization of search.

## Territoriality and Search Pattern

The significance of the spacing patterns of ant colonies and the division of foraging area has chiefly been approached by measuring dispersion and/or by studying territorial interactions (51, 62). Cost/benefit explanations suggest an interplay of the economics of foraging and defense and allow predictions to be made about how colonies should allocate workers to related tasks in time and space (51, 58). Dispersion patterns may change seasonally and appear to be nonrandom in most species for which data are available (42, 45, 49, 62, 90, 115).

Trail systems provide a mechanism by which the search patterns of foragers are directed away from neighboring colonies (42, 49, 90, 113). They also reduce search or travel time by directly channeling foragers to a persistent resource (49, 94, 95). In some cases more than one function may be served. Colonies of *P. rugosus* and *P. barbatus* feed at seed patches, and foragers travel along spatially separated trunk trails; the average nearest-neighbor distance between nests is 17–18 m (49). In contrast, *Pogonomyrmex maricopa* takes evenly distributed seeds, and foragers seem to depart randomly from the nest and collect seeds individually; colonies of this species are on average 46 m apart, apparently owing to the absence of trail systems (49). In *C. bicolor*, the radius of an area in which 90% of all search activity occurs approximates the average internest distance (119). In *Pheidole militicida*, trunk routes branch into smaller trails, and the persistence of use of these components of the trail system varies, with the trunk being most stable (52). Trunk routes in *P. militicida* channel foragers to areas where seeds are available, and the routes are abandoned when the resources are depleted. The biochemical ecology of trail system formation has been studied in regard to level of specificity and forager orientation (51, 52, 108, 113). Trail pheromones may reduce search time by serving as initial guides to areas of high seed density and may also perhaps decrease homing time, although visual cues are also important in orientation (49, 113). In *L. neoniger* the colony-specific persistent components of trail pheromone appear to direct forager search paths away from neighboring colonies, which may be as close as 5 cm (108, 113, 115). The highly directional patterns of foraging become random following confrontation with foragers from an alien nest, which suggests that the trail systems of this species minimize aggressive confrontation with competitors (113).

## Geometry of Search at the Colony Level

In the desert harvester ant *V. pergandei* the direction of foraging columns is rotated in a clockwise or counterclockwise manner between consecutive foraging periods, and the size of a foraging column is adjusted to seed production and patch distance (85). As food density decreases, columns increase in length and rotate more frequently. Column rotation seems to represent a system of sectorial search with a radial shift following resource depletion. Consecutive raids of the army ant *E. burchelli* are systematically rotated an average of 123° during the statary phase (35). Raiding also appears to be systematic in the nomadic phase, again seemingly for avoidance of redundant search. The geometry of search appears to be determined solely by prior raid direction, perhaps with modifications to avoid contact of foragers with other *E. burchelli* colonies (35).

Such systematic search during the statary and nomadic phases does not, however, appear to be the rule among army ants. In *Neivamyrmex nigrescens* the directions and distances of emigrations vary unpredictably during the nomadic phase (71). The striking differences in the search and raiding patterns of *E. burchelli* and *N. nigrescens* may be related to the distribution and abundance of prey or to nest sites and climatic factors (71). However, the density of prey species of ants in the southwest desert or grassland environment of *N. nigrescens* is approximately 3000 colonies/ha (71), and the density of ant species actually or potentially taken by *E. burchelli* is on average 6600 ± 2200 colonies/ha (range 4000–9000 colonies/ha) (61). The highly organized raid pattern of *E. burchelli* may therefore reflect selection for efficient prey harvesting given the greater abundance of its prey, but the role of prey distribution is not clear. Limited study of the nomadic obligate termite predator *Pachycondyla* (=*Termitopone*) *laevigata* has shown no predictable raiding pattern (30).

## Search Patterns of Individual Foragers

Forager decisions may include where to search, how long to search at a given site, and whether or not to return to a site where search was previously conducted. Some species, such as *Pachycondyla* (=*Neoponera*) *apicalis*, show strong route fidelity, each worker restricting its foraging activity to a given site (36). In *C. bicolor* search effort is concentrated in sectors of approximately 50° and 30-m radius, and average search path length differs among populations according to food availability (119). The search pattern of a colony contains random components as evidenced in the exponential distribution of search times of foragers, increasing average distance of search in proportion to the square root of time, and selection of initial foraging direction. Thus the colony-level search pattern approximates a random diffusion process, which may be an efficient solution to harvesting food that is unpredictably distributed (41). In *F. schaufussi*, search pattern is related to food distribution; collecting a single crop load of carbohydrate food produces a more spatially restricted and accurate search during a return trip than collecting a single load of insect prey (J. F. A. Traniello & A. J. Kozol, unpublished).

SEARCH DURATION AND PERSISTENCE    Search by *C. bicolor* foragers shows distinctive temporal patterns that are correlated with food availability. Colonies in Greece collect less prey biomass than colonies in Tunisia, and individuals in Greece make roughly double the number of trips, although trip duration is shorter (119). Trip duration is also dependent upon the type of food collected; it is longer for ants collecting plant secretions (39).

The tendency of a forager to return to and search at a site where it has found food has been studied in *C. bicolor* (91) and *F. schaufussi* (111, 112). In both species foragers differ in their tendency to restrict their search to the site of a prior find. Persistent foragers show intensive local search, whereas nonpersistent foragers often completely bypass a prior reward site. In *C. bicolor* these individually different search strategies appear to categorize foragers throughout their lives and reflect foraging specializations. Search pattern in *F. schaufussi* appears to be adjusted in response to food type, and search persistence seems to be correlated with the spatial and temporal persistence of the insect prey and homopteran excretions that compose this species' diet (112). Consecutive rewards of carbohydrate food tend to produce a shift from nonpersistent to persistent searching and increase search duration (112). The degree of spatial accuracy of search is greater in response to feeding on carbohydrate than on insect prey, but it does not improve with successive multiple rewards of either type of food, which indicates that spatial patterning of search and information acquired to control search are fixed resource-dependent parameters. However, search duration may be conditioned and may at least have a local effect on persistence. In *C. bicolor*, search persistence increases with the distance of the food to the nest (91), and in *V. pergandei* the frequency of seed sampling in a patch increases with the distance of the patch from the nest (84). Increasing search time in *V. pergandei* results in the sampling of a larger number of seeds in a patch.

# SOCIAL REGULATION OF FORAGING

## Age Polyethism and Development of Foraging Behavior

Workers of many ant species sequentially change tasks as they mature, beginning with brood care following worker eclosion and culminating with foraging (59, 75). Several factors affect the onset of foraging behavior. The completion of physiological processes related to cuticular sclerotization and the acquisition of sensory receptor competence may be required before foraging can begin. A callow phase during which tanning takes place may be necessary to provide time for the completion of physiological maturation; these processes occur during pupal development in some species (107). Feeding habits could affect the pattern of age polyethism if brood care is directly associated with foraging. This may be true for species such as *Amblyopone pallipes* that directly provision larvae with prey (107). In the temporal polyethism of most species, tasks with high associated risks, such as foraging, are scheduled as the ultimate worker contribution to a colony's economy (75). Ergonomically, loss is minimized if forager mortality is high because foragers hold fewer energy reserves in the form of lipid stores than younger workers (67).

The percentage of foragers in a colony varies across species (67, 78). The theory of adaptive demography predicts that the proportion of workers allocated to foraging should differ among populations in different ecological circumstances if worker age determines the transition to foraging performance (75). Recent field-based empirical tests suggest that age caste ratios are highly variable, and the proportion of older workers in a colony shows no clear correlation with ecology. Censuses of whole, field-collected *Pheidole dentata* colonies from two habitats showing significant differences in food resource biomass, ant species diversity, predation, and competitive environment revealed that older workers, which are typically foragers, make up a varying proportion of the total worker force (P. Calabi & J. F. A. Traniello, unpublished). Although temporal caste ratios vary, colonies with very different demographic profiles have very similar labor profiles. This suggests that changing foraging requirements are met by rapid changes in the task performance patterns of individual workers in response to colony need rather than by changes in age demography. This intercolony consistency in the proportion of total labor involved in foraging suggests a flexible switching of the behavior of individual workers in response to immediate foraging requirements. A worker's shift from another task to foraging is therefore related not only to age but also to the perception of social stimuli carrying information about the current nutritional status of a nest and/or the activities of other colony members. In *Novomessor albisetosus,* callow workers in colonies in which older workers were experimentally removed matured behaviorally in one third the time of normal development and began foraging at an earlier age than workers in unmanipulated control colonies (69). This type of socially mediated behavioral development appears to be common in ants (59). Flexibility in temporal polyethism appears to represent a major mechanism of social compensation of forager number in response to short-term ecological fluctuations.

Studies of trail system organization in *Formica* spp. have illustrated the role of age polyethism in the development of foraging behavior (15, 87). Foragers captured and marked in autumn and recaptured in the spring retained the use of their initial route through a period of winter inactivity. These topographic traditions result in a transfer of spatial foraging information through several worker generations. Routes have a spatiotemporal polyethic structure; mature (pioneer) foragers head the foraging column, whereas younger workers (recruits) are found close to the nest. The behavior of young foragers gradually changes to that of mature foragers, as evidenced from changes in their location along a foraging trail.

Because foraging behavior is often coordinated through chemical communication, maturational, social, or perceptual abilities may be associated with the onset of foraging. Callow army ants *(N. nigrescens)* travel along chemical trails with other colony members during emigration on the first nomadic day but do not begin foraging (raiding) until they are 3–7 days old. The absence of foraging behavior is not due to the sensory inability of callows, however, as these immature workers follow artificial trails as well as adults; this suggests that other social stimuli affect the development of foraging (102). That callows remain in the nest while mature workers conduct raids seems to be due to the absence of mechanical stimuli that are part of the mass-recruitment communication signals that accompany emigration and are required by callows to induce trail following (103).

## LEARNING AND FORAGING BEHAVIOR

The unpredictability of the resource environment of a colony may require that foraging be fine-tuned through experience-based change in the behavior of individual workers to enhance efficiency. Simple and complex types of learning have been described in ants, which may reflect selection for learning traits that have evolved in response to dietary habits and foraging ecology under the constraints of life span, sociality, and caste and the temporal, spatial, and size variability of food resources. Recent studies of ant learning have emphasized either ecology (81) or theoretical modeling (27, 28); additional empirical studies on the natural history of ant learning will be useful in defining the limits of individual flexibility and measuring forager competence (44). *Veromessor pergandei* and *P. rugosus* foragers develop preferences for harvesting the seeds of certain plants, which results in an individual foraging specialization that may persist for up to two weeks or longer (81). Learning rate and memory retention may be correlated with spatial variability in seed species in these ants (R. Johnson, personal communication). Similar forager specializations appear to occur in *Chelaner* spp. (25). The mechanism of specialization may be a learning process that permits a forager to identify qualitative properties of seeds and to narrow search image quickly to fix on a specific foraging task. This specialization may increase the net rate of energy returned to the colony. Learning may also have a role in the development of foraging in other species. In omnivorous species such as *Formica* spp. and *Cataglyphis* spp., workers tend homopterans or lick plants to collect carbohydrate secretions and also forage for arthropod prey (111, 112, 119). These two resources differ in their spatial and temporal patterns of distribution and therefore provide different patterns of reward. Workers of *C. bicolor* tend to specialize on collecting either type of food (39). In *F. schaufussi,* the ability of workers to modify their search behavior seems to depend on a food resource–dependent learning process. Local search duration is more easily conditioned by carbohydrate food rewards (sucrose) than by protein (freshly killed insects) (111, 112). The conditioning of search duration by carbohydrate food seems to be independent of resource quality. Repeated rewards of insect prey do not produce a more accurate search pattern, nor do they result in increased search effort. Search accuracy is greater after a reward of a single crop load of 1-M sucrose than after a reward of insect prey. However, for a given food type, there is no significant change in search accuracy over four or eight consecutive rewards (J. F. A. Traniello & A. J. Kozol, unpublished). Further studies are needed to determine if foraging specializations develop through resource-related learning processes. Resource-related learning processes may also be involved in the persistence of route fidelity and memory shown in *Formica* spp. (18, 88). At the colony level, learning may underlie competitor recognition systems and may mediate community interactions (13).

## CONCLUSIONS

The principal ecological determinants of ant foraging strategy are the distribution of food resources in size, time, space, and quality; competition with sympatric ant species; and predation. Because a colony is generally sessile, the resource and competitive environment is in essence defined by its location. A colony must therefore spatially pattern foraging to harvest food efficiently and minimize competition. This is accomplished through individual and colony-wide foraging.

Elements of social organization are linked to virtually every aspect of foraging strategy. Although patterns of physical and temporal caste evolution have been shown to be integral components of some foraging systems, more sensitive empirical tests are required to provide detailed explanations of the origin and ecological significance of forager size and age-related behavior. Similarly, forager functional morphology and biomechanical characteristics associated with loading capacity should be examined to describe precisely the role of form as well as size and behavior in ant foraging ecology. The overall conceptualization of ant foraging strategy should therefore encompass theory and analyses on subjects ranging from the major elements of colony organization such as caste and division of labor to the time and energy budgets of foragers. Although in this review I have suggested a components approach, foraging systems must be analyzed as whole entities. The pitfalls of extreme adaptationist reasoning that might develop from fragmenting the system should and can be avoided through an emphasis on social integration. A prudent application of foraging models, theories on caste evolution and ecology, and basic knowledge of natural history and social regulatory mechanisms can provide the necessary theoretical and empirical framework. Finally, it will be possible to evaluate ant foraging strategy fully when the energetics of foraging and the caloric and nutritional benefits of selected resources are quantitatively measured and used to estimate the conversion of food intake into alates. This analysis will remove the inaccuracies of assumptions of the energetic costs of foraging and will permit colony foraging economics to be quantified in terms of reproductive output.

ACKNOWLEDGMENTS

I thank many colleagues for their reprints and unpublished manuscripts, and regret that a lack of space prohibited me from including the results of a larger number of studies. The manuscript benefited from critical readings by Samuel Beshers, Michelle Scott, and Steve Rissing. I gratefully acknowledge the support of the National Science Foundation and the Whitehall Foundation.

*Literature Cited*

1. Adams, E. A., Traniello, J. F. A. 1981. Chemical interference competition by *Monomorium minimum*. *Oecologia* 51: 270–75
2. Andersen, A. N. 1986. Patterns of ant community organization in mesic southeastern Australia. *Aust. J. Ecol.* 11:87–97
3. Bartholomew, G. A., Lighton, J. R. B., Feener, D. H. 1988. Energetics of trail running, load carriage, and emigration in the column-raiding army ant *Eciton hamatum*. *Physiol. Zool.* 61:57–68
4. Bernstein, R. 1974. Seasonal food abundance and foraging activity in some desert ants. *Am. Nat.* 108:490–98
5. Bernstein, R. 1975. Foraging strategies of ants in response to variable food density. *Ecology* 56:213–19
6. Bowers, M. A., Porter, S. D. 1981. Effect of foraging distance on water content of substrates harvested by *Atta columbica* (Guerin). *Ecology* 62:273–75
7. Breise, D. T. 1982. Patterning of resources amongst seed-harvesting ants in an ant community in semi-arid Australia. *Aust. J. Ecol.* 7:299–307

8. Breise, D. T., Macauley, B. J. 1981. Food collection within an ant community in semi-arid Australia, with special reference to seed harvesters. *Aust. J. Ecol.* 6:1–19
9. Brian, M. V. 1978. *Production Ecology of Ants and Termites*. Cambridge, Mass: Cambridge Univ. Press
10. Brian, M. V. 1983. *Social Insects: Ecology and Behavioural Biology*. London: Chapman & Hall
11. Buckley, R. C. 1982. *Ant-Plant Interactions in Australia*. The Hague: Junk
12. Burton, J. L., Franks, N. R. 1985. The foraging ecology of the army ant *Eciton rapax*: an ergonomic enigma? *Ecol. Entomol.* 10:131–41
13. Carlin, N. F., Johnston, A. 1984. Learned enemy specification in the defense recruitment system of an ant. *Naturwissenschaften* 71:156–57
14. Carroll, C. R., Janzen, D. H. 1973. Ecology of foraging by ants. *Ann. Rev. Ecol. Syst.* 4:231–57
15. Cherix, D. 1987. Relation between diet and polyethism in *Formica* colonies. See Ref. 76, pp. 93–115

16. Chew, R. M., Chew, A. E. 1980. Body size as a determinant of small-scale distribution of ants in evergreen woodland in southeastern Arizona. *Insectes Soc.* 27:189–20

17. Chew, R. M., DeVita, J. 1980. Foraging characteristics of a desert ant assemblage: functional morphology and species separation. *J. Arid Environ.* 3:75–83

18. Cosens, D., Toussaint, N. 1985. An experimental study of the foraging strategy of the wood ant *Formica aquilonia*. *Anim. Behav.* 33:541–52

19. Crawford, D. L., Rissing, S. W. 1983. Regulation of recruitment by individual scouts in *Formica oreas* Wheeler (Hymenoptera; Formicidae). *Insectes Soc.* 30:177–83

20. Culver, D. 1974. Species packing in Caribbean and north temperate ant communities. *Ecology* 55:974–88

21. Davidson, D. W. 1977. Species diversity and community organization in desert seed-eating ants. *Ecology* 58:711–24

22. Davidson, D. W. 1977. Foraging ecology and community organization in seed-eating ants. *Ecology* 58:725–37

23. Davidson, D. W. 1978. Experimental tests of the optimal diet in two social insects. *Behav. Ecol. Sociobiol.* 4:35–41

24. Davidson, D. W. 1978. Size variability in the worker caste of a social insect (*Veromessor pergandei* Mayr) as a function of the competitive environment. *Am. Nat.* 112:523–32

25. Davison, E. A. 1982. Seed utilization by harvester ants. See Ref. 11, pp. 1–6

26. Davison, E. A. 1987. Respiration and energy flow in two Australian species of desert harvester ants, *Chelaner rothsteini* and *Chelaner whitei*. *J. Arid. Environ.* 12:61–82

27. Deneubourg, J. L. 1985. The role of errors and memory in the efficiency of ant foraging. *Ann. Soc. R. Zool. Belg.* 115:109–10

28. Deneubourg, J. L., Goss, S., Pasteels, J. M., Fresneau, D., Lachaud, J. P. 1987. Self-organizing mechanisms in ant societies (II): Learning in foraging and division of labor. See Ref. 76, pp. 177–96

29. DeVita, J. 1979. Mechanisms of interference and foraging among colonies of the harvester ant *Pogonomyrmex californicus* in the Mojave Desert. *Ecology* 60:729–37

30. Downing, H. 1978. *Foraging and migratory behavior of the ponerine ant* Termitopone laevigata. BA thesis. Smith Coll., Northampton, Mass. 121 pp.

31. Fellers, J. H. 1987. Interference and exploitation in a guild of woodland ants. *Ecology* 68:1466–78

32. Fewell, J. 1988. Energetic and time costs of foraging in harvester ants, *Pogonomyrmex occidentalis*. *Behav. Ecol. Sociobiol.* 22:401–8

33. Franks, N. R. 1985. Reproduction, foraging efficiency, and worker polymorphism in army ants. In *Experimental Behavioral Ecology and Sociobiology*, ed. B. Hölldobler, M. Lindauer, pp. 81–108. Stuttgart: Fischer

34. Franks, N. R. 1986. Teams in social insects: group retrieval of prey by army ants (*Eciton burchelli*, Hymenoptera: Formicidae). *Behav. Ecol. Sociobiol.* 18:425–29

35. Franks, N. R., Fletcher, C. A. 1983. Spatial patterns in army ant foraging and migration: *Eciton burchelli* on Barro Colorado Island, Panama. *Behav. Ecol. Sociobiol.* 12:261–70

36. Fresneau, D. 1985. Individual foraging and path fidelity in a ponerine ant. *Insectes Soc.* 32:109–16

37. Gentry, J. B. 1974. Response to predation by colonies of the Florida harvester ant *Pogonomyrmex badius*. *Ecology* 55:1328–38

38. Hansen, S. E. 1978. Resource utilization and coexistence of three species of *Pogonomyrmex* ants in an upper Sonoran grassland community. *Oecologia* 35:105–17

39. Harkness, M. L. R., Harkness, R. D. 1976. Functional differences between individual ants *Cataglyphis bicolor*. *J. Physiol.* 25:124–25

40. Harkness, R. D. 1978. The speed of walking of *Cataglyphis bicolor* (F.) (Hym., Formicidae). *Entomol. Mon. Mag.* 114:203–10

41. Harkness, R. D., Maroudas, N. G. 1985. Central place foraging by an ant (*Cataglyphis bicolor* Fab.): a model of searching. *Anim. Behav.* 33:916–28

42. Harrison, J. S., Gentry, J. B. 1981. Foraging pattern, colony distribution, and foraging range of the Florida harvester ant, *Pogonomyrmex badius*. *Ecology* 62:1467–73

43. Heinrich, B. 1978. The economics of insect sociality. In *Behavioural Ecology*, ed. J. Krebs, N. Davies, pp. 97–128. Sunderland, Mass: Sinauer

44. Herbers, J. M. 1981. Reliability theory and foraging by ants. *J. Theor. Biol.* 89:175–89

45. Herbers, J. M. 1985. Seasonal structuring of a north temperate ant community. *Insectes Soc.* 32:224–40

46. Hespenheide, H. A. 1973. Ecological inferences from morphological data. *Ann. Rev. Ecol. Syst.* 4:213–30

47. Hespenheide, H. A. 1975. Prey characteristics and predator niche width. In *Ecology and Evolution of Communities*, ed. M. L. Cody, J. M. Diamond, pp. 158–80. Cambridge, Mass: Belknap

48. Holder Bailey, K., Polis, G. A. 1987. Optimal and central place foraging theory applied to a desert harvester ant, *Pogonomyrmex californicus*. *Oecologia* 72:440–48

49. Hölldobler, B. 1976. Recruitment behavior, home range orientation and territoriality in harvester ants, *Pogonomyrmex*. *Behav. Ecol. Sociobiol.* 1:3–44

50. Hölldobler, B. 1984. Evolution of insect communication. In *Insect Communication*, ed. T. Lewis, pp. 349–77. London: Academic

51. Hölldobler, B., Lumsden, C. 1980. Territorial strategies in ants. *Science* 210:732–39

52. Hölldobler, B., Moglich, M. 1980. The foraging system of *Pheidole militicida* (Hymenoptera: Formicidae). *Insectes Soc.* 27:237–64

53. Hölldobler, B., Stanton, R. C., Markl, H. 1978. Recruitment and food-retrieving behavior in *Novomessor* (Formicidae: Hymenoptera). I. Chemical signals. *Behav. Ecol. Sociobiol.* 4:163–81

54. Hölldobler, B., Traniello, J. F. A. 1980. The pygidial gland and chemical recruitment communication in *Pachycondyla* (= *Termitopone*) *laevigata*. *J. Chem. Ecol.* 6:883–93

55. Howard, J. J. 1987. Leafcutting ant diet selection: the role of nutrients, water, and secondary chemistry. *Ecology* 68:503–15

56. Hunt, J. H. 1983. Foraging and morphology in ants: the role of vertebrate predators as agents of natural selection. See Ref. 57, 2:83–103

57. Jaisson, P. 1983. *Social Insects in the Tropics*, Vol. 1, 2. Paris: Presses Univ. Paris XIII

58. Johnson, L. K., Hubbell, S. P., Feener, D. H. 1987. Defense of food supply by eusocial colonies. *Am. Zool.* 27:347–58

59. Lenoir, A. 1987. Factors determining polyethism in social insects. See Ref. 76, pp. 219–40

60. Leonard, J. G., Herbers, J. M. 1986. Foraging tempo in two woodland ant species. *Anim. Behav.* 34:1172–81

61. Levings, S. C., Franks, N. R. 1982. Patterns of nest dispersion in a tropical ground ant community. *Ecology* 63:338–44

62. Levings, S. C., Traniello, J. F. A. 1982. Territoriality, nest dispersion, and community structure in ants. *Psyche* 88:265–319

63. Lighton, J. R. B., Bartholomew, G. A., Feener, D. H. 1987. Energetics of locomotion and load carriage and a model of the energy cost of foraging in the leaf-cutting ant *Atta colombica* Guer. *Physiol. Zool.* 60:524–37

64. Longhurst, C., Johnson, R. A., Wood, T. G. 1978. Predation by *Megaponera foetens* (Fabr.) (Hymenoptera: Formicidae) on termites in the Nigerian southern guinea savanna. *Oecologia* 32:101–7

65. Lynch, J. F., Balinksy, E. C., Vah, S. G. 1980. Foraging patterns in three sympatric forest ant species, *Prenolepis impuris*, *Paratrechina melanderi*, and *Aphaenogaster rudis* (Hymenoptera: Formicidae). *Ecol. Entomol.* 5:353–71

66. MacKay, W. P. 1982. The effect of predation of western widow spiders *Latrodectus hesperus* (Araneae:Theridiiae) on harvester ants *Pogonomyrmex rugosus* (Hymenoptera:Formicidae). *Oecologia* 53:406–11

67. MacKay, W. P. 1985. A comparison of the energy budgets of three species of *Pogonomyrmex* harvester ants (Hymenoptera: Formicidae). *Oecologia* 66:484–94

68. Maschwitz, U., Schonegge, P. 1983. Forage communication, nest moving recruitment and prey specialization in the oriental ponerine *Leptogenys chinensis*. *Oecologia* 57:175–82

69. McDonald, P., Topoff, H. 1985. Social regulation of behavioral development in the ant *Novomessor albisetosus* (Mayr). *J. Comp. Psychol.* 99:3–14

70. Mirenda, J. T., Eakins, D. G., Gravelle, K., Topoff, H. 1980. Predatory behavior and prey selection by army ants in a desert-grassland habitat. *Behav. Ecol. Sociobiol.* 7:119–27

71. Mirenda, J. T., Topoff, H. 1980. Nomadic behavior of army ants in a desert-grassland habitat. *Behav. Ecol. Sociobiol.* 7:129–35

72. Morton, S. R. 1982. Granivory in the Australian arid zone: diversity of harvester ants and structure of their communities. In *Evolution of the Flora and Fauna of Arid Australia*, ed. W. R. Barker, P. J. M. Greenslade, pp. 257–62. Adelaide, Australia: Peacock

73. Nielsen, M. G., Jensen, T. F., Holm-Jensen, I. 1982. Effect of load carriage on the respiratory metabolism of running worker ants of *Camponotus herculeanus* (Formicidae). *Oikos* 39:137–42

74. Nonacs, P., Dill, L. M. 1988. Foraging

responses of the ant *Lasius pallitarsus* to food sources with associated mortality. *Insectes Soc.* In press

75. Oster, G., Wilson, E. O. 1978. *Caste and Ecology in the Social Insects.* Princeton, NJ: Princeton Univ. Press

76. Pasteels, J. M., Deneubourg, J. L. 1987. *From Individual to Collective Behaviour in Social Insects.* Basel: Birkhauser

77. Peakin, G. J., Josens, G. 1978. Respiration and energy flow. See Ref. 9, pp. 111–63

78. Porter, S. D., Jorgenson, C. D. 1981. Foragers of the ant, *Pogonomyrmex owyheei:* a disposable caste? *Behav. Ecol. Sociobiol.* 9:247–56

79. Rettenmeyer, C. W., Chadab-Crepet, R., Naumann, M. G., Morales, L. 1983. Comparative foraging by neotropical army ants. See Ref. 57, 2:59–73

80. Reyes Lopez, J. L. 1987. Optimal foraging in seed harvester ants: computer-aided simulation. *Ecology* 68:1630–33

81. Rissing, S. W. 1981. Foraging specializations of individual seed-harvester ants. *Behav. Ecol. Sociobiol.* 9:149–52

82. Rissing, S. W. 1982. Foraging velocity of seed-harvester ants, *Veromessor pergandei* (Hymenoptera: Formicidae). *Environ. Entomol.* 11:905–7

83. Rissing, S. W. 1987. Annual cycles in worker size of the seed-harvester ant *Veromessor pergandei* (Hymenoptera: Formicidae). *Behav. Ecol. Sociobiol.* 20:117–24

84. Rissing, S. W., Pollock, G. B. 1984. Worker size variability and foraging efficiency in *Veromessor pergandei* (Hymenoptera: Formicidae). *Behav. Ecol. Sociobiol.* 15:121–26

85. Rissing, S. W., Wheeler, J. 1976. Foraging responses of *Veromessor pergandei* to changes in seed production. *Pan-Pac. Entomol.* 52:63–72

86. Rockwood, L., Hubbell, S. P. 1987. Host-plant selection, diet diversity, and optimal foraging in a tropical leafcutting ant. *Oecologia* 74:55–61

87. Rosengren, R. 1977. Foraging strategy of wood ants, *Formica rufa* group. Part 1. Age polyethism and topographic traditions. *Acta Zool. Fenn.* 149:1–30

88. Rosengren, R., Fortelius, W. 1986. Orstreue in foraging ants of the *Formica rufa* group—hierarchy of orienting cues and long-term memory. *Insectes Soc.* 33:306–37

89. Rudolph, S. G., Loudon, C. 1986. Load size selection by foraging leaf-cutter ants *(Atta cephalotes).* *Ecol. Entomol.* 11: 401–10

90. Ryti, R. T., Case, T. J. 1986. Over-dispersion of ant colonies: a test of hypotheses. *Oecologia* 69:446–53

91. Schmid-Hempel, P. 1984. Individually different foraging methods in the desert ant *Cataglyphis bicolor* (Hymenoptera, Formicidae). *Behav. Ecol. Sociobiol.* 14:263–71

92. Schmid-Hempel, P. 1987. Foraging characteristics of the desert ant *Cataglyphis.* See Ref. 76, pp. 43–61

93. Schmid-Hempel, P., Schmid-Hempel, R. 1984. Life duration and turnover of foragers in the ant *Cataglyphis bicolor* (Hymenoptera, Formicidae). *Insectes Soc.* 31:345–60

94. Shepherd, J. D. 1982. Trunk trails and the searching strategy of a leaf-cutter ant *Atta colombica.* *Behav. Ecol. Sociobiol.* 11:77–84

95. Shepherd, J. D. 1985. Adjusting foraging effort to resources in adjacent colonies of the leaf-cutter ant, *Atta colombica.* *Biotropica* 17:245–52

96. Sorensen, A. A., Busch, T. M., Vinson, S. B. 1983. Behaviour of worker subcastes in the fire ant, *Solenopsis invicta,* in response to proteinaceous food. *Physiol. Entomol.* 8:83–92

97. Stephens, D. W., Krebs, J. R. 1986. *Foraging Theory.* Princeton, NJ: Princeton Univ. Press

98. Stradling, D. J. 1978. Food and feeding habits of ants. See Ref. 9, pp. 81–106

99. Sudd, J. H. 1987. Individual behavior and mixed diet strategy in ants. See Ref. 76, pp. 81–92

100. Taylor, F. 1976. Foraging behavior in ants: experiments with two species of myrmecine ants. *Behav. Ecol. Sociobiol.* 2:147–68

101. Taylor, F. 1978. Foraging behavior of ants: theoretical considerations. *J. Theor. Biol.* 71:541–65

102. Topoff, H., Boshes, N., Trakimas, W. 1972. A comparison of trail following between callow and adult workers of the army ant *(Neivamyrmex nigrescens)* (Formicidae: Dorylinae). *Anim. Behav.* 20:361–66

103. Topoff, H., Mirenda, J. 1978. Precocial behaviour of callow workers of the army ant *Neivamyrmex nigrescens:* importance of stimulation by adults during mass recruitment. *Anim. Behav.* 26: 698–706

104. Topoff, H., Mirenda, J. 1980. Army ants do not eat and run: influence of food supply on emigration behaviour in *Neivamyrmex nigrescens.* *Anim. Behav.* 28:1040–45

105. Topoff, H., Mirenda, J. 1980. Army

ants on the move: relation between food supply and emigration frequency. *Science* 207:1099–100

106. Traniello, J. F. A. 1977. Recruitment behavior, orientation, and the organization of foraging in the carpenter ant *Camponotus pennsylvanicus* DeGeer. *Behav. Ecol. Sociobiol.* 2:61–79

107. Traniello, J. F. A. 1978. Caste in a primitive ant: absence of age polyethism in *Ambylopone. Science* 202:770–72

108. Traniello, J. F. A. 1980. Colony specificity in the trail pheromone of an ant. *Naturwissenschaften* 67:360

109. Traniello, J. F. A. 1983. Social organization and foraging success in *Lasius neoniger* (Hymenoptera: Formicidae): behavioral and ecological aspects of recruitment communication. *Oecologia* 59:94–100

110. Traniello, J. F. A. 1987. Comparative foraging ecology of north temperate ants: the role of worker size and cooperative foraging in prey selection. *Insectes Soc.* 34:118–30

111. Traniello, J. F. A. 1987. Individual and social modification of behavior in response to environmental factors in ants. See Ref. 76, pp. 63–80

112. Traniello, J. F. A. 1988. Variation in foraging behavior among workers of the ant *Formica schaufussi:* ecological correlates of search behavior and the modification of search pattern. In *Interindividual Behavioral Variability in Social Insects*, ed. R. L. Jeanne, pp. 91–112. Boulder, Colo: Westview

113. Traniello, J. F. A. 1988. Chemical trail systems, orientation, and territoriality in the ant *Lasius neoniger. J. Insect Behav.* In press

114. Traniello, J. F. A., Fujita, M. S., Bowen, R. V. 1984. Ant foraging behavior: Ambient temperature influences prey selection. *Behav. Ecol. Sociobiol.* 15:65–68

115. Traniello, J. F. A., Levings, S. C. 1986. Intra- and intercolony patterns of nest dispersion in the ant *Lasius neoniger:* correlations with territoriality and foraging ecology. *Oecologia* 69:413–19

116. Waller, D. A. 1982. Leaf-cutting ants and avoided plants: defenses against *Atta texana* attack. *Oecologia* 52:400–3

117. Waller, D. A. 1982. Leaf-cutting ants and live oak: the role of leaf toughness in seasonal intraspecific host choice. *Entomol. Exp. Appl.* 32:145–50

118. Wehner, R. 1987. Spatial organization of foraging behavior in individually searching desert ants, *Cataglyphis* (Sahara Desert) and *Ocymyrmex* (Namib Desert). See Ref. 76, pp. 15–42

119. Wehner, R., Schmid-Hempel, P., Harkness, R. D. 1983. *Foraging Strategies in Individually Searching Ants* Cataglyphis bicolor. Stuttgart: Fischer

120. Whitford, W. G. 1978. Foraging by seed-harvester ants. See Ref. 9, pp. 107–110

121. Whitford, W. G., Bryant, M. 1979. Behavior of a predator and its prey: the horned lizard *(Phrynosoma cornutum)* and harvester ants *(Pogonomyrmex* spp.). *Ecology* 60:686–94

122. Wilson, D. S. 1975. The adequacy of body size as a niche indicator. *Am. Nat.* 109:769–84

*Ann. Rev. Entomol. 1989. 34:211–29*

# ENTOMOLOGY OF OILSEED *BRASSICA* CROPS[1]

## *Robert J. Lamb*

Agriculture Canada Research Station, Winnipeg, Manitoba R3T 2M9, Canada

## INTRODUCTION

Humans have used vegetable oil pressed from the seeds of the *Brassica* plants known as rape for thousands of years (29, 95), but only during the last 30 years have oilseed *Brassica* crops become internationally important. The small spherical seeds are harvested and crushed to separate the oil, which makes up approximately 40% of the seed by weight, from the remaining meal. The oil has been used as a lamp fuel, as a lubricant, and in the chemical industry, but now is most often used in cooking or the production of food. The meal contains approximately 40% protein by weight. It has been used as an organic fertilizer, but now is most often blended in animal feed. Sales of this byproduct contribute substantially to the value of the crop.

The expanded use of the crop, and the consequent increase in production, occurred when plant breeders transformed the chemistry of the seed. First the fatty acid composition of the oil was modified to reduce the level of erucic acid, and then the levels of glucosinolates in the meal were reduced. Erucic acid was a potential health hazard for humans consuming the oil (28). Glucosinolates in the meal were repellent or toxic for some farm animals (97). The transformed crop is commonly referred to in Europe as 00 rape, signifying the goal of reducing the levels of erucic acid and glucosinolates in the seed to zero. In Canada it is known as Canola®. Knowledge of the effects on herbivory of secondary plant compounds such as the glucosinolates is critical to our understanding of herbivore-plant interactions (54, 81, 111). Glucosinolates are antifeedants for many polyphagous herbivores, but most insect pests

---

[1]The author is employed by the Canadian Federal Government, and by reason of this fact, the copyright covering this paper is owned by the Crown in right of Canada.

of *Brassica* crops use glucosinolates or their fission products as attractants or feeding stimulants (32).

The dramatic change that occurred in the secondary plant compounds of oilseed *Brassica* crops interests entomologists and chemical ecologists. Also, the enormous increase in production offered a new environment for insects, which provided entomologists with the opportunity to determine the ecological characteristics of a new insect community. Nevertheless, entomologists are interested in oilseed rape primarily because they need to control the many insect pests that compete with humans for the crop. The genus *Brassica* is in the family Cruciferae, and many of the pests that attack the oilseed *Brassica* crops also attack other cruciferous vegetables such as cabbage and radish. The common pests of crucifers were described by Bonnemaison (12) before oilseed rape became an important crop in Europe and North America.

Rape is an agricultural analog of soybeans, although the two herbaceous crops are in different plant families. Both produce seeds that yield vegetable oil and high-protein meal, and these compete on international markets. Both are grown over large areas in North America, Europe, and other parts of the world, although the oilseed *Brassica* crops grow better than soybeans in cooler, shorter-season climates and less well at higher temperatures. The plants differ in architecture, but both are sown in dense stands and grow to a similar height. Both crops harbor complex insect communities. In general, insect communities in crops have received less attention than insect communities in natural habitats (111). As the entomology of the oilseed *Brassica* crops becomes better known, a comparison of the insect community of rape with the more thoroughly studied soybean community (58, 123) will be instructive.

This review emphasizes the relationships between insect pests and the crop and the agricultural impact of pests on the crop. Thus, it focuses on the plant, its growth, responses to damage (80), and the way insects interact with the plant to affect the agricultural commodity, rapeseed.

## The Taxonomy and Origins of the Oilseed Brassica *Crops*

Although agriculture and the food industries treat rapeseed as a single commodity, the crop is a composite of seed from two or three species (29, 95). In Asia, *Brassica campestris* and *Brassica juncea* are widely grown, but in Europe, Canada, New Zealand, and Australia, *Brassica napus* and to a lesser extent *B. campestris* are grown. These species are closely related. *Brassica juncea* probably arose as a natural hybrid between *B. campestris* and *Brassica nigra*, and *B. napus* arose as a natural hybrid of *B. campestris* and *Brassica oleracea*. The latter species is the source of most of our cruciferous vegetables. *Brassica campestris* is native throughout Europe, central Asia, and the Near East and was probably domesticated as a source of oil in central Asia or adjacent northwestern India. *Brassica napus* does not occur in wild populations and was probably domesticated in southern Europe. The species differ morphologically and chemically, and they grow at different rates; these differences complicate the study of pests associated with the crops. Furthermore, the annual forms of *B. napus* and *B. campestris* are sown in the spring in most of Canada and in northern Europe, but in central and southern Europe winter-dormant, biennial forms are sown in the late summer (29). The differing phenologies of the two forms and the three species affect the synchronies of insect life histories with the crop.

## THE PESTS

The most serious pests of the oilseed *Brassica* crops are members of the order Coleoptera, particularly in Europe and North America (Table 1). On the Indian subcontinent and in New Zealand aphids cause severe damage (67, 108), although in India pests from other orders are also important (10). In Asia, insecticides are applied to control aphids and the diamondback moth, *Plutella xylostella,* in oilseed *Brassica* crops (72), but relatively little has been published on their pest status. Similarly, little has been published on the pests of oilseed *Brassica* crops in South America, Australia, and Africa. Previously unrecognized pests are still being identified, For example, recent research on *Lygus* bugs in Canada indicates that these bud and seed feeders may cause significant losses, although the subtle nature of the damage has caused it to be overlooked (R. A. Butts & R. J. Lamb, unpublished data).

Many of the important pests are as cosmopolitan as the crucifers themselves (Table 1). They may not be important pests wherever they occur, however. The flea beetle *Phyllotreta cruciferae* is not usually a significant pest in Europe (7), although it is the dominant flea beetle pest of oilseed *Brassica* crops in Canada (19) and in India (10). Root maggots, *Delia* spp., are significant pests of rape only in parts of Alberta, Canada (47) and in Finland (118), although *Delia radicum* is a widespread pest of other crucifers throughout most of Europe and North America (12).

In Europe and Asia, where the use of cruciferous crops is an ancient practice, most of the insect pests of oilseed *Brassica* crops are crucifer specialists. Some of the pests in North America, such as flea beetles, root maggots, and diamondback moth, are also crucifer specialists; these pests were introduced from Europe or Asia and became pests of cruciferous vegetables before oilseed *Brassica* crops were introduced. In North America, native pests that are not crucifer specialists have also adapted quickly to the new crop, as has one native crucifer specialist. The noctuid *Mamestra configurata* was the first pest to devastate the crop over a wide area in Canada (129). It had previously been known only as an occasional pest of noncruciferous

**Table 1**  Important pests of oilseed *Brassica* crops

| Pest | Plant organ attacked | Plant form attacked | Area | Ref. |
|------|---------------------|---------------------|------|------|
| **Hemiptera** | | | | |
| *Lygus elisus* | Bud, flower, pod, seed | Annual | Canada | [a] |
| *Lygus lineolaris* | Bud, flower, pod, seed | Annual | Canada | [a] |
| *Bagrada cruciferarum* | Stem, leaf, pod | Annual | India | 10 |
| **Homoptera** | | | | |
| *Brevicoryne brassicae* | Leaf, stem, bud, flower, pod | Annual, biennial | NZ/Australia | 67 |
| *Lipaphis erysimi* | Leaf, stem, bud, flower, pod | Annual | India | 107, 108 |
| **Coleoptera** | | | | |
| *Meligethes aeneus* | Flower | Annual, biennial | Europe | 90, 122 |
| *Phyllotreta cruciferae* | Leaf, stem, pod, root | Annual | Canada, India | 10, 19, 65 |
| *Phyllotreta striolata* | Leaf, stem, pod, root | Annual | Canada, Europe | 7, 19 |
| *Phyllotreta undulata* | Leaf, stem | Annual | Europe | 7, 118 |
| *Psylliodes chrysocephala* | Stem, leaf | Biennial | Europe | 5, 112 |
| *Entomoscelis americana* | Leaf | Annual | Canada | 41, 42 |
| *Ceutorhynchus assimilis* | Pod | Annual, biennial | Europe | 37, 44 |
| *Ceutorhynchus quadridens* | Stem | Annual, biennial | Europe | 45, 98, 138 |
| **Lepidoptera** | | | | |
| *Pieris brassicae* | Leaf | Annual | India | 10 |
| *Diacrisia obliqua* | Leaf | Annual | India | 10 |
| *Mammestra configurata* | Leaf, pod | Annual | Canada | 14, 129 |
| *Plutella xylostella*[b] | Leaf | Annual | India, Canada, NZ/Australia | 10, 72, 93 |
| **Diptera** | | | | |
| *Dasineura brassicae* | Pod | Annual, biennial | Europe | 44, 112, 136 |
| *Phytomyza horticola*[c] | Leaf | Annual | India | 10, 101 |
| *Delia radicum* | Root | Annual, biennial | Canada, Europe | 47, 73, 118 |
| **Hymenoptera** | | | | |
| *Athalia proxima* | Leaf | Annual | India | 10, 53 |

[a]R. A. Butts & R. J. Lamb, unpublished data.
[b]*P. xylostella* = *P. maculipennis*.
[c]Also identified as *Phytomyza atricornis*.

crops (124). Similarly, *Lygus* bugs have a wide host range in North America (39, 109); although they are associated primarily with noncruciferous crops, they now also attack Canola (R. A. Butts & R. J. Lamb, unpublished data). Another native pest, the chrysomelid beetle *Entomoscelis americana,* expanded its host range from native crucifers to cruciferous vegetables and subsequently to Canola (124).

## Pest Damage in Relation to Crop Growth and Structure

Insects attack the roots, stems, leaves, flower beds, flowers, pods, and seeds of oilseed *Brassica* plants (Table 1). The impact on yield of the various kinds of damage is best understood in the context of plant growth and the components of yield.

PLANT GROWTH     Oilseed *Brassica* crops have small seeds (2.0–5.5 g per 1000 seeds), which are usually sown at a depth of 2–3 cm (29). A pair of cotyledons emerges 4–7 days after seeding. The seedling quickly produces the first pair of true leaves and then a rosette of broader leaves. At this stage, a biennial crop enters dormancy until spring. The annual crop continues to the next growth stage, which consists of rapid elongation of the stem, often to a height of 1 m or more (116). As the stem elongates, flower buds are produced first in a raceme at the tip of the main stem and subsequently on side branches, or axillary racemes. The pale-yellow, four-petaled flowers soon open, beginning at the base of each raceme (78). When a flower is fertilized, its petals drop and a cylindrical pod (silique) forms. Each mature pod contains 15–40 spherical seeds, which vary in color from yellow to dark brown depending on the cultivar. As pods form, the leaves begin to senesce, first at the base of the plant. Pods on different parts of a plant mature at different times, so the ripest may open and drop their seed prior to harvest. The duration of the growth period may be as short as 70–80 days for *B. campestris,* but this period varies considerably depending on the species, cultivar, growth form, and environment (25, 56, 79). Harper & Berkencamp (48) have provided a numerical key for the growth stages of *B. napus.*

The yield of the crop depends on plant density, the number of pods produced per plant, the number of seeds per pod, and the average weight of a seed.

PLANT DENSITY     Oilseed *Brassica* crops effectively compensate in seed yield for variation in plant density (40, 79). Reductions in plant density from 200 to 40 plants per m² result in a loss in yield of only 20% or less (79). Plants in sparse populations grow larger and produce more pods and more seeds per plant than densely seeded crops. Thus, pests that kill seedlings, such as the flea beetles, may not have as great an impact on yield as the plant mortality in a damaged field indicates. The yield loss resulting from seedling mortality depends on the proportion of the seedlings killed and also on the distribution of mortality. Damage to surviving seedlings, however, may have a greater impact on yield than the loss of seedlings. For example, an insecticide applied for flea beetle control, which permits a higher seedling survival, results in lower yields than one that is less effective at preventing mortality but protects the surviving seedlings for a longer time (62).

ROOTS    Minor insect damage to roots that does not interfere with their transport and anchoring functions may not be important. In Canada, larvae of *Phyllotreta* spp. are abundant root feeders (19), yet only about 5% yield loss results from densities of 1600 larvae per $m^2$ (17). Root maggots, *Delia* spp., commonly tunnel the surface of rape taproots, sometimes severing the root and killing plants (73), but the impact of these pests on yield has not been quantified. *Fusarium* fungi invade the tunnels and may be a factor in plant mortality (46).

STEMS    Stems provide an important photosynthetic surface, especially when seeds are developing (99), in addition to supporting leaves and transporting nutrients. Damage to leaf petioles and stems by *Psylliodes chrysocephala* reduces yield (30), but the nature of the damage in relation to the functions of stems has not been described. One consequence of damage to stems by *P. chrysocephala* is higher than normal levels of winter plant mortality in fields with heavy infestations (18).

LEAVES    The photosynthetic activity of leaves produces much of the assimilate needed by oilseed *Brassica* plants until they flower. Later, the stems and then the pods gradually replace the leaves as the important photosynthetic organs (99). Thus, defoliation at the beginning of flowering may be an important source of loss, but leaf loss becomes less significant as the crop matures (26, 94, 117). In Canada, the bertha armyworm, *M. configurata*, is the major defoliating pest. It prefers leaves to stems or pods and grows more rapidly and larger when fed leaves (13). However, larvae consume 94% of their food in the last two instars (8), which are reached at the end of the flowering period (15). By this stage, leaf feeding by larvae has little impact on yield; however, the larvae also attack pods, which does result in yield loss (13).

BUDS AND FLOWERS    Because oilseed *Brassica* plants may abort up to 50–60% of their flower buds and flowers, they can compensate for damage to these organs by retaining buds or flowers that would otherwise be aborted (78, 115, 135). Low or moderate levels of damage to buds and flowers caused by pollen beetles, *Meligethes* spp. (90, 113, 115, 122, 137), *E. americana* (41), and *Lygus* bugs (R. A. Butts & R. J. Lamb, unpublished data) have little impact on yield. In fact, such damage may benefit the plant by reducing shading of the photosynthetic surfaces by flowers (117). Although such a benefit has not been confirmed, low levels of bud damage by *Meligethes* spp. have been shown to increase yield slightly (90). There is a limit to the extent to which compensation can occur, however, and yield loss resulting from high levels of *Meligethes* damage to buds and flowers is common in Europe (90, 113, 122). Inability to compensate fully is in part a result of damage to the tips of the main shoot and side branches rather than just a loss of buds (90). The damage probably also causes a loss of assimilates destined for seed production, because some damaged plots yielded less than protected control plots although the former had more pods (90).

PODS    Damage to young pods may be compensated in the same way as damage to buds and flowers (135). However, as plants and pods mature and seeds develop, the plants' capacity to compensate for pod loss declines (78, 135). Furthermore, pods are important photosynthetic organs later in the life of the plant (99). Thus, loss of expanded pods as occurs when plants are attacked by bertha armyworm can cause a substantial yield loss (14).

SEEDS    Feeding damage to developing seeds may be compensated for by reduced natural abortion of seeds (94) and possibly by an increase in the size of surviving seeds, although neither has been documented. Seeds are sometimes larger in defoliated plants (26), but this type of compensation has not been observed consistently (94, 135). Plants do appear to compensate for seed damage by larvae of the weevil *Ceutorhynchus assimilis* and by the pod midge, *Dasineura brassicae*, but through an increase in the number of pods (115). However, this compensation may be for loss of pods resulting from pod damage caused by adult weevils rather than for seed loss to larvae. The level of compensation often is not sufficient to prevent yield loss. Determining the degree of compensation for seed loss is difficult because of the intimate relationship between seed growth and pod development (94). Furthermore, in Europe *D. brassicae* depends on pod damage by adult *C. assimilis* for oviposition sites (112).

Damage that kills plants or damage that has immediate effects on the number of racemes, flowers, pods, or seeds is more easily quantified than damage that has lasting effects on the growth or development of the crop. Many of the pests that cause well-defined damage to particular plant organs also have effects that cannot be explained in terms of the physiological characteristics of plant growth and components of yield. For example, the yield losses caused by flea beetles (62) are not all due to the seedling mortality they cause, nor do the losses of buds and flowers by *Meligethes* spp. fully account for the loss in yield (90). Similarly, aphid damage has a complex series of effects on the plant that are not yet fully understood (67).

From an agronomic point of view, the return to the producer depends not only on the yield, but also on the harvestability and quality of the seed. Pests such as *Phyllotreta* spp. (17, 62), *C. assimilis* (121), and *Meligethes* spp. (90) affect the harvestability and quality of the seed by delaying the maturation of damaged plants and causing the crop to mature unevenly. Slowing maturation delays harvest, which increases the probability of frost damage where the growing season is short (62). When the crop matures unevenly, seeds of the most mature plants are lost when pods ripen, open, and drop their seed before harvest. Furthermore, the amount of chlorophyll in the harvested seed increases because the late-maturing plants are harvested before all their pods are ripe, reducing the quality of the crop (14, 62, 90, 121). Insect damage can also reduce the oil content (90, 108, 121) and increase the glucosinolate content of seed (68).

## Brassica *Chemistry in Relation to Insect Pests*

Oilseed *Brassica* plants produce glucosinolates, a group of more than 70 compounds characteristic of the Cruciferae and a few other plant families (57, 97, 130). The glucosinolates are secondary plant compounds (81), and we know little of their metabolic function in the Cruciferae (97). They or their fission products, however, may function as antifungal agents (82), feeding deterrents, or toxins for some insect herbivores (11, 31) and as antifeedants or toxins for some wild vertebrate herbivores (103) and some farm animals (97). The glucosinolates are hydrolyzed by myrosinases to yield volatile isothyiocyanates, particularly when plant tissue is damaged (130). Most insects that attack the Cruciferae are attracted, stimulated to feed, or stimulated to oviposit by allyl isothiocyanate or its parent glucosinolate, sinigrin (33, 34, 51, 77, 86, 119, 120). Although they are attracted to and confined to crucifers as hosts, some insects find certain fission products of glucosinolates toxic (2). Cruciferous plants probably evolved glucosinolates and their fission products as defensive chemicals; insects that specialize in attacking crucifers are thought to have evolved ways to minimize the toxic effects of these chemicals and to have then adopted the glucosinolates as cues for the identification of their hosts (32).

Only a small proportion (approximately 12) of the known glucosinolates occur in the genus *Brassica* (50), but these may produce a complex array of volatile fission products (27, 59, 132). Interestingly, allyl glucosinolate, which yields the commonly used insect attractant allyl isothiocyanate (23, 34), is not found in *B. napus* and *B. campestris* (55, 59, 102). It occurs as a significant component of the glucosinolates in *B. juncea* and *B. oleracea* (102). Nevertheless, many of the insects that attack oilseed mustard and the cruciferous vegetables and that are attracted to allyl isothiocyanate also attack *B. napus* and *B. campestris* (Table 1; 12). Presumably other volatiles produced by *B. napus* and *B. campestris* serve as attractants and feeding stimulants, although these have not yet been identified. The volatile cues produced by oilseed *Brassica* crops and related crucifers are likely to be very complex. Not only are many volatile chemicals produced, but the composition of volatiles varies rapidly over time with the growth of the plant and to a lesser extent varies with the plant part assayed (27, 55, 59, 102). The analysis of volatiles emanating from *Brassica* plants has not yet revealed the diversity of compounds suggested by analysis of extracted plant tissues, but has clearly demonstrated that the amount and probably the type of volatiles varies considerably between intact and damaged foliage (35, 132). Because of their diversity, proven biological activity, and limited distribution in the plant kingdom, glucosinolates have been emphasized in research on the chemical cues for the insects that attack crucifers, but other plant chemicals such as flavonol glycosides also stimulate feeding, and cucurbitacins and cardenolides are feeding inhibitors for some of these insects (85, 87, 88).

Pests of oilseed *Brassica* crops such as the aphid *Myzus persicae* (84, 104), the bertha armyworm, *M. configurata* (129), and *Lygus* bugs (39, 109) are polyphagous. When feeding on *B. napus*, *M. persicae* excretes high concentrations of glucosinolates, whereas another aphid, *Brevicoryne brassicae*, which is a crucifer specialist, excretes lower concentrations of these chemicals and retains them in its body (133). *Brevicoryne brassicae* has a glucosinolase system, which hydrolyzes glucosinolates (75). The mechanisms that allow other generalist herbivores that do not exhibit these adaptations to exploit the oilseed *Brassica* crops are not known. Sinigrin and ethyl isothiocyanate are mild feeding deterrents for the polyphagous *Lygus lineolaris* (49), but this species attacks Canola in Canada (R. A. Butts & R. J. Lamb, unpublished data). Obviously, not all polyphagous herbivores are deterred from attacking oilseed *Brassica* plants by glucosinolates and their fission products.

The glucosinolates and their fission products are antifeedant or toxic for domesticated animals such as fowl and hogs (97). The presence of glucosinolates limited the utility of rapeseed meal until plant breeders produced cultivars with very low levels of the aliphatic glucosinolates in the meal (29). The cultivars with low glucosinolate levels in the seeds also have lower levels of the aliphatic glucosinolates in the young foliage, although the indolyl glucosinolates are still present at higher levels (131). Changing the chemical composition of oilseed *Brassica* crops could conceivably affect their attractiveness, palatability, or toxicity to crucifer specialists or potential polyphagous pests. However, although virtually the whole crop area in Canada (28) and a large part of the crop area in Europe have been converted to low-glucosinolate types (110), no changes in the insect pests have been reported that can be attributed to the introduction of the new cultivars. In studies of individual pests, no consistent differences in the feeding responses or damage levels have been detected for the low- and high-glucosinolate cultivars (1,43, 64, 71, 133). Although glucosinolate levels are now at very low levels in the seeds of some cultivars, levels in the foliage differ (102, 131), which may partly explain the continued attraction of crucifer specialists to the crop. Nevertheless, the low-glucosinolate seeds are no more suitable as food for polyphagous insects that attack seeds than seeds of the high-glucosinolate cultivars (74). One observed effect of the transition to low-glucosinolate cultivars has been an apparent increase in mortality of deer and hares that feed on rape foliage in Europe (103). The increased mortality in deer results from increased palatability and hence increased consumption of low-glucosinolate rape foliage, which still has an unidentified toxic factor (131b).

## THE INSECT COMMUNITY IN OILSEED *BRASSICA* CROPS

The insects living in oilseed *Brassica* crops represent a community that interacts with other agricultural and nonagricultural habitats. The most striking feature of the rape habitat, particularly in Europe and Canada, is the recent enormous increase in its size. In England in 1971 about 5000 ha were planted in the south-central part of the country, but by 1984 the crop covered 260,000 ha spread over much of the country (69, 134). In Canada in 1950 rape was grown on a few hundred hectares in Saskatchewan, but by 1979 a peak of 3.4 million hectares was reached, spreading across the three prairie provinces (65). The explosion in the number of *Brassica* plants has had a major impact on the insects associated with them. Before the oilseed *Brassica* crops became the dominant crucifers in these areas, the crucifer specialists fed on vegetable crops, weeds, or scattered native plants. The density of weed and wild crucifers in and around agricultural land is difficult to estimate, but the area devoted to cruciferous vegetables is known. In England the area devoted to rape is at least five times greater than that planted with *Brassica* vegetables (134), and in Canada the difference is much greater. Initially the oilseed *Brassica* crops probably formed a sink for cruciferous insects (134), but the densities of pests have responded to the increasing food resource; thus pest problems have become more and more serious on the oilseed *Brassica* crops and have overflowed onto other cruciferous crops as well (69, 134). Furthermore, the introduction of the oilseed *Brassica* crops has changed the temporal availability of cruciferous hosts (134), particularly in areas where both summer and winter forms are grown in adjacent fields (89, 122).

The spatial relationships between oilseed *Brassica* crops and adjacent communities affects the pests of rape, but this subject has received little attention. Some pests use host plants in habitats other than rape fields (21, 22, 38) and overwinter in other habitats (20, 128), but the importance of these alternative habitats in the pest dynamics is not known. Rape fields are usually much larger than those of vegetable crops, particularly in Canada, where Canola fields of 100 ha are common. In small-scale experimental studies, crucifer monocultures support higher densities of insect herbivores and suffer more damage per unit area than small patches of crucifers or polycultures (6, 100, 114). On the scale of commercial production, however, larger rape fields lose less yield per unit area to flea beetles than smaller ones (65).

Little is known about the diversity and structure of the arthropod communities in rape, although many of the individual pests have been studied and many of the predators and parasites of these insects have been identified. The biomass of arthropods was found to be lower in a winter rape field than in a meadow (83). The densities of general predators such as spiders are low in rape fields because they must reinvade the large fields each season (131a). Some specialized predators use the same olfactory cues as their hosts to locate the host habitat (96), and thus may be less affected by spatial characteristics of the crop or temporal variation in its availability. Interactions between herbivores are beginning to be identified. For example, as mentioned above, *D. brassicae* is dependent on pod damage by *C. assimilis* for oviposition sites (3, 112).

## PEST MANAGEMENT

With the recent increase in the area devoted to the oilseed *Brassica* crops in Europe and North America and the increasing importance of pests to crop production, much research has been conducted on methods of reducing losses in yield due to pests. Insecticides are applied to control the most important pests: aphids (24, 67), *M. configurata* (129), *Phyllotreta* spp. (65, 118), *P. chrysocephala* (4), *Meligethes aeneus* (90, 122), and *C. assimilis* (121). Pests that attack the early growth stages of the crop are usually controlled with systemic insecticides applied as seed dressings or in-furrow granules (4, 65, 67, 118). Pests that attack at the time of stem elongation or later require one (121, 129) or more (122) foliar sprays. In some cases pesticide applications for one pest also reduce the damage caused by secondary pests (7). Pesticide resistance, pest resurgence, and outbreaks of secondary pests have not yet been reported to result from insecticide applications for control in rape.

*Phyllotreta* spp. in Canada and *M. aeneus* in parts of Europe can cause severe damage annually over wide areas, and insecticides are applied routinely to control these pests. Other pests are more sporadic (66), and economic thresholds in conjunction with pest monitoring have been used to minimize the use of insecticides (105). For example, in the control of *M. configurata*, an economic threshold for larvae was developed in caged field trials (14, 15) and then verified in commercial fields (16). An area-wide pheromone-monitoring system was devised to warn of pest outbreaks (125), and larval sampling methods were tested that may enable farmers to assess pest densities efficiently in individual fields to determine if control is necessary (127). Economic thresholds have also been defined for *M. aeneus* (90), *C. assimilis* (36, 121), and flea beetles in Finland (7). Pest monitoring and economic thresholds are not effective tools for flea beetle control in Canada because the invasion of the newly emerged crops is sudden and weather dependent. Seedling mortality can reach high levels before farmers are able to assess pest densities or apply insecticides (61).

Well-planned insecticide applications are economically justified, but nevertheless costly. A number of alternative control measures have been investigated for the important pests of oilseed *Brassica* crops. Trap cropping is an effective control for *M. aeneus* in Finland (52). In Canada, fall cultivation contributes to the control of *E. americana* (42) and *M. configurata* (126). Biological control of *Phyllotreta* spp. has been attempted in Canada (139). Host-plant resistance is being investigated for aphids in India (9, 76) and for *Phyllotreta* spp. in Canada (60, 63).

## CONCLUSIONS

The entomology of the oilseed *Brassica* crops is in an early stage of development (70). The recent enormous expansion of the crop area has led to the establishment of a new agricultural ecosystem with a novel pest complex. Although many of the pests of this crop are also well-known pests of cruciferous vegetables, their status in rape was previously unknown. A consensus is only now being reached on which insects are the primary pests. Furthermore, the drastic difference in scale and in value per unit area between rape and the cruciferous vegetable crops assures that the ecology of the pests and the strategies adopted to control them will be substantially different. The oilseed *Brassica* crop has been qualitatively dynamic as it has become established and has expanded. The three *Brassica* species vary spatially and temporally in their dominance, as do the annual and biennial forms of the crop in certain parts of Europe. As the crop has become established there has been a major transformation in the composition of the secondary plant compounds of the crop, which is still under way. This dynamic crop has provided many challenges and opportunities for entomological research.

The future is sure to bring new entomological challenges as plant breeders and agronomists continue to transform the crop. At the moment, research is under way to develop hybrid Canola (106). If a suitable commercial method of producing hybrid seed becomes available, the yield per unit area will increase substantially and the economics of pest control will be transformed accordingly. Furthermore, the hybrid plants may grow more vigorously and may compensate better for insect damage. Research is also under way to develop *B. juncea* as a Canola crop for North America. Production of *B. juncea* could extend the crop into the hotter and drier areas of the North American prairies, a new habitat for many crucifer pests. The chemical characteristics of the crop will continue to change as plant breeders tailor cultivars to the oilseed markets. The ability to modify the glucosinolate content of foliage through plant breeding opens up the possibility of developing a rape crop lacking the attractants and feeding stimuli used by the cruciferous pests. However, much more needs to be learned about the orientation cues and feeding stimuli used by these insects before this goal can be achieved. Mixed cropping systems of field peas and Canola are being tested in Canada because they are thought to have agronomic advantages and because the seed of the two crops can be readily separated. This practice is likely to have a substantial effect on the pest complex in Canola if experimental results on pest densities in polycultures (6, 100, 114) are borne out.

The insect pests of oilseed *Brassica* crops are sufficiently important that continued research will be demanded to minimize the damage they cause. Because of the cropping area involved and the annual requirements for large amounts of insecticides to control the pests, there will be a concerted effort to discover alternative methods of control. Given the intimate relationship between many of the pests and the host plants, and given our increasing knowledge of the secondary compounds of the *Brassicas* plants, the chemical aspect of the insect-plant relationship will undoubtedly be the starting point for developing alternative control measures.

ACKNOWLEDGMENTS

I thank C. E. Shearer for helping prepare a bibliography on the crop and its pests and for checking the manuscript. L. Burgess provided many of the references and advice as the project developed. R. Bodnaryk, P. Pachagounder, and W. J. Turnock provided useful comments on the manuscript.

This paper is Contribution No. 1317, Agriculture Canada Research Station, Winnipeg.

*Literature Cited*

1. Åhman, I. 1982. A comparison between high and low glucosinolate cultivars of summer oilseed rape (*Brassica napus* L.) with regard to their levels of infestation by the brassica pod midge (*Dasineura brassicae* Winn.). *Z. Angew. Entomol.* 94:103–9

2. Åhman, I. 1986. Toxicities of host secondary compounds to eggs of *Brassica* specialist *Dasineura brassicae*. *J. Chem. Ecol.* 12:1481–88

3. Åhman, I. 1987. Oviposition site characteristics of *Dasineura brassicae* Winn. (Dipt., Cecidomyiidae). *Z. Angew. Entomol.* 104:85–91

4. Alford, D. V. 1977. Chemical control of the cabbage stem flea beetle, *Psylliodes chrysocephala*, on winter oil-seed rape. *Ann. Appl. Biol.* 85:369–74

5. Alford, D. V. 1979. Observations on the cabbage stem flea beetle, *Psylliodes chrysocephala*, on winter oil-seed rape in Cambridgeshire. *Ann. Appl. Biol.* 93:117–23

6. Altieri, M. A., Gliessman, S. R. 1983. Effects of plant diversity on the density and herbivory of the flea beetle, *Phyllotreta cruciferae* Goeze, in California collard *(Brassica oleracea)* cropping systems. *Crop Prot.* 2:497–501

7. Augustin, A., Tulisalo, U., Korpela, S. 1986. Flea beetles (Coleoptera, Chrysomelidae, Halticinae) on rapeseed and sugarbeet in Finland. *J. Agric. Sci. Finl.* 58:69–82

8. Bailey, C. G., Singh, N. B. 1977. An energy budget for *Mamestra configurata* (Lepidoptera: Noctuidae). *Can. Entomol.* 109:687–93

9. Bakhetia, D. R. C. 1980. Breeding for insect-pest resistance in oilseed crops. In *Breeding Oilseed Crops*, ed. K. S. Gill, pp. 120–37. Ludhiana, India: Punjab Agric. Univ. 479 pp.

10. Bakhetia, D. R. C., Labana, K. S. 1978. Insect resistance in *Brassica* crops. *Crop Improv.* 5:95–103

11. Blau, P. A., Feeny, P., Contardo, L., Robson, D. S. 1978. Allylglucosinolate and herbivorous caterpillars: a contrast in toxicity and tolerance. *Science* 200:1296–98

12. Bonnemaison, L. 1965. Insect pests of crucifers and their control. *Ann. Rev. Entomol.* 10:233–56

13. Bracken, G. K. 1984. Within plant preferences of larvae of *Mamestra configurata* (Lepidoptera: Noctuidae) feeding on oilseed rape. *Can. Entomol.* 116:45–49

14. Bracken, G. K. 1987. Relation between pod damage caused by larvae of bertha armyworm, *Mamestra configurata* Walker (Lepidoptera: Noctuidae), and yield loss, shelling, and seed quality in Canola. *Can. Entomol.* 119:365–69

15. Bracken, G. K., Bucher, G. E. 1977. An estimate of the relation between density of bertha armyworm and yield loss on rapeseed, based on artificial infestations. *J. Econ. Entomol.* 70:701–5

16. Bracken, G. K., Bucher, G. E. 1984. Measuring the cost-benefit of control measures for bertha armyworm (Lepidoptera: Noctuidae) infestations in rapeseed. *Can. Entomol.* 116:591–95

17. Bracken, G. K., Bucher, G. E. 1986. Yield losses in Canola caused by adult and larval flea beetles, *Phyllotreta cruciferae* (Coleoptera: Chrysomelidae). *Can. Entomol.* 118:319–24

18. Buhl, K. 1959. Observations and investigations on the bionomics of the rape flea-beetle *(Psylliodes chrysocephala* L.) in Schleswig-Holstein. *Z. Pflanzenkr. Pflanzenschutz* 66:321–38 (In German)

19. Burgess, L. 1977. Flea beetles (Coleoptera: Chrysomelidae) attacking rape crops in the Canadian prairie provinces. *Can. Entomol.* 109:21–32

20. Burgess, L. 1981. Winter sampling to determine overwintering sites and estimate density of adult flea beetle pests of rape (Coleoptera: Chrysomelidae). *Can. Entomol.* 113:441–47

21. Burgess, L. 1982. Occurrence of some flea beetle pests of parkland rapeseed crops in open prairie and forest in Saskatchewan (Coleoptera: Chrysomelidae). *Can. Entomol.* 114:623–27

22. Burgess, L. 1984. Changes in the relative abundance of the flea beetles *Phyllotreta striolata* and *Phyllotreta cruciferae* (Coleoptera: Chrysomelidae) in Saskatchewan parkland with increasing distance from the boreal forest. *Can. Entomol.* 116:653–56

23. Burgess, L., Wiens, J. E. 1980. Dispensing allyl isothiocyanate as an attractant for trapping crucifer-feeding flea beetles. *Can. Entomol.* 112:93–97

24. Butani, D. K. 1974. Effect of various insecticides on the mustard aphid, *Lipaphis erysimi* (Kalt.) and yield of rape, *Brassica campestris* Linn. *Indian J. Entomol.* 36:243–46

25. Chauhan, Y. S., Bhargava, S. C. 1984. Physiological analysis of growth and yield variation of rapeseed and mustard. *J. Agric. Sci.* 103:249–52

26. Clarke, J. M. 1978. The effects of leaf removal on yield and yield components of *Brassica napus*. *Can. J. Plant Sci.* 58:1103–5

27. Cole, R. A. 1980. Volatile components produced during ontogeny of some cultivated crucifers. *J. Sci. Food Agric.* 31:549–57

28. Daun, J. K. 1984. Composition and use of Canola seed, oil, and meal. *Cereal Foods World* 29:291–96

29. Downey, R. K. 1983. The origin and description of the *Brassica* oilseed crops. In *High and Low Erucic Acid Rapeseed Oils: Production, Usage, Chemistry, and Toxicological Evaluation*, ed. J. K. G. Kramer, F. D. Sauer, W. J. Pigden, pp. 1–20. New York: Academic. 582 pp.

30. Ebbe-Nyman, E. 1952. The rape flea beetle *Psylliodes chrysocephala* L. Contributions to the knowledge of its biology and control. *Statens Vaxtskyddsanst. Medd.* 63:96–103 (In Swedish)

31. Erickson, J. M., Feeny, P. 1974. Sinigrin: a chemical barrier to the black swallowtail butterfly, *Papilio polyxenes*. *Ecology* 55:103–11

32. Feeny, P. 1977. Defensive ecology of

the Cruciferae. *Ann. Mo. Bot. Gard.* 64:221–34

33. Feeny, P., Paauwe, K. L., Demong, N. J. 1970. Flea beetles and mustard oils: host plant specificity of *Phyllotreta cruciferae* and *P. striolata* adults (Coleoptera: Chrysomelidae). *Ann. Entomol. Soc. Am.* 63:832–41

34. Finch, S. 1977. Effect of secondary plant substances on host-plant selection by the cabbage root fly. *Colloq. Int. CNRS* 265:251–67

35. Finch, S. 1978. Volatile plant chemicals and their effect on host plant finding by the cabbage root fly *(Delia brassicae)*. *Entomol. Exp. Appl.* 24:150–59

36. Free, J. B., Ferguson, A. W., Winfield, S. 1983. Effect of various levels of infestation by the seed weevil *(Ceuthorrhynchus assimilis* Payk.) on the seed yield of oil-seed rape *(Brassica napus* L.). *J. Agric. Sci.* 101:589–96

37. Free, J. B., Williams, I. H. 1978. A survey of the damage caused to crops of oil-seed rape *(Brassica napus* L.) by insect pests in south-central England and their effect on seed yield. *J. Agric. Sci.* 90:417–24

38. Free, J. B., Williams, I. H. 1978. The responses of the pollen beetle, *Meligethes aeneus*, and the seed weevil, *Ceuthorrhynchus assimilis*, to oil-seed rape, *Brassica napus*, and other plants. *J. Appl. Ecol.* 15:761–74

39. Fye, R. E. 1982. Damage to vegetable and forage seedlings by the pale legume bug (Hemiptera: Miridae). *J. Econ. Entomol.* 75:994–96

40. Geisler, G., Stoy, A. 1987. Investigations on the effect of number of plants/area on the yield potential of rapeseed plants *(Brassica napus* L. var. *napus)*. *J. Agron. Crop Sci.* 159:232–40 (In German)

41. Gerber, G. H. 1976. Effects of feeding by adults of the red turnip beetle, *Entomoscelis americana* Brown (Coleoptera: Chrysomelidae), during late July and August on the yield of rapeseed (Cruciferae). *Manit. Entomol.* 10:31–35

42. Gerber, G. H. 1978. Effects of burying the eggs in soil on survival in the red turnip beetle, *Entomoscelis americana* (Coleoptera: Chrysomelidae). *Manit. Entomol.* 12:49–51

43. Gerber, G. H., Obadofin, A. A. 1981. Growth, development, and survival of the larvae of the red turnip beetle, *Entomoscelis americana* (Coleoptera: Chrysomelidae), on *Brassica campestris* and *B. napus* (Cruciferae). *Can. Entomol.* 113:395–406

44. Gould, H. J. 1975. Surveys of pest incidence on oil-seed rape in south central England. *Ann. Appl. Biol.* 79:19–26

45. Graham, C. W., Gould, H. J. 1980. Cabbage stem weevil *(Ceuthorrhynchus quadridens)* on spring oilseed rape in southern England and its control. *Ann. Appl. Biol.* 95:1–10

46. Griffiths, G. C. D. 1986. Phenology and dispersion of *Delia radicum* (L.) (Diptera: Anthomyiidae) in Canola fields at Morinville, Alberta. *Quaest. Entomol.* 22:29–50

47. Griffiths, G. C. D. 1986. Relative abundance of the root maggots *Delia radicum* (L.) and *D. floralis* (Fallén) (Diptera: Anthomyiidae) as pests of Canola in Alberta. *Quaest. Entomol.* 22:253–60

48. Harper, F. R., Berkenkamp, B. 1975. Revised growth-stage key for *Brassica campestris* and *B. napus*. *Can. J. Plant Sci.* 55:657–58

49. Hatfield, L. D., Frazier, J. L., Ferreira, J. 1982. Gustatory discrimination of sugars, amino acids, and selected allelochemicals by the tarnished plant bug, *Lygus lineolaris*. *Physiol. Entomol.* 7: 15–23

50. Heaney, R. K., Fenwick, G. R. 1980. The glucosinolate content of *Brassica* vegetables. A chemotaxonomic approach to cultivar identification. *J. Sci. Food Agric.* 31:794–801

51. Hicks, K. L. 1974. Mustard oil glucosides: feeding stimulants for adult cabbage flea beetles, *Phyllotreta cruciferae* (Coleoptera: Chrysomelidae). *Ann. Entomol. Soc. Am.* 67:261–64

52. Hokkanen, H., Granlund, H., Husberg, G.-B., Markkula, M. 1986. Trap crops used successfully to control *Meligethes aeneus* (Col., Nitidulidae), the rape blossom beetle. *Ann. Entomol. Fenn.* 52:115–20

53. Hussain, T. 1980. Bionomics of mustard sawfly *Athalia proxima* Klug. (Hymenoptera: Tenthridinidae) on *Brassica campestris*. *Pak. J. Zool.* 12:127–30

54. Jermy, T. 1984. Evolution of insect/host plant relationships. *Am. Nat.* 124:609–30

55. Josefsson, E. 1967. Distribution of thioglucosides in different parts of *Brassica* plants. *Phytochemistry* 6:1617–27

56. Kasa, G. R., Kondra, Z. P. 1986. Growth analysis of spring-type oilseed rape. *Field Crops Res.* 14:361–70

57. Kjaer, A. 1976. Glucosinolates in the Cruciferae. In *The Biology and Chemistry of the Cruciferae*, ed. J. G. Vaughan, A. J. MacLeod, B. M. G. Jones, pp. 207–19. New York: Academic. 355 pp.

58. Kogan, M., Turnipseed, S. G. 1987.

Ecology and management of soybean arthropods. *Ann. Rev. Entomol.* 32:507–38

59. Kondo, H., Kawaguchi, T., Naoshima, Y., Nozaki, H. 1985. Changes in volatile components of rape seeds (*Brassica napus* L.) during germination. *Agric. Biol. Chem.* 49:217–19

60. Lamb, R. J. 1980. Hairs protect pods of mustard (*Brassica hirta* 'Gisilba') from flea beetle feeding damage. *Can. J. Plant Sci.* 60:1439–40

61. Lamb, R. J. 1983. Phenology of flea beetle (Coleoptera: Chrysomelidae) flight in relation to their invasion of Canola fields in Manitoba. *Can. Entomol.* 115:1493–502

62. Lamb, R. J. 1984. Effects of flea beetles, *Phyllotreta* spp. (Chrysomelidae: Coleoptera), on the survival, growth, seed yield and quality of Canola, rape and yellow mustard. *Can. Entomol.* 116:269–80

63. Lamb, R. J. 1988. Assessing the susceptibility of crucifer seedlings to flea beetle (*Phyllotreta* spp.) damage. *Can. J. Plant Sci.* 68:85–93

64. Lamb, R. J. 1988. Susceptibility of low- and high-glucosinolate oilseed rapes to damage by flea beetles, *Phyllotreta* spp. (Coleoptera: Chrysomelidae). *Can. Entomol.* 120:195–96

65. Lamb, R. J., Turnock, W. J. 1982. Economics of insecticidal control of flea beetles (Coleoptera: Chrysomelidae) attacking rape in Canada. *Can. Entomol.* 114:827–40

66. Lamb, R. J., Turnock, W. J., Hayhoe, H. N. 1985. Winter survival and outbreaks of bertha armyworm, *Mamestra configurata* (Lepidoptera: Noctuidae), on Canola. *Can. Entomol.* 117:727–36

67. Lammerink, J., Banfield, R. A. 1980. Effect of aphid control by disulfoton on seed yield components and seed quality of oilseed rape. *NZ J. Exp. Agric.* 8:45–48

68. Lammerink, J., MacGibbon, D. B., Wallace, A. R. 1984. Effect of the cabbage aphid (*Brevicoryne brassicae*) on total glucosinolate in the seed of oilseed rape (*Brassica napus*). *NZ J. Agric. Res.* 27:89–92

69. Lane, A. B. 1983. Benefits and hazards of new crops: oilseed rape in the U. K. *Agric. Ecosyst. Environ.* 10:299–309

70. Lane, A. B. 1984. An inquiry into the response of growers to attacks by insect pests in oilseed rape (*Brassica napus* L.), a relatively new crop in the United Kingdom. *Prot. Ecol.* 7:73–78

71. Larsen, L. M., Nielsen, J. K., Plöger, A., Sørensen, H. 1985. Responses of some beetle species to varieties of oilseed rape and to pure glucosinolates. In *Advances in the Production and Utilization of Cruciferous Crops,* ed. H. Sørenson, pp. 230–44. Dordrecht, the Netherlands: Nijhoff/Junk. 317 pp.

72. Lee, H. S. 1968. Evaluation of some granulated insecticides against the aphids and diamond-back moth on rapeseed by soil treatment. *Plant Prot. Bull.* 10:69–70 (In Chinese)

73. Liu, H. J., Butts, R. A. 1982. *Delia* spp. (Diptera: Anthomyiidae) infesting Canola in Alberta. *Can. Entomol.* 114:651–53

74. Loschiavo, S. R., Lamb, R. J. 1985. Food preferences, survival, and development of four stored-product pests (Coleoptera) on rapeseed and Canola (*Brassica* spp.). *Can. Entomol.* 117:575–80

75. MacGibbon, D. B., Allison, R. M. 1968. A glucosinolase system in the aphid *Brevicoryne brassicae. NZ J. Sci.* 11:440–46

76. Malik, R. S. 1981. Morphological, anatomical and biochemical basis of aphid, *Lipaphis erysimi* Kalt., resistance in cruciferous species. *Sver. Utsädesfören Tidskr.* 91:25–35

77. Matsumoto, Y. 1970. Volatile organic sulphur compounds as insect attractants with special reference to host selection. In *Control of Insect Behavior by Natural Products,* ed. D. L. Wood, R. M. Silverstein, M. Nakajima, pp. 133–60. New York: Academic. 345 pp.

78. McGregor, D. I. 1981. Pattern of flower and pod development in rapeseed. *Can. J. Plant Sci.* 61:275–82

79. McGregor, D. I. 1987. Effect of plant density on development and yield of rapeseed and its significance to recovery from hail injury. *Can. J. Plant Sci.* 67:43–51

80. McNaughton, S. J. 1983. Compensatory plant growth as a response to herbivory. *Oikos* 40:329–36

81. Metcalfe, R. L. 1987. Plant volatiles as insect attractants. *CRC Crit. Rev. Plant Sci.* 5:251–301

82. Mithen, R. F., Lewis, B. G., Heaney, R. K., Fenwick, G. R. 1987. Glucosinolates of wild and cultivated *Brassica* species. *Phytochemistry* 26:1969–73

83. Nabiałczyk-Karg, J. 1980. Density and biomass of soil inhabiting insect larvae in a rape field and in a meadow. *Pol. Ecol. Stud.* 6:305–16

84. Nault, L. R., Styer, W. E. 1972. Effects of sinigrin on host selection by aphids. *Entomol. Exp. Appl.* 15:423–37

85. Nielsen, J. K. 1978. Host plant dis-

crimination within Cruciferae: feeding responses of four leaf beetles (Coleoptera: Chrysomelidae) to glucosinolates, cucurbitacins and cardenolides. *Entomol. Exp. Appl.* 24:41–54

86. Nielsen, J. K., Dalgaard, L., Larsen, L. M., Sørensen, H. 1979. Host plant selection of the horse-radish flea beetle *Phyllotreta armoraciae* (Coleoptera: Chrysomelidae): feeding responses to glucosinolates from several crucifers. *Entomol. Exp. Appl.* 25:227–39

87. Nielsen, J. K., Larsen, L. M., Sørenson, H. 1977. Cucurbitacin E and I in *Iberis amara:* feeding inhibitors for *Phyllotreta nemorum*. *Phytochemistry* 16:1519–22

88. Nielsen, J. K., Larsen, L. M., Sørensen, H. 1979. Host plant selection of the horseradish flea beetle *Phyllotreta armoraciae* (Coleoptera: Chrysomelidae): identification of two flavonol glycosides stimulating feeding in combination with glucosinolates. *Entomol. Exp. Appl.* 26:40–48

89. Nillson, C. 1975. Damage of brassica pod midge to winter and summer oil seed crops in central Sweden. *Statens Vaxtskyddsanst. Medd.* 16:61–71 (In Swedish)

90. Nilsson, C. 1987. Yield losses in summer rape caused by pollen beetles (*Meligethes* spp.). *Swed. J. Agric. Res.* 17:105–11

91. Deleted in proof

92. Deleted in proof

93. Palaniswamy, P., Gillott, C., Slater, G. P. 1986. Attraction of diamondback moths, *Plutella xylostella* (L.) (Lepidoptera: Plutellidae), by volatile compounds of Canola, white mustard, and faba bean. *Can. Entomol.* 118:1279–85

94. Pechan, P. A., Morgan, D. G. 1985. Defoliation and its effects on pod and seed development in oil seed rape (*Brassica napus* L.) *J. Exp. Bot.* 36:458–68

95. Prakash, S., Hinata, K. 1980. Taxonomy, cytogenetics and origin of crop brassicas, a review. *Opera Bot.* 55:1–57

96. Read, D. P., Feeny, P. P., Root, R. B. 1970. Habitat selection by the aphid parasite *Diaeretiella rapae* (Hymenoptera: Braconidae) and hyperparasite *Charips brassicae* (Hymenoptera: Cynipidae). *Can. Entomol.* 102:1567–78

97. Röbbelen, G., Thies, W. 1980. Variation in rapeseed glucosinolates and breeding for improved meal quality. In *Brassica Crops and Wild Allies, Biology and Breeding*, ed. S. Tsunoda, K. Hinata, C. Gomez-Campo, pp. 285–99. Tokyo: Jpn Sci. Soc. 354 pp.

98. Roivainen, S. 1957. On the distribution

of *Ceuthorrhynchus* species (Col., Curculionidae) occurring as pests of cruciferous plants in Finland. *Ann. Entomol. Fenn.* 23:111–14

99. Rood, S. B., Major, D. J., Charnetski, W. A. 1984. Seasonal changes in $^{14}CO_2$ assimilation and $^{14}C$ translocation in oilseed rape. *Field Crops Res.* 8:341–48

100. Root, R. B. 1973. Organization of a plant-arthropod association in simple and diverse habitats: the fauna of collards *(Brassica oleracea)*. *Ecol. Monogr.* 43:95–124

101. Sandhu, G. S., Brar, K. S. 1975. Evaluation of techniques for the screening of different varieties of *Brassica* species to pea leaf-miner *Phytomyza atricornis* (Meigen) Agromyzidae: Diptera. *Oilseeds J.* 4:14–15

102. Sang, J. P., Minchinton, I. R., Johnstone, P. K., Truscott, R. J. W. 1984. Glucosinolate profiles in the seed, root and leaf tissue of cabbage, mustard, rapeseed, radish and swede. *Can. J. Plant Sci.* 64:77–93

103. Schellner, H.-P. 1987. Rape poisoning as a possible cause of death in hares and roe deer. *Tieraerztl. Umsch.* 42:1–6 (In German)

104. Schoonhoven, L. M., Derksen-Koppers, I. 1976. Effects of some allelochemics on food uptake and survival of a polyphagous aphid, *Myzus persicae*. *Entomol. Exp. Appl.* 19:52–56

105. Schott, H. 1961. On the methodology for prognoses for the occurrence of the cabbage stem flea beetle (*Psylliodes chrysocephala* L.). *Nachrichtenbl. Dtsch. Pflanzenschutzdienst Berlin* 15:234–39 (In German)

106. Sernyk, J. L., Stefansson, B. R. 1983. Heterosis in summer rape (*Brassica napus* L.). *Can. J. Plant Sci.* 63:407–13

107. Singh, O. P., Dhamdhere, S. V., Nema, K. K. 1983. Effect of different parts of mustard plant on the development of mustard aphid, *Lipaphis erysimi* (Kalt.). *Agric. Sci. Dig.* 3:5–7

108. Singhvi, S. M., Verma, N. D., Yadava, T. P. 1973. Estimation of losses in rapeseed (*Brassica campestris* L. var. *Toria*) and mustard (*Brassica juncea* Coss.) due to mustard aphid (*Lipaphis erysimi* Kalt.). *Horyana Agric. Univ. J. Res.* 3:5–7

109. Snodgrass, G. L., Scott, W. P., Smith, J. W. 1984. Host plants and seasonal distribution of the tarnished plant bug (Hemiptera: Miridae) in the delta of Arkansas, Louisiana, and Mississippi. *Environ. Entomol.* 13:110–16

110. Stefansson, B. R. 1983. The develop-

ment of improved rapeseed cultivars. See Ref. 29, pp. 143–59

111. Strong, D. R., Lawton, J. H., Southwood, T. R. E. 1984. *Insects on Plants, Community Patterns and Mechanisms.* Cambridge, Mass: Harvard Univ. Press. 313 pp.

112. Sylvén, E., Svenson, G. 1975. Relationship between density of *Ceuthorrhynchus assimilis* Payk. (Col.) and damage by *Dasineura brassicae* Winn. (Cec.) in a cage experiment in summer turnip rape. *Statens Vaxtskyddsanst. Medd.* 16: 53–60

113. Sylvén, E., Svenson, G. 1976. Effect on yield of damage caused by *Meligethes aeneus* F. (Col.) to winter rape. As indicated by cage experiments. *Ann. Agric. Fenn.* 15:24–33

114. Tahvanainen, J. O., Root, R. B. 1972. The influence of vegetational diversity on the population ecology of a specialized herbivore, *Phyllotreta cruciferae* (Coleoptera: Chrysomelidae). *Oecologia* 10:321–46

115. Tatchell, G. M. 1983. Compensation in spring-sown oil-seed rape (*Brassica napus* L.) plants in response to injury to their flower buds and pods. *J. Agric. Sci.* 101:565–73

116. Tayo, T. O., Morgan, D. G. 1975. Quantitative analysis of the growth, development and distribution of flowers and pods in oil seed rape (*Brassica napus* L.). *J. Agric. Sci.* 85:103–10

117. Tayo, T. O., Morgan, D. G. 1979. Factors influencing flower and pod development in oil-seed rape (*Brassica napus* L.). *J. Agric. Sci.* 92:363–73

118. Tiittanen, K., Varis, A.-L. 1960. The treatment of seeds of swede, turnip, and turnip rape in the control of flea beetles (*Phyllotreta* spp.) and cabbage root flies (*Hylemyia brassicae* Bouché and *H. floralis* Fall.). *Valt. Maatalouskoetoiminnan Julk.* 181:1–11

119. Traynier, R. M. M. 1967. Effect of host plant odour on the behaviour of the adult cabbage root fly, *Erioischia brassicae.* *Entomol. Exp. Appl.* 10:321–28

120. Traynier, R. M. M. 1967. Stimulation of oviposition by the cabbage root fly *Erioischia brassicae.* *Entomol. Exp. Appl.* 10:401–12

121. Tulisalo, U., Korpela, S., Pohto, A. 1976. The yield loss caused by the seed-pod weevil *Ceuthorrhynchus assimilis* Payk. (Col., Curculionidae) on summer turnip rape in cage experiments. *Ann. Entomol. Fenn.* 42:98–102

122. Tulisalo, U., Wuori, T. 1986. Blossom beetle (*Meligethes aeneus* Fab.) as a yield factor in turnip rape (*Brassica*

*campestris* L.). *J. Agric. Sci. Finl.* 58:221–37

123. Turnipseed, S. G., Kogan, M. 1976. Soybean entomology. *Ann. Rev. Entomol.* 21:247–82

124. Turnock, W. J. 1977. Adaptability and stability of insect pest populations in prairie agricultural ecosystems. In *Insect Ecology—Papers Presented in the A. C. Hodson Lectures, Tech. Bull. 310,* ed. H. M. Kulman, H. C. Chiang, pp. 89–101. Minneapolis, Minn: Univ. Agric. Exp. Stn.

125. Turnock, W. J. 1987. Predicting larval abundance of the bertha armyworm, *Mamestra configurata* Wlk., in Manitoba from catches of male moths in sex attractant traps. *Can. Entomol.* 119: 167–78

126. Turnock, W. J., Bilodeau, R. J. 1984. Survival of pupae of *Mamestra configurata* (Lepidoptera: Noctuidae) and two of its parasites in untilled and tilled soil. *Can. Entomol.* 116:257–67

127. Turnock, W. J., Bilodeau, R. J. 1985. A comparison of three methods of examining the density of larvae of the bertha armyworm, *Mamestra configurata,* in fields of Canola (*Brassica* spp.). *Can. Entomol.* 117:1065–66

128. Turnock, W. J., Lamb, R. J., Bilodeau, R. J. 1987. Abundance, winter survival, and spring emergence of flea beetles (Coleoptera: Chrysomelidae) in a Manitoba grove. *Can. Entomol.* 119:419–26

129. Turnock, W. J., Philip, H. G. 1977. The outbreak of bertha armyworm *Mamestra configurata* (Noctuidae: Lepidoptera), in Alberta, 1971 to 1975. *Manit. Entomol.* 11:10–21

130. Underhill, E. W. 1980. Glucosinolates. In *Encyclopedia of Plant Physiology,* Vol. 8, *Secondary Plant Products,* eds. E. A. Bell, B. V. Charlwood, pp. 493–511. New York: Springer-Verlag. 674 pp.

131. Uppström, B. 1983. Glucosinolate pattern in different growth stages of high and low glucosinolate varieties of *Brassica napus.* *Sver. Utsädesfören. Tidskr.* 93:331–36

131a. Von Nyffeler, M., Benz, G. 1979. Studies on the ecological importance of spider populations for the vegetation of cereal and rape fields. *Z. Angew. Entomol.* 87:348–76 (In German)

131b. Von Onderscheka, K., Tataruch, F., Steineck, T., Klansek, E., Vodnansky, M., Wagner, J. 1987. Increased losses of roe deer after consumption of rape. *Z. Jagdwiss.* 33:139–42 (In German)

132. Wallbank, B. E., Wheatley, G. A. 1976. Volatile constituents from cauli-

flower and other crucifers. *Phytochemistry* 15:763–66

133. Weber, G., Oswald, S., Zöllner, U. 1986. Suitability of rape cultivars with a different glucosinolate content for *Brevicoryne brassicae* (L.) and *Myzus persicae* (Sulzer) (Hemiptera, Aphididae). *Z. Pflanzenkr. Pflanzenschutz* 93:113–24 (In German)

134. Wheatley, G. A., Finch, S. 1984. Effects of oilseed rape on the status of insect pests of vegetable brassicas. In *Pests and Diseases, Br. Crop Prot. Conf. Brighton, UK,* 2:807–14. Croydon, UK: Br. Crop. Prot. Counc.

135. Williams, I. H., Free, J. B. 1979. Compensation of oil-seed rape (*Brassica napus* L.) plants after damage to their buds and pods. *J. Agric. Sci.* 92:53–59

136. Williams, I. H., Martin, A. P., Kelm, M. 1987. The phenology of the emergence of brassica pod midge (*Dasineura brassicae* Winn.) and its infestation of winter oil-seed rape (*Brassica napus* L.). *J. Agric. Sci.* 108:579–89

137. Winfield, A. L. 1961. Field observations on the control of blossom beetles (*Meligethes aeneus* F.) and cabbage-seed weevils (*Ceuthorrhynchus assimilis* Payk.) on mustard-seed crops in East Anglia. *Ann. Appl. Biol.* 49:539–55

138. Winfield, A. L. 1963. Pests of *Brassica* seed crops. *Agriculture London* 70:228–32

139. Wylie, H. G. 1988. Release in Manitoba, Canada of *Townesilitus bicolor* (Hym.: Braconidae) a European parasite of *Phyllotreta* spp. (Col.: Chrysomelidae). *Entomophaga* 33:25–32

Ann. Rev. Entomol. 1989. 34:231–45

# THE LOCK-AND-KEY HYPOTHESIS: EVOLUTIONARY AND BIOSYSTEMATIC INTERPRETATION OF INSECT GENITALIA

*Arthur M. Shapiro and Adam H. Porter*

Department of Zoology, University of California, Davis, California 95616

> If I see a number of keys, of very minute and elaborate workmanship, all different, I cannot doubt that every one is intended to fit some special lock . . . and this conviction is the stronger, the more varied and the more complex are the keys. We cannot withhold a hearty assent to the conclusion of one of the most eminent of modern physiologists, who, speaking of these organs in the class of Insects generally, says, "They prevent allied species from producing bastards by adulterous connexions; for the hard parts of the male correspond so exactly with those of the female, that the organs of one species cannot fit those of another" (Siebold). And Léon Dufour speaks of them as "a guarantee of the conservation of Types, and a safeguard for the legitimacy of species." But I should like to see these axioms demonstrated.
>
> P. H. Gosse, 1883 (40, p. 280)

## PERSPECTIVES AND OVERVIEW

Evolutionary biology, like ecology, has suffered throughout its history as a discipline from a shortage of rigorously defined and explicitly testable hypotheses. The related field of systematics, according to some of its critics, has until recently had no testable hypotheses at all. In this paper we discuss the lock-and-key hypothesis, one of the most venerable explicit hypotheses in both evolution and systematics. Although it was proposed nearly a century and a half ago, and although its truth or falsity bears on matters of great theoretical importance in both evolution and systematics, no one can yet say whether it is true or false. The failure to produce an answer is grounded in the

0066-4170/89/0101-0231$02.00

structure of biology as a science and of evolution and systematics as branches
of biology. To understand the history and current status of the lock-and-key
hypothesis is to understand why these disciplines are so often exercises in
frustration. Whether such understanding makes chronic frustration any more
bearable is a psychological, not a biological or philosophical, matter.

Loehle (51) recently analyzed the role of hypothesis testing in ecology,
echoing some of the general themes voiced earlier by Bronowski (13).
Although ecology and evolutionary biology are not isomorphically organized,
they are sufficiently similar that Loehle's observations are likely to bear on
evolution in at least a general way. He has stated that not all interesting ideas
in ecology lend themselves, at least initially, to hypothesis testing. Description and cataloging may be necessary before suitable data can be acquired for
hypothesis testing; as Lakatos (49) has observed, as a theory matures it
increases in empirical content and its predictions become more distinct. By
the mid-nineteenth century a suitable data base had accumulated, so the
lock-and-key hypothesis began its long life as a quite explicit hypothesis. It
emerged as an inductive generalization: Genitalic morphology tends to be
species specific. This was followed immediately by a functional explanation
for the perceived pattern: reproductive isolation. Although the initial claim of
explanation (29) was made outside a Darwinian framework (indeed, 15 years
before the *Origin of Species*), it fit so easily into a Darwinian context that it
was quickly adopted by many Darwinians. Later (with reservations) it was
admitted to the neo-Darwinian synthesis, which afforded it added credibility.
The first to attempt to harmonize the lock-and-key hypothesis with Darwinism
was Karl Jordan, who stated (43, p. 428): "Before Darwin brought forward
his theory of evolution, naturalists had to solve one question: which are the
differences found to exist among the various forms of animals and plants? The
question which is put to naturalists nowadays is, however, twofold: (1) Which
kinds of divergency do we find to exist in nature? (2) How has this divergency
come about?"

Jordan's point remains well taken. Darwinian and neo-Darwinian explanations typically invoke an adaptive historical process (natural selection) to
account for the attributes of organisms. All such explanations run the risk of
using contemporary functions to explain the attribute at issue when purely
historical functions may be the true explanation. (Contemporary functions
may indeed account for the persistence of the attribute, which must be
distinguished at least conceptually from its origin.) Evolutionary biologists
are well aware of this problem, particularly in discussions of preadaptation
(41); nonetheless, they often fail to disentangle contemporary from historical
aspects of function. The rigorous definition of function given by Wright (80),
"The function of $X$ is $Z$ means ($a$) $X$ is there because it does $Z$ and ($b$) $Z$ is a
consequence of $X$'s being there," is rarely applicable to the real world because

knowledge of past function (part *a*) is so often irretrievable (compare 11). Thus, as Jordan realized, the Darwinized version of the lock-and-key hypothesis is intrinsically harder to test than the pre-Darwinian version, which dealt only with purported function in the here and now (9).

Eberhard (31) has recently rejuvenated the topic of genitalic morphology. He has discussed not only the lock-and-key hypothesis, but the full range of explanations suggested to date for patterns of variation in animal genitalic morphology. Eberhard has evaluated the lock-and-key hypothesis vis-à-vis competing hypotheses, but not in a philosophical framework. We limit ourselves in this review to the historical evolution of the lock-and-key hypothesis and to the tests, both theoretical and empirical, that have been advanced during its convoluted career. We refer to competing hypotheses only insofar as it is necessary to note ambiguities in the logic used to discriminate between the lock-and-key hypothesis and others. We limit ourselves to the entomological literature (except with reference to genetics) and to works whose argument is derived from or dependent on the lock-and-key hypothesis or to works that explicitly attempt to test it, contain data that by implication are useful in testing it, or reinterpret or extend it in creative ways. At the end we return to its philosophical underpinnings to ask whether the lock-and-key hypothesis is ultimately rigorously testable.

## HISTORICAL EVOLUTION OF THE HYPOTHESIS

### The Problem

The lock-and-key hypothesis purports to explain the observation that the morphology of animal genitalia is often highly species specific. Some circularity is inherent here, since the genitalia have been used routinely to characterize morphospecies (particularly in the Insecta); but enough species were initially delimited on other grounds and only subsequently characterized genitalically to assure that the circularity is not absolute.

Genitalic morphology has been used in taxonomy in tandem with pure morphospecies concepts and with their descendants of broader scope, the phenetic species concepts. It has also been combined with the biological species concept. It has been heavily weighted, with or without an explicit rationale, in those schools of taxonomy that practice weighting. When rationales have been advanced they have varied from purely practical to highly theoretical. The latter have been grounded usually in claims of functional significance, but sometimes in lack of plasticity or sensitivity to environment (45) or even lack of function (R. Owen, quoted in 26). Even taking circularity into account, few insect taxonomists would dispute the value of genitalia, especially male genitalia, for species-level determinations. Obviously their value is not universal, and it varies among taxonomic groups (68). Tuxen's

(74) extensive coverage of genitalic morphology is evidence of its penetration into taxonomic practice. Genitalic morphology is routinely used in species-level determination of fossil Coleoptera, allowing inferences about rates of morphological evolution and the dynamics of geographic range in geologic time (22). Eberhard (31) has made it clear that the overall pattern of genitalic differentiation at the species level is too pervasive to be dismissed as trivial or artifactual. It demands explanation.

Dufour (29) advanced the lock-and-key hypothesis in a treatise on dipteran anatomy; his declaration that "l'armure copulatrice . . . est la garantie de la conservation des types, la sauvegarde de la légitimité de l'espèce" (29, p. 253) must be one of the most quoted sentences in systematics. Dufour assigned a function to species-specific genitalic morphology, and his hypothesis was echoed by several nineteenth-century writers (reviewed in 43). Darwin (26) did not mention the hypothesis explicitly. The first major commentator seems to have been Gosse (40), who carried out extensive morphological studies of both male and female genitalia in the Papilionidae in the hope of testing the lock-and-key hypothesis. Unfortunately, he had only a vague and largely incorrect idea of how the male and female abdomens articulate. He found much more morphological diversity in male than in female genitalic anatomy and was confounded by this, but no conclusions can be drawn from his superficial study of the female. Pérez (58) largely followed Dufour's line.

Jordan (43) followed up on Gosse's work (40), extending it to a broader sample of Papilionidae. He showed that the differences in male genitalia among species and even geographic subspecies were often pronounced; the females often differed also, seemingly to a lesser degree. His paper was remarkable in that Jordan asked sophisticated and essentially quantitative questions about the relative variability of the genitalia versus other characters, the degree of independence of genitalic versus nongenitalic variation, the susceptibility of the genitalia to environmental influences, and most importantly, the relative degree of intra- versus interpopulational variability in genitalic characters. He found a bewildering diversity even within one family. Some of his observations were surprising: "The individual variation within the Indo-Malayan subspecies of *Papilio sarpedon* (nominate *sarpedon*), of which a large material has been examined, is so great that the difference between every two nearest-allied subspecies is small compared with the differences exhibited by the extreme individuals [of *P. sarpedon sarpedon*]" (43, p. 483). "The individual variation within each subspecies is such that the differences between the harpes of several individuals [of *P. depilis*] from the same place are more obvious than those of the subspecies *inter se*" (43, p. 473). Such discoveries did not support the general validity of the lock-and-key hypothesis. Despite overwhelming evidence that genitalic morphology was largely refractory to environmental influences, and hence under strong genetic control, Jordan remained cautiously equivocal. Nine years later he published a second major paper (44), in which he surveyed 698 species of sphingid moths. He found 48 species with morphologically similar male genitalia, along with plenty of intraspecific variation both within and among populations. This time he was moved to a negative verdict on the lock-and-key hypothesis.

Boulangé (12) disagreed with Dufour (29) and Pérez (58). He was quickly seconded by Richards (63), who showed that in the apid genus *Bombus* the males exhibit excellent genitalic species characters and the females vary but little (and mainly at the level of species groups, not individual species). He also showed that there is little correspondence between the parts that vary and the parts that are actually engaged in copulation. Despite some cases of apparently coupled variation between the sexes (32, 42), Richards joined with Robson in 1936 (64) in a broadside attack on the lock-and-key hypothesis. Their historical review included some important and rather obscure references, but it was spotty and selective and did not explicitly justify all their conclusions. Nonetheless, those conclusions had an important role in defining subsequent discourse and identifying potential falsifiers of the hypothesis. They are as follows (64, p. 155): (*a*) The male armature differs specifically much more often and usually more markedly than that of the female. (*b*) There is often, perhaps usually, no close specific correlation between the male and female structures. At least such correlation has not been established. (*c*) The numerous interspecific crosses, mostly artificial but some natural, between species with very different genitalia show that the male and female armatures do not impose an insuperable barrier. (*d*) Most species with different genitalia probably do not try to interbreed. They are in fact separated by other types or combinations of isolating factors. (*e*) Large groups of species exist in which the female genitalia differ only slightly from species to species. There is no evidence that such species hybridize more readily than those in which the differences are marked. (*f*) There does not appear to be a high correlation between degree of differences in genitalia and the fertility of hybrids if pairing does take place.

This is not the most rigorously logical organization for the objections to the lock-and-key hypothesis. For example, not only is lack of evidence a poor argument (when it is not explicit that anyone has looked for such evidence), but if in fact species with similar female genitalia do not hybridize more often than others (*e*), this could simply be because they have the same frequency of prior isolating factors (*d*). The following discussion, in which tests of the lock-and-key hypothesis are enumerated and evaluated, thus does not follow Robson & Richards' (64) organization closely, though echoes of it are evident.

## Attempts to Test the Hypothesis

ARE MALES MORE VARIABLE THAN FEMALES?    Males are more variable
than females in a great variety of insect groups. Eberhard (31, pp. 30–31)
updated Robson & Richards' (64) bibliography very convincingly on this
point. His most interesting bit of data, however, comes from his own survey
of taxonomic papers published in four volumes of the *Journal of the Kansas
Entomological Society*. He found five papers that stated explicitly that male,
but not female, genitalia were diagnostic at the species level in the groups
discussed; 35 of 57 additional papers discussed only male genitalia, without
explicitly dismissing those of the female. This is an unambiguous demonstration of a mind-set among taxonomists. As Eberhard himself noted, however,
that demonstration need not mirror biological reality, since the female genitalia are soft, involuted, harder to study, and generally both less well known
and less often studied (30, 54, 66, 78). The data are not cheering, however, to
anyone seeking a precise one-to-one morphological match between the sexes
as mandated by the lock-and-key hypothesis.

ARE THERE FUNCTIONAL CORRELATES BETWEEN MALE AND FEMALE
GENITALIA?    To answer this question fully, one needs detailed anatomical
knowledge of the sort generally obtainable only from preparations of animals
killed instantaneously *in copula*. Such studies are rare. Many claims have
been made of inability to mate based on studies of males and females
examined individually (e.g. 36, 77), but all such claims should be considered
suspect (cf 53).

The most frequently cited papers are not based on dissections of mating
pairs, but rather on successful mating between animals superficially unable to
do so [e.g. ladybird beetles (69), hemipterans (47)]. Gering (37) reported in
minute detail on the copulatory apparatus in agelenid spiders, concluding that
"the lock-and-key concept is poorly supported, if at all . . . There appears to
be little if any mechanical preclusion of cross-mating within [the genus
*Agelenopsis*]." Gering, however, obtained no interspecific copulations because of differences in species-specific precopulatory behavior. Fennah (33)
found little support for the lock-and-key concept in homopterans.

The most thorough study of the functional anatomy of insect genitalia is
that by Kunze (48) on the ciccadellid genera *Euscelis* and *Typhlocyba*. This
study encompassed both inter- and intraspecific (seasonal) variation. Kunze's
exacting methods make his strongly negative conclusions particularly com-
pelling (48, pp. 384, 360–61):

> Only a loose morphological correlation exists between the male and female reproductive
> organs, albeit these organs do bear species-specific characteristics. A narrow, shape-
> determined correspondence, such as that between a complicated lock and its key, was not
> demonstrable. . . . As anatomical study demonstrated, spring and summer generations of

*Euscelis plebejus* differ considerably in the form and width of the penile shaft; they are largely similar, on the other hand, with respect to the formation and size of the female genital space. . . . the dimensions of the male and female copulatory organs . . . can vary within certain limits without reducing their functional capacity . . . Supposing it to be the case that the morphological correspondence between the penis and the female genital space were as precise as that between a key and its appropriate lock, a successful copulation between different seasonal forms would be impossible. . . . the following conclusions must be drawn: the shape of the penis in *Euscelis plebejus* is not linked to its function in every detail; rather it can vary in rather large measure, without impairing successful copulation.

Sengün (67) cataloged instances of interspecific, intergeneric, and even interfamilial hybridizations in Lepidoptera to demonstrate the inefficacy of genitalic anatomy as a barrier to mating. The most important part of the study, however, was the experimental alteration and mutilation of the genitalia in the domestic silkworm, *Bombyx mori,* which generally failed to prevent copulation and fertilization unless the penis itself were amputated. Although this suggests the jocular proof that fleas hear with their hind legs because a trained flea with hind legs amputated fails to jump on command, it certainly demonstrates that the precise configuration of the male armature is not a precondition for copulatory success.

ARE THERE GEOGRAPHIC PATTERNS IN GENITALIC MORPHOLOGY THAT SUGGEST REPRODUCTIVE CHARACTER DISPLACEMENT? If genitalic morphology diverges at the species level as an adaptive defense against hybridization, one would expect to find patterns of differential genitalic divergence in sympatric versus allopatric species pairs. In keeping with the general concept of character displacement (14) and the conventional model of reinforcement of reproductive isolating mechanisms (RIMs) in secondary sympatry (27, 50; but see also 57), one would predict that closely related taxa would be more similar genitalically in allopatry than in sympatry. Relevant data are few, and they are complicated by the presence of earlier acting RIMs, discussed later. Eberhard (31) found no cases of convincing character displacement in genitalic morphology, nor did we in preparing this review. Burns (17) took an indirect approach to the question. The hesperiid species *Atrytonopsis ovinia* and *Atrytonopsis edwardsi* are completely allopatric, with no indications of any previous sympatry since their differentiation. There is considerable variation in both habitus and genitalia (both sexes) within each species, but no geographic pattern is evident in this variation even among disjunct segments of their ranges. Burns (17) claimed that "this essential 'uniformity' through each sister-species suggests differentiation in a small isolate followed by dispersal and expansion of range without further important genetic changes." Burns proceeded to use this case to critique panselectionism and to endorse a "punctuated equilibrium" scenario of speciation. Although the example in *Atrytonopsis* does not support a claim that the genitalic differences function (or ever functioned) in reproductive isolation, the possibility of previous sympatry, however remote, can never be entirely foreclosed. Such cases cannot be powerful tests of the hypothesis.

Burns (15) demonstrated his negative point of view more convincingly in his landmark revision of the difficult hesperiid genus *Erynnis*. In these insects the genitalia are laterally asymmetrical, the particular configuration being species specific, except in rare aberrants that conserve the uniquely derived morphology of a species but reproduce it bilaterally (16). Thus, for example, a right-sided male should be unable to copulate with a left-sided (allo-specific) female. Closely related species, however, are same-sided; the right-left difference comes into play only at the species-group level and thus presumably does not function in reproductive isolation. Moreover, there are no detectable genitalic differences between partly sympatric very close relatives *(E. lucilius* and *E. baptisiae,* separated by host-plant and habitat), while there are more or less extensive genitalic differences among related but parapatric or fully allopatric taxa in which these differences could not be functioning as RIMs today *(E. horatius–E. tristis, E. propertius–E. meri-dianus, E. brizo brizo–E. brizo burgessi–E. brizo lacustra).* Related patterns can be seen in the genus *Autochton* (18). Very complex genera with geographically variable species richness, such as the hesperiid genus *Pyrgus,* which has many sympatric or parapatric sibling species in Europe but only a handful of species in North America, offer good opportunities for further work along these lines.

ARE GEOGRAPHICALLY OR TAXONOMICALLY ISOLATED SPECIES MORE VARIABLE THAN MEMBERS OF SPECIOSE GROUPS?    This question is closely allied to the preceding ones. Kosminsky (46) published a classic paper on a moth that is taxonomically isolated and has very variable genitalia. No one seems to have searched out such cases methodically. Eberhard (31, pp. 33–47) surveyed cases in which there was a low probability of interspecific encounter owing to geographic, taxonomic, or ecological isolation; several examples were not insects. The results, while equivocal, do not support the lock-and-key hypothesis.

IS GENITALIC DIFFERENTIATION CORRELATED WITH HYBRID INFERIOR-ITY?    If genitalic divergence occurs as a part of reproductive character displacement, there must be some disadvantage to hybridizing or selection would not occur. Such a disadvantage could be at the level of fertility, fecundity, sex ratio, survivorship, or other factors. A number of studies have documented at least in part an apparent lack of such handicaps once prezy-gotic RIMs are overcome (7, 69). Given the difficulty in documenting hybrid disadvantage and the spectrum of hybrid breakdown phenomena, which may not be manifest for three or more generations after first contact, this literature is not very convincing. One of the largest bodies of data of this sort is on swallowtail butterflies, Papilionidae, which are hybridized easily by hand-pairing (summarized in 2); the data suggest that there is little or no correlation of genitalic morphology and hybrid success except at higher (subgeneric or generic) levels. The entire model of reinforcement of RIMs has been under attack recently (57), perhaps because its logical consistency preadapts it to facile use. Insect genitalia, however, often resemble the pollinator-specific reproductive organs of orchids, which indisputably function as effective prezygotic RIMs, but often in the absence of any postzygotic RIMs even at the intergeneric level (1, 70, 79). Thus the genitalia are unlikely to have evolved to reduce hybridization frequency per se.

IS MECHANICAL ISOLATION UNECONOMICAL?    Mayr (52) argued that lock-and-key isolation would be unlikely to function often in nature because most potential hybridizers would never proceed far in courtship. Eberhard (31, pp. 24, 27) cited doubts expressed by Alexander (3) and supported by references in Alexander & Otte (4):

> Natural selection should favor females who are able to determine male species identity early in courtship sequences rather than late, since both courtship and copulation are often somewhat costly and dangerous for a female [see 25]. Thus species discrimination by genitalia, while feasible, would be less advantageous than discrimination based on stimuli received prior to copulation. . . . selection on genitalia in a species-isolation context would be expected to be only transitory at best.

The consequences of such selection, however, might not be transitory.

Tinkham & Rentz (72) found consistent premating RIMs in sympatric Jerusalem crickets with no species-specific genitalic differences. Again, there is an extensive, scattered literature of similar observations (e.g. 20, 69, 76). Rentz (62) studied multiple levels of reproductive isolation in shield-backed katydids and concluded that lock-and-key mechanisms did come into play when earlier barriers were lacking. His final argument was that (62)

> in species bearing morphologically similar genitalia, it is apparent that evolution has proceeded in other directions. In the case of the slant-faced and band-winged grasshoppers, behavior has become an important isolating mechanism, and little differentiation of the genitalia has been necessary. . . . The above-mentioned studies . . . can be seen in the proper light. They are not examples that discredit the lock-and-key theory. The species in these studies have merely utilized other isolating mechanisms to maintain species integrity.

This argument bears on our concluding analysis.

# IS THE LOCK-AND-KEY HYPOTHESIS TESTABLE?

*Old soldiers never die; they just fade away.*

General Douglas MacArthur

## Philosophical Considerations

In its 145-year history the lock-and-key hypothesis has provoked many confirmation and refutation attempts. It is remarkable that explicit tests were attempted as early as the late nineteenth century (40, 43) and continue to the present. The bulk of the evidence is unfavorable to the hypothesis. W. G. Eberhard (personal communication) clearly views it as unsalvageable. But the problem remains far from resolved. Indeed, it is not clear what "resolved" means in this context.

The pre-Darwinian version of the lock-and-key hypothesis merely asserts that genitalic morphology is effective in preventing interspecific hybridization today. Dufour (29) considered this the general explanation of species-specific genitalic morphology. Such a rigorous version of the hypothesis could be falsified by a single contrary case, and there are several. In addition, numerous potential hybridizers do not try out their genitalia because courtship is interrupted by earlier RIMs, so the claim of function is also falsified. Of course, a less rigorous version of the hypothesis would not be falsified by such evidence. The role postulated by Dufour might occur; it might even be important; it is just not universal.

The Darwinized version of the lock-and-key hypothesis asserts that the morphological differences evolved as a form of reproductive character displacement. Thus, failure of genitalic morphology to function as an RIM today does not falsify the hypothesis, because it might have functioned thus in the past. Nor do cases in which genitalic morphology functions as an RIM today particularly strengthen the hypothesis. The hypothesis has not yet been supported convincingly, but this failure is not tantamount to global falsification.

The reinforcement model predicts the successive replacement of effective RIMs by ever earlier ones. By this reasoning, species pairs hybridizing in secondary contact (with adverse consequences) should move, on average, from mechanical isolation to some form(s) of behavioral isolation; effective mechanical isolation, being transitory, should rarely be encountered in nature. This situation is directly analogous to the "ghost of competition past" (21) in evolutionary ecology. It is unclear what the implications for genitalic morphology would be if selection were relaxed after the genitalia had ceased to function as an RIM. Inductively, morphological stasis seems more probable than reversion to an earlier condition. One might envision the accumulation of successive layers of sequential RIMs forming a time series with decreasing probability of actual use. This succession is analogous to the "arms race escalation" postulated in the evolution of defensive phytochemistry (8).

The absence of genitalic differences between close relatives might be interpreted as prima facie evidence that the lock-and-key mechanism is irrelevant in that case, perhaps because selection of earlier prezygotic RIMs occurred (or they arose in allopatry, coming into play immediately upon secondary contact) before genitalic differentiation could occur. The presence of genitalic differences between two species, however, should not be presumed to imply that the differences have any function. This is an expansion of Rentz's argument, but it is even more pessimistic regarding the possibility of a definitive test of the lock-and-key hypothesis.

We conclude that there is no prospect of resolving the controversy in naive falsificationist terms (59, 60, 75). If we take a sophisticated approach to hypothesis testing (49), taking account of multiple layers of disconfirmation and "theory tenacity" (51), the best we can hope for is to restrict the generality of the lock-and-key hypothesis step by step. The logic of the Darwinian argument for lock-and-key evolution is so appealing that one may be reluctant to abandon it in toto, and one need not do so. In retrospect, it is striking that the concept was so reluctantly admitted into neo-Darwinism (27, 28, 38, 52) and that so seemingly coherent a notion did not generate more ideological bias than it did. Lakatos (49) would have argued that a more "progressive" research program will replace the lock-and-key hypothesis without its being definitively falsified; as its domain becomes more and more restricted, fewer and fewer people will care about it.

## Desiderata

In researching this review we have come to appreciate how far we are from a convincing causal chain connecting genes, genitalic morphology, and fitness. Our involvement in the issue grew out of a study (A. M. Shapiro & A. H. Porter, in preparation) of the role of genitalic morphology in the hybrid zone between two butterfly taxa in northwestern Patagonia. These nominal species differ genitalically in ways suggestive of, but ineffective in, mechanical isolation. The distribution of genitalic character states may be studied afield and compared with other species-specific character states in typical hybrid-zone studies of differential introgression (6). We produced artificial hybrid swarms in the laboratory and found that the distribution of genitalic character states converged to the situation in the field. Assuming a role for these characters in reproductive isolation, one could generate testable hypotheses about their diffusion. We looked in vain for a published precedent.

There is little but interesting literature on the genetic control of genitalic morphology, some of it involving vertebrates. This literature clearly bears on the susceptibility of this control to the rapid selection envisioned in the Darwinized lock-and-key hypothesis (and some alternatives) (5, 23, 24, 34, 35, 69, 73 for insects; 39, 65 for fish). There is also a small but significant body of evidence bearing on environmentally controlled (often seasonal) variation in genitalic morphology (48, 55, 61). This evidence not only undermines one classical rationale for the heavy weighting of genitalia in taxonomy, but gives implicit support to the notion that genitalic morphology may often be a by-product of other processes rather than a direct target of selection. These fundamental questions bear not only on the viability of the lock-and-key concept, but on most if not all of the proposed alternatives. Our understanding of the genetic and developmental controls on genitalia has been too long hampered by the methodological problems of quantifying complex three-dimensional variation; it is time to take advantage of modern computerized methods of morphometrics (10).

We note in closing that the lock-and-key mechanism as an explanation grades insensibly into Eberhard's sexual selection hypothesis (31); there is a gradient from mechanical inability to copulate, to increased probability of rejection or early termination of copulation due to poor fit. In many cases, without detailed dissections of the Kunze type (48) of animals taken *in flagrante delicto*, it will remain impossible to distinguish between inability to join and unwillingness to do so. Both hypotheses, then, potentially depend upon the existence of a neurobiological basis for receiving and evaluating the stimuli generated during copulation. Little is known of genitalic neuroanatomy; the existence of stretch receptors in butterfly bursae copulatrices has been indirectly demonstrated (56, 71), but it is too early to do more than speculate what other receptors and wiring we may find. Genitalic neuroanatomy may, as Eberhard (31, p. 185) hinted, open up new frontiers of hypothesis testing for the lock-and-key hypothesis and its rivals.

## SUMMARY

The lock-and-key hypothesis, which purports to explain species-specific genitalic morphology in terms of mechanical reproductive isolation, has fared poorly in many attempts to test it over a century, but it remains unfalsified in a definitive sense. We argue that it is unfalsifiable globally in either its unrigorous pre-Darwinian form or its Darwinized form. We predict that as it is falsified for more and more specific cases and its potential domain is ever more restricted, systematists and evolutionary biologists will lose interest in it, and it will be invoked only in those exceptional cases in which it appears to be supported by good data.

ACKNOWLEDGMENTS

We thank Bill Eberhard for bearing with us and Adrienne R. Shapiro for help in dense thickets of scientific German. John Burns, who has written (19) that "genitalia deserve all the respect and attention they can get," has been extremely helpful and in the process has demonstrated that he knows whereof he speaks.

## Literature Cited

1. Adams, H., Anderson, E. 1958. A conspectus of hybridization in the Orchidaceae. *Evolution* 12:512–18
2. Ae, S. A. 1979. The phylogeny of some *Papilio* species based on interspecific hybridization data. *Syst. Entomol.* 4:1–16
3. Alexander, R. D. 1962. Evolutionary change in cricket acoustical communication. *Evolution* 16:443–67
4. Alexander, R. D., Otte, D. 1967. The evolution of genitalia and mating behavior in crickets (Gryllidae) and other Orthoptera. *Misc. Publ. Mus. Zool. Univ. Mich.* 133:1–62
5. Bacheler, J. S., Habeck, D. H. 1974. Biology and hybridization of *Apantesis phalerata* and *A. radians* (Lepidoptera: Arctiidae). *Ann. Entomol. Soc. Am.* 67:971–75
6. Barton, N., Hewitt, G. M. 1985. Analysis of hybrid zones. *Ann. Rev. Ecol. Syst.* 16:113–48
7. Beheim, D. 1942. Uber den taxonomischen und isolierenden Wert der Forcepovariationen einiger Caraboidea. *Z. Morphol. Oekol. Tiere* 39:21–46
8. Berenbaum, M. 1983. Coumarins and caterpillars: a case for coevolution. *Evolution* 37:163–79
9. Bock, W. J. 1988. The nature of explanations in morphology. *Am. Zool.* 28:205–15
10. Bookstein, F., Chernoff, B., Elder, R., Humphries, J., Smith, G., Strauss, R. 1985. *Morphometrics in Evolutionary Biology. Spec. Publ. Acad. Nat. Sci. Philadelphia* Vol. 15, 277 pp.
11. Boorse, C. 1984. Wright on functions. In *Conceptual Issues in Evolutionary Biology*, ed. E. Sober, pp. 347–68. Cambridge, Mass: MIT Press. 725 pp.
12. Boulangé, H. 1924. Recherches sur l'appareil copulateur des Hyménoptères et spécialement des Chalastrogastres. *Mem. Trav. Fac. Cathol. Lille* 28:1–444
13. Bronowski, J. 1953. *The Common Sense of Science*. Cambridge, Mass: Harvard Univ. Press. 152 pp.
14. Brown, W. L., Wilson, E. O. 1956. Character displacement. *Syst. Zool.* 5:49–64
15. Burns, J. M. 1964. Evolution in skipper butterflies of the genus *Erynnis. Univ. Calif. Berkeley Publ. Entomol.* 37:1–216
16. Burns, J. M. 1970. Secondary symmetry of asymmetric genitalia in males of *Erynnis funeralis* and *E. propertius* (Lepidoptera: Hesperiidae). *Psyche* 77:430–35
17. Burns, J. M. 1983. Superspecies *Atrytonopsis ovinia* (*A. ovinia* plus *A. edwardsii*) and the nonadaptive nature of interspecific genitalic differences (Lepidoptera: Hesperiidae). *Proc. Entomol. Soc. Wash.* 85:335–58
18. Burns, J. M. 1984. Evolutionary differentiation: differentiating gold-banded skippers—*Autochton cellus* and more (Lepidoptera: Hesperiidae: Pyrginae). *Smithson. Contrib. Zool.* 405:1–38
19. Burns, J. M. 1987. The big shift: *nabokovi* from *Atalopedes* to *Hesperia* (Hesperiidae). *J. Lepid. Soc.* 41:173–86
20. Cardé, R. T., Shapiro, A. M., Clench, H. K. 1970. Sibling species in the *eurydice* group of *Lethe* (Lepidoptera: Satyridae). *Psyche* 77:70–103
21. Connell, J. 1980. Diversity and the coevolution of competitors, or the ghost of competition past. *Oikos* 35:131–38
22. Coope, G. R. 1979. Late Cenozoic fossil Coleoptera: evolution, biogeography and ecology. *Ann. Rev. Ecol. Syst.* 10:247–68
23. Coyne, J. A. 1983. Genetic basis of differences in genital morphology among three sibling species of *Drosophila. Evolution* 37:1101–8
24. Coyne, J. A., Kreitmann, M. 1986. Evolutionary genetics of two sibling species, *Drosophila simulans* and *D. sechellia. Evolution* 40:673–91
25. Daly, M. 1978. The cost of mating. *Am. Nat.* 112:771–74
26. Darwin, C. 1859. *On the Origin of Species by Means of Natural Selection*. London: Murray. 1st ed. 513 pp.
27. Dobzhansky, Th. 1951. *Genetics and the Origin of Species*. New York: Columbia Univ. Press. 3rd ed. 364 pp.
28. Dobzhansky, Th., Ayala, F. J., Stebbins, G. L., Valentine, J. W. 1977. *Evolution*. San Francisco: Freeman. 572 pp.
29. Dufour, L. 1844. Anatomie Générale des Diptères. *Ann. Sci. Nat.* 1:244–64
30. Dugdale, H. S. 1974. Female genital configuration in the classification of Lepidoptera. *NZ J. Zool.* 1:127–46
31. Eberhard, W. G. 1985. *Sexual Selection and Animal Genitalia*. Cambridge, Mass: Harvard Univ. Press. 244 pp.
32. Edwards, F. W. 1929. British non-biting midges (Diptera, Chironomidae). *Trans. Entomol. Soc. London* 77:279–430
33. Fennah, R. G. 1945. The Cixiini of the lesser Antilles (Homoptera: Fulgoroidea). *Proc. Biol. Soc. Wash.* 58:133–46
34. Foot, K., Strobell, E. C. 1914. Results

of crossing *Euschistus variolarius* and *Euschistus servus* with reference to the inheritance of an exclusively male character. *J. Linn. Soc. London Zool.* 32: 337–73

35. Foot, K., Strobell, E. C. 1915. Results of crossing two hemipterous species, with reference to the inheritance of two exclusively male characters. *J. Linn. Soc. London Zool.* 32:457–93

36. Freitag, R. 1934. Selection for a non-genitalic mating structure in female tiger beetles of the genus *Cicindela* (Coleoptera: Cicindelidae). *Can. Entomol.* 106:561–68

37. Gering, R. L. 1953. Structure and function in the genitalia in some American agelenid spiders. *Smithson. Misc. Collect.* 121:1–84

38. Goldschmidt, R. 1940. *The Material Basis of Evolution.* New Haven, Conn: Yale Univ. Press. 436 pp.

39. Gordon, M., Rosen, D. E. 1951. Genetics of species differences in the morphology of the male genitalia of Xiphophorin fishes. *Bull. Am. Mus. Nat. Hist.* 95: 413–64

40. Gosse, P. H. 1882. On the clasping-organs ancillary to generation in certain groups of the Lepidoptera. *Trans. Linn. Soc. London Zool.* 2:265–345

41. Gould, S. J., Vrba, E. 1982. Exaptation: a missing term in the science of form. *Paleobiology* 8:4–15

42. Harnisch, W. 1915. Über den männlichen Begattungsapparat einiger Chrysomeliden. *Z. Wiss. Zool.* 114:1–94

43. Jordan, K. 1896. On mechanical isolation and other problems. *Novit. Zool.* 3:426–525

44. Jordan, K. 1905. Der Gegensatz zwischen geographischer and nichtgeographischer Variation. *Z. Wiss. Zool.* 83:151–210

45. Klots, A. B. 1931. A generic revision of the Pieridae (Lepidoptera) together with a study of the male genitalia. *Entomol. Am.* 12:139–242

46. Kosminsky, P. 1912. Zur Frage über die Unbestandigkeit der morphologischen Merkmale bei *Abraxas grossulariata* L. *Rev. Russ. Entomol.* 12:313–28

47. Kullenberg, B. 1947. Der Kopulationsapparat der Insekten aus phylogenetischen Gesichtspunkt. *Zool. Bidr. Uppsala* 25:79–90

48. Kunze, L. 1959. Die funktionsanatomischen Grundlagen der Kopulation der Zwergzikaden, untersucht an *Euscelis plebejus* (Fall.) und einigen Typhlocybinen. *Dtsch. Entomol. Z.* (NF) 6:322–87

49. Lakatos, I. 1970. Falsification and the methodology of scientific research programs. In *Criticism and the Growth of Knowledge,* ed. I. Lakatos, A. Musgrave, pp. 91–195. Cambridge, UK: Cambridge Univ. Press. 282 pp.

50. Levin, D. A. 1978. The origin of isolating mechanisms in flowering plants. *Evol. Biol.* 11:185–317

51. Loehle, C. 1987. Hypothesis testing in ecology: psychological aspects and the importance of theory maturation. *Q. Rev. Biol.* 62:397–409

52. Mayr, E. 1963. *Animal Species and Evolution.* Cambridge, Mass: Belknap. 797 pp.

53. Mickel, C. E. 1924. An analysis of a bimodal variation of the parasite *Dasymutilla bioculata* (Hymenoptera: Mutillidae). *Entomol. News* 35:236–42

54. Mitter, C. 1988. Taxonomic potential of some internal reproductive structures in *Catocala* (Lepidoptera: Noctuidae) and related genera. *Ann. Entomol. Soc. Am.* 81:10–18

55. Müller, H. J. 1957. Die Wirkung exogener Faktoren auf die zyklische Formenbildung der Insekten, insebesondere der Gattung *Euscelis. Zool. Jahrb. Abt. Syst. Oekol. Geogr. Tiere* 85:317–430

56. Obara, Y., Tateda, H., Kuwabara, M. 1975. Mating behavior of the cabbage butterfly, *Pieris rapae crucivora* Boisduval. V. Copulatory stimuli inducing changes of female response patterns. *Zool. Mag. Jpn.* 84:71–76

57. Paterson, H. E. H. 1982. Perspective on speciation by reinforcement. *S. Afr. J. Sci.* 78:53–57

58. Pérez, J. 1894. De l'organe copulateur mâle des Hymenoptères et sa valeur taxonomique. *Ann. Soc. Entomol. Fr.* 63: 74–81

59. Popper, K. R. 1959. *The Logic of Scientific Discovery.* London: Hutchinson. 479 pp.

60. Popper, K. R. 1963. *Conjectures and Refutations: The Growth of Scientific Knowledge.* New York: Harper & Row. 431 pp.

61. Reinhardt, R. 1969. Über die Einfluss der Temperatur auf den Saisondimorphismus von *Araschnia levana* L. (Lepidoptera: Nymphalidae) nach photoperiodischer Diapause-Induktion. *Zool. Jahrb. Abt. Allg. Zool. Physiol. Tiere* 75:41–75

62. Rentz, D. C. 1972. The lock and key as an isolating mechanism in katydids. *Am. Sci.* 60:750–55

63. Richards, O. W. 1927. The specific characters of British bumble-bees (Hymenoptera). *Trans. R. Entomol. Soc. London* 75:233–65

64. Robson, G. C., Richards, O. W. 1936. *The Variation of Animals in Nature*. London: Longmans, Green. 425 pp.

65. Rosen, D. E., Gordon, M. 1953. Functional anatomy and evolution of male genitalia in poeciliid fishes. *Zoologica NY* 38:1–47

66. Scudder, G. G. E. 1971. Comparative morphology of insect genitalia. *Ann. Rev. Entomol.* 16:379–406

67. Sengün, A. 1944. Experimente zur sexuell-mechanischen Isolation. *Rev. Fac. Sci. Istanbul Ser. B* 9:239–53

68. Shapiro, A. M., Cardé, R. T. 1970. Habitat selection and competition among sibling species of satyrid butterflies. *Evolution* 24:48–54

69. Shull, A. F. 1946. Inheritance in lady beetles. IV. The form of the chitinous male genitalia in crosses of the species *Hippodamia quinquesignata* and *H. convergens*. *Genetics* 31:291–303

70. Stebbins, G. L., Ferlan, L. 1956. Population variability, hybridization, and introgression in some species of *Ophrys*. *Evolution* 10:32–46

71. Sugawara, T. 1979. Stretch reception in the bursa copulatrix of the butterfly *Pieris rapae crucivora*, and its role in behavior. *J. Comp. Physiol.* 130:191–99

72. Tinkham, E. R., Rentz, D. C. 1969. Notes on the bionomics and distribution of the genus *Stenopelmatus* in central California with the description of a new species. *Pan-Pac. Entomol.* 45:4–14

73. Turner, J. R. G., Clarke, C. A., Sheppard, P. M. 1961. Genetics of a difference in the male genitalia of East and West African stocks of *Papilio dardanus* (Lepidoptera). *Nature* 191:935–36

74. Tuxen, S. L. 1970. *Taxonomist's Glossary of Genitalia in Insects*. Copenhagen: Monksgaard. 2nd 3d. 359 ed.

75. Tweney, R. D., Doherty, M. E., Mynatt, C. R. 1981. *On Scientific Thinking*. New York: Columbia Univ. Press

76. Warren, B. C. S. 1926. Monograph of the tribe Hesperidi (European species) with a revised classification of the subfamily Hesperiinae (Palearctic species) based on the genital armature of the male. *Trans. Entomol. Soc. London* 74:1–170

77. Watson, J. A. L. 1966. Genital structure as an isolating mechanism in Odonata. *Proc. R. Entomol. Soc. London A* 41: 171–74

78. Weidner, H. 1934. Beiträge zur Morphologie und Physiologie des Genitalapparats der weiblichen Lepidopteren. *Z. Angew. Entomol.* 21:239–89

79. Withner, C., ed. 1959. *The Orchids*. New York: Ronald. 604 pp.

80. Wright, L. 1973. Functions. *Philos. Rev.* 82:139–68

Ann. Rev. Entomol. 1989. 34:247–71
Copyright © 1989 by Annual Reviews Inc. All rights reserved

# REMOTE SENSING IN ENTOMOLOGY

## J. R. Riley

Overseas Development Natural Resources Institute, Radar Entomology Unit, Royal Signals and Radar Establishment, Malvern, Worcestershire WR14 1LL, United Kingdom

## INTRODUCTION

The term "remote sensing" has come to be strongly associated with techniques of observation from earth-orbiting satellites, and in this guise would seem to have relatively little to offer entomologists. However, the subject more properly includes all methods of "observation of a target by a device some distance from it" (9), and this broader definition encompasses a large number of techniques that have been of benefit in entomological studies (3, 14, 19, 70, 80, 83, 109, 130, 150). There have been three distinct areas of application: the observation of insects themselves, the detection of the effects that insects produce (usually plant damage), and the monitoring of environmental factors likely to influence insect behavior. This review begins with an outline of remote sensing techniques used in entomology and then decribes how they have been applied in these three areas. The limited space available has meant that the descriptions are necessarily somewhat abbreviated.

## REMOTE SENSING TECHNIQUES USED IN ENTOMOLOGY

Remote sensing techniques used in entomology include photography and videography from aircraft and from the ground; satellite-borne photography, multispectral scanning, and thermal imaging; ground-based and airborne radar; and acoustic sounding and low-light optical methods. With the exception of acoustic sounding, all entomological applications of remote sensing depend on the propagation of electromagnetic waves between the target being observed and a sensing device. The range of wavelengths employed extends

247

0066-4170/89/0101-0247$02.00

over six orders of magnitude, from 0.4 $\mu$m at the ultraviolet end of the visible spectrum to tens of centimeters, characteristic of the longer-wavelength radars. Although the techniques used to exploit the different spectral regions are radically different, each with its own nomenclature and technology, they all have the same basic objective: to derive useful information about a target from the way in which it reflects or emits wave energy.

## Photography and Videography from the Air

Photography is the oldest and most well-known method of instrumental remote sensing. Images of a target are cast by a lens system onto a light-sensitive film, which once developed, may be analyzed to yield data about the target's shape, contrast, texture, and reflectivity. Quantitative interpretation of the image in terms of target properties is not, however, a trivial matter, because the image characteristics depend not only on the target, but also on the quality of the light illuminating it, the angles of illumination and view, the properties of the lens system, the characteristics of the film and the developing process, the distance of the target from the camera, and the clarity of the atmosphere. Slater et al (149) have described the large and multidisciplinary field that has evolved to deal with the complex and interactive factors of photographic remote sensing; I outline here only those aspects that are particularly relevant to insect-related observations.

AERIAL PHOTOGRAPHY    The main entomological use of photographic methods for remote sensing has been in aerial photography, with particular attention having been paid to methods of detecting the activity of insects from the changes they produce in the appearance of plant foliage (19, 70). Foliage may be changed if insects leave deposits on leaves (70, 71, 155) or if they induce alterations of leaf color, shape, or density (108) as a direct (18, 20, 64, 114, 166) or indirect (85, 108) result of feeding. More unusually, another indicator of insect activity that has been successfully detected by aerial photography is the mound structure excavated by certain ant species (57, 87, 88).

In principle, aerial photography offers two main advantages: the rapid survey of very large or inaccessible habitats (83) and the accurate delineation and recording of insect-affected areas (70, 83, 86). In practice, however, it seems that large habitats are most economically surveyed by airborne observers equipped with binoculars, at least in the case of forest surveys (86), and that photography is best reserved for quantifying the type, degree, and area of vegetation damage once it has been found.

*Film type*    Because foliage reflects light strongly in the near-infrared region (108), films sensitive to these wavelengths tend to highlight vegetation cover

(96). However, the combination of differential response to both the visible spectrum (0.35–0.70 $\mu$m) and the near infrared (0.79–0.90 $\mu$m) that is found in false color (or color infrared) film has proved most effective in detecting plant stress (67, 70, 74, 86, 108, 109). This type of film effectively transposes into the visible region the spectral range in which the most important changes in plant reflectance occur (108); thus it allows the color sensitivity of the human eye to be used to best effect during image analysis.

*Scale and altitude*    The ratio of the size of the image on a film to the object being photographed is known as the photographic scale and is determined by dividing the range to the target by the camera's focal length. Typically large, medium, and small scales would be 1:600 to 1:2000, 1:4000 to 1:15,000, and 1:20,000 to 1:60,000, respectively (73). Small-scale photography allows a larger area to be recorded on a single frame of film, but also reduces the capacity to register small features (148). The minimum useful scale in any investigation is thus dictated by the type and extent of the foliage changes being investigated (86, 108); the detection of individual defoliated trees would require a much larger scale than the detection of a defoliated forest area (76). Heller (73, 76) has given a useful assessment of the relative effectiveness of different scales and film types in detecting the damage caused by a variety of forest insects. Once a scale has been selected for a given application, the altitude of flight is determined by the focal length of the cameras available.

*Identification*    Insect identification is a major problem in almost all entomological remote sensing experiments. In the present context it is particularly serious because the photographic method generally relies on the detection of plant stress or damage, which in many cases could have a variety of causes quite unrelated to insect activity (108). In these instances identification is dependent on circumstantial evidence such as the previous history of insect activity in the area, the timing and pattern of the observed symptoms (76, 107), and the vulnerability of identified plants to particular insect species (61, 107); ultimately, identification depends on on-the-spot visual inspection by entomologists. There are a few possible exceptions to this general rule (53, 70); for example, the degree of sooty mold discoloration on the leaves of citrus trees was found to be a potential indicator of the insect species infesting the trees (53). In the main, however, identification difficulties remain a major limitation in entomological applications of aerial photography.

*Ecological surveys*    Although most entomologically related aerial photography has been concerned with the detection of damage to plants, a number of experiments have been carried out to assess the distribution of the principal host plants or habitats of some insect species (31, 66, 154, 165). The crucial factor in ecological studies based on aerial photography is the ability to identify these plants and habitats in the photographs. This may be done with some success by using a combination of methods, including inspection visits to enough representative sites to establish an interpretive key, the recognition of cultivation patterns (31), knowledge of the altitudinal distribution of plant types (31), and in the case of mosquito habitats, the detection of wetlands (105, 154).

AERIAL VIDEOGRAPHY    In contrast to photography, aerial videography has only recently been applied to entomological studies. It is, however, rapidly becoming popular (52, 152) as specialized systems for airborne use become available (90, 104, 111). In a video camera the image of remote objects is cast onto a photo-conductive surface rather than onto film and forms a charge distribution that duplicates the optical image. This distribution is read by a scanning electron beam and is converted into electrical signals, which are stored on a video tape recorder. The signals can also simultaneously generate a copy of the image on a television monitor. This ability to display imagery while it is actually being acquired is one of the main advantages of videography. Other advantages (reviewed in 52) include high sensitivity to light, which allows the use of narrow spectral band filters; good sensitivity in the near infrared, with no special processing required; a high image repetition rate (30 or 25 Hz); suitability for digitization and analysis by computer; the ability to obtain false color composites when multiple sensor heads are used; and the low price and reusable nature of video recording tape. A serious disadvantage of commercial video systems is their low resolution, typically 240 lines per format, compared with 720 lines across the long axis of a 35-mm slide (52). However, specialized video recording systems that match the resolution achievable with photography have been developed for military reconnaissance tasks (151), and it seems likely that high-resolution video equipment will gradually become widely available.

## Multispectral Scanning

Multispectral scanning (MSS) can form images over a much wider range of electromagnetic wavelengths (0.4–14 $\mu$m) than photographic methods. The improvement is gained, however, at the expense of a considerable increase in technical complexity, and MSS systems are in consequence far more expensive and much less common than photographic systems.

Instead of generating an instantaneous image of a whole scene, multispectral scanners use sensors with very narrow fields of view (1.5–2.5 mrad) to scan an area of interest systematically, and an image is built up as the scan progresses (96). The sensors are responsive to different segments of the electromagnetic spectrum, so each forms an image that represents the reflectance of the scene in its particular waveband. The widely used multispectral scanners in the Landsat series of satellites provide images in four bands (0.5–0.6, 0.6–0.7, 0.7–0.8, and 0.8–1.1 $\mu$m), with a ground resolution of 57 × 79 m (32). Landsats 4 and 5 carried in addition a more powerful scanner called a thematic mapper (TM), which had improved spatial and spectral resolution (148). This device also formed images in the thermal infrared region (10.2–12.2 $\mu$m), i.e. it responded to infrared radiation emitted by the scene, and thus provided a measure of temperature. MSS images can be stored on magnetic recording media and subsequently corrected for geometric distortion (96) prior to digital processing, display on an image tube, and transfer to film as color composites.

MSS FROM SATELLITES    The major source of MSS data for entomological studies has been earth-orbiting satellites. This is partly because the only MSS data routinely available has been that generated by satellite-borne equipment, and partly because satellite-based remote sensing intrinsically provides regular coverage of large land areas. Efforts have been concentrated into two principal areas: the detection of changes in vegetation and the measurement of meteorological data.

*Vegetation changes*    Surveys have been made to detect insect-induced defoliation of forest areas (32, 73, 76) and to delineate the ephemeral vegetation that might support locust breeding in normally desiccated habitats (16, 116, 150). A parameter called the ratio vegetation index (32), formed by dividing the near-infrared by the red response (MSS band 4/MSS band 2), was found to provide a good indicator in both types in study. Because reflectance in the red region is inversely proportional to chlorophyll density, and reflectance in the near-infrared region is proportional to leaf density, a large index is indicative of a high green-leaf biomass. Furthermore, because the index is a ratio of luminance in adjacent bands, it is relatively insensitive to changes in natural illumination (16, 150). The normalized vegetation index, (band 4 − band 2)/(Band 4 + band 2), is currently considered a somewhat better indicator of biomass than the simple band 4/band 2 ratio used in earlier studies.

Landsat MSS data is not well suited to the detection of dying vegetation because the yellow to orange color (0.58–0.62 $\mu$m) characteristic of stress is added to the top of the green band (0.5–0.6 $\mu$m) and to the bottom of the red band (0.6–0.7 $\mu$m), partially compensating for the drop in reflectance in the green region (76).

*Meteorological observations*    Two environmental factors of particular interest to field entomologists are rainfall and air temperature, both of which may be estimated by multispectral imaging systems on meteorological satellites.

Radiation from the scene is measured (usually at several wavelengths within the 15-$\mu$m band), and air temperature profiles are computed using radiative transfer equations (96). Rainfall is estimated indirectly from the extent, appearance, and life history of clouds imaged in the visible and infrared regions (9, 96).

## Radar

Radar has arguably a more obvious connection with entomology than any of the remote sensing methods discussed so far because it has been used effectively in direct observations of insects (130). Following the pioneering demonstration by Schaefer (140) that radar could be successfully used as a powerful entomological tool, application of the technique has led to spectacular advances in the study of long-distance migration (43, 132, 142) and other aspects of flight behavior in the field (128, 134). Insects cannot perceive radar waves, at least at the power levels used in normal entomological studies (127), so the technique allows observations of undisturbed, natural behavior. Radar has also been used to investigate windfields affecting airborne insects (30, 58, 119).

COMMON ENTOMOLOGICAL RADARS    Most of the radars used to date in entomological studies have been small, mobile, incoherent pulse systems using a wavelength of 3.2 cm and based on commercially available marine systems (36, 127, 140, 160). These radars transmit from their antennas a narrow, conical beam of short (typically 0.1 or 0.05 $\mu$s) pulses of electromagnetic waves. Any object illuminated by a pulse reflects or scatters some of the pulse energy, and a part of this scattered energy (the "echo") is returned in the direction of the radar. If the echo is strong enough, it is detected and amplified when it reaches the radar receiver, and the presence of a target is registered on a display device. Target direction can be deduced from the orientation and directive properties of the radar antenna, and target range from the time elapsing between transmission of the illuminating pulse and reception of the echo. Typical maximum detection ranges would be 1.5–2.5 km for individual medium-sized ($\sim$100 mg) insects and up to several tens of kilometers for dense concentrations.

*Ground-based scanning radars*    Large volumes of space (typically $10^7$ m$^3$) can be conveniently scanned by causing a radar antenna to rotate about a vertical axis so that the sensing beam sweeps out a circle. Targets at the appropriate elevations all around the radar are then briefly illuminated once per revolution, and tilting the antenna up and down allows measurements at different altitudes. The range and bearing of targets detected with this type of scan are usually displayed on a conventional circular radar display, called a plan position indicator (PPI) (127). Sequential registration of the position of a particular target on the PPI provides a record of its flight trajectory. Comparison of these trajectories with wind velocity (derived from the trajectories of small hydrogen or helium-filled balloons carrying pieces of aluminium foil) allows accurate estimates of the target's flying speed and heading. Aerial densities may be estimated by counting the number of targets on the screen (129) or, in the case of higher densities when insects are not separately resolved, by altering the radar threshold of detection. Drake (34, 35) has documented particularly well both the theoretical basis for threshold density measurements and the general observational procedures for PPI-based entomological radars.

Two stages of target recognition are required in radar entomology. First it is vital to discriminate the echoes returned by insects from those of birds, bats, and precipitation; secondly, it is often necessary to identify the species of insect detected. The first stage is usually accomplished easily, at least at close range, because the diffuse echo returns from precipitation are very different from discrete insect echoes and because the flying speeds and radar-derived wingbeat frequencies of insects usually differ clearly from those of other animals (130). On the other hand, recognition of individual insect species from their radar returns is made difficult by intraspecies spread and interspecies overlap of wingbeat frequency (130, 132). In most successful radar studies identification has relied heavily on supplementary evidence of species composition acquired, for example, by aerial trapping (42–44, 58, 78, 136) or ground sampling in the takeoff area (43, 135).

*Vertical-looking radars*    Attempts have been made to improve the identification capacity of entomological radars by using supplementary vertical-looking radars (VLR) equipped with rotating polarization (132, 134). These radars respond differently to insects with different body shapes and thus potentially supply an extra identification feature. They also yield insect wingbeat frequency (132) and displacement speed (4). Because of the limited dynamic range of the receivers used in equipment to date, the identification potential of VLR systems has not yet been fully realized. These systems have, however, been used to great advantage in measuring the heading distributions of migrating insects (132, 134).

A further development of the rotating polarization principle, which promises to be even more useful, produces a slight wobbling (nutation) of the radar beam as the polarization rotates. In principle, this feature permits estimation of the size of insects flying overhead as well as of their shape and displacement velocity (12).

*Airborne radar*    Although the range of radar greatly exceeds that of all other methods of observing insects, it is still very limited when compared to the flight range of many species. To overcome this limitation Schaefer (143) mounted a downward-looking rotating polarization radar in an aircraft. Vertical profiles of insect-density and of average alignment could then be measured over the hundreds of kilometers of the aircraft's flight track (58). This powerful technique is undergoing further development (W. W. Wolf, personal communication; S. E. Hobbs, in preparation) and can be expected to find widespread application in the future.

OTHER TYPES OF RADAR    In addition to the 3.2-cm pulse radars that have been widely used for insect flight studies, a number of other types of radar have provided useful entomological observations from time to time. These are discussed below.

*Bistatic and Doppler radars for low-altitude studies*    Because ground features and vegetation reflect radar waves quite strongly, the echos from insects flying at low altitude may be obscured by ground returns. Except over very flat terrain, this problem of "ground clutter" limits the minimum altitude accessible to scanning radars to 20–30 m above ground level. In experiments where it is desirable to monitor flight activity below this height, it is possible to use a bistatic, intercepting-beam type of radar that will register the presence of flying insects down to 2–3 m above the ground (J. R. Riley, in preparation). Another, less quantitative method is to use a hand-held Doppler radar of the type used to enforce vehicle speed restrictions (161). These Doppler radars are able to detect moving targets against returns from a static background, but they normally have ranges of only a few tens of meters for insects.

*Millimetric-wavelength radars*    The strength of the echoes returned from insects varies aproximately as the square of their mass (131, 142), so small insects give very weak returns and are hard to detect except at very close range. One way to increase detection range is to use a radar with a wavelength in the millimetric region, which strongly increases the echo and hence the insect's detectability. This strategy has worked effectively in radar studies of planthoppers, whch weigh only a few milligrams (136). The practical drawback to this approach is that unlike the 3.2-cm marine radars, which are cheap and can be readily adapted for entomological work, millimetric systems are not generally available and may have to be custom built.

*Large radar installations*    One of the first demonstrations that insects can be followed by radar was made with large, powerful tracking systems (56). Although large radars offer a substantial range advantage, in practice this tends to be offset by their lack of mobility, by the uncertainty about the degree to which the sensing beam is occupied by insect targets, and particularly by the very high costs of constructing and running a large radar. The entomological use of such systems appears to have been limited to a few instances in which meteorological radar installations were fortuitously available (78, 120).

*Frequency-modulated continuous-wave (FMCW) radars*    Highly specialized FMCW radars have been built to study the fine-scale structure of the atmosphere (10, 125). These radars have exceptionaly high sensitivity and spatial resolution and are capable of measuring insect motion in atmospheric waves (4, 5, 126). This intriguing area, which may yield clues to insect orientation mechanisms (134), has not yet been explored.

*Tracking radars*    Tracking radars with 3-cm wavelength that were designed for military purposes have been used extensively in ornithological studies (1, 15, 91, 139), but have only rarely been applied to the observation of insect flight (78), perhaps because only one target may be tracked at a time. In contrast, a PPI scanning system can simultaneously record segments of trajectories from tens of targets while registering the presence of hundreds of other insects. This high data rate makes it possible to keep up with rapidly changing events that a simple tracking radar would be unable to resolve. Nevertheless, small tracking radars seem to be suitable for applications in which detailed information about short flight trajectories of individual large insects is required.

*Harmonic radars*    If a conductor with nonlinear characteristics (e.g. a small diode) is illuminated with radar waves, currents at harmonic frequencies of the illuminating signal will be generated. These currents will in turn reradiate the harmonics of the illuminating signal. Detection of these harmonics can be used to locate the diode and therefore anything to which it is attached. Attempts to tag screwworm flies, *Cochliomyia hominivorax,* with harmonic-generating diodes have been curtailed by lack of funds and also because of concern that the wire antennas that must be attached to the diode might interfere with insect flight (D. J. K. Krause, personal communication). Wire attachments are less of an impediment to walking insects, and the technique has apparently been successfully used to follow the movement of walking and burrowing carabid beetles (100). The range of detection is small (a few meters), but the equipment is portable, and surveillance of typical field areas seems to be practicable.

## Acoustic Sounding or Sodar

Sodar works in the same way as pulse radar, except that high-frequency sound waves are transmitted rather than electromagnetic waves. Most acoustic sounding equipment projects a narrow beam vertically upward and is used to monitor aspects of the temperature and wind structure in the lower atmosphere, typically up to an altitude of 1 km (82). Although sounders register the presence of airborne insects (25), there have been no reports of flight studies with this type of equipment, perhaps because of concern that insect behavior might be seriously perturbed by the high-intensity sound pulses. However, a low-power, short-range sounding device has been used to count *Heliothis virescens* moths approaching pheromone traps (79). Acoustic sounders have the potential to make indirect but useful contributions to radar studies of flight behavior by providing concurrent records of atmospheric structure and (with supplementary radar) vertical temperature profiles (99).

The use of sounders is not confined to the aerial environment. For example, the underwater nocturnal migration of aquatic larvae of the dipteran *Chaoborus flavicans* has been monitored with echo-sounding equipment (50).

## Ground-Based Optical Methods

A number of optical devices have been employed to monitor insect flight in the field, sometimes to provide coverage at altitudes below those accessible to scanning radars. These devices include daylight photographic and video systems (6, 7, 29, 54, 153); image-intensifier night-vision binoculars (58, 93, 95) and low-light photographic and video cameras (94, 106), both supplemented by infrared illuminators; crossed-beam infrared detectors (49, 135); and a gated image intensifier system linked to an infrared-filtered xenon flash illuminator (144). The gated intensifier system will work in daylight, albeit over a very small volume ($<1$ $m^3$); all of the other night-sensitive devices require a dark background and become ineffective during dusk and dawn.

## Remote Sensing Methods in the Laboratory

A number of devices for monitoring insect activity have found application in the laboratory, including microwave sensors (17), infrared emitter/detector matrices (118), video recording equipment (8, 27, 28, 164), and electrostatic charge detectors (13). Only the video and infrared matrix methods are useful in experiments in which perception of the direction, rather than just the occurrence, of movement is required.

## REMOTE SENSING OF INSECTS

Individual desert locusts (*Schistocerca gregaria*) standing or walking on open ground can be accurately counted on aerial photographs taken at a scale of 1:100 (137), but the technique has not found any practical application. Attempts to use Landsat imagery to detect the presence of large numbers of late-instar hopper bands of the Australian plague locust (*Chortoicetes ter-*

*minifera*) have been unsuccessful (102), and it appears that the only remote sensing techniques to have been profitably used in direct field observations of insects are radar and specialized optical methods.

## Radar Studies

In 1976 Schaefer (142) reported an eight-year series of radar investigations of the nocturnal flight of locusts, grasshoppers, and moths. Although he did not present any detailed case studies, Schaefer's work clearly constitutes a spectacular advance in knowledge about insect migratory flight. In particular, he described mass takeoff at dusk, rates of climb and mean altitudes of flight, layering, common orientation to geographical or wind directions, and the concentration of insects in areas of wind convergence. Schaefer's work in the Sudan is described in more detail in Reference 141.

MOTH FLIGHT    Drake et al (45) have used radar to show that a number of noctuid moth species (including *Persectania ewingii, Heliothis punctigera,* and *Agrotis munda*) regularly migrate from Australia to Tasmania, and other studies in Australia (40, 42, 44) have convincingly demonstrated the propensity for noctuids to make long-distance migratory flights over land as well. Drake found that the tendency for moths and other insects to be concentrated by wind convergence could be exploited in studies of atmospheric waves (39, 41) and sea-breeze effects (34, 39, 41, 44a). Radar observations of East African armyworm moths, *Spodoptera exempta,* showed that this species, at least in its gregarious phase, is an obligate windborne migrant and that the concentration of flying moths, particularly by storm-front outflows, is a major factor in outbreaks of caterpillars (116a, 135, 136b). By contrast, the Old World bollworm, *Heliothis armigera,* in India shows little tendency to undertake long-range migration above its entomological boundary layer, i.e. above the altitude at which wind speed exceeds flying speed (J. R. Riley, D. R. Reynolds, N. J. Armes, A. B. S. King, unpublished). This result is consistent with Joyce's (89) radar observation that only a small fraction (<10%) of *H. armigera* moths in flight climbed beyond their boundary layer. Wolf et al (163) have found evidence for northerly movements of corn earworm moths, *Heliothis zea,* and tobacco budworm moths, *Heliothis virescens,* from Mexico to the United States. In a novel shipborne study, he also demonstrated that moths and other insects can cross the widest part of the Gulf of Mexico on favorable winds (162). A large-scale Canadian study with ground-based and airborne radars (58, 30) showed that long-range migration was an important factor in the population dynamics of spruce budworm (*Choristoneura fumiferana*) and that moth arrival was often associated with frontal systems. Most authors report that moth headings are often nonrandom and are distributed about a common mean direction.

GRASSHOPPERS AND LOCUSTS    Riley & Reynolds have studied the migration of grasshoppers in the Niger flood plain in West Africa (132) and in the Sahel (124, 133). In both cases evidence was found for flights of over 100 km per night. Drake & Farrow (43) reported comparable displacements of *C. terminifera*. The Sahelian grasshoppers and *C. terminifera* both exhibited some collective orientation (38, 128, 133), and in at least one case there was evidence of bimodal heading distributions (132). *Chortoicetes terminifera* has also been studied in daylight flight (121).

OTHER INSECTS    Hendrie et al (78), using a very powerful 10-cm radar, have detected concentrations of airborne aphids up to 1200 m above ground level. On a smaller scale, aphids have also been observed with 3-cm radar during an experiment designed to measure their density above a suction trap (145). In other studies of small insects, Riley et al (136) used a millimetric radar in the Philippines to show that planthoppers and leafhoppers associated with rice engage predominantly in only short-range flight, and Downing & Frost (33) demonstrated that 1.87-cm radar could be used to detect low-altitude mosquito swarms.

Careful siting of standard entomological radar equipment allowed Loper et al (97) to observe honey bee (*Apis mellifera*) drone congregations at an altitude as low as 15 m. Radar has also been used to describe the low-altitude concentration of insects above a river in West Africa (123).

## Optical Studies

Sayer (138) and later Waloff (153) used vertical-looking photography to investigate the orientation of locusts to the wind in day-flying swarms. Subsequently Baker et al (6, 7) developed the technique further, using high-speed cinephotography.

In an investigation of the mechanisms of pheromone-influenced flight, David et al (29) mounted a downward-looking video camera on a 27-m tower to record simultaneously the movement of gypsy moths, *Lymantria dispar*, and windborne soap bubbles. In another study of flight to pheromone sources, Hollingsworth (81) employed a pair of synchronized video cameras to produce three-dimensional flight tracks of Lepidoptera. Gibson & Brady (54) have used a video method to record the flight of tsetse flies in host odor plumes in the field, and Schaefer & Bent (144) applied their sophisticated video detection and analysis system to calibrate a large suction trap and to observe aphid flight above a crop canopy.

Murlis & Bettany (106, 106a) fitted an image intensifier to a cinecamera to study nocturnal flight maneuvers by moths in pheromone plumes. Lingren et al (94) used similar arrangements with video cameras to observe moth behavior in and above the crop canopy. Lingren et al (93) have also evaluated several night-viewing devices for entomological studies. Peloquin & Olson (117) used a video camera that was sensitive in the near infrared and an infrared illuminator to record flying swarms of male mosquitoes. In a novel experiment, Okubo & Chiang (113) found that by filming cecidomyid midges against a white background it was possible to combine shadow and image movement to reconstruct three-dimensional flight trajectories.

## REMOTE SENSING OF THE EFFECTS PRODUCED BY INSECTS

### Forest Insects

Remote sensing methods were first used by forest entomologists in the 1930s, when airborne observers made sketch maps of defoliation caused by hemlock looper (*Lambdina fiscellaria fiscellaria*) infestations (19). Sketch mapping of forest damage was supplemented, but not replaced, by increasingly sophisticated aerial photography from the 1950s onward (11, 19, 23, 76, 157). More recently, airborne videography (98, 110) and MSS and photographic data from satellites (32, 122, 157) have been introduced. These methods are used for rapid and economical identification of areas of forest being attacked by various insects so that the status and trend of these pest species can be evaluated and appropriate action can be taken (21). Many of these photographic and satellite imagery studies have been summarized elsewhere (22, 75, 76).

AIRBORNE SYSTEMS    Color and color-infrared aerial photography with conventional cameras can be used effectively to delineate damage caused by a number of serious pests, including the defoliators hemlock looper (86, 166) and spruce budworm (*C. fumiferana*) (86, 101, 166) and several destructive bark beetles (*Dendroctonus* spp.) (76, 84, 92, 159). Hall et al (60) have found that it is possible to detect beetle attack of douglas fir before any changes in foliage are visually apparent. Ward et al (156) recently summarized hardwood defoliation detection techniques. The power of aerial photography has been greatly increased by the advent of nonconventional camera systems that can quickly photograph very large areas at high resolution [e.g. 4400 square miles in 16 min (23)], so that it is now technically feasible to conduct rapid, extensive and detailed photographic surveys for damage caused by forest pests (2, 23, 46, 158). The United States Department of Agriculture has conducted three such surveys for gypsy moth (*L. dispar*) defoliation over large areas of the northeastern United States. However, substantial costs and uncertainties are involved in organizing major collaborative high-altitude reconnaissance missions within the narrow time windows when defoliation damage is most perceptable and not obscured by cloud cover (23). In many instances the use of experienced airborne observers and limited-area photography remains the best economic option, especially now that low-cost, accurate aerial navigation equipment (Loran-C) (59, 98) and computer-aided photograph registration methods have made damage location more precise and efficient (59). The statement made by Jano (86) in 1980 that "there is as yet nothing better than the keen eye of the airborne observer in the initial detection of damage" is probably still valid.

Initial experiments with airborne video recording of defoliation by larvae of the gypsy moth (98), defoliation by forest tent caterpillars, *Malacosoma disstria* (110), and leaf discoloration caused by mountain pine beetle, *Dendroctonus ponderosae* (110), have produced promising interim results; it seems likely that this inexpensive and flexible medium will find increasing use as a valuable supplement to visual observations. Recording and displaying Loran-C data on video imagery (110) should provide a convenient and time-saving solution to the awkward problem of geographically registering the images.

SATELLITE SYSTEMS    The coarse spectral and spatial resolution available from the Landsat MSS, combined with data loss due to cloud cover and the lengthy data turnaround time, severely limits the utility of satellite MSS imagery for forest entomologists (73, 86, 122, 147, 166), and it is now rarely used (75). Dottavio & Williams (32) claimed that the combination of MSS imagery with a statewide mosaic data base allowed automated assessment of gypsy moth defoliation in Pennsylvania. Only heavy defoliation could be registered accurately, however, and the authors conceded that problems with cloudiness would make reliance on the Landsat data "operationally risky."

High-altitude airborne simulations of thematic mapper (TM) performance have demonstrated that unlike MSS, this device should provide acceptable estimates of pine beetle infestation (122) and that its superior spectral and spatial resolution should also give better measurements of defoliation (32). Similarly, the three-spectral-band, high-resolution visible (HRV) scanners carried by the SPOT-1 satellite (148) offer the prospect of improved defoliation detection. Ciesla & Eav (24) concluded however, that although both Landsat 5 and SPOT-1 data are potential cost-effective alternatives to high-altitude aerial photographs, the probability of either satellite acquiring suitable data for hardwood defoliation surveys is low. Their applicability will increase if more earth resources satellites are launched.

## Insect Pests of Fruit Trees

Remote sensing of insect pests of fruit trees was pioneered in 1968 by Hart et al (62, 70), who exploited the fact that a number of insect pests of fruit trees produce honeydew that serves as a host medium for a sooty mold fungus. This mold blackens foliage and thus provides a clue of insect presence that is readily detectable by aerial photography. The initial studies were on brown soft scale, *Coccus hesperidum,* infesting citrus groves. The technique was later said to be sensitive enough to detect trees with less than 1% of their leaves infected (67, 109). The cited literature (65), however, did not explicitly make this claim.

Another honeydew-producing insect, the citrus mealybug, *Planococcus citri,* was also readily detectable; moreover, it could be specifically identified because its distribution within tree crowns was such as to produce a characteristic blotchy pattern of mold (67).

Infestations of citrus blackfly, *Aleurocanthus woglumi,* were less evident on aerial photographs, although localized heavy and medium attacks could be detected (67, 69). Payne et al (115) have reported that infestations of the European red mite, *Panonychus ulmi,* can be detected in low-altitude color infrared photographs of peach orchards. They concluded that reduction in leaf chlorophyll content caused by the mites may account for the different appearance of healthy and infected trees. Photographic evidence of attack on pecan trees by root borers (*Prionus* spp.) was also reported (115). In later experiments, Harris et al (61) showed that defoliation of pecans by walnut caterpillars, *Datana integerrima,* is clearly visible on color infrared film.

The motivation for these aerial studies was to promote the early detection of potentially damaging infestations and to evaluate the efficiency of control measures by comparing treated with untreated areas. Aerial photography was considered less expensive than the ground surveys required to accomplish the same tasks (70). In a 1975 examination of SKYLAB color infrared photographs, Hart et al (66) concluded that heavy infestations of honeydew-producing insects would have been detectable if any had been present. No other satellite-based studies of fruit trees have been reported.

## Insect Pests of Crops

Sooty mold has also been used to indicate the presence of corn leaf aphids, *Rhopalosiphum maidis* (155), and sweetpotato whiteflies, *Bemisia tabaci* (112). In the aphid study, photographs were taken from 2000 m, and different levels and areas of infestation were successfully measured with the aid of photographic enhancement and computerized area-estimation methods. The whitefly-induced sooty mold could be detected on cotton from 3000 m, and photographs from 2000 m yielded good resolution of mold growth patterns. It was also discovered that lettuce infectious yellow virus, which is transmitted by whitefly and causes leaf yellowing, could be detected from low altitude on normal color film (112).

# REMOTE SENSING OF THE INSECTS' ENVIRONMENT

Entomologists have used remote sensing methods to investigate three aspects of the environment that are of particular importance to insects: host-plant distribution, rainfall, and air temperature.

## Host Plants and Rainfall

Hart and co-workers (68) and Diez et al (31) have used aerial photography to investigate the distribution of host plants of tropical fruitflies in Hawaii, El Salvador, and Mexico. Zalucki et al (165) in Australia used the same technique to map areas of milkweed (*Asclepias* spp.), a major host of the monarch butterfly (*Danaus plexippus*). Because cotton left in the field after harvesting is an important host for overwintering boll weevils, *Anthonomus grandis*, in the lower Rio Grande valley in Texas, postharvest plant removal has been made mandatory. Aerial photography is now routinely used in annual surveys to detect infringements of this regulation (63). Researchers at the Overseas Development Natural Resources Institute (ODNRI) in London are currently developing a method based on SPOT and Landsat 5 high-resolution imagery to identify rice production areas in northern Luzon, the Philippines, that could act as hosts for the brown planthopper, *Nilaparvata lugens*. They are also using high-resolution imagery to investigate the possibility of creating a buffer zone against *A. grandis* in Paraguay (J. Pender, personal communication).

DESERT LOCUST HABITATS    The satellite-based studies initiated by Pedgley (116) in 1973, which sought to identify areas of ephemeral vegetation capable of supporting desert locusts (*S. gregaria*), have been continued by the United Nations Food and Agriculture Organization (FAO) (80a, 150). It has proved impractical and too expensive to use Landsat data for real-time locust forecasting, but the FAO researchers have found that the more frequent imagery of vegetation available from the advanced very high resolution radiometer (AVHRR) on United States National Oceanic and Atmospheric Administration (NOAA) meteorological satellites makes a valuable contribution to their operational desert locust forecasting procedure. These data are especially useful when combined with estimates of rainfall from infrared cloud imagery (J. Roffey, personal communication).

AUSTRALIAN PLAGUE LOCUST HABITATS    Mitchell grass (*Astrebla* spp.) tends to be the dominant vegetation in habitats favored by *C. terminifera* in the Channel Country of southwest Queensland, so it could be used as an indicator of areas where these insects might breed. This grass produces relatively sparse cover and is unfortunately not detectable on Landsat imagery (103). Bryceson & Wright (16) concluded nevertheless that a judicious combination of ground surveys, rainfall data, and imagery of other, more readily detectable vegetation makes it feasible to detect areas that have become suitable for locust beeding.

RAINFALL AND THE EAST AFRICAN ARMYWORM    Flying moths of the East African armyworm, *S. exempta,* are concentrated by convergent winds associated with rainstorms, and the subsequent mass laying of the concentrated moths leads to dense and destructive outbreaks of caterpillars (136a). Remote sensing of rainstorms in the appropriate areas thus offers the prospect of rapidly locating potential outbreaks. Garland (51) investigated this idea using thermal images of cloud tops from Meteosats 1 and 2, with promising results. The method is now being used by the armyworm forecasting service of the Desert Locust Control Organization for Eastern Africa (DLCOEA) in Nairobi to supplement its ground survey data (D. Pedgley, personal communication).

MOSQUITO HABITATS    Aerial photographs have proved extremely useful in surveying for mosquito breeding areas in western and southern Africa (105) and the United States (154). Attempts to use MSS data from Landsats 1 and 2 have been less successful because the satellite data lacked the resolution required to discern the complex patterns of mosquito habitats (72).

## Air Temperature

The screwworm (*C. hominivorax*), a serious pest of cattle, is particularly sensitive to variations in temperature. Thus its ingress into the United States from Mexico is controlled by the northward progression of the appropriate temperature window of 25–30°C each summer. United States National Aeronautics and Space Administration (NASA) scientists analyzed data from NOAA meteorological satellites to monitor this progression, and the results were used in coordinating a sterile-male release program (55). Although the method worked successfully, the daily handling and reducing of satellite data proved to be a major and expensive task, and the experiment was not repeated. Giddings (55) noted, however, that a more logical and more economical approach would be to use satellite data to create a temperature model that could thereafter be routinely used for interpolation of data acquired by surface stations.

## SUMMARY

Entomological remote sensing using radar and specialized optical techniques is a very active field and has already made major contributions to the study of insect flight. There is considerable promise that techniques currently under development will greatly improve the identification capabilities of entomological radars, and this will facilitate the study of more species. Automatic, remote-control entomological radars for routine monitoring will probably be available within the next few years.

In spite of much early optimism and effort, satellite-based MSS observations have proved at best only partially effective for routine insect damage surveys. Although higher-resolution data of the type available from Landsat 5 and SPOT have probably made useful damage surveys technically feasible, the cost of launching enough satellites specifically to obtain the necessary regularity of coverage would be prohibitive. Further applications in this field thus seem unlikely in the immediate future. In contrast, hybrid systems that incorporate data from meteorological satellites into advanced biogeographical models can be expected to continue to improve forecasting methods for at least two major pest species. The use of high-resolution imagery to build up cropping pattern data bases for subsequent use in planning pest control stategies also holds considerable promise.

Very high-altitude, panoramic aerial photography provides an effective and elegant means of conducting large-area surveys for insect damage. On the other hand, visual observations from light aircraft, supplemented by videography and photography, can be made with a much higher degree of local control and at much shorter notice. Choice between the two techniques is thus likely to be made on the basis of operational and economic considerations.

ACKNOWLEDGMENTS

I thank my colleagues at the Overseas Development Natural Resources Institute for granting me access to their records and for their advice. Particular thanks are due to Dr. D. R. Reynolds, who was responsible for identifying and gathering together much of the literature on which this review is based.

*Literature Cited*

1. Able, K. P. 1977. The flight behaviour of individual passerine nocturnal migrants: A tracking radar study. *Anim. Behav.* 25:924–35
2. Acciavatti, R. E., Ward, J. G. D. 1987. *Final report—optical bar panoramic-photography for mapping gypsy moth defoliation in portions of the northeastern United States.* US Dep. Agric. For. Serv., Morgantown, W. Va. 17 pp.
3. Aldrich, R. C. 1980. Satisfying multisource information needs for national assessment—a challenge to remote sensing. See Ref. 77, pp. 5–13
4. Atlas, D., Harris, F. I., Richter, J. H. 1970. Measurement of point target speeds with incoherent non-tracking radar: insect speeds in atmospheric waves. *J. Geophys. Res.* 75:7588–95
5. Atlas, D., Metcalf, J. I., Richter, J. H., Gossard, E. E. 1970. The birth of "CAT" and microscale turbulence. *J. Atmos. Sci.* 27:903–13
6. Baker, P. S., Gewecke, M., Cooter, R. J. 1981. The natural flight of the migratory locust, *Locusta migratoria* L. *J. Comp. Physiol.* 141:233–37
7. Baker, P. S., Gewecke, M., Cooter, R. J. 1984. Flight orientation of swarming *Locusta migratoria*. *Physiol. Entomol.* 9:247–52
8. Baker, T. C., Haynes, K. F. 1987. Manoeuvres used by flying male oriental fruit moths to relocate a sex pheromone

plume in an experimentally shifted wind field. *Physiol. Entomol.* 12:263–79

9. Barrett, E. C., Curtis, L. F. 1982. *Introduction to Environmental Remote Sensing.* London/New York: Chapman & Hall. 352 pp. 2nd ed.

10. Bean, B. R., McGavin, R. E., Chadwick, R. B. Warner, B. D. 1971. Preliminary results of utilizing the high resolution FM radar as a boundary-layer probe. *Boundary-Layer Meteorol.* 1: 466–73

11. Beaubien, J., Jobin, L. 1974. Forest insect damage and cover types from high altitude color-IR photographs and ERTS-1 imagery. *Proc. Symp. Remote Sensing Photo Interpretation, Banff, Canada,* pp. 449–54. Banff: Can. Cent. Remote Sensing

12. Bent, G. A. 1984. Developments in detection of airborne aphids with radar. *Proc. Br. Crop Prot. Conf. Brighton, UK,* 2:665–74. Croydon, UK: Br. Crop Prot. Counc.

13. Berry, I. L. 1973. Improved system for measuring flying activity of insects by detecting static charges. *J. Econ. Entomol.* 66(3):820–22

14. Blazquez, C. H. 1980. Remote sensing and its potential role in pest management. *Insectic. Acaric. Tests* 5:2–4

15. Bruderer, B., Steidinger, P. 1972. Methods of quantitative and qualitative analysis of bird migration with a tracking radar. In *Animal Orientation and Navigation,* ed. S. R. Galler, K. Schmidt-Koenig, G. J. Jacobs, R. E. Belleville, pp. 151–67. Washington, DC: NASA. 606 pp.

16. Bryceson, K. P., Wright, D. E. 1986. An analysis of the 1984 locust plague in Australia using multitemporal Landsat multispectral data and a simulation model of locust development. *Agric. Ecosyst. Environ.* 16:87–102

17. Buchan, P. B., Moreton, R. B. 1981. Flying and walking of small insects (*Musca domestica*) recorded differentially with a standing-wave radar actograph. *Physiol. Entomol.* 6:149–55

18. Chiang, H. C., Latham, R., Meyer, M. P. 1973. Aerial photography: use in detecting simulated insect defoliation in corn. *J. Econ. Entomol.* 66:779–84

19. Chiang, H. C., Meyer, M. P. 1974. Remote sensing applications to agricultural monitoring. *EPPO Bull.* 4(3):309–17

20. Chiang, H. C., Meyer, M. P., Jensen, M. S. 1976. Armyworm defoliation in corn as seen on IR aerial photographs. *Entomol. Exp. Appl.* 20:301–3

21. Ciesla, W. M. 1980. Operational remote sensing for forest insect and disease management—some challenges and opportunities. See Ref. 77, pp. 39–47

22. Ciesla, W. M. 1983. Aerial mapping of forest insect and disease image. *Proc. Symp. Appl. Remote Sensing Resour. Manage.* pp. 362–80. Falls Church, Va: Am. Soc. Photogramm. Remote Sensing

23. Ciesla, W. M. 1984. Mission: track the gypsy from 65,000 feet. *Am. For.* 90(7):30–33, 54–56

24. Ciesla, W. M., Eav, B. B. 1987. *Satellite imagery versus aerial photos for mapping hardwood defoliation: a preliminary evaluation of cost and acquisition feasibility. USDA forest pest management/methods application group Rep.* 87–2, Fort Collins, Colo. 16 pp.

25. Cronenwett, W. T., Walker, G. B., Inman, R. L. 1972. Acoustic sounding of meteorological phenomena in the planetary boundary layer. *J. Appl. Meteorol.* 11:1351–58

26. Danthanarayana, W., ed. 1986. *Insect Flight Dispersal and Migration.* Berlin/Heidelberg: Springer-Verlag. 289 pp.

27. David, C. T. 1982. Compensation for height in the control of groundspeed by *Drosophila* in a new "barber's pole" wind tunnel. *J. Comp. Physiol.* 147:485–93

28. David, C. T. 1985. Visual control of the partition of flight force between lift and thrust in free-flying *Drosophila. Nature* 313:48–50

29. David, C. T., Kennedy, J. S., Ludlow, A. R. 1983. Finding of a sex pheromone source by gypsy moths in the field. *Nature* 303:804–6

30. Dickison, R. B. B., Haggis, M. J., Rainey, R. C., Burns, L. M. D. 1986. Spruce budworm moth flight and storms: further studies using aircraft and radar. *J. Clim. Appl. Meteorol.* 25(11):1600–8

31. Diez, J. A., Hart, W. G., Ingle, S. J., Davis, M. R., Rivera, S. 1980. The use of remote sensing in detection of host plants of Mediterranean fruit flies in Mexico. *Proc. Int. Symp. Remote Sensing Environ. 14th, San Jose, Costa Rica,* pp. 675–83. Ann Arbor, Mich: Environ. Res. Inst. Mich.

32. Dottavio, C. L., Williams, D. L. 1983. Satellite technology: an improved means for monitoring forest insect defoliation. *J. For.* 81(1):30–34

33. Downing, J. D., Frost, E. L. 1972. Recent radar observations of diurnal insect behavior. *Proc. Annu. Meet. NJ Mosq. Exterm. Soc., 59th, Atlantic City, NJ,* pp. 114–31

34. Drake, V. A. 1981. Target density estimation in radar biology. *J. Theor. Biol.* 90:545–71

35. Drake, V. A. 1981. *Quantitative observation and analysis procedures for a manually operated entomological radar. CSIRO Div. Entomol. Tech. Pap. No. 19,* Melbourne, Australia. 41 pp.

36. Drake, V. A. 1982. The CSIRO entomological radar: a remote-sensing instrument for insect migration research. In *Scientific Instruments in Primary Production,* ed. L. A. Wisbey, pp. 63–73. Melbourne: Aust. Sci. Ind. Assoc. 181 pp.

37. Drake, V. A. 1982. Insects in the sea-breeze front at Canberra: a radar study. *Weather* 37(5):135–43

38. Drake, V. A. 1983. Collective orientation by nocturnally migrating Australian plague locusts *Chortoicetes terminifera* (Walker) (Orthoptera: Acrididae): a radar study. *Bull. Entomol. Res.* 73: 679–92

39. Drake, V. A. 1984. A solitary wave disturbance of the marine boundary layer over Spencer Gulf revealed by radar observations of migrating insects. *Aust. Meteorol. Mag.* 32:131–35

40. Drake, V. A. 1984. The vertical distribution of macro-insects migrating in the nocturnal boundary layer: a radar study. *Boundary-Layer Meteorol.* 28: 353–74

41. Drake, V. A. 1985. Solitary wave disturbances of the nocturnal boundary layer revealed by radar observations of migrating insects. *Boundary-Layer Meteorol.* 31:269–86

42. Drake, V. A. 1985. Radar observations of moths migrating in a nocturnal low-level jet. *Ecol. Entomol.* 10:259–65

43. Drake, V. A., Farrow, R. A. 1983. The nocturnal migration of the Australian plague locust, *Chortoicetes terminifera* (Walker) (Orthoptera: Acrididae): quantitative radar observations of a series of northward flights. *Bull. Entomol. Res.* 73:567–85

44. Drake, V. A., Farrow, R. A. 1985. A radar and aerial-trapping study of an early spring migration of moths (Lepidoptera) in inland New South Wales. *Aust. J. Ecol.* 10:223–35

44a. Drake, V. A., Farrow, R. A. 1988. The influence of atmospheric structure and motions on insect migration. *Ann. Rev. Entomol.* 33:183–210

45. Drake, V. A., Helm, K. F., Readshaw, J. L., Reid, D. G., 1981. Insect migration across Bass Strait during spring: a radar study. *Bull. Entomol. Res.* 71: 449–66

46. Eav, B. B., Dillman, R. D., White, W. B. 1984. High-altitude photography for inventories of mountain pine beetle damage. *J. For.* 82:175–77

47. Estes, J. E., Thorley, G. A., Colwell, R. N., eds. 1983. *Manual of Remote Sensing,* Vol. 2, *Interpretation and Applications.* Falls Church, Va: Am. Soc. Photogramm. 1218 pp.

48. Everitt, J. H., Nixon, P. R., eds. 1987. *Proc. Bienn. Workshop Color Aerial Photogr. Plant Sci. 11th, Weslaco, Tex.* Falls Church, Va: Am. Soc. Photogramm. Remote Sensing. 343 pp.

49. Farmery, M. J. 1981. *Optical studies of insect flight at low altitude.* PhD thesis. Univ. York, UK. 141 pp.

50. Franke, C. 1987. Detection of transversal migration of larvae of *Chaoborus flavicans* (Diptera, Chaoboridae) by the use of a sonar system. *Arch. Hydrobiol.* 109(3):355–66

51. Garland, A. C. 1985. *The use of Meteosat in the analysis of African armyworm outbreaks in East Africa.* PhD thesis. Univ. Bristol, UK. 289 pp.

52. Gausman, H. W. 1987. Photographic and video dramas of remote sensing with plants as thespians. See Ref. 48, pp. 5–20

53. Gausman, H. W., Hart, W. G. 1974. Reflectance of sooty mold fungus on citrus leaves over the 2.5 to 40 micrometer wavelength interval. *J. Econ. Entomol.* 67(4):479–80

54. Gibson, G., Brady, J. 1985. "Anemotactic" flight paths of tsetse flies in relation to host odour: a preliminary video study in nature of the response to loss of odour. *Physiol. Entomol.* 10:395–406

55. Giddings, L. E. 1976. *Extension of surface data by use of meteorological satellites. Lockheed Corp. Tech. Mem. LEC 8377,* Lockheed Electron. Corp., Houston, Tex. 32 pp.

56. Glover, K. M., Hardy, K. R., Konrad, T. G., Sullivan, W. N., Michaels, A. S. 1966. Radar observations of insects in free flight. *Science* 154:967–72

57. Green, L. R., Olson, J. K., Hart, W. G., Davis, M. R. 1977. Aerial photographic detection of imported fire ant mounds. *Photogramm. Eng. Remote Sensing* 43(8):1051–57

58. Greenbank, D. O., Schaefer, G. W., Rainey, R. C. 1980. *Spruce budworm (Lepidoptera: Tortricidae) moth flight and dispersal: new understanding from canopy observations, radar and aircraft. Mem. Entomol. Soc. Can.* Vol. 110. 49 pp.

59. Hain, F. P. 1983. Sampling and predicting population trends. In *The Southern*

Pine Beetle, *US Dep. Agric. For. Serv. Bull. 1631,* ed. R. C. Thatcher, J. L. Searcy, J. E. Coster, G. D. Hertel, pp. 107–35. Washington, DC: US Dep. Agric. For. Serv. 267 pp.

60. Hall, P. M., McClean, J. A., Murtha, P. A. 1983. Dye-layer density analysis identifies new Douglas-fir beetle attacks. *Can. J. For. Res.* 13:279–82

61. Harris, M. K., Hart, W. G., Davis, M. R., Ingle, S. J., Van Cleave, H. W. 1976. Aerial photographs show caterpillar infestation. *Pecan Q.* 10(2):12–18

62. Hart, W. G. 1978. Remote sensing in horticulture. *Proc. Int. Soc. Citricult.* 3:168–71

63. Hart, W. G. 1986. *Evaluation of ecological factors that contribute to insect population increase.* Presented at Parasitis '86, Geneva

64. Hart, W. G., Gausman, H. W., Rodriguez, R. R. 1976. Citrus blackfly (Hemiptera: Aleyrodidae) feeding injury and its influence on the spectral properties of citrus foliage. *Rio Grande Valley Hortic. Soc. J.* 30:37–43

65. Hart, W. G., Ingle, S. J. 1969. Detection of arthropod activity on citrus foliage with aerial infrared color photography. *Proc. Workshop Aerial Color Photogr. Plant Sci. Gainesville, Fla.,* pp. 85–88

66. Hart, W. G., Ingle, S. J., Davis, M. R. 1975. The use of skylab data to study the early detection of insect infestations and density and distribution of host plants. *NASA Earth Resour. Surv. Symp. Houston,* 1-A:203–19. Houston: NASA

67. Hart, W. G., Ingle, S. J., Davis, M. R. 1977. Remote sensing of insect populations attacking citrus. *Proc. Int. Soc. Citricult.* 2:485–87

68. Hart, W. G., Ingle, S. J., Davis, M. R. 1978. The use of color infrared aerial photography to detect plants attacked by tropical fruit flies. *Proc. Int. Symp. Remote Sensing Environ. 12th, Ann Arbor, Mich.,* 2:1409–13. Ann Arbor, Mich: Environ. Res. Inst. Mich.

69. Hart, W. G., Ingle, S. J., Davis, M. R., Mangum, C. 1973. Aerial Photography with Infrared Color Film as a Method of Surveying for Citrus Blackfly. *J. Econ. Entomol.* 66(1):190–94

70. Hart, W. G., Ingle, S. J., Davis, M. R., Mangum, C., Higgins, A., et al. 1971. Some uses of infrared aerial color photography in entomology. *Proc. Bienn. Workshop Color Aerial Photogr. Plant Sci. 3rd, Gainesville, Fla.,* pp. 99–113

71. Hart, W. G., Myers, V. I. 1968. Infrared aerial color photography for detection of populations of brown soft scale in citrus groves. *J. Econ. Entomol.* 61(3):617–24

72. Hayes, R. O., Maxwell, E. L., Mitchell, C. J., Woodzick, T. L. 1985. Detection, identification, and classification of mosquito larval habitats using remote sensing scanners in earth-orbiting satellites. *Bull. World Health Organ.* 63(2):361–74

73. Heller, R. C. 1975. Remote sensing to detect forest diseases and insects. *Proc. World Tech. Consult. For. Dis. Insects, 2nd, New Delhi.* Rome: Food Agric. Organ. 8 pp.

74. Heller, R. C. 1978. Case applications of remote sensing for vegetation damage assessment. *Photogramm. Eng. Remote Sensing* 44(9):1159–66

75. Heller, R. C. 1985. Remote sensing: its state of the art in forestry. In *Remote Sensing in Forest and Range Resource Management. Proc. Pecora Symp. 10th,* pp. 18–29. Falls Church, Va: Am. Soc. Photogramm. Remote Sensing

76. Heller, R. C., Ulliman, J. J., Aldrich, R. C., Hildebrandt, G., Hoffer, R. M., et al. 1983. Forest resource assessments. See Ref. 47, pp. 2229–324

77. Heller, R. C., Ulliman, J. J., Hall, W. B., eds. 1980. *Proc. Remote Sensing Nat. Resour., Moscow, Idaho, 1979*

78. Hendrie, L. K., Irwin, M. E., Liquido, N. J., Ruesink, W. G., Mueller, E. A., et al. 1986. Conceptual approach to modelling aphid migration. In *The Movement and Dispersal of Agriculturally Important Biotic Agents,* ed. D. R. MacKenzie, C. S. Barfield, G. G. Kennedy, R. D. Berger, D. J. Taranto, pp. 541–82. Baton Rouge, La: Claitor's. 611 pp.

79. Hendricks, D. E. 1980. Low-frequency sodar device that counts flying insects attracted to sex pheromone dispensers. *Environ. Entomol.* 9:452–57

80. Henninger, J., Hildebrandt, G. 1980. *Bibliography of Publications on Damage Assessment in Forestry and Agriculture by Remote Sensing Techniques.* Freiburg, W. Germany: Univ. Freiburg

80a. Hielkema, J. U., Roffey, J., Tucker, C. J. 1986. Assessment of ecological conditions associated with the 1980/81 desert locust plague upsurge in West Africa using environmental satellite data. *Int. J. Remote Sensing* 7:1609–22

81. Hollingsworth, T. S. 1986. *The influence of local wind effects upon the approach behaviour of some male Lepidoptera to field pheromone sources.*

PhD thesis. Cranfield Inst. Technol., Bedford, UK. 270 pp.

82. Holmes, N. E., Lunken, A. C., McIlvenn, J. F. R. 1976. Acoustic sounding of the atmosphere. *Weather* 31(7):218–22, 227–35

83. Hooper, A. J. 1980. Aerial photography as an aid to pest forecasting. *Aeronaut. J.* 1980:136–40

84. Hostetler, B. B., Young, R. W. 1979. *A pilot survey to measure annual mortality of ponderosa pine caused by the mountain pine beetle in the Black Hills of South Dakota. US Dep. Agric. For. Serv. Reg. 2 Tech. Rep. R2–20,* Lakewood, Colo.

85. Ingle, S. J., Kreasky, J. B., Davis, M. R., Meyerdirk, D. E., Hart, W. G. 1982. The use of color infrared aerial photography to monitor the progress of lethal decline of date palms in the lower Rio Grande valley of Texas. *Proc. Fla. State Hortic. Soc.* 95:260–61

86. Jano, A. P. 1981. Practical applications of remote sensing for forest pest damage appraisal. *Proc. Sem. Uses Remote Sensing For. Pest Damage Appraisal, Rep. NOR-X-238,* pp. 30–39. Edmonton, Canada: North. For. Res. Cent.

87. Jonkman, J. C. M. 1976. Biology and ecology of the leaf cutting ant *Atta vollenweideri* Forel, 1893. *Z. Angew. Entomol.* 81:140–48

88. Jonkman, J. C. M. 1979. Population dynamics of leaf cutting ant nests in Paraguayan pasture. *Z. Angew. Entomol.* 87:281–93

89. Joyce, R. J. V. 1983. Aerial transport of pests and pest outbreaks. *EPPO Bull.* 13:111–19

90. King, D. L., Vleck, J., Yuan, X. 1987. A 4-camera video sensor: its performance and application with special regards to forestry. See Ref. 48, pp. 286–94

91. Larkin, R. P., Thompson, D. 1980. Flight speeds of birds observed with radar; evidence for two phases of migratory flight. *Behav. Ecol. Sociobiol.* 7:301–17

92. Lee, Y. J., Alfaro, R. I., Van Sickle, G. A. 1983. Tree-crown defoliation measurement from digitised photographs. *Can. J. For. Res.* 13:956–61

93. Lingren, P. D., Raulston, J. R., Henneberry, T. J., Sparks, A. N. 1986. Night-vision equipment, reproductive biology and nocturnal behavior: importance to studies of insect flight, dispersal, and migration. See Ref. 26, pp. 253–64

94. Lingren, P. D., Raulston, J. R., Sparks, A. N., Wolf, W. W. 1982. Insect monitoring technology for evaluation of suppression via pheromone systems. In *Insect Suppression With Controlled Release Pheromone Systems,* ed. A. F. Kydonieus, M. Beroza, G. Zweig, 1:171–93. Boca Raton, Fla: CRC. 274 pp.

95. Lingren, P. D., Wolf, W. W. 1982. Nocturnal activity of the tobacco budworm and other insects. In *Biometeorology in Integrated Pest Management,* ed. J. L. Hatfield, I. J. Thomason, pp. 211–28. New York/London: Academic. 491 pp.

96. Lo, C. P. 1986. *Applied Remote Sensing.* New York: Longman. 393 pp.

97. Loper, G. M., Wolf, W. W., Taylor, O. R. 1987. Detection and monitoring of honeybee drone congregation areas by radar. *Apidologie* 18:163–72

98. Lusch, D. P., Sapio, F. J. 1987. Mapping gypsy moth defoliation in Michigan using airborne CIR video. See Ref. 48, pp. 261–69

99. Marshall, J. N., Peterson, A. M., Barnes, A. A. 1972. Combined radar-acoustic sounding system. *Appl. Optics.* 11:108–12

100. Mascanzoni, D., Wallin, H. 1986. The harmonic radar: a new method of tracing insects in the field. *Ecol. Entomol.* 11:387–90

101. McCarthy, J., Olson, C. E., Witter, J. A. 1983. Assessing spruce budworm damage with small-format aerial photographs. *Can. J. For. Res.* 13:395–99

102. McCulloch, L. 1983. Detection of nymphal bands from satellite imagery. In *Australian Plague Locust Commission Annual Report 1981–82,* p. 20. Canberra: Aust. Gov. Publ. Serv. 21 pp.

103. McCulloch, L., Hunter, D. M. 1983. Identification and monitoring of Australian plague locust habitats from Landsat. *Remote Sensing Environ.* 13: 95–102

104. Meisner, D. E., Lindstrom, O. M. 1985. Design and operation of a color infrared aerial video system. *Photogramm. Eng. Remote Sensing* 51(5): 555–60

105. Muirhead-Thomson, R. C. 1973. The use of vertical aerial photographs in rural yellow fever mosquito surveys. *Mosq. News* 33(2):241–43

106. Murlis, J., Bettany, B. W. 1977. Night flight towards a sex pheromone source by male *Spodoptera littoralis* (Boisd.) (Lepidoptera, Noctuidae). *Nature* 268: 433–35

106a. Murlis, J., Bettany, B. W., Kelly, J., Martin, L. 1982. The analysis of flight paths of male Egyptian cotton leafworm moths, *Spodoptera littoralis*, to a sex pheromone source in the field. *Physiol. Entomol.* 7:435–41

107. Murtha, P. A. 1978. Remote sensing and vegetation damage: a theory for detection and assessment. *Photogramm. Eng. Remote Sensing* 44(9):1147–58

108. Murtha, P. A. 1982. Detection and analysis of vegetation stress. In *Remote Sensing for Resource Management*, ed. C. J. Johannsen, J. L. Sandus, pp. 141–57. Ankeny, Iowa: Soil Conserv. Soc. Am. 688 pp.

109. Myers, V. I., Bauer, M. E., Gausman, H. W., Hart, W. G., Heilman, J. L., et al. 1983. Remote sensing applications in agriculture. See Ref. 47, pp. 2182–85

110. Myhre, R. J., Munson, A. S., Meisner, D. E., Dewhurst, S. 1987. Assessment of a color infrared aerial video system for forest insect detection and evaluation. See Ref. 48, pp. 244–51

111. Nixon, P. R., Escobar, D. E., Bowen, R. L. 1987. A multispectral false color-video imaging system for remote sensing applications. See Ref. 48, pp. 295–305

112. Nuessly, G. S., Meyerdirk, D. E., Hart, W. G., Davis, M. R. 1987. Evaluation of color-infrared aerial photography as a tool for the identification of sweetpotato whitefly induced fungal and viral infestations of cotton and lettuce. See Ref. 48, pp. 141–48

113. Okubo, A., Chiang, H. C. 1974. An analysis of the kinematics of swarming *Anarete pritchardi* (Kim) (Diptera: Cecidomyiidae) *Res. Popul. Ecol.* 16:1–42

114. Olfert, O. O., Gage, S. H., Mukerji, M. K., 1980. Aerial photography for detection and assessment of grasshopper (Orthoptera: Acrididae) damage to small grain crops in Saskatchewan. *Can. Entomol.* 112:559–66

115. Payne, J. A., Hart, W. G., Davis, M. R., Jones, L. S., Weaver, D. J., et al. 1971. Detection of peach and pecan pests and diseases with color infrared aerial photography. See Ref. 70, pp. 216–30

116. Pedgley, D. E. 1973. *Testing feasibility of detecting locust breeding sites by satellite. Final Report to NASA on ERTS-1 Experiment Period: July 1972 to August 1973*. London: Dep. Trade Ind. 19 pp.

116a. Pedgley, D. E., Reynolds, D. R., Riley, J. R., Tucker, M. R. 1982. Flying insects reveal small-scale wind systems. *Weather* 37:295–306

117. Peloquin, J. J., Olson, J. K. 1985. Observations on male swarms of *Psorophora columbiae* in Texas ricelands. *J. Am. Mosq. Control Assoc.* 1(4):482–88

118. Pickard, R. S., Hepworth, D. 1979. A method for electronically monitoring the ambulatory activity of honeybees under dark conditions. *Behav. Res. Methods Instrum.* 11(4):433–36

119. Rainey, R. C., Joyce, R. J. V. 1972. The use of airborne Doppler equipment in monitoring windfields for airborne insects. In *Preconf. Pap. 7th Int. Aerosp. Instrum. Symp., Cranfield, UK*, pp. 8.1–8.4. London: Peregrinus. 147 pp.

120. Ramana Murty, B. V., Roy, A. K., Biswas, K. R., Khemani, L. T. 1964. Observations of flying locusts by radar. *J. Sci. Ind. Res.* 23:289–96

121. Reid, D. G., Wardaugh, K. G., Roffey, J. 1979. *Radar studies of insect flight at Benalla, Victoria in February 1974. CSIRO Div. Entomol. Tech. Pap. No. 16*, Melbourne, Australia. 21 pp.

122. Rencz, A. N., Nemeth, J. 1985. Detection of mountain pine beetle infestation using Landsat MSS and simulated thematic mapper data. *Can. J. Remote Sensing* 11(1):50–58

123. Reynolds, D. R, Riley, J. R. 1979. Radar observations of concentrations of insects above a river in Mali, West Africa. *Ecol. Entomol.* 4:161–74

124. Reynolds, D. R., Riley, J. R. 1988. A migration of grasshoppers, particularly *Diabolocatantops axillaris* (Thumberg) (Orthoptera: Acrididae), in the West African Sahel. *Bull. Entomol. Res.* 78: 251–71

125. Richter, J. H. 1969. High resolution tropospheric radar sounding. *Radio Sci.* 12:1261–68

126. Richter J. H., Jensen, D. R., Noonkester, V. R., Kreasky, J. B., Stimmann, M. W., et al. 1973. Remote radar sensing: atmospheric structure and insects. *Science* 180:1176–78

127. Riley, J. R. 1974. Radar observations of individual desert locusts (*Schistocerca gregaria* (Forsk.) (Orthoptera, Locustidae)). *Bull. Entomol. Res.* 64:19–32

128. Riley, J. R. 1975. Collective orientation in night-flying insects. *Nature* 253:113–14

129. Riley, J. R. 1979. Quantitative analysis of radar returns from insects. In *Proc. Workshop Radar, Insect Popul. Ecol. Pest Manage., Wallops Island, Va. NASA Conf. Publ. 2070*, pp. 131–58. Wallops Island, Va: NASA

130. Riley, J. R. 1980. Radar as an aid to the study of insect flight. In *A Handbook on Biotelemetry and Radio Tracking*, ed. C. J. Amlaner, D. W. MacDonald, pp. 131–40. Oxford/New York: Pergamon. 804 pp.

131. Riley, J. R. 1985. Radar cross section of insects. *Proc. IEEE* 73(2):228–32

132. Riley, J. R., Reynolds, D. R. 1979. Radar-based studies of the migratory flight of grasshoppers in the middle Niger area of Mali. *Proc. R. Soc. London Ser. B* 204:67–82

133. Riley, J. R., Reynolds, D. R. 1983. A long-range migration of grasshoppers observed in the Sahelian zone of Mali by two radars. *J. Anim. Ecol.* 52:167–83

134. Riley, J. R., Reynolds, D. R. 1986. Orientation at night by high-flying insects. See Ref. 26, pp. 71–87

135. Riley, J. R., Reynolds, D. R., Farmery, M. J. 1983. Observations of the flight behaviour of the armyworm moth *Spodoptera exempta*, at an emergence site using radar and infra-red optical techniques. *Ecol. Entomol.* 8:395–418

136. Riley, J. R., Reynolds, D. R., Farrow, R. A. 1987. The migration of *Nilaparvata lugens* (Stal) (Delphacidae) and other Hemiptera associated with rice during the dry season in the Philippines: a study using radar, visual observations, aerial netting and ground trapping. *Bull. Entomol. Res.* 77:145–69

136a. Rose, D. J. W., Dewhurst, C. F., Page, W. W., Fishpool, L. D. C. 1987. The role of migration in the life-system of the African armyworm, *Spodoptera exempta*. *Insect Sci. Appl.* 8:561–69

136b. Rose, D. J. W., Page, W. W., Dewhurst, C. F., Riley, J. R., Reynolds, D. R., et al. 1985. Downwind migration of the African armyworm moth *Spodoptera exempta*, studied by mark-and-capture and by radar. *Ecol. Entomol.* 10:299–313

137. Rossetti, C. 1970. *Aerial photographic detection of low-density desert locust populations in north-west Muritania.* FAO Rep. AGP: DL/70/3, Rome. 19 pp.

138. Sayer, H. J. 1956. A photographic method for the study of insect migration. *Nature* 177:226

139. Schaefer, G. W. 1966. *The study of bird echoes using a tracking radar synopsis of recent experiments.* Presented at Int. Ornithol. Congr., 14th, Oxford, UK

140. Schaefer, G. W. 1969. Radar studies of locust, moth and butterfly migration in the Sahara. *Proc. R. Entomol. Soc. London* 34:33, 39–40

141. Schaefer, G. W. 1975. *Radar studies of the flight activity and the effects of windfields on the dispersal of Sudan Gezira insects.* Presented at Sem. Strategy Cotton Pest Control Sudan Gezira, Wad Medani, Sudan

142. Schaefer, G. W. 1976. Radar observations of insect flight. *Symp. R. Entomol. Soc. London* 7:157–97

143. Schaefer, G. W. 1979. An airborne radar technique for the investigation and control of migrating pest insects. *Philos. Trans. R. Soc. London Ser. B* 287:459–65

144. Schaefer, G. W., Bent, G. A. 1984. An infra-red remote sensing system for the active detection and automatic determination of insect flight trajectories (IRADIT). *Bull. Entomol. Res.* 74:261–78

145. Schaefer, G. W., Bent, G. A., Allsopp, K. 1985. Radar and opto-electronic measurements of the effectiveness of Rothamstead Insect Survey suction traps. *Bull. Entomol. Res.* 75:701–15

146. Simonett., D. S., Ulaby, F. T., Colwell, R. N., eds. 1983. *Manual of Remote Sensing*, Vol. 1, *Theory and Techniques*. Falls Church, Va: Am. Soc. Photogramm. 1232 pp.

147. Skidmore, A. K., Wood, G. R., Shepherd, K. R. 1987. Remotely sensed data in forestry, a review. *Aust. For.* 50(1):40–53

148. Slater, P. N. 1985. Survey of multispectral imaging systems for earth observations. *Remote Sensing Environ.* 17:85–102

149. Slater, P. N., Doyle, F. J., Fritz, N. L., Welch, R. 1983. Photographic systems for remote sensing. See Ref. 146, pp. 231–91

150. Tucker, C. J., Hielkema, J. U., Roffey, J. 1985. The potential of satellite remote sensing of ecological conditions for survey and forecasting desert-locust activity. *Int. J. Remote Sensing* 6(1):127–38

151. Uttley-Moore, W. J., Carew-Jones, R., Morgan, T., Williams, P. J., Fulton, R. 1986. *Proc. IEE Meet. Adv. Infra-red Detect. Syst., London,* pp. 103–9. London: IEE

152. Vlcek, J. 1983. Videography: some remote sensing applications. *Proc. Annu. Meet. Am. Soc. Photogramm.* 49:63–69

153. Waloff, Z. 1972. Orientation of flying locusts *Schistocerca gregaria* (Forsk.) in migrating swarms. *Bull. Entomol. Res.* 62:1–72

154. Wagner, V. E., Hill-Rowley, R., Narlock, S. A., Newson, H. D. 1979. Remote sensing: a rapid and accurate method of data aquisition for a newly formed mosquito control district. *Mosq. News* 39(2):283–87

155. Wallen, V. R., Jackson, H. R., MacDiarmid, S. W. 1976. Remote sensing of corn aphid infestation, 1974 (Hemiptera: Aphididae). *Can. Entomol.* 108:751–54

156. Ward, J. D., Acciavatti, R. E., Ciesla, W. M. 1986. *Mapping insect defoliation in eastern hardwood forests with color-*

*IR aerial photos—a photo interpretation guide. US Dep. Agric. For. Pest Manage./Methods Appl. Group Rep. 86–2,* Fort Collins, Colo. 25 pp.

157. Weber, F. P. 1976. Forest stress detection. In *Evaluation of SKYLAB (EREP) data for forest and rangeland surveys, US Dep. Agric. For. Serv. Res. Pap. PSW-113,* pp. 55–63. Berkeley, Calif: Pac. Southwest For. Range Exp. Stn.

158. Weber, F. P. 1980. Large area forest pest impact surveys using high resolution panoramic photography from suborbital reconnaissance platforms. See Ref. 77, pp. 467–74

159. Wert, S. L., Roettgering, B. 1968. Douglas-fir beetle survey with color photos. *Photogramm. Eng.* 34:1243–48

160. Wolf, W. W. 1978. Entomological radar studies in the United States. In *Movement of Highly Mobile Insects: Concepts and Methodology in Research,* ed. R. L. Rabb, G. G. Kennedy, pp. 263–66. Raleigh, NC: Univ. Graphics. 456 pp.

161. Wolf, W. W., Loper, G. M., Greneker, G. 1987. *Doppler radar for detecting insects.* Presented at Summer Meet. Am. Soc. Agric. Eng., Baltimore, Md.

162. Wolf, W. W., Sparks, A. N., Pair, S. D., Westbrook, J. K., Truesdale, F. M. 1986. Radar observations and collections of insects in the Gulf of Mexico. See Ref. 26, pp. 221–34

163. Wolf, W. W., Westbrook, J. K., Sparks, A. N. 1986. Relationship between radar entomological measurements and atmospheric structure in south Texas during March and April 1982. In *Long-Range Migration of Moths of Agronomic Importance to the United States and Canada: Specific Examples of Occurrence and Synoptic Weather Patterns Conducive to Migration,* ed. A. N. Sparks, pp. 84–97. Beltsville, Md; US Dep. Agric. Agric. Res. Serv.

164. Young, S., Getty, C. 1987. Visually guided feeding behaviour in the filter feeding cladoceran, *Daphnia magna. Anim. Behav.* 35:541–48

165. Zalucki, M. P., Chandica, A., Kitching, R. L. 1981. Quantifying the distribution and abundance of an animals resource using aerial photography. *Oecologia* 50:176–83

166. Zsilinszky, V. G. 1980. A user's notes on remote sensing application. See Ref. 77, pp. 493–510

*Ann. Rev. Entomol. 1989. 34:273–92*
*Copyright © 1989 by Annual Reviews Inc. All rights reserved*

# ECOLOGY AND BEHAVIOR OF *NEZARA VIRIDULA*

## J. W. Todd

Department of Entomology, Coastal Plain Experiment Station, University of Georgia, Tifton, Georgia 31793

## INTRODUCTION

*Nezara viridula* is one of the most important pentatomid insect pests in the world. It is cosmopolitan and highly polyphagous on many important food and fiber crops. It has been referred to in the literature and throughout the warmer regions of the world by various common names: the green stink bug, the green vegetable bug, the green pumpkin bug, the cotton green bug, the green tomato bug, the green soldier bug, the southern green plant bug, the tomato and bean bug, and the green bug of India (31). Recently, however, it has been most commonly referred to as the southern green stink bug, particularly in the papers published from the Americas and Asia. The literature pertaining to the species is vast; a bibliography by DeWitt & Godfrey (29) lists over 690 references. Many excellent papers concerning *N. viridula* have been published in the last 10 years; however, only a few review articles are among them, and those have not adequately covered all aspects of the bug's biology. The reader who desires more detailed information is referred to References 18, 62, 122, 129, 144, and 152.

## GEOGRAPHICAL AND HOST RANGE

The worldwide distribution of *N. viridula* has been discussed by several authors (20, 21, 29, 31, 35, 37, 62, 66, 75, 93, 129, 143, 152). The bug occurs throughout the tropical and subtropical regions of the Americas, Africa, Asia, Australasia, and Europe between latitudes 45° N and 45° S, and it is still spreading to new areas (51, 106, 129, 162). *Nezara viridula* is a strong flier and undoubtedly has utilized wind and weather frontal systems as

273

0066-4170/89/0101-0273$02.00

well as the lanes of commerce in its wide dispersal (4, 9, 30, 68, 76, 146, 162).

Several authors have postulated that the ancestral home of *N. viridula* is the eastern Palearctic or Indo-Malayan region of southeast Asia (70, 76, 162). However, based on Freeman's (37) 1940 review of the known *Nezara* spp., Jones & Powell (63) suggested that *N. viridula* more likely originated in the Ethiopian region of eastern Africa. Hokkanen (53) and Jones (62) both subsequently presented evidence supporting this proposal. The same four color morphs have been reported from India (5, 143) and Japan (76, 88, 162), while five different color morphs occur in different provinces in Italy (140). Hokkanen (53) reasoned that the increased frequency and variety of color morphs in Italy constitutes evidence that *N. viridula* originated in the Mediterranean or Ethiopian region. Jones (62) concluded that since 13 of the 20 recognized species of *Nezara* are confined to Africa and since nearly all of them possess each of the East Asian color morphs, the Ethiopian region of eastern Africa is the most likely point of origin of *N. viridula*. Jones (62) suggested that a historical, biogeographical analysis would help locate the origin of the genus *Nezara* and would help identify areas where host-specific parasitoids were most likely to have coevolved.

The Commonwealth Institute of Entomology prepared a distribution map in 1953 (20, 29) and updated it in 1970 and 1980 (21, 152). However, the map does not include areas of the United States (California) and South America (Argentina, Paraguay, Uruguay, and Brazil) where the insect now occurs (51, 129).

The host range of *N. viridula* encompasses over 30 families of di-cotyledonous plants and a number of monocots (31, 33, 52, 60, 84, 135, 142, 152, 157). Although a comprehensive host list is currently unavailable, *N. viridula* seems to have a strong preference for certain legumes (22, 24, 31, 52, 65, 130, 152). Preference varies somewhat with the host maturity and time of year (28, 107, 121, 122). The most highly preferred stage of plant development is that of fruit or pod formation. As plants senesce, bugs move to more succulent hosts (23, 65, 111, 139). Noncultivated host species have a key role in maintaining populations when preferred crops are not available (65).

Since *N. viridula* is highly polyphagous and is already prevalent in many areas suitable for further agricultural expansion, *N. viridula* population densities are likely to increase with greater quantities of suitable host crops. In the absence of established natural enemies, *N. viridula* may subsequently attain more serious pest status. Several authors have pointed out that the pest status of *N. viridula* may be related to climatic factors and small changes in the environment over time in addition to geographical expansion of a highly preferred legume or other host (22, 24, 62, 72, 75, 78, 129, 149, 152). In addition, as agriculture expands in the developing countries and to a lesser degree in the developed countries, conditions are created that allow pest species to increase their geographical range and possibly change their pest status in a relatively unstable environment over a long period of time. Kiritani (77) detailed such circumstances surrounding the rise of *N. viridula* to major pest status in Japan.

# BIONOMICS AND SEASONAL PHENOLOGIES

## *Life History*

The life history of *N. viridula* has been studied in many parts of the world (22, 26, 31, 35, 45, 60, 71, 78, 143). In Japan, Kiritani and coworkers (70–88) have comprehensively studied the population ecology of *N. viridula* on rice and companion crops. Basic biological studies from Louisiana (60, 95–100), Florida (31), Georgia (41–50), and South Carolina (61, 62, 64, 65) provide a broad data base for *N. viridula* in the southern United States.

In the early spring, as temperatures rise, adults emerge from overwintering quarters in search of food and begin to mate almost immediately (31). Females have been observed dispersing up to 1000 m per day by flight in search of feeding or oviposition sites (88). The duration of stay of egg-laying females is shorter than that of feeding adults, which indicates that females redisperse to food plants soon after egg-laying is completed (84).

Eggs are pale yellow or cream colored at oviposition and are generally deposited in polygonal clusters. Individual eggs are closely packed in regular rows and are firmly glued together and to the substrate. Incubation averages 5 days in the summer (45), but may reach 2–3 wk in early spring and late fall (31).

*Nezara viridula,* like other stink bug species, develops through five nymphal instars. First instar nymphs cluster on or near the egg mass and apparently do not feed (71). At approximately the first molt, the nymphs begin to disperse slightly and feed (31). Aggregation is reported to affect development rate and mortality (71, 85, 86). It occurs through the third stage but disappears during the fourth stage (84, 128). Nymphal aggregation may impart a measure of protection from predators such as ants owing to the pooling of chemical defense (97).

The nymphs, like the adults, are usually found upon those portions of the plant on which they prefer to feed—the tender growing shoots and especially the developing fruit or seed. During the summer the period from egg to adult is 35–37 days, depending on temperature (31, 45, 60). Kariya (67) reported that development from egg hatch to adult took 58.4, 34.2, and 23.2 days at 20, 25, and 30°C, respectively; the optimal temperature for development is about 30°C. Four generations per year have commonly been observed in

Florida and Louisiana, and a fifth generation may occur at the latitude of southern Florida (31, 60).

Generally, published reports on the life cycle and generation times of *N. viridula* are similar. Variations in development time and number of eggs and egg masses laid are due to differential seasonal temperatures at the locations under study (42, 43, 45, 67, 70, 71, 76–80, 82–85, 89).

Laboratory rearing of *N. viridula* has been described (46, 59, 61) along with successful foodstuff combinations (12, 33, 46, 68). Efficient laboratory rearing procedures, including the use of a standardized synthetic diet, will make possible many of the needed studies concerning the relationships of *N. viridula* with biotic and abiotic components of its environment. Meridic, Parafilm®-encapsulated diets developed for rearing lepidopterous larvae, mirid bugs, and aphids have been modified and tested for suitability in rearing *N. viridula* (12, 59), but these have not proven to be entirely satisfactory.

Nutrition of *N. viridula* during the last nymphal stadium is particularly important because of energy demands during the final molting, flight, and reproduction. Although diet switching during the nymphal stages is unusual, it commonly occurs with adults under certain circumstances. For example, a newly molted adult may move from its nymphal host plant to feed on a new host prior to ovipositing on a different host, which will subsequently serve as the food source for the next generation (68). When fifth-stage nymphs of *N. viridula* were switched from a diet of green beans and peanuts to various inadequate foods, the flight capacity, growth, and fecundity of adult females were irreversibly reduced. Conversely, adult females reared on a diet containing mature soybean or peanut seeds exhibited increased fat-body deposition and capacity for extended flight compared to those switched to diets of immature green beans (68). Fecundity, however, was greater in adult females reared on a combination of green beans and peanuts than in those reared on peanuts alone. Further research on the nutritional aspects of seasonal host sequence and possible interactions among nutrient components from various hosts is needed to determine the role of host plants in major population outbreaks of *N. viridula* (68, 69).

*Nezara viridula* is particularly susceptible to inbreeding depression while restricted to laboratory culture, and care must be taken to avoid sibling mating (42, 61). Recolonization with field-collected specimens is usually necessary after four to six laboratory-reared generations or at least annually (61).

## Diapause and Overwintering

*Nezara viridula*, like most pentatomids, overwinters in the adult stage, mainly under litter, bark, and other objects that offer protection (31, 136, 160). The overwintering habits and survival of *N. viridula* have been investigated in several widely separated areas of its range and under various extremes of winter conditions (8, 59, 64, 74, 78, 81, 82, 87, 132). Initially males were reported to overwinter in a state of quiescence, while females were thought to overwinter in a diapausing state characterized by arrested ovarian development, accumulation of the fat body, and a reddish-brown or russet coloration (8, 70). Subsequently, it was reported that diapause in both sexes is characterized by a cessation of feeding and mating and by a reddish-brown (diapausing) coloration (11). Overwintering survival is reportedly greater for females than for males, for larger individuals than for smaller ones, and for those with reddish-brown coloration (87).

Near the northern limits of the bugs' distribution in Japan, survival of adults in winter was dependent partly on the physiological state of the adults (81, 83) and partly on the degree of protection and food provided by the habitat (75, 77, 81). When the severity of winter (primarily low temperature) exceeded the physiological tolerance of the insect, heterogeneity of hibernacula was the main safety valve that prevented total overwintering mortality (84). Overall, winter mortality is one of the major limiting factors of *N. viridula* populations (72, 73, 75, 81, 83).

The role of photoperiod on induction of diapause in *N. viridula* has been studied (3, 113, 132, 145). Pitts (132) demonstrated that the reception of photoperiodic stimuli for diapause induction was most critical in fifth-stage nymphs. Diapause was induced by photophases of 10, 11, and 12 hr, but was suppressed by a daily 13-hr photophase. Diapause in *N. viridula* reared for three generations in the laboratory under a 14-hr photophase has also been reported (49). The last observation is perplexing, since the experimental insects had never been exposed to environmental conditions or stimuli normally associated with diapause induction. One possible explanation is that entry into reproductive diapause may be under endogenous genetic control, rather than strictly environmentally (photoperiodically) induced. Conclusive evidence concerning this aspect of the biology of *N. viridula* is needed to elucidate the rather precarious relationship of the species with its environment during winter.

Despite the occurrence of reproductive diapause in *N. viridula,* overwinter survival is enhanced by the presence of succulent food (31, 122). Adults often feed during mild periods in the winter. Mustard (*Brassica* spp.) and wild radish (*Raphanus* sp.) provide excellent hibernacula and a constant source of highly preferred food (31, 49, 132), thus increasing survival of *N. viridula.*

## Polyphagous Nature and Seasonal Host Sequence

As temperatures warm in spring, adults move out of winter cover and begin feeding and oviposition in clover, small grains, early spring vegetables (such as mustard, turnip, beet, and radish), corn, tobacco, and weed hosts such as showy crotalaria and coffee senna (31, 64, 65, 120, 152). Resulting nymphs and adults constitute the first generation. In temperate regions tomatoes, certain leguminous weeds, vegetables and row crops, cruciferous vegetables, and okra are highly suitable, typical midsummer hosts (123–125, 143, 149, 151, 161). Different hosts may be just as suitable in other regions. Third generation adults migrate into soybean when it becomes attractive to *N. viridula* during the onset of bloom and podset, in late summer. The fourth and fifth generations also occur in soybean.

Synchronization of host plant and pest phenologies is extremely important for pest populations (65, 82, 85, 86, 152). *Nezara viridula* and most if not all phytophagous Heteroptera are primarily pod, seed, or fruit feeders. Therefore, population peaks almost always coincide with or lag slightly behind the development of the reproductive stages (fruiting) of the primary host species (13, 32–34, 64, 65, 68, 80, 94, 111, 139, 143, 155).

Dispersal of *N. viridula* in relation to feeding and oviposition has been studied on several hosts in Japan (84). Census data taken in a wide variety of crops indicate that rape, radish, wheat, and barley are primary sites for feeding and mating. Gravid females then disperse to other kinds of host plants, e.g. potato, beans, and rice, for oviposition.

The population dynamics of *N. viridula* in relation to soybean phenology have been studied in the United States and Brazil (13, 23, 25, 101, 111, 139). Oviposition was initiated during the late vegetative stages of soybean and peaked between R3 (beginning pod development) and early R5 (beginning podfill) (36, 139). More females than males were found in the earlier stages of soybean development. Third-generation *N. viridula* developed in early-maturing cultivars, while fourth and fifth generations developed in the later-maturing cultivars. Although cultivar and planting date affected *N. viridula* population density, soybean maturity stage had a more significant impact (139).

## ASPECTS OF BEHAVIOR

### Immature Stages

GREGARIOUSNESS, DISTRIBUTION, FEEDING, AND DEFENSIVE BEHAVIOR    Distribution of egg masses in a field is primarily determined by the degree of activity and movement patterns of ovipositing females (139). *N. viridula* deposits eggs in masses containing 30–130 eggs (22, 31, 45, 60, 61, 78, 143), with a mean of $60 \pm 24$ eggs per mass (41, 45, 46). Ovipositing females and egg masses are randomly distributed in most fields (41, 44, 55, 84). Females oviposit in the upper portions of canopied crops; the undersurfaces of leaves or the pods and other fruiting structures are the most preferred substrate. Females leave the plant soon after oviposition is complete (84).

Lockwood & Story (95–100) investigated the effects of endogenous rhythms and environmental and sibling influences on egg hatch in *N. viridula*. Hatching was found to be a function of a photoperiodically entrained rhythm. It was temperature dependent, but it was not thermoperiodically entrained. Also, the synchronization of hatching within an egg mass was due to the interaction of emerging nymphs stimulating hatching in adjacent eggs, and not to the synchrony of the eclosion of each egg.

As eggs mature they become deep yellow, then pinkish-yellow, and finally bright orange at eclosion. A red crescent-shaped spot or pattern appears beneath the operculum 3 days after oviposition, delineating the embryo's eyes and facial features (98). The crescent permits the orientation of the embryos within the egg to be determined before hatching. Lockwood & Story (98) utilized this phenomenon to study the mechanisms and functions of embryonic orientation in relation to nymphal behavior in *N. viridula* and other pentatomids. They determined that orientation toward the center of the egg mass by outer-row embryos was not a function of oviposition pattern or photic cues, but rather was mediated by the glue binding the egg mass. Physical and chemical orientation cues associated with the glue were present only on the inner surfaces of outer-row eggs. These cues assure that upon hatching, outer-row nymphs receive the necessary stimulation from the egg mass and siblings to contain these first-stage nymphs in the aggregation.

*N. viridula* nymphs do not use visual cues to aggregate, and they do not possess an aggregation pheromone until 2 days postemergence (96). Lockwood & Story (95–98) concluded that emerging nymphs remain and aggregate during this 2-day period using tactile cues from the egg mass and emerging siblings. These authors also presented interesting work on the orientation of outer-margin nymphs arising from outer-margin embryos. They found that nymphs that hatched from eggs in the interior of a mass oriented at random with respect to the center, whereas nymphs forming the perimeter of an aggregation oriented inward at an angle of less than 90° (90). This orientation scheme may facilitate cohesiveness and/or communication in the aggregation or may be related to pooled chemical defense as referred to later in this section.

First-stage nymphs apparently do not feed on plant tissue and seldom leave the egg mass unless disturbed (31, 71). Second-stage nymphs begin feeding on plant tissue but remain congregated on or near the egg mass. Third-stage nymphs disperse as a group from the egg mass, but usually remain clustered. The clumped distribution of the first- through third-stage nymphs is maintained because of this tendency to aggregate, and the degree of aggregation is diminished only through mortality or disturbance. Aggregation decreases abruptly in fourth-stage nymphs because of their increased ability and tendency to disperse. By the fifth stage, populations approach a random or Poisson distribution (55). Nymphs move more along the rows in a soybean field than across rows, with maximum movement of 12 and 7.2 m, respectively (128).

On warm, sunny mornings, fourth- and fifth-stage nymphs and adults tend to bask at the canopy surface of hosts such as soybean, beans, peas, and okra (22, 158). They return to the protection of the canopy as the direct noonday sunshine becomes too hot for them. On crops such as wheat, sorghum, linseed, and safflower where feeding sites are situated above or at the canopy surface, the bugs move around the seed head, basking when conditions are suitable and retreating to the shaded side when radiation becomes too intense (158).

In Louisiana, Russin and coworkers (138) studied within-plant distribution of stink bug feeding damage by natural populations of *N. viridula*, *Acrosternum hilare*, and *Euschistus* spp. Bugs preferentially fed upon seeds in the upper halves of plants until high population densities or reduction in food quality or quantity forced some bugs to feed in lower levels (138). Feeding and damage potential of fifth-stage nymphs and adults are comparable (155). The damage potential of third- and fourth-stage nymphs is much less than that of adults (108). At high population densities of *N. viridula* in combination with *Anticarsia gemmatalis*, interspecific competition or interference caused the damage and yield reduction to be lower than when one or the other species was present alone (154). Also, *N. viridula* appeared to interfere more with *A. gemmatalis* than the reverse.

Nymphs of *N. viridula* use *n*-tridecane as a bifunctional pheromone, which causes dispersal at high concentrations and aggregation at low concentrations (96). Aggregated nymphs suffer less predation than isolated nymphs (56, 97), which indicates that *n*-tridecane may serve as a defensive secretion at high concentrations.

## Adult Behavior

SPATIAL PATTERNS OF DISPERSION AND AGGREGATION    Dispersion of stink bug species has been researched on rice in Japan (80) and soybean in the United States (137). Under most conditions, invading females of *N. viridula* apparently disperse at random into rice fields (119) and vegetative soybean (139) or into trap crops of early-planted, early-maturing soybean (107, 121). In other cases invading adults first appear in a corner or edge of a field, exhibiting an "edge effect" similar to that reported by Nakasuji et al (119). This pattern is probably due to directional movement among a sequence of hosts in an area in response to deteriorating suitability of the host or other unsuitable conditions of surrounding crops or native vegetation. Since fruiting structures, the preferred feeding site, are not present in vegetative soybean, the initial invasion of adults probably involves the search for an oviposition site.

The presence of adult *N. viridula* males and females in a host crop with ample feeding sites may give quite a different scenario. One to several males are often found in close proximity to the females, which results in a clumped distribution of males and of the adult population as a whole (44). Clumped distributions of males can also result from physiological processes related to mating behavior (44, 119).

COURTSHIP AND MATING BEHAVIOR    Mitchell & Mau (115) reported that *N. viridula* males are the source of intraspecific attraction over considerable distances. These authors considered the attraction to be due to a sex pheromone, which appeared to be attractive only to females of *N. viridula* and the tachinid parasitoid *Trichopoda pennipes*. Harris & Todd (44) suggested that a more likely behavioral scenario involves aggregation of the sexes. They demonstrated that male *N. viridula* and fifth-stage nymphs were as strongly attracted as females to males in the field. Those results supported the hypothesis that males of *N. viridula* release an olfactory substance that, rather than being directly sexual in nature, acts as an aggregation pheromone for males, females, and fifth instars of the species; it also acts as a kairomone for *T. pennipes*. The resultant aggregation may then, at least for adults, facilitate mate-finding (44).

Since the initial report of the presence of a sex attractant in *N. viridula* (115), further investigations of premating and mating behavior have been published (6, 10, 11, 44, 45, 50, 103–105, 114). Courtship and mating behavior can be divided into two phases: long-range mate location and short-range courtship, which results in mating (10). Long-range behavioral activity includes aggregation resulting in the arrival of females in the vicinity of males (10, 44, 115). Arrival of the sexes in the same general vicinity marks the transition between long-range attraction and courtship behavior.

Earlier reports that geographically isolated populations of *N. viridula* were differentially parasitized by *T. pennipes* (18) led to a suspicion that pheromone strains of *N. viridula* existed. Chemically different, male-released attractant pheromones have recently been isolated from European (131), Brazilian (7), and southern United States (2) populations of *N. viridula*. These are sesquiterpene isomers ranging in molecular weight from 220 to 224. Field and laboratory work revealed that there are probably two pheromone strains of *N. viridula*, which are distinguished by the presence or absence of *cis*-(Z)-α-bisabolene epoxide (2). Baker et al (7) reported that *Nezara* males from Brazil produce *trans*-(Z)-α-bisabolene epoxide, but not the *cis*-isomer. In contrast, he found that *Nezara* males from southern France produce a 2:1

ratio of *trans/cis*-(Z)-α-bisabolene, while male bugs from the southern United States produce the *trans-* and *cis*-isomers in a 3:1 ratio. Aldrich et al (2) reported that males from the southern United States do not possess the 224-dalton sesquiterpene isomer components reported in European males (131) and that all *N. viridula* male-specific blends also include (Z)-α-bisabolene, *n*-nonadecane, and several less abundant alkanes and sesquiterpenes. Aldrich et al (2) further stated that these findings should allow delineation of the global distribution of *Nezara* pheromone strains and may give additional evidence of the biogeographical relationships of *N. viridula* with many if not all of its natural enemies (2).

A likely sequence of courtship behavior once the sexes arrive in the same general vicinity has been described (10, 45, 114). Although the male pheromone has been shown to attract males and females to its source, it has not been demonstrated to elicit courtship behavior (10). After the close-range courtship sequence is initiated, visual (17, 102), acoustic (11, 50, 92), and tactile (11, 22, 45, 50, 104) stimuli probably mediate progress toward copulation. Studies in Yugoslavia revealed a repertoire of six stridulatory calls involved with mating behavior in *N. viridula* (19). Three male and two female songs must alternate in a specific order for mating to ensue, and the sixth call is a male rivalry song. Subsequent work with *N. viridula* in Georgia and Florida (50) has identified an extensive acoustic repertoire that mediates a complex mating behavior. Three female songs, all acoustically and behaviorally distinct, have been identified. Males produce seven behaviorally distinct songs, but three of these are very similar acoustically. Mating success varies with different acoustic strategies and combinations of tactile and acoustic behavioral sequences (summarized in 41, 50). Also, significant differences were noted in the acoustic repertoire of the geographical races studied from the southern United States and Yugoslavia (19, 50). Much work remains to be done in the characterization of the specific behavioral sequences involved in the acoustic and tactile communication used as close-range sexual communication mechanisms in *N. viridula*. However, the literature concerning these areas of behavioral biology clearly indicates that these modes of communication are widely utilized in the Heteroptera (10, 19, 44, 45, 50, 92, 103–105).

DEFENSIVE BEHAVIOR    The defensive secretion of *N. viridula* adults is a complex blend of at least 19 components (38, 39, 57, 159). One of these components, E-2-hexanal, is present in both the metathoracic glands (38) and dorsal abdominal glands III–IV (1). This compound has been studied to determine its role in pheromonal, allomonal, and kairomonal functions (1, 57, 58, 100). *E*-2-Hexanal may be the first exocrine product to have a primary role in all three of the behavioral responses mentioned above (100). This compound is present in the defensive secretion of at least 13 species of pentatomids and can provide defense against predators by eliciting alarm behavior. It may also be repellent to ants and possibly other predators (147, 159).

## BIOLOGICAL CONTROL

With the spread of *N. viridula* throughout the world and its subsequent emergence as a serious pest, there have been numerous attempts to establish parasitoids in newly invaded areas (15, 62). Several of the programs have been highly successful and are listed among landmark examples of successes in classical biological control (15, 18, 27).

In his recent worldwide reviews of *N. viridula* parasitoids, Jones (62) stated that even though 57 parasitoids are recorded from *N. viridula,* many are incidental and are more closely associated with other hosts or habitats. The most successful biological control agents employed thus far are the scelionid egg parasitoid *Trissolcus basalis* and the tachinid flies *T. pennipes* and *Trichopoda pilipes,* which attack adults and large nymphs of *N. viridula* (16, 22, 47, 54, 62, 109, 110, 112, 117, 141, 150, 153). Of the egg parasitoids, *T. basalis* is both the most important and the most widely distributed species (14, 31, 62, 127). Although *T. basalis* has frequently been recorded from several other pentatomids, it is clearly most closely associated with *N. viridula* (62, 133, 156). This parasitoid's utility may be somewhat limited geographically, however, since it appears to be restricted to the coastal areas of the pentatomid's range (62).

The early nymphal stages of *N. viridula* are generally free of significant attack by parasitoids (62), although fourth- and fifth-stage nymphs have been commonly reported to sustain attack by *Trichopoda* spp. Jones (62) listed 13 species of Tachinidae that are the only known parasitoids that attack adult *N. viridula.* He further stated that six of the New World tachinid species are well adapted to adult *N. viridula* and should be considered for establishment in new areas.

Predation is also an important mortality factor for *N. viridula* (31, 60, 71, 118, 148, 149). Ragsdale et al (134) quantitatively assessed egg and nymphal predation using an enzyme-linked immunosorbent assay (ELISA). They found a complex of nonspecific predators associated with *N. viridula* and calculated predator efficiency ratings for each. Detailed studies of predation on *N. viridula* in a soybean ecosystem in Louisiana (148, 149) showed 18 insect and 6 spider species representing 6 orders and 19 families to be important. *Solenopsis invicta* was the dominant egg predator during the vegetative stages of soybean development, while grasshoppers were more important egg predators during the reproductive stages of plant development.

*Solenopsis invicta* is also effective as a predator on early-stage nymphs and to a lesser extent on late-stage nymphs and adults (91). The impact of *S. invicta* as a major regulator of *N. viridula* populations is unclear, however, since *N. viridula* has difficulty maintaining its population during the vegetative and early reproductive stages of soybean growth even in the absence of predation (149). Life-table analyses illustrated that survival to adult increased with plant maturity. Survival rates range from 0.5% during the vegetative and flowering stages to 4.5% during podset (R4 and R5) (36) and 38.2% during seed formation (149).

## NATURAL MORTALITY AND ENVIRONMENTAL CONSIDERATIONS

Several studies of the effects of various abiotic and biotic stresses on population dynamics and mortality of *N. viridula* have been published (31, 33, 34, 60, 62, 64, 67, 71, 78, 81, 83, 87, 91, 110, 111, 116–118, 126, 129, 132, 134, 149, 161).

Kiritani and coworkers (72, 73, 75, 78, 81, 83) developed over a dozen life tables that showed that total mortality from egg to third instar varied from 70 to 90%. Key-factor analysis revealed that egg mortality due to parasitism was the main mortality factor; this factor accounted for the majority of the total mortality sustained by all stages during the spring and summer growing seasons. Mortality factors are predominantly stage specific: Parasites and predators work against eggs (73, 78, 116, 134, 149), weather (excessive rainfall, low humidity) against first-stage nymphs (73, 78, 134, 149), predators (especially spiders and ants) against second-stage nymphs (73, 78, 91, 134, 148, 149), direct or indirect deaths associated with molting against the third stage to adult, and parasitism against adults (78, 81, 91, 109, 110, 148, 149). First-stage nymphs appear to be particularly vulnerable to desiccation during periods of low humidity and high temperatures. This weather combination may be associated with long periods of drought and food shortages (availability of necessary fruiting structures), which contribute to nymphal mortality.

Research on rearing methods has shown that although the first-stage nymphs do not feed, the addition of a green bean to a rearing chamber aids in survival (46, 61). Presumably the increased humidity enhanced the survival of *N. viridula* neonates (46, 59, 61).

## CONCLUSIONS

The pest status of *N. viridula* is well documented over a substantial portion of the world's agricultural land. Despite the extensive knowledge on the population ecology of the species, ecologically based pest management programs are in their infancy in most localities, if they exist at all. Many natural control factors have been shown to exert substantial impact on population densities of the species. Certain efforts to establish biological control programs have worked relatively well, and a few recent attempts seem to have good chances for success. Additional research in the areas proposed by Jones (62) to further assess several additional parasitoids would be highly desirable.

Progress has also been made in identification of soybean germplasm with resistance to *N. viridula,* but very little progress has been made in the development of agronomically acceptable resistant cultivars (40). More importantly, the impact of the use of resistant varieties on the population ecology of *N. viridula* has not been determined.

The use of trap crops as a control strategy for management of *N. viridula* has been adopted only in very limited areas, but it has excellent potential for reducing crop damage (107). Other methods of cultural control that offer substantial benefit but that are still underutilized are (*a*) selection and location of alternate crops that may be less attractive to *N. viridula;* (*b*) the use of highly attractive late-season hosts as trap crops in combination with chemical insecticides in a manner similar to that used in diapause control for other insects such as the boll weevil, *Anthonomus grandis grandis;* and (*c*) the elimination of preferred hibernacula. Combinations of certain of these control tactics may be particularly effective while contributing far fewer environmental problems than conventional control with pesticides.

## ACKNOWLEDGMENTS

The author would like to thank M. H. Bass, D. J. Isenhour, J. R. Chamberlin, R. M. McPherson, W. A. Jones, V. E. Harris, and Ann Rice for helpful discussions and reviewing the manuscript. Special thanks go to Sheran Thompson for typing of the manuscript, help in gathering relevant literature, and other assistance.

## *Literature Cited*

1. Aldrich, J. R., Blum, M. S., Lloyd, H. A., Fales, H. M. 1978. Pentatomid natural products: chemistry and morphology of the III–IV dorsal abdominal glands of adults. *J. Chem. Ecol.* 4(2):161–72
2. Aldrich, J. R., Oliver, J. E., Lusby, W. R., Kochansky, J. P., Lockwood, J. A. 1987. Pheromone strains of the cosmopolitan pest, *Nezara viridula* (Heteroptera: Pentatomidae). *J. Exp. Zool.* 244:171–75
3. Ali, M., Eweiss, M. A. 1977. Photoperiodic and temperature effects on rate of development and diapause in the green stink bug, *Nezara viridula* L. (Heteroptera: Pentatomidae). *Z. Angew. Entomol.* 84(3):256–64
4. Asahina, S., Turuoka, Y. 1970. Records of the insects visited a weather-ship located at the ocean weather station 'Tango' on the Pacific. V. Insects captured during 1968. *Konchu* 38:318–30
5. Azim, M. N., Shafee, S. A. 1978. Indian species of the genus *Nezara* Amyot and Serville (Hemiptera: Pentatomidae). *J. Bombay Nat. Hist. Soc.* 75(2):507–11
6. Azmy, N. M. 1976. Sexual activity, fecundity and longevity of *Nezara viri-*

*dula* (L.). *Bull. Soc. Entomol. Egypte* 60:323–30

7. Baker, R., Borges, M., Cooke, N. G., Herbert, R. H. 1987. Identification and synthesis of (Z)-1'S,3'R,4'S)(−)-2(3',4'-epoxy-4'-methylcyclohexyl)-6-methol-hepta-2,5-diene, the sex pheromone of the southern green stinkbug, *Nezara viridula* (L.). *J. Chem. Soc. Chem. Commun.* 1987:414–16

8. Banerjee, T. C., Chatterjee, M. 1985. Seasonal changes in feeding, fat body and ovarian conditions of *Nezara viridula* L. (Heteroptera: Pentatomidae). *Insect Sci. Appl.* 6(6):633–35

9. Baust, J. G., Benton, H. B., Aumann, G. D. 1981. The influence of off-shore platforms on insect dispersal and migration. *Bull. Entomol. Soc. Am.* 27(1):23–25

10. Borges, M., Jepson, P. C., Howse, P. E. 1987. Long-range mate location and close-range courtship behaviour of the green stink bug, *Nezara viridula* and its mediation by sex pheromones. *Entomol. Exp. Appl.* 44(3):205–12

11. Brennan, B. M., Chang, F., Mitchell, W. C. 1977. Physiological effects on sex pheromone communication in the southern green stink bug, *Nezara viridula*. *Environ. Entomol.* 6(1):169–73

12. Brewer, F. D., Jones, W. A. Jr. 1985. Comparison of meridic and natural diets on the biology of *Nezara viridula* (Heteroptera: Pentatomidae) and eight other phytophagous Heteroptera. *Ann. Entomol. Soc. Am.* 78(5):620–25

13. Buschman, L. L., Pitre, H. N., Hodges, H. F. 1984. Soybean cultural practices: effects of populations of geocorids, nabids, and other soybean arthropods. *Environ. Entomol.* 13:305–17

14. Buschman, L. L., Whitcomb, W. H. 1980. Parasites of *Nezara viridula* (Hemiptera: Pentatomidae) and other Hemiptera in Florida. *Fla. Entomol.* 63(1):154–62

15. Caltagirone, L. E. 1981. Landmark examples in classical biological control. *Ann. Rev. Entomol.* 26:213–32

16. Capeluto, R. 1949. Some studies on the parasitic relationship between the feather-legged tachina fly, *Trichopoda pennipes* Fab. and the southern green stink bug, *Nezara viridula* Linne. *J. Newell Entomol. Soc.* 4:31–33

17. Cardé, R. T., Baker, T. C. 1984. Sexual communication with pheromone. In *Chemical Ecology of Insects,* ed. W. J. Bell, R. T. Cardé, pp. 335–83. London: Chapman & Hall

18. Clausen, C. P. 1978. *Introduced Parasites and Predators of Arthropod Pests and Weeds: A World Review. US Dep. Agric. Agric. Handb.* No. 480. 545 pp.

19. Čokl, A. M., Gogala, M., Jež, M. 1972. Analiza zvočnih signalov stenice *Nezara viridula* (L.) (The analysis of the acoustic signals of the bug *Nezara viridula*). *Biol. Vestn.* 20:47–53

20. Commonwealth Institute of Entomology. 1953. *Distribution Maps of Insect Pests,* Ser. A, Map 27, *Nezara viridula.* London: Commonw. Inst. Entomol. 4 pp.

21. Commonwealth Institute of Entomology. 1970. *Distribution Maps of Insect Pests,* Ser. A, Map 27, *Nezara viridula.* London: Commonw. Inst. Entomol. 4 pp. (Revised)

22. Corpuz, L. R. 1969. The biology, host range and natural enemies of *Nezara viridula* L. (Pentatomidae: Hemiptera). *Philipp. Entomol.* 1:225–39

23. Corrêa-Ferreira, B. S., Panizzi, A. R. 1982. Percevejos-pragas da soja no norte do Paraná: abundância em relação a fenologia da planta e hospedeiros intermediários. *An. Sem. Nac. Pesqui. Soja* 2:140–51

24. Corrêa-Ferreira, B. S., Panizzi, A. R., Newman, G. C., Turnipseed, S. G. 1977. Geographical distribution and seasonal abundance of the major soybean insects of Brazil. *An. Soc. Entomol. Bras.* 6(1):40–50

25. Costa, E. C., Link, D. 1982. Adult dispersion of *Piezodorus guildinii* and *Nezara viridula* (Hemiptera: Pentatomidae) on soybean (Brazil). *Rev. Cent. Cienc. Rurais* 12(1):51–57

26. Cumber, R. A. 1949. The green vegetable bug *Nezara viridula. NZ J. Agric.* 79:563–64

27. DeBach, P. 1962. An analysis of success in biological control of insects in the Pacific area. *Proc. Hawaii. Entomol. Soc.* 18:69–79

28. DeWitt, J. R., Armburst, E. J. 1978. Feeding preference studies of adult *Nezara viridula* (Hemiptera: Pentatomidae) morphs from India and the United States. *Great Lakes Entomol.* 11(1):67–69

29. DeWitt, N. B., Godfrey, G. L. 1972. A bibliography of the southern green stink bug *Nezara viridula* (Linnaeus)(Hemiptera: Pentatomidae). *Ill. Nat. Hist. Surv. Biol. Notes* No. 78. 23 pp.

30. Distant, W. L. 1880. Insecta. Rhynchota. Hemiptera—Heteroptera. *Biol. Cent. Am. London* 1(14):78–79

31. Drake, C. J. 1920. The southern green stink bug in Florida. *Fla. State Plant Board Q. Bull.* 4:41–94

32. Dutcher, J. D., Todd, J. W. 1983.

Hemipteran kernel damage of pecan. *Misc. Publ. Entomol. Soc. Am.* 13(2):1–11

33. Egwuatu, R. I. 1981. Food plants in the survival and development of *Nezara viridula* L. (Hemiptera, Pentatomidae). *Beitr. Trop. Landwirtsch. Veterinarmed. Leipzig, Karl Marx Univ.* 19(1):105–12

34. Egwuatu, R. I., Ani, A. C. 1986. Some aspects of the effects of temperature, rearing density, and food sources on the biology of *Nezara viridula* L. (Hemiptera: Pentatomidae). *Beitr. Trop. Landwirtsch. Veterinarmed. Leipzig, Karl Marx Univ.* 24(1):71–83

35. Everett, P. 1950. Spread of green vegetable bug. *NZ J. Agric.* 80:145–46

36. Fehr, W. R., Caviness, C. E., Burmood, D. T., Pennington, J. S. 1971. Stage of development descriptions for soybeans, *Glycine max* (L.) Merrill. *Crop Sci.* 11(6):929–31

37. Freeman, P. 1940. A contribution to the study of the genus *Nezara* Amyot and Serville (Hemiptera, Pentatomidae). *Trans. R. Entomol. Soc. London* 80: 351–71

38. Gilby, A. R., Waterhouse, D. F. 1965. The composition of the scent of the green vegetable bug, *Nezara viridula*. *Proc. R. Entomol. Soc. London Ser. B* 162:105–20

39. Gilby, A. R., Waterhouse, D. F. 1967. Secretions from the lateral scent glands of the green vegetable bug, *Nezara viridula*. *Nature* 216:90–91

40. Gilman, D. F., McPherson, R. M., Newsom, L. D., Herzog, D. C., Williams, C. 1982. Resistance in soybeans to the southern green stink bug. *Crop Sci.* 22(3):573–76

41. Harris, V. E. 1980. *Basic biology, intraspecific communication and reproductive behavior of the southern green stink bug*, Nezara viridula *(L.) (Hemiptera: Pentatomidae)*. PhD thesis. Univ. Ga., Athens. 188 pp.

42. Harris, V. E., Todd, J. W. 1980. Comparative fucundity, egg fertility and hatch among wild-type and three laboratory-reared generations of the southern green stink bug. *Nezara viridula* (L.) (Hemiptera: Pentatomidae). *J. Ga. Entomol. Soc.* 15:245–53

43. Harris, V. E., Todd, J. W. 1980. Duration of immature stages of the southern green stink bug, *Nezara viridula* (L.), with a comparative review of previous studies. *J. Ga. Entomol. Soc.* 15(2): 114–24

44. Harris, V. E., Todd, J. W. 1980. Male-mediated aggregation of male, female

and 5th instar southern green stink bugs and concomitant attraction of a tachinid parasite, *Trichopoda pennipes*. *Entomol. Exp. Appl.* 27(2):117–26

45. Harris, V. E., Todd, J. W. 1980. Temporal and numerical patterns of reproductive behavior in the southern green stink bug, *Nezara viridula* (Hemiptera: Pentatomidae). *Entomol. Exp. Appl.* 27 (2):105–16

46. Harris, V. E., Todd, J. W. 1981. Rearing the southern green stink bug, with relevant aspects of its biology. *J. Ga. Entomol. Soc.* 16(2):203–10

47. Harris, V. E., Todd, J. W. 1981. Validity of estimating percent parasitization of *Nezara viridula* populations by *Trichopoda pennipes* using parasite-egg presence on host cuticle as the indicator. *J. Ga. Entomol. Soc.* 16(4):505–10

48. Harris, V. E., Todd, J. W. 1982. Longevity and reproduction of the southern green stink bug, *Nezara viridula*, as affected by parasitization by *Trichopoda pennipes*. *Entomol. Exp. Appl.* 31(4): 409–12

49. Harris, V. E., Todd, J. W., Mullinix, B. G. 1984. Color change as an indicator of adult diapause in the southern green stink bug, *Nezara viridula*. *J. Agric. Entomol.* 1(1):82–91

50. Harris, V. E., Todd, J. W., Webb, J. C., Benner, J. C. 1982. Acoustical and behavioral analysis of the songs of the southern green stink bug, *Nezara viridula*. *Ann. Entomol. Soc. Am.* 75(3):234–49

51. Hoffman, M. P., Wilson, L. T., Zalom, F. G. 1987. Control of stink bugs in tomatoes. *Calif. Agric.* 41(5/6):4–6

52. Hoffman, W. E. 1935. The food plants of *Nezara viridula* (L.) (Hem., Pent.). *Proc. 6th Int. Congr. Entomol., Madrid*, pp. 811–16

53. Hokkanen, H. 1983. *Interspecific homeostasis, pest problems, and the principle of classical biological pest control*. PhD thesis. Cornell Univ., Ithaca, NY. 157 pp.

54. Hokkanen, H. 1986. Polymorphism, parasites, and the native area of *Nezara viridula* (Hemiptera, Pentatomidae). *Ann. Entomol. Fenn.* 52(1):28–31

55. Hokyo, N., Kiritani, K. 1962. Sampling design for estimating the population of the southern green stink bug, *Nezara viridula* (Pentatomidae, Hemiptera) in the paddy field. *Jpn. J. Ecol.* 12:228–35

56. Hokyo, N., Kiritani, K. 1963. Mortality process in relation to aggregation in the southern green stink bug. *Res. Popul. Ecol.* 5:23–30

57. Ishiwatari, T. 1974. Studies on the scent

of stink bugs (Hemiptera: Pentatomidae). I. Alarm pheromone activity. *Appl. Entomol. Zool.* 9(3):153–58

58. Ishiwatari, T. 1976. Studies on the scent of stink bugs (Hemiptera: Pentatomidae). II. Aggregation pheromone activity. *Appl. Entomol. Zool.* 11:38–44

59. Jensen, R. L., Gibbens, J. 1973. Rearing the southern green stink bug on an artificial diet. *J. Econ. Entomol.* 66:269–71

60. Jones, T. H. 1918. The southern green plant bug. *USDA Bull.* 689:1–27

61. Jones, W. A. Jr. 1985. *Nezara viridula.* In *Handbook of Insect Rearing,* ed. P. Singh, R. F. Moore, 1:339–44. Amsterdam: Elsevier-Sci.

62. Jones, W. A. Jr. 1988. World review of the parasitoids of the southern green stink bug, *Nezara viridula* (L.) (Heteroptera: Pentatomidae). *Ann. Entomol. Soc. Am.* 81(2):262–73

63. Jones, W. A. Jr., Powell, J. E. 1982. Potential for biological control of the southern green stink bug on soybeans. *Miss. Entomol. Assoc. Proc.* 1(1):21–22 (Abstr.)

64. Jones, W. A. Jr., Sullivan, M. J. 1981. Overwintering habitats, spring emergence patterns and winter mortality of some South Carolina Hemiptera. *Environ. Entomol.* 10(3):409–14

65. Jones, W. A. Jr., Sullivan, M. J. 1982. Role of host plants in population dynamics of stink bug pests of soybean in South Carolina. *Environ. Entomol.* 11(4):867–75

66. Kamal, M. 1937. The cotton green bug, *Nezara viridula,* L. and its important egg-parasite, *Microphanurus megacephalus* (Ashmead) (Hymenoptera: Proctotrupidae). *Bull. Soc. Entomol. Egypte* 21:175–207

67. Kariya, H. 1961. Effect of temperature on the development and mortality of the southern green stink bug, *Nezara viridula* and the oriental green stink bug, *N. antennata. Jpn. J. Appl. Entomol. Zool.* 5:191–96

68. Kester, K. M., Smith, C. M. 1984. Effects of diet on growth, fecundity and duration of tethered flight of *Nezara viridula. Entomol. Exp. Appl.* 35(1):75–81

69. Kester, K. M., Smith, C. M., Gilman, D. F. 1984. Mechanisms of resistance in soybean (*Glycine max* [L.] Merrill) Genotype PI171444 to the southern green stink bug, *Nezara viridula* (L.) (Hemiptera: Pentatomidae). *Environ. Entomol.* 13(5):1208–15

70. Kiritani, K. 1963. The change in reproductive system of the southern green stink bug, *Nezara viridula,* and its application to forecasting of the seasonal history. *Jpn. J. Appl. Entomol. Zool.* 7:327–37

71. Kiritani, K. 1964. The effect of colony size upon the survival of larvae of the southern green stink bug, *Nezara viridula.* Jpn. J. Appl. Entomol. Zool. 8:45–54

72. Kiritani, K. 1964. Natural control of populations of the southern green stink bug, *Nezara viridula. Res. Popul. Ecol.* 6:88–98

73. Kiritani, K. 1965. The natural regulation of the population of the southern green stink bug, *Nezara viridula* L. *Proc. 12th Int. Congr. Entomol. London,* p. 375

74. Kiritani, K. 1969. The differences in biological and ecological characteristics between neighboring populations in the southern green stink bug *Nezara viridula. Jpn. J. Ecol.* 19(5):177–84

75. Kiritani, K. 1972. Distribution and abundance of the southern green stink bug *Nezara viridula. Jpn. Pestic. Inf.* 10:117–41

76. Kiritani, K. 1970. Studies on the adult polymorphism in the southern green stink bug, *Nezara viridula* (Hemiptera: Pentatomidae). *Res. Popul. Ecol.* 12(1):19–34

77. Kiritani, K. 1971. Distribution and abundance of the southern green stink bug *Nezara viridula. Proc. Symp. Rice Insects. Trop. Agric. Res. Ser.* 5:235–48. Tokyo: Jpn. Minist. Agric. For.

78. Kiritani, K., Hokyo, N. 1962. Studies on the life table of the southern green stink bug, *Nezara viridula. Jpn. J. Appl. Entomol. Zool.* 6:124–40

79. Kiritani, K., Hokyo, N. 1965. Variation of egg mass size in relation to the oviposition pattern in Pentatomidae. *Konchu* 33:427–32

80. Kiritani, K., Hokyo, N., Iwao, S. 1966. Population behavior of the southern green stink bug, *Nezara viridula,* with special reference to the developmental stages of early-planted paddy. *Res. Popul. Ecol.* 8:133–46

81. Kiritani, K., Hokyo, N., Kimura, K. 1962. Differential winter mortality relative to sex in the population of the southern green stink bug, *Nezara viridula* (Pentatomidae, Hemiptera). *Jpn. J. Appl. Entomol. Zool.* 6:242–46

82. Kiritani, K., Hokyo, N., Kimura, K. 1963. Survival rate and reproductivity of the adult southern green stink bug, *Nezara viridula,* in the field cage. *Jpn. J. Appl. Entomol. Zool.* 7:113–24

83. Kiritani, K., Hokyo, N., Kimura, K. 1966. Factors affecting the winter mortality in the southern green stink

bug, *Nezara viridula* L. *Ann. Soc. Entomol. Fr.* 2:199–207

84. Kiritani, K., Hokyo, N., Kimura, K., Nakasuji, F. 1965. Imaginal dispersal of the southern green stink bug, *Nezara viridula*, in relation to feeding and oviposition. *Jpn. J. Appl. Entomol. Zool.* 9:291–97

85. Kiritani, K., Kimura, K. 1965. The effect of population density during nymphal and adult stages on the fecundity and other reproductive performances. *Jpn. J. Ecol.* 15:233–36

86. Kiritani, K., Kimura, K. 1966. A study of the nymphal aggregation of the cabbage stink bug, *Eurydema rugosum* Motschulsky (Heteroptera: Pentatomidae). *Appl. Entomol. Zool.* 1:21–28

87. Kiritani, K., Nakasuji, F., Hokyo, N. 1966. The survival rate of eggs and larvae in relation to group size in the southern green stink bug, *Nezara viridula* L. *Jpn. J. Appl. Entomol.* 10:205–11

88. Kiritani, K., Sasaba, T. 1969. The difference in bio- and ecological characteristics between neighbouring populations in the southern green stink bug, *Nezara viridula* L. *Jpn J. Ecol.* 19 (5):177–84

89. Kobayashi, T. 1959. The developmental stages of some species of the Japanese Pentatomidae (Hemiptera). VII. Developmental stages of *Nezara* and its allied genera. *Jpn. J. Appl. Entomol. Zool.* 3:221–31

90. Kogan, M., Turnipseed, S. G., Shepard, M., De Oliveira, E. B., Borgo, A. 1977. Pilot insect pest management program for soybean in southern Brazil. *J. Econ. Entomol.* 70(5):659–63

91. Krispyn, J. W., Todd, J. W. 1982. The red imported fire ant as a predator of the southern green stink bug in Georgia. *J. Ga. Entomol. Soc.* 17(1):19–26

92. Leston, D., Pringle, J. W. S. 1963. Acoustical behaviour of Hemiptera. In *Acoustical Behaviour of Animals*, ed. R. G. Busnel, pp. 391–410. Amsterdam/London/New York: Elsevier. 933 pp.

93. Lethierry, L., Severin, G. 1893. *Catalogue général des Hémiptères. Tomo 1. Hétéroptères Pentatomidae. F. Hayez.* Brussels: Acad. R. Belg. 286 pp.

94. Lim, G. S. 1970. The importance and control of *Nezara viridula* Linnaeus on the rice crop in West Malaysia. *Malays. Agric. J.* 47(4):465–82

95. Lockwood, J. A., Story, R. N. 1985. Bifunctional pheromone in the first instar of the southern green stink bug, *Nezara viridula* (L.) (Hem.: Pent.): Its characterization and interaction with other stimuli. *Ann. Entomol. Soc. Am.* 78(4):474–79

96. Lockwood, J. A., Story, R. N. 1985. Photic, thermic and sibling influences on the hatching rhythm of the southern green stink bug, *Nezara viridula* (L.). *Environ. Entomol.* 14(5):562–67

97. Lockwood, J. A., Story, R. N. 1986. Adaptive functions of nymphal aggregation in the southern green stink bug, *Nezara viridula* (L.) (Hemiptera: Pentatomidae). *Environ. Entomol.* 15:739–49

98. Lockwood, J. A., Story, R. N. 1986. Embryonic orientation in pentatomids: its mechanism and function in southern green stink bug (Hem: Pent). *Ann. Entomol. Soc. Am.* 79(6):963–70

99. Lockwood, J. A., Story, R. N. 1986. The diurnal ethology of the southern green stink bug, *Nezara viridula* (L.) in cowpeas. *J. Entomol. Sci.* 21(2):175–84

100. Lockwood, J. A., Story, R. N. 1987. Defensive secretion of the southern green stink bug (Hemiptera: Pentatomidae) as an alarm pheromone. *Ann. Entomol Soc. Am.* 80(5):686–91

101. Marsolan, N. F., Rudd, W. G. 1976. Modeling and optimal control of insect pest populations of *Nezara viridula* (Linnaeus), a major pest of soybeans. *Math. Biosci.* 30(3/4):231–44

102. Matthews, R. W., Matthews, J. R. 1978. *Insect Behavior.* New York: Wiley. 507 pp.

103. McLain, D. K. 1980. Female choice and the adaptive significance of prolonged copulation in *Nezara viridula* (Hemiptera: Pentatomidae). *Psyche* 87(3/4): 325–36

104. McLain, D. K. 1981. Sperm precedence and prolonged copulation in the southern green stink bug, *Nezara viridula*. *J. Ga. Entomol. Soc.* 16(1):70–77

105. McLain, D. K. 1985. Male size, sperm competition and the intensity of sexual selection in the southern green stink bug, *Nezara viridula* (Hemiptera: Pentatomidae). *Ann. Entomol. Soc. Am.* 78(1):86–89

106. McPherson, J. E., Cuda, J. P. 1974. The first record in Illinois of *Nezara viridula* (Hemiptera: Pentatomidae). *Trans. Ill. State Acad. Sci.* 67(4):461–62

107. McPherson, R. M., Newsom, L. D. 1984. Trap crops for control of stink bugs in soybean. *J. Ga. Entomol. Soc.* 19(4):470–80

108. McPherson, R. M., Newsom, L. D., Farthing, B. F. 1979. Evaluation of four stink bug species from three genera affecting soybean yield and quality in

Louisiana. *J. Econ. Entomol.* 72(2): 88–94

109. McPherson, R. M., Pitts, J. R., Newsom, L. D., Chapin, J. B., Herzog, D. C. 1982. Incidence of tachinid parasitism of several stink bug (Heteroptera: Pentatomidae) species associated with soybean. *J. Econ. Entomol.* 75(5):783–86

110. Menezes, E. B., Herzog, D. C., D'Almada, P. J. 1985. A study of parasitism of the southern green stink bug, *Nezara viridula* (L) (Hemiptera: Pentatomidae), by *Trichopoda pennipes* (F) (Diptera: Tachinidae). *An. Soc. Entomol. Bras.* 14(1):29–35

111. Menezes, E. B., Herzog, D. C., Teare, D. C., Sprenkel, R. K. 1985. Phenological events affecting southern green stink bug on soybean. *Proc. Soil Crop Sci. Soc. Fla.* 44:227–31

112. Michael, P. J. 1981. *Trichopoda*: A tricky parasite. *J. Agric. West. Aust.* 22(2):56–57

113. Michieli, S., Žener, B. 1968. Der sauerstoffverbrauch verschiedener farbstadien bei der wanze *Nezara viridula* (L.). *Z. Vgl. Physiol.* 58:223–24

114. Mitchell, W. C., Mau, R. F. L. 1969. Sexual activity and longevity of the southern green stink bug, *Nezara viridula.* *Ann. Entomol. Soc. Am.* 62(6):1246–47

115. Mitchell, W. C., Mau, R. F. L. 1971. Response of the female southern green stink bug and its parasite, *Trichopoda pennipes*, to male stink bug pheromones. *J. Econ. Entomol.* 64(4):856–59

116. Moeira, G. R. P., Becker, M. 1986. Mortality of *Nezara viridula* Linnaeus 1785 (Heteroptera: Pentatomidae) in the egg stage in a soybean field. I. All causes of mortality. *An. Soc. Entomol. Bras.* 15(2):271–90

117. Moeira, G. R. P., Becker, M. 1986. Mortality of *Nezara viridula* Linnaeus 1785 (Heteroptera: Pentatomidae) in the egg stage in a soybean field. II. Parasitoids. *An. Soc. Entomol. Bras.* 15(2): 291–308

118. Moeira, G. R. P., Becker, M. 1986. Mortality of *Nezara viridula* Linnaeus 1785 (Heteroptera: Pentatomidae) in the egg stage in a soybean field. III. Predators. *An. Soc. Entomol. Bras.* 15(2): 309–26

119. Nakasuji, F., Hokyo, N., Kiritani, K. 1965. Spatial distribution of three plant bugs in relation to their behavior. *Res. Popul. Ecol.* 4(7):99–108

120. Negron, J. F., Riley, T. J. 1987. Southern green stink bug, *Nezara viridula* (Heteroptera: Pentatomidae), feeding in corn. *J. Econ. Entomol.* 80(3):666–69

121. Newsom, L. D., Herzog, D. C. 1977. Trap crops for control of soybean pests. *La. Agric.* 20:14–15

122. Newsom, L. D., Kogan, M., Miner, F. D., Rabb, R. L. Sr., Turnipseed, S. G., et al. 1980. General accomplishments toward better pest control in soybean. In *New Technology of Pest Control*, ed. B. B. Huffaker, pp. 51–78. New York: Wiley-Intersci.

123. Nilakhe, S. S., Chalfant, R. B., Singh, S. V. 1981. Damage to southern peas by different stages of the southern green stink bug. *J. Ga. Entomol. Soc.* 16 (4):409–14

124. Nilakhe, S. S., Chalfant, R. B., Singh, S. V. 1981. Evaluation of southern green stink bug damage to cowpeas. *J. Econ. Entomol.* 74(5):589–92

125. Nilakhe, S. S., Chalfant, R. B., Singh, S. V. 1981. Field damage to lima beans by different stages of southern green stink bug. *J. Ga. Entomol. Soc.* 16 (3):392–96

126. Nishida, T. 1966. Behavior and mortality of the southern green stink bug *Nezara viridula* in Hawaii. *Res. Popul. Ecol.* 8:78–88

127. Orr, D. B., Russin, J. S., Boethel, D. J., Jones, W. A. Jr. 1986. Stink bug (Hemiptera: Pentatomidae) egg parasitism in Louisiana soybeans. *Environ. Entomol.* 15(6):1250–54

128. Panizzi, A. R., Galileo, M. H. M., Gastal, H. A. O., Toledo, J. F. F., Wild, C. H. 1980. Dispersal of *Nezara viridula* and *Piezodorus guildinii* nymphs in soybeans. *Environ. Entomol.* 9(3):293–97

129. Panizzi, A. R., Slansky, F. Jr. 1985. Review of phytophagous pentatomids (Hemiptera: Pentatomidae) associated with soybean in the Americas. *Fla. Entomol.* 68(1):184–214

130. Passlow, T., Waite, G. K. 1971. Green vegetable bug as a soybean pest, *Nezara viridula.* *Queensl. Agric. J.* 97(7):491–93

131. Pavis, C., Malosse, C. 1986. A sex pheromone produced by mature males in the southern green stink bug, *Nezara viridula* (L.) (Heteroptera: Pentatomidae). *C. R. Acad. Sci. Ser. III* 303(7):273–76

132. Pitts, J. R. 1977. Effect of temperature and photoperiod on *Nezara viridula* L. MS thesis. La. State Univ., Baton Rouge. 89 pp.

133. Powell, J. E., Shepard, M. 1982. Biology of Australian and United States

strains of *Trissolcus basalis*, a parasitoid of the green vegetable bug, *Nezara viridula. Aust. J. Ecol.* 7:181–86

134. Ragsdale, D. W., Larson, A. D., Newsom, L. D. 1981. Quantitative assessment of the predators of *Nezara viridula* eggs and nymphs within a soybean agroecosystem using an ELISA. *Environ. Entomol.* 10(3):402–5

135. Rizzo, H. F. E. 1968. Aspectos morfologicos y biologicos de *Nezara viridula* (L.) (Hemiptera: Pentatomidae). *Agron. Trop. Maracay* 18:249–74

136. Rosenfeld, J. H. 1911. Insects and spiders in Spanish moss. *J. Econ. Entomol.* 4:398–409

137. Rudd, W. G., Jensen, R. L. 1977. Sweep net and ground cloth sampling for insects in soybeans. *J. Econ. Entomol.* 7:301–4

138. Russin, J. S., Layton, M. B., Orr, D. B., Boethel, D. J. 1987. Within-plant distribution of, and partial compensation for, stink bug (Heteroptera: Pentatomidae) damage to soybean seeds. *J. Econ. Entomol.* 80:215–20

139. Schumann, F. W., Todd, J. W. 1982. Population dynamics of the southern green stink bug (Heteroptera: Pentatomidae) in relation to soybean phenology. *J. Econ. Entomol.* 75(4):748–53

140. Servadei, A. 1967. *Fauna d'Italia*, Vol. 9, *Rhyncota (Heteroptera, Homoptera, Auchenorrhyncha)*. Bologna: Calderini. 851 pp.

141. Shahjahan, M., Beardsley, J. W. Jr. 1975. Egg viability and larval penetration in *Trichopoda pennipes pilipes* Fabricius (Diptera: Tachinidae). *Proc. Hawaii. Entomol. Soc.* 22(1):133–36

142. Silva, A. G. d'A., Goncalves, C. R., Galvo, D. M., Goncalves, A. J. L., Gomes, J., et al. 1968. *Quarto Catálogo dos Insetos que Vivem nas Plantas do Brasil—Seus Parasitas e Predadores*, Parte II, Vol. 1. Rio de Janeiro: Minist. Agric.

143. Singh, Z. 1972. *Bionomics of the southern green stink bug*, Nezara viridula *(Linn.) (Hemiptera: Pentatomidae) in Central India*. PhD thesis. Univ. Ill., Urbana-Champaign. 136 pp.

144. Singh, Z. 1973. *Southern Green Stink Bug and its Relationship to Soybeans. Bionomics of the Southern Green Stink Bug,* Nezara viridula *(Linn.) (Hemiptera: Pentatomidae) in Central India*. Delhi, India: Metropolitan. 105 pp.

145. Sisli, M. N., Bosgelmez, A. 1973. Effect of photoperiod on the biology of *Nezara viridula* f. *smaragdula* (F.) (Hemiptera: Pentatomidae). *Commun.* *Fac. Sci. Univ. Ankara Ser.* C 17:201–11

146. Sparks, A. N., Jackson, R. D., Carpenter, J. E., Muller, R. A. 1986. Insects captured in light traps in the gulf of Mexico. *Ann. Entomol. Soc. Am.* 79: 132–39

147. Staddon, B. W. 1979. The scent glands of Heteroptera. *Adv. Insect Physiol.* 14: 351–418

148. Stam, P. A. 1978. *Relation of predators to population dynamics of* Nezara viridula *(L.) in a soybean ecosystem*. PhD thesis. La. State Univ., Baton Rouge. 220 pp.

149. Stam, P. A., Newsom, L. D., Lambremont, E. N. 1987. Predation and food as factors affecting survival of *Nezara viridula* (L.) (Hemiptera: Pentatomidae) in a soybean ecosystem. *Environ. Entomol.* 16(6):1211–16

150. Su, T. H., Tseng, H. K. 1984. The introduction of an egg parasite, *Trissolcus basalis* (Wollaston), for control of the southern green stink bug, *Nezara viridula* (L.) in Taiwan. *J. Agric. For.* 33(2):49–54

151. Talekar, N. S., Chen, B. S. 1983. Seasonality of insect pests of soybean and mungbean in Taiwan. *J. Econ. Entomol.* 76(1):34–37

152. Todd, J. W., Herzog, D. C. 1980. Sampling phytophagous Pentatomidae in soybean. In *Sampling Methods in Soybean Entomology*, ed. M. Kogan, D. C. Herzog, pp. 438–78. New York: Springer-Verlag. 587 pp.

153. Todd, J. W., Lewis, W. J. 1976. Incidence and oviposition patterns of *Trichopoda pennipes* (F.), a parasite of the southern green stink bug, *Nezara viridula* (L.). *J. Ga. Entomol. Soc.* 11 (1):50–54

154. Todd, J. W., Mullinix, B. G. 1985. Effects of insect-pest complexes on soybean. In *World Soybean Res. Conf. III Proc., Ames, Iowa*, pp. 624–34. Boulder/London: Westview. 1262 pp.

155. Todd, J. W., Turnipseed, S. G. 1974. Effects of southern green stink bug damage on yield and quality of soybeans. *J. Econ. Entomol.* 67(3):421–26

156. Turner, J. W. 1983. Influence of plant species on the movement of *Trissolcus basalis* Wollaston (Hymenoptera: Scelionidae)—a parasite of *Nezara viridula* L. *J. Aust. Entomol. Soc.* 22(3):271–72

157. Velez, J. R. 1974. Observaciones sobre la biologia de la chinche verde, *Nezara viridula* (L.) en el Valle del Fuerte Sin. *Folia Entomol. Mex.* 28:5–12

158. Waite, G. K. 1980. The basking behaviour of *Nezara viridula* (L.) (Pentatomidae: Hemiptera) on soybeans and its implication in control. *J. Aust. Entomol. Soc.* 19(2):157–59

159. Waterhouse, D. F., Forss, D. A., Hackmann, R. H. 1961. Characteristic odour components of the scent of stink bugs. *J. Insect Physiol.* 6:113–21

160. Watson, J. R. 1918. Insects of a citrus grove. *Fla. Agric. Exp. Stn. Bull.* 148:165–67

161. Wuensche, A. L. 1976. *Relative abundance of seven pest species and three predaceous genera in three soybean ecosystems in Louisiana.* MS thesis. La. State Univ., Baton Rouge. 384 pp.

162. Yukawa, J., Kiritani, K. 1965. Polymorphism in the southern green stink bug. *Pac. Insects* 7:639–42

*Ann. Rev. Entomol. 1989. 34:293–313*

# ECONOMICS OF AGRICULTURAL PESTICIDE RESISTANCE IN ARTHROPODS

## Alan L. Knight

Department of Entomology, The Pennsylvania State University, Biglerville, Pennsylvania 17307

## George W. Norton

Department of Agricultural Economics, Virginia Polytechnic Institute and State University, Blacksburg, Virginia 24061

## PERSPECTIVE AND OVERVIEW

Pesticide resistance among arthropods poses a severe threat to agricultural productivity in the United States and in other areas of the world. While humans have had the ingenuity to develop thousands of pesticides to protect foodstuffs, livestock, and health, the use of these poisons has triggered a rapid evolutionary response in many target species and some nontarget species of invertebrates, microorganisms, higher plants, and other vertebrates. Exposure to pesticides has led to genetic selection for individuals with the biochemistry or behavior necessary to nullify their toxic effects (31, 70, 76). Despite the frequent development of resistance, the tremendous initial success of agricultural chemicals has created an intractable dependency on their use.

Continued development of new compounds to replace old ineffective chemistries has placed agriculture on a pesticide treadmill. The trend toward greater use and higher costs of pesticides as inputs into agricultural production continues both despite and because of exponential increases in resistance (30). Moreover, the threat of pesticide resistance to agricultural productivity is accelerating as the identification and commercial development of new pesticides slows. For some pests the availability of efficacious pesticides is already

0066-4170/89/0101–0293$02.00

low to none. Fueled by public concern over health and environmental effects of pesticide use, increased regulatory action is reducing the number of compounds available and increasing the threat of resistance to the remaining compounds.

Reduction in pesticide efficacy from pest resistance has major economic, environmental, and human health implications. Understanding the economics of resistance at the farm and beyond-farm levels will allow us to achieve optimal use of pesticides over time. Solutions to the resistance problem will require coordinated efforts among farmers, pest management firms, the chemical industry, government policymakers, and researchers (23, 63).

Several reviews have previously discussed the economics of pesticide use in pest management (15, 37, 52, 54, 60, 64, 66). Yet no work has specifically examined the major economic issues associated with pesticide resistance in agriculturally important arthropods. In our review we deliberately do not focus on specific economic evaluations of crop-pest-pesticide systems. Instead, we choose to draw a broader picture of the economics of pesticide resistance by linking the range of biological facets of resistance with their economic impacts. First, we consider the nature of resistance impacts both on farm and off, and briefly review economic theory and the methods used to analyze the impacts. Secondly, we discuss implications for resistance management for both the private and public sectors. Finally, we suggest important areas for future research.

## RESISTANCE IMPACTS

Pesticide resistance can influence pesticide costs and crop yields, affecting both the level and stability of farm income. Because of the mobility of pests, the pest-control actions of one farmer affect other farmers, and pest-control activities have significant environmental and distributional implications for society as a whole. We identify first the nature of field or farm-level impacts and then consider effects beyond the farm gate.

### Field or Farm-Level Impacts

Arthropod damage to crops is significant, perhaps representing a loss of 13% of potential crop production (68). For many years, increased insecticide usage appeared to be the least expensive means of insect control and a major source of productivity gains in agriculture, particularly for cotton, corn, fruits, and vegetables. Currently, over one billion pounds of pesticides are applied each year on farms in the United States alone (5). Crop production per pound of insecticide applied for two of the most heavily treated crops (cotton and corn) has dropped precipitously since the 1920s (54). Although many factors influence the level of insect damage and pesticide usage, evidence suggests that increased insecticide resistance over time has had an important effect on both (29). Pimentel and coworkers (67) estimated that resistance from increased pesticide use costs United States agriculture at least $118 million annually.

The cost of resistance is also affected by the need to substitute newer and almost always higher-priced insecticides to replace ineffective ones. For example, the more complex chemical syntheses needed to produce synthetic pyrethroids and growth regulators made these compounds approximately 100-fold more expensive by weight than DDT at the time of their introduction (54). For species resistant to a wide variety of pesticides, growers commonly spend hundreds of dollars per hectare to control a single pest (7, 8, 26).

Increased use of insecticides due to resistance magnifies the development of secondary pest outbreaks and pest resurgences. For example, cotton production in Mexico and the Rio Grande Valley was devastated by *Helicoverpa (Heliothis) zea* and *Heliothis virescens* following field sterilization with broad-spectrum insecticides during the late 1960s (1). These species were previously considered secondary pests, and in most seasons they were adequately controlled by their natural enemies. Increasing pesticide use to compensate for the destruction of the biological control agents eventually led to the cotton growers' inability to manage these now-resistant pests, and hundreds of thousands of acres went out of production.

The development of cross and multiple resistance to a class or classes of pesticides greatly accelerates the depletion of pest susceptibility and multiplies the cost of resistance. Because DDT and the synthetic pyrethroids have similar modes of action, the prior widespread use of DDT and the development of resistance to it (*kdr* gene) threatens the efficacy of the synthetic pyrethroids (70). The increasing number of pests with genes conferring resistance to several classes of pesticides is a major threat to successful pest control (76).

Unfortunately, the potential for reversion of pesticide resistance in the field to prolong the life of a compound is generally considered low under continued pesticide usage (54). Yet the presence and interactions of a myriad of genetic factors (e.g. recessive resistance genes, polygenic mechanisms, lower fitness levels associated with resistance), ecological factors (e.g. high immigration rates of susceptibles, presence of refugia), and operational factors [e.g. low pesticide selection under integrated pest management (IPM) programs] have maintained the usefulness of some pesticides for over 20 years.

Conversely, the augmentation of resistance in arthropods has been used advantageously. Management of tolerant or resistant arthropod natural enemies (predators and parasites of pest species) has allowed successful integration of biological and chemical control tactics and has thus reduced pesticide use and presumably the development of resistance in a number of cropping systems (19). For example, the conservation and augmentation of phytoseiid mites to manage herbivorous mite populations is widespread for deciduous

tree fruit and nut crops (41). Hoyt (42) estimated that the use of the predaceous mite *Typhlodromus occidentalis* on apple in Washington state alone
saves growers over $5 million per year on pesticide material and application
costs.

Genetic selection for improved strains of natural enemies has increased
these arthropods' ability to establish, reproduce, and disperse under pesticide
pressure in field conditions (41). Commercially bred resistant natural enemies
have been distributed in a few areas for high value crops. For example, an
integrated mite management program using genetically selected strains of the
phytoseiid *Metaseiulus occidentalis* has been adopted by nearly 60% of
California almond growers. Economic analyses have estimated that growers
who adopt the program can save $60–110 per hectare. The annual return on
the almond growers' investment in this program is anticipated to be 280–
370% (39).

Farm-level decisions to apply insecticides are also influenced by the perceived income risk (variability) associated with alternative methods of pest
control. The farmer's aversion to risk is a private incentive to overapply
insecticides as a form of insurance. For example, Moffitt and coworkers (58)
found that while cotton yields in the Imperial Valley would on average be
lower, the expected net profits would be greater, but more variable, with the
adoption of short-season production practices to control the pink bollworm,
*Pectinophora gossypiella*. Despite tremendous resistance problems, a 30%
decline in productivity due to this pest, and pesticide expenditures of $400–
700 per hectare for some growers (7), along with an 80% decline in acreage
planted in the valley, growers only recently adopted the short-season production program in this region (35).

## Impacts Beyond the Farm Gate

Because of pest mobility, pesticide resistance for certain pests cannot easily
be managed by individual farmers. Pests are not unique to a single farm, and
consequently pest control can be considered a communal problem (47). For
species such as spider mites that disperse passively, movement of resistant
pests may be limited to between adjacent fields and crops (56). For somewhat
more mobile species, such as pear psylla, resistance levels may be
characteristic of a region (28). Movement of resistant beetles and moths,
however, may cross several state or national borders, such as the movement
of resistant *Heliothis* spp. from Mexico into the United States (91). Movement of resistant pests suggests that analysis of optimal pest control levels and
of means for implementing those levels must consider the possibilities for
collective action within the agricultural sector (57). That collective action
may involve the agrochemical industry or groups of farmers themselves.

The cost and difficulty of developing new pesticides has increased rapidly
during the last decades. It is estimated that, on average, over 15,000 com-

pounds are screened, 8–10 yr of research and development are required, and costs range from $20–40 million to isolate, test, develop, and market a new synthetic pesticide (45, 51, 54). Within this atmosphere of tremendous risk and uncertainty, agrochemical companies tend to develop products with a large potential market (ignoring pest problems of minor crops) and promote their pesticides in a way that recovers their large investment quickly (heavy reliance on chemical inputs in crop production).

The international agrochemical industry has recently organized to deal with resistance problems (44). The Pyrethroid Efficacy Group, formed by pyrethroid manufacturers in 1979, and the Insecticide Resistance Action Committee, formed in 1984, are part of the International Group of National Associations and Manufacturers of Agrochemical Products, which provides advice and coordinates industry efforts to prolong the life of pesticides. Both organizations are specifically involved in monitoring suspected cases of resistance and advise governments and growers' associations on developing strategies to manage resistance.

The agrochemical industry, however, has moved more slowly in modifying its various market strategies to combat resistance. Concern about antitrust laws has allegedly limited cooperation among companies in the United States (R. Roush, personal communication). Nevertheless, growers have been organized through state, regional, and national programs to manage resistance outbreaks. Collective voluntary programs have been developed to restrict the use of pyrethroids on cotton to manage resistance in *Heliothis armigera* in Queensland, Australia (21) and the Egyptian cotton leafworm, *Spodoptera littoralis,* in Egypt (77).

Unfortunately, resistance management is complicated by the fact that pesticide use can result in water pollution, food residues, and dangers to human health and nontarget species, which can cause the socially optimum level of pesticide use to differ from the private optimum level. Private decision makers have limited incentives to include such environmental and social costs in their pesticide application decisions. Pimentel and coworkers (67) estimated the social and environmental costs of the use of pesticides at $839 million annually in the United States. These added costs are exacerbated if resistance induces higher levels of pesticide use.

The environmental costs of pesticide use provide a rationale for pesticide regulatory action by environmental protection agencies. The divergence of social and private costs must be estimated if appropriate (from society's point of view) pesticide regulatory decisions are to be made. The analysis, however, is complicated by the potential impact of pesticide regulatory actions on the management of pesticide resistance. The use of a variety of pesticides is a promising tactic to combat developing resistance to any particular compound. Conversely, the banning of an effective pesticide may lead to more rapid development of resistance to the other chemicals. For example, the discontinuation of aldicarb use due to ground water contamination on Long Island, New York (40) has been a major factor leading to a significant drop in potato production in this area and has increased the selection pressure on the few remaining effective chemicals (26). Chlordimeform has been used on cotton to improve control of *Heliothis* spp. and to reduce their development of resistance to pyrethroids (9). Yet owing to potential worker exposure and associated health risks, the registration of this compound has been canceled in California (35), and it may soon be withdrawn from the marketplace (R. Frisbie, personal communication). The selective acaricide cyhexatin, which enabled entomologists to develop highly successful integrated mite management programs for fruit crops following years of repeated acaricide failures (20), has recently been removed by the manufacturer because of potential human health risks.

Another pesticide resistance issue is related to the widespread subsidization of pesticide use by international agencies and governments. Repetto (75) found that on average 44% of the total real cost of pesticides was subsidized in the nine countries he surveyed in Central America, Africa, and Southeast Asia. Subsidies encourage growers to use more chemicals and not to use other pest control methods. Promoting pesticides obviously increases the risk of resistance and environmental degradation. High rates of agricultural pesticides, especially on rice and cotton, may have been a factor in reversing the initial successes of malaria control in the Third World because of the development of anopheline resistance (16).

Finally, there is a large class of distributional effects generated by increased pesticide resistance. When yields decrease or costs increase, prices can be affected, as well as the net economic benefits of consumers and producers. Producers in one region can benefit relative to those in another depending on the effects on yield, costs, and prices in the different regions. The structure of the farm sector can be affected if costs or new technologies stimulated by pest resistance affect one size farm more than another or one commodity more than another. In addition, effects of resistance on agriculture have multiple effects on employment and income outside the farm sector, particularly in the chemical industry.

Clearly the effects of pesticide resistance can be widespread and diverse. In the next two sections we examine these effects in more detail.

## ECONOMIC ANALYSIS OF FARM-LEVEL ISSUES

Farm-level economic issues associated with pesticide use and with resistance in particular have been assessed both theoretically and empirically over the past 15 years. Many of the studies have explored the optimal use of pesticides in light of the dynamic nature of resistance. Others have examined the choice of alternative pest management strategies. Both types of studies are concerned with the effects of resistance on pesticide productivity (efficacy). Because of the potential for resistance, future as well as current productivity must be considered.

## Effects of Resistance on Pesticide Productivity

The productivity of a new pesticide immediately following its introduction is generally high. The use of DDT in Wisconsin to control the Colorado potato beetle, *Lepinotarsa decemlineata,* and the potato leafhopper, *Empoasca fabae,* for example, generated an initial return of $29 per dollar invested (55). But as pests lose their susceptibility to pesticides, the productivity of pesticides declines over time through either reduced yields or higher costs for the same yields. The effect of pesticide resistance on productivity is illustrated graphically in Figure 1. If nonpesticide inputs are held constant, for a given pest density the effect of the pesticide input on agricultural output can be represented with the hypothetical response $Z_1$. With partial resistance, the efficacy of the pesticide is reduced as represented by the response curve $Z_2$. To achieve the same level of production $(A)$, more pesticide must be applied $(B_2$ rather than $B_1)$.

Early attempts to measure the productivity of pesticides, irrespective of resistance, used static models (often simple production functions) (10, 27, 36, 49). Headley (36), for example, estimated the marginal product of pesticide use by fitting a Cobb-Douglas production function with 1963 state data for 59

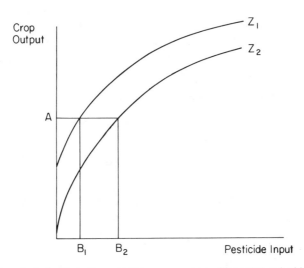

*Figure 1*   Hypothetical effect of a pesticide on crop output without ($Z_1$) and with ($Z_2$) partial pesticide resistance in insects.

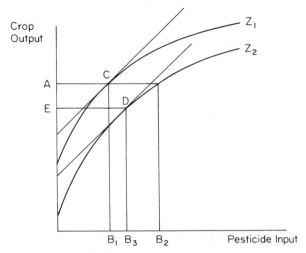

*Figure 2*  Hypothetical optimum amount of a pesticide to apply ($B_3$) against insects that exhibit partial pesticide resistance.

The optimal timing dimension involves two important but separate factors. First, pesticide productivity depends on the number and timing of applications within a season and across seasons irrespective of pesticide resistance, because pesticide productivity depends on both the population dynamics of the pest and the stage of plant growth. Resistance then alters the pest population dynamics. The second important factor is related to the time value of money. A dollar received today is worth more to a farmer than a dollar received tomorrow because the farmer could invest the dollar received today and earn interest. For this reason, economic models discount future benefits by an interest rate that represents the value that could be earned on alternative investments.

Much of the literature on economics of pesticide resistance includes dynamic optimizing models in which the optimal allocation of pesticides implies management of both the pest and its associated stock of susceptibility. Shoemaker (78–80) and Hueth & Regev (43) presented dynamic agricultural pest control models incorporating resistance. Their theoretical models contained no empirical estimation, but laid the basis for subsequent work in this area. Shoemaker's model (78–80) incorporated resistance along with many other dimensions of pest management. Hueth & Regev (43) investigated the effect of increasing pest resistance to insecticides on the optimal control of a pest population by constructing a single-pest, single-crop management model. Their simplified model excluded pest migration. The model's conditions for optimal use of pesticides at any time assumed that the marginal profits, excluding susceptibility, are equated to a user cost resulting from depletion of the stock of susceptibility. In addition, the optimal time path of pesticide use depended on a variable economic threshold during the course of the season. Only under very restrictive assumptions did the neglect of resistance result in overuse of chemicals based on this analysis.

These dynamic models offer optimal actions (for example, pesticide applications) given a set of state variables in the system, such as potential plant product, pest population density, and the stock of pest susceptibility. The problem is that of choosing time paths for control variables, which in turn imply, via a set of differential equations, time paths for the state variables. The time paths for the control variables are chosen to maximize a given function depending on the time paths of the control and state variables. The time paths can be both within seasons and across seasons. Empirically the model can be solved using a dynamic programing or control theory approach.

The inherent complexity of the biological processes, and hence the mathematical complexity of the models, has constrained the empirical application of these approaches. Studies that include theoretical or empirical dynamic optimizing models incorporating resistance, however, have become more common over time, particularly as computer capabilities have increased (3, 35, 47, 69, 73, 86). Taylor & Headley (86) presented a control theory model for use in dynamic evaluation of pest control strategies, although they did not analyze any data. Regev and coworkers (73) discussed the theoretical implications from their control theory model but then applied a related but simplified version of it using nonlinear programing. They explored the optimal policy for pesticide application given resistance, solving the model for a closed region rather than a farm because of pest mobility. Lazarus & Dixon (47) also utilized a nonlinear programing model to assess potential gains from internalizing resistance through regional coordination. Their model of an Illinois cash grain farm was used to estimate the gains from corn rootworm control; in this model, rotation to soybeans was considered as an alternative to insecticides. The authors computed taxes, subsidies, and environmental costs that would affect the regional optimum by enabling farmers to maximize individual profits. Harper (35) used a dynamic simulation model to explore the potential effects of build-up of resistance to chlordimeform by the pink bollworm on cotton in California.

Archibald's (3) economic analysis of resistance in California cotton production is one of the most ambitious empirical attempts to date to analyze optimal pesticide use with resistance. She used a dynamic programing model to determine optimal input given resistance under alternative pesticide regulatory policy and pest management strategies. Her study indicated that while insecticide usage has increased in cotton, the probability of *H. virescens* surviving all insect controls at the levels of materials used has increased, indicating that resistance has developed. Additionally, her data suggested that newer insecticides that meet regulatory requirements are more variable in their effectiveness and lose their effectiveness more rapidly. She estimated that economic losses from resistance in the early 1980s were in the range of $45–120 per hectare of cotton.

## Choice of Alternative Pest Control Strategies

Appropriate use of pesticides, given actual or potential pesticide resistance, depends on the efficacy of alternative means of pest control. One of the motivations behind the development of IPM strategies has been the growing pesticide resistance problem (89). These IPM strategies have been assessed using a variety of approaches varying in complexity from simple budgeting to the dynamic programing approach mentioned above. Many of these analyses of pest management alternatives have included an assessment of net income risk. Others have not.

Perhaps the most widely used method for evaluating alternative pest control strategies has been budgeting. Many examples of budgeting are found in Environmental Protection Agency and United States Department of Agriculture studies evaluating the possible impacts of regulatory actions, as well as in the scientific literature (e.g. 32, 81, 93). Most of these studies have not explicitly considered resistance even if they have alluded to its importance, although some have included resistant crop varieties as an alternative pest control measure (93).

Other studies have used simulation (74), linear programing (87), and dynamic programing (92) to evaluate pest management alternatives. Again, most of these have not considered resistance except as related to resistant cultivars.

Studies that have explicitly incorporated net income risk have employed Bayesian decision theory (11), simulation (48), and stochastic dominance (6, 18, 32, 33, 53, 59, 61). Feder (24) and Carlson (14) have presented theoretical discussions of the effects of risk on pest management strategies.

Unfortunately, owing to the dynamic nature of resistance it is difficult to evaluate alternative pest management strategies meaningfully, with or without risk, unless a dynamic model is employed that captures the effects of resistance on these alternatives over time. Thus a stochastic dynamic programing model is the preferred approach for in-depth analysis of both the optimal level of pesticide use and the choice of pest management alternatives given the existence or potential for pest resistance to pesticides.

## ECONOMIC ANALYSIS OF IMPACTS BEYOND THE FARM

The influence of pest resistance to pesticides extends beyond the farm primarily because (*a*) pests are mobile, (*b*) pesticides can enter the off-farm environment, and (*c*) resistance-induced changes in crop productivity can affect prices. In this section we consider the implications of these three factors on optimal pesticide use and on the distribution of benefits and costs of pesticide use with resistance.

## Mobility of Pests

The communal nature of the pest resistance problem has been considered in several studies. Regev and associates (72) pointed out the importance of analyzing the gap between private and social benefits (which arises, in part, because of pesticide resistance) when developing pest control policies. More recently, several authors have examined, theoretically or empirically, the implications of pesticide resistance on pesticide use, given the fact that pests can be mobile (17, 47, 57, 71).

Because of pest mobility, the decisions made by farmers that affect pest populations directly or indirectly are not optimal from society's point of view. Pests travel from one farmer's field to another's; thus the resistance level in each field is not the result of the individual farmer's application of pesticides, but is rather the result of cumulative action taken by all farmers in the area (71). As a result, the individual farmer has no incentives to consider future resistance when he or she applies a pesticide, but considers only the current level of pesticide resistance for the pests in his or her field. There is incentive to apply large quantities of pesticides to maximize current-period profits. The farmer will not consider future profits because they are independent of the current pesticide application.

Pest susceptibility is a renewable resource, but repeated use of a specific pesticide or class of pesticides can reduce the stock of susceptibility and eventually exhaust it (43, 57). If pest susceptibility declines as more of a pesticide is applied, irrespective of who applies the pesticide, then each farmer has an incentive to apply the amount needed to maximize only current-period profits because even if he or she applies less pesticide now in an effort to conserve the stock of susceptibility, other farmers will apply pesticides and deplete the stock. Attempts to analyze public pesticide policies under resistance therefore require a model that considers collective action on the part of farmers to optimally conserve the stock of pest susceptibility over time.

## Environmental Effects

Environmental effects of pesticides include water pollution, residues left on food, and impacts on human health and on nontarget species. A number of papers have discussed the divergence of private and social optima created by these effects (25, 35, 46, 71). This divergence exists regardless of resistance, but increases if higher levels of pesticides are applied to compensate for resistance.

One of the difficulties in incorporating environmental effects, either physical or economic, into pesticide regulatory decisions is the difficulty of measuring them. For example, water pollution is difficult to trace back to individual farms. Food residues are somewhat easier to measure because foods can be sampled. However, human exposure beyond the farm requires careful monitoring, and it is difficult to measure its economic consequences. Hospital costs, value of wages foregone, and increases in insurance premiums have been used to place a value on the cost of human exposure (67). These are crude approximations and apply to acute pesticide poisoning more than to the effects of long-term exposure. Harper (35) has discussed in depth the difficulties in assessing the human health implications of pesticide use.

Valuation of pesticide impacts on nontarget species requires, in many cases, valuation of nonmarket goods. Resource economists have attempted such valuations, but few studies have examined the cost of resistance as it affects nontarget species.

## Levels and Distribution of Economic Benefits

We noted earlier that pesticide resistance can lower agricultural productivity. This productivity change can have a variety of distributional impacts, many of them resulting from induced price effects. Lower production of a commodity means an increase in its per unit cost of production. Economists often represent this effect as an upward or backward shift in the supply curve for the affected good, e.g. from $S_1$ to $S_0$ in Figure 3. At the aggregate or market level, producers supply less ($Q_0$ compared to $Q_1$) at a higher price ($P_0$ compared to $P_1$). The amount of the price and quantity change depends on the slope of the demand curve. That slope, in turn, depends on the availability of substitute goods, the importance of exports of the good, and other factors.

Figure 3 can be used to illustrate the distributional effect of resistance on producers versus consumers. Resistance causes a loss to consumers that can be represented by the area $P_0ABP_1$. This area is called the loss in consumer surplus because consumers pay a higher price for fewer goods. The net effect to the producer is represented by $P_0ACP_1 - CBE$. This net effect, called the change in producer surplus, may be positive or negative. It will be negative if producers sell fewer goods that cost more to produce, increasing $CBE$. It will be positive if goods that are sold receive a higher price, resulting in a gain in $P_0ACP_1$. Net economic losses to consumers and producers as a group are represented by $ABE$.

Several formulas are available for calculating the changes in consumer surplus, producer surplus, and net economic welfare. The decision about which formula to use in a particular study depends on the specification of the supply and demand curves, the nature of the shifts in supply and demand curves (e.g. whether the curves shift out more at high prices than at low

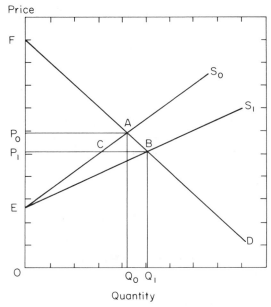

*Figure 3*  Graphical illustration of changes in consumer and producer surplus. See text for discussion.

prices), whether the commodity is traded internationally, and the nature of price policies (65).

Consumer and producer surplus analysis with respect to pesticide use or alternative pest management practices has been employed in several studies (62, 84, 85, 87). Some authors have applied economic surplus formulas directly, while others have incorporated consumer and producer surplus into the objective function of a mathematical (usually quadratic) programing model.

The concept of calculating consumer and producer surplus can be extended and applied to various income groups within society because the quantity demanded of different foods depends on income levels. Furthermore, the concept can be extended to consider the relative effects of resistance on different regions or countries.

Differential effects of resistance on regions or countries arise because of productivity and price differences. If resistance lowers productivity of a commodity in one region of a country but not another (perhaps because the pest does not exist in the other region), it can alter the comparative advantage of the two regions in producing the commodity. The unaffected region will benefit relative to the affected region both because its productivity is now relatively higher and because its product may bring a higher price even though its production did not drop. Over time, the differential advantage can cause a shift in production toward the region without the resistance problem. The same principle can hold across countries.

The structure of the farm sector also can be altered over time if farms of particular sizes are better able to adopt new technologies developed to control resistance than are others. However, these effects may be relatively small because of pest mobility, unless farm size varies greatly by region. Production may become more capital or labor intensive if new pest management practices (e.g. scouting, trap-cropping, rotational changes) require more capital or labor than before.

Finally, there can be distributional effects on the chemical industry and other sectors of the economy. In addition, pesticide manufacturers have varying degrees of control over the speed of resistance development. Decisions to develop new chemical control alternatives and the pricing of products influence depletion of the stock of pest susceptibility. For example, the competitive structure of the chemical industry may affect the stock of pesticide susceptibility over time (57). If there are no close substitutes for a pesticide and if a particular firm has a monopoly on its production (often provided by patent protection), the chemical firm will have an incentive to produce less of it and charge a higher price than it would if there were market competition. Thus, monopolistic control of a pesticide might retard resistance development. If the pesticide is available cheaply, resistance might develop more quickly (57). This may be very important for the management of resistance in minor pests or minor crops, which tend to be neglected by pesticide manufacturers and to have a limited arsenal of compounds available.

The economic effects of resistance extend beyond the farm not only to the chemical industry, but also to suppliers of other agricultural inputs, to the transportation sector, and to other sectors in the economy that demand agricultural products. These effects are called multiplier effects, and they can be as large as the direct on-farm impacts of pesticide resistance. Input-output analysis is commonly used to estimate these impacts. For example, Bernat & Norton (4) used this approach to calculate the effects of a pesticide used on apples.

## IMPLICATIONS FOR RESISTANCE MANAGEMENT

The diversity of economic impacts of pesticide resistance, the dynamic nature of resistance, pest mobility, and environmental effects from pesticide use all have important implications for resistance management by the private sector. Public policies are needed to provide incentives for collective action to control the rate of depletion of the stock of pest susceptibility by pesticides.

## Private-Sector Resistance Management

Miranowski & Carlson (57) have suggested a variety of conditions under which various elements of the private sector (farms, groups of farms, and chemical companies) have incentives to manage the build-up of pesticide resistance. If pests have little mobility (as some do), if substitute controls are more costly or nonexistent, and especially if the crop has a high value and is subject to serious pest damage in the absence of pest control, farmers have an incentive to seek to retard resistance.

If pests are mobile but can be confined to a region so that collective action within a region is justified, and if the costs of coordinating farms are relatively low and most farmers can be encouraged to cooperate, then incentives exist for multifarm resistance cooperatives or other group efforts.

A single pesticide manufacturer has an incentive to manage resistance if it possesses a highly profitable pesticide with no actual or potential close substitutes, if it can monitor resistance at a relatively low cost, if it has a monopoly in marketing the pesticide, and if pests are mobile so that incentives for farm-level management of resistance are noneconomic. Collective action on the part of chemical firms is also possible if resistance can be managed by mixtures of compounds owned by several firms, if resistance monitoring is valuable because of potential cross-resistance, and if coordinated rotation of compounds over time can reduce resistance (57).

A major point is that private incentives for resistance management (irrespective of public policies and regulations) depend on the particular set of conditions prevailing with respect to pest mobility, market economics, and other factors. Any procedure developed to estimate optimal pesticide use over time with developing resistance must consider these conditions, or private agents will have little incentive to act on the resulting recommendations. Because many pests are common to many or all of the farms in an area, individual farmers have little incentive to control resistance; yet collective actions of farmers will only be undertaken if the added administrative and coordination costs are outweighed by the benefits.

## Public Policy Implications

If pests are mobile, pest management coordination among farms is costly, the market structure of the chemical industry discourages resistance management, and pest resistance is becoming or is likely to become a serious problem, public policies can be designed to provide incentives for resistance management. First, user charges or subsidies can be enacted to control the rate of use of a particular pesticide over time (22, 23, 90). Revenue generated from user charges could be used to develop a national program to finance resistance-related research and the implementation of resistance management projects (90). Yet this proposal has been rejected by a National Academy of Sciences task force on implementing management of resistance to pesticides (63). Instead, this distinguished group has suggested that user charges imposed by individual pest management districts may be more appropriate for combating localized pest resistance problems (63).

Secondly, regulatory agencies can restrict (without banning) the use of current compounds and can allow special registration of new compounds for use in resistance management programs. In addition, the United States Environmental Protection Agency through Section 18 of the Federal Insecticide, Fungicide, and Rodenticide Act and the Cooperative Extension Service can take leadership roles by facilitating improved documentation of resistance and encouraging field implementation of resistance management strategies.

Thirdly, additional support can be provided to increase our understanding of the mechanisms through which pests develop resistance, to develop improved methods for monitoring resistance, and to discover ways of integrating field evaluations of resistance with economic models in designing optimal resistance management strategies.

## RESEARCH IMPLICATIONS

More than one third of public agricultural research expenditures are allocated to work on maintaining previous yield gains (2). Yet very little money is directed specifically at combating pesticide resistance. We anticipate that as resistance problems become more severe, the priority of funding for these types of studies will increase. We feel there are two broad research areas that need to be further addressed. The first includes additional studies on the genetics, biochemistry, physiology, and population biology of resistant organisms, development of improved methods to detect and monitor resistance, and field evaluation of tactics to prevent or slow resistance. These studies, while biological in nature, will improve subsequent economic analyses of resistance.

The second broad research area would include economic analyses of several types:

1. Assessment of alternative resistance management strategies for individual farmers, groups of farmers, and agrochemical companies, given the pest mobility, pest and crop dynamics, and industry structure associated with the particular pesticides.
2. Assessment of the economic impacts associated with pesticide resistance that would result from alternative regulatory actions by public agencies, such as the United States Environmental Protection Agency.
3. Assessment of the environmental implications of pesticide use, given the projected increases in pesticide resistance.

4. Assessment of interregional and international economic and environmental implications associated with differential resistance control activities by region and country.

The purpose of these analyses would be threefold: first, to design institutional arrangements to stimulate resistance management activities by farmers and agrochemical companies; second, to assist public agencies to design appropriate regulations in light of resistance; third, to help design national and international institutions or policies to control resistance development.

## CONCLUSIONS

Pesticide resistance among agriculturally important arthropod pests has become a growing problem in recent years. Some progress has been made in assessing its economic implications, but the nature of the problem may dictate collective actions, which must pay for themselves.

The complexity of the mechanisms through which resistance develops and the myriad economic impacts of resistance complicate its management. Unless a simultaneous assault is undertaken at the farm, agrochemical company, and public policy levels, agriculture and the environment (and hence people) will suffer the consequences of a sharply escalating resistance problem in the immediate future.

ACKNOWLEDGMENTS

We thank William Brindley, Gerald Carlson, Brian Croft, Michael Dover, Raymond Frisbie, Carolyn Harper, J. C. Headley, William Lazarus, William Plapp, Edward Rajotte, Steve Riley, Richard Roush, and Bernie Smale for their critical reviews of an earlier draft of this manuscript.

*Literature Cited*

1. Adkisson, P. L., Niles, G. A., Walker, J. K., Bird, L. S., Scott, H. B. 1982. Controlling cotton's insect pests: a new system. *Science* 216:19–22
2. Adusei, E. 1988. *Evaluation of the importance and magnitude of agricultural maintenance research in the United States.* PhD thesis. Va. Polytech. Inst. State Univ., Blacksburg. 226 pp.
3. Archibald, S. O. 1984. *A dynamic analysis of production externalities: pesticide resistance in California cotton.* PhD thesis. Univ. Calif., Davis. 325 pp.
4. Bernat, A. G., Norton, G. W. 1983. *Economic impacts beyond the farm gate from pesticide use on apples in Virginia.* *Staff Pap. SP-7-83.* Dep. Agric. Econ.,

Va. Polytech. Inst. State Univ., Blacksburg. 11 pp.
5. Berry, J. H. 1978. *Pesticides and energy utilization.* Presented at Annu. Meet. Am. Assoc. Adv. Sci., Washington, DC
6. Boggess, W. G., Cardelli, D. J., Barfield, C. S. 1985. A bioeconomic simulation approach to multispecies insect management. *South. J. Agric. Econ.* 24: 13–23
7. Burrows, T. M., Sevacherian, V., Browning, H., Baritelle, J. 1982. History and cost of the pink bollworm (Lepidoptera: Gelechiidae) in the Imperial Valley. *Bull. Entomol. Soc. Am.* 28:286–90
8. California Celery Research Advisory

Board. 1984. *Economic impact report: the leafminer in California celery.* Dinuba, Calif. 15 pp.

9. Campanhola, C., Plapp, F. W. 1987. Toxicity of pyrethroids and other insecticides against susceptible and resistant tobacco budworm larvae and synergism by chlordimeform. *Proc. Beltwide Cotton Prod. Res. Conf., Dallas, Tex.,* pp. 326–29. Dallas, Tex: Natl. Cotton Counc. Am.

10. Campbell, H. F. 1976. Estimating the marginal productivity of agricultural pesticides: the case of tree-fruit farms in the Okanagan Valley. *Can. J. Agric. Econ.* 24:23–30

11. Carlson, G. A. 1970. A decision theoretic approach to crop disease prediction and control. *Am. J. Agric. Econ.* 52:216–33

12. Carlson, G. A. 1977. Long-run productivity of insecticides. *Am. J. Agric. Econ.* 59:543–48

13. Carlson, G. A. 1979. Pest control risks in agriculture. In *Risk, Uncertainty, and Agricultural Development,* ed. J. A. Roumasset, J. Boussard, I. Singh, pp. 199–209. New York: Agric. Dev. Counc.

14. Carlson, G. A. 1984. Risk reducing inputs related to agricultural pests. In *An economic analysis of risk management strategies for agricultural production firms: concepts, information requirements and policy issues. Proc. South. Reg. Proj. S-180,* New Orleans, La. 13 pp.

15. Carlson, G. A., Main, C. E. 1976. Economics of disease-loss management. *Ann. Rev. Phytopathol.* 14:381–403

16. Chapin, G., Wasserstrom, R. 1983. Pesticide use and malaria resurgence in Central America and India. *Soc. Sci. Med.* 17:273–90

17. Clark, J. S., Carlson, G. A. 1987. *Econometrically distinguishing between private and common property: the case of pesticide resistance.* Presented at Annu. Meet. Am. Agric. Econ. Assoc., Lansing, Mich.

18. Cochran, M., Lodwick, W., Jones, A., Robison, L. 1982. *Selection of apple scab pest management strategies under uncertainty: an application of various stochastic dominance technologies. Staff Pap. 10–82–34.* Dept. Agric. Econ., Mich. State Univ., East Lansing

19. Croft, B. A., Brown, A. W. A. 1975. Responses of arthropod natural enemies to insecticides. *Ann. Rev. Entomol.* 20: 285–335

20. Croft, B. A., Hoyt, S. C., Westigard, P. H. 1987. Spider mite management on pome fruits, revisited: organotin and acaricide resistance management. *J. Econ. Entomol.* 80:304–11

21. Daly, J. C., McKenzie, J. A. 1986. Resistance management strategies in Australia: the *Heliothis* and "Wormkill" programmes. *Proc. Br. Crop Prot. Conf.* 3:951–59

22. Dover, M. 1985. Getting off the pesticide treadmill. *Tech. Rev.* 88:52–63

23. Dover, M., Croft, B. 1984. *Getting Tough: Public Policy and the Management of Pesticide Resistance.* Washington, DC: World Resources Inst. 80 pp.

24. Feder, G. 1979. Pesticides, information, and pest management under uncertainty. *Am. J. Agric. Econ.* 61:97–103

25. Feder, G., Regev, U. 1975. Biological interactions and environmental effects in the economics of pest control. *J. Environ. Econ. Manage.* 2:75–91

26. Ferro, D. N. 1985. Pest status and control strategies of the Colorado potato beetle. In *Proc. Colorado Potato Beetle. Mass. Agric. Exp. Stn. Res. Bull.* 704:1–8

27. Fischer, L. A., 1970. The economics of pest control in Canadian apple production. *Can. J. Agric. Econ.* 16:90–100

28. Follett, P. A., Croft, B. A., Westigard, P. H. 1985. Regional resistance to insecticides in *Psylla pyricola* from pear orchards in Oregon. *Can. Entomol.* 117:565–73

29. Forgash, A. J. 1984. History, evolution, and consequences of insecticide resistance. *Pestic. Biochem. Physiol.* 22:178–86

30. Georghiou, G. P. 1986. The magnitude of the resistance problem. See Ref. 63, pp. 14–46

31. Gould, F. 1984. Role of behavior in the evolution of insect adaptation to insecticides and resistant host plants. *Bull. Entomol. Soc. Am.* 30:34–41

32. Greene, C. R., Kramer, R. A., Norton, G. W., Rajotte, E. G., McPherson, R. M. 1985. An economic analysis of soybean integrated pest management. *Am. J. Agric. Econ.* 67:567–72

33. Greene, C. R., Rajotte, E. G., Norton, G. W., Kramer, R. A., McPherson, R. M. 1985. Revenue and risk analysis of soybean pest management options in Virginia. *Am. J. Agric. Econ.* 78:10–18

34. Hall, D. C., Norgaard, R. B. 1973. On the timing and application of pesticides. *Am. J. Agric. Econ.* 55:198–201

35. Harper, C. R. 1986. *Optimal regulation of agricultural pesticides: a case study of chlordimeform in the Imperial Valley.* PhD thesis. Univ. Calif., Berkeley. 148 pp.

36. Headley, J. C. 1968. Estimating the productivity of agricultural pesticides. *Am. J. Agric. Econ.* 50:13–23
37. Headley, J. C. 1972. Economics of agricultural pest control. *Ann. Rev. Entomol.* 17:273–86
38. Deleted in proof
39. Headley, J. C., Hoy, M. A. 1987. Benefit/cost analysis of an integrated mite management program for almonds. *J. Econ. Entomol.* 80:555–59
40. Holden, P. W. 1986. *Pesticides and Groundwater Quality: Issues and Problems in Four States.* Washington, DC: Natl. Acad. Press. 124 pp.
41. Hoy, M. A. 1985. Recent advances in genetics and genetic improvement of the Phytoseiidae. *Ann. Rev. Entomol.* 30: 345–70
42. Hoyt, S. C. 1982. Summary and recommendations for future research and implemetation. In *Recent Advances in Knowledge of the Phytoseiidae*, ed. M. A. Hoy, pp. 90–92. Berkeley, Calif: Univ. Calif.
43. Hueth, D., Regev, U. 1974. Optimal agricultural pest management with increasing pest resistance. *Am. J. Agric. Econ.* 56:543–52
44. Jackson, G. J. 1986. Insecticide resistance—what is industry doing about it? *Proc. Br. Crop Prot. Conf.* 3:943–49
45. Kinoshita, G. B. 1985. The economics of entomological effort: viewpoint of the pesticide industry in Canada. *Can. Entomol.* 117:909–21
46. Langham, M., Edwards, W. F. 1969. Externalities in pesticide use. *Am. J. Agric. Econ.* 51:1195–201
47. Lazarus, W. F., Dixon, B. F. 1984. Agricultural pests as common property; control of the corn rootworm. *Am. J. Agric. Econ.* 66:456–65
48. Lazarus, W. F., Swanson, E. 1983. Insecticide use and crop rotation under risk: rootworm control in corn. *Am. J. Agric. Econ.* 65:738–47
49. Lee, J. Y., Langham, M. R. 1973. A simultaneous equations model of the economic-ecological system in citrus groves. *South. J. Agric. Econ.* 5:175–80
50. Lichtenberg, E., Zilberman, D. 1986. The econometrics of damage control: why specification matters. *Am. J. Agric. Econ.* 68:261–73
51. Marmet, J. P. 1977. Problems faced by the pesticide manufacturer in relation to social and economic aspects of crop production. *Pestic. Sci.* 8:380–88
52. McCarl, B. A. 1981. *Economics of integrated pest management: an interpretative review of the literature. Spec.*
53. McGuckin, T. 1983. Alfalfa management strategies for a Wisconsin dairy farm—an application of stochastic dominance. *North Cent. J. Agric. Econ.* 5:43–49
54. Metcalf, R. L. 1980. Changing role of insecticides in crop protection. *Ann. Rev. Entomol.* 25:219–56
55. Metcalf, R. L., Luckman, W. H. 1975. *Introduction to Insect Pest Management.* New York: Wiley. 587 pp.
56. Miller, R. W., Croft, B. A., Nelson, R. D. 1985. Effects of early season immigration on Cyhexatin and Formetanate resistance of *Tetranychus urticae* (Acari: Tetranychidae) on strawberry in central California. *J. Econ. Entomol.* 78: 1379–88
57. Miranowski, J. A., Carlson, G. A. 1986. Economic issues in public and private approaches to preserving pest susceptibility. See Ref. 63, pp. 436–48
58. Moffitt, L. J., Burrows, T. W., Baritelle, J., Sevacherian, V. 1984. Risk evaluation of early termination for pest control in cotton. *West. J. Agric. Econ.* 9:145–51
59. Moffitt, L. J., Tanigoshi, L. K., Baritelle, T. L. 1983. Incorporating risk in comparisons of alternate pest control methods. *Environ. Entomol.* 12:1003–11
60. Mumford, J. D., Norton, G. A. 1984. Economics of decision making in pest management. *Ann. Rev. Entomol.* 29: 157–74
61. Musser, W. W., Tew, B. V., Epperson, J. V. 1981. An examination of an integrated pest management production system with a contrast between E-V and stochastic dominance analysis. *South. J. Agric. Econ.* 13:119–23
62. Napit, K. 1986. *Economic impacts of extension integrated pest management programs in several states.* MS thesis. Va. Polytech. Inst. State Univ., Blacksburg. 150 pp.
63. National Academy of Sciences. 1986. *Pesticide Resistance: Strategies and Tactics for Management.* Washington, DC: Natl. Acad. Press. 471 pp.
64. Norgaard, R. B. 1976. The economics of improving pesticide use. *Ann. Rev. Entomol.* 21:45–60
65. Norton, G. W., Ganoza, V. G., Pomareda, C. 1987. Potential benefits of agricultural research and extension in Peru. *Am. J. Agric. Econ.* 69:247–57
66. Osteen, C. D., Bradley, E. B., Moffitt, L. J. 1981. *The economics of agricul-*

*Rep. 636. Agric. Exp. Stn., Int. Plant Prot. Cent., Oreg. State Univ. Agric. Res. Econ., Corvallis. 146 pp.*

tural pest control: an annotated bibliography, 1960–80. USDA Bibliogr. Lit. Agric. No. 14. Washington, DC: US Dep. Agric. 53 pp.

67. Pimentel, D., Andow, D., Dyson-Hudson, R., Gallahan, D., Jacobson, S., et al. 1980. Environmental and social costs of pesticides: a preliminary assessment. Oikos 34:126–40

68. Pimentel, D., Krummel, J., Gallahan, D., Hough, J., Merrill, A., et al. 1978. Benefits and costs of pesticide use in U. S. food production. BioScience 28:777–84

69. Plant, R. E., Mangel, M., Flynn, L. E. 1985. Multiseasonal management of an agricultural pest II: the economic optimization problem. J. Environ. Econ. Manage. 12:45–61

70. Plapp, F. W. 1976. Biochemical genetics of insecticide resistance. Ann. Rev. Entomol. 21:179–98

71. Regev, U. 1984. An Economic Analysis of Man's Addiction to Pesticides in Pest and Pathogen Control: Strategic, Tactical and Policy Models. New York: Wiley. 384 pp.

72. Regev, U., Gutierrez, A. P., Feder, G. 1976. Pests as a common property resource: a case study of alfalfa weevil control. Am. J. Agric. Econ. 58:186–97

73. Regev, U., Shalit, H., Gutierrez, A. P. 1983. On the optimal allocation of pesticides with increasing resistance: the case of alfalfa weevil. J. Environ. Econ. Manage. 10:86–100

74. Reichelderfer, K. H., Bender, F. E. 1979. Application of a simulative approach to evaluating alternative methods for the control of agricultural pests. Am. J. Agric. Econ. 61:258–67

75. Repetto, R. 1985. Paying the Price: Pesticide Subsidies in Developing Countries. Washington, DC: World Resources Inst.

76. Roush, R. T. McKenzie, J. A. 1987. Ecological genetics of insecticide and acaricide resistance. Ann. Rev. Entomol. 32:361–80

77. Sawicki, R. M., Denholm, I. 1986. Evaluation of existing resistance-management strategies against arthropod pests of cotton. Proc. Br. Crop Prot. Conf. 3:933–41

78. Shoemaker, C. 1973. Optimization of agricultural pest management I: biological and mathematical background. Math. Biosci. 16:143–73

79. Shoemaker, C. 1973. Optimization of agricultural pest management II: for-mulation of a control model. Math. Biosci. 17:357–65

80. Shoemaker, C. 1973. Optimization of agricultural pest management III: results and extension of a model. Math. Biosci. 18:1–22

81. Steiner, H. 1973. Cost-benefit analysis in orchards where integrated control is practiced. OEPP/EPPO Bull. 3:27–36

82. Talpaz, H., Borosh, I. 1974. Strategy for pesticide use: frequency and applications. Am. J. Agric. Econ. 56:769–75

83. Talpaz, H., Curry, G. L., Sharpe, P. J., DeMichele, D. W., Frisbie, R. E. 1978. Optimal pesticide application for controlling the boll weevil on cotton. Am. J. Agric. Econ. 60:470–75

84. Taylor, C. R. 1980. The nature of benefits and costs of use of pest control methods. Am. J. Agric. Econ. 62:1007–11

85. Taylor, C. R., Frohberg, K. K. 1977. The welfare effects of erosion controls banning pesticides and limiting fertilizer applications in the corn belt. Am. J. Agric. Econ. 59:25–36

86. Taylor, C. R., Headley, J. C. 1975. Insecticide resistance and the evaluation of control strategies for an insect population. Can. Entomol. 107:237–42

87. Taylor, C. R., Lacewell, R. D. 1977. Bollweevil control strategies: regional benefits and costs. South. J. Agric. Econ. 9:129–35

88. Deleted in proof

89. van den Bosch, R., Stern, V. M. 1962. The integration of chemical and biological control of arthropod pests. Ann. Rev. Entomol. 7:367–86

90. Whalon, M. 1987. A means of funding pesticide resistance management. Presented at Annu. Meet. West. Reg. Coord. Comm., Proj. 68, New Orleans, La.

91. Wolfenbarger, D. A., Lukefahr, M. J., Graham, H. M. 1973. $LD_{50}$ values of methyl parathion and endrin to tobacco budworms and bollworms collected in the Americas and hypothesis on the spread of resistance in these lepidopterans to these insecticides. J. Econ. Entomol. 66:211–16

92. Zacharias, T. P., Grube, A. H., 1986. Integrated pest management strategies for approximately optimal control of corn rootworm and soybean cyst nematode. Am. J. Agric. Econ. 68:704–15

93. Zavaleta, L., Ruesink, W. 1980. Expected benefits from nonchemical methods of alfalfa weevil control. Am. J. Agric. Econ. 62:801–5

*Ann. Rev. Entomol. 1989. 34:315–50*

# ECOLOGICAL AND EVOLUTIONARY ASPECTS OF LEARNING IN PHYTOPHAGOUS INSECTS

*Daniel R. Papaj and Ronald J. Prokopy*

Department of Entomology, University of Massachusetts at Amherst, Amherst, Massachusetts 01003

## PERSPECTIVES AND OVERVIEW

To date, learning has been demonstrated in six orders of insects with phytophagous habits. Most examples of learning in these orders have come to light relatively recently [all but a few since the last review of learning in the *Annual Review of Entomology* 17 years ago (3)], and many more will surely be uncovered. Still, the study of learning in other insects such as the social Hymenoptera remains far more advanced than in phytophagous insects. In this review, work on learning in phytophagous insects is placed into the broader perspective of learning in a variety of insects including bees and parasitic wasps. Although historically the study of learning in phytophagous insects has emphasized the acquisition of host resources for feeding and oviposition, we believe that, as in bees and wasps, learning impinges on almost every aspect of the natural history of a phytophagous insect. Following discussion of the definition of learning, the natural history of learning, and the host stimuli involved in learning, we review the ecological significance of learning with respect to both the traditional categories erected by behaviorists and the more recent concept of programed learning developed by ethologists. Finally, we evaluate critically hypotheses about the evolution of learning and suggest approaches for future research.

315

0066-4170/89/0101-0315$02.00

# DEFINITION OF LEARNING

## Properties of Learning

Learning eludes a rigorous and all-encompassing definition (10, 66, 103, 171). Some definitions are too broad and fail to exclude phenomena intuitively not considered to be learning. The definition "a change in behavior with experience" (164, 192), for example, does not exclude the phenomena collectively termed "event-dependent responsiveness," i.e. growth, maturational, and other processes that are triggered by events such as mating or feeding (132). More restrictive definitions, by contrast, exclude phenomena intuitively considered to be learning. The definition "a reversible change in behavior with experience" (129) excludes phenomena in which the modification caused by some experience is fixed and resistant to further change. Kimble's oft-cited definition of learning as "a more or less permanent change in behavior that occurs as a result of practice" (90) skirts this problem but engenders another: Both the phrase "more or less permanent" and the term "practice" are somewhat ambiguous (3, 103). Like others before us (66, 103, 171), we sidestep an attempt at an all-embracing definition of learning. Instead, we summarize below some properties characteristic of learning and recommend their use (with caution!) as criteria by which to specify learning.

1. *The individual's behavior changes in a repeatable way as a consequence of experience.* While this property is rarely explicitly acknowledged and may seem trivially obvious, this is often the only criterion used for acceptance of publications on insect learning. The extent to which the change is repeatable is judged typically by the outcome of statistical analysis of the results of a learning experiment. This criterion effectively excludes changes in behavior that, through chance alone, accompany experience.

2. *Behavior changes gradually with continued experience,* often following a so-called "learning curve" to an asymptote (163; cf 95). This criterion effectively excludes motor programs that are simply switched on or off by particular stimuli. For example, a fly that increases its turning rate or decreases walking speed after an encounter with a food resource has not necessarily learned anything. If, however, turning rate increases or walking speed decreases progressively as resource items are successively encountered (cf 39), the fly may well be learning to restrict the area over which it searches. This criterion unfortunately excludes single-trial learning, i.e. learning in which behavior changes after one experience and changes little with further experience.

3. *The change in behavior accompanying experience wanes in the absence of continued experience of the same type or as a consequence of a novel experience or trauma* (180, 189). To use familiar terminology, in the absence of continued experience the insect forgets what was learned. In response to a

novel experience, the insect often switches from one learned behavior to another. This criterion effectively excludes maturational processes that are triggered by an event such as mating or feeding (132, 180), but also excludes certain irreversible effects of experience sometimes considered to be learning [e.g. imprinting (50, 180)]. In practice, forgetting can be difficult to distinguish from effects of resource deprivation on behavior. If, for example, feeding on a host plant induces a change in behavior and the insect is then deprived of the opportunity to feed, it may revert back to its original behavior either because the experience has been forgotten or because deprivation of food itself has enforced a change in behavior that counters the change with experience (10, 131).

An effect of experience on behavior that meets all three criteria is almost certainly learning. Yet a change in behavior due to experience may exhibit just the first of these properties and still be considered learning. Likewise, effects of experience on behavior not usually considered to be learning may meet one or more of these criteria. Learning is ultimately characterized by the exclusion of phenomena that are not learned (66, 171).

## Learning Defined as Adaptation

When insects are trained to select among alternative kinds of resources such as host plants or flowers, researchers typically anticipate reciprocity in the results. Individuals exposed to different host species, for example, are expected to show reciprocal changes in behavior toward alternative hosts, usually an increase in the tendency to feed or oviposit on the host species with which they have had experience. It is important to realize that this expectation often betrays an adaptationist perspective toward learning.

The adaptationist perspective has strong historical roots in the study of insect learning. Thorpe (180) defined learning as "that process which manifests itself by adaptive change in individual behavior as the result of experience." By including under the rubric of learning only changes in behavior with experience that are adaptive, Thorpe essentially undermined efforts to examine the evolution and adaptive significance of learning (57); learning was, by definition, adaptive. In fact, rigorous application of the standard of adaptiveness to changes in behavior would probably force us to dismiss most phenomena traditionally designated as learning, in part because adaptation is perhaps as difficult to define as learning and almost certainly more difficult to test (41, 52, 98).

There are a number of ways (e.g. genetic drift, mutation, pleiotropy, and linkage) in which maladaptive behavior may be maintained in natural populations (41, 52). So too, the plasticity of behavior (i.e. learning) may be maladaptive and yet still be preserved in a population. While it may be true that experience almost always affects behavior in a way that appears adaptive at least in the short term (101), there are exceptions even among phytophagous insects. Larvae of some lepidopterous species will die of starvation before they eat food other than that to which they have become accustomed, even if the novel food is adequate and acceptable under other circumstances (59, 102, 162). This so-called "starving-to-death-at-Lucullian-banquets" phenomenon is not easily reconciled with an adaptationist point of view (78). Malfeasant effects of experience might conceivably be repeatable, cumulative, subject to decay, and reversible, and it would be inappropriate not to include them as examples of learning. Despite the reasonableness of these arguments, learning is still often defined in terms of its own adaptive significance (cf 2, 50, 101, 115).

## WHAT IS LEARNED?

Animals in various phyla learn attributes of their natural history associated with acquisition of food and mates, care for kin or offspring, interaction among conspecifics, escape from or defense against enemies, and navigation to and from home (45, 50). In phytophagous insects, the primary research focus has been on learning in relation to acquisition of food, with scattered attention given to learning in mate acquisition, recognition of conspecific competitors, and establishment of home range.

### Mate Acquisition, Recognition of Competitors, and Establishment of Home Range

While the process of acquiring a mate is often assumed to be fixed and unsusceptible to change through experience, recent work on the apple maggot fly, *Rhagoletis pomonella*, suggests otherwise. Male apple maggot flies spend considerable time on host fruit, where they copulate with arriving females. Prior exposure to host fruit of a given species (either apple or hawthorn fruit) alters the time males spend on that fruit (141). Males of a saprophagous *Puliciphora* fly species learn the location of favorable oviposition sites and "airlift" females to those sites while copulating (118). In addition to learning characteristics of sites of mating and oviposition, male phytophagous insects may conceivably learn to assess characteristics of females. Males of saprophagous *Drosophila* fly species, for example, learn to respond to females in a particular physiological state (58). Male bees (*Lasioglossum* spp.) learn characteristics of individual females (204).

Although phytophagous insects have not been shown to learn to recognize mates, learning is involved, in at least one instance, in recognition of conspecifics competing for an essential resource: Female apple maggot flies learn to recognize the presence of conspecific eggs in host fruit. After oviposition into a host fruit, females deposit a trail of pheromone on the surface of the fruit, which deters oviposition by flies landing subsequently. Females require prior contact with the mark in order to be subsequently deterred (156).

Numerous hymenopterans learn the geography of their home range and learn to navigate an area encompassed by that range (46, 49, 157). Such learning has not been demonstrated rigorously for any phytophagous insect, but is implied strongly by field observations on butterflies. *Heliconius* butterflies occupy relatively large home ranges within which they travel repeatedly over certain routes at specific times of day to visit feeding, oviposition, and roosting sites (105, 190). While direct evidence has not been forthcoming, these butterflies (as well as other phytophagous insects that exhibit site fidelity) might learn landmarks as do bees (49 and references within) and might even form cognitive maps (48). Fidelity to a particular feeding site is achieved through experience in one phytophagous insect in a unique way: Eastern tent caterpillars *(Malacosoma americanum)* are more likely to trailmark in response to the plant species on which they were reared (either sweet cherry or scarlet oak) than in response to an alternative species. The implication is that feeding experience affects the tendency of the caterpillar to lay down a trail from sites in the foliage and, by doing so, affects the tendency of the caterpillar (and conspecific recruits) to return to those sites (136).

## Food Acquisition

A phytophagous or saprophagous insect acquires food associated with living or decaying plant material by a catenary process (89) involving perhaps four phases: finding the habitat, finding a plant within the habitat, examining a plant, and using a plant for food or oviposition (62, 117). We know of no instance in which experience affects the first phase, in which phytophagous insects select the habitat in which plants are found. Such examples are well known among entomophagous parasitoids, which learn to find the microhabitat (typically living or decaying plant material) with which their larval hosts are associated (97, 192, 193). Examples will likely be found for phytophagous insects as well. Experience has been shown to affect each of the last three phases of food selection in phytophagous insects.

With respect to finding plants that furnish food or egg-laying sites (and hence food for larval progeny), prior experience with a plant or plant stimuli has been shown to have a role in adult *Heliconius* (175), *Pieris* (95, 96, 185–188), *Battus* (124–127, 147), and *Colias* (173) butterflies, *Manduca* moth larvae (158), *Melanoplus* grasshopper and *Schistocerca* locust nymphs (8, 94), and adult *Drosophila* flies (69, 70, 75, 76). With respect to examining and using a plant for food or oviposition after arrival, prior experience with a plant or plant stimuli has been shown to have a role in adult *Pieris* butterflies (95, 96), *Haltica, Callosobruchus,* and *Deloyala* beetles (106,

137, 150), adult *Rhagoletis, Ceratitis,* and *Dacus* tephritid flies (22, 129–131, 140, 143, 145), adult *Drosophila* flies (68, 70, 72, 73, 75), adult and nymphal *Carausius* stick insects (19), *Schistocerca* and *Locusta* locust nymphs (10, 79, 176), *Melanoplus* grasshopper nymphs (8), and various species of lepidopterous larvae (see reviews in 78 and 129).

In most cases, phytophagous insects learn properties specific to a plant species; learning results in some modification of preference (preference sensu 168) for particular plant species. Sometimes insects learn characteristics below the species level. Ovipositing *Pieris* butterflies, for example, learn to approach and land on particular cultivars of cabbage (186). Similarly, ovipositing *Rhagoletis* fruit flies learn to accept particular cultivars of apples (144). Female *Battus* butterflies appear to learn to discriminate not only between host species, but also among individual conspecific plants according to the size and phenological age of the plant (134) or the presence of a leaf bud (127).

## HOST STIMULI INVOLVED IN LEARNING

Compared with what is known for other insects such as honey bees (46, 114) or entomophagous parasitoids (97, 192, 193), little is known about the stimuli involved in learning in phytophagous insects (see 78 and 129 for reviews). Prior to an insect's arrival on a plant or plant part, experience can modify its host-selection response to both visual and phytochemical stimuli. Host experience enhances upwind responses to host plant odor in *Leptinotarsa* beetles (197) and *Schistocerca* nymphs (94). Color and/or light intensity cues are learned by *Heliconius* and *Pieris* butterflies in finding nectar sources or oviposition sites (96, 175, 185–188) and by *Melanoplus* nymphs in finding food (8). Experience with host fruit of a particular size increases the tendency of female *Ceratitis* fruit flies to land on fruit of that size (R. J. Prokopy, T. A. Green & T. T. Y. Wong, unpublished). Ovipositing *Battus* butterflies find suitable hosts sometimes by learning the shape of the leaves of a host (124, 126, 147) and sometimes by learning visual cues associated with the terminal leaf bud (127). It remains to be seen whether phytophagous insects store pictorial images of resource items, as has been suggested for bees (47).

After arrival on a plant or plant part, experience principally affects host-selection responses to phytochemicals. Experience alters feeding responses to contact chemicals in *Locusta* nymphs (10) and a number of species of lepidopterous larvae (11; reviewed in 78, 129). In female *Rhagoletis* and *Ceratitis* flies, prior fruit experience modifies oviposition responses to both fruit size and chemistry after arrival on a fruit (128, 129). The stimuli involved in assessment of fruit size are presently unknown.

Responses of insects to stimuli of one modality are not necessarily independent of responses to stimuli of another modality (61, 139). Experience can modify not just responses to different kinds of stimuli, but also the interaction between different sensory modalities. Such higher-order effects of experience on behavior are well known in bees. Bees fed at a target of a particular color and odor learn to associate the combination of color and odor with a sucrose reward and assess potential targets according to their color, odor, and a composite representation of color and odor (23).

The interaction between sensory modalities may be modified by experience in phytophagous insects as well. Adult experience affects the interaction between responses to fruit size and chemical stimuli in *Ceratitis* flies (128). Female flies attempt to oviposit into a test fruit similar in size to the fruit with which they are experienced; experience "closes the window" of sizes that are acceptable to ovipositing females. While the window also closes with experience when wax models are tested in place of real fruit, experienced females generally show a lower response to such models than inexperienced females. The learned response to model size is more strongly expressed if models are first treated with surface chemicals from the fruit with which females had experience; hence experience affects the interaction between the insect's responses to size and surface chemistry.

## KINDS OF LEARNING AND THEIR ECOLOGICAL FUNCTIONS

To date, at least two distinct kinds of learning have been identified in phytophagous insects: habituation and associative learning (including aversion learning). Another kind, sensitization, is likely to underlie many examples of learning in phytophagous insects. Many authors have placed their work on effects of experience on host-selection behavior in a fourth category, induction of preference.

### Habituation

Habituation is characterized by a waning of response to a stimulus with repeated exposure to that stimulus (111, 135, 180). Habituation is specific to a particular stimulus and is more persistent than fatigue or sensory adaptation (66, 135, 179, 180). Prior experience alters feeding (79, 159, 176) or oviposition (72, 73) behavior in phytophagous insects in ways that are consistent with habituation, specifically habituation to chemical deterrents in the host medium. In practice, however, it is often difficult to distinguish habituation from effects of deprivation on behavior (79, 129, 176).

PRESUMED ECOLOGICAL FUNCTION OF HABITUATION    Widespread among animals (179), habituation is thought to be a general mechanism by which certain default behavior patterns are turned off when their expression is no longer functional (103, 135, 180). The time over which entomophagous parasitoids foraging for oviposition sites remain in host patches, for example, may be governed by the rate at which they habituate to host stimulants (198, 199). Waage (198, 199) postulated that as female *Nemeritis* wasps forage unsuccessfully for unparasitized host larvae within an infested patch of grain, they habituate to the kairomone of the host and become progressively less likely to turn back into the patch when they arrive at its edge. Habituation thus provides a mechanism by which the forager emigrates when the patch becomes depleted of unparasitized hosts. Whether habituation represents a mechanism for regulating dispersal of phytophagous insects from host plants or patches of host plants is not known. Habituation to a deterrent in the diet may conceivably reduce the insects' tendency to disperse from an initially unpalatable but adequately nutritious food plant. This notion, while reasonable, is somewhat impervious to testing in the sorts of laboratory assays in which habituation has usually been observed.

The argument that habituation enables insects to use food that is unpalatable but suitable for growth and survival (78) has at least two assumptions: (*a*) Diets that are unpalatable are not always toxic or nutritionally deficient, and (*b*) insects cannot or do not evolve appropriate responses to unpalatable but nutritious host species other than habituation. The first assumption is supported by experiments in which insects deprived of the organs by which they detect deterrent stimuli consumed and thrived on previously unpalatable plants (200). The second assumption is more difficult to support. What, for example, has prevented the insect from evolving the sensory apparatus and processing equipment that renders the plant palatable? Possibly, a given phytochemical stimulus occurs in both toxic and nontoxic plant species. Natural selection might conceivably maintain deterrent responses to that phytochemical even if some nutritionally adequate plants are avoided as a consequence. The likelihood that a phytochemical is shared by both toxic and nontoxic species may increase as more and more host species are included in the diet. If so, polyphagous herbivores should be more likely to habituate to deterrents than herbivores with more restricted diets. This prediction is tentatively supported by the available evidence (78).

## Associative Learning and Sensitization

Associative learning involves the establishment, through experience, of an association between two stimuli or between a stimulus and a response. Typically, the association is formed through the close temporal and spatial pairing of the stimuli or the stimulus and response. Associative learning has been proposed for a number of phytophagous insects, including butterflies (96, 124, 126, 186–188), grasshoppers (8, 94), locusts (10, 12), and lepidopterous larvae (11; see also 180). Studies with locusts (10, 12) illustrate how difficult it can be to distinguish between associative learning and nonassociative processes such as sensitization.

Sensitization, often considered the counterpart of habituation (28, 77), is characterized by a gradual increase in response to a stimulus with repeated exposure to that stimulus even when it has not been paired with any other stimulus (103, 111). *Locusta* nymphs progressively reject unpalatable leaf material at an earlier stage in the food-selection sequence as they gain experience with such leaf material. Whereas at early meals they reject food primarily after biting into it, in later meals they often reject the food after palpation but before biting. The change in rejection behavior is host specific: When locusts that had learned to reject *Senecio vulgaris* at the palpation stage were offered *Brassica oleracea,* they initially rejected it primarily at the biting stage (10, 12).

Blaney and colleagues (10, 12) postulated that locusts [and *Spodoptera* larvae, in which a similar pattern was found (11)] learned to associate sensations at palpation with deterrent sensations at biting. However, these results could also be explained if the locusts became sensitized to a host-specific deterrent stimulus perceived by the palps. The appropriate control for sensitization entails exposing locusts only to the stimuli perceived by the palps (and not to the stimuli perceived during biting) and examining whether rejection at the palpation stage still increases with prolonged exposure to those stimuli. Because of the speed at which these events occur, however, it is impractical to perform this experiment (10). Similar caveats about associative learning apply to learning in apple maggot flies (learning described as associative in References 111 and 140) and, to a lesser extent, to almost every supposed example of associative learning in phytophagous insects to date (129). In general, sensitization may account for effects of experience on behavior much more often than has been appreciated (103).

CLASSICAL AND INSTRUMENTAL CONDITIONING     Usage of the word "conditioning" by evolutionary biologists has often differed from its usage by ethologists and psychologists. Evolutionary biologists (and especially those who study plant-insect interactions) have used the word synonymously with "learning," often to describe any change in behavior with experience. Most contemporary behaviorists (66, 103, 104, 154), however, refer to conditioning only in the context of two experimental paradigms by which associative learning is commonly studied: classical and instrumental conditioning.

In classical conditioning [sometimes referred to as Pavlovian conditioning (154)], a neutral stimulus, the conditioned stimulus (CS), is paired with a nonneutral stimulus, the unconditioned stimulus (US), such that the response elicited previously only by the US is subsequently elicited also by the CS.

Following Pavlov's original design, classical conditioning paradigms require that presentation of the CS just precedes presentation of the US. In instrumental conditioning [sometimes referred to as operant conditioning (104)], an outcome such as access to food (often called a "reward") is contingent upon performance of some response; this contingency subsequently causes the frequency of that response to change.

While examples of classical conditioning are known among insects (3, 23, 116, 120), we know of no unambiguous example among phytophagous forms. If the learning by feeding locusts described above is associative in nature, it may well be conditioning of the classical type (10). In that case, the CS (whose exact nature is unknown) is presumably detected during palpation just prior to detection of the US (also uncharacterized) during biting. In subsequent feeding episodes, the CS elicits a rejection response by the locust similar to that elicited by the US. More than two stimuli may well be involved in locust learning; simple one-to-one pairing of a single CS and US probably occurs only rarely during Pavlovian conditioning (154).

Since selection of food or oviposition sites involves a catenary sequence of stimulus-response pairs in which performance of some response at a prior stage permits the animal to perceive and respond to stimuli at the next stage, it is not surprising that associative learning in phytophagous insects often resembles instrumental conditioning. Nectar-extraction learning by pierid butterflies, for example, probably involves instrumental conditioning. As in bees (46, 64, 92), the time required for *Pieris* butterflies to find and extract nectar from a flower of a given species decreases over successive encounters with flowers of that species (95). Perhaps, as postulated for bees (46, 92), spontaneous performance of an undirected motor pattern at a novel flower results occasionally in more rapid access to nectar. The association between the previously undirected behavior and more effective feeding presumably increases the likelihood that the same behavior will be expressed at subsequent flowers.

In practice, it can be extraordinarily difficult to distinguish rigorously between classical and instrumental conditioning (104), especially when insects are learning under natural conditions (116). Either classical or instrumental conditioning, for example, might explain learning by ovipositing butterflies. Adult experience with particular cabbage varieties increased the tendency of *Pieris* butterflies females to approach and alight on those varieties or on paper discs of similar color and light intensity (186–188). Experience with moist paper discs treated with sinigrin (a constituent of cruciferous hosts that elicits oviposition upon contact) likewise conditioned responses to color and/or light intensity. A single landing by a female on a disc treated with sinigrin solution induced a preference for the color of the disc, even if the landing did not result in oviposition (186). Similarly, female *Battus* butterflies were conditioned by a single contact with phytochemical stimuli from their

*Aristolochia* hosts to approach and land on leaves of shapes associated with those stimuli, even when contact was not followed by successful oviposition (124). Which does the female butterfly associate with the phytochemical stimulants, the color and leaf shape of a host or the act of approaching and landing on a host with that color and leaf shape?

Most natural learning probably involves elements of both classical and instrumental conditioning. Butterflies learning to manipulate a flower, for example, presumably associate some spontaneous and random movements with successful extraction of nectar (instrumental conditioning). At the same time, motor patterns previously triggered by nonneutral stimuli are gradually elicited by previously neutral stimuli (classical conditioning). Finally, butterflies may also, through trial and error, assemble component motor patterns into an order that enables them to extract nectar effectively (50). Associative learning may thus affect the catenary sequence of food selection by (*a*) adding de novo responses to food stimuli; (*b*) changing the likelihood of preexisting responses (including their excision from the sequence); and (*c*) changing the order in which different responses are arranged in the sequence.

PRESUMED ECOLOGICAL FUNCTION OF ASSOCIATIVE LEARNING    Associative learning is the neural mechanism by which the causal structure of the real world is extracted and used to predict the occurrence of future events (113, 183). Associative learning is judged to be almost always adaptive (cf 2, 101) because cause is associated with effect in a manner that usually appears to enhance survival or fecundity. While examples of associative learning (or sensitization) are usually consistent with an adaptationist perspective, we offer two caveats. First, negative results might not be published. Second, what appears to benefit fitness in the short term may have no effect or even negative effects on fitness measured over the life of the animal. Acquisition of a strong preference for a particular host plant, for example, might increase the growth rate and survival of an herbivore while it is on that plant (and might thus seem adaptive), but it may be fatal if the herbivore, when it disperses, fails to find another plant of that species and refuses other perfectly nutritious species (78). Thorpe, anticipating this objection to his characterization of learning as adaptive, commented that " . . . by 'adaptive change' we mean a change which is 'economical' over a considerable stretch of the animal's life" (180).

## Food Aversion Learning

Food aversion learning, sometimes considered a kind of associative learning, is defined as an association of an ingested food with a postingestional malaise that results in subsequent avoidance of ingestion of that food. Dethier (32, 35, 36) was the first to investigate food aversion learning in phytophagous insects. Two arctiid caterpillars, *Diacrisia virginica* and *Estigmene congrua,* ate proportionately less of the initially favored *Petunia* sp. than of an alternative plant following acute illness caused by ingestion of *Petunia* sp. (32). Dethier's laboratory results were recently confirmed under more natural conditions (35).

*Schistocerca* grasshoppers exhibit a steady decline in meal length over successive feeding bouts on spinach, which is consistent with aversion learning and not due to sensory feedback from the gut (93). Unequivocal evidence for aversion learning was provided by experiments in which a specific noxious stimulus was artificially linked to feeding on plant material (7). Immediately after feeding on either spinach or one of two types of broccoli, the grasshoppers were injected with a toxic alkaloid, nicotine hydrogen tartrate (NHT). Grasshoppers fed on spinach avoided it after NHT poisoning, but those fed one type of broccoli did not avoid the broccoli even though feeding on it had also been followed by NHT poisoning. The extent to which aversion learning was manifested apparently depended on the extent to which the plant was initially acceptable, broccoli being much more acceptable than spinach. This dependence of aversion learning on the initial palatability of the food is consistent with findings in vertebrates, including humans (references in 7, 56, 207).

PRESUMED ECOLOGICAL FUNCTION OF AVERSION LEARNING    The ecological function of aversion learning has been described as obvious (35, 78): prevention of continued consumption of poisonous food. What is not obvious is why, given that phytophagous insects as a group are capable of extreme behavioral specialization, insects have not evolved the capacity to identify and reject toxic food before ingestion. As stated previously, a given phytochemical stimulus may occur in both toxic and nontoxic plant species. Natural selection might not produce rejection responses to phytochemicals in the toxic plant if, by doing so, nutritionally suitable plants would be avoided as a consequence. The likelihood that particular phytochemicals are represented in both toxic and nontoxic species may increase as more and more host species are included in the diet, which may buttress assertions (26, 43) that generalists should be more likely to exhibit aversion learning than specialists.

## Induction of Preference

Some, if not most, effects of experience on feeding or oviposition preference in phytophagous insects cannot be assigned to any traditional category of learning. Induction of preference (80) is the catchall category to which such phenomena have been consigned. Dethier (33, 35) and Jermy (78), have argued that induction of feeding preference in lepidopterous larvae (where induction has been most studied) does not fit the usual definitions of learning.

Induction of preference, however, at least sometimes fits the criteria presented above; effects of experience termed "induction," for example, can be cumulative (35) and reversible (19, 206). Many examples of induction of preference may eventually be accounted for by habituation, sensitization, or even associative learning.

PRESUMED ECOLOGICAL FUNCTION OF INDUCTION    Induction of feeding preference may permit a larva's preference to track physiological changes in the utilization of alternative host species. Lepidopterous larvae, for example, adjust metabolically to the plant on which they are feeding through the activation of mixed-function oxidases (16). If feeding experience increases the efficiency with which a host is utilized in some way that enhances survival or fecundity on that host relative to other hosts, and if physiological adjustment to a novel host is costly in terms of time or energy, it might benefit a larva to modify its preference if, by doing so, the larva could avoid switching from one host species to another.

Evidence that induction of preference in Lepidoptera permits tracking of changes in the relative efficiency with which alternative host species are converted to biomass (and corresponding changes in relative growth and survival on those species) is somewhat limited. While the efficiency with which nutrients were processed sometimes depended on previous feeding history in southern armyworm larvae *(Spodoptera eridania)* (162), the relative growth rate of individuals fed on one host species was little different from that of individuals forced to switch from one species to another (161). Even in instances where feeding experience affected the efficiency with which alternative hosts were utilized (160) or the survival of insects on alternative hosts (60), it was not clear if the effect reflected physiological adjustment, behavioral induction, or both (53). However, in a study of *Colias* butterflies, larvae clearly became both behaviorally and physiologically specialized as a consequence of experience with particular fabaceous plants. Larvae fed on alfalfa, for example, had a higher growth rate and pupal weight on alfalfa than larvae fed on white sweet clover or crown vetch owing to changes in both consumption rate (behavioral induction) and the efficiency with which digested food was converted to biomass (physiological adjustment) (88). Since relative growth rate and pupal weight are highly correlated with fecundity in *Colias* butterflies, induction of feeding preference at least in these butterflies seems to be a mechanism for avoiding the adverse physiological effects associated with switching from one host species to another.

There are other possible hypotheses for the function of induction of larval preference, which draw upon lepidopteran natural history and which are better examined in a field setting. In the only published field study of induction of larval preference, Dethier (35) postulated that induction of preference, in concert with effects of deprivation, minimizes interruptions in feeding that might have deleterious consequences (53). Alternatively, the complex life history of the Lepidoptera may have promoted the evolution of flexibility of feeding preference in the larval stage in response to flexibility in oviposition preference in the adult stage. In many lepidopterous species, the ovipositing female chooses the site at which the larva initially feeds. The female's preference, after modification by effects of deprivation (166) and/or experience, often reflects both the availability of alternative host species for oviposition and the suitability of those species for juvenile survival (167). If young larvae exhibit a congenital preference for alternative host species that is correlated roughly with early growth and survival on those hosts, and if hatching on a relatively unsuitable plant implies that more suitable plants are relatively scarce, larvae may do well to acquire a preference for the less suitable host. This induction of preference may render larvae less likely to disperse from the plant on which they hatched in search of a more suitable but rare host or, when they do disperse, less likely to reject plants that are less suitable but more frequently encountered. In this scheme, induction of preference confers an advantage on the larva by reducing losses in fitness associated with dispersal.

The notion that the evolution of learning at one life history stage may depend on behavioral flexibility (including learning) at another stage is speculative (see also 81). Nevertheless, it illustrates the need to know more about individual feeding histories and dispersal patterns in nature. Except in Dethier's recent study (35), induction of larval preference has only been observed in the laboratory. We suspect that, for this reason, discussions of its ecological significance have generated only laboratory-testable hypotheses that do not depend on detailed knowledge of the natural history of the studied organisms.

## Integration of Learning Mechanisms

Recent neurobiological studies suggest that associative learning is mechanistically related to habituation and sensitization (18, 58). Likewise, exposure to a given stimulus activates both habituation and sensitization processes simultaneously (28 and references within); the insect's change in behavior with experience is a balance between these two processes. It may thus be inappropriate to invoke separate ecological functions for different kinds of learning as though each acts independently of the other. Some kinds of learning, e.g. aversion learning and habituation to deterrents, make more ecological sense if they act in concert to regulate feeding by phytophagous insects. Consider a grasshopper arriving at a potential host plant. The grasshopper begins to feed when the stimulant:deterrent ratio reaches a certain internal acceptance threshold (33, 117), at first taking short meals with long intermeal intervals. If the plant on which it cautiously begins to feed does not cause acute illness, the grasshopper habituates to one or more deterrents in the plant and continues to feed, taking longer meals with fewer interruptions (79, 176). If the plant causes acute illness, however, the grasshopper acquires an aversion for the plant (7), disperses, and refuses any plants of that species that are subsequently encountered.

This particular integration of learning mechanisms would be especially useful if unpalatability is often but not always an indication of toxicity. Indeed, a significant but less than perfect correlation between deterrency and toxicity would be predicted on theoretical grounds and seems to be supported by what little evidence is available (34, 40). Grasshoppers and other polyphagous insects may cope with the loose correlation between unpalatability and toxicity by employing a homeostatic combination of learning routines that simultaneously minimizes both their likelihood of rejecting nutritious plants and their likelihood of repeatedly consuming toxic plants.

## PROGRAMED ELEMENTS OF LEARNING

Each kind of learning discussed above may be manifested only with respect to certain cues or motor responses or only in certain contexts (45, 50, 114, 182). Bees, for example, are "prepared to learn" (sensu 114) certain stimuli associated with a nectar reward faster than other stimuli. Honeybees learn violet more rapidly than other colors and learn flower-like odors more rapidly than other odors (114 and references therein). Bumblebees in enclosure arrays switch more readily from yellow to blue artificial flowers than from blue to yellow (65). Analogous cue biases for phytophagous insects are likely to be uncovered. Responses of lepidopterous larvae to host-specific stimuli, for example, are modified by experience, but responses to general stimuli such as sucrose and inositol are not (references in 60, 129). *Pieris* butterflies oviposit readily on paper wetted with sinigrin solution if the paper is yellow, green, blue, or white, but not if it is violet, red, or black (187). Although the experiment remains to be done, these females are presumably less able to associate the latter group of colors with sinigrin than the former.

Insects that learn associatively are biased not only with respect to the stimuli to which they are conditioned, but also with respect to the stimuli that evoke conditioning. Certain so-called "sign stimuli" often trigger learning (182). *Battus* butterflies learn or relearn a leaf shape only upon contact with an *Aristolochia* host plant (124). Learning is triggered by contact with one or a few host stimulants. Learning of color and/or light intensity by female *Pieris* butterflies is similarly triggered by contact with the host stimulant sinigrin (186, 187) and is unaffected by simultaneous contact with deterrent stimuli such as chlorogenic acid (188).

Insects also exhibit biases in the motor patterns that are susceptible to modification. Probing of flowers by naive bees learning to extract nectar efficiently, for example, is not random (92). Apparently certain motor patterns are more likely to be modified by experience than others. Similar motor biases remain to be defined for phytophagous insects.

Learning is sometimes restricted to certain contexts. Imprinting is a kind of programed learning in which experience exerts specific and irreversible effects on behavior only during early sensitive periods in an animal's ontogeny (107). Few examples of imprinting as dramatic as those in vertebrates are known in insects (180), although the efficacy of learning varies with age in ants (77) and hymenopteran parasitoids (201) and with developmental stage in one phytophagous insect (19). In both ant and parasitoid, young insects are more susceptible to effects of experience on habitat selection than old insects. By contrast, adult *Carausius* stick insects are more labile in their feeding preferences than nymphs. Such contextual biases may themselves be learned. Bees learn to visit particular flower species at the time of day at which flowers are most rewarding, apparently by linking the time of day to particular floral stimuli (references in 46). While alternative explanations are possible, such time-linked learning could account for the tendency of *Heliconius* butterflies to visit host plants or nectar sources in the same sequence time after time (i.e. traplining).

The restriction of learning to a limited set of stimuli and motor patterns in highly specific contexts is thought to have an adaptive function. Learning may be confined to situations that are complex or unpredictable for preprogramed responses. In other situations, congenitally fixed responses may be sufficient and costs associated with learning in terms of time or errors might be avoided (45, 46, 50). Localized learning, i.e. learning triggered by specific sign stimuli (182), may be an efficient way to store complex images or databases (46). If the brain or ganglia are limited in capacity, it may pay to have congenitally programed responses to a few key identifier stimuli and to learn and remember details of the image associated with those identifier stimuli only while the details are required. For example, a newly eclosed beetle might respond congenitally to only one or a few volatile compounds that unambiguously identify its host, but might eventually learn to respond to more compounds and even to a specific blend (which might conceivably increase the rate at which it finds hosts). The costs in terms of occasional mistakes or time required to learn an odor blend would presumably be offset by the advantages of being able to store those details temporarily. When the beetle no longer needs information about host odor, e.g. when it enters a migratory phase and is refractory to host stimuli (132), it can replace this information, perhaps with information relating to migratory stimuli.

## LIMITS TO MEMORY: CONSTRAINT OR ADAPTATION?

The notion that insects can store only a limited amount of information is consistent with the frequent observation among vertebrates that a novel experience appears to interfere with recall of an earlier experience (107). *Colias* butterflies, for example, land progressively more often on host plants and less often on nonhost plants over the course of an oviposition bout; they apparently learn to recognize host plants before landing (173). If the butterflies engage in nectar foraging, however, erroneous landings on nonhosts increase significantly at the beginning of the next oviposition bout. Nectar foraging (and possibly learning of floral cues during nectar foraging) in some way interferes with retention or retrieval of stored information about host plants.

*Pieris* butterflies cannot remember the methods for extracting nectar from more than one species of flower (95). Learning to handle a second species interferes with recall of handling the first species, which suggests perhaps that butterflies are somehow neurally restricted in their ability to store or tap such information. *Pieris* butterflies also fail to learn to avoid colors that are not associated with a reward, perhaps because recall of all unrewarding species would exceed the insect's memory capacity (96). The inability of flower-foraging insects such as bees and butterflies to remember simultaneously more than one or a few flower types may account for the remarkable tendency of these insects to restrict their visits to flowers of one or several species while bypassing other species that are equally or even more rewarding (27, 95, 202).

The number of host species that phytophagous insects can remember may be similarly limited. Under enclosure conditions, female *Battus* butterflies searching in arrays containing a single *Aristolochia* host species adopted a search mode for the leaf shape of that host (124). In arrays containing three *Aristolochia* species females appeared confused, switching back and forth between search modes for broad and narrow leaves as they landed on hosts with those leaf shapes (125). Associated with this frequent switching was a higher number of erroneous landings on nonhost leaves in multihost arrays than in single-host arrays (123). In *Aristolochia macrophylla* and *Aristolochia reticulata* arrays, for example, landings on nonhosts comprised 55 and 61% of all landings in one experiment. In an array of the same overall host density but in which three host species that differed in leaf shape *(A. macrophylla, A. reticulata,* and *Aristolochia serpentaria)* were represented, the corresponding percentage was significantly higher (71%).

Although it is intuitively appealing to attribute interference effects to physiological constraints on information storage or retrieval (95, 202), memory limits may conceivably have adaptive value (113, 164). Memory in insects is a two-step process involving a short-term phase and a consolidation phase (113). Short-term memory fades within a few minutes of an experience, and the insect's behavior is briefly very susceptible to alteration by a new and contradictory experience. While short-term memory was originally thought to be a means of temporarily storing information until slower biochemical processes were able to consolidate storage into long-term memory, it may actually be an adaptive mechanism enabling the insect to respond to new and contradictory information (113 and references within). The vulnerability of short-term memory to alteration by intervening experiences may protect a bee or butterfly, for example, from making ecologically rash decisions based on just a few successive rewarding visits to a new flower species. Since bees can reach peak responsiveness to a rewarding flower in just a few visits and can retain that responsiveness for life (45; 114 and references within), the transience of short-term memory may guard against entrainment (by a short succession of lucky strikes) to a novel floral resource that is actually relatively unprofitable.

## DOES LEARNING EVOLVE UNDER SELECTION?

Virtually all of the adaptive functions of learning and memory discussed thus far are embraced by two non–mutually exclusive hypotheses for the evolution of learning: the neural economy hypothesis and the environmental unpredictability hypothesis. Despite the intense attention devoted to learning processes by psychologists, neither hypothesis has ever been properly evaluated for any animal species (81).

### Neural Economy Hypothesis

The neural economy hypothesis supposes that the world is too complex for insects to have all appropriate recognition and response behavior programed into the nervous system or into the DNA encoding the nervous system (81). In Mayr's words (110), "The great selective advantage of a capacity for learning is, of course, that it permits storing far more experiences, far more detailed information about the environment, than can be transmitted in the DNA of the fertilized zygote." Gould (46) has speculated that processing of object shape and spatial information by honey bees is ultimately limited by the number of neurons in the brain. This argument can easily be extended to phytophagous insects. If information storage is under physiological constraints, learning would be favored over congenitally fixed behavior if it were the more economical in terms of the amount of useful information stored over the lifetime of the animal. Williams (205) suggested that, to the contrary, learning may be costly in terms of the extra genetic material necessary to encode both a trait and its plasticity. Learning may also involve costs in terms of additional nerve cells and connections between nerve cells (81). While there is currently little direct support for either position (181), a growing body of foraging theory supposes that animals do not have the capacity to remember indefinitely all incoming information about the quality and abundance of their resources (e.g. 24, 121, 155).

An appropriate test of the neural economy hypothesis would ask whether differences among species or populations in the storage capacity of the nervous system are associated with differences in the complexity of the environment. Since so little is known about the neural mechanisms underlying learning and memory, this hypothesis is somewhat resistant to testing (81, 181).

## Environmental Unpredictability Hypothesis

The environmental unpredictability hypothesis for the evolution of learning supposes that the environment is too unpredictable within the life of an individual and/or over successive generations for natural selection to preset all appropriate recognition and response behaviors (81, 180). This hypothesis also supposes that the environment is not so unpredictable that the individual cannot or should not behaviorally track its change. Recent foraging models, for example, support the idea that tracking of changes in the quality of variable resources is favored so long as the changes are not too unpredictable; extreme unpredictability in a resource may lead to a fixed preference for a mediocre but predictable alternative (174 and references within).

An appropriate test of the environmental unpredictability hypothesis would ask whether differences in learning ability among species or populations are associated with differences in the predictability of the environment. The expectation is that learning would be more common in moderately unpredictable environments, less common in extremely predictable environments, and no more common in extremely unpredictable environments than fixed behavior. While this hypothesis seems more amenable to testing than the neural economy hypothesis, evidence for or against it is remarkably scarce.

## Between-Species Approach

Studies of the evolution of learning have traditionally employed comparisons among species across major taxa (cf 9). As species were commonly chosen according to the outmoded concept of scala naturae, early comparisons not surprisingly yielded few patterns of significance (37, 67, 86). The occurrence of learning in insects is not well correlated with level of phylogenetic advancement even when modern systematic criteria are applied (4). For example, the Hymenoptera, a relatively advanced order, are quite sophisticated with respect to learning while the Diptera, also an advanced order, are apparently not (31). In phytophagous insects no patterns among orders are obvious yet. Learning has been found in each of the orders in which it has been examined [i.e. Hemiptera (including both Homoptera and Heteroptera), Coleoptera, Phasmatodea, Orthoptera, Lepidoptera, and Diptera] (78) and remains to be examined in the Collembola (but see reference in 78), in the

Thysanoptera, and ironically, in phytophagous forms within the Hymenoptera.

Aside from phylogenetic analysis, species comparisons have also been applied to the proposition that species with generalist habits will be more plastic in their host-selection behavior than specialist species (32, 35, 43, 50, 78, 192). Polyphagous herbivores, for example, may be more likely to habituate to feeding deterrents than herbivores with more restricted diets (78). The three polyphagous species tested thus far exhibited aversion learning, while the single oligophagous species tested did not (7, 32, 35, 36). In contrast, the degree to which lepidopterous larvae can be induced to prefer a host plant is not related to the degree of specialization (29; but see 59). Fruit acceptance in tephritid flies is affected by oviposition experience in the specialist species *Rhagoletis pomonella* as well as in the generalist species *Ceratitis capitata* and *Dacus tryoni* (22, 128, 131, 140, 143, 145).

In short, the prediction that generalists are more likely to learn than specialists has received only limited support in scattered comparisons among phytophagous insect species. Problems familiar to those who have pondered the evolutionary underpinnings of host specialization in phytophagous insects may account in part for these conflicting results. For instance, a species that is classified as a generalist over its entire range may consist of populations specialized on different host species (38), yet most species comparisons are made using insects collected from just one or a few localities. Furthermore, a specialist with respect to host species may be a generalist with respect to individual plants within a host population or with respect to different parts of a particular plant (7). Learning of preference at the level of host species may simply be an epiphenomenon (sensu 52) of learning of preference at the level of conspecific host individuals or host-plant parts (125).

The between-species approach to analyzing the evolution of learning has a more general drawback: Species may differ in learning for reasons other than differences in the set of selective factors presumed to shape learning (37, 41). First, differences in learning between species could be due to differences in the methods used to assay learning (9). Second, such differences could be due to differences in selective factors other than the ones supposed. Third, even if selection on learning is similar between species, differences in learning could arise from differences in the genetic background against which genes encoding learning are selected (e.g. differences between species in pleiotropy). Finally, the differences in learning between species could be caused by factors other than natural selection (e.g. differences in gene flow between or genetic drift within populations). The between-species approach has yet another pitfall—species may not differ in learning even in the face of observed differences in selection on learning. A lack of difference could be rooted in history if selective factors operating in the past were different from those currently observed.

Comparisons of closely related species minimize these problems (37, 86). The more similar are the test species, the more similar can be the methods used to assay learning, and the lower is the likelihood that any observed differences in learning will be due to differences in ecological factors other than those under scrutiny.

## Within-Species Approach

Whether the genetic mechanisms (i.e. mutation, natural selection, and genetic drift) that lead to differences between species are the same as those that give rise to differences among populations within a species or among individuals within a population is a matter of some controversy (25, 41). Differences among species, for example, can vary in sign as well as magnitude from corresponding differences among populations within a species (51). Comparisons among closely related species may thus permit evaluation of genetic differentiation in behavior after speciation but do little to elucidate the evolutionary forces that brought about the original differences leading to speciation (194). Complementing the traditional comparative approach, the microevolutionary approach examines sources of variation in behavior within a species, either among populations or among individuals within a population. Knowledge about genetically based differences between populations or between individuals within populations can be used to construct hypotheses about the mechanisms that gave rise to differences among species as well as to predict the course of future evolutionary change. A significant methodological advantage of the within-species approach is that different populations or families of insects can usually be trained and tested under almost identical experimental regimes, so that differences among groups are not likely to be due to variations in the methodology by which learning is assayed. A possible disadvantage is that differences within species may be smaller than those between species and may thus be harder to detect.

LEARNING IN VARIABLE ENVIRONMENTS: A WITHIN-SPECIES TEST    A variable environment is not necessarily unpredictable. When the environment is variable in a regular and predictable way (e.g. when the resources change seasonally in abundance or quality), insects could respond adaptively either through behavioral plasticity (e.g. learning) or through genetic tracking. Genetic tracking means that the gene frequency in the population at polymorphic loci controlling behavioral traits changes over successive generations in a way that tracks the change in the environment. Bradshaw (13) proposed that phenotypic plasticity would be a more common adaptation to regularly fluctuating environments than genetic tracking (see 133 for theoretical justification).

Bradshaw's hypothesis was supported by a study of an east Texas population of *Battus philenor* butterflies that shifted seasonally in oviposition preference for two *Aristolochia* host species (133). Although two broods emerging successively in the spring preferred different host species in the field, stocks originating from each brood did not differ genetically in their propensity to find and lay eggs on either host. Moreover, individuals from either stock were able to learn or relearn the leaf shape of either host species, showing no apparent preparedness to learn (sensu 114) one or the other leaf shape. The difference in host preference of the two broods in the field was apparently solely the result of differences in the outcome of learning to search in early versus late spring environments.

The shift in host preference in east Texas accompanied well-defined seasonal changes in the relative abundance of the two host species and their suitability as larval food (148). Learning and switching thus appeared to be a local adaptation that enabled the population to track those changes (124, 133). This notion was tested by comparing the learning abilities of butterflies in the east Texas population with those of butterflies in a Virginia population that used just one host species, *A. macrophylla* (125). Contrary to expectations, learning did not depend on the variability of the host environment: Both populations were equally capable of learning leaf shape.

Several hypotheses can account for the lack of differentiation in learning ability between *Battus* populations. For example, there may be no genetic variation in learning within monophagous populations on which selection can act. Alternatively, selection on learning of leaf shape may be genetically constrained by selection on other learning traits such as learning to forage for nectar sources. Which of these hypotheses explains the absence of local adaptation in learning ability is not yet clear. Nevertheless, it is interesting that the only other explicit test of the environmental variability hypothesis also yielded a negative result (54).

# CAN LEARNING EVOLVE UNDER SELECTION?

As we have pointed out, there is at present little evidence, either within or between species, that learning has undergone adaptive evolutionary change in phytophagous insects. We can still ask whether or not learning has potential to evolve in an adaptive manner. The notion that natural selection can alter learning supposes (*a*) that genetically based variation in learning exists and (*b*) that differences in learning are associated with differences in fitness.

## Genetic Variation in Learning Ability

Insect populations clearly harbor genetic variation in learning. Existing genetic variation within strains has been used to analyze the physiology of learning in *Drosophila* and *Phormia* flies (111, 112). Variation in learning in honey bees has a heritable component (15), and bidirectional selection in a thelytokous parthenogenic strain of honey bees has produced strong differences in learning after a single generation (14). In phytophagous insects, little information on genetically based variation in learning is available. A natural population of *Deloyala* beetles exhibited no genetic variation in the extent to which feeding and oviposition experience with their convolvulaceous hosts altered oviposition preference (150). Laboratory and wild strains of the Mediterranean fruit fly differed in the extent to which oviposition experience affected fruit acceptance (131), but maternal effects were not ruled out, so the differences were not necessarily genetically based.

The best evidence for genetic variation in learning in phytophagous insects has come from analyses of the effects of larval diet on adult feeding and oviposition preference in agromyzid flies (194). Individual *Liriomyza* leafminers varied in the effect of larval food plant on adult feeding and oviposition preference, and the variation had a genetic basis. Interestingly, as in most studies examining larval induction of adult preference (reviewed in 78 and 129), larval diet did not affect mean adult preference in the study populations. Despite the sparseness of available data, evidence for genetic variation in host preference is accumulating rapidly (42, 71, 74, 195, 203). This variation may reflect at least in part genetic variation in the extent to which host preference is modified by experience (99).

## Effect of Learning on Components of Fitness

Given that so little is known about genetic differences in learning among individuals, it is not surprising that nothing is known about the extent to which such differences are associated with differences in fitness. We might still ask whether or not phenotypic variation in learned behavior (whether genetically based or not) is associated with variation in expected survival or reproductive success. Even in this limited sense, almost nothing is known about the relative costs and benefits of learning for either invertebrates or vertebrates (6, 81, 122). Components of fitness should be easier to measure in phytophagous insects than in, say, eusocial insects or long-lived vertebrates, but we still have little idea whether differences in level of experience among individual phytophages are associated with differences in survival and reproductive success. In particular, it is not clear how currencies such as time or energy (in which costs and benefits of learning are typically measured) relate to fitness. Does the time saved by learning to handle a flower species provide some benefit to the female *Pieris* butterflies in terms of rate of energy uptake? If so, is the survival or reproductive success of nectar-foraging butterflies affected by the rate of energy uptake? Does the time saved by locusts and caterpillars in rejecting a host plant at an earlier stage in the food-selection sequence (10, 11) confer any increase in the rate at which suitable hosts are consumed? If so, are increases in consumption rate directly related to increases in survival or fecundity?

Pulliam & Dunford (146) remarked succinctly, "The obvious problem with trial-and-error learning is error." But do errors [such as inaccurate searching by egg-laying butterflies (173)] really translate into losses in fitness and, if so, how? As noted above, the relative frequency of erroneous landings (i.e. landings on nonhost plants) declined with experience for ovipositing *Battus* butterflies. Since the relative frequency of errors was inversely correlated with the rate at which females alighted on host plants (123), a reduction in errors is undoubtedly associated with an increase in the rate at which hosts are discovered. Does the oviposition rate (and potentially the lifetime fecundity) of butterflies depend on the rate at which they find host plants? In fact, differences in host discovery rate due to learning are not accompanied by differences in oviposition rate (123, 147). Increases in discovery rate might still enhance reproductive success by permitting females to be more selective about the plants on which eggs are deposited and to place more eggs on plants that are highly suitable for larval survival (151). Indeed, in an enclosure array featuring equal numbers of low-quality and high-quality plants, *Battus* females that landed at higher rates deposited proportionately more egg clusters on high-quality plants than females that landed on hosts at lower rates (123).

A different pattern was observed in female apple maggot flies foraging for oviposition sites in trees within field cages (130). Experience with host fruit of a given species (either apple or hawthorn fruit) increased the rate at which *Rhagoletis* females oviposited in trees bearing such fruit but not the rate at which they found the fruit. Experience enhances oviposition rates by increasing the likelihood that, once having landed on a fruit, a female will oviposit. It is unknown whether increases in oviposition rate measured within a tree over a relatively short period (~1 hr) translate into increases in expected lifetime fecundity. In summary, phenotypic differences in learning are tentatively thought to be associated with differences in fitness. Yet the consequences of experience for individual fitness are qualitatively different for the only two phytophagous insect species examined outside the laboratory, which illustrates the need for detailed field studies on more species.

## EVOLUTION OF PROGRAMED ELEMENTS OF LEARNING

Increasing attention is being devoted not to whether learning per se is adaptive, but to whether the way in which learning is programed with respect to cues, motor patterns, and context is adaptive. Bee species, for example, differ not in their ability to learn color per se but in the selectiveness of their learning, specifically in their preparedness to learn those colors that elicit the strongest congenital responses (114). A bee species that shows a relatively strong congenital response to violet, for example, also learns to associate violet with a reward faster than other bee species; this difference is potentially adaptive. While evidence is lacking thus far, phytophagous insect species will undoubtedly also exhibit potentially adaptive differences in the cues that they are prepared to learn. An essentially unlimited number of functional questions concerning programed learning emerges readily from considerations of the lifestyles of animals. We mention just a few in the following paragraphs.

Parasitoids that specialize on generalist host larvae seem to respond in a fixed manner to the invariant cues of such larvae (97; O. Zanen, W. J. Lewis & J. H. Tumlinson, unpublished) but learn by association to attend to the cues of the variable host microhabitat (97, 193). Are responses of phytophagous insects to unpredictable cues more flexible than responses to predictable cues? Do generalists learn cues that vary among host species, but have congenitally fixed responses to those cues that are common to all host species?

In rats, olfactory or gustatory stimuli are more likely to be associated with illness during aversion learning than are visual stimuli. The reverse is true in quail. The difference is linked to species differences in the cues used in foraging behavior (56). Do phytophagous insect species show similar differences? Horses, which graze continuously on a mixed diet, show less tendency toward aversion learning than rats, which forage sporadically with long pauses (207). Grasshoppers and caterpillars, which, like rats, forage nomadically and with digestive pauses, exhibit aversion learning (7, 32, 35). Do closely related species that forage more continuously on mixed diets show less tendency toward aversion learning?

Animals learn better when the temporal pattern of reinforcement resembles the pattern of encounter with resources (83, 86, 164). Host resources of phytophagous insects tend to be patchy in distribution (172). Do phytophagous insects learn fastest when the temporal pattern of reinforcement most closely resembles the pattern of encounter with spatially clumped resources? Birds that forage for rapidly diminishing resource patches such as nectar in flowers tend to avoid revisiting depleted flowers (20, 84). Moreover, hummingbirds can be trained more readily to avoid a recently visited position (shift learning) than they can be trained to revisit that position (stay learning) (20). Do phytophagous insects similarly adopt a win-shift strategy when foraging for nectar or pollen? Do the same insects adopt a win-stay strategy when foraging for slowly diminishing resources such as host plants in large patches? Birds that cache food, wasps that provision burrows, and bees that trapline exhibit astounding abilities to remember locations (45, 84, 164). Is the spatial memory of phytophagous insects that trapline (cf 105, 190) superior to that of phytophagous insects that do not?

# LEARNING AND THE EVOLUTION OF HOST USE BY PHYTOPHAGOUS INSECTS

Whether or not learning evolves under the influence of natural selection, learning can theoretically influence the adaptive evolution of other traits (109,

181). Maynard Smith (108), for instance, suggested that genetic subdivision within a population in a heterogeneous environment would be facilitated if individuals tended to breed in the same habitat in which they developed. Maynard Smith and other authors since (17, 42, 169, 177) noted that effects of experience during the juvenile stage on habitat-selection behavior by adult animals is one mechanism by which such habitat preference might arise. For a variety of reasons, phytophagous insects are considered likely candidates for genetic subdivision within populations (17, 42). Although juvenile experience almost never affects adult preference in phytophagous insects (78, 132; but see 5, 194), Jaenike (72, 76) pointed out that learning during the adult stage (which is common in phytophagous insects) can also promote the establishment within a population of genotypes specialized to survive and breed on particular hosts.

Studies on learning in the apple maggot fly [whose populations are thought by some to have diverged sympatrically in historical times in specialization for apple or hawthorn fruit (17, 142; but see 42)], suggest that prior adult experience could promote genetic subdivision within populations in more complex ways than previously imagined. Under semifield conditions, experience with fruit of a given species increased both the tendency of *Rhagoletis* females to accept that fruit for oviposition and the length of their residence in trees bearing that fruit (130). In addition, because experience extended the proportion of time that a female spends on fruit (130) and because virtually all matings are initiated on the surface of fruit (141), females should be more likely to mate on fruit of the species with which they are experienced. Males also remain longer on fruit of the species with which they have had experience and should also be more likely to mate on such fruit (141).

Each consequence of learning should promote sympatric divergence as long as individuals are more likely to gain experience with the host for which they are genotypically specialized than with the alternative host. This is not unlikely, since apple maggot flies eclose near the host on which they have developed (17) and since individuals originating from apple and hawthorn hosts differ in fruit acceptance (differences that are probably genetically based) (142). In summary, learning could promote sympatric divergence in apple maggot fly populations by increasing the flies' acceptance of the fruit for which they are behaviorally specialized and by increasing the fidelity of flies to trees bearing that fruit. In addition, experience should increase the likelihood that males and females will mate assortatively on fruit of the species on which they are genotypically specialized (see also 21).

The extent to which adult experience promotes genetic subdivision within apple maggot fly populations depends on whether or not adults disperse from their larval hosts to alternative hosts before any learning takes place (69, 76). Early adult experience of the sort that might affect genetic subdivision has been shown for saprophagous flies. When the length of time over which emerging *Drosophila* flies were exposed to host resources at the site of eclosion was determined by the flies themselves, early experience at the site of larval development at least sometimes significantly increased preference for the larval food resource (69, 76). Investigators were circumspect about the extent to which such effects could promote genetic divergence within *Drosophila* populations. Hoffman (69) observed that the effect of experience on host fidelity was transient and disappeared once flies had dispersed a few meters. Jaenike (76) cautioned that the specific impact of early experience depends on the natural distribution of host resources, about which little is known. Clearly, further investigation of rates of learning in relation to adult dispersal capability is essential for determining whether effects of experience on host preference have any role in promoting genetic differentiation within populations of phytophagous insects.

## FOOD FOR THOUGHT

Constraints on the amount of information stored in our review did not permit discussion of many worthwhile ecological and evolutionary aspects of learning in phytophagous insects. Below we summarize three areas worthy of further research with respect to learning: host and herbivore population dynamics, vegetational complexity and search image formation, and plant-herbivore coevolution.

Learning can theoretically influence the dynamics of host and herbivore populations (165), principally by affecting the functional response of the herbivore to the density of its host (82, 178, 191). Learning and switching of host preference by individuals can have an impact on host-herbivore dynamics when it results in a shift in the equilibrium preference of the population for the currently most abundant host. Population-level switching (not to be confused with individual switching) between host species can potentially produce stable equilibria in host and herbivore populations (63, 119). Studies of the dynamics of switching by individuals within insect populations (cf 64, 126, 152, 153) will be enormously helpful in predicting how fast populations will respond to changes in relative plant abundance and how likely the plant-insect system is to reach a stable equilibrium.

Learning can either create or destroy behavioral variation in a population and can thus affect population dynamics in yet another way. By analogy with arguments about genetic variation and population dynamics (100), variation created by learning could stabilize insect populations if experience caused differences in competitive ability among individuals. If effects of experience conceal genetically based variation (e.g. 170) in competitive ability among individuals, however, learning might have the opposite effect. The extent to which experience homogenizes or diversifies a population and the ramifications for population growth and stability remain to be explored both theoretically and empirically.

Learning can also alter the responses of an insect population to the structure of the plant community. Vegetational diversity, for instance, commonly influences the susceptibility of hosts to herbivore attack (87); an herbivore's foraging success may depend on the extent to which it can learn to detect a host plant against a diverse nonhost background. The concept of the search image (defined as a change in the ability to detect a cryptic host or prey as a consequence of recent experience with that host or prey) is currently on shaky ground (55) and continues to be misused in the plant-insect literature (132). Still, herbivores do respond to changes in vegetational complexity in ways that suggest that plants are cryptic (149). It has even been suggested that host species are under selection to resemble nonhost species (and to reduce thereby their visual apparency to ovipositing female insects) (44, 149). Like other animals (55), phytophagous insects may well learn to forage in one way when their host is cryptic and in another when it is conspicuous.

Finally, learning may be a force in plant-insect coevolution. Over a decade ago, Gilbert (44) postulated that the diversity in leaf shape among *Passiflora* species was a consequence of selection exerted by ovipositing *Heliconius* butterflies. While direct evidence for selection on leaf shape by herbivory is still lacking, support for elements of Gilbert's hypothesis has emerged in studies of other plant-insect systems. First, butterflies use leaf shape as a host-finding cue (124, 126, 147). Second, foraging butterflies can remember only a limited number of items and thus cannot search as efficiently for a range of items as they do for a single item (95). Third, host species differ in leaf shape when sympatric and when under attack by a butterfly that learns leaf shape (147). It is not inconceivable that plants evolve defenses aimed at limitations in the memory of their insect enemies.

## CONCLUDING REMARKS

A number of factors make phytophagous insects exemplary organisms for inquiry into the ecology and evolution of learning. A major component of the habitat of the phytophagous insect, the plant itself, is readily accessible to experimental manipulation. Insect behavior can often be observed directly under natural or seminatural conditions and is generally easier to quantify than vertebrate behavior. Components of fitness can be measured more directly than in eusocial insects or long-lived vertebrates. Genetic variation in behavioral traits such as host preference is relatively well known, and the framework for assessing variation in learning with respect to such traits is already in place. Finally, assessment of the impact of learning on the dynamics and coevolution of plant-herbivore systems is within reach of established theory and techniques. In spite of these advantages, much remains to be done. The potential for future investigation is almost as overwhelming as it is enticing. For this reason, we might be grateful that phytophagous insects are capable of learning and equally grateful that plants are not (but see 184).

ACKNOWLEDGMENTS

We are grateful to M. Aluja, Y. Drost, D. Karowe, L. Vet, S. Yoerg, and O. Zanen for comments on earlier drafts. We also thank E. Bernays, D. Karowe, J. Lee, and A. Lewis for access to unpublished manuscripts. The senior author is particularly indebted to L. Vet, J. Schmidt, R. Cardé, and L. Schoonhoven for stimulating discussions that contributed greatly to the development of the ideas expressed here. Special thanks to T. Green for providing a quiet place in which to write. Funds were provided by the USDA under Competitive Research Grant 8603158.

## Literature Cited

1. Ahmad, S., ed. 1983. *Herbivorous Insects: Host-Seeking Behavior and Mechanisms.* New York: Academic. 257 pp.
2. Alcock, J. 1984. *Animal Behavior: An Evolutionary Approach.* Sunderland, Mass: Sinauer. 596 pp. 3rd ed.
3. Alloway, T. M. 1972. Learning and memory in insects. *Ann. Rev. Entomol.* 17:43–56
4. Alloway, T. M. 1972. Learning in insects except Apoidea. In *Invertebrate Learning,* ed. W. C. Corning, J. A. Dyal, A. O. D. Willows, 1:131–71. New York: Plenum. 313 pp.
5. Alloway, T. M. 1972. Retention of learning through metamorphosis in the grain beetle *(Tenebrio molitor).* Am. Zool. 12:471–77
6. Bateson, P. P. G. 1984. Genes, evolution, and learning. See Ref. 107, pp. 75–88.
7. Bernays, E. A., Lee, J. C. 1988. Food aversion learning in the polyphagous grasshopper *Schistocerca americana.* Physiol. Entomol. 13:131–37
8. Bernays, E. A., Wrubel, R. P. 1985. Learning by grasshoppers: association of colour and light intensity with food. *Physiol. Entomol.* 10:359–69
9. Bitterman, M. E. 1965. Phyletic differences in learning. *Am. Psychol.* 20:396–410
10. Blaney, W. M., Simmonds, M. S. J. 1985. Food selection by locusts: the role of learning in rejection behaviour. *Entomol. Exp. Appl.* 39:273–78

11. Blaney, W. M., Simmonds, M. S. J. 1987. Experience: a modifier of neural and behavioral sensitivity. See Ref. 91, pp. 237–41
12. Blaney, W. M., Winstanley, C., Simmonds, M. S. J. 1985. Food selection by locusts: an analysis of rejection behaviour. *Entomol. Exp. Appl.* 38:35–40
13. Bradshaw, A. D. 1965. Evolutionary significance of phenotypic plasticity in plants. *Adv. Genet.* 13:115–55
14. Brandes, C. 1987. Effects of bidirectional selection on learning behaviour in honeybees *(Apis mellifera capensis).* In *Chemistry and Biology of Social Insects,* ed. J. Eder, H. Rembold, pp. 192–93. Munich: Peperny. 757 pp.
15. Brandes, C. 1988. Estimation of heritability of learning behavior in honeybees *(Apis mellifera capensis). Behav. Genet.* 18:119–32
16. Brattsen, L. B. 1979. Biochemical defense mechanisms in herbivores against plant allelochemicals. In *Herbivores: Their Interaction With Secondary Plant Metabolites,* ed. G. A. Rosenthal, D. H. Janzen, pp. 199–270. New York: Academic. 718 pp.
17. Bush, G. L. 1975. Sympatric speciation in phytophagous parasitic insects. In *Evolutionary Strategies of Parasitic Insects and Mites,* ed. P. W. Price, pp. 187–206. New York: Plenum. 236 pp.
18. Byrne, J. H. 1987. Cellular analysis of associative learning. *Physiol. Rev.* 67:329–439

19. Cassidy, M. D. 1978. Development of an induced food plant preference in the Indian stick insect, *Carausius morosus*. *Entomol. Exp. Appl.* 24:87–93

20. Cole, S., Hainsworth, F., Kamil, A. C., Mercier, T., Wolf, L. L. 1982. Spatial learning as an adaptation in hummingbirds. *Science* 217:655–57

21. Colwell, R. K. 1985. Community biology and sexual selection: lessons from hummingbird flower mites. In *Community Ecology*, ed. J. R. Diamond, T. Case, pp. 406–24. New York: Harper & Row. 704 pp.

22. Cooley, S. S., Prokopy, R. J., McDonald, P. T., Wong, T. T. Y. 1986. Learning in oviposition site selection by *Ceratitis capitata* flies. *Entomol. Exp. Appl.* 40:47–51

23. Couvillon, P. A., Bitterman, M. E. 1987. Compound-component and conditional discrimination of colors and odors by honeybees: Further tests of a continuity model. *Anim. Learn. Behav.* 16:67–74

24. Cowie, R. J. 1977. Optimal foraging in great tits *(Parus major)*. *Nature* 268:137–39

25. Coyne, J. A., Lande, R. 1985. The genetic basis of species differences in plants. *Am. Nat.* 126:141–45

26. Daly, M., Rauschenberger, J., Behrends, P. 1982. Food aversion learning in kangaroo rats: a specialist-generalist comparison. *Anim. Learn. Behav.* 10:314–20

27. Darwin, C. 1876. *On the Effects of Cross and Self Fertilisation in the Vegetable Kingdom*. London: Murray. 482 pp.

28. Davis, M., File, S. E. 1984. Intrinsic and extrinsic mechanisms of habituation and sensitization: implications for the design and analysis of experiments. In *Habituation, Sensitization, and Behavior*, ed. H. V. S. Peeke, L. Petrinovich, pp.287–323. New York: Academic. 471 pp.

29. de Boer, G., Hanson, F. E. 1984. Food-plant selection and induction of feeding preference among host and non-host plants in larvae of the tobacco hornworm, *Manduca sexta*. *Entomol. Exp. Appl.* 35:177–93

30. Denno, R. F., McClure, M. S., eds. 1983. *Variable Plants and Herbivores in Natural and Managed Systems*. New York: Academic. 717 pp.

31. Dethier, V. G. 1976. *The Hungry Fly: A Physiological Study of the Behavior Associated With Feeding*. Cambridge, Mass: Harvard Univ. Press. 489 pp.

32. Dethier, V. G. 1980. Food-aversion learning in two polyphagous caterpillars, *Diacrisia virginica* and *Extigmene congrua*. *Physiol. Entomol.* 5:321–25

33. Dethier, V. G. 1982. Mechanism of host-plant recognition. *Entomol. Exp. Appl.* 31:49–56

34. Dethier, V. G. 1984. Analyzing proximate causes of behavior. See Ref. 71, pp. 319–28

35. Dethier, V. G. 1988. Induction and aversion-learning in polyphagous arctiid larvae (Lepidoptera) in an ecological setting. *Can. Entomol.* 120:125–31

36. Dethier, V. G., Yost, M. T. 1979. Oligophagy and the absence of food-aversion learning in tobacco hornworms, *Manduca sexta*. *Physiol. Entomol.* 4:125–30

37. Domjan, M., Galef, B. G. Jr. 1983. Biological constraints on instrumental and classical conditioning: retrospect and prospect. *Anim. Learn Behav.* 11:151–61

38. Fox, L. R., Morrow, P. A. 1981. Specialization: species property or local phenomenon? *Science* 211:887–93

39. Fromm, J. E., Bell, W. J. 1987. Search orientation of *Musca domestica* in patches of sucrose drops. *Physiol. Entomol.* 12:297–307

40. Futuyma, D. J. 1984. The role of behavior in host-associated divergence in herbivorous insects. See Ref. 71, pp. 295–302

41. Futuyma, D. J. 1986. *Evolutionary Biology*. Sunderland, Mass: Sinauer. 600 pp. 2nd ed.

42. Futuyma, D. J., Peterson, S. C. 1985. Genetic variation in the use of resources by insects. *Ann. Rev. Entomol.* 30:217–38

43. Gelperin, A., Forsythe, D. 1975. Neuroethological studies of learning of mollusks. In *Simpler Networks and Behavior*, ed. J. C. Fentress, pp. 239–50. Sunderland, Mass: Sinauer. 403 pp.

44. Gilbert, L. E. 1975. Ecological consequences of a coevolved mutualism between butterflies and plants. In *Coevolution of Animals and Plants*, ed. L. E. Gilbert, P. R. Raven, pp. 100–40. Austin, Texas: Univ. Texas Press. 246 pp.

45. Gould, J. L. 1982. *Ethology: The Mechanisms and Evolution of Behavior*. New York: Norton. 544 pp.

46. Gould, J. L. 1984. Natural history of honey bee learning. See Ref. 107, pp. 149–80

47. Gould, J. L. 1986. Pattern learning by honey bees. *Anim. Behav.* 34:990–97

48. Gould, J. L. 1986. The locale map of honey bees: Do insects have cognitive maps? *Science* 232:861–62

49. Gould, J. L. 1987. Landmark learning

by honey bees. *Anim. Behav.* 35:26–34

50. Gould, J. L., Marler, P. 1984. Ethology and the natural history of learning. See Ref. 107, pp. 47–74

51. Gould, S. J. 1966. Allometry and size in ontogeny and phylogeny. *Biol. Rev.* 41:587–640

52. Gould, S. J., Lewontin, R. C. 1979. The spandrels of San Marcos and the Panglossian paradigm: a critique of the adaptionist programme. *Proc. R. Soc. London Ser. B* 205:581–98

53. Grabstein, E. M., Scriber, M. 1982. Host-plant utilization by *Hyalophora cecropia* as affected by prior feeding experience. *Entomol. Exp. Appl.* 32:262–68

54. Greenberg, R. 1985. A comparison of foliage discrimination learning in a specialist and generalist species of migrant wood warbler (Aves: Parulidae). *Can. J. Zool.* 63:773–76

55. Guilford, T., Dawkins, M. S. 1987. Search images not proven: a reappraisal of recent evidence. *Anim. Behav.* 35:1838–45

56. Gustavson, C. R. 1977. Comparative and field aspects of learned food aversions. In *Learning Mechanisms in Food Selection,* ed. L. M. Barker, M. R. Best, M. Cornjan, pp. 23–43. Waco, Texas: Baylor Univ. Press. 632 pp.

57. Hailman, J. P. 1985. Historical notes on the biology of learning. See Ref. 138, pp. 27–57

58. Hall, J. C. 1985. Genetic analysis of behavior in insects. In *Comprehensive Insect Physiology, Biochemistry, and Pharmacology,* ed. G. A. Kerkut, L. I. Gilbert, 9:287–373. Oxford: Pergamon. 735 pp.

59. Hanson, F. E. 1976. Comparative studies on induction of food choice preferences in lepidopterous larvae. *Symp. Biol. Hung.* 16:71–77

60. Hanson, F. E. 1983. The behavioral and neurophysiological basis of food plant selection by lepidopterous larvae. See Ref. 1, pp. 3–23

61. Harris, M. O., Miller, J. R. 1983. Color stimuli and oviposition behavior of the onion fly, *Delia antiqua* (Meigen) (Diptera: Anthomyiidae). *Ann. Entomol. Soc. Am.* 76:766–71

62. Hassell, M. P., Southwood, T. R. E. 1978. Foraging strategies of insects. *Ann. Rev. Ecol. Syst.* 9:75–98

63. Hassell, M. P., Waage, J. K. 1984. Host-parasitoid population interactions. *Ann. Rev. Entomol.* 29:89–114

64. Heinrich, B. 1976. The foraging specializations of individual bumblebees. *Ecol. Monogr.* 46:105–28

65. Heinrich, B., Mudge, P., Deringus, P. 1977. A laboratory analysis of flower constancy in foraging bumblebees: *Bombus ternarius* and *B. terricola. Behav. Ecol. Sociobiol.* 2:247–66

66. Hinde, R. A. 1970. *Animal Behavior: A Synthesis of Ethology and Comparative Psychology.* New York: McGraw-Hill. 876 pp. 2nd ed.

67. Hodos, W., Campbell, C. B. G. 1969. Scala naturae: Why there is no theory in comparative psychology. *Psychol. Rev.* 76:337–50

68. Hoffmann, A. A. 1985. Conditioning in *Drosophila:* Comparing apples and oranges. *Am. Nat.* 126:41–51

69. Hoffmann, A. A. 1988. Early adult experience in *Drosophila melanogaster. J. Insect Physiol.* 34:197–204

70. Hoffmann, A. A., Turelli, M. 1985. Distribution of *Drosophila melanogaster* on alternative resources: effects of experience and starvation. *Am. Nat.* 126:662–79

71. Huettel, M. D., ed. 1984. *Evolutionary Genetics of Invertebrate Behavior. Progress and Prospects.* New York: Plenum

72. Jaenike, J. 1982. Environmental modification of oviposition behavior in *Drosophila. Am. Nat.* 119:784–802

73. Jaenike, J. 1983. Induction of host preference in *Drosophila melanogaster. Oecologia* 58:320–25

74. Jaenike, J. 1985. Genetic and environmental determinants of food preference in *Drosophila tripunctata. Evolution* 39:362–69

75. Jaenike, J. 1986. Intraspecific variation for resource use in *Drosophila. Biol. J. Linn. Soc.* 27:47–56

76. Jaenike, J. 1988. Effects of early adult experience on host selection in insects: some experimental and theoretical results. *J. Insect Behav.* 1:3–15

77. Jaisson, P. 1980. Environmental preference induced experimentally in ants (Hymenoptera: Formicidae). *Nature* 286:388–89

78. Jermy, T. 1987. The role of experience in the host selection of phytophagous insects. In *Perspectives in Chemoreception and Behavior,* ed. R. F. Chapman, E. A. Bernays, J. G. Stoffolano, Jr., pp. 143–57. New York: Springer-Verlag. 207 pp.

79. Jermy, T., Bernays, E. A., Szentesi, A. 1982. The effect of repeated exposure to feeding deterrents on their acceptability to phytophagous insects. See Ref. 196, pp. 25–32

80. Jermy, T., Hanson, F. E., Dethier, V. G. 1968. Induction of specific food pref-

erence in lepidopterous larvae. *Entomol. Exp. Appl.* 11:211–30

81. Johnston, T. D. 1982. Selective costs and benefits in the evolution of learning. *Adv. Study Behav.* 12:65–106

82. Kacelnik, A., Krebs, J. R. 1985. Learning to exploit patchily distributed food. See Ref. 165, pp. 189–206

83. Kamil, A. C. 1983. Optimal foraging theory and the psychology of learning. *Am. Zool.* 23:291–302

84. Kamil, A. C., Roitblat, H. L. 1985. The ecology of foraging behavior: Implications for animal learning and memory. *Ann. Rev. Psychol.* 36:141–69

85. Kamil, A. C., Sargent, T. D., eds. 1981. *Foraging Behavior: Ecological, Ethological and Psychological Approaches.* New York: Garland. 534 pp.

86. Kamil, A. C., Yoerg, S. I. 1982. Learning and foraging behavior. In *Perspectives in Ethology*, ed. P. P. G. Bateson, P. H. Klopfer, 5:325–64. New York: Plenum. 536 pp.

87. Karieva, P. 1983. Influence of vegetation texture on herbivore populations: resource concentration and herbivore movement. See Ref. 30, pp. 259–89

88. Karowe, D. N. 1989. Facultative monophagy as a consequence of prior feeding experience: behavioral and physiological specialization in individual *Colias philodice* larvae. *Oecologia* In press

89. Kennedy, J. S. 1965. Mechanisms of host plant selection. *Ann. Appl. Biol.* 56:317–22

90. Kimble, G. A. 1961. *Hilgard and Marquis' Conditioning and Learning.* New York: Appleton-Century-Crofts

91. Labeyrie, V., Fabres, G., Lachaise, D., eds. 1987. *Insects–Plants. Proc. 6th Int. Symp. Insect-Plant Relat., PAU 1986.* Dordrecht, the Netherlands: Junk. 459 pp.

92. Laverty, T. M. 1980. The flower-visiting behaviour of bumble bees: floral complexity and learning. *Can. J. Zool.* 58:1324–35

93. Lee, J. C., Bernays, E. A. 1988. Declining acceptability of a food plant for the polyphagous grasshopper *Schistocerca americana* (Drury) (Orthoptera: Acrididae): the role of food aversion learning. *Physiol. Entomol.* In press

94. Lee, J. C., Bernays, E. A., Wrubel, R. P. 1987. Does learning play a role in host location and selection by grasshoppers? See Ref. 91, pp. 125–27

95. Lewis, A. C. 1986. Memory constraints and flower choice in *Pieris rapae. Science* 232:863–65

96. Lewis, A. C., Lipani, G. A. 1988. Learning of visual flower cues in the cabbage butterfly, *Pieris rapae. Physiol. Entomol.* In press

97. Lewis, W. J., Tumlinson, J. H. 1988. Host detection by chemically mediated associative learning in a parasitic wasp. *Nature* 331:257–59

98. Lewontin, R. C. 1978. Adaptation. *Sci. Am.* 239:212–30

99. Lofdahl, K. 1984. A genetic analysis of habitat selection in the cactophilic species, *Drosophila mojavensis.* See Ref. 71, pp. 153–62

100. Lomnicki, A. 1982. Individual heterogeneity and population regulation. In *Current Problems in Sociobiology*, ed. King's Coll. Sociobiol. Group, pp. 153–67. Cambridge, UK: Cambridge Univ. Press. 400 pp.

101. Lorenz, K. Z. 1981. *The Foundations of Ethology.* New York: Springer-Verlag. 380 pp.

102. Ma, W. C. 1972. Dynamics of feeding responses in *Pieris brassicae* Linn. as a function of chemosensory input: a behavioural, ultrastructural and electrophysiological study. *Meded. Landbouwhogesch. Wageningen* 11:1–162

103. Mackintosh, N. J. 1983. General principles of learning. In *Animal Behavior. Genes, Development and Learning*, ed. T. R. Halliday, P. J. B. Slater, 3:149–77. Oxford: Blackwell. 246 pp.

104. Mackintosh, N. J. 1983. *Conditioning and Associative Learning.* Oxford: Oxford Univ. Press. 350 pp.

105. Mallet, J., Longino, J. T., Murawski, D., Murawski, A., De Gamboa, A. S. 1987. Handling effects in *Heliconius*: where do all the butterflies go? *J. Anim. Ecol.* 56:377–86

106. Mark, G. A. 1982. Induced oviposition preference, periodic environments, and demographic cycles in the bruchid beetle *Callosobruchus maculatus* Fab. *Entomol. Exp. Appl.* 32:155–60

107. Marler, P., Terrace, H. S., eds. 1984. *The Biology of Learning. Dahlem Konf. Life Sci. Res. Rep. No. 29.* Berlin: Springer-Verlag. 739 pp.

108. Maynard Smith, J. 1966. Sympatric speciation. *Am. Nat.* 100:637–50

109. Maynard Smith, J. 1987. When learning guides evolution. *Nature* 329:761–62

110. Mayr, E. 1974. Behavior programs and evolutionary strategies. *Am. Sci.* 62:650–59

111. McGuire, T. R. 1984. Learning in three species of Diptera: the blow fly, *Phormia regina*, the fruit fly, *Drosophila melanogaster*, and the house fly, *Musca domestica. Behav. Genet.* 14:479–526

112. McGuire, T. R., Tully, T. 1987. Characterization of genes involved with classical conditioning that produce differences between bidirectionally selected strains of the blow fly *Phormia regina. Behav. Genet.* 17:97–107

113. Menzel, R. 1983. Neurobiology of learning and memory: The honeybee as a model system. *Naturwissenschaften* 70: 504–11

114. Menzel, R. 1985. Learning in honeybees in an ecological and behavioral context. In *Experimental Behavioral Ecology and Sociobiology*, ed. B. Holldobler, M. Lindauer, pp. 55–74. Sunderland, Mass: Sinauer. 500 pp.

115. Menzel, R., Bicker, G., Carew, T. J., Fischbach, K.-F., Gould, J. L., et al. 1984. Biology of invertebrate learning: group report. See Ref. 107, pp. 249–70

116. Menzel, R., Bitterman, M. E. 1983. Learning by honeybees in an unnatural situation. In *Neuroethology and Behavioural Physiology*, ed. F. Huber, H. Markl, pp. 206–15. New York: Springer-Verlag. 440 pp.

117. Miller, J. R., Strickler, K. L. 1984. Finding and accepting host plants. In *Chemical Ecology of Insects*, ed. W. Bell, R. Cardé, pp. 128–57. Sunderland, Mass: Sinauer. 550 pp.

118. Miller, P. L. 1984. Alternative reproductive routines in a small fly, *Puliciphora borinquenensis* (Diptera: Phoridae). *Ecol. Entomol.* 9:293–302

119. Murdoch, W. W., Oaten, A. 1975. Predation and population stability. *Adv. Ecol. Res.* 9:1–130

120. Nelson, M. C. 1971. Classical conditioning in the blowfly. *J. Comp. Physiol. Psychol.* 77:353–68

121. Ollason, J. G. 1987. Learning to forage in a regenerating patchy environment: Can it fail to be optimal? *Theor. Popul. Biol.* 31:13–32

122. Orians, G. H. 1981. Foraging behavior and the evolution of discrimination abilities. See Ref. 85, pp. 389–405

123. Papaj, D. R. 1984. *Causes of variation in host discrimination behavior in the butterfly*, Battus philenor: PhD dissertation. Duke Univ. Durham, NC

124. Papaj, D. R. 1986. Conditioning of leaf-shape discrimination by chemical cues in the butterfly, *Battus philenor. Anim. Behav.* 34:1281–88

125. Papaj, D. R. 1986. Interpopulation differences in host preference and the evolution of learning in the butterfly *Battus philenor. Evolution* 40:518–30

126. Papaj, D. R. 1986. Shifts in foraging behavior by a *Battus philenor* population: field evidence for switching by individual butterflies. *Behav. Ecol. Sociobiol.* 19:31–39

127. Papaj, D. R. 1986. Leaf buds: A factor in host selection by *Battus philenor* butterflies. *Ecol. Entomol.* 11:301–7

128. Papaj, D. R., Opp, S. B., Prokopy, R. J., Wong, T. T. Y. 1988. Cross-induction of host fruit acceptance in medfly: the role of fruit size and chemistry. *J. Insect Behav.* In press

129. Papaj, D. R., Prokopy, R. J. 1986. Phytochemical basis of learning in *Rhagoletis pomonella* and other herbivorous insects. *J. Chem. Ecol.* 12:1125–43

130. Papaj, D. R., Prokopy, R. J. 1988. The effect of prior adult experience on components of habitat preference in the apple maggot fly *(Rhagoletis pomonella). Oecologia* In press

131. Papaj, D. R., Prokopy, R. J., McDonald, P. T., Wong, T. T. Y. 1987. Differences in learning between wild and laboratory *Ceratitis capitata* flies. *Entomol. Exp. Appl.* 45:65–72

132. Papaj, D. R., Rausher, M. D. 1983. Individual variation in host location by phytophagous insects. See Ref. 1, pp. 77–124

133. Papaj, D. R., Rausher, M. D. 1987. Genetic differences and phenotypic plasticity as causes of variation in oviposition preference in *Battus philenor. Oecologia* 74:24–30

134. Papaj, D. R., Rausher, M. D. 1987. Components of conspecific host plant discrimination by *Battus philenor* (Papilionidae). *Ecology* 68:245–53

135. Peeke, H. V. S. 1984. Habituation and the maintenance of territorial boundaries. See Ref. 28, pp. 393–421

136. Peterson, S. C. 1986. Host specificity of trail marking to foliage by eastern tent caterpillars, *Malacosoma americanum. Entomol. Exp. Appl.* 42:91–96

137. Phillips, W. M. 1977. Modification of feeding "preference" in the flea-beetle *Haltica lythri* (Coleoptera: Chrysomelidae). *Entomol. Exp. Appl.* 21:71–80

138. Pietrewicz, A. T., Johnston, T. D., eds. 1985. *Issues in the Ecological Study of Learning*. Hillsdale, NJ: Erlbaum. 451 pp.

139. Prokopy, R. J. 1986. Visual and olfactory stimulus interaction in resource finding by insects. In *Mechanisms in Insect Olfaction*, ed. T. L. Payne, M. C. Birch, C. E. J. Kennedy, pp. 81–89. Oxford: Oxford Univ. Press. 364 pp.

140. Prokopy, R. J., Averill, A. L., Cooley, S. S., Roitberg, C. A. 1982. Associative learning in egglaying site selection by apple maggot flies. *Science* 218:76–77

141. Prokopy, R. J., Cooley, S. S., Opp, S.

B. 1988. Prior experience influences fruit residence time of male apple maggot flies. *J. Insect Behav.* In press

142. Prokopy, R. J., Diehl, S. R., Cooley, S. S. 1988. Behavioral evidence for host races in *Rhagoletis pomonella* flies. *Oecologia* 76:138–47

143. Prokopy, R. J., Fletcher, B. S. 1987. The role of adult learning in the acceptance of host fruit for egglaying by the Queensland fruit fly, *Dacus tryoni*. *Entomol. Exp. Appl.* 45:259–63

144. Prokopy, R. J., Papaj, D. R. 1988. Learning of apple fruit biotypes by apple maggot flies. *J. Insect Behav.* 1:67–74

145. Prokopy, R. J., Papaj, D. R., Cooley, S. S., Kallet, C. 1986. On the nature of learning in oviposition site acceptance by apple maggot flies. *Anim. Behav.* 34:98–107

146. Pulliam, H. R., Dunford, C. 1980. *Programmed to Learn.* New York: Columbia Univ. Press

147. Rausher, M. D. 1978. Search image for leaf shape in a butterfly. *Science* 200:1071–73

148. Rausher, M. D. 1980. Host abundance, juvenile survival, and oviposition preference in *Battus philenor*. *Evolution* 34:342–55

149. Rausher, M. D. 1981. The effect of native vegetation on the susceptibility of *Aristolochia reticulata* (Aristolochiaceae) to herbivore attack. *Ecology* 65:1187–95

150. Rausher, M. D. 1983. Conditioning and genetic variation as causes of individual variation in oviposition behaviour of the tortoise beetle, *Deloyala guttata*. *Anim. Behav.* 31:743–47

151. Rausher, M. D. 1983. Alteration of oviposition behavior by *Battus philenor* butterflies in response to variation in host-plant density. *Ecology* 64:1028–34

152. Rausher, M. D. 1986. Variability for host preference in insect populations: mechanistic and evolutionary models. *J. Insect Physiol.* 31:873–99

153. Rausher, M. D., Odendaal, F. J. 1987. Switching and the pattern of host use by *Battus philenor* butterflies. *Ecology* 68:869–77

154. Rescorla, R. A. 1988. Pavlovian conditioning: it's not what you think it is. *Am. Psychol.* 3:151–60

155. Roitberg, B. D., Mangel, M. 1989. Dynamic information and host acceptance by a tephritid fruit fly. *Ecol. Entomol.* In press

156. Roitberg, B. D., Prokopy, R. J. 1981. Experience required for pheromone recognition by the apple maggot fly. *Nature* 292:540–41

157. Rosenheim, J. A. 1987. Host location and exploitation by the cleptoparasitic wasp *Argochrysis armilla:* the role of learning (Hymenoptera: Chrysididae). *Behav. Ecol. Sociobiol.* 21:401–6

158. Saxena, K. N., Schoonhoven, L. M. 1982. Induction of orientational and feeding preference in *Manduca sexta* larvae for different food sources. *Entomol. Exp. Appl.* 32:173–80

159. Schoonhoven, L. M. 1969. Sensitivity changes in some insect chemoreceptors and their effects on food selection behavior. *Proc. K. Ned. Akad. Wet. Ser. C* 72:491–98

160. Schoonhoven, L. M., Meerman, J. 1978. Metabolic cost of changes in diet and neutralization of allelochemics. *Entomol. Exp. Appl.* 24:689–93

161. Scriber, J. M. 1981. Sequential diets, metabolic cost, and growth of *Spodoptera eridanea* feeding upon dill, lima bean and cabbage. *Oecologia* 51:175–80

162. Scriber, J. M. 1982. The behavior and nutritional physiology of southern armyworm larvae as a function of plant species consumed in earlier instars. *Entomol. Exp. Appl.* 31:359–69

163. Shaw, R. E., Ailey, T. R. 1985. How to draw learning curves: Their use and justification. See Ref. 138, pp. 275–304

164. Shettleworth, S. J. 1984. Learning and behavioural ecology. In *Behavioural Ecology. An Evolutionary Approach.* ed. J. R. Krebs, N. B. Davies, pp. 170–94. Sunderland, Mass: Sinauer. 493 pp.

165. Sibly, R. M., Smith, R. H., eds. 1985. *Behavioural Ecology: Ecological Consequences of Adaptive Behaviour. 25th Symp. Br. Ecol. Soc.* Oxford: Blackwell Sci.

166. Singer, M. C. 1982. Quantification of host preference by manipulation of oviposition behavior in the butterfly *Euphydryas editha*. *Oecologia* 52:230–35

167. Singer, M. C. 1984. Butterfly-hostplant relationships: host quality, adult choice and larval success. *Symp. R. Entomol. Soc. London* 11:81–88

168. Singer, M. C. 1986. The definition and measurement of oviposition preference in plant-feeding insects. In *Insect-Plant Interactions*, ed. T. A. Miller, J. Miller, pp. 65–94. New York: Springer-Verlag. 342 pp.

169. Smith, M. A., Cornell, H. V. 1979. Hopkins host-selection in *Nasonia vitripennis* and its implications for sympatric speciation. *Anim. Behav.* 27:365–70

170. Soliman, M. H., Hay, D. A. 1978. Interaction of genotype and learning in the food preference of the flour beetle, *Tri-*

*bolium castaneum. Experientia* 34:329–31

171. Staddon, J. E. R. 1983. *Adaptive Behavior and Learning.* Cambridge, UK: Cambridge Univ. Press. 640 pp.

172. Stanton, M. 1983. Spatial patterns in the plant community and their effects upon insect search. See Ref. 1, pp. 125–52

173. Stanton, M. L. 1984. Short-term learning and the searching accuracy of egg-laying butterflies. *Anim. Behav.* 32:33–40

174. Stephens, D. W. 1987. On economically tracking a variable environment. *Theor. Popul. Biol.* 32:15–25

175. Swihart, C. A., Swihart, S. L. 1970. Colour selection and learned feeding preferences in the butterfly, *Heliconius charitonius* Linn. *Anim. Behav.* 18:60–64

176. Szentesi, A., Bernays, E. A. 1984. A study of behavioral habituation to a feeding deterrent in nymphs of *Schistocerca gregaria. Physiol. Entomol.* 9:329–40

177. Taylor, C. E. 1986. Habitat choice by *Drosophila pseudoobscura:* the roles of genotype and experience. *Behav. Genet.* 16:271–79

178. Taylor, R. J. 1974. Role of learning in insect parasitism. *Ecol. Monogr.* 44:89–104

179. Thompson, R. F., Spencer, W. A. 1966. Habituation: a model phenomenon for the study of neuronal substrates of behaviour. *Psychol. Rev.* 73:16–43

180. Thorpe, W. H. 1963. *Learning and Instinct in Animals.* London: Methuen. 558 pp.

181. Tierney, A. J. 1986. The evolution of learned and innate behavior: contributions from genetics and neurobiology to a theory of behavioral evolution. *Anim. Learn. Behav.* 14:339–48

182. Tinbergen, N. 1951. *The Study of Instinct.* London: Oxford Univ. Press. 228 pp.

183. Tolman, E. C., Brunswick, E. 1935. The organism and the causal texture of the environment. *Psychol. Rev.* 42:43–47

184. Tompkins, P., Bird, C. 1973. *The Secret Life of Plants.* New York: Harper & Row. 402 pp.

185. Traynier, R. M. M. 1979. Long-term changes in the oviposition behavior of the cabbage butterfly *Pieris rapae* induced by contact with plants. *Physiol. Entomol.* 4:87–96

186. Traynier, R. M. M. 1984. Associative learning in the ovipositional behavior of the cabbage butterfly, *Pieris rapae. Entomol. Exp. Appl.* 9:465–72

187. Traynier, R. M. M. 1986. Visual learning in assays of sinigrin solution as an oviposition releaser for the cabbage butterfly, *Pieris rapae. Entomol. Exp. Appl.* 40:25–33

188. Traynier, R. M. M. 1987. Learning without neurosis in host finding and oviposition by the cabbage butterfly, *Pieris rapae.* See Ref. 91, pp. 243–47

189. Tully, T. 1984. *Drosophila* learning: behavior and biochemistry. *Behav. Genet.* 14:527–57

190. Turner, J. R. G. 1971. Experiments on the demography of tropical butterflies. II. Longevity and home-range behavior in *Heliconius erato. Biotropica* 3:21–31

191. van Alphen, J. J. M., van Harsel, H. H. 1982. Host selection by *Asobara tabida* Nees (Braconidae; Alysiinae), a larval parasitoid of fruit inhabiting *Drosophila* species. III. Host species selection and functional response. In *Foraging behaviour of* Asobara tabida, *a larval parasitoid of Drosophilidae,* ed. J. J. M. van Alphen, pp. 61–93. PhD thesis, Univ. Leiden, the Netherlands. 165 pp.

192. van Alphen, J. J. M., Vet, L. E. M. 1986. An evolutionary approach to host finding and selection. *Symp. R. Entomol. Soc. London* 13:23–61

193. Vet, L. E. M., van Opzeeland, K. 1984. The influence of conditioning on olfactory microhabitat and host location in *Asobara tabida* (Nees) and *A. rufescens* (Foerster) (Braconidae: Alysiinae) larval parasitoids of Drosophilidae. *Oecologia* 63:171–77

194. Via, S. 1984. Quantitative genetic analysis of feeding and oviposition behavior in the polyphagous leafminer *Liriomyza sativae.* See Ref. 71, pp. 185–96

195. Via, S. 1986. Genetic covariance between oviposition preference and larval performance in an insect herbivore. *Evolution* 40:778–85

196. Visser, J. H., Minks, A. K., eds. 1982. *Proc. 5th Int. Symp. Insect-Plant Relat., Wageningen.* Wageningen, the Netherlands: Pudoc. 464 pp.

197. Visser, J. H., Thiery, D. 1986. Effects of feeding experience on odour-conditioned anemotaxes of Colorado potato beetle. *Entomol. Exp. Appl.* 42:198–200

198. Waage, J. K. 1978. Arrestment responses of the parasitoid *Nemeritis canescens* to a contact chemical produced by its host, *Plodia interpunctella. Physiol. Entomol.* 3:339–51

199. Waage, J. K. 1979. Foraging for patchily distributed hosts by the parasitoid *Nemeritis canescens. J. Anim. Ecol.* 48:353–71

200. Waldbauer, G. P., Fraenkel, G. 1961. Feeding on normally rejected plants by maxillectomized larvae of the tobacco hornworm, *Protoparce sexta* (Johan.). *Ann. Entomol. Soc. Am.* 54:477–85
201. Wardle, A. R., Borden, J. H. 1985. Age-dependent associative learning by *Exeristes roborator* (F.) (Hymenoptera: Ichneumonidae). *Can. Entomol.* 117: 605–16
202. Waser, N. M. 1986. Flower constancy: definition, cause, and measurement. *Am. Nat.* 127:593–603
203. Wasserman, S. S. 1986. Genetic variation in adaptation to foodplants among populations of the southern cowpea weevil, *Callosobruchus maculatus:* evolution of oviposition preference. *Entomol. Exp. Appl.* 42:201–12
204. Wcislo, W. T. 1987. The role of learning in the mating biology of a sweat bee *Lasioglossum zephyrum* (Hymenoptera: Halictidae). *Behav. Ecol. Sociobiol.* 20:179–85
205. Williams, G. C. 1966. *Adaptation and Natural Selection.* Princeton, NJ: Princeton Univ. Press. 307 pp.
206. Yamamoto, R. T. 1974. Induction of host plant specificity in the tobacco hornworm, *Manduca sexta. J. Insect Physiol.* 20:641–50
207. Zahorik, D., Houpt, K. A. 1981. Species differences in feeding strategies, food hazards, and the ability to learn food aversions. See Ref. 85, pp. 289–311

*Ann. Rev. Entomol. 1989. 34:351–72*

# EXPRESSION OF FOREIGN GENES IN INSECTS USING BACULOVIRUS VECTORS

*Susumu Maeda*

Department of Entomology, University of California, Davis, California 95616

## PERSPECTIVES AND OVERVIEW

Recent advances in biotechnology have made it possible to express foreign genes in heterologous living organisms. However, expressions of foreign genes are limited and are still far from satisfactory in most cases. The goals of current work can be divided into two categories: (*a*) to develop methods for efficient production of large amounts of purified, fully active proteins, e.g. for use as pharmaceuticals, and (*b*) to study the basic mechanisms of gene expression and the biological effects of the expressed products in cells or organisms.

*Escherichia coli,* together with plasmid or phage vectors, has been widely used for the expression of foreign genes encoding polypeptides with simple structures. Complex polypeptides, however, are sometimes difficult to produce in biologically active forms, and if produced, they are also difficult to refold to the natural form. Furthermore, posttranslational modifications, such as glycosylation, phosphorylation, and C-terminus amidation, do not occur properly in prokaryotic systems owing to the lack of the necessary enzymatic machinery. To overcome these problems, many hosts, especially higher eukaryotic organisms with appropriate vectors, have been explored as alternatives.

In studies of gene expression in living organisms, especially the multicellular higher organisms, two different types of experimental strategies have been used to introduce foreign genes. One approach involves the stable integration of the foreign genetic material into chromosomal DNA; the second requires only the physical transportation of the foreign DNA into the cell

0066-4170/89/0101-0351$02.00

nucleus, which gives transient expression. Both approaches have advantages and disadvantages in the length of time needed to obtain the expected transformant and the efficiency of introduction. Although with both approaches it is also possible to introduce DNA fragments directly into cells, e.g. by electroporation or injection, the efficiency of transformation is not high. On the other hand, viruses that are extraordinarily efficient in transporting their own genome into susceptible cells in nature have been utilized as vectors. Examples of such viral vectors are polyoma virus (23), bovine papilloma virus (45), adenoviruses (66), herpesviruses (97), vaccinia virus (73), and retroviruses (55). In recent years insect baculoviruses have been commonly recognized as very efficient vectors, and they receive the most attention in this review.

The fruit fly *Drosophila melanogaster* has a unique and efficient gene integration system using the P element; such an efficient system is exceptional in insects (79). *Drosophila melanogaster* is also one of the most thoroughly studied animals with regard to gene expression, including differentiation. There has been one report of successful integration of a foreign gene into the genome of the mosquito *Anopheles gambiae* (61), but no other system has yet been established in other insects. Many types of viruses infect insects, and one of them, the baculovirus, has recently become an extremely useful vector in the successful expression of foreign genes at high levels to produce biologically active products. Both the *Autographa californica* nuclear polyhedrosis virus (AcNPV) (74, 86) and *Bombyx mori* NPV (BmNPV) (52, 53) have been used for this purpose. The BmNPV vector has an additional advantage of having a natural host, the silkworm *B. mori,* that can be used for in vivo expression (53). This in vivo expression system has several attractive features: First, the advanced system for mass culture of silkworms, developed by continuous improvements in sericulture for silk production, provides an abundant supply of insect hosts. Second, studies of foreign gene expression will provide basic information on the biochemistry and physiology of the silkworm, one of the most thoroughly studied lepidopterans (25). Third, the baculovirus expression system will provide additional information that can be applied in the mass production of useful polypeptides, such as pharmaceuticals, and active viral insecticides.

This review is primarily concerned with the expression of foreign genes in insect cells, especially in silkworm larvae, using BmNPV vectors. Applications of baculoviruses to production of pharmaceuticals and genetically engineered viral insecticides are also discussed.

## BACULOVIRUSES AS VECTORS

Insect-pathogenic viruses are classified into seven families (60). The most common group is the family Baculoviridae, which is characterized by a circular double-stranded DNA genome and rod-shaped enveloped virion. The Baculoviridae are divided into three subgroups: nuclear polyhedrosis viruses (NPV), granulosis viruses, and nonoccluded-type viruses. NPVs are found in several orders of insects, mainly lepidopterans, and have the unique property of producing proteinacious nuclear occlusion (inclusion) bodies in which progeny virions are embedded at a late stage of infection. Two other insect virus groups, the cytoplasmic polyhedrosis viruses and insect poxviruses, also produce viral occlusion bodies. The proteinacious occlusion bodies are believed to have an important role in the horizontal transmission of these insect viruses.

## Life Cycle of Baculoviruses

After the insect host dies from NPV infection, occlusion bodies containing mature virions are released from the disintegrating larval remains into the environment and are horizontally transmitted to other larvae. The proteinacious coating protects progeny virions from inactivating agents in the environment. When the occlusion bodies are ingested by the host larva, they are dissolved by the gut juices. The enveloped virions are released and invade the midgut cells by fusion. Secondary infection is caused by virions budded from the initially infected cells. At a late stage of infection, numerous occlusion bodies containing mature virions are produced in the infected cell nuclei (18).

## Structure of Baculovirus Genomes

Studies on baculovirus replication advanced rapidly after the establishment of an in vitro replication system using AcNPV and susceptible established cell lines. Most knowledge at the molecular level has been obtained with AcNPV, the type baculovirus, but other baculoviruses and cell lines have added further information (18).

The genome of baculoviruses consists of 90–160 kb of closed circular double-stranded DNA (60). The genome is believed to contain more than 100 different genes. Sequencing and hybridization experiments reveal that it consists largely of unique sequences that may encode structural or nonstructural polypeptides. Coding sequences are localized tightly in the genome and are spaced with short sequences of a junction containing a proposed promoter and poly(A) signal(s) (5, 14, 70). Four or five dispersed repeated sequences (1, 41), which have enhancing activity to downstream promoters (21), have been found in the viral genome. The arrangement of the genes is conserved throughout the various NPVs, even among NPVs having relatively low DNA sequence homology (43).

Very little is known about the functions of most of the viral genes. Two late genes, polyhedrin (24, 78) and p10 (42, 44), have been mapped and sequenced. Polyhedrin genes, which have an important role in the transmission of the virus in nature, are well conserved in many NPVs (78). The gene encoding an arginine-rich probable core protein has also been identified and sequenced (98). In other cases, open reading frames in NPVs have been identified by sequencing (5, 14, 70). However, the expression and function of the gene product have not yet been studied.

Expression of viral genes is sequentially controlled in a characteristic manner during the baculovirus infection process. The signals for initiation and termination of transcription change during the viral replication process; this variation is believed to be a strategy for the temporal regulation of baculovirus gene expression (13, 47, 54, 70, 76). An immediate early expressed gene, which *trans*-activates other promoters in the viral genome, has been identified (22). The activation of the late gene promoter is considered to be under the control of several early genes; however, there has been a report that the late polyhedrin gene promoter is activated in mammalian BHK cells infected with adenovirus (37). Although splicing is a common phenomenon in the expression of eukaryotic genes, including the viral DNA genome in its host, there is no evidence to indicate the existence of introns in the genes of the baculovirus genome (46).

## Characteristics of Baculoviruses as Vectors

As previously discussed by others (62, 84), baculoviruses possess several unique characteristics that are extremely advantageous for a useful expression vector. The virus has (a) a circular double-stranded genome, which is easily modified by commonly used biotechnological techniques such as specific cleavage and ligation; (b) a rod-shaped capsid, which allows the virus to contain extra DNA fragments; (c) a cell line that is susceptible to viral infection; and (d) the polyhedrin gene. This gene is suitable for the insertion of foreign genes because it is nonessential for viral reproduction, it is a late gene with a strong promoter, and it is a marker, detected easily by light microscopy. Furthermore, with regard to the safety of recombinant DNA experiments, NPVs lacking the polyhedrin gene have the advantage of being easily inactivated in the field. Another advantageous characteristic of the BmNPV expression system is that it can be easily expanded to an in vivo system using silkworm larvae, as discussed earlier.

## Construction of Recombinant Viruses

Since the baculovirus genome is very large (about 130 kb for BmNPV), recombinant viruses containing foreign genes can be obtained only by recombination in host cells cotransfected with the wild-type viral DNA and a recombinant transfer plasmid vector containing the foreign gene. Six different transfer vectors for general expression and 13 different transfer vectors for the expression of fused genes are available for the construction of recombinant BmNPV (51). These vectors generally possess an appropriate length (about 3 kb) of each of the 5' and 3' fragments for homologous recombination, a

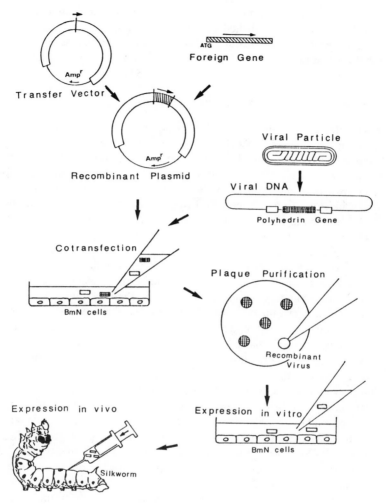

*Figure 1*  The strategy for the insertion of a foreign gene into BmNPV.

linker between these fragments for the insertion of foreign genes, and an antibiotic (ampicillin) resistance gene and a replication origin from a bacterial plasmid. Details of the construction of the transfer vectors have been described elsewhere (51, 53). Similar transfer vectors derived from AcNPV have been constructed (48).

The strategy used for the insertion of a foreign gene into BmNPV is shown in Figure 1 (53). A foreign gene is introduced to a transfer vector to produce recombinant plasmids that contain the 3' and 5' flanking region of the polyhedrin gene, including the promoter. Recombinant viruses are produced by the calcium-mediated cotransfection of the recombinant plasmid and wild-type viral DNA in an established cell line of *B. mori*. Recombinant viruses that do not produce polyhedral occlusion bodies are isolated by plaque purification. Foreign genes are expressed either in the established cell line or in an in vivo system with the silkworm larva (53).

## Characteristics of the Silkworm as a Host

We have selected the silkworm as a host for the mass production of useful foreign gene products (53) because it has many advantageous characteristics as discussed earlier. Humans have used the silkworm for silk production for several thousand years. Its normal development features extraordinarily rapid, efficient growth. It increases its body weight $10^4$-fold within 20 days, from 0.5 mg at hatching to 5 g at the fifth instar, by eating only 20 g of mulberry leaves. Although the body weight is much less than that of many vertebrates, it is nonetheless $10^3$ times that of *D. melanogaster*. Because the efficiency of silkworm protein synthesis is extremely high during the larval stage, several milligrams of a specific protein can be synthesized within a day (50).

An insect larva can be considered an extremely sophisticated factory for the production and storage of polypeptides, which change dramatically during metamorphosis. The thin but strong cuticle layer surrounds a sterile space filled with hemolymph and organs. Oxygen is supplied by direct gas exchange through tracheal tubes attached to all tissues, and nourishment is supplied through the midgut epithelium. Storage proteins are secreted from the fat body and stored in the hemolymph before use (91).

With respect to safety considerations in recombinant DNA experiments, the silkworm is a prime host because it lacks the ability to survive in the field. The silkworm adult cannot fly, and the larvae cannot migrate far to find new food sources. The silkworm has been used in studies of genetics, biochemistry, and physiology as a model of the Lepidoptera. If the BmNPV can introduce the genes of interest into the silkworm, much more knowledge can be accumulated in these areas.

The silkworm is easily mass-cultured at low cost, even under sterile conditions, by an automatic feeding machine and advanced artificial diets (32). It is now possible to obtain essentially unlimited quantities of silkworm larvae.

# EXPRESSION OF FOREIGN GENES IN ESTABLISHED CELL LINES AND IN LARVAE OF THE SILKWORM

Many recombinant baculoviruses have been constructed and have successfully expressed different kinds of inserted foreign genes derived from a variety of sources, from bacteria to human beings. With the BmNPV vector, the genes encoding the following proteins have been expressed: chloramphenicol acetyltransferase (S. Maeda, unpublished results), human alpha interferon (IFN-$\alpha$) (26, 50, 52, 53), insulin-like growth factor-II (IGF-II) (57), human interleukin 3 (A. Miyajima & S. Maeda, unpublished results), mouse interleukin 3 (64), human and mouse interleukin 4 (A. Miyajima & S. Maeda, unpublished results), human T-lymphotropic virus (HTLV-1) p40$^x$ (69, 100), HTLV-1 *env* (H. Nyunoya & K. Shimotono, personal communication), human immunodeficiency virus (HIV-1) *gag, pol, sor,* gp41, and gp120 (N. Kobayashi & S. Maeda, unpublished results), adenovirus E1a (H. Handa, personal communication), Japanese encephalitis virus *env* (N) (Y. Aira, A. Igarashi & S. Maeda, unpublished results), bovine papillomavirus 1 (BPV1) E2 (90), HPV6b E2 (82, 90), BPV1 E6 (A. Fuse, S. Maeda, & B. Simizu, unpublished results), and human apolipoproteins A and E (M. Iwano, personal communication). The AcNPV vector (reviewed in 48, 63a) has produced chloramphenicol acetyltransferase (9), $\beta$-galactosidase (74, 95), hepatitis B surface antigen (34a), HIV-1 *env* (28, 80), HIV-1 *gag* (49), HTLV-1 p40$^x$ (33), human IFN-$\beta$ (86), human interleukin 2 (85), c-*myc* (65), *D. melanogaster* Kruppel gene product (71), bluetongue virus VP2 (31) and VP3 (30), human parainfluenza virus hemagglutinin (HA) (94), influenza polymerases PA, PB1, and PB2 (87), influenza virus HA (40, 75), lymphocytic choriomeningitis virus (LCMV) GPC and N proteins (58, 59), *Neurospora crassa* activator protein (3), polyomavirus T antigen (77), simian virus 40 (SV40) small t antigen (33), SV40 large T antigen (71a), Punta Toro phlebovirus N and Ns proteins (72), simian rotavirus VP6 (12), CD4 (T4) (29), human erythropoietin (99), Hantaan virus structural protein (81), human epidermal growth factor (EGF) receptor (19), human insulin receptor (10a, 23a), human B lymphotrophic virus 130-kd protein (14a), hepatitis A virus VP1 (22a), human tyrosine hydroxylase (15a), human glucocerebrosidase (56a), and mouse p53 (71a).

In addition to the polyhedrin gene, the p10 genes of baculoviruses have also been replaced by foreign genes [encoding neomycin (17) and $\beta$-galactosidase (95)], and the inserted genes have been expressed by the recombinant viruses.

## Characteristics of Recombinant Viruses

Since the polyhedrin gene is not required for the production of progeny virus, the recombinant viruses are expected to replicate like the wild-type virus in established cell lines. The recombinant viruses contain foreign genes at the site formerly occupied by the polyhedrin gene, and they replicate normally without the help of the wild-type virus. After several passages in vitro no significant changes had occurred in the patterns of viral DNA and the inserted genes. Released nonoccluded wild-type and recombinant viruses did not differ in heat stability (50).

The most reliable silkworm infection procedure at present is the inoculation of the virions into the hemocoel, since no polyhedra are produced for use in per os infection (53). Symptoms of recombinant virus infections are the same as those of wild-type virus infections except that the hemolymph does not turn whitish because of the absence of released polyhedra.

## Efficiency of Production

Efficiency of expression is the most important factor in obtaining foreign gene products. The original viral products, i.e. the polyhedral proteins of BmNPV or AcNPV, generally make up almost 20% of the total protein in infected cells. This proportion subsequently increases to more than 50% after the degradation of other proteins at a very late stage of infection. About 20 mg of polyhedrin is produced in one fifth-instar larva of the silkworm. In contrast, the quantities of foreign gene products are relatively low, usually less than a few percent of the original polyhedrin. The reasons for the reduced expression of foreign genes are complex; several different factors appear to be involved.

One factor affecting the efficiency of the production of foreign gene products is the sequence between the promoter and the translation start. A 71-base sequence is highly conserved in AcNPV and BmNPV (53). Its importance was indicated in deletion experiments with BmNPV in which the removal of more than 20 bases upstream from the translational start in this sequence resulted in dramatic reduction of the expression of the foreign gene for human IFN-$\alpha$, both in the established cell line and in the silkworm larvae (26). Similar results were obtained when the corresponding conserved sequence of the recombinant AcNPV was removed (59).

The efficiency of expression of foreign gene products seems to depend primarily on the characteristics of the foreign protein. On average, nuclear localized or nonstructural proteins are most highly expressed, secreted proteins are intermediate, and enveloped proteins are least expressed. This dependence on the type of gene product may explain the importance of the N-terminal sequence of the polyhedrin for high-level expression of fused foreign genes. If the foreign gene, e.g. p40$^x$ of HTLV-1, is inserted at a site downstream from the translational start of the polyhedrin gene so as to produce a fusion protein (containing the N-terminus of the polyhedrin), the fused gene is highly expressed compared to a foreign gene starting from the original translational start (69). However, this effect does not increase the production rate of enveloped proteins, even though some of the mature genes are highly expressed (S. Maeda, unpublished results). In most cases the expression rate in established cell lines depends strongly on the details of the construction of the vectors and the characteristics of the gene products, but not on the species of NPV, i.e. BmNPV or AcNPV.

The site of gene activity in cells in vitro or in vivo is of interest because little is known of the expression of foreign genes in the higher eukaryotic organisms. It is possible, although not easy, to compare the production of foreign gene products in the silkworm and in the established cell lines, e.g. by determining the concentration of foreign gene products in the hemolymph or culture medium. By such estimation, expression of some proteins, such as interleukin 3 (64), is extremely high in silkworm larvae, more than 500-fold that achieved in the established cell line (almost 1 mg of this polypeptide is synthesized in one silkworm larva). In the case of human IFN-$\alpha$, 20 times more (about 50 $\mu$g) is produced in the silkworm (53). Such high levels of expression indicate that these polypeptides are well suited for expression in the silkworm organs.

## Fusion Proteins

As described earlier, polyhedrin fusion genes are, in many cases, expressed at higher levels than the original mature genes. For example, fusion proteins of p40$^x$ of HTLV-1 (69) or of E2 of papilloma viruses (82, 90) produced in cells of *B. mori* appear, like the polyhedral protein, as abundant bands on polyacrylamide gels of cell homogenates. If antibodies directed against the foreign gene products are not available, the fusion gene products can be identified with antibody against polyhedrin. For example, Tada et al (90) constructed a fusion gene containing 156 bp derived from the polyhedrin gene linked to 1230 bp from the E2 gene of BPV1. The polyhedrin-fused E2 protein is able to induce specific antibody against E2 (82). This antibody has been found useful for the diagnosis of viral infection and for detecting and localizing the expressed and accumulated products in infected tissues (82).

The silkworm also produces a large amount of E2 fused protein; however, there is little difference between the production rate in vitro and in vivo because even in the established cell line the protein produced accounts for up to 10%, i.e. close to the maximum, of the total cellular protein (90). The polyhedrin-fused p40$^x$ protein of HTLV-1 is highly expressed; however, the enhancing activity is quite low compared to that of the mature gene product prepared in the same BmNPV expression system (H. Nyunoya & K. Shimoto-no, personal communication).

The copiously produced polyhedrin-fused IGF-II protein is useful for mass production of small peptides. The mature IGF-II peptide is cleaved with CNBr, and then the biologically active peptide is purified by column chromatography (57).

## Cleavage of Signal Peptides and Secretion of Polypeptides

Proteins secreted from cells are synthesized as precursor molecules containing hydrophobic N-terminal signal peptides. The signal peptides are then cleaved by a peptidase on the membrane of the endoplasmic reticulum when the proteins pass through the membrane (96).

To examine whether signal peptides of mammalian proteins are recognized and cleaved at the correct sites, the mammalian genes encoding secreted proteins have been introduced into BmNPV. The N-terminal amino acid sequences of the secreted polypeptides human IFN-$\alpha$ (53) and mouse interleukin 3 (64) synthesized by cells of *B. mori* were identical with those produced in the original mammalian cells. When silkworms were infected with the recombinant viruses, all foreign gene products were secreted into the hemolymph. The N-termini of these molecules were identical with those of mature polypeptides cleaved in mammalian cell lines (53, 64). The results show that the signal sequences for the secretion of mammalian genes are recognized in the silkworm and are cleaved at the correct sites. These results also indicate that the mechanisms involved in the recognition and cleavage signal sequences are conserved throughout many different species in the animal kingdom.

To confirm the importance of the signal peptides for secretion, a recombinant virus carrying the human IFN-$\alpha$ gene without the signal sequence has been constructed. The translational start codon is inserted before the mature interferon sequence, and the sequence corresponding to the signal peptide is deleted. When the virus infects cells of *B. mori,* most of the interferon activity is found within the cell, and little is secreted. Furthermore, the total activity is quite low. Much of the interferon produced in the cytoplasm is probably degraded by endogenous proteinases (52).

Using AcNPV as a vector, mammalian genes for human interleukin 2 (85), human glucocerebrosidase (56a), and CD4 (30) have been expressed in cells of *Spodoptera frugiperda.* The N-terminal amino acid sequences of these polypeptides were identical with those produced in the original mammalian system.

## Glycosylation

Glycosylation is a posttranslational modification found abundantly in membrane-bound and secreted proteins in eukaryotes. Although many proteins are glycosylated, the significance of the oligosaccharide for protein structure and function is in most cases still unclear. Some evidence suggests that some of the mechanisms of glycosylation in insects differ from those in mammals (39). Glycosylation in mosquito cells is limited; conversion to complex N-linked oligosaccharides does not occur, and the structure of the oligosaccharides is simple (27). A charged sugar residue that is usually found at the end of oligosaccharides in mammalian cells, sialic acid, is absent in mosquito cells (8). Experiments aimed at producing glycosylated proteins in heterologous host cells may help to clarify the mechanisms and functions of glycosylation.

Several genes for mammalian glycoproteins have been expressed using the

BmNPV expression vector, both in cultured insect cells and in silkworm larvae. Mouse interleukin 3 was expressed in silkworm larvae and purified by ion exchange and reverse-phase chromatography (64). In silkworm larvae three distinct species of interleukin 3 were produced, of 22, 20, and 18 kd. The silkworm-produced interleukin 3 polypeptides were all much smaller than those produced in mammals. These polypeptides were fractionated by a second round of reverse-phase high-pressure liquid chromatography (HPLC). Since N-terminal sequencing indicated that the three glycoproteins were identical in polypeptide sequence, the variation in molecular weights appeared to result from variation in glycosylation. This proposal was confirmed in $N$-glycanase digestion experiments, which converted all species to the same $\sim$ 15-kd core peptide. Thus, interleukin 3 produced in the silkworm is variably glycosylated and is glycosylated to a much lesser extent than in mammals. The recombinant and mammalian interleukin 3 exhibited equal binding affinity and equal potency in cell proliferation and colony formation assays. The same heterogeneity of produced interleukin 3 was observed when the established cell line was used (64).

When the glycoprotein gp41 of HIV-1 was expressed by the BmNPV vector, the single gp41 polypeptide was synthesized and fully glycosylated in both the established cell line and the silkworm, so the molecular weight was similar to that of the polypeptide produced in human cells. When another glycoprotein of HIV-1, gp120, was expressed by the BmNPV vector, it was not glycosylated at all, either in the silkworm or in the established cell line (N. Kobayashi & S. Maeda, unpublished results). As these polypeptides are expressed as fusion proteins, it is very difficult to conclude whether the glycosylation of these peptides occurs normally in insect cells or not. Another possible cause of the difference in glycosylation may be inappropriate localization of produced fused polypeptides resulting from an unknown change in their sorting characteristics within the cell.

The AcNPV vector has produced the following $N$-glycosylated polypeptides: human IFN-$\beta$ (86), human parainfluenza virus HA (94), influenza virus HA (40, 75), LCMV GPC protein (58), Punta Toro phlebovirus N and Ns proteins (72), human erythropoietin (99), human EGF receptor (19), and human glucocerebrosidase (56a). These polypeptides, like those obtained using the BmNPV vector, are generally less glycosylated than those produced by mammalian cells.

An interesting result has shown that oligosaccharides with lower molecular weights produced in insect cells have the same role in biological activity as their counterparts in mammalian cells. When N-linked oligosaccharides of human erythropoietin produced in insect cells were enzymatically removed by $N$-glycanase, the biological activity, i.e. differentiation of red blood cells, was reduced (99). Thus, in this case where glycosylation was required for functional activity, the incompletely glycosylated oligosaccharide produced in insect cells exhibited full biological activity.

When interleukin 2 was expressed in cells of *S. frugiperda* by the AcNPV vector, O-linked glycosylation did not occur (85).

## Phosphorylation

When the phosphorylated proteins c-*myc* (65), p40[x] (33), EGF receptor (19), insulin receptor (10a, 23a), SV40 large T antigen (71a), mouse p53 (71a), and *D. melanogaster* Kruppel gene product (71) were expressed in cells of *S. frugiperda* using AcNPV vectors, the proteins were also found to be phosphorylated.

## C-Terminus Amidation

C-terminus amidation, another common posttranslational modification, is important in protecting against digestion by carboxypeptidase and also provides biological activity. Cecropin, a bactericidal peptide that is secreted from the fat body in *Hyalophora cecropia* in response to bacterial infection, is amidated at its C-terminus (88). Recent cDNA cloning experiments showed that glycine is involved in this modification, as suggested by many examples of amidation of mammalian peptides (93). Furthermore, *B. mori* and other insects also produce similar peptides whose C-termini are amidated (6). These results indicate that the insect fat body contains an adequate amount of the enzyme involved in C-terminus amidation. Since little amidation activity has been found in established cell lines (16), the silkworm may be the preferred system for the production of these peptides. We have expressed an amidated peptide gene using the BmNPV vector. The peptide expressed in the silkworm is highly active in functional assays and therefore seems to be amidated (S. Maeda, unpublished results).

## Internal Cleavage of Polypeptides

Proteins and peptides are sometimes produced as large precursors and become biologically active forms by cleavage after transportation to an appropriate location, i.e. a targeted cellular organelle or tissue. The polypeptides are cleaved either by their own proteinase activity (autonomous cleavage) or by endogenous proteinases found in the target cells. There has been a report of successful cleavage of the HIV-1 *gag-pol* product expressed in cells of *S. frugiperda* by the AcNPV vector. Comparison of cleavage of polypeptides produced by recombinant viruses that contained different areas in the *gag-pol* genome indicated the possible involvement of cellular proteinase rather than autonomous cleavage (49).

When influenza hemagglutinins and the envelope protein of HIV-1 were expressed in cells of *S. frugiperda* by the AcNPV vector, no cleavage of these precursors was detected on SDS gels (40, 75, 80). The absence of cleavage indicates the lack of appropriate specific endogenous proteinase in insect cells. It has not been tested whether the silkworm possesses this specific proteinase or not. With this in mind, we have initially cleaved the genes encoding gp120 and gp41 from HIV-1 and introduced them separately into the BmNPV. Both proteins were successfully expressed in the established cell line and in silkworm larvae (N. Kobayashi & S. Maeda, unpublished results).

## Biological Activity

In the *E. coli* expression system, some proteins are synthesized at high levels, but in an insoluble and biologically inactive form. In the baculovirus expression system, all the expressed polypeptides possess biological properties such as DNA-binding, *trans*-activating, hormonal, immunogenetic, and immunoreactive activities equal per amount of protein to those of the original products (48, 63a). Proteins produced in the silkworm using BmNPV vectors are also all biologically active (26, 53, 64, 82, 90).

## Lipid Uptake

Lipids are essential for living organisms and are major components of the membranous structures of cell walls and intracellular organelles. Lipid transportation is conducted in a special manner, since the organs of living organisms are composed mainly of water-soluble compounds. The mechanisms of lipid transportation in insects differ from those of mammals (83). The expression experiment is a unique system for investigating the polypeptides involved in lipid transportation.

Apolipoproteins in humans carry lipids in the blood vessels. When a BmNPV with inserted human apolipoprotein A or E genes infected the silkworm, the apolipoproteins were produced in large quantities. The preliminary experiments showed that the produced proteins contained lipids. The human apolipoproteins were found in the hemolymph in a complex formed with the original lipoproteins produced in the silkworm (M. Iwano, personal communication). Accordingly, the expression system can be used to obtain basic information on the mechanism and function of lipoproteins.

## Localization of Produced Proteins

Proteins have been shown to contain signals that ensure their direction to appropriate target organelles in the eukaryotic cell (15). Foreign gene products produced using baculovirus vectors can be appropriately targeted within the heterologous insect cell. For example, influenza HA (40, 75), LCMV glycoprotein GPC (59), and Punta Toro phlebovirus N and Ns proteins (72) are directed to the cell surface; p40$^x$ (33), *D. melanogaster* Kruppel gene product (71), SV40 large T antigen (71a), mouse p53 (71a), and c-*myc* (65) are found in the nuclei of insect cells.

The silkworm has controlling mechanisms for the transportation of proteins between organs (92). For example, the major hemolymph proteins, storage proteins 1 and 2, are produced by the fat body at the fifth instar and are secreted into the hemolymph. Stored storage proteins are then transferred back into the fat body at the pupal stage and are degraded as a source of amino acids. Another major group of hemolymph proteins, called 30K proteins for their molecular weight, are secreted by the fat body, stored in the hemolymph at the fifth instar, transferred into the ovary, and deposited in the eggs. Vitellogenin in the silkworm has the same transport pathway as the 30K proteins, but its expression is activated mainly in the spinning and pupal stages.

When a recombinant BmNPV carrying the human IFN-$\alpha$ gene infected the silkworm, the interferon was secreted by the fat body and stored in the hemolymph without the degradation observed in the case of the hemolymph proteins (53, 64). Based on this observation, I predict that the silkworm together with recombinant BmNPVs containing genes encoding functional proteins and appropriate promoters will provide a useful system for the study of the mechanisms and dynamics of protein transportation in vivo.

## Gene Splicing

As described earlier, there is no indication of gene splicing in the expression of the baculovirus genome (46). Since some RNA molecules possess auto-catalytic activity for splicing, it has been assumed that the gene-containing exons would be processed in the infected cells. When the DNA fragment containing the large T and the small t antigen genes of SV40 was inserted into the AcNPV vector, the produced mRNAs were found to be spliced, and at least the mature product of the small t antigen was produced (34).

# USE IN PHARMACEUTICAL PRODUCTION

Many organizations have already started to produce pharmaceutical products using recombinant DNA technology. The pharmaceuticals produced by many of the different procedures have to be tested carefully and may require considerable time before they are ready for commercial use. However, this is not the case with products produced in the silkworm with the BmNPV expression vector system, owing in part to the many advantages described above. Since the silkworm has been used as food by many animals, including humans, it is a safe factory for the production of pharmaceuticals. Furthermore, the silkworm does not seem as prone to induce allergy as some insects.

The purification of the expressed products is important. The foreign gene products secreted into insect hemolymph, e.g. human IFN-$\alpha$ (53) and mouse interleukin 3 (64), are easily purified by methods commonly used in in vitro culture. In the case of interleukin 3, purification from the hemolymph is much easier than from transfected mammalian cultures containing fetal calf serum.

For pharmaceutical use, it is required that the biologically active form will not cause an unexpected effect or antibody production. The results obtained to date show that the products produced with the baculovirus vector in insect cells and in the silkworm are almost identical with those produced in human beings. The only difference detected is in glycosylation. Although the biological activity is the same as that of the mature products in mammalian cells, different oligosaccharides may cause specific antibody production against the products in insects. However, since the difference seems to be caused not by the addition of sugar residues that differ from those found in mammals, but only by the incomplete addition of some sugar residues (27), this may not be a factor of concern in the immunological response by humans.

The BmNPV system can also be easily applied for diagnostic work or the production of vaccines. We have successfully expressed five different polypeptides coded on the HIV genome. When one of the products, gp41, produced by the BmNPV system was immunologically tested with the serum of 109 AIDS patient and 50 healthy individuals, all AIDS patients could be completely distinguished (N. Kobayashi & S. Maeda, unpublished results). This system can be used not only in diagnosis, but also in vaccination. The envelope proteins of HIV-1 produced in cells of S. frugiperda by the AcNPV vector have now been approved for testing in humans for AIDS vaccination (2).

## USE FOR PHYSIOLOGICAL STUDIES IN INSECTS

For studies in physiology, biochemistry, and endocrinology of insects it is very helpful to obtain purified small peptides, such as neuropeptides, and large polypeptides that are usually found in small amounts. Small, simple peptides can be produced by automated peptide synthesis. For complex peptides [e.g. prothoracicotropic hormone (PTTH) (68) or eclosion hormones (35, 38, 56)] recombinant DNA methods are preferable. A large amount of biologically active peptides can be produced using baculovirus vectors.

Several genes encoding insect polypeptides have been isolated from lepidopterans. The expression of many genes is considered to be under the control of insect hormones. If the promoters connected to an easily detectable gene, such as chloramphenicol acetyltransferase, can be introduced into BmNPV, the resulting recombinant BmNPVs will be useful for studies on the controlling mechanisms of interesting silkworm genes. If the effect of the virus can be reduced in the silkworm by some means, the BmNPV expression system will provide opportunities for endocrinological studies and also for studies of gene expression in lepidopterans.

## USE IN INSECT PEST CONTROL

Problems with the use of chemical pesticides, e.g. mammalian toxicity and the appearance and spread of resistant insect pest species, are widely known. Awareness of these problems has resulted in increased interest in the use of baculoviruses, some of which have been registered and are commercially available, as viral insecticides (4). The baculoviruses generally are of low virulence and cause larval death after relatively long periods of infection. Some workers have suggested that by introducing appropriate foreign genes into the baculovirus genomes pathogenicity and insecticidal effectiveness may be increased (36, 63, 89). If this new technology is to be introduced into agriculture, it will be necessary to demonstrate the safety of these engineered viruses.

Six of seven families of insect-pathogenic viruses have counterparts with similar physicochemical characteristics among the noninsect animal and plant viruses (60). The Baculoviridae are the only exception. They infect only arthropods, mainly Lepidoptera. In vitro, baculoviruses can penetrate into mammalian cells, but only into the cytoplasm and not into the nucleus, where virogenesis occurs in insect cells. Furthermore, the virus is not known to persist in mammalian cells (7, 20). The lack of infectivity for mammals was clearly shown with AcNPV recombinant viruses possessing the chloramphenicol acetyltransferase gene under the control of mammalian-activated LTR promoter and the $\beta$-galactosidase gene under the control of the polyhedrin gene promoter (9, 10). When this recombinant virus infected susceptible cells, both gene products were produced. In a nonpermissive $D$. *melanogaster* cell line, only the chloramphenicol acetyltransferase was activated, which indicated that penetration into the cell nucleus had taken place. On the contrary, when a mammalian cell was infected, neither gene was activated. These results suggest that the baculovirus lacks the ability to transfer the genomic DNA into the mammalian cell nucleus even if an established cell line is used. The unusual physicochemical properties and limited host range suggest that the baculovirus would be safe as a viral insecticide.

The first approach in the formation of a baculovirus insecticide is to make a recombinant baculovirus of increased toxicity by introducing appropriate foreign genes into the highly expressed polyhedrin gene site. Polypeptide toxins that block neuronal function and the important peptide hormones are currently being studied. Most of these toxins are toxic only to insects. Infection with such recombinant viruses may cause direct toxicity, alter behavior, or arrest development in insects. The onset of these effects may be much faster than the pathology caused by the original parent virus. Recently, we have obtained an exciting result showing the addition of toxicity by the introduction of this type of gene (S. Maeda, unpublished results). This result indicates that the gene product produced by the baculovirus has promise for pest-insect control.

The stability of baculoviruses under field conditions is dependent on the occlusion bodies made up of polyhedrin. A recombinant virus that retains its ability to produce polyhedrin can be created by introducing the foreign gene into the polyhedrin gene site without removing the polyhedrin gene itself. For example, the polyhedrin gene of *Spodoptera litura* NPV was introduced into the recombinant BmNPV at a site quite distant from the original polyhedrin gene location, which now contained the interferon gene. The heterologous polyhedrin gene was expressed with high efficiency together with the human interferon gene (S. Maeda, unpublished results). With the use of the AcNPV vector, the introduced LCMV glycoprotein and the original polyhedrin were expressed by the recombinant AcNPV with insertions of two different genes and the same polyhedrin gene promoters (11).

Furthermore, if the promoter of the polyhedrin gene, used for the gene for insect toxin, is replaced with one of the early gene promoters of NPV or with a wild pest insect promoter that is highly activated in the target organs, such a virus may express the gene for insect toxin at an early stage of infection. If the right gene for insect toxin and the promoter can be isolated and introduced into a baculovirus, the resulting recombinant virus will likely be an efficient viral insecticide.

## CONCLUSIONS

Recombinant DNA technology has been applied in entomology. Insects can be made to serve as miniature factories for the production of foreign gene products. Baculoviruses have served as vectors for the introduction of foreign genes into insects. Some of the expressed products may be pharmaceuticals, others may serve as insecticides, and still others (e.g. hormones) may be of value in basic studies on the physiology and metabolism of insects. There is every indication that this new technology will be used widely in entomology.

ACKNOWLEDGMENTS

I would like to thank E. Cooper, C. L. Ecale, B. D. Hammock, S. J. Kramer, and Y. Tanada for critical review of the manuscript and Y. Aira, A. Fuse, H. Handa, A. Igarashi, M. Iwano, N. Kobayashi, L. K. Miller, A. Miyajima, H. Nyunoya, K. Shimotono, and B. Simizu for providing unpublished results.

## Literature Cited

1. Arif, B. M., Doerfler, W. 1984. Identification and localization of reiterated sequences in the *Choristoneura fumiferana* MNPV genome. *EMBO J.* 3:525–29
2. Barnes, D. M. 1987. AIDS vaccine trial OKed. *Science* 237:973
3. Baum, J. A., Geever, R., Giles, N. H. 1987. Expression of qa-1F activator protein: identification of upstream binding sites in the *qa* gene cluster and localization of the DNA-binding domain. *Mol. Cell. Biol.* 7:1256–66
4. Betz, F. S. 1986. Registration of baculoviruses as pesticides. In *The Biology of Baculoviruses*, Vol. 2, *Practical Application for Insect Control*. ed. R. R. Granados, B. A. Federici, pp. 203–22. Boca Raton, Fla: CRC. 275 pp.
5. Bicknell, J. N., Leisy, D. J., Rohrmann, G. F., Beaudreau, G. S. 1987. Comparison of the p26 gene of two baculoviruses. *Virology* 161:589–92
6. Boman, H. G., Hultmark, D. 1987. Cell-free immunity in insects. *Ann. Rev. Microbiol.* 41:103–26
7. Brusca, J., Summers, M., Couch, J., Courtney, L. 1986. *Autographa californica* nuclear polyhedrosis virus efficiently enters but does not replicate in poikilothermic vertebrate cells. *Intervirology* 26:207–22
8. Butters, T. D., Hughes, R. C. 1981. Isolation and characterization of mosquito cell membrane glycoproteins. *Biochim. Biophys. Acta* 640:655–71
9. Carbonell, L. F., Klowden, M. J., Miller, L. K. 1985. Baculovirus-mediated expression of bacterial genes in dipteran and mammalian cells. *J. Virol.* 56:153–60
10. Carbonell, L. F., Miller, L. K. 1987. Baculovirus interaction with nontarget organisms: A virus-borne reporter gene is not expressed in two mammalian cell lines. *Appl. Environ. Microbiol.* 53:1412–17
10a. Ellis, L., Levitan, A., Cobb, M. H., Ramos, P. 1988. Efficient expression in insect cells of a soluble, active human insulin receptor protein-tyrosine kinase domain by use of a baculovirus vector. *J. Virol.* 62:1634–39
11. Emery, V. C., Bishop, D. H. L. 1987. The development of multiple expression vectors for high level synthesis of eukaryotic proteins: expression of LCMV-N and AcNPV polyhedrin protein by a recombinant baculovirus. *Protein Eng.* 1:359–66
12. Estes, M. K., Crawford, S. E., Penaranda, M. E., Petrie, B. L., Burns, J. W., et al. 1987. Synthesis and immunogenicity of the rotavirus major capsid antigen using a baculovirus expression system. *J. Virol.* 61:1488–94
13. Friesen, P. D., Miller, L. K. 1985. Temporal regulation of baculovirus RNA: overlapping early and late transcripts. *J. Virol.* 54:392–400
14. Friesen, P. D., Miller, L. K. 1987. Divergent transcription of early 35- and 94-kilodalton protein genes encoded by the *Hind*III K genome fragment of the baculovirus *Autographa californica* nuclear polyhedrosis virus. *J. Virol.* 61:2264–72
14a. Fung, M.-C., Chiu, K. Y. M., Weber, T., Chang, T.-W., Chang, N. T. 1988. Detection and purification of a recombinant human B lymphotropic virus (HHV-6) in the baculovirus expression system by limiting dilution and DNA dot-blot hybridization. *J. Virol. Methods* 19:33–42
15. Garoff, H. 1985. Using recombinant DNA techniques to study protein targeting in the eucaryotic cell. *Ann. Rev. Cell Biol.* 1:403–45
15a. Ginns, E. I., Rehavi, M., Martin, B. M., Weller, M., O'Malley, K. L., et al. 1988. Expression of human tyrosine hydroxylase cDNA in invertebrate cells using a baculovirus vector. *J. Biol. Chem.* 263:7406–10
16. Glembotski, C. C. 1984. The α-amidation of α-melanocyte stimulating hormone in intermediate pituitary requires ascorbic acid. *J. Biol. Chem.* 259:13041–48
17. Gonnet, P., Devauchelle, G. 1987. Neomycin-resistant insect cells infected with a p10 recombinant baculovirus. *C. R. Acad. Sci. Ser. III* 305:111–14 (In French)
18. Granados, R. R., Federici, B. A., eds. 1986. *The Biology of Baculoviruses*, Vol. 1, *Biological Properties and Molecular Biology*. Boca Raton, Fla: CRC. 275 pp.
19. Greenfield, C., Patel, G., Clark, S., Jones, N., Waterfield, M. D. 1988. Expression of the human EGF receptor with ligand-stimulatable kinase activity in insect cells using a baculovirus vector. *EMBO J.* 7:139–46
20. Groner, A., Granados, R. R., Burand, J. P. 1984. Interaction of *Autographa californica* nuclear polyhedrosis virus with two nonpermissive cell lines. *Intervirology* 21:203–9
21. Guarino, L. A., Gonzalez, M. A.,

Summers, M. D. 1986. Complete sequence and enhancer function of the homologous DNA regions of *Autographa californica* nuclear polyhedrosis virus. *J. Virol.* 60:224–29

22. Guarino, L. A., Summers, M. D. 1987. Nucleotide sequence and temporal expression of a baculovirus regulatory gene. *J. Virol.* 61:2091–99

22a. Harmon, S., Johnston, J. M., Ziegelhoffer, T., Richards, O. C., Summers, D. F., et al. 1988. Expression of hepatitis A virus capsid sequences in insect cells. *Virus Res.* 10:273–80

23. Hassell, J. A., Mueller, C., Mes, A.-M., Featherstone, M., Naujokas, M., et al. 1982. The construction of polyoma virus vectors: functions required for gene expression. In *Eukaryotic Viral Vectors*, ed. Y. Gluzman, pp. 71–77. Cold Spring Harbor, NY: Cold Spring Harbor Lab. 221 pp.

23a. Herrera, R., Lebwohl, D., de Herreros, A., Kallen, R. G., Rosen, O. M. 1988. Synthesis, purification, and characterization of the cytoplasmic domain of the human insulin receptor using a baculovirus expression system. *J. Biol. Chem.* 263:5560–68

24. Hooft van Iddekinge, B. J. L., Smith, G. E., Summers, M. D. 1983. Nucleotide sequence of the polyhedrin gene of *Autographa californica* nuclear polyhedrosis virus. *Virology* 131:561–65

25. Horie, Y., Watanabe, H. 1980. Recent advances in sericulture. *Ann. Rev. Entomol.* 25:49–71

26. Horiuchi, T., Marumoto, Y., Saeki, Y., Sato, Y., Furusawa, M., et al. 1987. High-level expression of the human-α-interferon gene through the use of an improved baculovirus vector in the silkworm, *Bombyx mori*. *Agric. Biol. Chem.* 51:1573–80

27. Hsieh, P., Robbins, P. W. 1984. Regulation of asparagine-linked oligosaccharide processing: oligosaccharide processing in *Aedes albopictus* mosquito cells. *J. Biol. Chem.* 259:2375–82

28. Hu, S.-L., Kosowski, S. G., Schaaf, K. F. 1987. Expression of envelope glycoproteins of human immunodeficiency virus by an insect virus. *J. Virol.* 61:3617–20

29. Hussey, R. E., Richardson, N. E., Kowalski, M., Brown, N. R., Chang, H.-C., et al. 1988. A soluble CD4 protein selectively inhibits HIV replication and syncytium formation. *Nature* 331:78–81

30. Inumaru, S., Ghiasi, H., Roy, P. 1987. Expression of bluetongue virus group–specific antigen VP3 in insect cells by a baculovirus vector: its use for the detection of bluetongue virus antibodies. *J. Gen. Virol.* 68:1627–35

31. Inumaru, S., Roy, P. 1987. Production and characterization of the neutralization antigen VP2 of bluetongue virus serotype 10 using a baculovirus expression vector. *Virology* 157:472–79

32. Ito, T., Kobayashi, M. 1978. Rearing of the silkworm. In *The Silkworm: An Important Laboratory Tool*, ed. Y. Tazima, pp. 83–102. Tokyo: Kodansha. 307 pp.

33. Jeang, K.-T., Giam, C.-Z., Nerenberg, M., Khoury, G. 1987. Abundant synthesis of functional human T-cell leukemia virus type I p40ˣ protein in eucaryotic cells by using a baculovirus expression vector. *J. Virol.* 61:708–13

34. Jeang, K.-T., Holmgren-Konig, M., Khoury, G. 1987. A baculovirus vector can express intron-containing genes. *J. Virol.* 61:1761–64

34a. Kang, C. Y., Bishop, D. H. L., Seo, J.-S., Matsuura, Y., Choe, M. 1987. Secretion of particles of hepatitis B surface antigen from insect cells using a baculovirus vector. *J. Gen. Virol.* 68:2607–13

35. Kataoka, H., Troetschler, R. G., Kramer, S. J., Cesarin, B. J., Schooley, D. A. 1987. Isolation and primary structure of the eclosion hormone of the tobacco hornworm, *Manduca sexta*. *Biochem. Biophys. Res. Commun.* 146:746–50

36. Kirshbaum, J. B. 1985. Potential implication of genetic engineering and other biotechnologies to insect control. *Ann. Rev. Entomol.* 30:51–70

37. Knebel, D., Doerfler, W. 1987. Activation of an insect baculovirus promoter in mammalian cells by adenovirus functions. *Virus Res.* 8:317–26

38. Kono, T., Nagasawa, H., Isogai, A., Fugo, H., Suzuki, A. 1987. Amino acid sequence of eclosion hormone of the silkworm, *Bombyx mori*. *Agric. Biol. Chem.* 51:2307–8

39. Kornfield, R., Kornfield, S. 1985. Assembly of asparagine-linked oligosaccharides. *Ann. Rev. Biochem.* 54:631–64

40. Kuroda, K., Hauser, C., Rott, R., Klenk, H.-D., Doerfler, W. 1986. Expression of the influenza virus haemagglutinin in insect cells by a baculovirus vector. *EMBO J.* 5:1359–65

41. Kuzio, J., Faulkner, P. 1984. Regions of repeated DNA in the genome of *Choristoneura fumiferana* nuclear polyhedrosis virus. *Virology* 139:185–88

42. Kuzio, J., Rohel, D. Z., Curry, C. J., Krebs, A., Carstens, E. B., et al. 1984. Nucleotide sequence of the p10 poly-

peptide gene of *Autographa californica* nuclear polyhedrosis virus. *Virology* 139:414–18

43. Leisy, D. J., Rohrmann, G. F., Beaudreau, G. S. 1984. Conservation of genome organization in two multicapsid nuclear polyhedrosis viruses. *J. Virol.* 52:699–702

44. Leisy, D. J., Rohrmann, G. F., Nesson, M., Beaudreau, G. S. 1986. Nucleotide sequencing and transcriptional mapping of the *Orgyia pseudotsugata* multicapsid nuclear polyhedrosis virus p10 gene. *Virology* 153:157–67

45. Lowy, D. R., Dvoretzky, I., Schober, R., Law, M.-F., Engel, L., et al. 1980. In vitro tumorigenic transformation by a defined sub-genomic fragment of bovine papilloma virus DNA. *Nature* 287:72–74

46. Lubbert, H., Doerfler, W. 1984. Mapping of early and late transcripts encoded by the *Autographa californica* nuclear polyhedrosis virus genome: Is viral RNA spliced? *J. Virol.* 50:497–506

47. Lubbert, H., Doerfler, W. 1984. Transcription of overlapping sets of RNAs from the genome of *Autographa californica* nuclear polyhedrosis virus: a novel method for mapping RNAs. *J. Virol.* 52:255–65

48. Luckow, V. A., Summers, M. D. 1988. Trends in the development of baculovirus expression vectors. *Biotechnology* 6:47–55

49. Madisen, L., Travis, B., Hu, S.-L., Purchio, A. F. 1987. Expression of the human immunodeficiency virus *gag* gene in insect cells. *Virology* 158:248–50

50. Maeda, S. 1987. Expression of human interferon α gene in silkworms with a baculovirus vector. In *Biotechnology in Invertebrate Pathology and Cell Culture*, ed. K. Maramorosch, pp. 221–33. San Diego, Calif: Academic. 511 pp.

51. Maeda, S. 1989. Gene transfer vectors of a baculovirus, *Bombyx mori* nuclear polyhedrosis virus, and their use for expression of foreign genes in insect cells. In *Invertebrate Cell System and Applications*, ed. J. Mitsuhashi. Boca Raton, Fla: CRC. In press

52. Maeda, S., Kawai, T., Obinata, M., Chika, T., Horiuchi, T., et al. 1984. Characteristics of human interferon-α produced by a gene transferred by a baculovirus vector in the silkworm, *Bombyx mori*. *Proc. Jpn. Acad.* 60:423–26

53. Maeda, S., Kawai, T., Obinata, M., Fujiwara, H., Horiuchi, T., et al. 1985. Production of human α-interferon in silkworm using a baculovirus vector. *Nature* 315:592–94

54. Mainprize, T. H., Lee, K., Miller, L. K. 1986. Variation in the temporal expression of overlapping baculovirus transcripts. *Virus Res.* 6:85–99

55. Mann, R., Mulligan, R. C., Baltimore, D. 1983. Construction of a retrovirus packaging mutant and its use to produce helper-free defective retrovirus. *Cell* 33:153–59

56. Marti, T., Takio, K., Walsh, K. A., Terzi, G., Truman, J. W. 1987. Microanalysis of the amino acid sequence of the eclosion hormone from the tobacco hornworm. *Manduca sexta*. *FEBS Lett.* 219:415–18

56a. Martin, B. M., Tsuji, S., LaMarca, M. E., Maysak, K., Eliason, W., et al. 1988. Glycosylation and processing of high levels of active human glucocerebrosidase in invertebrate cells using a baculovirus expression vector. *DNA* 7:99–106

57. Marumoto, Y., Sato, Y., Fujiwara, H., Sakano, K., Saeki, Y., et al. 1987. Hyperproduction of polyhedrin-IGF II fusion protein in silkworm larvae infected with recombinant *Bombyx mori* nuclear polyhedrosis virus. *J. Gen. Virol.* 68:2599–606

58. Matsuura, Y., Possee, R. D., Bishop, D. H. L. 1986. Expression of the S-coded genes of lymphocytic choriomeningitis arenavirus using a baculovirus vector. *J. Gen. Virol.* 67:1515–29

59. Matsuura, Y., Possee, R. D., Overton, H. A., Bishop, D. H. L. 1987. Baculovirus expression vectors: the requirements for high level expression of proteins, including glycoproteins. *J. Gen. Virol.* 68:1233–50

60. Matthews, R. E. F. 1982. Classification and nomenclature of viruses. *Intervirology* 17:1–199

61. Miller, L. H., Sakai, R. K., Romans, P., Gwadz, R. W., Kantoff, P., et al. 1987. Stable integration and expression of a bacterial gene in the mosquito *Anopheles gambiae*. *Science* 237:779–81

62. Miller, L. K. 1981. A virus vector for genetic engineering in invertebrates. In *Genetic Engineering in the Plant Sciences*, ed. N. J. Panapoulos, pp. 203–24. New York: Praeger. 271 pp.

63. Miller, L. K. 1987. Expression of foreign genes in insect cells. See Ref. 50, pp. 295–303

63a. Miller, L. K. 1988. Baculoviruses as gene expression vectors. *Ann. Rev. Microbiol.* 42:177–99

64. Miyajima, A., Schreurs, J., Otsu, K., Kondo, A., Arai, K., et al. 1987. Use of

the silkworm, *Bombyx mori,* and an insect baculovirus vector for high-level expression and secretion of biologically active mouse interleukin-3. *Gene* 58:273–81

65. Miyamoto, C., Smith, G. E., Farrell-Towt, J., Chizzonite, R., Summers, M. D., et al. 1985. Production of human c-*myc* protein in insect cells infected with a baculovirus expression vector. *Mol. Cell. Biol.* 5:2860–65

66. Morin, J. E., Lubeck, M. D., Barton, J. E., Conley, A. J., Davis, A. R., et al. 1987. Recombinant adenovirus induces antibody response to hepatitis B virus surface antigen in hamsters. *Proc. Natl. Acad. Sci. USA* 84:4626–30

67. Deleted in proof

68. Nagasawa, H., Kataoka, H., Isogai, A., Tamura, S., Suzuki, A., et al. 1986. Amino acid sequence of a prothoracicotropic hormone of the silkworm *Bombyx mori. Proc. Natl. Acad. Sci. USA* 83:5840–43

69. Nyunoya, H., Maeda, S. 1988. Baculovirus vectors. *Tissue Cult.* 14:112–17 (In Japanese)

70. Oellig, C., Happ, B., Muller, T., Doerfler, W. 1987. Overlapping sets of viral RNAs reflect the array of polypeptides in the *Eco*RI J and N fragments (map position 81.2–85.0) of the *Autographa californica* nuclear polyhedrosis virus genome. *J. Virol.* 61:3048–57

71. Ollo, R., Maniatis, T. 1987. *Drosophila* Kruppel gene product produced in a baculovirus expression system is a nuclear phosphoprotein that binds to DNA. *Proc. Natl. Acad. Sci. USA* 84:5700–4

71a. O'Reilly, D. R., Miller, L. K. 1988. Expression and complex formation of simian virus 40 large T antigen and mouse p53 in insect cells. *J. Virol.* 62:3109–19

72. Overton, H. A., Ihara, T., Bishop, D. H. L. 1987. Identification of the N and NS$_s$ proteins coded by the ambisense S RNA of Punta Toro phlebovirus using monospecific antisera raised to baculovirus expressed N and NS$_s$ proteins. *Virology* 157:338–50

73. Panicali, D., Paoletti, E. 1982. Construction of poxviruses as cloning vectors: insertion of the thymidine kinase gene from herpes simplex virus into the DNA of infectious vaccinia virus. *Proc. Natl. Acad. Sci. USA* 79:4927–31

74. Pennock, G. D., Shoemaker, C., Miller, L. K. 1984. Strong and regulated expression of *Escherichia coli* β-galactosidase in insect cells with a baculovirus vector. *Mol. Cell. Biol.* 4:399–406

75. Possee, R. D. 1986. Cell-surface expression of influenza virus haemagglutinin in insect cells using a baculovirus vector. *Virus Res.* 5:43–59

76. Rankin, C., Ladin, B. F., Weaver, R. F. 1986. Physical mapping of temporally regulated, overlapping transcripts in the region of the 10K protein gene in *Autographa californica* nuclear polyhedrosis virus. *J. Virol.* 57:18–27

77. Rice, W. C., Lorimer, H. E., Prives, C., Miller, L. K. 1987. Expression of polyomavirus large T antigen by using a baculovirus vector. *J. Virol.* 61:1712–16

78. Rohrmann, G. F. 1986. Polyhedrin structure. *J. Gen. Virol.* 67:1499–513

79. Rubin, G. M., Spradling, A. C. 1982. Genetic transformation of *Drosophila* with transposable element vectors. *Science* 218:348–53

80. Rusche, J. R., Lynn, D. L., Robert-Guroff, M., Langlois, A. J., Lyerly, H. K., et al. 1987. Humoral immune response to the entire human immunodeficiency virus envelope glycoprotein made in insect cells. *Proc. Natl. Acad. Sci. USA* 84:6924–28

81. Schmaljohn, C. S., Sugiyama, K., Schmaljohn, A. L., Bishop, D. H. L. 1988. Baculovirus expression of the small genome segment of Hantaan virus and potential use of the expressed nucleocapsid protein as a diagnostic antigen. *J. Gen. Virol.* 69:777–86

82. Sekine, H., Fuse, A., Tada, A., Maeda, S., Simizu, B. 1988. Expression of human papillomavirus type 6b E2 gene product with DNA binding activity in insect *(Bombyx mori)* cells using a baculovirus expression vector. *Gene* 65:187–93

83. Shapiro, J. P., Law, J. H., Wells, M. A. 1988. Lipid transport in insects. *Ann. Rev. Entomol.* 33:297–318

84. Smith, G. E., Fraser, M. J., Summers, M. D. 1983. Molecular engineering of the *Autographa californica* nuclear polyhedrosis virus genome: deletion mutations within the polyhedrin gene. *J. Virol.* 46:584–93

85. Smith, G. E., Ju, G., Ericson, B. L., Moschera, J., Lahm, H.-W., et al. 1985. Modification and secretion of human interleukin 2 produced in insect cells by a baculovirus expression vector. *Proc. Natl. Acad. Sci. USA* 82:8404–8

86. Smith, G. E., Summers, M. D., Fraser, M. J. 1983. Production of human β interferon in insect cells infected with a baculovirus expression vector. *Mol. Cell. Biol.* 3:2156–65

87. St. Angelo, C., Smith, G. E., Summers, M. D., Krug, R. M. 1987. Two of the

three influenza viral polymerase proteins expressed by using baculovirus vectors form a complex in insect cells. *J. Virol.* 61:361–65

88. Steiner, H., Hultmark, D., Engström, A., Bennich, H., Boman, H. G. 1981. Sequence and specificity of two antibacterial proteins involved in insect immunity. *Nature* 292:246–48

89. Summers, M. D., Smith, G. E. 1985. Genetic engineering of the genome of the *Autographa californica* nuclear polyhedrosis virus. In *Genetically Altered Viruses and the Environment,* ed. B. Fields, M. A. Martin, D. Kamely, pp. 319–39. Cold Spring Harbor, NY: Cold Spring Harbor Lab. 234 pp.

90. Tada, A., Fuse, A., Sekine, H., Simizu, B., Kondo, A., et al. 1988. Expression of the E2 open reading frame of papillomaviruses BPV1 and HPV6b in silkworm by a baculovirus vector. *Virus Res.* 9:357–67

91. Tojo, S., Betchaku, T., Ziccardi, V. J., Wyatt, G. R. 1978. Fat body protein granules and storage proteins in the silkmoth, *Hyalophora cecropia. J. Cell Biol.* 78:823–38

92. Tomino, S. 1985. Major plasma proteins of *Bombyx mori. Zool. Sci.* 2:293–303

93. Van Hofsten, P., Faye, I., Kockum, K., Lee, J.-Y., Xanthopoulos, K. G., et al. 1985. Molecular cloning, cDNA sequencing, and chemical synthesis of cecropin B from *Hyalophora cecropia. Proc. Natl. Acad. Sci. USA* 82:2240–43

94. Van Wyke Coelingh, K. L., Murphy, B. R., Collins, P. L., Lebacq-Verheyden, A.-M., Battey, J. F. 1987. Expression of biologically active and antigenically authentic parainfluenza type 3 virus hemagglutinin-neuraminidase glycoprotein by a recombinant baculovirus. *Virology* 160:465–72

95. Vlak, J. M., Klinkenberg, F. A., Zaal, K. J. M., Usmany, M., KlingeRoode, E. C., et al. 1988. Functional studies on the p10 gene of *Autographa californica* nuclear polyhedrosis virus using a recombinant expressing a p10-β-galactosidase fusion gene. *J. Gen. Virol.* 69:765–76

96. Wickner, W. T., Lodish, H. F. 1985. Multiple mechanisms of protein insertion into and across membranes. *Science* 230:400–8

97. Wigler, M., Silverstein, S., Lee, L.-S., Pellicer, A., Cheng, Y.-C., et al. 1977. Transfer of purified herpes virus thymidine kinase gene to cultured mouse cells. *Cell* 11:233–32

98. Wilson, M. E., Mainprize, T. H., Friesen, P. D., Miller, L. K. 1987. Location, transcription, and sequence of a baculovirus gene encoding a small arginine-rich polypeptide. *J. Virol.* 61: 661–66

99. Wojchowski, D. M., Orkin, S. H., Sytkowski, A. J. 1987. Active human erythropoietin expressed in insect cells using a baculovirus vector: a role for N-linked oligosaccharide. *Biochim. Biophys. Acta* 910:224–32

Reference Added in Proof

100. Nyunoya, H., Akagi, T., Ogura, T., Maeda, S., Shimotohno, K. 1989. Evidence for phosphorylation of *trans*-activator p40$^x$ of human T-cell leukemia virus type I produced in insect cells with a baculovirus expression vector. *Virology* In press

*Ann. Rev. Entomol. 1989. 34:373–81*

# INSECT CONTROL WITH GENETICALLY ENGINEERED CROPS

## Ronald L. Meeusen

Sandoz Crop Protection Corporation, 975 California Avenue, Palo Alto, California 94304-1104

## Gregory Warren

Rohm and Haas Company, 727 Norristown Road, Spring House, Pennsylvania 19477

## INTRODUCTION

For decades geneticists have bred crops with improved characteristics, among them resistance to insect pests. At the same time chemists and entomologists have developed a stream of improved chemical and biological insecticides successful enough to amount to a worldwide business of some $4 billion annually. Recently molecular biologists have combined the two in a third approach to insect pest control, genetic engineering of crops that produce insecticidal or antifeedant proteins continuously in the field.

The first published report of success with this approach is barely two years old (20), and the literature includes only three other reports at the time of this writing (3, 9, 13). Yet already the first prototype products, caterpillar-resistant tobacco and tomato plants that produce the delta endotoxin of *Bacillus thuringiensis,* are undergoing field trials.

The speed with which this new tool is developing and its potential influence on the way our society protects its food and fiber crops from insect predation make this a timely topic for consideration even at this early stage.

## TECHNOLOGY

The methodology for inserting genes into crop plants has been reviewed thoroughly in the scientific and popular press (see 14 for an excellent in-

373

0066-4170/89/0101-0373$02.00

troduction). Central to this work is the common crown gall bacterium, *Agrobacterium tumefaciens*, a ubiquitous soil organism with the remarkable ability to move a portion of its own DNA into a plant cell during infection. In the early 1980s a number of groups in the United States and Europe found ways to use this bacterium to insert other genes into plant cells; they were then able to recover whole plants from the transformed cells (4, 11, 12). The resulting plants were found to have incorporated the inserted genes into their own genomes and thus had acquired new, stable, and inherited traits. These pioneering studies used bacterial genes for resistance to antibiotics or for production of unusual amino acids. While such genes were useful in the development of the technique, they imparted no agronomically beneficial traits to the plants.

Even before this milestone had been attained many public and private research groups had begun planning to use gene insertion techniques to improve crops. One of the first genes to draw attention was from the biocontrol organism *Bacillus thuringiensis*.

## Bacillus thuringiensis *Delta Endotoxin*

The history, biology, and insecticidal properties of *Bacillus thuringiensis* (Bt) have been extensively reviewed elsewhere (1, 2, 5, 6). Early efforts to engineer useful genes into crops focused on genes for the Bt endotoxin because of a number of advantages: (*a*) The Bt endotoxin had been safely used in the United States as a registered pesticide since 1960. (*b*) The safety, biodegradability, and selectivity of the endotoxin had made it a preferred material for use in environmentally sensitive applications. (*c*) The Bt endotoxin was known to be the product of a single gene, which had already been isolated and partially characterized (17, 18). (*d*) The endotoxin was known to be active against the larvae of lepidopteran pests, perhaps the most destructive insect pests worldwide.

The safety, efficacy, and relative simplicity of this protein insecticide made it a nearly ideal candidate for testing the concept of engineering crops to produce protective materials.

## *The* Bacillus thuringiensis *Endotoxin Gene in Plants*

The first successful use of the technology was reported by a Belgian biotechnology company, Plant Genetic Systems, in July of 1987 (20). Using a gene from *B. thuringiensis* this group developed tobacco plants that produced enough of the endotoxin to kill first-instar *Manduca sexta* larvae. Insects placed on the leaves of these plants displayed the same response as insects placed on leaves sprayed with commercial *B. thuringiensis* products—feeding suppression after about 18 hr and death within 3 days. Levels of the endotoxin as low as 30 ng per gram of leaf protein provided complete protection against

*Manduca* neonates, and the production of endotoxin was shown to be inherited like a simple dominant trait.

The first report was followed in August of 1987 by similar results from a research group at the Monsanto Company using tomato. Shortly thereafter the Agracetus Company reported another success in tobacco (3, 9). Concerted efforts are now underway in these and other laboratories to duplicate these results in other crops with important lepidopteran pests, such as cotton, corn, and vegetables.

The rapid progress of this field is driven home even more forcefully by the fact that prototype insect-resistant varieties of tomato and tobacco have already undergone one and two seasons of field trials, respectively. The Rohm and Haas Company, Northrup King Company, and Monsanto Company conducted trials in 1986 and 1987 after review by the United States Department of Agriculture. If we assume that field testing and seed production for these novel varieties will follow the same path as for traditionally bred crop lines, we can expect the first genetically engineered insect-resistant seed to reach the marketplace between 1992 and 1995.

## The Trypsin Inhibitor Gene in Plants

Genes for the endotoxins of *B. thuringiensis* are not the only insect resistance genes under development for crops. In late 1987 a group in the United Kingdom reported success with a gene from the cowpea *Vigna unguiculata* (13). This gene encodes a serine protease inhibitor, which is believed to impart resistance to a beetle herbivore of the cowpea by interfering with its digestion.

When introduced into tobacco, the trypsin inhibitor gene imparted resistance to *Heliothis virescens,* a pest of tobacco. Growth reduction in the insects clearly correlated with levels of the inhibitor in plant leaves. As with the endotoxin genes, the new trait proved both stable and heritable.

## PROSPECTS

These recent successes have engendered sharp debate over the role of genetically engineered crop varieties in the future of insect control. Proponents (largely molecular biologists) envision a day when traditional insecticide use will be largely replaced by genetically modified crops that will stand impervious to the onslaught of insect pests. Skeptics (largely those in the agrochemical industry) remind us that breeders have long manipulated the genetics of crops to provide insect resistance, and that despite many striking successes the need for chemical insecticides today remains as pressing as ever. It is certainly too early to predict where between these extremes the pendulum will come to rest. However, it is not too early to consider the factors, both technical and nontechnical, that will determine the outcome.

The degree to which this technology will be adopted in the future depends on its relative advantages compared to currently deployed insect control methods. The technology offers advantages in performance and application costs, cost of development, and environmental and safety areas. Adoption of the technology will also be influenced by uncertainties yet to be addressed such as insect resistance, regulatory barriers, and availability of patent protection. Crops containing the *Bacillus thuringiensis* endotoxin genes provide a useful test case for consideration of each of these elements.

## Potential Advantages

PERFORMANCE AND APPLICATION COSTS    Control of caterpillar pests with a Bt endotoxin–producing crop variety might be expected to offer four immediate advantages over traditional approaches from the grower's perspective. First, the grower would be freed from dependence on weather. His crop would be protected even when his fields are too muddy to pull spray equipment through or when the weather is too severe to allow aerial application of insecticides. A related advantage would be the protection of plant parts that are difficult to reach with sprays, such as roots, shaded lower leaves, and new growth that emerges between insecticide applications. Thirdly, the control agent is in the field continuously, freeing the grower from the need to scout his field to time his spray applications properly. Conceptually it should not matter when caterpillar eggs are laid or hatch. Since the endotoxin is in the plant tissues continuously, insects will be "treated" as soon as they start to feed. For caterpillars this is typically the time they are most sensitive to the endotoxin. Finally, in crops that require heavy insecticide use, such as cotton, the application costs now rival the chemical costs. Equivalent control offered by a resistant seed would not only provide savings on insecticide expenditures, but would also reduce the grower's outlays for spray equipment, labor, and fuel (or aerial application fees).

COST OF DEVELOPMENT    The four potential advantages discussed thus far are of greatest value to the grower, since they bear directly on improved performance, reduced production costs, and hence greater production and profits. But the new technology may offer broader economic environmental advantages as well.

The cost to discover, develop, register, and produce a new chemical insecticide today is well over $25 million (16), and it can easily run higher if a new chemical production plant must be built. If we assume that about $75,000 are spent annually per employee (on salary, benefits, overhead, and other costs), the equivalent of over 300 man-yr of effort is necessary to add each new chemical to our insect control arsenal.

In contrast, development of a new crop variety typically costs under $1 million. Even allowing for an additional research charge of several million dollars for the expensive molecular research involved in genetic engineering, a novel insect-resistant crop line could be adopted for a fraction of the investment that traditional chemical products require.

ENVIRONMENTAL AND SAFETY BENEFITS    From an environmental and human-health perspective, the use of engineered insect-resistant crops promises other attractive benefits. Because the active materials are produced directly in the crop tissue, such concerns as spray drift and groundwater contamination are obviated. The two resistance factors already in development, the Bt endotoxin and the trypsin inhibitor, are both natural proteins of high molecular mass (60 and 13 kd, respectively). Thus they should prove completely biodegradable to nontoxic products (amino acids).

Toxicity to nontarget organisms should also be obviated, since the only organisms in the field environment able to receive a dose of the active elements are the pest organisms that feed on the crop.

Finally, monitoring the safety of crops for human consumption should be easier, as the added materials are fully defined in advance. Assessing risks from traditional chemical insecticide use entails expensive residue analysis and toxicology. In contrast, before a new gene can be inserted into a crop it must be fully characterized. Problems, if any, are expected to be less frequent.

It has been argued that the insertion of novel genes by molecular methods is even safer than traditional crop breeding, which has until now been essentially unregulated. In traditional breeding resistance genes are generally used without any attempt to determine the insecticidal product produced within the plant. No such lack of information should haunt genetically engineered crops.

## Potential Disadvantages and Uncertainties

The foregoing considerations argue persuasively that this technology will become a major component of our future insect control strategies. However, a proper perspective of its potential also demands a close look at the limitations and uncertainties, both technical and otherwise, that may reduce its impact on agriculture.

RESISTANCE    A concern already voiced has to do with the rate at which resistance will arise in the target insect populations. For crops containing the Bt endotoxin, history provides some comfort that it will not be rapid. Products based on the endotoxin have been in use for over a quarter century without reports of significant resistance in the field. Few chemical insecticides have displayed such longevity. However, the degree to which one can extrapolate from experience with sprays, to which the insect population undergoes intermittent exposure, to the situation in which the endotoxin is continuously present in the food supply remains an open question. Indeed, the sole report of resistance to a Bt product involved just such a continuous exposure, through spiking of the food supply (15). More recent work has confirmed this result in another species, but suggests that much of the resistance is not to the endotoxin itself (19), but to the bacterium containing it. Thus resistance to endotoxin-containing plants, while a concern, may not arise significantly faster than to sprays.

Modeling studies would be helpful in predicting whether an insecticide produced within the plant constitutes a greater, lesser, or equivalent selection for resistance. This question will be of keen interest to the companies developing these crops, as pest resistance may dictate product lifetime. Too short a lifetime could seriously limit the usefulness of this approach.

A factor that at least temporarily limits the usefulness of genetic engineering for insect control is our inability to insert genes into many major crops, particularly grains such as wheat, corn, and rice. This difficulty is widely expected to be overcome in the next few years.

REGULATORY UNCERTAINTIES    When is a tomato not a tomato? A possible future answer is "When it is an insecticide." A tomato plant producing Bt endotoxin in its tissues poses a difficult legal dilemma for regulatory agencies. The United States Environmental Protection Agency (EPA) could potentially require its registration as a pesticide, since it falls within that definition in the Federal Insecticide, Fungicide, and Rodenticide Act. If the EPA does not require registration, then the Food and Drug Administration (FDA) could require food additive petitions under the Pure Food, Drug, and Cosmetics Act. In addition, the United States Department of Agriculture (USDA) enjoys regulatory authority over plants that are potentially injurious to agriculture under the Federal Noxious Weed and Plant Pest Acts. Thus a modified crop could well be a pesticide to one agency, a food additive to another, and a plant pest to a third.

Clearly such regulatory overkill could preclude any use of genetically engineered insect-resistant crops. How could a seed company ever recover the $10 million pesticide registration and an even greater food additive registration from a crop variety that cost under $1 million to develop?

To prevent expensive overlaps and redundancies in the registration process, the Reagan administration undertook in 1983 to coordinate its regulation of the products of biotechnology. A major goal was to minimize multiple review. In December of 1984 the three agencies mentioned above published a proposal for coordinated regulation, which designated one lead agency to handle each type of product (7). Responsibility for plants primarily lay with the USDA, which in June of 1987 published its formal regulations (8). At that time the immediate need was for oversight of initial field trials with prototype products such as the endotoxin-containing tobacco and tomato plants. For this purpose the USDA's regulations, operating under the coordinated framework (7), seem to have served adequately. The successful field tests in 1986 and 1987 by the Rohm and Haas Company, the Northrup King Company, and the Monsanto Company were reviewed without undue delay by the USDA with input from the EPA.

Small-scale field trials are only a first step toward commerical use, however. The role of the EPA and FDA in this stage of product development is currently the topic of a lively debate, whose outcome could determine the future of such products.

If the EPA decides to assert jurisdiction over insect-resistant crops it must resolve two types of dilemmas, legal and practical. From the legal (or even scientific) perspective, there is little basis to distinguish between insect-resistant crops developed wtih molecular techniques and insect-resistant crops developed by traditional breeding. For the agency to assert jurisdiction over the former would require either regulating the entire seed industry as well or drawing a logically and legally indefensible line based on the process by which a crop was developed. The agency has already publicly promised to avoid such "process-based" regulation in its policy statement (7).

From a practical perspective, the development of crop varieties simply cannot bear the regulatory costs associated with chemical insecticides. To superimpose a $7–10 million registration burden onto a crop that costs under $1 million to develop would clearly preclude any significant adoption of this technology in commerce, regardless of the environmental benefits or advantages to the grower. On the other hand, the Bt endotoxin is currently a registered pesticide, and the agency might find itself equally hard pressed to legally justify exempting it from oversight.

The agencies must arrive at a regulatory framework that assures the public that salient questions of food safety and environmental suitability are addressed. At the same time this framework must avoid multiple review by the three agencies and keep total costs well under 10% of the current cost of registering an insecticide. Although efforts to date are encouraging, the ultimate ability of these agencies to develop such a framework remains an open question.

PATENT AVAILABILITY    The final major element of uncertainty affecting the future of genetically engineered insect-resistant crops is the degree to which those supporting this research can protect their inventions or investments. Although development of a resistant crop variety with molecular techniques is far less expensive than development of a chemical insecticide, it is currently far more expensive than traditional crop breeding.

No seed company would invest in the new technology if its competitors were entitled to buy a bag of the new seed, cross the new gene into their varieties, and sell it without compensation for the developers. Yet until recently the Plant Variety Protection Act (PVPA) provided the only intellectual property rights for breeders, and this act contained a research exemption permitting free use of proprietary varieties by competing breeders for development of their own new varieties.

In 1985, however, the United States Circuit Court of Appeals ruled in *Ex parte* Hibberd (5a) that genetically engineered plants were indeed patentable subject matter under the Utility Patent Act. This ruling logically extended the famous 1980 Chakrabarty Supreme Court decision (16a) to plants. In theory this ruling now provides a choice of protections for a plant invention; inventors can seek variety protection certificates or utility patents. Utility patent coverage offers a second advantage in that it does not contain the troublesome "farmer's exemption," a provision of the PVPA that permits growers to reproduce proprietary varieties and sell the seed in unlabeled bags to neighbors. The widespread abuse of this exemption is a serious concern to the seed industry.

While in theory the Hibberd ruling is a most promising development and should spur continued investment in research on novel crop varieties, the picture in practice is still very murky. Fundamental questions about the scope of patent coverage need to be answered. Can a single patent cover all crop plants containing the Bt endotoxin, or is each modified crop a separate invention? Who holds the dominant patent, the inventor of the endotoxin gene or the inventor of the endotoxin-containing plant, or must each cross-license the other? Can a patent holder control what a grower does with the seed he harvests from his own field? Will public breeders, accustomed to free access to any germplasm for breeding purposes under the old PVPA, accept the new practices?

An interesting aspect of these questions is that without new legislation they cannot be answered until courts have the chance to adjudicate actual cases. These cannot be initiated until products reach the market and infringement suits are filed. Assuming that the first genetically engineered products will reach the market by the mid-1990s and allowing several years for trials and appeals, we probably cannot expect answers much before the turn of the millennium. In the meantime the field must progress in the presence of these uncertainties.

## CONCLUSION

The ability to incorporate genes into crops to confer insect resistance has already been demonstrated. Potentially this approach offers dramatic advantages over conventional methods, and these advantages ought to drive its adoption as a major new technology for the control of insect pests. However, uncertainties about the rates at which insects will overcome the new resistances, the rapidly evolving regulatory systems, and the more slowly evolving patent coverage deserve careful attention. Each holds the potential to limit widespread adoption of this potential control strategy, however desirable the technology.

## Literature Cited

1. Andrews, R. E., Faust, R. M., Wabiko, H., Roymond, K. C., Bulla, L. A. 1987. The biotechnology of *Bacillus thuringiensis*. *CRC Crit. Rev. Biotechnol.* 6:163–232

2. Aronson, A. I., Backman, W., Dunn, P. 1986. *Bacillus thuringiensis* and related insect pathogens. *Microbiol. Rev.* 50:1–24

3. Barton, K. A., Whiteley, H. R., Yang, N. 1987. *Bacillus thuringiensis* delta-endotoxin expressed in transgenic *Nicotiana tabacum* provides resistance to lepidopteran insects. *Plant Physiol.* 85:1103–9

4. Bevan, M., Flavell, R. 1983. A chimeric antibiotic resistance gene as a selectable marker for plant cell transformation. *Nature* 304:18–187

5. Dean, D. H. 1984. Biochemical genetics of the bacterial insect-control agent *Bacillus thuringiensis:* Basic principles and prospects for genetic engineering. *Biotechnol. Genet. Eng. Rev.* 2:341–63

5a. *Ex parte* Hibberd et al. 1985. *US Pat. Q.* 227:443–48

6. Fast, P. G. 1981. The crystal toxin of *Bacillus thuringiensis.* In *Microbial Control of Pests and Plant Diseases 1970–1980*, ed. H. D. Burges, pp. 223–48. London: Academic

7. *Federal Register.* 1984. Proposal for a coordinated framework for regulation of biotechnology. 49(252):50856–907

8. *Federal Register.* 1987. Plant pests: introduction of genetically engineered organisms or products; final rule. 52(115):22892–915

9. Fischhoff, D. A., Bowdish, K. S., Perlack, F. J., Marrone, P. G., McCormick, S. M., et al. 1987. Insect tolerant transgenic tomato plants. *Biotechnology* 5:807–13

10. Deleted in proof

11. Fraley, R., Rogers, R., Horsch, R., Sanders, R., Flick, J., et al. 1983. Expression of bacterial genes in plant cells. *Proc. Natl. Acad. Sci. USA* 80:4803–7

12. Herrera-Estrella, L., De Block, M., Messens, E., Hernalsteens, J. P., Van Montagu, M., Schell, J. 1983. Chimeric genes as dominant selectable markers in plant cells. *EMBO J.* 2:987–95

13. Hilder, V. A., Gatehouse, A. M. R., Sheerman, S. E., Barker, R. F., Boulter, D. 1987. A novel mechanism of insect resistance engineered into tobacco. *Nature* 330:160–63

14. Hooykaas, P. J., Schilperoort, R. A. 1985. The Ti-plasmid of *Agrobacterium tumefaciens:* a natural genetic engineer. *Trends Biochem. Sci.* 10:307–9

15. McGaughy, W. H. 1985. Insect resistance to the biological insecticide *Bacillus thuringiensis*. *Science* 229:193–94

16. Metcalf, R. 1980. Changing role of insecticides in crop production. *Ann. Rev. Entomol.* 25:219–56

16a. Parker v. Bergey. 1979. *US Pat. Q.* 201:352–95

17. Schnepf, H. E., Whiteley, H. R. 1981. Cloning and expression of the *Bacillus thuringiensis* crystal protein gene in *E. coli. Proc. Natl. Acad. Sci. USA* 78:2893–97

18. Schnepf, H. E., Whiteley, H. R. 1985. Delineation of a toxin-encoding segment of a *Bacillus thuringiensis* crystal protein gene. *J. Biol. Chem.* 260:6273–80

19. Stone, T. B., Sims, S. R., Marrone, P. G. 1988. Selection of tobacco budworm for resistance to a genetically engineered *Pseudomonas fluorescens* containing the delta-endotoxin of *Bacillus thuringiensis* subsp. Kurstaki. *J. Invertebr. Pathol.* In press

20. Vaeck, M., Reynaerts, A., Hofte, H., Jansens, S., De Beukeleer, M., et al. 1987. Transgenic plants protected from insect attack. *Nature* 328:33–37

*Ann. Rev. Entomol. 1989. 34:383–400*

# BIONOMICS OF THE NABIDAE

*John D. Lattin*

Systematic Entomology Laboratory, Department of Entomology, Oregon State University, Corvallis, Oregon 97331–2907

## INTRODUCTION

The family Nabidae, or damsel bugs, is a small family of the true bugs (Hemiptera: Heteroptera) containing 31 genera and approximately 380 species (70, 72). While all known species are terrestrial, some species are found in moist areas on the ground or at the edge of streams, ponds, and marshes (fresh and saline) (77, 107). Nabidae prey on a variety of small invertebrates, chiefly arthropods (7, 27, 61, 75). Some plant feeding by nabids may occur, but no development follows (141), and it is likely that moisture is the chief objective (25, 116). The predaceous habit, together with the widespread occurrence of some species in a variety of ecosystems, particularly agroecosystems, has attracted the attention of entomologists.

Much of the world literature on the Nabidae is taxonomic (e.g. 51, 65, 67, 70, 107, 113–115, 135, 139). There have been few summaries of the family, but there are two notable exceptions. Kerzhner (70) dealt primarily with the nabids of the USSR, but he summarized morphology, life stages, parasites, biogeography, fossils, and especially importantly, his own work on the classification and phylogeny of the Nabidae. Pericart (107) dealt with the Nabidae of the western Palearctic, treating each species in detail. Both books have extensive bibliographies. Although many taxa found outside the Holarctic region remain unknown to us biologically, much is now known about the Palearctic species and, to a lesser extent, the Nearctic species. As our knowledge grows, it is certain that we will discern considerable biotic diversity in the habits and habitats of the Nabidae.

0066-4170/89/0101–0383$02.00

# SYSTEMATICS

## Position Within the Heteroptera

Leston et al (86) placed the family Nabidae within the superfamily Cimicoidea in the group Cimicomorpha, one of the two subdivisions of the Geocorisae, or land bugs. Earlier, Carayon (16) had compared internal genital organs and adult scent glands of the Nabidae and Reduviidae and concluded that these two families were not closely related as had been suggested by earlier workers (e.g. 114). China & Miller (23) considered the Nabidae close to the Velocipedidae, a taxon sometimes included as a subfamily of the Nabidae. Carayon (20) reviewed the status of the Nabidae and included both the Velocipedinae and the Medocostinae as subfamilies within the family.

Stys (146) considered the Medocostidae a distinct family. Stys & Kerzhner (147) considered both the Medocostidae and the Velocipedidae as families separate from the Nabidae. Kerzhner (70) placed the Nabidae within the Cimicoidea in the Cimicomorpha and retained the Velocipedinae and Medocostinae within the family. Pericart (106) removed the Velocipedinae from the Nabidae and elevated it to family status, but restored it to subfamily status in 1987. Schuh (121) reviewed the impact of cladistics on the classification of the Heteroptera. He presented the current classification of the Heteroptera and considered the cladistic work that had been done on various subgroups. Schuh placed the Nabidae in the Cimicomorpha, and he removed both the Velocipedinae and the Medocostinae and considered them distinct families. He reviewed the Cimicomorpha and presented a cladogram of the group, based in part upon the work of Kerzhner (70).

## Family Classification

Stål (135) included three subfamilies in the Nabidae: Nabina, Coriscina, and Pachynomina. Carayon (16) considered the Pachynominae as a taxon closer to the Reduviidae and removed it from the Nabidae. He included the Scotomedinae (=Velocipedinae) as a subfamily of the Nabidae, following Blöte (11). Leston et al (86) concurred with Carayon. China & Miller (23) recognized five subfamilies in the Nabidae: Nabinae, Prostemminae, Arachnocorinae, Gorpinae, and Carthasinae. They considered the Velocipedidae and Pachynomidae as separate families. Carayon (20) included in the Nabidae four subfamilies, the Nabinae, Prostemmatinae, Velocipedinae, and Medocostinae, with the last subfamily intermediate between the Velocipedinae and the Nabinae. Kerzhner (70) recognized the same four subfamilies, with the subfamily Nabinae divided into four tribes (Nabini, Arachnocorini, Carthasini, and Gorpini). Schuh (121) questioned the status of the Velocipedinae and Medocostinae, and felt that additional information is needed to clarify their position within the Cimicomorpha. He considered them as distinct families of uncertain position. Pericart (107) retained the Velocipedinae and Medocostinae in the family Nabidae, along with the Nabinae and Prostemmatinae. Additional work is needed to clarify the uncertainties associated with the major subdivisions of the family.

The chromosomes of a number of species have been characterized (84, 89, 95, 96, 103). The type number for the family is $2n = 18$ ($16A + X + Y$) (84). While most species that have been investigated show this number, several have $2n = 20$, and *Aptus mirmicoides* is known to have $2n = 34$ (84). Supernumerary sex chromosomes have been reported from several species (95). There is considerable disagreement over the mechanisms involved in the increase in chromosome number (see e.g. 152, 153). Thomas (152) suggested that the most likely explanation of autosome increase is aneuploidy.

# BIONOMICS

## Movement

Southwood (130) defined two types of animal movement, trivial and migratory. Trivial movements occur within the habitat of the population, in contrast to migratory movements, which carry the animal away (140). Most available information on movement in the nabids is confined to members of the subfamily Nabinae (15, 30, 32, 39, 75, 100, 157).

TRIVIAL MOVEMENT    Nabids are active shortly after hatching and begin feeding immediately, often on prey considerably larger than themselves (100). Koschel (75) found that early instar nymphs of *Himacerus apterus* lived on forbes and grasses but moved to shrubs and trees later in their life. This species is one of the few that occurs regularly in trees (7, 15, 27, 75, 134, 145), where they have been reported to feed on a variety of prey. Scudder (123) reported *Nabicula vanduzeei* from *Pinus contorta* in North America. On cotton, adult nabids were found higher on the plant than the nymphs (157). The adults also displayed a stronger preference for the fruits than the nymphs. Diel movement occurs in nabids in grasslands; bugs were found in the upper portions of grass at night (39). This movement might be a response to microclimatic conditions rather than a nocturnal activity, since there was no evidence that the nabids were not active during the day. Soybeans sampled by sweeping at different times during the day showed no significant change in numbers (32). Night activity has been documented for *Himacerus mirmicoides* in the United Kingdom (30).

FLIGHT    Adult Nabidae are winged insects, and some species are able to move considerable distances (71). *Nabis capsiformis* is the most widespread species and is pantropical in distribution. There is ample evidence that

Nabidae fly. McPherson & Weber (93), working on the seasonal flight patterns of Hemiptera in North Carolina, collected six species of nabids from window traps set at different heights above the ground (1–7 m). The three common species included *Nabis americoferus, Nabis roseipennis,* and *N. capsiformis.* Johnson & Southwood (64) collected *Nabis ferus* 50 ft above the ground in aerial nets. Edwards (36), reporting on arthropods from alpine aeolian ecosystems, mentioned *Nabis alternatus* as one of the five species of insects taken on the summit of Mount Rainier, Washington. A number of genera of nabids are commmonly taken in light traps (128); specimens are always macropterous. Southwood (128) demonstrated Heteroptera showing high levels of flight activity were associated with temporary habitats. A high level of brachyptery may restrict nabids to stable habitats.

Some nabids may be found high in the air. Glick (43) reported *N. roseipennis* from 2000 ft and *Hoplistoscelis sordidus* from 1000 ft in Louisiana. Specimens of *N. capsiformis* have been collected many miles from the nearest land in the Atlantic, Pacific, and Indian Oceans (71). This species lives in open areas, including coastal habitats, and is a generalist in its feeding habits. Leston (83) considered the spread potential and island colonization of the Heteroptera. While the family Miridae ranked highest, the Nabidae ranked eighth of 15 families considered.

WING POLYMORPHISM    Although many adult Nabidae have completely developed (macropterous) wings, a number of species show considerable wing polymorphism (microptery and brachyptery) (51, 81, 82, 137). Wing reduction may be slight [e.g. *Nabis brevis* (107)] or greatly reduced [e.g. *Phorticus brevipennis* (68)]. In a few instances the wing reduction is unilateral rather than bilateral (82), although this is an aberrant condition. Nonetheless, wing reduction of some sort is common in the Nabidae. Of the 32 species of nabids found in Europe and the Maghreb (107), 25 (78%) display some form of wing reduction in some individuals. Seven (22%) are known in the macropterous state only. Of the North American taxa in which sufficient specimens were available to constitute an adequate sample, nine (32%) species are known only from macropterous forms and 19 (68%) species have some individuals showing some type of wing reduction. These percentages will change as more information becomes available.

Brachyptery is common in insects on islands, including the Nabidae (66, 158). Zimmerman (158) discussed 26 species of nabids from the Hawaiian Islands, and at least an equal number remain to be described. All known species save one *(N. capsiformis)* are endemic to the islands. Twelve of the 26 species have at least partly modified wings. When the full study on this family has been completed in the Hawaiian Islands, we will have a much better understanding of the role of wing reduction in speciation and habitat utilization. Considerably more species of Nabidae are known from the Hawaiian Islands than from North America or Europe.

Southwood (129) proposed a hormonal theory to explain wing polymorphism in the Heteroptera. He suggested that shortness of wings is a character of the juvenile and is determined by the concentration of juvenile hormone. He considered the reduction of wings in mountain forms of *N. ferus* (136, 137) as a case of metathetely, whereas he cited the occurrence of only four nymphal instars in *Nabicula limbata* (most Heteroptera have five) as an example of prothetely. *Nabicula limbata* is most commonly found in the brachypterous state, but according to Stehlik (136) the long-winged form is found in cooler, mountainous circumstances (in contrast to the short-winged form of *N. ferus*).

The degree of wing development influences dispersal capabilities (53). Many of the species that show strong wing reduction are found in stable habitats, especially on the ground (51, 59, 68, 75, 145). Fully winged individuals (e.g. *N. capsiformis*) are often associated with temporary habitats, including agroecosystems (37, 56, 128, 130, 132). In some instances the brachypterous forms might be considered typical K-strategists and the macropterous species r-strategists. Even though ready dispersal is normally associated with r-strategists, Hamilton & May (48) suggested that dispersal may be expected not only from species occupying patchy environments, but also from those species found in uniform, permanent habitats. McPherson & Weber (93) reported the capture of *Pagasa fusca* in aerial samples taken at 1–3 m in North Carolina. This species is a ground inhabitant and is usually found in the brachypterous state.

Dispersal forms may occur in otherwise immobile populations. For example, Freeman (41) reported a very rare macropterous female of a mirid, *Mecomma dispar*, taken at 300 ft. Normally, only the male of this species is macropterous (134). I collected over 1000 specimens of a species of *Rhagovelia* (Veliidae) on a large river in Oregon. All were apterous except for a single male and female.

INTRODUCTIONS    Commercial activities influence the movement of insects, and many species have been introduced into areas well removed from their native range (78). Since the time of Columbus the North American continent has received at least 1683 species, mostly from the western Palearctic region (87, 119, 125); remarkably few species have gone in the opposite direction (some notable exceptions include the grape phylloxera, fall webworm, and Colorado potato beetle). At least two species of introduced Nabidae have been found in North America: *Nabis brevis* and *Anaptus major*. Parshley (104) reported that *N. brevis* was found "on nursery stock imported from England." McAtee (91) objected to the inclusion of this record by Blatchley (10) on the grounds that many insects are detected by agricultural inspectors, but Parshley's citation does provide a specific record of a means of introduction. *Nabis brevis* is usually slightly brachypterous, lives in moist environments, deposits its eggs in grasses, and hibernates as an adult (107). The species does not appear to have become established in North America (57).

*Anaptus major* is established in North America and is known from California, Oregon, Washington, and British Columbia on the Pacific Coast and from New York and Pennsylvania in the eastern United States (5, 51, 57, 77, 122, 156). There is reasonable evidence that *A. major* was introduced (at least in the Pacific Northwest) via ballast materials brought from Europe, particularly the United Kingdom (77, 122). Southwood & Fewkes (133) reported this species as ". . . common on wastelands" in England. Specimens have been collected on disturbed sites around ports in Oregon, including areas close to known ballast dump sites (J. D. Lattin, personal observations). *Anaptus major* is fully winged, but its favored habitats are the ground and low vegetation (107). In Europe this species may occur along the water's edge together with *Nabicula lineata* (107, 134). A North American analog of *N. lineata, Nabicula propinqua*, was collected with *A. major* in a coastal salt marsh in Oregon (77).

*Nabis ferus* is a common, widespread species in the Palearctic region (70, 107). Before 1960 it was regarded as one of the most widespread species in North America and was thought to represent an introduction from the Old World (51). Then Carayon (19) demonstrated that the North American species was actually distinct from *N. ferus;* he named the species *N. americoferus*. It is curious that only *A. major* appears to have become established in North America, and that none of the many Old World species of the genus *Nabis* have done so. Careful attention to disturbed environments along both coasts of North America may yet disclose exotic species.

*Nabis capsiformis* is a special case. This pantropical species is the most widespread nabid known (52, 70, 71). It occurs only in the macropterous state (107). Kerzhner (71) reported it as the only species recovered from aerial samples taken over the Atlantic, Indian, and Pacific Oceans. Its status as an immigrant is less certain, but Zimmerman (158) considered it so in Hawaii. Collection of this species in Central Europe has been reported only occasionally (69, 107, 138). While human activities may have had little or no role in the dispersal of *N. capsiformis,* its ability to disperse is well established.

## Mortality Factors

PARASITES    A variety of organisms parasitize the eggs, nymphs, and adults of Nabidae. Carayon (17) found protozoans and bacteria in the digestive tract of nabids. He considered both groups to be parasitic. Further, he felt that the bacteria were not of the symbiotic type found in some other Heteroptera (usually phytophagous taxa). Morrill (98) reported the presence of nematode parasites in *N. alternatus* in the United States. Rubtsov (118) described a new species of Mermithidae parasitizing a nabid from the far eastern Soviet Union. Both parasitic (phytoseiid) and phoretic (trombidiiform) mites have been found associated with Nabidae (22, 51, 70, 100). Chant & Lindquist (22) described the otopheidominid (76) mite *Nabiseius duplicisetus* from the dorsal surface of the abdomen of a nabid from Chile. Mundinger (100) observed a trombidiid mite attached to a second-instar nymph of *N. roseipennis*. Pericart (107) reported that only about 1% of the specimens of nabids that he examined had ectoparasitic mites associated with them.

Two families of parasitic flies have been reported from nabids, the Chloropidae (70, 127) and Tachinidae (e.g. 70, 107). Smith (127) observed a species of *Elachipteron* on the back of a nabid, inserting its ovipositor beneath the hemelytra. Representatives of at least three genera of Tachinidae have been reported parasitizing Nabidae: *Leucostoma simplex,* a Holarctic species, has been reported from *A. mirmicoides* (58), *N. ferus* (35), *N. alternatus* (24, 98, 143, 150, 151, 154), *N. americoferus* (2, 55, 143), and *N. capsiformis* (143); *Hyalomya aldrichii,* a species that sometimes parasitizes the lygaeid *Geocoris* (24), was commonly encountered in *N. alternatus, N. americoferus,* and *N. capsiformis* in Arizona (143); and *Athrycia cinerea* was reared from *N. alternatus* or *N. americoferus* (2). Parasitization by the last species was 3% (2). This is quite low compared to parasitization rates recorded for *L. simplex* of 24% in nymphs and 44% in adults of *N. alternatus* and *N. americoferus* (24) and 6–28% in *N. alternatus* (98). Stoner et al (143) recorded parasitism varying from 0 to 48.2% in Arizona for both *L. simplex* and *H. aldrichii* on *N. alternatus* and *N. americoferus*. Percent parasitism was high in January, low in the spring, high in midsummer, low in the fall, and then high again in the winter. The nabids had two peaks in their populations, one in May and another in November. The degree of parasitism lagged behind these population peaks.

Hendrick & Stern (55) provided a detailed study of the development of *L. simplex* on *N. americoferus* in California. The egg is inserted through the intersegmental membrane of the abdomen, usually between the convexium and the sternites. Both nymphs and adult bugs are attacked. The eggs hatch three days later (at 80°F). There are three larval instars, and the total larval period is about 26 days. Although superparasitism did occur in the laboratory, it was rarely encountered in the field. The mature larva exits the bug quickly, usually through the dorsal surface between the terminal abdominal tergites. The host generally dies shortly thereafter. Pupation occurs quickly outside the host and lasts for about 9 days before the adult emerges.

Three families of Hymenoptera parasitize Nabidae: Scelionidae and Mymaridae on eggs and Braconidae on nymphs and adults. Southwood &

Leston (134) reported a scelionid, *Telenomus* sp., attacking the eggs of *Anaptus major* and *Nabicula flavomarginata* in England. They also reported the mymarid *Polynema gracile* parasitizing the eggs of *A. major* and *Nabicula limbata*. Benedict & Cothran (8) found that as many as 70% of *N. americoferus* eggs in alfalfa were parasitized by *Polynema boreum* in California. *P. boreum* also parasitized *N. americoferus* and *N. alternatus* in Arizona, where 35% parasitism occurred (45). The percentages of *Nabis* eggs parasitized were lower in eggs deposited in the apical portions of host-plant stems. Although *Anaphes ovijentatus* has been reported on *Nabis* species in the laboratory (24, 144), attempts to establish this well-known parasite of *Lygus* species on various species of *Nabis* have not proven very successful (62). This lack of success suggests that releases of this species of wasp would have little impact on these important predators of *Lygus*.

The braconid *Wesmaelia pendula* parasitizes several species of Nabidae in North America and Europe (70, 85, 88, 99, 107). Although the wasp has been widely known on both continents, the host association is recent (85, 99). Leston (85) used the presence of euphorine parasites in the Miridae and the Nabidae as partial evidence of the evolutionary derivation of the former family from the latter. Hendrick & Stern (55) reported a rate of parasitism of less than 1% for *W. pendula* on nymphal and adult *N. americoferus* in southern California. Stoner (142), however, reported rates of parasitism as high as 40% for male *N. alternatus* in southern Arizona. The parasite was also common in males of *N. americoferus* and *N. capsiformis,* although the percent parasitism was lower for both species. *Wesmaelia pendula* occurs only rarely in females of *N. alternatus* and *N. americoferus* (143). The *Wesmaelia* life cycle has a two-brood pattern in eastern Canada; the nabid nymphs are attacked in the fall, and the parasite larvae overwinter in the adult hosts (88).

PREDATORS    While it is clear that nabids are predators that occupy a specific trophic level, little specific information is available about the next trophic level up, i.e. their predators. Cannibalism is common throughout the life cycle of nabids (63, 75, 100); thus members of their own trophic level may contribute to their mortality. Hutchinson (60) briefly discussed the possible influence of cannibalism on food-chain length, particularly as the predator grows; *H. apterus* ranges in size from 2 mm (first instar) to 11.5 mm (female adult) (75, 107). Some predators have been reported, such as sphecid wasps (33, 34) and pentatomid (150) and reduviid (150) bugs. Spiders have also preyed on nabids (75), although Southwood & Leston (134) reported a situation in which the attacking spider was killed by the nabid. Koschel (75) also reported the predation of another nabid species, *N. limbata,* upon *H. apterus*. Further, he mentioned possible predation by vertebrate predators, particularly birds. The size of nabids makes them potential prey for many invertebrate and vertebrate predators.

DISEASE    The entomophagous fungus *Verticillium lecanii* has a wide geographical range and attacks a variety of insects although the fungus controls aphids, it is also a potent killer of *N. alternatus,* a major predator of aphids (50). The mortality rate of nabids dipped in a spore suspension was 82% (50).

## Feeding

PLANTS    Although the basic feeding strategies of Heteroptera have been the subject of some debate (25, 26, 148), the nabids are known to be largely predaceous. Some plant feeding has been observed, however (73, 116, 141). Ridgway & Jones (116) used cotton plants labeled with $^{32}$P to demonstrate plant feeding. Their experiments indicated that nabids could be exposed to systemic insecticides applied to cotton plants. Indeed, systemic chemicals have reduced *Nabis* populations (80). Stoner (141) noted plant feeding but indicated that, while longevity increased, no nymphal growth occurred. The plant-feeding habits of *N. alternatus* have been implicated in the possible spread of a yeast *(Nematospora coryli)* found in *Brassica juncea* seed grown in Canada (13). The yeast was recovered from the nabids during the summer, but overwintering individuals tested negative; however, the test sample of bugs was small. The relationship between the bug and the yeast is not clear; the bug may acquire the yeast by feeding on insects that have fed on the plants.

ANIMAL PREY    The mandibles of the Nabidae project ahead of the maxillae during penetration of animal prey (25). Both mouthparts have serrated tips. Arnold (3) observed nabids touching their potential prey with the foreleg several times prior to inserting the mouthparts; he proposed that this action accustomed the prey to the predator's touch.

Feeding is essential to growth and development, but the insect is able to survive short periods without food (75). At 17°C, the average survival time without food for nymphs of *H. apterus* was 15.9 days (75); younger nymphs died earlier than older nymphs. Most nabids overwinter in the adult stage and thus are able to go long periods without food.

Most of the prey of Nabidae are small invertebrates, chiefly insects (15, 70, 107). Collyer (27) reported *H. apterus,* a tree-inhabiting nabid, as an occasional predator of the red spider mite in orchards. The same nabid is a known predator of many insects in trees (75, 145). Most types of insect prey of nabids are plant-feeding species, but nabids sometimes attack predaceous insects, including members of their own species (63, 75). *Nabis alternatus* feeds on the lygaeid *Geocoris punctipes,* an important predator in some agroecosystems (4). Several genera within the nabid subfamily Prostemmatinae (e.g. *Prostemma, Alloeorhynchus,* and *Pagasa*) are known predators of other Lygaeidae, particularly members of the Blissinae, Geocorinae, and Rhyparochrominae (20, 42, 51, 70, 107, 112). These particular nabids and lygaeids live on the ground. Froeschner (42) and Reinert (112) reported two *Pagasa* species feeding on two different species of chinch bugs, both pests of considerable importance. A number of nabid species are predators of Miridae, and especially the genus *Lygus* (108, 109, 149, 155). Smaller nymphal nabids can handle smaller prey; the larger damsel bugs can handle all life stages of *Lygus.* Immunological techniques have been used to demonstrate that *N. americoferus* (like *N. ferus*) is a predator of *Lygus* spp. (155).

Nabids are confirmed predators of leafhoppers in all stages, including the beet leafhopper and the potato leafhopper (40, 74, 90). Nabids also prey upon aphids of many different species, including a number of species of economic importance (9, 21, 47, 63, 79, 111, 117). The polyphagous feeding habits of the nabids make them less effective than species-specific predators against specific prey species.

Among the Coleoptera, species of the Chrysomelidae have been reported as prey for nabids, including the Colorado potato beetle (38, 44, 94, 105). Two nabid species were considered major predators on two species of goldenrod leaf beetles (94). Eggs and larvae of Lepidoptera constitute common prey for many Nabidae, with feeding usually concentrated on the smaller stages (31, 92, 120, 126). Several native nabids attack the introduced *Hyphantria curea* in the Soviet Union (7). One of these, *H. apterus* (F.), one of the few tree-inhabiting nabids, was reported as a predator of the gypsy moth *(Lymantria dispar)* in the Ukraine (29) and on a variety of geometrid larvae in Poland (145). As was the case with other prey, younger stages of nabids were able to handle smaller prey, and the larger damsel bugs attacked most prey sizes.

Polyhedral inclusion bodies of a nuclear polyhedrosis virus of the moth *Heliothis punctigera* were found in the feces of *Nabis tasmanicus,* one of the predators of *H. punctigera* in Australia (6). A bioassay showed no loss of infectivity during the passage through the nabid's gut. Cooper (28) reported that 39% of field-collected *N. tasmanicus* adults contained polyhedra in their feces. He considered this predator and a predatory pentatomid as potential agents for disseminating infective polyhedra in the field.

## Functional Role in Ecosystems

NATURAL ECOSYSTEMS    The Nabidae are members of a guild of arthropod predators found in different terrestrial ecosystems. They are often associated with representatives of several other predaceous families of Heteroptera such as Anthocoridae, Lygaeidae *(Geocoris* spp.), Miridae (Deraeocorinae and others), Pentatomidae (Asopinae), and Reduviidae. Although the Nabidae is a small family of about 380 species (72) displaying rather modest ecological diversity, there is some apparent ordering of habitat preferences. Southwood (131) discussed the influence of the habitat on ecological strategies and provided a good discussion of the r-K selection spectrum. Most known nabids seem to be K strategists, particularly in regard to dispersal, since a high percentage show some sort of wing reduction. Further, some species display restricted niche breadth. For example, the Prostemmatinae are found on the ground and in the litter layer, usually closely associated with their prey, ground-inhabiting Lygaeidae (20, 107). *Himacerus apterus* lives in the grass-forbe zone in the first several nymphal instars and then moves up into trees (75). *Arachnocoris albomaculatus* is adapted to live in spider webs, where it apparently feeds on trapped insects (101). The usually brachypterous *Nabicula subcoleoptrata* is a predator of several species of Chrysomelidae within its range (38, 94). As discussed above, most of the prey species belong to certain insect taxa, although nabids appear to have the capacity to attack a number of other arthropods. Indeed, some species are generalist feeders and can be considered r-strategists.

AGROECOSYSTEMS    Much of what we know about the habits of nabids has come from a large number of studies involving their functional role as predators in various agroecosystems (e.g. 12, 14, 46, 102, 110, 120, 143). The number of species is quite small, and most belong to the genus *Nabis*. Most, if not all, would be considered r-strategists according to Southwood's definition (131, 132). He posited that polyphagous predators are likely to be important in the temporary habitats often created by agriculture, where habitats may not be stable enough for K-strategists to become established. Ehler & Miller (37) provided support for the idea that r-selected natural enemies can provide control for r-selected pests; they cited *N. americoferus* as an example of such an enemy.

There is considerable evidence that nabids respond to habitat manipulation such as that occurring in agroecosystems. Intercropped fields of bean and maize supported more nabids than monocultures (97). Although differing maturity dates, locations, and row spacing in soybeans had little impact on predator populations, grassy soybean fields had more nabids than broadleaf-weed and weed-free fields (124). In conservation tillage systems in soybean fields, nabids were most numerous in no-till fields previously planted with soybeans that had not received an insecticide application (49).

Carayon (18) proposed that predatory Heteroptera (including Nabidae) could have an important role in the natural control of pest species, but felt it would be difficult to utilize them because of their polyphagy, cannibalism, and their variation in numbers over time and space. He recognized two size classes of predaceous Heteroptera. Since nabids are generally part of a group of predators, it is the summed effort of this predator guild that is important. The size classes of the Anthocoridae, Lygaeidae, Miridae, Nabidae, Reduviidae, and Pentatomidae provide a continuum able to cope with wide size range of prey (61).

## RESEARCH NEEDS

Any review discloses potential areas for productive research. For the Nabidae, important areas include the compilation of comprehensive manuals on the fauna of most areas of the world. Such publications would stimulate biological studies. One region that deserves special attention is the Hawaiian Islands; a detailed study of the fauna of this region would be likely to enrich our ideas about insular evolution. Biological studies of almost any tropical species of nabid would be of special interest, since most existing information is for temperate species. The role of nabids in natural and manipulated ecosystems, including their varying feeding strategies, needs much additional attention. Special features of the nabids, such as the common occurrence of reduced wing development, might have potential value in manipulated ecosystems; the reduced mobility of the adults might be of value in reducing pest populations (54). There seems to be an interesting series of interactions between the ground-inhabiting Prostemmatinae and the ground-dwelling Lygaeidae, which include some pest taxa (e.g. *Blissus*). Finally, much progress is being made on the chemical ecology of many insect groups, but we know almost nothing about the importance of semiochemicals in the family Nabidae (1).

ACKNOWLEDGMENTS

Special thanks are due to the late R. L. Usinger, who expanded my horizons in the Heteroptera; the late R. H. Cobben, Wageningen, friend and colleague, whose superb volumes on the Heteroptera have provided us with a rich source of information; J. Carayon, Paris, who has produced elegant work on the functional morphology of the Heteroptera, and especially the Nabidae; I. M. Kerzhner, Leningrad, for his high-quality systematics work on the Nabidae; J. Pericart, Montereau, for his recent volume on the Nabidae; T. J. Henry, Washington, and C. W. Schaeffer, Storrs, for loan of important literature; and Sir Richard Southwood, Oxford, long-time friend and colleague, who provided me with a haven for five months where much of this was written. He too has produced a rich literary legacy in the Heteroptera and in ecology, where his examples have always been drawn liberally from the Heteroptera. I also thank K. A. Bennett, Corvallis, who patiently dealt with the preparation of this manuscript, and A. Asquith, Corvallis, for valuable assistance in the acquisition of pertinent literature.

## Literature Cited

1. Aldrich, J. R. 1988. Chemical ecology of the Heteroptera. *Ann. Rev. Entomol.* 33:211–38
2. Angalet, G. W., Stevens, N. A. 1977. The natural enemies of *Brachycolus asparagi* in New Jersey and Delaware. *Environ. Entomol.* 6:97–100
3. Arnold, J. W. 1971. Feeding behavior of a predaceous bug (Hemiptera: Nabidae). *Can. J. Zool.* 49:131–32
4. Atim, A. B., Graham, H. M. 1984. Predation of *Geocoris punctipes* by *Nabis alternatus*. *Southwest. Entomol.* 9:227–31
5. Barber, H. G. 1932. Two Palearctic Hemiptera in the Nearctic fauna (Heteroptera–Pentatomidae : Nabidae). *Proc. Entomol. Soc. Wash.* 34:65–66
6. Beekman, A. G. B. 1980. The infectivity of polyhedra of nuclear polyhedrosis virus after passage through gut of an insect predator. *Experientia* 36:858–59
7. Bel'skaya, E. A., Sharov, A. A., Izhevskii, S. S. 1985. Predators of the American white web-worm moth *(Hyphantria cunea)* in the south of the European part of the USSR. *Zool. Zh.* 64:1384–91 (In Russian)
8. Benedict, J. H., Cothran, W. R. 1975. Identification of the damsel bugs, *Nabis alternatus* Parshley and *N. americoferus* Carayon (Heteroptera: Nabidae). *Pan-Pac. Entomol.* 51:170–71
9. Berest, Z. L. 1981. Predators of cereal aphids in wheat fields in the steppe zone of the right bank region of the Ukraine. In *Ekologo-Morfologicheskie Osobennosti Zhivotnykh i Sreda ikh Obitaniya,* ed. G. L. Topchii, pp. 88–90. Kiev: Nauk. Dumka (In Russian)
10. Blatchley, W. S. 1926. *Heteroptera of Eastern North America.* Indianapolis, Indiana: Nature. 1116 pp.
11. Blöte, H. C. 1945. On the systematic position of *Scotomedes* (Heteroptera–Nabidae). *Zool. Meded. Rijks Mus. Nat. Hist. Leiden* 25:321–24
12. Braman, S. K., Sloderbeck, P. E., Yeargan, K. V. 1984. Effects of temperature on the development and survival of *Nabis americoferus* and *N. roseipennis* (Hemiptera: Nabidae). *Ann. Entomol. Soc. Am.* 77:592–96
13. Burgess, L., Dueck, J., McKenzie, D. L. 1983. Insect vectors of the yeast *Nematospora coryli* in mustard, *Brassica juncea,* crops in southern Saskatchewan. *Can. Entomol.* 115:25–30
14. Buschman, L. L., Pitre, H. N., Hodges, H. F. 1984. Soybean cultural practices: effects on populations of geocorids, nabids, and other soybean arthropods. *Environ. Entomol.* 13:305–17
15. Butler, E. A. 1923. *A Biology of the British Hemiptera–Heteroptera.* London: Witherby. 682 pp.
16. Carayon, J. 1950. Caractères anatomiques et position systématique des Hémiptères Nabidae (Note préliminier). *Bull. Mus. Natl. Hist. Nat.* 22:95–101
17. Carayon, J. 1951. Les organes génitaux males des Hémiptères Nabidae: absence de symbiontes dans ces organes. *Proc. R. Entomol. Soc. London. Ser. A* 26:1–10
18. Carayon, J. 1961. Quelques remarques sur les Hémiptères–Hétéroptères: leur importance comme insectes auxiliaires et les possibilités de leur utilisation dans la lutte biologique. *Entomophaga* 6:133–41
19. Carayon, J. 1961. Valeur systématique des voies ectodermiques de l'appareil génital femelle chez les Hémiptères Nabidae. *Bull. Mus. Natl. Hist. Nat.* 33:183–96
20. Carayon, J. 1970. Étude des Alloeorhynchus d'Afrique Centrale avec quelques remarques sur la classification des Nabidae (Hemiptera). *Ann. Soc. Entomol. Fr.* (NS) 6:899–931
21. Carroll, D. P., Hoyt, S. C. 1984. Natural enemies and their effects on apple aphid, *Aphis pomi* DeGeer (Homoptera: Aphididae), colonies on young apple trees in central Washington. *Environ. Entomol.* 13:469–81
22. Chant, D. A., Lindquist, E. E. 1965. *Nabiseius duplicisetus,* a new genus and species of Otopheidomeninae (Acarina: Phytoseiidae) from nabid bugs. *Can. Entomol.* 97:515–21
23. China, W. E., Miller, N. C. E. 1959. Check list and keys to the families and subfamilies of the Hemiptera–Heteroptera. *Bull. Br. Mus. Nat. Hist. Entomol.* 8:1–45
24. Clancy, D. W., Pierce, H. D. 1966. Natural enemies of some *Lygus* bugs. *J. Econ. Entomol.* 59:853–58
25. Cobben, R. H. 1978. Evolutionary trends in Heteroptera. Part II: Mouthpart structures and feeding strategies. *Meded. Landbouwhogesch. Wageningen.* No. 289. 407 pp.
26. Cobben, R. H. 1979. On the original feeding habits of the Hemiptera (Insecta): a reply to Merrill Sweet. *Ann. Entomol. Soc. Am.* 72:711–15
27. Collyer, E. 1953. Biology of some predatory insects and mites associated with

the fruit tree red spider mite (*Metatetranychus ulmi* (Koch)) in south-eastern England. III. Further predators of the mite. *J. Hortic. Sci.* 28(2):98–113

28. Cooper, D. J. 1981. The role of predatory Hemiptera in disseminating a nuclear polyhedrosis virus of *Heliothis punctiger*. *J. Aust. Entomol. Soc.* 20: 145–50

29. Dei, E. A., Nikitenko, G. N. 1980. The predators and symbionts of the gypsy moth *Porthetria dispar* L. (Lepidoptera) in the lower cis Dneprov'ya. *Vestn. Zool.* 3:91–92 (In Russian)

30. Dicker, G. H. L. 1946. Nocturnal activity of *Nabis lativentris* Boh. (Hem.–Het., Nabidae). *Entomol. Mon. Mag.* 82:236

31. Donahoe, M. C., Pitre, H. N. 1977. *Reduviolus roseipennis* behavior and effectiveness in reducing numbers of *Heliothis zea* on cotton. *Environ. Entomol.* 6:872–76

32. Dumas, B. A., Boyer, W. P. Whitcomb, W. H. 1962. Effect of time of day on surveys of predaceous insects in field crops. *Fla. Entomol.* 45(3):121–28

33. Dupuis, C. 1947. Les proies des Sphégides chasseurs d'Hémiptères. *Feuille Nat.* (NS) 2:111–13

34. Dupuis, C. 1959. Notes, remarques et observations diverses sur les Hémiptères. Quatrieme série: notes IX–XII. *Cah. Nat.* (NS) 15:45–52

35. Dupuis, C. 1963. Essai monographique sur les Phasiinae. *Mém. Mus. Natl. Hist. Nat. Ser. A Zool.* (NS) 26:1–461

36. Edwards, J. S. 1987. Arthropods of alpine aeolian ecosystems. *Ann. Rev. Entomol.* 32:163–79

37. Ehler, L. E., Miller, J. C. 1978. Biological control in temporary agroecosystems. *Entomophaga* 23:207–12

38. Eickwort, K. R. 1977. Population dynamics of a relatively rare species of milkweed beetle (*Labidomera*). *Ecology* 58:527–38

39. Fewkes, D. W. 1961. Diel vertical movements in some grassland Nabidae (Heteroptera). *Entomol. Mon. Mag.* 97: 128–30

40. Flinn, P. W., Hower, A. A., Taylor, R. A. J. 1985. Preference of *Reduviolus americoferus* (Hemiptera: Nabidae) for potato leafhopper nymphs and pea aphids. *Can. Entomol.* 117:1503–8

41. Freeman, J. A. 1945. Studies in the distribution of insects by aerial currents. The insect population of the air from ground level to 300 feet. *J. Anim. Ecol.* 14:128–54

42. Froeschner, R. C. 1944. Contributions to a synopsis of the Hemiptera of Missouri, Pt. III. Lygaeidae, Pyrrhocoridae, Piesmidae, Tingidae, Enicocephalidae, Phymatidae, Ploiariidae, Reduviidae, Nabidae. *Am. Midl. Nat.* 31:638–83

43. Glick, P. A. 1939. The distribution of insects, spiders, and mites in the air. *US Dep. Agric. Tech. Bull.* 673:1–150

44. Goeden, R. D., Ricker, D. W. 1979. Life history of *Zygogramma tortuosa* Rogers on the ragweed, *Ambrosia eriocentra* (Gray) Payne, in southern California (Coleoptera: Chrysomelidae). *Pan-Pac. Entomol.* 55:261–66

45. Graham, H. M., Jackson, C. G. 1982. Distribution of eggs and parasites of *Lygus* spp. (Hemiptera: Miridae), *Nabis* spp. *(Hemiptera: Nabidae)*, and *Spississtilus festinus* (Say) (Homoptera: Membracidae) on plant stems. *Ann. Entomol. Soc. Am.* 75:56–60

46. Guppy, J. C. 1986. Bionomics of the damsel bug, *Nabis americoferus* Carayon (Hemiptera: Nabidae), a predator of the alfalfa blotch leafminer (Diptera: Agromyzidae). *Can. Entomol.* 118:745–51

47. Hagen, K. S., Van Den Bosch, R. 1968. Impact of pathogens, parasites, and predators on aphids. *Ann. Rev. Entomol.* 13:325–84

48. Hamilton, W. D., May, R. M. 1977. Dispersal in stable habitats. *Nature* 269: 578–81

49. Hammond, R. B., Stinner, B. R. 1987. Soybean foliage insects in conservation tillage systems: effects of tillage, previous cropping history, and soil insecticide application. *Environ. Entomol.* 16:524–31

50. Harper, A. M., Huang, H. C. 1986. Evaluation of the entomophagous fungus *Verticillium lecanii* (Moniliales: Moniliaceae) as a control agent for insects. *Environ. Entomol.* 15:281–84

51. Harris, H. M. 1928. A monographic study of the hemipterous family Nabidae as it occurs in North America. *Entomol. Am.* (NS) 9:1–98

52. Harris, H. M. 1931. Nabidae from the State of Paraná. *Ann. Mus. Zool. Pol.* 9(14):179–85

53. Harrison, R. G. 1980. Dispersal polymorphisms in insects. *Ann. Rev. Ecol. Syst.* 11:95–118

54. Hass, H. G. 1978. A remedy for flighty insects. *Agric. Res.* 26(7):8–9

55. Hendrick, R. D., Stern, V. M. 1970. Biological studies of three parasites of *Nabis americoferus* (Hemiptera: Nabidae) in southern California. *Ann. Entomol. Soc. Am.* 63:382–91

56. Henry, T. J., Lattin, J. D. 1987. Taxonomic status, biological attributes,

and recommendations for future work on the genus *Lygus* (Heteroptera: Miridae). In *Economic Importance and Biological Control of* Lygus *and* Adelphocoris *in North America. ARS-64*, ed. R. C. Hedlund, H. M. Graham, pp. 54–68. Washington, DC: US Dep. Agric. Agric. Res. Serv. 95 pp.

57. Henry, T. J., Lattin, J. D. 1988. Family Nabidae. In *Catalog of the Heteroptera of Canada and the Continental United States*, ed. R. C. Froeschner, T. J. Henry, pp. 500–12. Leiden: Brill

58. Horváth, G. 1885. Poloskákban élüsködó legiek. *Rovartani Lapok* 2(11):238–39 (In Hungarian)

59. Hussey, R. F. 1953. Some new and little-known American Hemiptera. *Occas. Pap. Mus. Zool. Univ. Mich.* 550:1–12

60. Hutchinson, G. E. 1959. Homage to Santa Rosalia or why are there so many kinds of animals? *Am. Nat.* 93:145–59

61. Irwin, M. E., Shepard, M. 1980. Sampling predaceous Hemiptera on soybean. In *Sampling Methods in Soybean Entomology*, ed. M. Kogan, D. C. Herzog, pp. 505–31. New York/Heidelberg/Berlin: Springer-Verlag. 587 pp.

62. Jackson, C. G., Graham, H. M. 1983. Parasitism of four species of *Lygus* (Hemiptera: Miridae) by *Anaphes ovijentatus* (Hymenoptera: Mymaridae) and an evaluation of other possible hosts. *Ann. Entomol. Soc. Am.* 76:772–75

63. Jessep, C. T. 1964. A note on the feeding habits of *Nabis capsiformis* Germar. *NZ Entomol.* 3(3):23

64. Johnson, C. G., Southwood, T. R. E. 1949. Seasonal records in 1947 and 1948 of flying Hemiptera–Heteroptera, particularly *Lygus pratensis* L., caught in nets 50 ft. to 3,000 ft. above the ground. *Proc. R. Entomol. Soc. London Ser. A* 24:128–30

65. Kerzhner, I. M. 1963. Beitrag zur Kenntnis der Unterfamilie Nabinae (Heteroptera Nabidae). *Acta Entomol. Mus. Natl. Pragae* 35:5–61

66. Kerzhner, I. M. 1968. Insects of the Gálapagos Islands (Heteroptera: Nabidae). *Proc. Calif. Acad. Sci.* (4th Ser.) 36(4):85–91

67. Kerzhner, I. M. 1968. Novye i maloizvestnye palearktitsheskie poluzhestkokrylye semejstva Nabidae (Heteroptera). *Entomol. Obozr.* 47:848–63

68. Kerzhner, I. M. 1970. Heteroptera Nabidae from Rennell and Guadalcanal Islands. In *The Natural History of Rennell Island, British Solomon Islands*, ed. T. Wolff, 6:39–41. Copenhagen: Dan. Sci. Press

69. Kerzhner, I. M. 1977. Zufällige Einflüge als Ursache der Funde von *Nabis capsiformis* (Heteroptera, Nabidae) in Mitteleuropa. *7te Int. Symp. Entomofaun. Mitteleur. Zusammenfassungen.* pp. 47–48. Leningrad: Nauka

70. Kerzhner, I. M. 1981. *Fauna SSSR. Nasekomye Khobotnye,* t. 13, vyp. 2, *Poluzhestkokrylye Semejstva Nabidae.* Leningrad: Akad. Nauk SSSR. 326 pp.

71. Kerzhner, I. M. 1983. Airborne *Nabis capsiformis* (Heteroptera: Nabidae) from the Atlantic, Indian and Pacific Oceans. *Int. J. Entomol.* 25(4):273–75

72. Kerzhner, I. M. 1986. Neotropical Nabidae (Heteroptera), I: a new genus, some new species, and notes on synonymy. *J. NY Entomol. Soc.* 94:180–93

73. Kiman, Z. B., Yeargan, K. V. 1985. Development and reproduction of the predator *Orius insidiosus* (Hemiptera: Anthocoridae, reared on diets of selected plant material and arthropod prey. *Ann. Entomol. Soc. Am.* 78:464–67

74. Knowlton, G. F. 1935. Beet leafhopper insect predator studies. *Proc. Utah Acad. Sci. Arts Lett.* 12:255–60

75. Koschel, H. 1971. Zur Kenntnis der Raubwanze *Himacerus apterus* F. (Heteroptera, Nabidae). Teil I, II. *Z. Angew. Entomol.* 68:1–24, 113–37

76. Krantz, G. W. 1978. *A Manual of Acarology.* Corvallis, Ore: Ore. State Univ. Book Store. 508 pp. 2nd ed.

77. Lattin, J. D. 1966. *Stalia major* (Costa) in North America (Hemiptera: Nabidae). *Proc. Entomol. Soc. Wash.* 68:314–18

78. Lattin, J. D., Oman, P. 1983. Where are the exotic insect threats? In *Exotic Plant Pests and North American Agriculture*, ed. C. L. Wilson, C. L. Graham, pp. 93–137. New York/London: Academic. 522 pp.

79. Leathwick, D. M., Winterbourn, M. J. 1984. Arthropod predation on aphids in a lucerne crop. *NZ Entomol.* 8:75–80

80. Lentz, G. L., Chambers, A. Y., Hayes, R. M. 1983. Effects of systemic insecticide-nematicides on midseason pest and predator populations in soybean. *J. Econ. Entomol.* 76:836–40

81. Leston, D. 1951. Alary dimorphism in *Nabis apterus* F. (Hem., Nabidae) and *Coranus subapterus* Deg. (Hem., Reduviidae). *Entomol. Mon. Mag.* 87:242–44

82. Leston, D. 1955. Unilateral microptery in *Nabis ericetorum* Scholtz (Hem., Nabidae). *Entomol. Mon. Mag.* 91:90–91

83. Leston, D. 1957. Spread potential and

the colonization of islands. *Syst. Zool.* 6(1):41–46

84. Leston, D. 1957. Cytotaxonomy of Miridae and Nabidae (Hemiptera). *Chromosoma* 8:609–16

85. Leston, D. 1961. Testis follicle number and the higher systematics of Miridae (Hemiptera-Heteroptera). *Proc. Zool. Soc. London* 137:89–106

86. Leston, D., Pendergrast, J. G., Southwood, T. R. E. 1954. Classification of the terrestrial Heteroptera (Geocorisae). *Nature* 174:91–92

87. Lindroth, C. H. 1957. *The Faunal Connections Between Europe and North America.* New York: Wiley. 344 pp.

88. Loan, C. 1983. Host and generic relations of the Euphorini (Hymenoptera: Braconidae). *Contrib. Am. Entomol. Inst. Ann Arbor* 20:388–97

89. Makino, S. 1951. *An Atlas of the Chromosome Numbers in Animals.* Ames, Iowa: Iowa State Univ. Press. 290 pp.

90. Martinez, D. G., Pienkowski, R. L. 1982. Laboratory studies on insect predators of potato leafhopper eggs, nymphs, and adults. *Environ. Entomol.* 11:361–62

91. McAtee, W. L. 1927. Notes on "Heteroptera or True Bugs of Eastern North America." *Brooklyn Entomol. Soc.* 22:267–81

92. McCarty, M. T., Shepard, M. Turnipseed, S. G. 1980. Identification of predaceous arthropods in soybeans by using auto radiography. *Environ. Entomol.* 9:199–203

93. McPherson, J. E., Weber, B. C. 1981. Seasonal flight patterns of Hemiptera in a North Carolina black walnut plantation. 4. Cimicoidea. *Great Lakes Entomol.* 14:19–22

94. Messina, F. J. 1982. Comparative biology of the goldenrod leaf beetles, *Trirhabda virgata* Leconte and *T. borealis* Blake (Coleoptera: Chrysomelidae). *Coleopt. Bull.* 36(2):255–69

95. Mikolajski, M. 1965. Chromosome numbers in *Nabis* Lt. (Heteroptera, Nabidae). *Experientia* 21(8):445

96. Mikolajski, M. 1967. Chromosome studies in the genus *Nabis* Lt. (Heteroptera, Nabidae). *Zool. Pol.* 17:323–34

97. Milanez, J. M. 1984. Ocorrência de artrópondos em um sistema de consórcio feijão-milho, comparado aos respectivos monocultivos. *Resumos IX Congr. Bras. Entomol. Londrina, Brasil,* p. 38. Londrina: Soc. Entomol. Bras.

98. Morrill, W. L. 1969. Parasitization of

*Nabis alternatus* by *Leucostoma simplex* in eastern South Dakota. *Ann. Entomol. Soc. Am.* 62:240

99. Muesebeck, C. F. W. 1963. Host relationships of the Euphorini (Hymenoptera: Braconidae). *Proc. Entomol. Soc. Wash.* 65:306

100. Mundinger, F. G. 1922. The life history of two species of Nabidae (Hemiptera: Heteroptera). *NY State Coll. For. Tech. Publ.* 16:149–67, plates 12–19

101. Myers, J. G. 1925. Biological notes on *Arachnocoris albomaculatus* Scott (Hemiptera: Nabidae). *J. NY Entomol. Soc.* 33:136–45

102. Nadgauda, D., Pitre, H. N. 1986. Effects of temperature on feeding, development, fecundity, and longevity of *Nabis roseipennis* (Hemiptera: Nabidae) fed tobacco budworm (Lepidoptera: Noctuidae) larvae and tarnished plant bug (Hemiptera: Miridae) nymphs. *Environ. Entomol.* 15:536–39

103. Nokkala, S., Nokkala, C. 1984. Achiasmatic male meiosis in the heteropteran genus *Nabis* (Nabidae, Hemiptera). *Hereditas* 101:31–35

104. Parshley, H. M. 1917. Fauna of New England. 14. List of the Hemiptera–Heteroptera. *Occas. Pap. Boston Soc. Nat. Hist.* 7:1–125

105. Parshley, H. M. 1923. Records of Nova Scotia Hemiptera–Heteroptera. *Proc. Acadian Entomol. Soc.* 8:102–8

106. Pericart, J. 1983. *Hémiptères Tingidae euro-méditerranéens. Faune Fr.* Vol. 69. 618 pp.

107. Pericart, J. 1987. *Hémiptères Nabidae d'Europe Occidentale et du Maghreb. Faune Fr.* Vol. 71. 185 pp.

108. Perkins, P. V., Watson, T. F. 1972. Biology of *Nabis alternatus* (Hemiptera: Nabidae). *Ann. Entomol. Soc. Am.* 65:54–57

109. Propp, G. D. 1982. Functional response of *Nabis americoferus* to two of its prey, *Spodoptera exigua* and *Lygus hesperus*. *Environ. Entomol.* 11:670–4

110. Putchkov, A. V. 1980. Distribution of species of *Nabis ferus* L. (Heteroptera) in natural biotopes and agrocenoses of certain districts in the chernozem zone (Ukranian SSR). *Vestn. Zool.* 4:89–92 (In Russian)

111. Racz, V., Visnyovszky, É. 1985. Changes in the abundance of aphidophagous Heteroptera and syrphids occurring in maize stands of different management types. *Acta Phytopathol. Acad. Sci. Hung.* 20:193–200

112. Reinert, J. A. 1978. Natural enemy complex of the southern chinch bug in

Florida. *Ann. Entomol. Soc. Am.* 71: 728–31

113. Remane, R. 1964. Weitere Beiträge zur Kenntnis der Gattung *Nabis* Latr. (Hemiptera Heteroptera Nabidae). *Zool. Beitr.* 10:253–314

114. Reuter, O. M. 1908. Bemerkungen über Nabiden nebst Beschreibung neuer Arten. *Mem. Soc. Entomol. Belg.* 15: 87–130

115. Reuter, O. M., Poppius, B. 1909. Monographia Nabidarum orbis terrestris. I. *Acta Soc. Sci. Fenn.* 37(2):1–62

116. Ridgway, R. L., Jones, S. L. 1968. Plant feeding by *Geocoris pallens* and *Nabis americoferus. Ann. Entomol. Soc. Am.* 61:232–33

117. Rohitha, B. H., Pottinger, R. P., Firth, A. C. 1985. Population monitoring studies of lucerne aphids and their predators in the Waikato. *Proc. NZ Weed Pest Control Conf.* 38:31–34

118. Rubtsov, I. A. 1977. New species of mermithids from the Far East. *Tr. Biol. Pochv. Inst. Dal'nevost. Nauchn. Tsentr Akad. Nauk SSSR* (NS) 46:149–54 (In Russian)

119. Sailer, R. I. 1983. History of insect introductions. See Ref. 78, pp. 15–38

120. Samson, P. R., Blood, P. R. B. 1980. Voracity and searching ability of *Chrysopa signata* (Neuroptera: Chrysopidae), *Micromus tasmaniae* (Neuroptera: Hemerobiidae), and *Tropiconabis capsiformis* (Hemiptera: Heteroptera: Nabidae). *Aust. J. Zool.* 28:575–80

121. Schuh, R. T. 1986. The influence of cladistics on heteropteran classification. *Ann. Rev. Entomol.* 31:67–93

122. Scudder, G. G. E. 1961. Some Heteroptera new to British Columbia. *Proc. Entomol. Soc. BC* 58:26–29

123. Scudder, G. G. E. 1985. Heteroptera new to Canada. *J. Entomol. Soc. BC* 82: 66–71

124. Shelton, M. D., Edwards, C. R. 1983. Effects of weeds on the diversity and abundance of insects in soybeans. *Environ. Entomol.* 12:296–98

125. Simberloff, D. 1986. Introduced insects: a biogeographic and systematic perspective. In *Ecology of Biological Invasions of North America and Hawaii,* ed. H. A. Mooney, J. A. Drake, pp. 1–26. New York/Berlin: Springer-Verlag. 321 pp.

126. Sloderbeck, P. E., Yeargan, K. V. 1983. Comparison of *Nabis americoferus* and *Nabis roseipennis* (Hemiptera: Nabidae) as predators of the green cloverworm (Lepidoptera: Noctuidae). *Environ. Entomol.* 12:161–65

127. Smith, F. 1855. Note on *Elachipteron brevipennis. Proc. Entomol. Soc. London.* 1855:108

128. Southwood, T. R. E. 1960. The flight activity of Heteroptera. *Trans. R. Entomol. Soc. London* 112:173–220

129. Southwood, T. R. E. 1961. A hormonal theory of the mechanism of wing polymorphism in Heteroptera. *Proc. R. Entomol. Soc. London. Ser. A* 36:63–66

130. Southwood, T. R. E. 1962. Migration of terrestrial arthropods in relation to habitat. *Biol. Rev.* 37:171–214

131. Southwood, T. R. E. 1977. Habitat, the templet for ecological strategies? *J. Anim. Ecol.* 46:337–65

132. Southwood, T. R. E. 1977. The relevance of population dynamic theory to pest status. In *Origins of Pest, Parasite, Disease and Weed Problems. 18th Symp. Br. Ecol. Soc., Bangor, Wales, 1976,* pp. 35–54. Oxford: Blackwell Sci.

133. Southwood, T. R. E., Fewkes, D. W. 1961. The immature stages of the commoner British Nabidae (Heteroptera). *Trans. Soc. Br. Entomol.* 14:147–66

134. Southwood, T. R. E., Leston, D. 1959. *Land and Water Bugs of the British Isles.* London: Warne. 436 pp.

135. Stål, C. 1873. Enumeratio Hemipterorum 3. *K. Sven. Vetenskapsakad. Handl.* 11(2):1–167

136. Stehlik, J. 1952. Fauna Heteropter Hrubeho Jeseniku. *Acta Mus. Moraviae* 37:132–248

137. Stehlik, J. 1954. Príspêvek pozání pterygo polymorphismu u Heteropter. *Acta Mus. Moraviae* 39:127–32

138. Stehlik, J. 1970. Contribution to the knowledge of Heteroptera of Moravia and Slovakia. *Acta. Mus. Moraviae* 55:209–32

139. Stichel, W. 1959–1960. Nabidae. *Illustrierte Bestimmungstabellen der Wanzen. II. Europa,* 3:185–206, 381–84. Berlin: Moller. 428 pp.

140. Stoltz, R. L., McNeal, C. D. Jr. 1982. Assessment of insect emigration from alfalfa hay to bean fields. *Environ. Entomol.* 11:578–80

141. Stoner, A. 1972. Plant feeding by *Nabis,* a predaceous genus. *Environ. Entomol.* 1:557–58

142. Stoner, A. 1973. Incidence of *Wesmaelia pendula* (Hymenoptera: Braconidae), a parasite of male *Nabis* species in Arizona. *Ann. Entomol. Soc. Am.* 66(2):471–73

143. Stoner, A., Metcalfe, A. M., Weeks, R. E. 1975. Seasonal distribution, reproductive diapause, and parasitization of

three *Nabis* spp. in southern Arizona. *Environ. Entomol.* 4:211–14

144. Stoner, A., Surber, D. E. 1969. Notes on the biology and rearing of *Anaphes ovijentatus*, a new parasite of *Lygus hesperus* in Arizona. *J. Econ. Entomol.* 62:501–2

145. Strawinski, K. 1937. Przyczynek do badan nad biologia *Nabis apterus* Fabr. *Pol. Pismo Entomol.* 14–15:349–63

146. Stys, P. 1967. Medocostidae: a new family of cimicomorphan Heteroptera based on a new genus and two new species from tropical Africa. I. Descriptive part. *Acta Entomol. Bohemoslov.* 64(6):439–65

147. Stys, P., Kerzhner, I. M. 1975. The rank and nomenclature of higher taxa in recent Heteroptera. *Acta Entomol. Bohemoslov.* 72(2):65–79

148. Sweet, M. H. 1979. On the original feeding habits of the Hemiptera (Insecta). *Ann. Entomol. Soc. Am.* 72:575–79

149. Tamaki, G., Olsen, D. P., Gupta, R. K. 1978. Laboratory evaluation of *Geocoris bullatus* and *Nabis alternatus* as predators of *Lygus. J. Entomol. Soc. BC* 75:35–37

150. Taylor, E. J. 1949. Biology of the damsel bug *Nabis alternatus*. *Proc. Utah Acad. Sci. Arts Lett.* 26:132–33

151. Taylor, E. J. 1949. A life history study of *Nabis alternatus*. *J. Econ. Entomol.* 42:991

152. Thomas, D. B. Jr. 1987. Chromosome evolution in the Heteroptera (Hemiptera): agmatoploidy versus aneuploidy. *Ann. Entomol. Soc. Am.* 80:720–30

153. Ueshima, N. 1979. Hemiptera II: Heteroptera. In *Animal Cytogenetics*, Vol. 3, *Insecta*, ed. B. John, pp. 1–117. Berlin: Borntraeger

154. Werner, F. G., Butler, G. D. Jr. 1957. The reduviids and nabids associated with Arizona crops. *Univ. Ariz. Agric. Exp. Stn. Tech. Bull.* 133:1–12

155. Whalon, M. E., Parker, B. L. 1978. Immunological identification of tarnished plant bug predators. *Ann. Entomol. Soc. Am.* 71:453–56

156. Wheeler, A. G. Jr. 1976. *Anaptus major* established in eastern North America (Hemiptera: Nabidae). *Proc. Entomol. Soc. Wash.* 78:382

157. Wilson, L. T., Gutierrez, A. P. 1980. Within-plant distribution of predators on cotton, *Gossypium hirsutum:* comments on sampling and predator efficiencies. *Hilgardia* 48(2):3–11

158. Zimmerman, E. C. 1948. *Insects of Hawaii*, Vol. 3, *Heteroptera*. Honolulu: Univ. Hawaii Press. 254 pp.

*Ann. Rev. Entomol. 1989. 34:401–21*

# CHEMICAL ECOLOGY AND BEHAVIORAL ASPECTS OF MOSQUITO OVIPOSITION

*Michael D. Bentley*

Department of Chemistry, University of Maine, Orono, Maine 04469

*Jonathan F. Day*

Florida Medical Entomology Laboratory, IFAS—University of Florida, Vero Beach, Florida 32962

## PERSPECTIVES AND OVERVIEW

Location and selection of an oviposition site is an essential part of the life history of all mosquito species. The remarkable number of oviposition behaviors range from the common, e.g. deposition of eggs on or near the water surface, to the unusual, e.g. egg brooding by adult females (73, 74). The initiation of an ovipositional flight is linked with environmental factors, especially rainfall, relative humidity, temperature, and wind speed. The location and selection of an oviposition site involves visual, olfactory, and tactile responses.

Since pathogen acquisition by vector mosquitoes usually requires the taking of at least one blood meal, disease transmission usually requires the completion of at least one oviposition cycle before pathogen transfer can occur with a subsequent blood meal. Oviposition is thus an important component of most mosquito-borne diseases.

A tremendous amount has been written concerning mosquito oviposition. A large proportion of the studies have dealt with laboratory investigations on mosquito response to chemical and physical aspects of the oviposition site; relatively few field studies have been attempted. In this review we discuss general aspects of mosquito oviposition in nature, as well as the interplay

0066-4170/89/0101-0401$02.00

of chemical, physical, and physiological factors in oviposition site selection. Considering the number of published papers in the literature on oviposition, it is not possible for us to cite all of the important work. Rather, we attempt a broad survey of selected aspects of the chemical ecology and the oviposition behavior of gravid mosquitoes.

## OVIPOSITION BEHAVIOR IN NATURE

There are many similarities between mosquito host-seeking and oviposition behaviors. Both require complex integration of physical and chemical cues by searching mosquitoes. Long-range cues, probably involving vision, allow mosquitoes to identify different habitats (14) and specific host and oviposition site characteristics. As mosquitoes approach a host or an oviposition site other cues become important. For example, olfactory cues help mosquitoes to identify $CO_2$ odor plumes from a host or volatile factors at the oviposition site. Once a host or an oviposition site has been identified, short-range cues become increasingly important. Short-range cues include temperature and chemical signals received by contact chemoreceptors. Electrophysiological studies have demonstrated that as the blood meal is digested in *Aedes aegypti,* neurons sensitive to host-produced cues, such as lactic acid, become less sensitive, while neurons sensitive to oviposition site attractants, such as methyl butyrate, become more sensitive (25).

Certain mosquito species exhibit a great deal of specialization in both host and oviposition site selection, while others are completely opportunistic with respect to both behaviors. For example, *Culex nigripalpus* mosquitoes oviposit in almost any aquatic habitat from salt marsh to artificial containers such as discarded jars, cans, and tires. Egg rafts of this species have also been found in wheel ruts and leaf axils of plants. This mosquito is also an opportunistic blood-feeder that accepts a wide range of hosts (26). Among more selective species, *Culiseta melanura* shows a strong preference for avian hosts, and its oviposition site is usually restricted to cedar or bay-head swamp potholes. This is not to suggest that all mosquito species that are ovipositional specialists can also be classified as blood-feeding specialists. Some species are extremely specific in their ovipositional requirements (e.g. *Deinocerites cancer* in flooded crab holes) but are quite opportunistic with respect to their blood-feeding behavior. Species that are opportunistic and able to oviposit in most aquatic habitats would be expected to have a distinct advantage over species that depend on specialized or rare habitats. In fact, the distributions of species with specialized oviposition sites tend to be restricted by those sites.

Once a blood meal large enough to initiate ovarian development has been obtained and the temperature-dependent egg maturation period has been completed, mosquitoes must locate suitable oviposition sites. Mosquito egg-laying has been defined as two distinct behaviors: preoviposition, which includes all of the behaviors involved in attraction to an oviposition site, and oviposition, the actual deposition of eggs on the substrate. Klowden & Blackner (62) have shown that preoviposition behavior is induced by a hemolymph-borne substance in *A. aegypti*. Preoviposition behavior may involve extensive searching flights in species such as *Aedes taeniorhynchus* that can migrate and blood-feed miles inland from the salt marsh habitat in which they oviposit. Other mosquitoes, including *Wyeomyia* spp., *Deinocerites* spp., *A. aegypti,* and *Aedes triseriatus,* are less mobile and remain in more localized habitats to seek-hosts and to feed on blood. These species use specialized oviposition sites including epiphytic bromeliads, crab holes, and natural or artificial containers. Because of their more restricted habitat their preoviposition searches are briefer, but they may rely on specialized visual cues to help identify potential oviposition sites. Egg characteristics and general oviposition strategies of mosquito subfamilies have been reviewed (81).

Oviposition flights of many species occur during twilight (36, 90). Mosquito species that display two crepuscular biting peaks (one in the evening followed by one in the morning) often also have two peaks of ovipositional flight (15). Like female mosquitoes in other physiological stages, such as those that have had a blood meal or those seeking hosts, gravid mosquitoes of species that are normally active at night fly more during moonlight than under cloudy or moonless conditions (13).

Mosquitoes use four broad ovipositional strategies in nature. Females of some species deposit individual eggs on the water surface, usually while hovering above the water. Often, these species make no contact with the water. Genera that exhibit this type of behavior include *Anopheles, Sabethes, Toxorhynchites,* and *Wyeomyia.* Mosquitoes of the genera *Coquillettidia, Culex,* and *Culiseta* deposit egg rafts directly on the water surface. These females use contact stimuli to evaluate water chemistry and other factors prior to oviposition. Some *Aedes* spp. and *Psorophora* spp. deposit individual eggs on a substrate at or above the water line. The final strategy involves the complex attachment of an egg raft to vegetation, usually below the water surface. This behavior is exhibited by members of the tropical subgenera *Mansonioides* and *Mansonia* (72), by species of *Aedeomyia,* and by some *Culex* and *Anopheles* species.

Eggs laid on a substrate above the water line are resistant to desiccation, especially since they are usually shaded from direct sunlight. The eggs can survive for many months or even years until flooding provides the hatching stimulus and a suitable habitat for completion of larval development (63). Since mosquitoes that oviposit above the water line do not require an immediate larval habitat, they can oviposit as soon as development of the egg batch is complete and an oviposition site is located. In general, females of these species are short lived. For example, in southern Florida C. *nigripalpus* and *Aedes vexans* oviposit in flooded citrus furrows during the dry winter months. Adult female C. *nigripalpus* can be collected up to 10 wk after groves are flooded, while female A. *vexans* disappear within 2 wk of the flooding. Likewise, mosquitoes that oviposit in bodies of permanent water such as sewage retention ponds and swamps can deposit their eggs as soon as egg development has been completed. On the other hand, females that lay individual eggs or rafts on the surface of temporary pools or that attach eggs to specific plants must locate, evaluate, and select oviposition sites that will be immediately habitable by the larvae. In the absence of suitable larval habitats or vegetation on which to oviposit, these mosquitoes must retain their eggs until sites become available. Thus, these species may have an extended lifetime, which may be important with respect to disease incubation and transmission by vector species (7, 26, 111).

Physiological factors such as nutritional status and insemination are known to influence mosquito ovipositional behavior. Prolonged exposure to sugar inhibits oviposition in the laboratory (27, 41). Insemination also affects oviposition in the laboratory. Matrone, a substance produced by the male accessory gland, is transferred to the female during copulation, rendering her refractory to further insemination (20). This substance is known to be an ovipositional stimulant in several mosquito species (40, 69). The influence of both carbohydrate feeding and insemination on the oviposition behavior of A. *vexans* has been examined in the laboratory (111). Inseminated females oviposited more often during the gonotrophic cycle than virgins. Carbohydrate-starved females laid an average of 49 eggs per oviposition, while sugar-fed females laid only 22. In nature, many mosquito species mate soon after adult emergence or around host animals at the time of blood-feeding, and it is likely that most females have been inseminated by the time they become gravid. However, sugar availability in the field may be important in directing the ovipositional behavior of gravid females. It is possible that depleted nutritional reserves may force gravid females to oviposit in poor or overcrowded habitats.

The oviposition behavior of several mosquito species in combination with human agricultural and industrial practices has resulted in a variety of unique and important mosquito control problems. For example, the recent introduction of *Aedes albopictus* from Asia into the United States is a direct consequence of the oviposition behavior exhibited by this species. Females often lay eggs in artificial containers, especially discarded tires. The eggs, which resist desiccation and hatch after flooding, have been transported in tires from the temperate Orient to several important port cities in the Western Hemisphere. This species has become well established and is a potential threat as a human disease vector (37, 100).

Changes in agricultural practices of the southern Florida citrus industry created an extensive new habitat for the pest mosquito *A. vexans*. This species prefers to deposit its drought-resistant eggs in grove furrows at the drip line of mature citrus trees. Under natural conditions, only exceptionally heavy rains produce enough water to flood furrows to the depth necessary to hatch eggs. In the early 1960s, growers began routine flooding of groves during periods of drought. This resulted in many more broods of *A. vexans* each year and a dramatic increase in their population in southern Florida, where at times they become one of the most important nuisance species (22).

## SELECTION OF AN OVIPOSITION SITE

Oviposition site selection by mosquitoes is a critical factor in both survival and population dynamics and has important implications with regard to mosquito control. Oviposition sites in nature cover practically the range of available aquatic niches, including artificial containers, treeholes, ponds, marshes, and plant axils (33). Intensive field studies have shown that mosquitoes are quite discriminating in selecting sites for egg deposition (75), and considerable evidence points to this site discrimination by ovipositing females as a key factor in determining larval distribution (5, 6, 39, 57, 103, 118). Although species overlap in habitat preference, oviposition site selectivity is considerably species dependent.

Oviposition site selection is the net result of the interaction of a complex array of both chemical and physical factors (reviewed in 77). In the last two decades the advent of increasingly sensitive chemical instrumentation has made possible the separation and identification of low concentrations of substances in complex mixtures. The availability of this technology has resulted in a surge of investigations on insect chemical ecology, including chemical aspects of mosquito oviposition site selection. In this review attention is focused on those studies that present evidence for a chemical role in the selection of an oviposition site. Physical factors long recognized to be extremely important in site choice include color and optical density of the site, site texture, temperature, and reflectance (19). We discuss below the interaction of these physical factors with the chemical factors involved in site selection.

To date, many of the reported studies have involved oviposition experiments in the laboratory. These studies, of course, do not necessarily reflect mosquito behavior in the field. Many of the reported behaviors may be laboratory artifacts and may have no relevance to behavior in nature. It is hoped that more field studies, with their intrinsically greater ecological significance, will be forthcoming.

Unfortunately, there is much confusion in the literature with regard to terms used to describe mosquito response in oviposition bioassays. Substances are frequently referred to as "attractants" or "stimulants" and "repellents" or "deterrents" even though experiments appropriate for providing the basis of meaningful definitions have not been conducted. Dethier et al (28) proposed that an oviposition attractant may be defined as a chemical that causes insects to make oriented movement toward the source, while an oviposition stimulant is a chemical that elicits oviposition. An oviposition repellent is a chemical that causes insects to make oriented movements away from the source, and a deterrent inhibits oviposition when present in a place where insects would, in its absence, oviposit. This terminology requires careful analysis of behavioral sequences (58–60), for which experimental design has not been adequate in most mosquito oviposition studies. Often, site selectivity has been determined by comparing egg numbers in test and control sites. In these experiments the observed response may be to olfactory or gustatory stimuli or to both. In several cases an olfactometer has been used or the substances have been assayed in screened containers where only olfactory detection was possible. Forced oviposition techniques have also been used to compare rates of oviposition in response to contact of specific mosquito parts in test substances with ovipositional responses to tap or distilled water. A few electrophysiological and organ extirpation experiments have been conducted to pinpoint the location of specific receptors.

## OVIPOSITION ATTRACTANTS AND STIMULANTS

For many mosquito species, oviposition site selection is influenced by the presence of chemical substances. The origins of these substances are wide-ranging and their molecular structures, where known, are highly variable. Since the literature on this subject is extensive, we limit our discussion to selected studies that demonstrate the diversity of these phenomena.

### Interaction of Chemical and Physical Factors

Chemicals at the oviposition site are not isolated and independent factors, but are components in a web of external influences that together may elicit oviposition site selection. The interaction of factors has been demonstrated for the container-breeding Queensland strain of A. *aegypti* in a well-designed series of combination laboratory experiments that linked humidity and visual, tactile, chemotactile, and olfactory stimuli (91). Similar complexity was demonstrated for A. *triseriatus* (120), which prefers to oviposit in rot cavities of hardwood trees but which has also been observed to oviposit in a variety of other containers, including discarded tires (84). Wilton (120) examined the effects of six environmental factors on oviposition site selection by A. *triseriatus* in the laboratory. The mosquitoes showed preferences for containers with horizontal openings, rough-textured and dark-colored walls, and dark backgrounds. Water of high optical density also proved important, and experiments with treehole water demonstrated the importance of both olfactory and contact chemoreception.

In laboratory investigations of oviposition by *A. triseriatus* in sites containing water that had previously held conspecific larvae, McDaniel et al (85) observed a strong correlation of chemical attraction with color of the site. Williams (119) and Wilton (120) had explored the effect of site color on oviposition in earlier work, but had not noted this correlation. McDaniel et al (85) observed little discrimination between sites containing attractants of larval origin and those containing only water if both were presented in highly attractive amber vessels. In the visually less attractive green or colorless containers, however, more than 95% of the eggs were deposited in sites containing larva-produced attractants. These results may indicate that visual cues are important for the location of dark treeholes, while olfactory attractants are less important for oviposition site location by this species.

Physical factors associated with the oviposition site are known to be important components of site selection. Substrate moisture is particularly important for *A. taeniorhynchus* and *Aedes sollicitans* (63), *Psorophora howardii* (92), *Psorophora columbiae* (87), and *A. vexans* (21). Surrounding vegetation is also carefully evaluated prior to oviposition (104). Many of the physical aspects of a potential oviposition site are probably evaluated visually before the mosquito initiates oviposition behavior or makes contact with water or substrate. Pool brightness is an important factor in the oviposition behavior of *Culex restuans* and *Anopheles gambiae* (8, 83). Surface reflectance is undoubtedly an important ovipositional cue for species that must locate standing water, and it is especially important for species that deposit eggs while hovering above the oviposition site (57). It is likely that species that oviposit in containers, both natural and artificial, rely on some sort of visual analysis for site location and subsequent oviposition (29, 30, 71).

## Substances of Larval Origin

Considerable evidence has accumulated for the existence of larva-produced oviposition attractants and stimulants. In laboratory experiments *Aedes atropalpus* preferentially oviposited in water that had previously held conspecific larvae (56). The active substance was stable and could be stored in solution for several weeks. The aqueous solution could be evaporated to dryness and reconstituted to restore the original activity. Although the authors referred to the substance as an oviposition attractant, appropriate experiments to differentiate olfactory and gustatory modes were not conducted, and the low volatility of the substance suggests that it may actually be a contact stimulant. *Aedes atropalpus* is a rock-pool breeder that utilizes the same temporary sites repeatedly in nature. The apparent stability and low volatility of the active substance would be advantageous in this habitat; the substance can remain active over long periods and can be reconstituted after the drying and subsequent reflooding of the rock pools.

*Aedes togoi* occurs along rocky shores in East Asia and southwestern Canada and preferentially oviposits in rock pools above the high-tide mark. For oviposition this species preferred water that had previously held larvae over water that had not (116). The larva-produced activity remained both when the larvae were first purged with kaolin and when the water containing larvae was microfiltered. The active substance resulted in greater egg deposition only when the females came in contact with the solution; it was thus a contact stimulant rather than an attractant.

With *A. triseriatus,* water in which fourth-instar larvae had been held for 48 hr induced significantly greater oviposition than distilled water (10, 85, 86). Water that had held larvae whose gut contents had been flushed by kaolin exhibited similar activity. The active substance retained activity when enclosed in a screened container within the oviposition site. When water that had held larvae was distilled under reduced pressure, the distillate, but not the reconstituted residue, displayed activity. The substance was thus an olfactory attractant. Similar oviposition attractant and stimulant activity of water in which larvae had been held has been reported for a number of other species. The phenomenon is not, however, characteristic of all species. For example, the concurrent presence of *A. gambiae* larvae reduced oviposition, while turbid water from natural breeding sites increased oviposition selectivity (83).

The origin of the larva-produced attractants and stimulants remains an important question requiring further exploration. These substances may, in some cases, be pheromones produced by the larvae, but the possibility that activity is due to fecal material or microbial metabolites must also be considered. Neither microfiltration of the oviposition medium nor flushing of the gut contents prior to site selection experiments is adequate to eliminate the latter possibility (116). In the most definitive experiment in this area, *A. atropalpus* reared under axenic conditions preferred larval rearing water over distilled water for oviposition (78). This experiment indicated that the attractive substance was probably a pheromone.

Larval density also affects site selectivity. Rearing water of higher larval density (900 larvae per liter) was repellent to ovipositing *A. atropalpus* females reared under axenic conditions (78). *Culex tritaeniorhynchus* preferred water that had held uncrowded rather than crowded conspecific larvae (99). For *Anopheles stephensi,* no such preference was evident. It was not determined whether the factor influencing *Culex* site selection was an attractant, a stimulant, or an effect of larval density, and its origin was not elucidated. This overcrowding factor, whether an inhibitor produced by the larvae or a result of excess excretory production, could have evolved as a mechanism regulating oviposition in less favorable, overcrowded sites.

## Substances of Pupal Origin

*Culex tarsalis* females preferentially oviposited in water containing conspecific pupae, exuviae, and emerging adults (42). The active substance was of low volatility and was heat stable, as evidenced by the observation that both boiled emergence water and reconstituted residue from evaporation treatments retained activity. The response was probably initiated by contact chemoreception, and the substance was probably an oviposition stimulant. The substance was species specific, since *C. tarsalis* preferred water in which conspecifics had emerged to that in which *Culiseta inornata*, *A. aegypti*, or *Culex pipiens* had emerged. Osgood (93), however, found these results (42) unreproducible in 24-hr bioassays with *C. tarsalis*. He attributed the earlier results to bacterial contamination caused by the longer assay period or to a factor added to the sites by ovipositing mosquitoes. The latter explanation would seem more consistent with the high species specificity observed.

Ovipositional preference for water that contained pupae has been demonstrated for *Culex salinarius* (2), *A. aegypti* (112), *A. atropalpus* (19), *Aedes caspius* (1), and *A. togoi* (116). However, the origin of the active substance has not been proved in any of these studies. In the case of *A. aegypti*, suspensions of bacteria cultured from water that contained pupal exuviae or incubated pupal exuviae induced increased laboratory oviposition site selectivity (101). On the other hand, *A. togoi* females showed no preference for water containing microorganisms filtered from water that had held pupae (116).

## Substances of Egg Origin

Ether washings of *Culex pipiens fatigans* (=*quinquefasciatus*) egg rafts elicited preferential oviposition by this subspecies in laboratory experiments (114). The active fraction was reported to be a mixture of 1,3-diglycerides of acetylated monohydroxy and dihydroxy fatty acids. The principal dihydroxy fatty acid methyl ester obtained by methanolysis of the 1,3-diglyceride mixture corresponded to 5,6-dihydroxyhexadecanoic acid, while the major monohydroxy methyl ester from the methanolysis corresponded to 3-hydroxytetradecanoic acid.

A later study reported that the increased oviposition response of *C. pipiens fatigans* to the presence of its own egg rafts was due to a substance associated with the egg apical droplet (17). The substance was nonspecific, since this species responded in an identical manner to egg rafts of *Culex pipiens molestus* and *C. tarsalis* (17).

Using gas chromatography–mass spectrometry (GC-MS), microchemical methods, and synthesis, Laurence & Pickett (67) demonstrated that the major component of the active substance in the *C. pipiens fatigans* apical droplet was erythro-6-acetoxy-5-hexadecanolide. Laboratory assays of the synthetic compound showed it to be as active as egg rafts containing an equivalent amount of the compound. Application of chiral stationary-phase gas chromatography demonstrated that the absolute configuration of the pheromone was (−)-(5R,6S) (66). Bioassays of the synthetic enantiomers were consistent with this stereochemical assignment. All four stereoisomers have been subjected to bioassays, which have shown the (−)-(5R,6S) isomer to be the most active (44, 105). A number of groups have synthesized the optically active pheromone (32, 70, 76, 79, 89, 106). A short synthesis of the racemic pheromone has also been reported (54). The 6-trifluoroacetoxy analog exhibited activity similar to that of the natural pheromone (16). This analog is more volatile than the natural substance, which is only effective over short distances (68); thus it may have potential for increasing the attraction range.

An oviposition pheromone has also been found associated with egg rafts of *C. tarsalis* (93). Olfactometer experiments demonstrated an attraction, but egg deposition ratios did not indicate the presence of oviposition stimulants associated with the egg rafts (94). The substance was ether soluble and of somewhat low volatility, since activity was retained in the residue when it was distilled at 40°C under vacuum. The oviposition response of *C. tarsalis* to conspecific egg rafts has been further confirmed (17). An active fraction has been chromatographically isolated from the ether extract of *C. tarsalis* egg rafts (113). The nuclear magnetic resonance (NMR) spectrum was consistent with the presence of 1,3-diglycerides of acetylated hydroxy fatty acids. Degradation experiments demonstrated that the fatty acids associated with the 1,3-diglycerides were 3-hydroxytetradecanoic acid, 3-hydroxyhexadecanoic acid, 3-hydroxyoctadeca-*cis*-11-enoic acid, and erythro-5,6-dihydroxyhexadecanoic acid. Starratt & Osgood (113) postulated that the monohydroxy acid residues were esterified with one of the positions of the glycerol moiety and the dihydroxy acid residues with the other. Acid hydroxy functions were acetylated in the diglycerides.

A chemical relationship has been noted between the oviposition attractant fraction from *C. tarsalis* egg rafts (113, 114) and the more recently reported *C. pipiens fatigans* apical droplet pheromone (67). The *C. pipiens fatigans* apical droplet pheromone, if subjected to the methanolysis conditions used by Starratt & Osgood (113), would lead to the formation of methyl-5,6-dihydroxyhexadecanoate, a product found in the 1,3-diglyceride methanolysis product mixture. Considering the apparent nonspecificity of the apical droplet attractant, it seems possible that the lactone might have been present at low levels in the *C. tarsalis* active fraction studied by Starratt & Osgood (113). If present in low concentration, it may not have been detectable in the 1,3-diglyceride fraction by 60 MHz NMR, and methanolysis would have converted it to methyl-5,6-dihydroxyhexadecanoate. The lactone as the actual active compound would be more consistent with olfactometer experiments that showed that the active compound in *C. tarsalis* egg raft washings functioned as an attractant (94). The apical droplet lactone would be more volatile than the larger 1,3-diglycerides, which contained more than 35 carbons. The case for 1,3-diglycerides as oviposition attractants for several *Culex* species, although speculative, warrants reevaluation in light of the analysis of the apical droplet pheromone of *C. pipiens fatigans* (67).

## Substances From Other Sources

Perry & Fay (96) reported that gravid *A. aegypti* exhibited an olfactory response to fatty acid esters in the laboratory. The authors concluded that site choice was mediated primarily by stimulation of olfactory receptors. The lower–molecular weight members of the acetate, propionate, and butyrate homologous series proved most active, possibly because of their higher vapor pressures. Oviposition experiments on antennectomized females indicated probable loss of olfactory orientation to chemicals. The ecological significance of these findings is not clear, since the esters were not demonstrated to be present in water from field oviposition sites. In a subsequent electrophysiological study, Davis (23) reported that the short, blunt-tipped sensilla trichodea Type II (A2II) of female *A. aegypti* responded to methyl propionate, one of the esters active as an oviposition attractant in the earlier study (96).

The terpenes carvacrol, citral, eugenol, and farnesol have also elicited preferential oviposition responses by *A. aegypti* in laboratory studies (107). Each of these terpenes was also found to inhibit egg hatching. Although these substances are all naturally occurring phytochemicals, their association with field oviposition sites has not been established.

Capric acid, a larvicide, first repelled gravid *C. restuans* and later became attractive (82). The attraction was attributed to the presence of bacteria utilizing the capric acid as a food source. In laboratory experiments, water from pools containing capric acid in the attractive phase was also effective in inducing increased oviposition by *A. aegypti* as well as by other *Culex* species. Later studies (50, 52) reported that cultures of the bacterium *Pseudomonas aeruginosa* produced an *A. aegypti* oviposition attractant, which was attributed to a metabolite, 7,11-dimethyloctadecane. Hwang & Schultz (45), however, were unable to demonstrate activity of this compound with the same assay. Various species of bacteria, including *Pseudomonas maltophilia, Escherichia coli, Enterobacter agglomerans,* and *Enterobacter aerogenes,* have acted as oviposition attractants for *Culex pipiens pipiens* in the laboratory (102). Substances that stimulated oviposition by *C. pipiens fatigans* were produced in cultures of *Pseudomonas reptilivora* isolated from the field oviposition sites of this mosquito species (53).

The synthetic insect growth regulator methoprene elicited increased oviposition by *A. aegypti* in field and laboratory experiments (18). In field experiments with methoprene briquets, test sites received an average of about twice as many eggs as untreated controls. This result has interesting implications for control applications, since areas treated for larvae may then become attractive to gravid females.

The response of gravid *A. triseriatus* to decayed wood infusions, their associated chemicals, and synthetic analogs has been reported (11, 12). A distillate of an aqueous infusion of decayed paper birch elicited increased oviposition, while extracts of freshly cut wood of the same species had no effect. A volatile, active material identified as *p*-cresol exhibited activity similar to that of the original wood distillate at 3 ppm. It also proved highly attractive when enclosed in screened vials within the oviposition sites; this experiment demonstrated the importance of an olfactory response in oviposition site selection by *A. triseriatus*. In laboratory trapping experiments significantly more mosquitoes visited sites containing *p*-cresol as bait than sites containing distilled water. Interestingly, no significant difference was noted between numbers of males and females in the traps. Whether this aggregation phenomenon has ecological significance is not known, but it is noteworthy that known oviposition attractants evoked similar electrophysiological responses in both male and female *A. aegypti* (24).

Synthetic analogs of *p*-cresol also evoked oviposition responses by *A. triseriatus* (12). Some of the analog acted as olfactory attractants, while others were contact stimulants. In electrophysiological studies, Bentley et al (9) found that one analog (4-methylcyclohexanol) and its individual *cis-trans* isomers were detected by short, pointed A2 sensilla, while *p*-cresol was detected by short, blunt A2 sensilla. An important aspect of this study is the reminder that when investigating structure-activity relationships, one must be sure that the same receptor is involved in detection of the various compounds for the correlation between molecular parameters and receptor site fit to have any significance. It is also important to note that in the studies described above, *p*-cresol was detected in a wood infusion, but the other compounds were simply structurally related. Whether ecological significance can be attached to the results will not be known until the phenols in a variety of treehole environments in the field have been investigated.

Of 150 chemicals screened in laboratory oviposition bioassays with *C. pipiens fatigans,* only two, beechwood creosote and *N*-butyl-*N*-ethyl-*o*-veratrylamine, elicited higher oviposition than the distilled water control (34). Further studies with this species (35) showed that grass infusions, methane-saturated water, and furfural elicited increase oviposition. Hay infusions also resulted in increased oviposition by *C. pipiens fatigans* (38). Tests using an olfactometer showed that an olfactory response to volatile bacterial metabolites was responsible for the observed effect.

Ikeshoji & Mulla (51) reported the results of *C. pipiens fatigans* oviposition assays using 124 compounds including alkyl carbonyls and compounds from several other classes. A positive response was recorded when mosquitoes detected the compound vapor in a stream of air. Of the compounds assayed, only eight were attractive. There was some correlation between activity and molecular structure; most of the attractive compounds were nine-carbon systems with an $\alpha$-carbon ethyl branch.

Water that previously held *C. pipiens fatigans* larvae was reported to induce increased oviposition activity in the laboratory (49). Egg-white suspension also increased oviposition, but mosquitoes from which the proboscis had been removed did not discriminate between the suspension and tap water. Ikeshoji (49) concluded that labial receptors responded to the egg-white protein. Using a forced oviposition technique in which the tarsi and proboscis of immobilized gravid females were placed in contact with various solutions, Ikeshoji (48) observed a high ovipositional response to septic tank and laboratory water and a low response to egg-white suspension. In a later report (49), the percentage of females that oviposited was higher with the forced oviposition method than in cage oviposition assays. Removal of various organs showed that detection by the proboscis tip was primarily responsible for the observed oviposition stimulation of breeding water in forced oviposition experiments; but stimulatory substances were apparently also detected by the tarsi and even by the antennae. If these observations are correct, volatile attractants might also act as stimulants after the arrival of the mosquito at the oviposition site.

## OVIPOSITION DETERRENTS AND REPELLENTS

### Salinity

Inorganic salts greatly influence mosquito oviposition behavior in the laboratory and in nature. Ovipositing mosquitoes can tolerate a variety of salt concentrations, and considerable work has focused on egg hatchability and larval survival in oviposition water of varying salinity (118, 121, 122). *Culiseta inornata* laid eggs in NaCl concentrations up to 0.1 M, but more than 50% of the larvae failed to develop in salt concentrations above 0.01 M (95). Eggs, however, were viable at the 0.1 M concentrations, which suggests that this species may choose an oviposition site based on optimal egg hatch rather than optimal larval survival (95).

*Aedes togoi* encounters a wide range of salinities in the marine rock pools that serve as its natural oviposition site. In the laboratory this species showed no preference between fresh water and water with salt concentrations up to 20 g NaCl per liter, but avoided higher concentrations (115). This behavior may be an adaptation for avoidance of media unfavorable to eggs or larvae. The high salt concentration also may alert the female that the pool is in late stages of evaporation and may not hold water long enough for the completion of larval development.

The larvae of *Anopheles albimanus* are sometimes found in estuaries that are periodically flooded with salt water (4). In the laboratory females of this species prefer to oviposit in fresh water, but they will oviposit in salt concentrations as high as 33 parts per thousand. Species normally found in salt-marsh habitats are more tolerant of salinity at ovipositional sites than species that oviposit in freshwater habitats (97, 98). In view of the negligible vapor pressure of NaCl, contact chemoreception must be the mechanism for site salinity discrimination. Consistent with this conclusion is the finding that *Toxorhynchites splendens,* an aerial ovipositor, showed no significant sensitivity to NaCl in the oviposition site even though the eggs of this species are viable only in water containing less than 1% salt (123).

## Fatty Acids

Infusions of lab chow were effective repellents to ovipositing *C. tarsalis* and *Culex quinquefasciatus.* This activity was attributed to the action of microorganisms (65). In a follow-up study, acetic, propionic, isobutyric, isovaleric, and caproic acids were identified as the active repellent compounds (43). The negative oviposition responses were quantitated (64), and the investigators suggested that where a negative oviposition response was observed, an orientational response to the source (taxis) was probably operative. Butyric acid proved to be highly toxic to first-instar *C. quinquefasciatus* at the concentration at which it occurred in the infusion. Since low–molecular weight carboxylic acids are probably formed by anaerobic fermentation of vegetation in some field sites, it is possible that the presence of these compounds betrays sites unsuitable for larval development (64).

Further work on $C_5$–$C_{13}$ carboxylic acids demonstrated that nonanoic acid is the most effective repellent to *A. aegypti, C. quinquefasciatus,* and *C. tarsalis* (46). This activity was further demonstrated under field conditions (108). Structure-activity studies have also been conducted on higher unsaturated $C_{14}$–$C_{24}$ carboxylic acids (47). Acids of Z configuration were most active. Among the acids assayed, *(Z)*-9-octadecenoic acid was the most active against *C. quinquefasciatus.*

## Phytochemicals

Phytochemicals are important mosquito oviposition attractants and stimulants. They can, however, have a negative or deterrent effect on oviposition. For example, *Lemna minor* extracts have deterred *A. aegypti* oviposition; it was demonstrated that the active substance was volatile (55). Other plant extracts have also exhibited deterrent activity against the same mosquito species (3, 109). Eucalyptol (1,8-cineole), a constituent of the volatile oil of the aster *Hemizonia fitchii* (61), and other terpenes such as citronellal and geraniol (107) act as ovipositional deterrents. The possible importance of compounds in any of these extracts in influencing mosquito egg distribution at field sites is not clear.

*Wyeomyia vanduzeei* oviposits in the leaf axils of the epiphytic bromeliad *Tillandsia utriculata*. These plants die soon after flowering, and the presence of a flower spike foreshadows the impending loss of that aquatic habitat. Gravid *W. vanduzeei* are able to identify flowering plants and avoid them as oviposition sites (31). The exact mechanism of this recognition is unknown. Laboratory experiments have ruled out vision, and it seems likely that a chemical change associated with the blooming plant repels gravid mosquitoes (G. A. Curtis, unpublished data).

## Insecticides

Several insecticides used for the control of mosquitoes repel gravid mosquitoes. Ovitraps containing 1125 ppm of Dursban® 2E repelled *A. triseriatus,* while traps containing the same amount of Abate® 5G or Dursban 1G displayed no significant repellency (80). Cypermethrin, fenvalerate, decamethrin, and permethrin have been tested with caged *Anopheles stephensi, A. aegypti,* and *C. quinquefasciatus* in laboratory oviposition experiments (117). Some repellency was observed for each of these substances against all of the species tested. In field tests with ovipositing *A. aegypti,* malathion was repellent at concentrations above 125 ppm, while Abate was repellent at concentrations above and including 50 ppm (88).

## CONCLUDING REMARKS

Substantial documentation has been presented here for the important roles of chemical substances and physical factors in influencing mosquito oviposition behavior. The investigations have in large part, however, been conducted in laboratory settings. Although continuing laboratory exploration of this type is important, future studies should also emphasize field evaluation of these factors. A number of the laboratory studies have involved screening of compounds for activity, and it is not known whether the active substances discovered actually exist in normal field oviposition sites. Microanalysis of the chemical composition of field sites should be the target of further investigation. In most cases, particularly in the work involving the influence of immature stages on oviposition behavior, the origins of the active substances have not been elucidated. For example, in the numerous studies of larva-produced attractants and stimulants, only one investigator has unambiguously defined the source of the attractive substance (78). Similar clarification of these phenomena is needed for other species. In addition, the isolation and identification of the associated active compounds should be pursued.

Another major gap is an understanding of the modes of action of substances that influence oviposition. Appropriate experimental design for differentiating attractants, stimulants, deterrents, and repellents as well as for exploring the interplay between these factors is important. A further goal of this research should be an understanding of the receptors and receptor protein involved in triggering ovipositional behavior.

Finally, observation and documentation of the basic components of oviposition behavior are needed. As computer-enhanced video equipment becomes more accessible, studies similar to those of Linley (71) and Lounibos & Linley (72) will lead to a better understanding of the behavioral components of oviposition.

ACKNOWLEDGMENTS

We thank J. D. Edman, G. A. Curtis, L. P. Lounibos, G. F. O'Meara, J. R. Rey, and D. A. Shroyer for helpful suggestions on an earlier draft of the manuscript. This review was supported in part by NIH Grant No. AI-20983. MDB thanks the Council for International Exchange of Scholars for a Fulbright Senior Research Scholar Award, under which a portion of this review was written.

Institute for Food and Agricultural Sciences, University of Florida Experiment Stations Journal Series No. 8956.

## Literature Cited

1. Adham, F. K. 1979. Studies in laboratory oviposition behavior of *Aedes caspius* (Diptera: Culicidae). *Acta Entomol. Bohemoslov.* 76:99–103
2. Andreadis, T. G. 1977. An oviposition attractant of pupal origin in *Culex salinarius*. *Mosq. News* 37:53–56
3. Angerilli, N. P. D. 1980. Influences of extracts of freshwater vegetation on the survival and oviposition by *Aedes aegypti*. *Can. Entomol.* 112:1249–52
4. Bailey, D. L. 1981. Effects of salinity on *Anopheles albimanus* ovipositional behavior, immature development and population dynamics. *Mosq. News* 41:161–67
5. Bates, M. 1940. Oviposition experiments with anopheline mosquitoes. *Am. J. Trop. Med.* 20:569–83
6. Beattie, M. V. G. 1932. The physico-chemical factors of water in relation to mosquito breeding in Trinidad. *Bull. Entomol. Res.* 23:477–96
7. Bellamy, R. E., Reeves, W. C., Scrivani, R. P. 1968. Experimental cyclic transmission of St. Louis encephalitis virus in chickens and *Culex* mosquitoes through a year. *Am. J. Epidemiol.* 87:484–95
8. Belton, P. 1967. Effect of illumination and pool brightness on oviposition by *Culex restuans* (Theobald) in the field. *Mosq. News* 27:66–68
9. Bentley, M. D., McDaniel, I. N., Davis, E. E. 1982. Studies of 4-methylcyclohexanol: an *Aedes triseriatus* (Diptera: Culicidae) oviposition attractant. *J. Med. Entomol.* 19:589–92
10. Bentley, M. D., McDaniel, I. N., Lee, H.-P., Stiehl, B., Yatagai, M. 1976. Studies of *Aedes triseriatus* oviposition attractants produced by larvae of *Aedes triseriatus* and *Aedes atropalpus*. *J. Med. Entomol.* 13:112–15
11. Bentley, M. D., McDaniel, I. N., Yatagai, M., Lee, H.-P., Maynard, R. 1979. *p*-Cresol: an oviposition attractant of *Aedes triseriatus*. *Environ. Entomol.* 8:206–9
12. Bentley, M. D., McDaniel, I. N., Yatagai, M., Lee, H.-P., Maynard, R. 1981. Oviposition attractants and stimulants of *Aedes triseriatus* (Say) (Diptera: Culicidae). *Environ. Entomol.* 10:186–89
13. Bidlingmayer, W. L. 1974. The influence of environmental factors and physiological stage on flight patterns of mosquitoes taken in the vehicle aspirator and truck, suction, bait and New Jersey light traps. *J. Med. Entomol.* 11:119–46
14. Bidlingmayer, W. L. 1975. Mosquito flight paths in relation to the environment. Effect of vertical and horizontal visual barriers. *Ann. Entomol. Soc. Am.* 68:51–57

15. Bidlingmayer, W. L., Franklin, B. P., Jennings, A. M., Cody, E. F. 1974. Mosquito flight paths in relation to the environment. Influence of blood meals, ovarian stage and parity. *Ann. Entomol. Soc. Am.* 67:919–27

16. Briggs, G. S., Cayley, G. R., Dawson, G. W., Griffiths, D. C., Macauley, E. D., et al. 1986. Some fluorine-containing pheromone analogs. *Pestic. Sci.* 17:441–48

17. Bruno, D. W., Laurence, B. R. 1979. The influence of the apical droplet of *Culex* egg rafts on oviposition of *Culex pipiens fatigans* (Diptera: Culicidae). *J. Med. Entomol.* 16:300–5

18. Carroll, M. K. 1979. Methoprene briquets as an attractant for gravid *Aedes aegypti* (L.). *Mosq. News* 39:680–81

19. Clements, A. N. 1963. *The Physiology of Mosquitoes*. New York: MacMillan. 393 pp.

20. Craig, G. B. 1967. Mosquitoes: female monogamy induced by male accessory gland substance. *Science* 156:1499–501

21. Curtis, G. A. 1985. Habitat selection strategies of mosquitoes inhabiting citrus irrigation furrows. *J. Am. Mosq. Control Assoc.* 1:169–73

22. Curtis, G. A., Frank, J. H. 1981. Establishment of *Aedes vexans* in citrus groves in southeastern Florida. *Environ. Entomol.* 10:180–82

23. Davis, E. E. 1976. A receptor sensitive to oviposition site attractants on the antennae of the mosquito, *Aedes aegypti*. *J. Insect Physiol.* 22:1371–76

24. Davis, E. E. 1977. Response of the antennal receptors of the male *Aedes aegypti* mosquito. *J. Insect Physiol.* 23:613–17

25. Davis, E. E., Takahashi, F. T. 1980. Humoral alteration of chemoreceptor sensitivity in the mosquito. *Proc. Int. Symp. Olfaction Taste,* 4:139–42. London: IRL

26. Day, J. F., Edman, J. D. 1988. Host location, blood-feeding and oviposition behavior of *Culex nigripalpus* (Diptera: Culicidae): their influence on St. Louis encephalitis virus transmission in southern Florida. *Misc. Publ. Entomol. Soc. Am.* 68:1–8

27. de Meillon, B., Sebastian, A., Khan, Z. H. 1967. Cane-sugar feeding in *Culex pipiens fatigans. Bull. WHO* 36:53–65

28. Dethier, V. G., Browne, L. B., Smith, C. N. 1960. The designation of chemicals in terms of the responses they elicit from insects. *J. Econ. Entomol.* 53:134–36

29. Focks, D. A., Sackett, S. R., Dame, D. A., Bailey, D. L. 1983. Ability of *Tox-*

*orhynchites amboinensis* (Doleschall) (Diptera: Culicidae) to locate and oviposit in artificial containers in an urban environment. *Environ. Entomol.* 12:1073–77

30. Frank, J. H. 1985. Use of an artificial bromeliad to show the importance of color value in restricting colonization of bromeliads by *Aedes aegypti* and *Culex quinquefasciatus. J. Am. Mosq. Control. Assoc.* 1:28–32

31. Frank, J. H., Curtis, G. A. 1981. Bionomics of the bromeliad-inhabiting mosquito *Wyeomyia vanduzeei* and its nursery plant *Tillandsia utriculata. Fla. Entomol.* 64:491–505

32. Fuganti, C., Grasselli, P., Servi, S. 1982. Synthesis of the two enantiomeric forms of erythro-6-acetoxy-5-hexadecanolide, the major component of a mosquito oviposition attractant pheromone. *J. Chem. Soc. Chem. Commun.* 1982(22):1285–86

33. Gillett, J. D. 1971. *Mosquitoes.* London: Weidenfeld & Nicolson. 274 pp.

34. Gjullin, C. M. 1961. Oviposition responses of *Culex pipiens quinquefasciatus* to waters treated with various chemicals. *Mosq. News* 21:109–13

35. Gjullin, C. M., Johnsen, J. O., Plapp, F. W., Jr. 1965. The effect of odors released by various waters on the oviposition sites selected by two species of *Culex. Mosq. News* 25:268–71

36. Haddow, A. J., Ssenkubuge, Y. 1962. Laboratory observations on the oviposition-cycle in the mosquito *Anopheles (Cellia) gambiae* Giles. *Ann. Trop. Med. Parasitol.* 56:352–55

37. Hawley, W. A., Reiter, P., Copeland, R. S., Pumpuni, C. B., Craig, G. B. Jr. 1987. *Aedes albopictus* in North America: probable introduction in used tires from northern Asia. *Science* 236:1114–16

38. Hazard, E. I., Mayer, M. S., Savage, K. E. 1967. Attraction and oviposition stimulation of gravid female mosquitoes by bacteria isolated from hay infusions. *Mosq. News* 27:133–36

39. Herms, W. B., Freeborn, S. B. 1921. The egg laying habits of Californian anophelines. *J. Parasitol.* 7:69–79

40. Hiss, E. A., Fuchs, M. S. 1972. The effect of matrone on oviposition in the mosquito, *Aedes aegypti. J. Insect Physiol.* 18:2217–27

41. Hudson, A. 1970. Factors affecting egg maturation and oviposition by autogenous *Aedes atropalpus. Can. Entomol.* 102:939–49

42. Hudson, A., McLintock, J. 1967. A chemical factor that stimultes oviposi-

tion by *Culex tarsalis* Coquillett (Diptera: Culicidae). *Anim. Behav.* 15:336–41

43. Hwang, Y. S., Kramer, W. L., Mulla, M. S. 1980. Oviposition attractants and repellents of mosquitoes: isolation and identification of oviposition repellents for *Culex* mosquitoes. *J. Chem. Ecol.* 6:71–80

44. Hwang, Y. S., Mulla, M. S., Chaney, J. D., Lin, G., Xu, H. 1987. Attractancy and species specificity of 6-acetoxy-5-hexadecanolide, a mosquito oviposition attractant pheromone. *J. Chem. Ecol.* 13:245–52

45. Hwang, Y. S., Schultz, G. W. 1983. Is 7,11-dimethyloctadecane an ovipositional attractant for *Aedes aegypti*? *J. Pestic. Sci.* 8:221–22

46. Hwang, Y. S., Schultz, G. W., Axelrod, H., Kramer, W. L., Mulla, M. S. 1982. Ovipositional repellency of fatty acids and their derivatives against *Culex* and *Aedes* mosquitoes. *Environ. Entomol.* 11:223–26

47. Hwang, Y. S., Schultz, G. W., Mulla, M. S. 1984. Structure-activity relationship of unsaturated fatty acids as mosquito ovipositional repellents. *J. Chem. Ecol.* 10:145–51

48. Ikeshoji, T. 1966. Studies of mosquito attractants and stimulants. Part II. A laboratory technique for obtaining mosquito eggs by forced oviposition. *Jpn. J. Exp. Med.* 36:61–65

49. Ikeshoji, T. 1966. Studies on mosquito attractants and stimulants. Part III. The presence in mosquito breeding waters of a factor which stimulates oviposition. *Jpn. J. Exp. Med.* 36:67–72

50. Ikeshoji, T., Ichimoto, I., Konishi, J., Naoshima, Y., Ueda, H. 1979. 7,11-Dimethyloctadecane: an ovipositional attractant for *Aedes aegypti* produced by *Pseudomonas aeruginosa* on capric acid substrate. *J. Pestic. Sci.* 4:187–94

51. Ikeshoji, T., Mulla, M. S. 1974. Attractancy and repellency of alkyl carbonyl compounds for mosquito oviposition. *Jpn. J. Sanit. Zool.* 25:89–94

52. Ikeshoji, T., Saito, K., Yano, A. 1975. Bacterial production of the ovipositional attractants for mosquitoes on fatty acid substrates. *Appl. Entomol. Zool.* 10:239–42

53. Ikeshoji, T., Umino, T., Hirakoso, S. 1967. Studies on mosquito attractants and stimulants. Part IV. An agent producing stimulative effects for oviposition of *Culex pipiens fatigans* in field water and the stimulative effects of various chemicals. *Jpn. J. Exp. Med.* 37:61–69

54. Jefford, C. W., Jaggi, D., Boukouvalas, J. 1986. A short, stereodivergent synthesis of the racemic erythro and threo diastereomers of 6-acetoxy-5-hexadecanolide. A mosquito oviposition attractant pheromone. *Tetrahedron Lett.* 27:4011–14

55. Judd, G. J. R., Borden, J. H. 1980. Oviposition deterrents for *Aedes aegypti* in extracts of *Lemna minor*. *J. Entomol. Soc. BC* 17:30–33

56. Kalpage, K. S. P., Brust, R. A. 1973. Oviposition attractant produced by immature *Aedes atropalpus*. *Environ. Entomol.* 2:729–30

57. Kennedy, J. S. 1942. On water-finding and oviposition by captive mosquitoes. *Bull. Entomol. Res.* 32:279–301

58. Kennedy, J. S. 1977. Olfactory responses to distant plants and other odor sources. See Ref. 110, pp. 67–91

59. Kennedy, J. S. 1977. Behaviorally discriminating assays of attractants and repellents. See Ref 110, pp. 215–29

60. Kennedy, J. S. 1978. The concepts of olfactory "arrestment" and "attraction." *Physiol. Entomol.* 3:91–98

61. Klocke, J. A., Darlington, M. V., Balandrin, M. F. 1987. 1,8-Cineole (eucalyptol), a mosquito feeding and ovipositional repellent from volatile oil of *Hemizonia fitchii* (Asteraceae). *J. Chem. Ecol.* 13:2131–41

62. Klowden, M. J., Blackmer, J. L. 1987. Humoral control of pre-oviposition behaviour in the mosquito, *Aedes aegypti*. *J. Insect Physiol.* 33:689–92

63. Knight, K. L., Baker, T. E. 1962. The role of the substrate moisture content in the selection of oviposition sites by *Aedes taeniorhynchus* (Wied.) and *A. Sollicitans* (Walk.). *Mosq. News* 22:247–54

64. Kramer, W. L., Hwang, Y. S., Mulla, M. S. 1980. Oviposition of mosquitoes: negative responses elicited by lower aliphatic acids. *J. Chem. Ecol.* 6:415–24

65. Kramer, W. L., Mulla, M. S. 1979. Oviposition attractants and repellents of mosquitoes: oviposition responses of *Culex* mosquitoes to organic infusions. *Environ. Entomol.* 8:1111–17

66. Laurence, B. R., Mori, K., Otsuka, T., Pickett, J. A., Wadhams, L. J. 1985. Absolute configuration of mosquito oviposition attractant pheromone, 6-acetoxy-5-hexadecanolide. *J. Chem. Ecol.* 11:643–48

67. Laurence, B. R., Pickett, J. A. 1982. Erythro-6-acetoxy-5-hexadecanolide, the major component of a mosquito oviposition attractant pheromone. *J. Chem.*

Soc. Chem. Commun. 1982(1):59–60
68. Laurence, B. R., Pickett, J. A. 1985. An oviposition attractant pheromone in Culex quinquefasciatus Say (Diptera: Culicidae). Bull. Entomol. Res. 75:283–90
69. Leahy, M. G., Craig, G. B. 1965. Accessory gland substance as a stimulant for oviposition in Aedes aegypti and Aedes albopictus. Mosq. News 25:448–52
70. Lin, G., Xu, H., Wu, B., Guo, G., Zhou, W. 1985. Studies on the identification and synthesis of insect pheromones. XXI. Stereoselective synthesis of all the possible optical isomers on the mosquito oviposition attractant pheromone. Tetrahedron Lett. 26:1233–36
71. Linley, J. R. 1987. Aerial oviposition flight of Toxorhynchites amboinensis (Diptera: Culicidae). J. Med. Entomol. 24:637–50
72. Lounibos, L. P., Linley, J. R. 1987. A quantitative analysis of underwater oviposition by the mosquito Mansonia titillans. Physiol. Entomol. 12:435–43
73. Lounibos, L. P., Machado-Allison, C. E. 1983. Oviposition and egg brooding by the mosquito Trichoprosonon digitatum in cacao husks. Ecol. Entomol. 8:475–78
74. Lounibos, L. P., Machado-Allison, C. E. 1987. Female brooding protects mosquito eggs from rainfall. Biotropica 19:83–85
75. Macan, T. T. 1961. Factors that limit the range of freshwater animals. Biol. Rev. 36:151–98
76. Machiya, K., Ichimoto, I., Kirihata, M., Ueda, H. 1985. A convenient synthesis of four stereoisomers of 6-acetoxy-5-hexadecanolide, the major component of the mosquito oviposition attractant pheromone. Agric. Biol. Chem. 49:643–49
77. Maire, A. 1983. Sélectivité des femelles de moustiques (Culicidae) pour leurs sites d'oviposition: état de la question. Rev. Can. Biol. Exp. 42:235–41
78. Maire, A. 1985. Effect of axenic larvae on the oviposition site selection by Aedes atropalpus. J. Am. Mosq. Control. Assoc. 1:320–23
79. Masaki, Y., Nagata, K., Kaji, K. 1983. Enantiospecific synthesis of (+)-erythro-(5S,6R)-6-acetoxy-5-hexadecanolide, an optically active form of the major component of a mosquito oviposition attractant pheromone. Chem. Lett. 12:1835–36
80. Mather, T. N., DeFoliart, G. R. 1983. Repellency and initial toxicity of Abate and Dursban formulations to Aedes

triseriatus in oviposition sites. Mosq. News 43:474–79
81. Mattingly, P. F. 1971. Ecological aspects of mosquito evolution. Parassitologia Rome 13:31–65
82. Maw, M. G. 1970. Capric acid as a larvicide and an oviposition stimulant for mosquitoes. Nature 227:1154–55
83. McCrae, A. W. R. 1984. Oviposition by African malaria vector mosquitoes. II. Effects of site tone, water type and conspecific immatures on target selection by freshwater Anopheles gambiae Giles, sensu lato. Ann. Trop. Med. Parasitol. 78:307–18
84. McCray, E. M., Wilton, D. P., Schoof, H. F. 1967. Preliminary observations concerning the absence of Aedes aegypti in Savannah, Georgia and nearby areas. Proc. 54th Annu. Meet. NJ Mosq. Exterm. Assoc., Atlantic City, pp. 152–58. New Brunswick, NJ: NJ Mosq. Exterm. Assoc.
85. McDaniel, I. N., Bentley, M. D., Lee, H.-P., Yatagai, M. 1976. Effects of color and larval-produced oviposition attractants on oviposition of Aedes triseriatus. Environ. Entomol. 5:553–56
86. McDaniel, I. N., Bentley, M. D., Lee, H.-P., Yatagai, M. 1979. Studies of Aedes triseriatus (Diptera: Culicidae) oviposition attractants. Evidence for attractant production by kaolin-treated larvae. Can. Entomol. 111:143–47
87. Meek, C. L., Williams, D. C. 1986. Diel periodicity of oviposition and soil moisture preference of Psorophora columbiae. J. Entomol. Sci. 21:185–90
88. Moore, C. G. 1977. Insecticide avoidance by ovipositing Aedes aegypti. Mosq. News 37:291–93
89. Mori, K., Otsuka, T. 1983. Pheromone synthesis. 59. Synthesis of both the enantiomers of erythro-6-acetoxy-5-hexadecanolide. The major component of a mosquito oviposition attractant pheromone. Tetrahedron 39:3267–69
90. Nielsen, E. T., Nielsen, A. T. 1953. Field observations on the habits of Aedes taeniorhynchus. Ecology 34:141–56
91. O'Gower, A. K. 1963. Environmental stimuli and the oviposition behavior of Aedes aegypti var. queenslandensis Theobald (Diptera: Culicidae). Anim. Behav. 11:189–97
92. Olson, J. K., Meek, C. L. 1977. Soil moisture conditions that are most attractive to ovipositing females of Psorophora columbiae in Texas ricelands. Mosq. News 37:19–26
93. Osgood, C. E. 1971. An oviposition pheromone associated with the egg rafts

of *Culex tarsalis. J. Econ. Entomol.* 64:1038–41

94. Osgood, C. E., Kempster, R. H. 1971. An air-flow olfactometer for distinguishing between oviposition attractants and stimulants of mosquitoes. *J. Econ. Entomol.* 64:1109–10

95. Pappas, L. G., Pappas, C. D. 1983. Laboratory studies on the significance of NaCl as an oviposition deterrent in *Culiseta inornata. Mosq. News* 43:153–55

96. Perry, A. S., Fay, R. W. 1967. Correlation of chemical constitution and physical properties of fatty acid esters with oviposition response of *Aedes aegypti. Mosq. News* 27:175–83

97. Petersen, J. J. 1969. Oviposition response of *Aedes sollicitans, Aedes taeniorhynchus,* and *Psorophora confinnis* to seven inorganic salts. *Mosq. News* 29:472–83

98. Petersen, J. J., Rees, D. M. 1966. Selective oviposition response of *Aedes dorsalis* and *Aedes nigromaculis* to soil salinity. *Mosq. News* 26:168–74

99. Reisen, W. K., Siddiqui, T. F. 1978. The influence of conspecific immatures on the oviposition preferences of the mosquitoes *Anopheles stephensi* and *Culex tritaeniorhynchus. Pak. J. Zool.* 10:31–41

100. Reiter, P., Darsie, R. F. Jr. 1984. *Aedes albopictus* in Memphis, Tennessee (USA): an achievement of modern transportation? *Mosq. News* 44:396–99

101. Roberts, D. R., Hsi, B. P. 1977. A method of evaluating ovipositional attractants of *Aedes aegypti* (Diptera: Culicidae) with preliminary results. *J. Med. Entomol.* 14:129–31

102. Rockett, C. L. 1987. Bacteria as ovipositional attractants for *Culex pipiens* (Diptera: Culicidae). *Great Lakes Entomol.* 20:151–55

103. Rudolfs, W., Lackey, J. B. 1929. The composition of water and mosquito breeding. *Am. J. Hyg.* 9:160–80

104. Russell, P. F., Rao, T. R. 1942. On the relation of mechanical obstruction and shade to ovipositing of *Anopheles culicifacies. J. Exp. Zool.* 91:303–29

105. Sakakibara, M., Ikeshoji, T., Machiya, K., Ichimoto, I. 1984. Activity of four stereoisomers of 6-acetoxy-5-hexanecanolide, the oviposition pheromone of culicene mosquito. *Jpn. J. Sanit. Zool.* 35:401–3

106. Sato, T., Watanake, M., Honda, N., Fujisawa, T. 1984. Synthesis of both enantiomers of erythro-5-acetoxy-5-hexadecanolide, the major component of

a mosquito oviposition attractant pheromone. *Chem. Lett.* 7:1175–76

107. Saxena, K. N., Sharma, R. N. 1972. Embryonic inhibition and oviposition induction in *Aedes aegypti* by certain terpenoids. *J. Econ. Entomol.* 65:1588–91

108. Schultz, G. W., Hwang, Y. S., Kramer, W. L., Axelrod, H., Mulla, M. S. 1982. Field evaluation of ovipositional repellents against *Culex* (Diptera: Culicidae) mosquitoes. *Environ. Entomol.* 11:968–71

109. Sharma, R. N., Joshi, V., Jadu, G., Bhosale, A. S., Gupta, A. S., et al. 1981. Oviposition deterrence activity in some Lamiaceae plants against insect pests. *Z. Naturforsch. Teil C* 36:122–25

110. Shorey, H. H., McKelvey, J. Jr., eds. 1977. *Chemical Control of Insect Behavior.* New York: Wiley. 414 pp.

111. Shroyer, D. A., Sanders, D. P. 1977. The influence of carbohydrate-feeding and insemination on oviposition of an Indiana strain of *Aedes vexans* (Diptera: Culicidae). *J. Med. Entomol.* 14:121–27

112. Soman, R. S., Reuben, R. 1970. Studies on the preference shown by ovipositing females of *Aedes aegypti* for water containing immature stages of the same species. *J. Med. Entomol.* 7:485–89

113. Starratt, A. N., Osgood, E. E. 1972. An oviposition pheromone of the mosquito *Culex tarsalis:* diglyceride composition of the active fraction. *Biochem. Biophys. Acta* 280:187–93

114. Starratt, A. N., Osgood, C. E. 1973. 1,3-Diglycerides from eggs of *Culex pipiens quinquefasciatus* and *Culex pipiens pipiens. Comp. Biochem. Physiol. B* 46:857–59

115. Trimble, R. M., Wellington, W. G. 1979. Effects of salinity on site selection by ovipositing *Aedes togoi* (Diptera: Culicidae). *Can. J. Zool.* 57:593–96

116. Trimble, R. M., Wellington, W. G. 1980. Oviposition stimulant associated with fourth-instar larvae of *Aedes togoi* (Diptera: Culicidae). *J. Med. Entomol.* 17:509–14

117. Verma, K. V. S. 1986. Deterrent effect of synthetic pyrethroids on the oviposition of mosquitoes. *Curr. Sci.* 55:373–75

118. Wallis, R. C. 1954. A study of oviposition activity of mosquitoes. *Am. J. Hyg.* 60:135–68

119. Williams, R. E. 1962. Effect of coloring oviposition media with regard to the mosquito *Aedes triseriatus* (Say). *J. Parasitol.* 48:919–25

120. Wilton, D. P. 1968. Oviposition site selection by the tree-hole mosquito, *Aedes triseriatus* (Say). *J. Med. Entomol.* 5:189–94

121. Woodhill, A. R. 1938. Salinity tolerances and pH range of *Culex fatigans* Wied. with notes on the anal papillae of salt water mosquitoes. *Proc. Linn. Soc. NSW* 63:273–81

122. Woodhill, A. R. 1941. The oviposition responses of three species of mosquitoes *Aedes (Stegomyia) aegypti* Linnaeus, *Culex (Culex) fatigans* Wiedemann, *Aedes pseudoskusea concolor* Taylor, in relation to the salinity of the water. *Proc. Linn. Soc. NSW* 66:287–92

123. Yap, H. H, Foo, A. E. S. 1984. Laboratory studies on the oviposition site preference of *Toxorhynchites splendens* (Diptera: Culicidae). *J. Med. Entomol.* 21:183–87

*Ann. Rev. Entomol. 1989. 34:423–51*

# GUILDS: THE MULTIPLE MEANINGS OF A CONCEPT

*Charles P. Hawkins*

Department of Fisheries and Wildlife and Ecology Center, Utah State University, Logan, Utah 84322–5210

*James A. MacMahon*

Department of Biology and Ecology Center, Utah State University, Logan, Utah 84322–5305

## INTRODUCTION

Root (161) defined a *guild* as "a group of species that exploit the same class of environmental resources in a similar way." With this intuitively appealing, seductively simple definition, he commenced the development of what has become a Gordian knot of ecological thinking. Root saw the term as a way to group "together species without regard to taxonomic position, that overlap significantly in their niche requirements." He also felt that one advantage to the guild concept was that it "focuses attention on all sympatric species involved in a competitive interaction, regardless of their taxonomic relationship" (161, p. 335).

Nearly 20 years later, Terborgh & Robinson (195, p. 90) cited the concept of guild as the new perspective in ecology: "Guilds will become the standard currency of ecologists in their efforts to understand community relationships of many kinds." Yet few authors have critically examined the implications of Root's definition or provided an operational definition for it in their studies. Also, many authors have developed meanings for the term that seem quite different from Root's emphasis and intent. For example, Balon (6) used the term to classify "reproductive styles" in fish, while Bambach (7) used charac-

423

0066-4170/89/0101-0423$02.00

teristics of basic body plans and physiological systems (Baupläne) of fossil marine invertebrates as a partial basis for his guilds.

One reason the concept of guild is so appealing and so often invoked is that after observing any natural system long enough, workers studying particular groups of organisms have difficulty placing their subjects into strictly defined subunits of communities or ecosystems, such as trophic levels. They also realize that their particular taxonomic group is not the only one using a particular resource. This observation leads to a resource-centered classification of community components rather than one based on taxonomic affinities, an emphasis consistent with Root's guild concept. This tendency is not new. For example, Elton (48, pp. 63–68) alluded to a guildlike niche in animal communities when he stated "we might take as a niche all the carnivores which prey upon small mammals, and distinguish them from those which prey upon insects." Interestingly, the literature of the period 1967–1973 contains several references to guildlike phenomena developed without knowledge of Root's definition. For example, Evans & Murdoch (50) stated, "We have been impressed by the evidence that the feeding activities of the organisms impose a pattern on the community which overrides taxonomic composition." Their studies dealt with terrestrial arthropods, but aquatic insects were the subject of a similar convergence in thinking (35).

When Root codified this common experience by adopting a familiar term, not surprisingly many community ecologists wholeheartedly embraced the term, but they flavored it with their own experiences and shades of personal meaning. Despite the apparent common need for a term such as *guild* to describe a resource-oriented, organizational unit within communities that transcends taxonomic boundaries, current opinion regarding the guild concept is mixed. Some view guilds as natural units worthy of detailed study, while others deny that guilds exist except in the minds of ecologists. We believe this multiplicity of viewpoints arises because few workers have critically addressed the dimensions of the word. Thus the literature is full of the self-assured use of the word *guild* but lacks a systematic account of the concept's implications.

In this chapter we review the various explicit and implicit meanings attributed to the term *guild*. Related terms, such as *functional group,* are compared with the guild concept. We also consider the ways that *guild* has been used to describe natural ecological units. Finally, we analyze the usefulness of the concept, its validity as a level of organization in nature, and its prospective use as a basis for management decisions.

Our review emphasizes studies of invertebrates, especially arthropods. Because the concept of guild is prominent in studies of vertebrates, however, we also refer to those vertebrate studies that provide general points about the guild concept.

# HISTORICAL USAGE, SHADES OF MEANING, AND SYNONYMS

To obtain some quantitative sense of the prevalence of the term *guild* in the ecological literature, we conducted a bibliographic search of the citation frequency of Root's original paper (161) (Figure 1A). The citation frequency has gradually but unevenly increased over the years. A noticeable decline occurred around 1983, followed by a spate of citations paralleling a reawakening of interest in community ecology (42, 53, 63, 80, 91, 192). Root's monograph discussed much more than guilds, but inspection of these 358 papers revealed that the usage of the term *guild* was most often the only reason for citation. In fact, the real scientific content of Root's manuscript seems to have been overshadowed by the new definition.

Because not all authors who discuss guilds cite Root, we conducted a search of titles and abstracts for the word *guild* (Figure 2). Of the 432 references found through 1986, only about 100 treated invertebrates. The infrequency of the term in the invertebrate literature in part stems from the alternative use of the term *functional group,* coined by Cummins (35) for

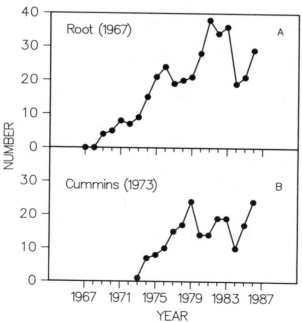

*Figure 1*    Number of times Root (161) *(A)* and Cummins (35) *(B)* were cited each year between 1967 and 1986. Data compiled from *Science Citation Index* (84a).

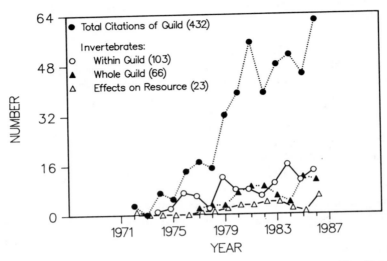

*Figure 2*  Number of papers published per year containing the word *guild* in either the title or abstract. Total citations include papers for any animal taxon. We placed papers on invertebrates into three categories, which are not mutually exclusive. Within-guild papers addressed patterns of resource partitioning or competition. Whole-guild papers addressed factors affecting overall abundance of one or more guilds. The third group of papers examined the effects of guilds on resource abundance or ecosystem processes. Data were compiled from a computer search of BIOSIS. Papers containing the phrase *functional group* but not the word *guild* are not included, because we would not have been able to extract guild-related papers from those citing chemical functional groups.

aquatic invertebrates (Figure 1*B*). MacMahon et al (106) attempted to differentiate these two terms because they seemed so similar, and by 1984 the terms were accepted as synonymous (38). In fact, some authors used functional groups as descriptors of guilds (154).

Problems with understanding of the guild concept occurred early. Therefore MacMahon et al (106) attempted to define *guild* explicitly in a general discussion of community and ecosystem theory. Later, Jaksic (85) took a similar approach and argued for a distinction between true (resource-based) community guilds and taxonomically based assemblage guilds. Although he indicated that workers should use quantitative techniques to define guilds, rather than a priori classifications, several attempts to do so have met with mixed success.

## CONCEPTS AND ECOLOGICAL IMPLICATIONS OF GUILDS

Ecologists have used the guild concept in three ways. Their different perspectives have strongly influenced the specific questions addressed and the avenues of inquiry used to consider them. One perspective views the guild as the ecologically appropriate context for the study of interspecific competition. In the extreme of this viewpoint, guild members are the only individuals likely to compete. A second viewpoint is strongly utilitarian and uses guild classifications to simplify the complex and numerous interactions among species that characterize real communities and ecosystems. A vast amount of species-specific information is thus condensed into a few "black boxes" that have generalized attributes. The third perspective views guilds as natural ecological units that recur across communities; this perspective is not always distinct from the other two. Root emphasized the first and third perspectives, whereas Cummins emphasized the last two. In this section we discuss each of these views and consider their ecological implications.

## Guilds as Arenas of Interspecific Competition

Ostensibly, a major advantage of a guild approach is that the importance of competition to community structure can be assessed only if interactions between the entire cast of participating characters are understood. Before the development of the guild perspective, studies of interspecific competition and resource partitioning generally focused on taxonomically related species (reviewed in 152). Insights generated from these studies catalyzed more than 50 years of research on the process of competition, but the emphasis on members of the same taxocene limited inferences regarding the importance of competition to whole communities or ecosystems. The demonstration that ants compete with rodents and birds as well as with other ants (17) illustrates the value of this idea.

Root's original idea is now so firmly established in the ecological literature that guilds are often considered "arenas of intense interspecific competition, with strong interactions within guilds but weak interactions between members of different guilds" (142). Many of the comparative and experimental studies of resource partitioning and interspecific competition now define the species of interest as members of the same guild (e.g. 41, 54, 87, 109, 146, 153, 155). The emphasis of many recent papers is on whether and how competition shapes guild structure—i.e. the composition, relative abundance, and number of species observed within guilds (e.g. 4, 22, 97, 107, 175, 176). Because of this emphasis the guild concept is used to discuss all of the aspects of competition and niche related to a wide variety of ecological phenomena including size-ratio differences between guilds (and the related Dyar's constant) (49, 145) and even global patterns of ecological equivalents (195).

## Community and Ecosystem Analysis

COMMUNITY STRUCTURE    If guilds represent the basic building blocks of ecosystems, communities should share similarities that transcend species composition. Elton (48) discussed this idea as early as 1927. He argued that "although the actual species of animals are different in different habitats, the ground plan [in terms of ecological types] of every animal community is much the same." Others extended Elton's basic idea by recognizing that the use of a natural, functional classification could facilitate comparison of communities that are either overwhelmingly complex in terms of number of species (147, 161, 162, 189) or poorly known taxonomically (5, 36).

Furthermore, if each guild exploits different types of resources, guild composition of communities should reflect the availability of resource types. This idea has permeated the literature on most ecosystem or habitat types [e.g. oceans (115, 189), fresh water (148, 201, 206), trees (127, 191), grasslands (50, 87), other plants (100], but it appears to have most strongly influenced the study of freshwater benthic invertebrate assemblages. The functional classification of Cummins (35; see also 120) is almost universally used by ecologists studying freshwater invertebrates and is apparently based on the hypothesis that "the relative dominance of invertebrate groups shifts with differences in available sources of energy" (37, pp. 147–148; see also 39, 74, 125, 199).

The idea that guilds are groups of functionally similar species also leads to the idea that ecosystems contain many functionally redundant species all capable of performing the same ecosystem function (132, 134). If so, guild structure may be more predictable and stable than either the abundance of individual species or species composition (50, 75, 134, 162, 196). Presumably, density compensation within guilds should maintain overall guild abundance at or near carrying capacity, while the fortunes of different species within a guild vary individually in response to factors other than resource availability—e.g. weather, predators, local conditions (162).

ECOSYSTEM ANALYSIS    Another important implication of the guild concept is that it provided a conceptual bridge that might unify community and ecosystem approaches to ecology. If different guilds exploit different kinds of resources, the processes of energy flow and material cycling that occur within ecosystems should closely correspond to community structure in terms of guild composition. Focusing on interactions between guilds and their resources offered a means of describing the structure and dynamics of food webs (5, 16, 36, 134) more realistically than was possible with traditional food-chain models (e.g. 103, 132). This last idea may largely explain why functional classifications are now used so frequently by ecosystem ecologists (3, 60, 106, 117, 121, 126, 171, 187).

## Guilds as Natural Ecological Units

The concept of guild has engendered the idea that communities are comprised of a limited number of relatively discrete functional subunits (16, 36, 118,

162). This idea, if true, has profound implications for the study and understanding of natural ecosystems. Root (161, p. 336) suggested that guilds represented major ecological categories that have been "molded by adaptation to the same class of resources." Root & Chaplin (164; see also 163) subsequently introduced the term *adaptive syndrome* to describe a "coordinated set of characteristics that are associated with . . . specialization on a . . . resource." They proposed that consumers have only a limited number of ways to obtain the resources they need, and that natural selection has molded the morphological and behavioral traits of species to fit one of these "occupations." Other ecologists have proposed similar ideas (108, 189). In general, the structure of the environment (habitat) is thought to impose selective pressures that cause sympatric species to evolve similar adaptive solutions for resource exploitation (see 108).

This type of reasoning is part of the basis for the guild classification that Root proposed (161). Others have subsequently constructed similar classifications. For example, Steneck & Watling (189; cf 7, 209) based a functional classification of marine invertebrate herbivores on the concept of adaptive zones (sensu 183, 188). They argued that relatively few adaptive zones exist in an evolutionary landscape and that one or more species occupies each of these zones.

Convergent evolution is also an important concept here, because it is often cited as the means by which similar adaptive syndromes or "ecological equivalents" arise among different taxa in geographically isolated ecosystems (26, 132, 135, 195). Morphological or behavioral equivalents are thought to represent the manifestation of similar evolutionary pressures to produce individuals that perform similar ecological roles.

## APPLICATIONS

In ecology, the fit between concept and reality is often poorly known because new ideas are generated faster than empirical information becomes available. Fortunately, however, the abundance of guild-related studies conducted during the past 20 years permits us to evaluate the utility of some of the ideas and concepts described above. We explore five related questions:

1. What a priori criteria have ecologists used to classify guilds?
2. Are invertebrate assemblages organized into a few natural, relatively discrete ecological groups?
3. Is the guild composition of natural communities predictable based on the resource availability?
4. Are guild structure and abundance more stable than species composition?
5. Have guild classifications facilitated understanding of ecosystem processes?

Although space limitations preclude an evaluation of whether guild theory has enhanced the understanding of competitive relationships (see 4, 29, 98, 101, 168 for review), we discuss in conclusion the adequacy of the guild concept as a framework for understanding such interactions.

## A Priori Classifications: Assigning Species to Investigator-Defined, Resource-Based Guilds

Table 1 lists some of the criteria used to define ecological groupings of invertebrate and other taxa. Although specific criteria for classification depended on author and situation, most classifications were based on presumed differences in the use of food resources. In some cases, groupings were largely based on the broad trophic categories of herbivore, detritivore, and predator (75, 127). In others, larger trophic groups were divided into guildlike groups based on feeding specializations or exact time or site of feeding. Regardless of the level of resolution, the classifications appear similar in that the taxa were grouped by differential use of a priori, investigator-defined sets of resources.

Guildlike classifications have frequently been used to classify invertebrate consumers over the past 20 years. For example, Evans & Murdoch (50) recognized that some herbivorous insects were specialized leaf-and-stem or flower feeders, whereas others belonged to a more general herbivore group. Root (162) formally introduced the term *guild* into the entomological literature when he divided folivores on collards into "three distinct guilds based on mode of attacking the plant" (i.e. pit feeders, strip feeders, and sap feeders). Several other presumed guilds of terrestrial insects have since been suggested, including chewers (strip feeders), miners, suckers (sap feeders), gall formers, seed feeders, predators, parasites, wood borers, stem borers, nectivores, omnivores (scavengers), fungivores, epiphyte grazers, and dung feeders (e.g. 19, 51, 57, 79, 82, 96, 100, 126, 127, 139, 155, 174, 191). Some authors even recognized a tourist (i.e. visitor) guild (127), even though such a grouping has no obvious basis in similarity of resource use.

Similar groupings have been erected for aquatic consumers. Cummins and coworkers (35, 37, 38) recognized six guilds (functional feeding groups) of freshwater invertebrates based on both food resource and mode of feeding: large particle shredders, small particle collectors, scrapers of algae and other attached organic material, macrophyte piercers, engulfers (predators), and parasites. Like Root, Cummins suggested that mode of feeding corresponds closely to the nature of the resource class. The resource class was defined based on the size of the material and its location (e.g. attached organic matter, deposits of fine or coarse detritus, suspended material). The mode of feeding was defined by the morphological structure of the mouthparts and the behavior of the consumer.

**Table 1** Examples of criteria used to define functional groupings in the entomological and ecological literature

| Consumers | Criteria[a] | Ref. |
|---|---|---|
| Grassland insects | Food habits | 50 |
| Model organisms | Energy source | 205 |
| Arthropods | Trophic position | 75 |
| Herbivorous insects | Exploitation of the same resource in the same way | 162 |
| Aquatic invertebrates | Morphobehavioral mechanisms of food acquisition | 35 |
| Flies, ants, dragonflies, drosophilids, bark beetles | Taxonomic relationship | 41 |
| | | 119 |
| | | 128 |
| | | 175 |
| | | 30 |
| | | 65 |
| Forest insects | Trophic level, feeding type, and reproductive rate | 33 |
| | | 171 |
| Marine polychaete worms | Set of relations among food particle size and composition, mechanism involved in food intake, and motility patterns associated with feeding | 89 |
| | | 52 |
| Phytophagous insects | Feeding types | 100 |
| Marine benthos | Ways in which [species] exploit their substratum environment and nature of their effects on the substratum | 209 |
| Grasshoppers | Coexisting species that use resources in a similar fashion | 87 |
| Spiders | Hunting manners | 137 |
| Marine herbivorous molluscs | Use of certain aspects of the environment in similar ways | 189 |
| Whole communities (40 studies) | Sets of species in a given ecosystem with the property that every pair in the set has some food resource in common | 210 |
| Marine benthos | Food source, space utilization, and Baupläne | 7 |
| Arthropods in trees | Feeding habits | 127 |
| | | 191 |
| Exopterygote insects | Trophic categories | 19 |
| Stream insects | Trophic and functional categories and habitat preferences | 54 |
| Mayfly larvae | Gut contents | 64 |
| | | 70 |
| Multispecies assemblages (29 studies) | Preference for a common set of key resources | 2 |
| Aquatic insects | Fatty acid composition | 67 |
| Protozoa | Type and size of food | 147 |
| | | 148 |

**Table 1**   (*Continued*)

| Consumers | Criteria[a] | Ref. |
|---|---|---|
| Cicadas | Closely related species (usually co-ordinal or confamilial) that are both sympatric and synchronously active and that forage on similar items in similar ways | 109 |
| Marine invertebrates | Mobility, size, and taxon | 15 |
| Soil arthropods | Principal food source, mode of feeding, reproductive rate, defense against predation, or distribution | 126 |

[a]Criteria are listed as closely as possible to that given in the original citations.

Unambiguous classification of species into guilds requires that species fit within some set of morphological or behavioral criteria that define an adaptive syndrome. However, explicit criteria have seldom been used. Species have often been assigned to groups based on either loosely defined criteria or assumptions regarding their feeding biology. Although a few authors clearly acknowledged these problems (e.g. 9, 23, 50, 127, 191), many others did not.

One reason for these assumptions is that the biology of most of these species is poorly known. As a consequence, many authors assumed that species were in the same guilds as taxonomically related species (e.g. 50, 127). These assumptions were often justified by the generalization that genera or families represent major ecological themes and species are minor variations on those themes (e.g. 206; see also functional group assignments in 120). The validity of such generalizations is largely untested, however, and species in at least some genera exhibit pronounced variation in resource use and clearly belong in different guilds (e.g. 64, 70, 156, 191, 197).

Operational placement of species into guilds has often been hampered by pronounced polyphagy and ontogenetic shifts in the diet of species (70, 191). Some studies have placed opportunistic taxa into a separate, generalist guild [e.g. ants (75) and snails (73)]. Species that are moderately opportunistic or that have different diets depending on life stage or sex (e.g. mosquitoes) are more problematic. The individuals of these species are logically members of two or more guilds.

If opportunistic feeding is common, at least two serious implications arise regarding the conceptual and pragmatic utility of the guild framework. First, the idea that the consumptive strategies of different species can be understood in terms of a few adaptive syndromes is compromised. Secondly, operational comparison of community structure becomes problematic because the percentage of the consumptive activity of opportunistic individuals that should be apportioned to each guild is seldom apparent (e.g. 74). This problem appears especially significant in freshwater systems. So many species of aquatic insects exhibit some degree of omnivory (e.g. 27, 64, 70, 130) that Hutchinson (83) termed them "selective omnivores." Extreme omnivory may not be as significant for terrestrial taxa. Nonetheless, variability in diet within terrestrial taxa can be sufficient to affect interpretations of the guild structure of terrestrial communities (e.g. 191, 204). These problems notwithstanding, the material actually assimiliated by consumers may differ sufficiently that taxa can be functionally distinguished from one another (e.g. 61, 67). If invertebrates generally exploit different sets of resources based on biochemical criteria, however, the criterion of similar way of using resources imposes severe practical difficulties for guild classification.

Adams (2) contended that groups of sympatric species were frequently referred to as a guild based on little or no objective evidence. To check the validity of assignments, he applied a technique called psychophysical unfolding to 38 published data sets (17 invertebrate assemblages). This technique determines whether the resource-use preferences of a group of species can be resolved to fit a single resource axis, a criterion Adams believed was necessary for guild membership. Based on his analyses, 28 sets of species appeared to be properly classified as guilds in the sense that species within groups partitioned a single resource dimension. Sets of species that did not fit his criterion included the guilds described by Moran & Southwood (127; see also 191).

Although Adams determined that many species groups were real guilds as defined by his criteria, he skirted the more important issue of what were the proper criteria for defining guilds. To Adams, guild membership depended on the criteria of sympatry and resource partitioning along a single niche axis. To Root (161), Moran & Southwood (127), and others, guilds were groups of species occupying similar functional roles within communities. The disparity between these definitions prompts several questions. Over what spatial scale, if any, should sympatry be a criterion for guild membership? In how many ways should species be functionally similar? Should guilds be defined in terms of a single resource dimension (e.g. food) or multiple dimensions? We believe that lack of consideration of these questions is the basis for significant confusion in the literature.

## A Posteriori Descriptions: Are Guilds Natural Ecological Groupings?

If guilds are naturally occurring ecological groups, a posteriori detection of groups of ecologically similar species within assemblages of consumer species should be possible. Over the past 10 years, a few studies have attempted to describe the guild composition of naturally occurring groups of consumers. With multivariate statistics, it is possible to represent the distributions of species in niche space and look for ecological groupings of species. Inger & Colwell (84) and Pianka (142) used nearest-neighbor analysis of pairwise species overlaps and principal component analysis (PCA) to examine amphibian and reptile assemblages. Others used cluster analysis and PCA to delineate guilds of birds (77, 78, 95) and assemblages of grasshoppers (86, 87) and mayflies (70).

Yodzis (210) used graph theory to determine whether 40 community food webs contained guildlike sets of species, i.e. cliques. He defined a clique as "a set of species with the property that every pair in the set has some food source in common." Cliques are therefore hierarchical in nature, but a dominant clique contains no other cliques and might therefore be regarded as a trophic guild that need not meet the criterion of similar way of using resources. Yodzis found that the number of dominant cliques increased with species richness and that cliques overlapped less than expected by chance, i.e. guildlike groupings of species occurred.

Although these analyses suggest that ecological aggregations of functionally similar species may exist in nature, interpretations are not usually straightforward. The cluster and ordination analyses, for example, implied that species were clumped in niche space. In all cases, however, the structure of these assemblages was clearly hierarchical, and the boundaries suggested by some authors seemed artificial (e.g. 78). Also, some analyses can produce artifacts and thus promote incorrect inferences. Based on an apparent relationship between food web connectance and the point correlation coefficient, for example, McNaughton (118) suggested that species of grass were organized into guilds (sensu 142). His conclusions, however, were based on an artifact created by his analysis (68, 99).

## Predictability and Stability of Guild Structure

If guilds are the basic building blocks of ecosystems, guild structure should be predictable in both space and time. This idea is conceptually appealing to many ecologists who either explicitly or implicitly accept the view that populations exist in dynamic balance with available resources. To date, however, the collective evidence that communities are so highly organized is equivocal. In this section we review studies that examined (a) whether guild composition is similar in systems with similar resources, (b) whether guild composition shifts predictably with spatially varying resource availability, and (c) whether guild composition shifts temporally.

SPATIAL PATTERNS    Moran & Southwood (127) compared the guild composition of arthropod assemblages on trees from Britain and South Africa. Stork (191) conducted a similar analysis of arthropods on trees in Borneo and compared his results with those of Moran & Southwood. These studies showed that in terms of species richness, the relative composition of the major guilds was similar across sites and species of broad-leaved trees. Guild composition varied considerably between sites, however, both in relative numbers and in biomass of individuals. For example, sap suckers represented 55, 18, and 12% and chewers represented 6, 9, and 14% of total individuals from trees in Britain, South Africa, and Borneo, respectively. Corresponding biomass values were 39 and 13% for sapsuckers and 15 and 32% for chewers from trees in Britain and South Africa. Stork did not report biomass values. Moran & Southwood also found that guild structure varied significantly between broad- and narrow-leaved trees; narrow-leaved trees had proportionately more chewers and sap suckers and fewer predators.

A few data sets have been used to compare convergence in restricted groups of species. Lawton (96) showed that convergence in resource use is not strong among the herbivore species on bracken from England and the southwestern United States. Bracken in Arizona and New Mexico was undersaturated with species. Some parts of bracken in the southwestern sites were never used, and no equivalent species existed. Lawton also observed no evidence for either niche expansion or density compensation in the assemblages he studied. In contrast, Morton & Davidson (129) expected the Australian ant fauna to be more abundant and diverse than the North American fauna because of the absence of competing rodents (cf 17). They found however, that the two faunas were similar in terms of species richness and abundance. Furthermore, the Australian ant fauna did not appear to be in equilibrium with resources. These results imply that similarities in the structure of insect faunas may stem from factors other than the use of a similar resource class and that faunas may differ strongly even when their resources are similar.

A much larger set of comparative studies on guild composition exists for invertebrate consumers in stream ecosystems. This abundance of studies appears to be due to acceptance by many stream ecologists of functional groups (35) as an appropriate means to describe consumer assemblages. In addition, the 1980 publication (199) of a general conceptual model for river ecosystems has prompted worldwide evaluation attempts.

Although many studies of guild structure in stream ecosystems have been published (e.g. 8, 12, 21, 24, 25, 34, 44, 45, 59, 71, 73, 74, 76, 92, 93, 105, 111, 112, 114, 122, 124, 125, 140, 141, 160, 166, 167, 185, 198, 208), their combined results are as equivocal as those for terrestrial systems. Some studies reported positive correlations between guild abundance and resource availability (8, 74, 140). A few reported patterns of guild abundance more or less consistent with presumed patterns of resource availability but did not quantitatively describe relationships between resource abundance and guild abundance (e.g. 12, 34, 44, 167). One study showed that the relative abundance of guilds could shift among sites even though food resources did not appear to do so (25). Several other studies showed that the relative abundance of different guilds did not vary as expected with respect to either presumed or measured differences in resource availability (45, 73, 111, 114). The strongest trend in these studies was for invertebrate consumers to respond rather similarly as a whole to variation in the resource spectrum. Hunt & Shure (82) also noted similar responses among terrestrial invertebrates.

A common problem in interpreting data from both terrestrial and freshwater studies was that resources were often more ambiguously defined than the presumed guilds (see above). Of the few quantitative studies, most used crude measurements of resource availability. Considering the importance of resource quality to growth and fitness of invertebrates (55, 61, 184), the quantification of resources used in many of these studies may not adequately characterize resource availability. It is therefore impossible to evaluate critically whether the observed differences in the relative abundance of guilds between sites was in fact related to differences in resource availability. It is probably significant that the strongest and most consistent associations between guild and resource abundance were observed when resource availability could be measured with reasonable confidence (e.g. for predators and prey).

Several factors other than food resources could influence guild composition and hence confound comparisons. The character of the foraging substrate (e.g. shape and size) appears to affect proportions of different guilds. For example, Moran & Southwood (127) showed that for tree arthropods the proportion of guilds was associated with shape of leaves. Coyle (32) found that the amount of litter affected the abundances of different spider guilds. Robinson (159) found that the three-dimensional architecture of the habitat also affected the proportion of spider guilds. Proportions of stream invertebrate guilds were found to vary with particle size of the stream bed (12, 185). In contrast, no differences in spider guild proportions could be attributed to the shape of leaf litter (190).

TEMPORAL DYNAMICS    If resource availability varies with time, guild structure might shift as resources change. Several authors have addressed seasonal shifts in guild structure, general stability of guild structure, or long-term successional patterns. Georgian & Wallace (54) concluded that the life histories and hence production dynamics of species in a periphyton-grazing guild tracked the seasonal availability of algae in streams. Similar arguments have been given for other groups. Cummins (36, 37) argued, for example, that a tendency for shredders to start growth in autumn evolved in response to the pulsed autumnal inputs of leaf material into streams. Other factors, including seasonal changes in temperature (e.g. 165) and durational stability of habitats (186) may also affect the seasonal dynamics of guilds, however. For example, Rader & Ward (153) suggested that the growth periods and production dynamics of a guild of "scraper-collectors" was best explained by life history adjustment to physical factors (presumably seasonal flooding); this interpretation was in direct contrast to that of Georgian & Wallace (54). Although the phenologies of some species within some guilds may be tightly coupled to temporal variation in resources, it is unclear how well guilds as a whole track resources. The answer to this question probably depends in part on how narrowly guilds are defined.

Few studies have explored successional shifts in guild composition of invertebrate faunas after disturbance. Some (18–20) showed that for arthropods associated with plants, most groups increased in density with successional stage. On the other hand, the relative abundance of different groups varied with time. Epiphyte grazers increased in relative abundance, while sap suckers declined. These trends appeared to be consistent with changes in resource availability. Minshall et al (123) also observed an apparent trophic reorganization in the invertebrate fauna of a stream after a catastrophic flood. Algal grazers appeared first, followed by detritus collectors and then predators. The appearance of different groups did not seem to track closely the availability of resources, however. Gore (56) observed a similar pattern of differential colonization after construction of a new stream channel, although he did not compare guild abundances to availability of resources.

Successional shifts in the guild composition of more narrowly defined taxocenes have also been noted. Hanski & Koskela (66) observed that dung specialists (coprophages) were the first beetle species to colonize droppings and that more generalist species (saprophages) appeared later. The arrival of saprophages apparently coincided with the appearance of fungal hyphae, a required food source. Parmenter & MacMahon (138) noted a significant difference in the guild structure of the beetle fauna across a successional gradient on reclaimed strip mines. Recently reclaimed plots had fewer herbivores, dung feeders, and predators and more fungivores and insect-carrion feeders than undisturbed plots. Trophic structure also appeared to track changes in resource availability(138).

Results of several other studies imply that functional equilibria may be characteristic of many natural communities. For example, the trophic composition of a grassland insect fauna did not change even though species composition shifted over time (50). Heatwole & Levins (75) also suggested consumer assemblages exist in trophic equilibrium. They reported that the arthropod assemblages on the defaunated mangrove islands studied by Simberloff & Wilson (179, 181, 182, 207) rapidly returned to an equilibrium structure in the proportion of species in different trophic groups.

Wallace et al (202) recently conducted a whole-stream manipulation similar to the mangrove experiments conducted by Simberloff & Wilson. They noted that two years after experimental poisoning, the guild composition of a stream had returned to predisturbance sturcture in terms of standing biomass even though strong differences in taxonomic composition persisted. Wallace et al concluded, as Heatwole & Levins (75) had, that functional equilibria had been established even though the original species composition had not. As a whole, these results suggest that invertebrate assemblages are strongly organized around functional roles and that guilds may be the basic building blocks of communities. The validity of these conclusions is clouded, however, by contrary observations in other systems and potential errors in interpretation caused by statistical artifact. For example, Root (162) tested the idea that compensatory shifts in arthropod species on collards would promote the kind of stability reported by Evans & Murdoch (50). He found little evidence that guild structure was more stable than species composition. In general, guild structure varied in concert with the abundances of a few dominant species, a result also observed for disturbed stream systems (71).

Simberloff (178, 180) questioned the conclusions of Heatwole & Levins (75) because the observed equilibrium could be explained as a passive consequence of random colonization from a large species pool. Cole (28) similarly criticized the conclusions of Evans & Murdoch (50). Teraguchi et al (194) also noted that although the trophic structure of insect assemblages of defaunated patches rapidly returned to that found in surrounding old-fields, the observed patterns could have resulted from random colonization. These problems in interpretation may also apply to the conclusions of Wallace et al (202). If colonization is largely a random process, the guild structure of a recovering community may simply reflect the overall guild composition of the surrounding species pool. The recolonization of several streams differentially disturbed by the eruption of Mount St. Helens (71) is consistent with this interpretation of the colonization process. Even though the streams differed in the types and amounts of resources available to consumers, their guild composition was nearly identical five years after disturbance. For these assemblages, overall species composition appeared to vary randomly, and the presence of a few dominant species strongly influenced overall guild structure.

Recently, the analysis of guild composition has also been applied to fossil assemblages. Bambach (7) assigned species from fossil marine faunas (mostly invertebrates) to guilds to examine changes in "ecospace utilization." Based on these data he concluded that the increase in species richness between the Paleozoic and Cenozoic eras resulted from an addition of guilds and that resource partitioning (species packing) within guilds generally remained unchanged. Presumably, community structure was tracking changes in resource availability via radiation of new species into new adaptive syndromes. Unfortunately, the absence of direct data on resource availability prevents critical evaluation of this idea.

## Ecosystem Analyses

Ecosystem science has gained prominence because the simulation modeling of ecosystem functions promises a quantitative predictability that is only now becoming common in ecology. Additionally, the creation of a model has heuristic value in the sense that an investigator must develop a blueprint of the system, and this activity reveals voids in knowledge of system components and processes.

A common finding during ecosystem analysis is that one cannot hope to model all of the species individually. Thus, the need to aggregate species into functional units, or guilds, emerges rapidly. Botkin (16) correctly asserted that during the modeling process, aggregating a variety of organisms into biologically based "black boxes" could reduce an ecosystem's bewildering complexity to "mathematically tractable proportions." This aggregation is implicit in ecosystem treatises that do not even mention the term *guild* (e.g. 133). The guild concept provides a convenient, supposedly biological basis for simplifying ecosystem models. Whether guilds have aided in the elucidation of ecosystem processes, however, remains unclear.

That guilds are often based on food resources makes them useful for ecosystem studies because of the universal emphasis on the flow of matter and energy. Even in the theoretical literature, the convenience of a group such as a guild is clear, e.g. in theories related to food web design and nutrient cycling (143), theories on diet selection as an ecosystem process (47), and in the application of ecological theories to insects (116). McNaughton (118), using cybernetic analysis, has even suggested that "organized blocks" of species can act as guilds.

Many examples illustrate the usefulness of a guildlike approach, especially for its heuristic value. Recent reviews of the role of arthropods in below ground detrital food webs have emphasized a functional group approach (172) and have led to the development of simulation models (126). Studies of detritus processing by aquatic macroinvertebrates have similarly used a functional group approach (3, 117, 121), as have analyses of secondary production in aquatic insects (11). Guildlike approaches were used to examine arthropod herbivory in forests (170) and streams (60) by dividing taxa into groups such as folivores, sap-feeders, and grazers and by emphasizing the plant as a heterogeneous resource in space and time (60, 170). Wallwork (203) similarly grouped oribatid mites and emphasized that any one species can participate in several food-based "feeding types" simultaneously (203). Recent studies of trophic interactions have criticized the reliance of the Lindeman model (103) on history of energy flow rather than on an assessment of present resource states; these studies imply the usefulness of guilds but do not require a guild approach per se (31). Despite this wealth of examples of the utility of the guild concept, guild theory does little to elucidate ecosystem structure and functioning, and most papers have not adequately addressed how to allocate the polyphagous activities of guild members among trophic levels. The utility of the guild concept does not prove that guilds exist in nature or that systems are organized around groups of organisms such as guilds. Given these shortcomings, the uncritical use of the guild concept in the literature of applied ecology is surprising. This adoption has implications for the care with which we transfer ecological information to management-oriented individuals and agencies.

## APPLIED ECOLOGY AND THE CONCEPT OF GUILD

As the guild concept caught on, it was so thoroughly embraced that it took the form of a theory or truism. The concept was subsequently suggested as a basis for assessing environmental impacts of anthropogenic disturbances as required by law in the United States (National Environmental Policy Act of 1969) and elsewhere.

Although invertebrates are not the usual objects of these applied studies and recommendations, the political pressure created by proposals to use guilds for management purposes, coupled with the poor understanding of the problems associated with the term's ambiguity, is instructive. In time, similar proposals will be made for the management of beneficial and appreciated invertebrates such as butterflies. The guild or functional group approach is already used for management-related studies by members of United States governmental agencies such as the National Park Service. Also, recent studies of the response of invertebrates to anthropogenic disturbances have used a guild-based approach (32, 73, 110, 136, 138).

The early applied work often attempted to develop fool-proof guild classification schemes. Johnson (88) presented a dichotomous key to ecological characteristics of plants that was to "provide a clear-cut placement of plant species into a specific guild thus eliminating much of the subjectivity of classification." The method, based on growth form, leaf persistence, ability to fix nitrogen, and other quantifiable components, identified 95 guilds. Simultaneously, Severinghaus (173) classified mammalian and avian guilds into 30 and 31 divisions, respectively. Short & Burnham (177) followed with a classification of wildlife guilds that relied strongly on the physical strata and vegetative structure used by vertebrates. They formally applied a variety of statistical analyses to a matrix of species and habitats and produced a system that was supposed to be useful in modeling the relationships between wildlife species and their habitats.

Severinghaus (173) strongly implied that guilds were objectively defined, real assemblages of organisms that interacted as natural units. Landres (94) responded by stating that the guild approach is useful for environmental impact assessements but may not represent a real level of organization in nature. Additionally, Landres suggested that a theoretical basis for guilds was lacking. About that time, Hairston (62) argued that, at least for the salamanders that he studied, guilds may not exist. In a recent monograph (63), however, he used a guild concept as a major organizing theme. Verner (200) railed at the suggestion that management techniques could be developed using a single indicator species for each guild rather than all guild members. He also appealed for the use of a matrix approach to define guilds, in which the rows and columns consisted of primary feeding and nesting zones. Note that the emphasis in the literature had changed since 1967. Initially guilds were discussed generally for their importance in dynamic processes such as competition. By the 1980s, however, guilds were defined on the basis of a series of habitat-characteristic variables, with the implied assumption that co-occurrence in one of these habitats assured competition. A further change from the original view of guilds occurred when Verner (200) coined the phrase *management guild* to classify a group of species that respond in a similar way to a variety of changes likely to affect their environme. t. In essence his definition relied on the summed response of the organisms to a change in their habitat, rather than on their role. This approach was challenged (113) because it masked individual species responses.

By 1985, despite the wide variety of usages and inconsistency of approaches in management, DeGraaf et al (40) concluded that the guild concept had become a useful tool for looking at competition, niche separation, and functional relationships within the community. These authors then produced a guild classification for North American birds based on food, feeding substrate, and foraging technique. In a sense this classification was the extreme application of previous methods that were used for subdividing guilds within local taxocenes (95).

Szaro (193) registered serious reservations about the wide variety of characteristics used to define guilds and postulated that guilds based on the analysis of one or two niche dimensions were too simplistic to describe nature. He decried the a priori, investigator-defined nature of guilds recommended for management purposes and urged extreme caution before wholehearted acceptance of the guild concept as a basis for management decisions. Roberts (158), apparently unaware of Szaro's paper, again suggested a matrix approach to guild definition, in which he included layers of vegetation and other features of the environment that are used for feeding and breeding. He admitted that the ease of use of guilds was previously exaggerated but stated that the concept is still worthwhile. Block et al (14) emphasized the use of guild-indicator species in resource management.

This series of papers highlights a problem that plagues the guild concept wherever it is used: Definitions are loose and are inconsistent among investigators (or even among studies by the same investigator), and guilds are seldom analyzed before they are accepted as existing in nature. The guild concept seems to have developed a life of its own, and its disciples forget that it rests on verbal arguments, not on experimental data.

## CONCLUDING REMARKS

Guilds are most useful when they include a variety of interacting taxa. This perspective in ecology has fostered great interest and is consistent with Root's (161) original intention. For a variety of reasons, however, guilds are still most often studied, we believe inappropriately, within a narrow taxocene. In recent explicit review of guilds and their utility in ecology, Terborgh & Robinson (195) suggested that the "special value of the concept as applied to ecology is that guilds can be more or less objectively defined independently of the particular species that comprise them." Nonetheless, they later cited examples to show that guild members share the same resources, occur in the same or overlapping microhabitats, and are taxonomically related. They stated that taxonomic relatedness has been assumed "because of limitations of time, methodology, and the expertise of the investigators." This series of statements suggests that the taxocene remains the first constraint on guild membership simply for practical reasons. We believe that such a constraint is biologically inappropriate.

Guilds are most useful when species co-occur in space and time. If they are offset in either dimension, a guild approach may yield heuristic value, but the farther the separation, the less likely it is that species or individuals will influence each other in measurable ways. Thus, we doubt that a fossil and a living species would truly be in the same guild, nor would a red oak, leaf-eating insect species in the Ozarks and one in Virginia.

A caveat is in order. Our approach emphasizes the interaction of guild components. The functional classification approach to the guild concept uses a broader community perspective, in which direct comparison of the components of similar communities in different places or at different times is desirable. Such a perspective does not use or require a detailed analysis of the effect of and participants in a specific interaction.

The crux of Root's (161) original definition, that guild members use the resource in the same manner, is not always useful. The problem of defining "similar manner" is complex. In spiders, for example, an assemblage of species that all capture prey in shrubs might be divided into separate guilds such as web-builders, stalkers, ambushers, and so on (1, 69, 146), even though they are all generalist predators of the same types of arthropod prey with, in fact, substantially overlapping diets. From another perspective, species with a wide variety of strategies for hunting arthropods may form a single guild specifically because they are all generalist predators, regardless of how they get their prey. For example, Polis & McCormick (144) defined a guild of scorpions, solpugids, and spiders on desert flats. In fact, a less parochial community viewpoint might also require the inclusion of lizards, some birds (169), and a variety of insects in a ground-dwelling, generalist, arthropod-eating guild. Regardless of the manner used, these disparate species interact as each obtains the same resource.

To state the question with a bit of hyperbole, does it matter that a particular insect species is captured by a silken spider web as opposed to a bird's beak? The ecosystem and community consequences are similar—one less insect of that species—and manner is irrelevant from that specific perspective.

The viewpoint stated above emphasizes that the definition of guilds should be related to the interaction of organisms in nature. If one uses guilds only as a classification scheme for community components, however, then a similar manner of resource use is an important criterion. Root's original intention seems to be related to classification. Recent use of the guild concept is related to community interactions, especially competition. We reiterate that we do not relate guilds solely to competition, nor do we infer that competition is only an intraguild phenomenon. Clear examples demonstrate interguild competition (e.g., 13, 72, 90, 209) and relationships other than competition, such as mutualism.

The "same resource" should be clearly defined; the resource definition probably depends on the specific ecological questions being addressed, rather than on some natural rule of definition. For example, all 10 arthropods living as phyllophages of beech *(Fagus silvatica)* have been termed a guild (131). This approach groups species that feed on one specific part of one plant species. In contrast, the parasites of insect species have been differentiated by the hosts' life history stage; each stage is said to support a different guild of parasites (43, 149–151). Is a life-cycle stage equivalent to a plant part? Perhaps for some questions it is and for others it is not. In a granivore guild the same part (seed) of several species is involved, and in a grazer guild several parts (leaves, flowers, stems) of numerous species may be the unit resource. Clearly, a single resource ranges from a specific morphological part of one species, to one life history stage, to nearly all parts of several species, depending on the specific question being asked and the particular organisms being studied.

The problems stated above show why *guild* is such an elusive term. The definition of a guild depends on many user-defined parameters that have no absolute guidelines. Even though there have been some recent theoretical excursions into the nature of guilds (2), their properties (175), and their basis as the foundation of communities (63), confusion (perhaps better stated as multiple viewpoints) remains. Ecologists should be mindful of the problems attendant to the use of this concept.

The heuristic value of the general concept of guild continues unabated, however; even general ecology textbooks organize whole sections (46, pp. 326–91) of community discussions around the elusive guild concept. The value of the concept stems partly from the intuitive appeal of guilds as a classification system based on the attributes of community components. We agree that the concept is useful and will use it ourselves, but we hope to make our assumptions and definitions explicit.

The concept of guild may be a perfect example of a problem that was recently elucidated by Loehle (104):

> Proper hypothesis testing is the subject of much debate in ecology. According to studies in cognitive psychology, confirmation bias (a tendency to seek confirming evidence) pervasively influences actual problem solving and hypothesis testing, often interfering with effective testing of alternative hypotheses. On the other hand, these psychological factors play a positive role in the process of theory maturation by helping to protect and nurture a new idea until it is suitable for critical evaluation. As a theory matures it increases in empirical content and its predictions become more distinct. Efficient hypothesis testing is often not possible when theories are in an immature state, as is the case in much of ecology. Problem areas in ecology are examined in light of these considerations, including failure to publish negative results, misuses of mathematical models, confusion resulting from ambiguous terms such as "diversity" and "niche," and biases against new ideas.

The guild concept has been subject to the sins of confirmation bias and theory tenacity. Although its empirical content has increased, we believe that the concept is a useful but artificial construct of the minds of ecologists. The immaturity of the guild hypothesis may be responsible, inevitably, for the lack of adequate testing of this notion.

To us, *guild* still describes all organisms that use the same investigator-defined resource; the usefulness of the concept depends more on the investigator's acuity and care than it does on the organisms and their interactions in nature.

Acknowledgments

Two "Rootlets," Ted Evans and Frank Messina, critiqued this manuscript, to its benefit. Research in the laboratory of CPH has been supported by NSF grants BSR-8306892 and BSR-8416127; that in the laboratory of JAM has been supported by NSF grants DEB-7904534, DEB-8101827, BSR-8317358, DEB-8022641, and DEB81-16914.

*Literature Cited*

1. Abraham, B. J. 1983. Spatial and temporal patterns in a sagebrush steppe spider community (Arachnida, Araneae). *J. Arachnol.* 11:31–50
2. Adams, J. 1985. The definition and interpretation of guild structure in ecological communities. *J. Anim. Ecol.* 54:43–59
3. Anderson, N. H., Sedell, J. R. 1979. Detritus processing by macroinvertebrates in stream ecosystems. *Ann. Rev. Entomol.* 24:351–77
4. Arthur, W. 1987. *The Niche in Competition and Evolution*. New York: Wiley. 175 pp.
5. Bahr, L. M. Jr. 1982. Functional taxonomy: an immodest proposal. *Ecol. Modeling* 15:211–33
6. Balon, E. K. 1975. Reproductive guilds of fishes: a proposal and definition. *J. Fish. Res. Board Can.* 32:821–64
7. Bambach, R. K. 1983. Ecospace utilization and guilds in marine communities through the Phanerozoic. In *Biotic In-*

teractions in Recent and Fossil Benthic Communities, ed. M. J. S. Tevesz, P. L. McCall, pp. 719–46. New York: Plenum. 837 pp.

8. Barmuta, L. A. 1988. Benthic organic matter and macroinvertebrate functional feeding groups in a forested upland stream in temperate Victoria. Verh. Int. Ver. Limnol. 23:In press

9. Barmuta, L. A., Lake, P. S. 1982. On the value of the river continuum concept. NZ J. Mar. Freshwater Res. 16:227–31

10. Barnes, J. R., Minshall, G. W. 1983. Stream Ecology: Application and Testing of General Ecological Theory. New York: Plenum. 399 pp.

11. Benke, A. C. 1984. Secondary production of aquatic insects. See Ref. 157, pp. 289–322

12. Benke, A. C., Van Arsdall, T. C. Jr., Gillespie, D. M., Parrish, F. K. 1984. Invertebrate productivity in a subtropical blackwater river: the importance of habitat and life history. Ecol. Monogr. 54:26–63

13. Blakely, N. R., Dingle, H. 1978. Competition: butterflies eliminate milkweed bugs from a Caribbean island. Oecologia 37:133–36

14. Block, W. M., Brennan, L. A., Gutierrez, R. J. 1987. Evaluation of guild-indicator species for use in resource management. Environ. Manage. 11:265–69

15. Bosman, A. L., Hockey, P. A. R., Siegfried, W. R. 1987. The influence of coastal upwelling on the functional structure of rocky intertidal communities. Oecologia 72:226–32

16. Botkin, D. B. 1975. Functional groups of organisms in model ecosystems. See Ref. 102, pp. 98–102

17. Brown, J. H., Davidson, D. W., Reichman, O. J. 1979. An experimental study of competition between seed-eating desert rodents and ants. Am. Zool. 19:1129–43

18. Brown, V. K. 1985. Insect herbivores and plant succession. Oikos 44:17–22

19. Brown, V. K., Southwood, T. R. E. 1983. Trophic diversity, niche breadth and generation times of exopterygote insects in a secondary succession. Oecologia 56:220–25

20. Brown, V. K., Southwood, T. R. E. 1987. Secondary succession: patterns and strategies. See Ref. 58, pp. 315–37

21. Bruns, D. A., Minshall, G. W., Brock, J. T., Cushing, C. E., Cummins, K. W., Vannote, R. L. 1982. Ordination of functional groups and organic matter parameters from the middle fork of the Salmon River, Idaho. Freshw. Invert. Biol. 1:2–11

22. Bultman, T. L., Faeth, S. H. 1985. Patterns of intra- and interspecific association in leaf-mining insects on three oak host species. Ecol. Entomol. 10:121–29

23. Bultman, T. L., Uetz, G. W., Brady, A. R. 1982. A comparison of cursorial spider communities along a successional gradient. J. Arachnol. 10:23–33

24. Bunn, S. E. 1986. Spatial and temporal variation in the macroinvertebrate fauna of streams of the northern Jarrah Forest, Western Australia: functional organization. Freshwater Biol. 16:621–32

25. Canton, S. P., Chadwick, J. W. 1983. Seasonal and longitudinal changes in invertebrate functional groups in the Dolores River, Colorado. Freshwater Invertebr. Biol. 2:41–47

26. Cody, M. L., Mooney, H. A. 1978. Convergence versus nonconvergence in Mediterranean-climate ecosystems. Ann. Rev. Ecol. Syst. 9:265–322

27. Coffman, W. P., Cummins, K. W., Wuycheck, J. C. 1971. Energy flow in a woodland stream ecology: 1. Tissue support trophic structure of the autumnal community. Arch. Hydrobiol. 68:232–76

28. Cole, B. J. 1980. Trophic structure of a grassland insect community. Nature 288:76–77

29. Connell, J. H. 1983. On the prevalence and relative importance of interspecific competition: evidence from field experiments. Am. Nat. 122:661–96

30. Coulson, R. N., Flamm, R. O., Pulley, P. E., Payne, T. L., Rykiel, E. J., Wagner, T. L. 1986. Response of the southern pine bark beetle guild (Coleoptera: Scolytidae) to host disturbance. Environ. Entomol. 15:850–58

31. Cousins, S. H. 1980. A trophic continuum derived from plant structure, animal size and a detritus cascade. J. Theor. Biol. 82:607–18

32. Coyle, F. A. 1981. Effects of clearcutting on the spider community of a southern Appalachian forest. J. Arachnol. 9:285–98

33. Crossley, D. A. Jr., Callahan, J. T., Gist, C. S., Maudsley, J. R., Waide, J. B. 1976. Compartmentalization of arthropod communities in forest canopies at Coweeta. J. Ga. Entomol. Soc. 11:44–49

34. Culp, J. M., Davies, R. W. 1982. Analysis of longitudinal zonation and the river continuum concept in the Oldman-South Saskatchewan river system. Can. J. Fish. Aquat. Sci. 39:1258–66

35. Cummins, K. W. 1973. Trophic rela-

tions of aquatic insects. *Ann. Rev. Entomol.* 18:183–206
36. Cummins, K. W. 1974. Structure and function of stream ecosystems. *BioScience* 24:631–41
37. Cummins, K. W., Klug, M. J. 1979. Feeding ecology of stream invertebrates. *Ann. Rev. Ecol. Syst.* 10:147–72
38. Cummins, K. W., Merritt, R. W. 1984. Ecology and distribution of aquatic insects. See Ref. 121, pp. 59–65
39. Cummins, K. W., Minshall, G. W., Sedell, J. R., Cushing, C. E., Petersen, R. C. 1984. Stream ecosystem theory. *Verh. Int. Ver. Limnol.* 22:1818–27
40. DeGraaf, R. M., Lilghman, N. G., Anderson, S. H. 1985. Foraging guilds of north American birds. *Environ. Manage.* 9:493–536
41. Denno, R. F., Cothran, W. R. 1975. Niche relationships of a guild of necrophagous flies. *Ann. Entomol. Soc. Am.* 68:741–54
42. Diamond, J., Case, T. J., 1986. *Community Ecology.* New York: Harper & Row. 665 pp.
43. Dowell, R. V., Horn, D. J. 1977. Adaptive strategies of larval parasitoids of the alfalfa weevil (Coleoptera: Curculionidae). *Can. Entomol.* 109:641–48
44. Dudgeon, D. 1984. Longitudinal and temporal changes in functional organization of macroinvertebrate communities in the Lam Tsuen River, Hong Kong. *Hydrobiologia* 111:207–17
45. Duncan, W. F., Brusven, M. A. 1985. Benthic macroinvertebrates in logged and unlogged low-order southeast Alaskan streams. *Freshwater Invertebr. Biol.* 4:125–32
46. Ehrlich, P. R., Roughgarden, J. 1987. *The Science of Ecology.* New York: MacMillan. 710 pp.
47. Ellis, J. E., Wiens, J. A., Rodell, C. L., Anway, J. C. 1976. A conceptual model of diet selection as an ecosystem process. *J. Theor. Biol.* 60:93–108
48. Elton, C. 1927. *Animal Ecology.* London: Sidgwick & Jackson. 207 pp.
49. Enders, T. 1976. Size, food-finding, and Dyar's constant. *Environ. Entomol.* 5:1–10
50. Evans, F. C., Murdoch, W. W. 1968. Taxonomic composition, trophic structure and seasonal occurrence in a grassland insect community. *J. Anim. Ecol.* 37:259–73
51. Faeth, S. H. 1985. Host leaf selection by leaf miners: interactions among three trophic levels. *Ecology* 66:870–75
52. Fauchald, K., Jumars, P. A. 1979. The diet of worms: a study of Polychaete feeding guilds. *Ann. Rev. Oceanogr. Mar. Biol.* 17:193–284
53. Gee, H. R., Giller, P. S., eds. 1987. *Organization of Communities Past and Present. 27th Symp. Br. Ecol. Soc., Aberstwyth, UK.* Oxford: Blackwell Sci. 576 pp.
54. Georgian, T., Wallace, J. B. 1983. Seasonal production dynamics in a guild of periphyton-grazing insects in a southern Appalachian stream. *Ecology* 64:1236–48
55. Gordon, H. T. 1984. Growth and development of insects. See Ref. 81, pp. 53–77
56. Gore, J. A. 1982. Benthic invertebrate colonization: source distant effects on community composition. *Hydrobiologia* 94:183–93
57. Grant, S., Moran, V. C. 1986. The effects of foraging ants on arboreal insect herbivores in an undisturbed woodland savanna. *Ecol. Entomol.* 11:83–93
58. Gray, A. J., Crawley, M. J., Edwards, P. J., eds. 1987. *Colonization, Succession and Stability. 26th Symp. Br. Ecol. Soc. Southampton, Hampshire, UK.* Palo Alto, Calif: Blackwell Sci. 482 pp.
59. Gregg, W. W., Rose, F. L. 1985. Influences of aquatic macrophytes on invertebrate community structure, guild structure, and microdistribution in streams. *Hydrobiologia* 128:45–56
60. Gregory, S. V. 1983. Plant-herbivore interactions in stream systems. See Ref. 10, pp. 157–89
61. Hagen, K. S., Dadd, R. H., Reese, J. 1984. The food of insects. See Ref. 81, pp. 79–112
62. Hairston, N. G. 1981. An experimental test of a guild: salamander competition. *Ecology* 62:65–72
63. Hairston, N. G. Sr. 1987. *Community Ecology and Salamander Guilds.* New York: Cambridge Univ. Press. 230 pp.
64. Hamilton, H. R., Clifford, H. F. 1983. The seasonal food habits of mayfly (Ephemeroptera) nymphs from three Alberta, Canada, streams, with special reference to absolute volume and size of particles ingested. *Arch. Hydrobiol. Suppl.* 65:197–234
65. Hanski, I. 1987. Colonization of ephemeral habitats. See Ref. 58, pp. 155–85
66. Hanski, I., Koskela, H. 1977. Niche relations among dung-inhabiting beetles. *Oecologia* 28:203–31
67. Hanson, B. J., Cummins, K. W., Cargill, A. S., Lowry, R. R. 1985. Lipid content, fatty acid composition, and the effect of diet on fats of aquatic insects. *Comp. Biochem. Physiol. B* 80:257–76

68. Harris, J. R. W. 1979. The evidence for species guild is artifact. *Nature* 279:350–51

69. Hatley, C. L., MacMahon, J. A. 1980. Spider community organization: seasonal variation and the role of vegetation architecture. *Environ. Entomol.* 9:632–39

70. Hawkins, C. P. 1985. Food habits of species of ephemerellid mayflies (Ephemeroptera: Insecta) in streams of Oregon. *Am. Midl. Nat.* 113:343–52

71. Hawkins, C. P. 1988. Effects of watershed vegetation and disturbance on invertebrate community structure in western Cascade streams: implications for stream ecosystem theory. *Verh. Int. Ver. Limnol.* 23:In press

72. Hawkins, C. P., Furnish, J. K. 1987. Are snails important competitors in stream ecosystems? *Oikos* 49:209–20

73. Hawkins, C. P., Murphy, M. L., Anderson, N. H. 1982. Effects of canopy, substrate composition, and gradient on the structure of macroinvertebrate communities in Cascade range streams of Oregon. *Ecology* 63:1840–56

74. Hawkins, C. P., Sedell, J. R. 1981. Longitudinal and seasonal changes in functional organization of macroinvertebrate communities in four Oregon streams. *Ecology* 62:387–97

75. Heatwole, H., Levins, R. 1972. Trophic structure stability and faunal change during recolonization. *Ecology* 53:531–34

76. Hildrew, A. G., Townsend, C. R., Francis, J. 1984. Community structure in some southern English streams: the influence of species interactions. *Freshwater Biol.* 14:297–310

77. Holmes, R. T., Bonney, R. E. Jr., Pacala, S. W. 1979. Guild structure of the Hubbard Brook bird community: a multivariate approach. *Ecology* 60:512–520

78. Holmes, R. T., Recher, H. F. 1986. Determinants of guild structure in forest bird communities: an intercontinental comparison. *Condor* 88:427–39

79. Holter, P. 1982. Resource utilization and local coexistence in a guild of scarabaeid dung beetles (*Aphodius* spp.). *Oikos* 39:213–27

80. Howe, H. F., Westley, L. C. 1988. *Ecological Relationships of Plants and Animals*. New York: Oxford Univ. Press. 273 pp.

81. Huffaker, C. B., Rabb, R. L. 1984. *Ecological Entomology*. New York: Wiley. 844 pp.

82. Hunt, E. J., Shure, D. J. 1980. Vegetation and arthropod responses to wastewater enrichment of a pine forest. *Oecologia* 47:118–24

83. Hutchinson, G. E. 1981. Thoughts on aquatic insects. *BioScience* 31:495–500

84. Inger, R. F., Colwell, R. K. 197. Organization of contiguous communities of amphibians and reptiles in Thailand. *Ecol. Monogr.* 47:229–53

84a. Institute for Scientific Information. 1967–1986. *Science Citation Index*. Philadelphia, Pa: ISI

85. Jaksic, F. M. 1981. Abuse and misuse of the term "guild" in ecological studies. *Oikos* 37:397–400

86. Joern, A., Lawlor, L. R. 1980. Food and microhabitat utilization by grasshoppers from arid grasslands: comparisons with neutral models. *Ecology* 61:591–99

87. Joern, A., Lawlor, L. R. 1981. Guild structure in grasshopper assemblages based on food and microhabitat resources. *Oikos* 37:93–104

88. Johnson, R. A. 1981. Application of the guild concept to environmental impact analysis of terrestrial vegetation. *J. Environ. Manage.* 13:205–22

89. Jumars, P. A., Fauchald, K. 1977. Between community contrasts in successful polychaete feeding strategies. In *Ecology of Marine Benthos*, ed. B. C. Coull, pp. 1–20. Columbia, SC: Univ. South Carolina

90. Karban, R. 1986. Interspecific competition between folivorous insects on *Erigeron glaucus*. *Ecology* 67:1063–72

91. Kikkawa, J., Anderson, D. J. 1986. *Community Ecology*. Palo Alto, Calif: Blackwell Sci. 432 pp.

92. Lake, P. S., Doeg, T. J. 1985. Macroinvertebrate colonization of stones in two upland southern Australian streams. *Hydrobiologia* 126:99–211

93. Lake, P. S., Doeg, T., Morton, D. W. 1985. The macroinvertebrate community of stones in an Australian upland stream. *Verh. Int. Ver. Limnol.* 22:2141–47

94. Landres, P. B. 1983. Use of the guild concept in environmental impact assessment. *Environ. Manage.* 7:393–98

95. Landres, P. B., MacMahon, J. A. 1980. Guilds on community organization: analysis of an oak woodland avifauna in Sonora, Mexico. *Auk* 97:351–65

96. Lawton, J. H. 1982. Vacant niches and unsaturated communities: a comparison of bracken herbivores at sites on two continents. *J. Anim. Ecol.* 51:573–95

97. Lawton, J. H. 1986. Predicting fruitfly guild sizes. *Nature* 323:398

98. Lawton, J. H., Hassell, M. P. 1984. Interspecific competition in insects. See Ref. 81, p. 451–95

99. Lawton, J. H., Rallison, S. P. 1979.

Stability and diversity in grassland communities. *Nature* 279:351

100. Lawton, J. H., Schroder, D. 1978. Some observations on the structure of phytophagous insect communities: the implications for biological control. *Proc. 4th Int. Symp. Biol. Control Weeds, Gainesville, Fla.*, pp. 57–73. Gainesville, Fla: Cent. Environ. Programs Inst. Food Agric. Sci.

101. Lawton, J. H., Strong, D. R. Jr. 1981. Community patterns and competition in folivorous insects. *Am. Nat.* 118:317–38

102. Levin, S. A. 1975. *Ecosystem Analysis and Prediction.* Philadelphia: Soc. Ind. Appl. Math. 337 pp.

103. Lindeman, R. L. 1942. The trophic-dynamic aspect of ecology. *Ecology* 23:401–18

104. Loehle, C. 1987. Hypothesis testing in ecology: psychological aspects and the importance of theory maturation. *Q. Rev. Biol.* 62:397–409

105. MacFarlane, M. B. 1983. Structure of benthic macroinvertebrate communities in a midwestern plains stream. *Freshwater Invertebr. Biol.* 2:147–53

106. MacMahon, J. A., Schimpf, D. J., Anderson, D. C., Smith, K. G., Bayn, R. L. Jr. 1981. An organism-centered approach to some community and ecosystem concepts. *J. Theor. Biol.* 88:287–307

107. MacNally, R. C. 1983. On assessing the significance of interspecific competition to guild structure. *Ecology* 64:1646–52

108. MacNally, R. C., Doolan, J. M. 1986. Patterns of morphology and behaviour in a cicada guild: a neutral model analysis. *Aust. J. Ecol.* 11:279–94

109. MacNally, R. C., Doolan, J. M. 1986. An empirical approach to guild structure: habitat relationships in nine species of eastern-Australian cicadas. *Oikos* 47:33–46

110. Majer, J. D., Sartori, M., Stone, R., Perriman, W. S. 1982. Recolonization by ants and other invertebrates in rehabilitated mineral sand mines near Eneabba, Western Australia. *Reclam. Reveg. Res.* 1:63–81

111. Malmqvist, B., Bronmark, C. 1985. Reversed trends in the benthic community structure in two confluent streams: one spring-fed, the other lake-fed. *Hydrobiologia* 124:65–71

112. Malmqvist, B., Nilsson, L. M., Svensson, B. S. 1978. Dynamics of detritus in a small stream in southern Sweden and its influence on the distribution of the bottom animal communities. *Oikos* 31:3–16

113. Mannan, R. W., Morrison, M. L., Meslow, E. C. 1984. Comment: The use of guilds in forest bird management. *Wildl. Soc. Bull.* 12:426–30

114. Marchant, R., Metzeling, L., Graesser, A., Suter, P. 1985. The organization of macroinvertebrate communities in the major tributaries of the LaTrobe River, Victoria, Australia. *Freshwater Biol.* 15:315–31

115. Maurer, D., Leathem, W. 1981. Polychaete feeding guilds from Georges Bank, USA. *Mar. Biol.* 62:161–71

116. May, R. M. 1978. The dynamics and diversity of insect faunas. In *Diversity of Insect Faunas*, ed. L. A. Mound, N. Waloff, pp. 188–204. Oxford: Blackwell Sci. 204 pp.

117. McIntire, C. D., Colby, J. A. 1978. A hierarchical model of lotic ecosystems. *Ecol. Monogr.* 48:167–90

118. McNaughton, S. J. 1978. Stability and diversity of ecological communities. *Nature* 274:251–53

119. Mehlhop, P., Scott, N. J. Jr. 1983. Temporal patterns of seed use and availability in a guild of desert ants. *Ecol. Entomol.* 8:69–85

120. Merritt, R. W., Cummins, K. W., eds. 1984. *An Introduction to the Aquatic Insects of North America.* Dubuque, Iowa: Kendall/Hunt. 722 pp. 2nd ed.

121. Merritt, R. W., Cummins, K. W., Burton, T. M. 1984. The role of aquatic insects in the processing and cycling of nutrients. See Ref. 157, pp. 134–63

122. Miller, C. 1985. Correlates of habitat favourability for benthic macroinvertebrates at five stream sites in an Appalachian Mountain drainage basin, U.S.A. *Freshwater Biol.* 15:709–33

123. Minshall, G. W., Andrews, D. A., Manuel-Faler, C. Y. 1983. Application of island biogeographic theory to streams: macroinvertebrate recolonization of the Teton River, Idaho. See Ref. 10, pp. 279–97

124. Minshall, G. W., Brock, J. T., LaPoint, T. W. 1982. Characterization and dynamics of benthic organic matter and invertebrate functional feeding group relationships in the upper Salmon River, Idaho (USA). *Int. Rev. Gesamten Hydrobiol.* 67:793–820

125. Minshall, G. W., Petersen, R. C., Cummins, K. W., Bott, T. L., Sedell, J. R., et al. 1983. Interbiome comparison of stream ecosystem dynamics. *Ecol. Monogr.* 53:1–25

126. Moore, J. C., Walter, D. E., Hunt, H. W. 1988. Arthropod regulation of micro- and mesobiota in below-ground detrital food webs. *Ann. Rev. Entomol.* 33:419–39

127. Moran, V. C., Southwood, T. R. E. 1982. The guild composition of arthropod communities in trees. *J. Anim. Ecol.* 51:289–306

128. Morin, P. J. 1984. Odonate guild composition: experiments with colonization history and fish predation. *Ecology* 65:1866–73

129. Morton, S. R., Davidson, D. W. 1988. Comparative structure of harvester ant communities in arid Australia and North America. *Ecol. Monogr.* 58:19–38

130. Muttkowski, R. A., Smith, G. M. 1929. The food of trout stream insects in Yellowstone National Park. *Roosevelt Wildl. Bull.* 2:241–63

131. Nielsen, B. O. 1978. Food resource partitioning in the beech leaf-feeding guild. *Ecol. Entomol.* 3:193–201

132. Odum, E. P. 1971. *Fundamentals of Ecology.* Philadelphia: Saunders. 574 pp. 3rd ed.

133. Odum, H. J. 1983. *Systems Ecology: An Introduction.* New York: Wiley. 644 pp.

134. O'Neil, R. V., DeAngelis, D. L., Waide, J. B., Allen, T. F. H. 1986. *A Hierarchical Concept of Ecosystems.* Princeton, NJ: Princeton Univ. Press. 253 pp.

135. Orians, G. H., Paine, R. T. 1984. Convergent evolution at the community level. In *Coevolution,* ed. D. J. Futuyma, M. Slatkin, pp. 431–58. Sunderland, Mass: Sinauer. 575 pp.

136. Osborne, L. L., Davies, R. W., Linton, K. J. 1979. Effects of limestone strip mining on benthic macroinvertebrate communities. *Water Res.* 13:1285–90

137. Otto, C., Svensson, B. S. 1982. Structure of communities of ground-living spiders along altitudinal gradients. *Holarct. Ecol.* 5:35–47

138. Parmenter, R. R., MacMahon, J. A. 1987. Early successional patterns of arthropod recolonization on reclaimed strip mines in southwestern Wyoming: the ground-dwelling beetle fauna (Coleoptera). *Environ. Entomol.* 16:168–77

139. Peck, S. B., Forsyth, A. 1982. Composition, structure and competitive behavior in a guild of Ecuadorian rain forest dung beetles (Coleoptera: Scarabeidae). *Can. J. Zool.* 60:1624–34

140. Peckarsky, B. L. 1980. Influence of detritus upon colonization of stream invertebrates. *Can. J. Fish. Aquat. Sci.* 37:957–63

141. Perry, J. A., Schaeffer, D. J. 1987. The longitudinal distribution of riverine benthos: a river dis-continuum? *Hydrobiologia* 148:257–68

142. Pianka, E. R. 1980. Guild structure in desert lizards. *Oikos* 35:194–201

143. Pimm, S. L. 1982. *Food Webs.* London: Chapman & Hall. 219 pp.

144. Polis, G. A., McCormick, S. J. 1986. Scorpions, spiders and solpugids: predation and competition among distantly related taxa. *Oecologia* 71:111–16

145. Pontin, A. J. 1982. *Competition and Coexistence of Species.* Bath, UK: Pitman. 102 pp.

146. Post, W. M. III, Riechert, S. E. 1977. Initial investigation into the structure of spider communities. I. Competitive effects. *J. Anim. Ecol.* 46:729–49

147. Pratt, J. R., Cairns, J. Jr. 1985. Functional groups in the protozoa: roles in differing ecosystems. *J. Protozool.* 32:415–23

148. Pratt, J. R., Horwitz, R., Cairns, J. Jr. 1987. Protozoan communities of the Flint River–Lake Blackshear ecosystem (Georgia, USA). *Hydrobiologia* 148:159–74

149. Price, P. W. 1970. Characteristics permitting coexistence among parasitoids of a sawfly in Quebec. *Ecology* 51:445–54

150. Price, P. W. 1971. Niche breadth and dominance of parasitic insects sharing the same host species. *Ecology* 52:587–96

151. Price, P. W. 1972. Parasitoids utilizing the same host: adaptive nature of differences in size and form. *Ecology* 53:190–95

152. Price, P. W. 1984. *Insect Ecology.* New York: Wiley. 607 pp. 2nd ed.

153. Rader, R. B., Ward, J. V. 1987. Resource utilization, overlap and temporal dynamics in a guild of mountain stream insects. *Freshwater Biol.* 18:521–28

154. Ratchke, B. J. 1976. Insect-plant patterns and relationships in the stem-boring guild. *Am. Midl. Nat.* 96:98–117

155. Rathcke, B. J. 1976. Competition and coexistence within a guild of herbivorous insects. *Ecology* 57:76–87

156. Resh, V. H. 1976. Life histories of coexisting species of *Ceraclea* caddisflies (Tricoptera: Leptoceridae): the operation of independent functional units in a stream ecosystem. *Can. Entomol.* 108:1303–18

157. Resh, V. H., Rosenberg, D. M., eds. 1984. *The Ecology of Aquatic Insects.* New York: Praeger. 625 pp.

158. Roberts, T. H. 1987. Construction of guilds for habitat assessment. *Environ. Manage.* 11:473–77

159. Robinson, J. V. 1981. The effect of architectural variation in habitat on a spider community: an experimental field study. *Ecology* 62:73–80

160. Rooke, B. 1986. Macroinvertebrates associated with macrophytes and plastic

imitations in the Eramosa River, Ontario, Canada. *Arch. Hydrobiol.* 106:307–25

161. Root, R. B. 1967. The niche exploitation pattern of the blue-gray gnatcatcher. *Ecol. Monogr.* 37:317–50

162. Root, R. B. 1973. Organization of a plant-arthropod association in simple and diverse habitats: the fauna of collards *(Brassica oleracea). Ecol. Monogr.* 43:95–124

163. Root, R. B. 1975. Some consequences of ecosystem texture. See Ref. 102, pp. 83–97

164. Root, R. B., Chaplin, S. J. 1976. The life-styles of tropical milkweed bugs, *Oncopeltus* (Hemiptera: Lygaeidae) utilizing the same hosts. *Ecology* 57:132–40

165. Ross, H. H. 1963. Stream communities and terrestrial biomes. *Arch. Hydrobiol.* 59:235–42

166. Rounick, J. S., Winterbourn, M. J. 1982. Benthic faunas of forested streams and suggestions for their management. *NZ J. Ecol.* 5:140–50

167. Scheiring, J. F. 1985. Longitudinal and seasonal patterns of insect trophic structure in a Florida sand-hill stream. *J. Kans. Entomol. Soc.* 58:207–19

168. Schoener, T. W. 1983. Field experiments on interspecific competition. *Am. Nat.* 122:240–85

169. Schoener, T. W., Toft, C. A. 1983. Spider populations: extraordinary high densities on islands without top predators. *Science* 219:1353–55

170. Schowalter, J. D., Hargrove, W. W., Crossley, D. A. Jr. 1986. Herbivory in forested ecosystems. *Ann. Rev. Entomol.* 31:177–96

171. Schowalter, T. D., Webb, J. W., Crossley, D. A. Jr. 1981. Community structure and nutrient content of canopy arthropods in clearcut and uncut forest ecosystems. *Ecology* 62:1010–19

172. Seastedt, L. R. 1984. The role of microarthropods in decomposition and mineralization processes. *Ann. Rev. Entomol.* 29:25–46

173. Severinghaus, W. D. 1981. Guild theory develop as a mechanism for assessing environmental impact. *Environ. Manage.* 5:187–90

174. Sholes, O. D. 1984. Responses of arthropods to the development of goldenrod inflorescences *(Solidago:* Asteraceae). *Am. Midl. Nat.* 112:1–14

175. Shorrocks, B., Rosewell, J. 1986. Guild size in drosophilids: a simulation model. *J. Anim. Ecol.* 55:527–41

176. Shorrocks, B., Rosewell, J., Edwards, K. 1984. Interspecific competition is not

a major organizing force in many insect communities. *Nature* 310:310–12

177. Short, H. L., Burnham, K. P. 1982. *Technique for structuring wildlife guilds to evaluate impacts on wildlife communities. US Fish Wildl. Serv. Spec. Sci. Rep. Wildl.* No. 244. 34 pp.

178. Simberloff, D. 1976. Trophic structure determination and equilibrium in an arthropod community. *Ecology* 57:395–98

179. Simberloff, D. S. 1969. Experimental zoogeography of islands: a model for insular colonization. *Ecology* 50:296–314

180. Simberloff, D. S. 1978. Colonization of islands by insects: immigration, extinction, and diversity. *Symp. R. Entomol. Soc. London* 9:139–53

181. Simberloff, D. S., Wilson, E. O. 1969. Experimental zoogeography of islands: the colonization of empty islands. *Ecology* 50:278–96

182. Simberloff, D. S., Wilson, E. O. 1970. Experimental zoogeography of islands: a two-year record of colonization. *Ecology* 51:934–37

183. Simpson, G. G. 1953. *The Major Features of Evolution.* New York: Columbia Univ. Press. 434 pp.

184. Slansky, F. Jr., Rodriquez, J. G. 1986. *Nutritional Ecology of Insects, Mites, and Spiders.* New York: Wiley. 1016 pp.

185. Smock, L. A., Gilinsky, E., Stoneburner, D. L. 1985. Macroinvertebrate production in a southeastern United States blackwater stream. *Ecology* 66:1491–503

186. Southwood, T. R. E. 1977. Habitat, the templet for ecological strategies? *J. Anim. Ecol.* 46:337–65

187. Sprules, W. G., Holtby, L. B. 1979. Body size and feeding ecology as alternatives to taxonomy for the study of limnetic zooplankton community structure. *J. Fish. Res. Board Can.* 36:1354–63

188. Stanley, S. M. 1979. *Macroevolution.* San Francisco: Freeman. 332 pp.

189. Steneck, R. S., Watling, L. 1982. Feeding capabilities and limitations of herbivorous molluscs: a functional group approach. *Mar. Biol.* 68:299–319

190. Stevenson, B. G., Dindal, D. L. 1982. Effect of leaf shape on forest litter spiders: community organization and microhabitat selection of immature *Enoplognatha avata* (Clerck) (Theridiidae). *J. Arachnol.* 10:165–78

191. Stork, N. E. 1987. Guild structure of arthropods from Bornean rain forest trees. *Ecol. Entomol.* 12:69–80

192. Strong, D. R., Simberloff, D., Abele, L. G., Thistle, A. B., eds. 1984. *Ecological Communities: Conceptual Issues and the Evidence.* Princeton; NJ: Princeton Univ. Press. 613 pp.

193. Szaro, R. C. 1986. Guild management: an evaluation of avian guilds as a predictive tool. *Environ. Manage.* 10:681–88

194. Teraguchi, S., Teraguchi, M., Upchurch, R. 1977. Structure and development of insect communities in an Ohio old-field. *Environ. Entomol.* 6:247–57

195. Terborgh, J., Robinson, S. 1986. Guilds and their utility in ecology. See Ref. 91, pp. 65–90

196. Thompson, J. N. 1984. Insect diversity and the trophic structure of communities. See Ref. 81, pp. 591–606

197. Towns, D. R. 1987. The mayflies (Ephemeroptera) of Great Barrier Island, New Zealand: macro- and micro-distributional comparisons. *J. R. Soc. NZ* 17:349–61

198. Townsend, C. R., Hildrew, A. G., Francis, J. 1983. Community structure in some southern English streams: the influence of physicochemical factors. *Freshwater Biol.* 13:521–44

199. Vannote, R. L., Minshall, G. W., Cummins, K. W., Sedell, J. R., Cushing, C. E. 1980. The river continuum concept. *Can. J. Fish. Aquat. Sci.* 37:130–37

200. Verner, J. 1984. The guild concept applied to management of bird populations. *Environ. Manage.* 8:1–14

201. Wallace, J. B., Merritt, R. W. 1980. Filter-feeding ecology of aquatic insects. *Ann. Rev. Entomol.* 25:103–32

202. Wallace, J. B., Vogel, D. S., Cuffney, T. F. 1986. Recovery of a head water from an insecticide-induced community disturbance. *J. North Am. Benthol. Soc.* 5:115–26

203. Wallwork, J. A. 1983. Oribatids in forest ecosystems. *Ann. Rev. Entomol.* 28:109–30

204. Wheeler, A. G. Jr. 1976. Lygus bugs as facultative predators. In *Lygus Bug: Host-Plant Interactions. Proc. Workshop, 15th Int. Congr. Entomol., Washington, DC,* pp. 28–35. Moscow, Idaho: Univ. Press Idaho. 38 pp.

205. Wiegert, R. G., Owen, D. F. 1971. Trophic structure, available resources and population density in terrestrial vs. aquatic ecosystems. *J. Theor. Biol.* 30:69–81

206. Wiggins, G. B., Mackay, R. J. 1978. Some relationships between systematics and trophic ecology in Nearctic aquatic insects, with special reference to trichoptera. *Ecology* 59:1211–20

207. Wilson, E. O., Simberloff, D. S. 1969. Experimental zoogeography of islands: defaunation and monitoring techniques. *Ecology* 50:267–78

208. Winterbourn, M. J., Rounick, J. S., Cowie, B. 1981. Are New Zealand stream ecosystems really different? *NZ J. Mar. Freshwater Res.* 15:321–28

209. Woodin, S. A., Jackson, J. B. C. 1979. Interphyletic competition among marine benthos. *Am. Zool.* 19:1029–43

210. Yodzis, P. 1982. The compartmentation of real and assembled ecosystems. *Am. Nat.* 120:551–70

Ann. Rev. Entomol. 1989. 34:453–76

# ENHANCED BIODEGRADATION OF INSECTICIDES IN SOIL: Implications for Agroecosystems

## A. S. Felsot

Section of Economic Entomology, Illinois Natural History Survey, 607 East Peabody, Champaign, Illinois 61820

## PERSPECTIVES AND OVERVIEW

A vast amount of research since the publication of *Silent Spring* (13) shows that soil, water, air, and the biota have become the ultimate sinks for all pesticides and their degradation products. During the late 1950s and throughout the 1960s great concern was expressed about the accumulation of the chlorinated hydrocarbon and cyclodiene insecticides in the biosphere. From the mid-1970s and throughout the 1980s these persistent insecticides have gradually been phased out by regulation and replaced by highly toxic but biodegradable insecticides. Biodegradation, a term used usually in reference to microbial metabolism of natural and synthetic chemicals in soil and water, is needed to reduce the accumulation of pesticide residues in the environment. Without biodegradation, the environmental sinks would rapidly fill with all kinds of chemical contaminants.

Most pesticides used for insect and weed control are applied directly to the soil. Pesticides are mainly dissipated from soil by microbiological metabolism. Paradoxically, biodegradation of pesticides, which has been viewed as a positive process for reducing environmental hazards, is a double-edged sword in regard to management of crop damage by soil-borne pests. Because prophylactic treatment is the strategy most frequently used to control soil pests, the efficacy of the pesticide depends on its persistence in the soil before the appearance of the pest. A growing body of literature has shown that some environmentally labile pesticides are degraded at accelerated rates in soils that are retreated with the same chemical. In several cases, this enhancement of biodegradation has been associated with ineffective pest control. Physiologi-

0066-4170/89/0101-0453$02.00

cally adapted or conditioned microorganisms are believed to have a key role in rapidly degrading pesticides in retreated soils.

The objective of this review is to examine the phenomenon known as enhanced biodegradation of pesticides (51) from the perspective of its implications for control of soil-borne insect pests. Interactions of soil microorganisms and pesticides have been reviewed elsewhere (3, 45, 54, 59, 61, 93, 108, 113). More recent reviews have focused specifically on enhanced pesticide biodegradation (51, 52) and its relationship to problems of insect (112), weed (43), and disease (112) control.

I explore development of enhanced biodegradation in agroecosystems and its subsequent effect on pesticide efficacy by reviewing case histories of carbamate and organophosphate insecticides. A discussion of the theoretical aspects of the development of enhanced biodegradation phenomena preludes a discussion of specific agronomic and environmental factors that have a bearing upon crop protection practices. Proposed methodologies for circumventing or coping with enhanced biodegradation are discussed. Finally, the positive aspects of enhanced biodegradation in decontamination of pesticide wastes are highlighted.

## CASE HISTORIES OF ENHANCED PESTICIDE BIODEGRADATION

Enhanced biodegradation is the accelerated dissipation of a chemical from soil after repeated exposure of the soil to that chemical. Enrichment of microorganisms that use the chemical as a nutrient or carbon source causes the observed increase in the degradation rate of the chemical (6). The first incidence of microbial enrichment (also called conditioning or adaptation) in response to the presence of a pesticide was reported for 2,4-dichlorphenoxyacetic acid (2,4-D) in soil perfusion columns about 40 years ago (5). Later studies confirmed the enhanced biodegradation of 2,4-D (66) and its monochloro analog 2-methyl-4-chlorophenoxyacetic acid (MCPA) in the field (55). The implications of the development of enhanced biodegration in the soil for crop protection practices, however, were not realized until the early 1970s (85). To elucidate how conditioning soils with pesticides affects pest control efficacy, I examine case histories based on a combination of field and laboratory observations for several pesticides. Although microbial adaptations for pesticide degradation have been receiving a great deal of attention, the purpose of this section is to link the phenomenon with field practices and insect control problems.

### Methyl Carbamate Insecticides

The relationship between enhanced biodegradation and pest control problems has been most thoroughly studied with carbamate pesticides, i.e. the methyl carbamate insecticides and the thiocarbamate herbicides. The failure of carbofuran to control the grape phylloxera was the earliest indication that rapid degradation of a carbamate insecticide could adversely affect control of a soil insect (116). Problems with efficacy of soil insecticides used to control corn rootworm species, however, have generated the most attention.

CARBOFURAN    During the last 20 years, several soil insecticides have shown inadequate or totally ineffective control of feeding damage by the western and northern corn rootworms, *Diabrotica virgifera virgifera* and *Diabrotica barberi,* in small- and large-plot studies conducted annually in the Corn Belt. All classes of insecticides have had efficacy problems at one time or another, but in Illinois aldrin, heptachlor, bufencarb, carbofuran, and phorate have proved most troublesome (57). Testing of adult cornworm susceptibility in the early 1960s indicated that both *D. virgifera virgifera* and *D. barberi* had developed resistance to aldrin with cross-resistance to heptachlor, both of which are environmentally persistent chlorinated cyclodienes (8, 11). When the methyl carbamate bufencarb failed to perform as expected in field efficacy tests in Nebraska and Illinois (7, 56), insect resistance was suspected, but the suspicion was never thoroughly tested or proven.

Carbofuran first failed to control corn rootworm feeding damage during the mid 1970s. There was little doubt that the efficacy of carbofuran in preventing corn rootworm damage had declined since its commerical introduction in 1970. During 1969–1974, corn root damage ratings in carbofuran-treated plots exceeded the economic injury level in 6% of the tests conducted by researchers from the University of Illinois (57). During 1976–1980, the economic injury level was exceeded in 28% of the test plots that were treated with carbofuran. In Iowa during 1975–1976, carbofuran failed to prevent root damage in test plots with a history of carbofuran use; however, carbofuran always performed well in test plots that had not previously been treated with insecticides or in plots that had been treated with organophosphate insecticides (107).

Based on the experiences with the chlorinated cyclodienes, it was logical to assume that the corn rootworms had developed resistance to carbofuran, especially considering that carbofuran was used on most of the acreage that was treated. Western corn rootworm beetles collected from different fields in Illinois seemed to exhibit low levels of resistance to carbofuran (25), but the primitive susceptibility level for field-collected beetles was uncertain. Furthermore, the western corn rootworm response to carbofuran was highly variable among populations in any one year, and the response of specific populations varied among years of collection (7, 29). Although there was a consensus that rootworm control failures associated with the use of cyclodiene insecticides were related to the development of high levels of insecticide resistance, no correlation was found between susceptibility of *D. virgifera*

*virgifera* to carbofuran and root damage ratings in the fields where the tested beetles were collected (29). Further testing showed that the susceptibility of corn rootworm larvae to carbofuran probably had not changed significantly from primitive levels (28).

Lack of carbofuran persistence was suggested as an alternative to the resistance hypothesis for explaining the failure of the insecticide to control corn rootworm feeding damage (24).This hypothesis rested primarily on three observations. First, soil insecticides were applied 3–5 wk before rootworm egg hatching; thus there was ample time for the chemical to dissipate before it was actually needed. Secondly, poor control was observed in fields where carbofuran had been used continuously for 2–4 yr, but control was acceptable in fields lacking a history of continuous use of carbofuran (58, 107). Since *D. virgifera virgifera* and *D. barberi* are univoltine and cornfields are not treated completely (thereby providing untreated refuges), it was unlikely that resistance high enough to cause control problems would have developed only 5 yr after the commercial introduction of carbofuran. Lastly, an association between rapid degradation of carbofuran and poor control of a soil pest had already been demonstrated (116).

The chemical persistence of carbofuran was examined during 1976–1977 in soils collected from fields where rootworm control problems had been reported in 1975 (24). Carbofuran completely dissipated within 30 days after application to soil taken from fields where carbofuran had been used for several consecutive years and failed to control corn rootworm feeding damage. In contrast, chemical persistence of carbofuran was much longer in soils taken from fields with either no previous use of an insecticide or previous use of an organophosphate. Prolonged persistence of carbofuran in sterilized soils indicated that a microbial factor was responsible for rapid carbofuran degradation (24, 51). From a practical perspective, these findings were significant because corn rootworm insecticides are usually applied at planting time, which normally extends from mid April to late May; therefore, a sufficient concentration of insecticide must be present 1–2 mo later to provide adequate root protection.

Earlier field studies had contradicted the laboratory observations of rapid carbofuran degradation in soils with a history of carbofuran use (2, 38). Other studies suggested that enhanced biodegradation of carbofuran could occur under field conditions (30, 37, 39). Differences in relative control of feeding damage by corn rootworms were associated with differences in carbofuran concentrations in the soil (30). Also, the 50%-disappearance time of carbofuran tended to be significantly shorter in fields with efficacy problems than in fields with no control problems (29).

The hypothesis of a relationship among repeated treatments of soil with carbofuran, enhancement of carbofuran degradation, and problems of insect control has been validated by several studies from agroecosystems other than corn. Rapid breakdown of carbofuran in soil from Prince Edward Island, Canada has been associated with failure to protect rutabagas from feeding damage by root maggots (77–79). In the United Kingdom, field performance of carbofuran against cabbage maggot, *Delia radicum,* and carrot rust fly, *Psila rosae,* in cabbage and carrots was significantly worse in soil previously treated with granular formulations of carbofuran than in previously untreated soil (103). Field data collected in Newfoundland, Canada since the mid 1960s has indicated that the efficacy of carbofuran against *D. radicum, P. rosae,* and the carrot weevil, *Listronotus oregonensis,* has declined over time (41). The diminished control could not be attributed to the development of insecticide resistance, but laboratory tests indicated that microbial populations had become adapted for the degradation of carbofuran. In Kansas, systemic control of chinch bugs, *Blissus leucopterus leucopterus,* and greenbug, *Schizaphis graminum,* in seedling sorghum was significantly lower in soils taken from fields that had histories of carbofuran use than in soils taken from adjacent fields that had had no prior carbofuran treatments (115).

Laboratory experiments in the United States (12, 51), Canada (77, 78), and the United Kingdom (10, 102) have shown that degradation of carbofuran is significantly more rapid in soil collected from previously treated fields than in soil collected from previously untreated portions of the same field or adjacent fields. The extent of prior carbofuran treatment in these studies ranged from one application the previous year (77) to two per year for seven years (102). Loss of carbofuran from the conditioned soils was greater than 90% within 20 days after treatment regardless of the number of prior applications to the field.

Without doubt, microbial activity is responsible for the rapid degradation of carbofuran in retreated soils. In laboratory studies, radiolabeled carbofuran (whether carbonyl- or ring-labeled) was more rapidly degraded to $^{14}CO_2$ in soils with a history of carbofuran use from the Corn Belt (denoted as "problem" soils) than in previously untreated soils of the same origin (51). Harris et al (40) developed laboratory methods for repeatedly treating soil with carbofuran and monitoring the effect on its degradation rate. Untreated soil incubated simultaneously with the treated soil served as a control. In a sandy loam soil from Ontario, Canada that had been pretreated only once with 10 ppm carbofuran, >90% of subsequently applied carbofuran was lost within one day; a comparable loss took 21 days in a control soil. Similar induction experiments using soils from South Carolina led to the same results, although degradation of carbofuran was not as rapid or as complete as that observed in the Ontario soil (12). Sterilization, freezing, or drying of the carbofuran-conditioned soils resulted in loss of carbofuran-degrading activity (40, 51).

Although enhanced biodegradation of carbofuran seems to develop easily in aerobic soils, unusual and somewhat ambiguous results have been reported for flooded rice soils. The efficacy of carbofuran for controlling brown planthopper on rice decreased after intensive use for two or three years (73). A bacterium was isolated from flooded rice soils that could degrade carbofuran within 40 days in pure culture (110), but repeated applications of carbofuran to flooded soil did not cause a rapid proliferation of microorganisms capable of decomposing carbofuran (91, 111). In contrast, another study (92) showed an accelerated degradation of carbofuran in retreated, flooded rice soils, which was ascribed to a buildup of carbofuran-decomposing microorganisms; the rate of degradation of carbofuran, however, was not as fast as the rate reported for aerobic soils. Enrichment cultures of bacteria from flooded soils previously treated with carbofuran rapidly degraded carbofuran when it was the sole source of carbon and nitrogen (73). Recently, the hydrolysis product of carbofuran, carbofuran phenol, was found to initiate enhanced biodegradation of carbofuran in flooded soils (74). Low temperatures (6°C) during six pretreatments of carbofuran inhibited the development of a carbofuran-adapted microflora in a flooded soil; incubations at 36°C were optimal for microbial adaptation (76).

OTHER METHYL CARBAMATES    Conditioning of soils for the enhanced biodegradation of other methyl carbamate insecticides has not been well studied. Most of the studies have indicated that other methyl carbamates are susceptible to enhanced biodegradation, because these chemicals have been degraded rapidly when added to soil previously conditioned with carbofuran. "Cross-conditioning" and "cross-adaptation" are terms used to describe the accelerated breakdown of a chemical after pretreatment of the soil with another chemical. Usually the two chemicals involved are structurally related analogs.

Three proinsecticide analogs of carbofuran, benfuracarb, carbosulfan, and furathiocarb, did not control cabbage maggot larvae when applied to soil treated the previous year with carbofuran (103). Pesticide efficacy in soil with a history of carbofuran use was at least an order of magnitude lower than efficacy in previously untreated soil. Persistence of furathiocarb was not affected by previous carbofuran treatments; however, once furathiocarb was hydrolyzed, the resulting carbofuran was rapidly degraded in the conditioned soil (40). The poor performance of the proinsecticide analogs against cabbage maggot probably resulted from rapid degradation of the newly formed carbofuran rather than from a direct microbial adaptation to the parent compounds (103).

In greenhouse tests with soils treated with benfuracarb or cloethocarb, significantly more chinch bugs and greenbugs survived on sorghum seedlings planted in soils that had a history of carbofuran use than in soil of the same type that had no prior insecticide use (115). Only 6% of the cloethocarb remained one day after application in a carbofuran-conditioned soil, but 91% remained in an unconditioned soil (40). These experiments have serious implications for the prospects for cloethocarb in the Corn Belt, where the pesticide has been under development for use as a corn rootworm soil insecticide.

Other methyl carbamate insecticides, including bufencarb, carbaryl, propoxur, and trimethacarb, have been shown to be degraded at significantly faster rates in soils previously treated with carbofuran than in soils previously untreated (40). Trimethacarb degraded significantly faster in a soil collected from a field with a history of efficacy problems following carbofuran use than in a soil collected from the untreated ditch surrounding the field (A. S. Felsot, unpublished). Direct pretreatment of a soil with formulated trimethacarb caused enhancement of biodegradation of technical trimethacarb and significantly lowered corn rootworm larval mortality in comparison to those in soil that was not pretreated (27). The efficacy of bufencarb against corn rootworm declined during the early 1970s, and eventually it was deleted from the recommendation lists of cooperative extension services (107).

Rodriguez & Dorough (82) have noted that carbaryl can condition soil for enhancement of its own biodegradation, but the practical effect on pest control is unknown because this compound is not usually used as a soil insecticide. Pretreatment of flooded soil with the hydrolysis product of carbaryl, 1-naphthol, induced enhanced biodegradation of carbaryl (74). Bacteria that utilize carbaryl as sole carbon and nitrogen source have been isolated from garden soils but paradoxically failed to metabolize carbaryl rapidly (60). Resting cell suspensions of a pure bacterial isolate adapted for carbofuran degradation also rapidly degraded carbaryl and propoxur (48).

## Oxime Carbamate Insecticides

Cross-conditioning experiments with carbofuran and aldicarb suggested that soil could be conditioned by aldicarb applications; one day after application, four times less aldicarb remained in a soil pretreated with carbofuran than in a soil that had not been pretreated (40). A pure bacterial culture adapted for carbofuran metabolism metabolized aldicarb extensively within 5 hr (48).

Under laboratory conditions, degradation of formulated aldicarb was enhanced in soil collected from field plots previously treated with aldicarb (80). Soil bioassays of cabbage maggot were used to determine aldicarb residues remaining in the soil. This experiment implied that aldicarb as a systemic for control of foliar feeders or as a contact agent for control of soil pests might be unsuccessful if used repeatedly in the same field. Indeed, aldicarb failed to control stem nematode, *Ditylenchus dipsaci*, on strawberries in the United Kingdom after several years of successful use (104). Aldicarb also failed to control greenhouse whitefly, *Trialeurodes vaporariorum*, on tomatoes grown under glass. Chemical assays showed that total aldicarb residues (i.e. aldicarb plus aldicarb sulfoxide and aldicarb sulfone) declined more rapidly in previously treated soils than in the corresponding control soils. The decline in total aldicarb residues was largely due to an accelerated degradation of aldicarb sulfoxide rather than of the parent compound (104). Furthermore, previous applications of carbofuran induced enhanced degradation of aldicarb more readily than pretreatment with aldicarb itself. Control of cabbage maggot with aldicarb, however, was not affected by previous carbofuran treatments (103). The efficacy of aldicarb for controlling chinch bugs and greenbugs on sorghum was also not affected by previous applications of carbofuran (115).

Nematode control in potato fields decreased after one to three seasons of aldicarb and oxamyl use in the Netherlands (95). Oxamyl residues were lost significantly faster from the soil of plots treated previously with oxamyl than from the soil of control plots. Although the rate of aldicarb loss was slightly higher in conditioned soil than in in control soil, the rate of aldicarb sulfoxide degradation was over five times faster in the conditioned soil. Oxamyl conditioned soil for enhanced biodegradation of aldicarb (95), and aldicarb (95) and carbofuran (40) conditioned soil for enhanced biodegradation of oxamyl.

## Organophosphate Insecticides

DIAZINON    The relationship between enhanced biodegradation and efficacy of insect control by organophosphate insecticides has not been as well documented as the relationship for carbamates. Diazinon was the first pesticide that failed to control a targeted pest as the result of rapid microbial degradation (85). After 3.5 yr the efficacy of diazinon against rice brown planthoppers, *Nilaparvata lugens,* in flooded rice culture had declined. Diazinon that was incubated in paddy water taken from treated fields degraded almost completely within 3 days. Upon five more additions of diazinon to the water, degradation was complete in only 6 hr. Soil sterilization studies showed that soil microbes had adapted for diazinon degradation. A *Flavobacterium* sp. that utilized diazinon as sole carbon source was isolated and found also to hydrolyze parathion (90).

In greenhouse studies using pots containing rice paddy soil and rice plants, the amount of diazinon recovered 10 days after a second application was less than half that recovered after a first application (87). These greenhouse results were corroborated with field studies in which efficacy was reduced and a buildup of rice brown planthoppers occurred after three successive applications of diazinon at 5-day intervals. Diazinon was normally applied every 20 days throughout the growing season (85), and such practices could have led to a rapid development of insect resistance. In field studies, however, reductions in planthopper populations were seen 10 days after the first applications of diazinon (87); thus, resistance to diazinon was probably not responsible for the subsequent decline in efficacy.

In subsequent studies, diazinon degradation was most rapid in paddy water, rhizosphere soils, and nonrhizosphere soils (in that order) collected from previously treated rice plots and much slower in water and soil from untreated plots of the same field (88). Mineralization of diazinon was also faster in soil containing a bean rhizosphere than in soil without a rhizosphere (47).

In soil treated repeatedly for 3 yr, diazinon could no longer protect lettuce from damage by the root aphid *Pemphigus bursarius* (32). Diazinon was degraded completely within 2 days in a soil conditioned by two previous treatments, but remained for 8 days in a previously untreated soil. An isolated *Flavobacterium* sp. similar to that reported from rice soils (90) utilized diazinon as a sole carbon source and also hydrolyzed parathion. Rate of parathion degradation, however, was similar in diazinon-conditioned soil and in the untreated control soil.

FENSULFOTHION    In Prince Edward Island, Canada, fensulfothion failed to control cabbage maggot damage to rutabagas in one retreated field (77). A long-term data base for fensulfothion efficacy in rutabagas showed adequate control (>90%) of root maggots for at least 12 yr prior to 1976 (79). Efficacy then began to drop sharply under both wet and dry growing conditions. Based on a cabbage maggot bioassay, fensulfothion bioactivity lasted twice as long in a previously untreated soil as in a soil treated the previous year (77). Similar results were observed even when the rate of the second fensulfothion treatment was doubled.

Fensulfothion bioactivity was lost more rapidly in soil collected from field plots previously treated with carbofuran than in soil from field plots without an insecticide history (77). Conversely, carbofuran was degraded more rapidly in soil that had initially been treated with fensulfothion than in soil without previous treatments. These observations were the first to indicate that cross-conditioning could occur between structurally unrelated insecticides.

Efficacy trials conducted in Newfoundland, Canada showed that control of cabbage maggot damage by fensulfothion has gradually declined since the 1960s (41). Tests for insect resistance showed little change from earlier levels. A laboratory test did reveal that fensulfothion dissipated more rapidly in nutrient broths inoculated with soil previously treated with fensulfothion than in nutrient broths inoculated with soil that was never treated. Cross-conditioning between carbofuran and fensulfothion was not observed.

ISOFENPHOS    The history of isofenphos use for control of corn rootworm feeding damage in the Corn Belt epitomizes the negative effects of enhanced biodegradation in agroecosystems. Isofenphos was commercially introduced in the corn-growing market in 1981. During the 1983 growing season there were numerous reports from many Corn Belt states that isofenphos failed to reduce corn rootworm feeding damage in fields where it had been used the previous year. By 1984, the manufacturer had decided to remove the formulation of isofenphos known as Amaze® from the market. Ironically, the manufacturer had advertised the product with the phrase "Amaze stays." Indeed, field and laboratory studies had shown that isofenphos was more persistent than the other soil insecticides registered for corn rootworm control (14, 22).

Evidence implicating the enhancement of biodegradation of isofenphos came from field studies during 1983–1985 in Iowa and Canada (1, 17). Isofenphos used in Iowa for a second year in corn plots under four different tillage treatments failed to reduce root damage significantly when compared to control in previously untreated plots (1). Percentage mortality of southern corn rootworm larvae in soils collected from the pretreated plots dropped to under 35% and to almost zero after 3 and 5 wk, respectively, following insecticide application.

In Canada, isofenphos dissipated much faster in field plots retreated during 1984 than in field plots treated for the first time during 1983 (17). Accelerated degradation of isofenphos in the field was corroborated by laboratory incubations in soil removed from treated and untreated plots just prior to renewed treatment in 1984. More than 90% of the isofenphos dissipated after 3 days in the previously treated soils, compared to 30% in the untreated controls (17).

Soil sterilization studies proved that the soil microflora was responsible for the rapid loss of isofenphos (1). A *Pseudomonas* sp. that utilized isofenphos as sole carbon source was isolated from conditioned soil in Iowa (69).

Isofenphos is also formulated as Oftanol®, which is used to control scarabaeid larvae in turf. Normally, it is applied in spring, and control can extend to the summer generation of larvae. In 1984, turfgrass managers in Ohio reported poor control of late-summer infestations of larvae following spring applications of isofenphos (67). Isofenphos degradation was examined in nutrient broth cultures inoculated with soils and thatch from golf courses with reported efficacy problems. Over 90% of the added isofenphos disappeared within 3 days. No degradation was observed in control broths containing inocula from previously untreated golf courses.

ETHOPROP    Ethoprop was ineffective against root-knot nematodes in corn and peas after 3 yr of prior use in Georgia (81). Residues declined by over 85% within 5 days of application, but 90% disappearance was expected to require 3 wk. The suggestion that ethoprop could condition soil for enhanced biodegradation was supported by observations of failed nematode control on potatoes after one to three growing seasons in the Netherlands (95). Chemical assays of samples collected from field plots proved that ethoprop was degraded more rapidly in soil from annually treated plots than in soil from untreated plots. Soil sterilization indicated the involvement of a microbial factor. There was no cross-conditioning effect from previous applications of the nematicides oxamyl or aldicarb.

PARATHION    The environmental chemistry of parathion has been intensively studied, and the compound serves as a general model for understanding pesticide interactions in the soil. Parathion is not normally used for the control of soil-inhabiting pests; therefore, enhanced biodegradation would not be detected by observation of a failed effort to control pests. Parathion, however, would be transported to the soil by runoff from foliage and deposition from sprays; thus microbial adaptation for accelerated breakdown of parathion would be beneficial in allaying adverse environmental effects. Enhanced biodegradation of parathion was first observed in flooded acid soils in India (86). Repeated additions of parathion to aerobic cranberry soils also resulted in accelerated loss of parathion as measured by carbon dioxide evolution (31). In contrast, repeated additions of parathion to a silt loam soil did not enhance the rate of parathion disappearance (72).

Inorganic and organic amendments can either inhibit or enhance the breakdown of parathion (31, 75, 89). Interestingly, after soil was repeatedly exposed to parathion, the metabolic pathway of the chemical changed from reduction to hydrolysis (101). Furthermore, pretreatment of soil with $p$-nitrophenol, the hydrolysis product of parathion, conditioned the soil for more rapid degradation of parathion (101).

OTHER ORGANOPHOSPHATE INSECTICIDES    All registered soil insecticides have failed to prevent corn rootworm feeding damage at one time or another, but no direct links between decreased corn rootworm control and rapid degradation rates have been reported for organophosphate insecticides other than isofenphos. Fonofos, however, was degraded more readily in soil collected from a field where it had been used previously than in soil collected from a field without its use (70). Previous applications of isofenphos did not seem to affect the degradation rate of fonofos.

Metabolism of terbufos was studied in soils collected from fields with reported problems of corn rootworm control and several years' prior use of terbufos (46). Terbufos was generally as persistent in soils with a prior history of its use as in soils collected from untreated fencerows. The primary environmental metabolite of terbufos, terbufos sulfoxide, was lost at significantly faster rates in previously treated soil than in untreated fencerow soil. The possible effect of such a change in degradation rate on corn rootworm control is open to speculation because the parent terbufos is toxic to third-instar $D$.

*virgifera virgifera* at concentrations below 0.1 ppm (105) and terbufos sulfoxide is almost 10 times less toxic to larvae than terbufos (23).

## THEORETICAL ASPECTS OF ENHANCED BIODEGRADATION

Knowledge of the mechanisms by which soil microflora become adapted for the biodegradation of insecticides is important for understanding the practical aspects of the relationship between enhanced biodegradation and control of insect pests. For example, prediction of the number of insecticide applications that will condition a soil for enhanced biodegradation requires a knowledge of the ecological physiology of microbial adaptation. Some knowledge of enyzme specificity for metabolism may allow the prediction of cross-conditioning between pesticides.

The enhancement of pesticide biodegradation can be viewed from ecological, enzymological, and genetic perspectives. The ecological aspects of microbial adaptation are understood fairly well. A pesticide in soil or water is one of many chemical substrates that microorganisms encounter. Microorganisms exhibit two ecological strategies of breaking down the pesticide: cometabolism and mineralization (3).

Metabolism of an absorbed substrate normally leads to energy generation, but some chemicals can only be partially broken down (i.e. cometabolized) by incidental enzymatic reactions. The metabolites that result from cometabolism do not enter into the cell's energy-generating metabolic pathways. As a consequence, the microbial population does not grow at the expense of the absorbed substrate, and the rate of change of substrate concentration in the medium remains relatively constant.

In the mineralization strategy (3), the absorbed substrate is broken down into smaller molecules, which are further metabolized via energy-generating pathways. In this case, the biomass of the population increases at the expense of the substrate. The rate of change in substrate concentration in the medium increases coincidentally with the expanding microbial population. As the substrate concentration in the medium quickly declines to a very low level, growth of the microbial population reaches a plateau at a higher cell density.

Enhanced pesticide biodegradation is due to microorganisms that derive energy or nutrients from metabolism of the pesticide. When a pesticide is added to soil, water, or a culture medium, the population density increases; thus the metabolic capacity for biodegradation is much greater than before. Subsequently added pesticide then disappears from the medium at an augmented rate.

The underlying principles of microbial adaptation to pesticides were best illustrated by classic experiments with the herbicide 2,4-D (5; discussed in 6).

After addition of 2,4-D to soil perfusion columns, a lag period was observed before the herbicide disappeared. The lag period was believed to result from the induction of pesticide-degrading enzymes followed by an increase in the microbial population. When the herbicide was cycled through the perfusion columns again, the lag period before complete loss of the pesticide was shorter. After a third treatment, the lag period had completely disappeared. The population of microorganisms had grown sufficiently from the two earlier treatments to cause the immediate degradation of 2,4-D. The phenomenon of microbial proliferation in response to new herbicide addition was termed "enrichment" (6).

Later studies with herbicides and insecticides gave credibility to the earlier hypotheses of microbial adaptation and enrichment. The population density of specific 2,4-D–degrading microorganisms was measured in soil treated twice in succession with 2,4-D (33). After the second treatment the lag period (or acclimation period) decreased coincidentally with an increase in the population of herbicide-specific degraders. "Ecocores" (water/sediment systems) previously treated with p-nitrophenol, the hydrolytic metabolite of parathion, harbored 1000-fold as many specific nitrophenol degraders as untreated ecocores (97). The larger degrader population was correlated with a faster loss of methyl parathion from pretreated ecocores than from control ecocores. Conditioned soils contained several thousand specific isofenphos-degrading microorganisms per gram of soil (69). Soils without prior isofenphos treatments contained no isofenphos-specific degraders. The lag period may simply be the time required for small populations of mineralizing populations to increase to the cell densities necessary to cause detectable changes in the substrate (114).

Both inducible and constitutive enzymes are involved in pesticide degradation. Studies of carbaryl metabolism by several bacterial strains isolated from a garden soil indicated that enzymes in the catabolic pathway for carbaryl may be inducible (60). In contrast, work with parathion-degrading microorganisms has shown that the initial hydrolysis to p-nitrophenol is cometabolic (i.e. no net energy is derived), but that further metabolism of the phenol is carried out by a constitutive enzyme (65), which has been called parathion hydrolase (63). A carbofuran hydrolase has been isolated from an *Achromobacter* sp. that survived on carbofuran as sole nitrogen source via the hydrolysis of carbofuran to methyl amine and carbofuran phenol (19, 48).

Knowledge of the enzymes involved in enhancement of biodegradation and an understanding of their specificity are important for the prediction of cross-conditioning in the field. The high hydrolytic activity of carbofuran hydrolase against carbaryl and aldicarb (19) might explain the cross-conditioning for enhanced aldicarb biodegradation (103, 104). Whether the lack of specificity characteristic among carbamate-hydrolyzing enzymes applies also to organophosphates is somewhat ambiguous. Parathion hydrolase exhibited hydrolytic activity toward diazinon and chlorpyrifos (64), but diazinon-adapted bacteria were reported to be unaffected by previous exposures to parathion (32, 85) or chlorpyrifos (85).

The biochemical genetics and molecular biology of microbial adaptation are beyond the scope of this review, but strong arguments have been made for studying these aspects to better understand the development of enhanced biodegradation (54). Plasmids, which are extrachromosomal pieces of DNA, carry the genome for synthesizing organophosphate-degrading enzymes (84). In relation to the applied aspects of enhanced biodegradation, the transfer and stability of pesticide-degrading ability in the soil are of interest.

## PRACTICAL ASPECTS OF ENHANCED BIODEGRADATION

Theories of the relation of microbial growth and substrate utilization and theories regarding the genetic potential for metabolism of synthetic organic chemicals indicate that the development of enhanced biodegradation of pesticides is an evolutionary phenomenon somewhat analogous to the development of insecticide resistance. The case histories presented earlier clearly show that this natural phenomenon has adversely affected crop protection practices in certain agroecosystems. Viewed in this manner, management of the phenomenon becomes a priority. Successful management, however, requires an understanding of the factors affecting the enhancement of biodegradation and its relationship with insect control.

Ecological and biochemical genetic research is now aimed at elucidating the number of times a field can be treated with the same insecticide before biodegradation becomes enhanced. Microbial enrichment can occur if a pesticide serves as a usable source of energy; thus, biodegradation can be enhanced after one field treatment. Rapid biodegradation of carbofuran was associated wth poor control of corn rootworm feeding damage only after a number of fields had been treated for at least 2 yr (24, 30). However, current research suggests that the phenomenon probably developed after only one treatment (e.g. 1, 16, 40, 102, 103).

Because the arsenal of soil insecticides is limited, the interval before a field can be retreated safely without rapid loss of bioactivity needs clarification. Is microbial enrichment stable in the absence of further inputs? Insecticide studies have not adequately addressed this issue. The enhanced biodegradation in carbofuran-conditioned soil remained for 13 wk after initial applications of the insecticide (40). A soil from a field in which fonofos was used for five consecutive years seemed to have lost the capacity to degrade the insecticide rapidly after the winter (70).

Studies with herbicides have indicated that the stability of enhanced pesticide degradation differs with the chemical. For example, acclimation of soil for enhanced 2,4-D degradation was stable for a year in the absence of added pesticide (6). Soils retained the ability to degrade MCPA 5 yr after the final field application of the herbicide (34). In contrast, the thiocarbamate herbicide S-ethyl dipropylthiocarbamate (EPTC) quickly conditioned soil for its enhanced biodegradation, but microbial adaptation did not persist for more than 1 yr in soil (42) or in culture (62) that had been stored in the absence of the pesticide. Spontaneous loss of a plasmid encoding EPTC metabolism was hypothesized to be responsible for the instability of enhanced biodegradation (62, 106).

Many structural analogs exist among the carbamate and organophosphate insecticides. Specificity of microbial enzymes would influence whether treatment of a field by one type of organophosphate or carbamate insecticide would cause enhanced biodegradation of a related insecticide. Cross-conditioning among carbamates is common (e.g. 27, 40, 71, 95, 103, 104). On the other hand, biodegradation of organophosphates is more specific (32, 70, 85), although one study suggested cross-conditioning between chlorpyrifos and terbufos (46). The only known cross-conditioning between a carbamate and an organophosphate was reported for carbofuran and fensulfothion (77); however, a streptomycete bacterium capable of growing on several commercial carbamate and organophosphate insecticides was isolated from a field sample previously treated with isofenphos (35). Thiocarbamate herbicides, such as EPTC and butylate, are commonly used in the Corn Belt, but cross-conditioning with carbofuran has not been observed (43, 94).

The influence of agronomic practices and environmental factors on enhancement of biodegradation has received some attention. Very low rates of application may delay or inhibit microbial enrichment of soil (16, 104). Both granular and flowable formulations of carbofuran can condition soil (15), but granular formulations degrade more slowly (102), especially at lower levels of soil moisture (16). Adapted organisms can be transported from one field to another on machinery carrying soil residues (79).

The amount of insecticide and moisture content necessary for conditioning soil for enhanced biodegradation differs among soil types (16). For example, higher doses of carbofuran are required to condition a muck than to condition a sandy loam. Carbofuran was not rapidly degraded in a pretreated sandy loam with less than 5% moisture, but it was rapidly degraded at higher moisture contents. In a clay loam, 20% moisture prevented the acceleration of carbofuran degradation. In another study, soils from fields with a natural pH below 5.8 did not tend to develop rapid breakdown of carbofuran (78).

Soil amendments can have a profound effect on rate of pesticide degradation. Rapid biodegradation of carbofuran was inhibited in a soil with a 7-yr manuring history (96). Inorganic fertilizers such as ammonium sulfate inhibited the breakdown of parathion (31). Fungicides and soil sterilants are also known to inhibit the rapid biodegradation of parathion (31) and carbofuran (103).

In sum, if a pesticide can be utilized as a substrate for microbial growth, then a soil can be easily conditioned for enhanced biodegradation by only one pesticide application. Whether enhancement of biodegradation persists under field conditions for longer than one year is not well characterized for insecticides, but herbicide studies have shown that stability depends on the pesticide. Finally, specific agronomic practices and environmental factors can inhibit or facilitate the microbial adaptations for biodegradation.

## COPING WITH ENHANCED BIODEGRADATION

Microbial adaptations for metabolism are natural processes that cannot be eliminated completely. Problems with this beneficial phenomenon arise only when crop protection practices are adversely affected. Some research suggests, however, that enhanced biodegradation may be managed to avoid these adverse effects. Two general kinds of strategies for coping with enhanced biodegradation have been proposed (26): operational and technological.

### Operational Strategies

Operational strategies for coping with enhanced biodegradation rely on management techniques based on biological principles. These strategies include conservation of pesticides, crop rotation, proper calibration of equipment, altered timing of applications, and chemical rotation.

Pesticides are conserved by adhering to the basic elements of integrated pest management: thorough understanding of the agroecosystem, monitoring of pest populations through scouting, and application of pesticides only when economic thresholds are reached. The goal is elimination of prophylactic pesticide treatments, which predominate in the Corn Belt. Strong arguments support the view that many of the insecticide treatments commonly used against corn rootworm are not needed (109). Research with aldicarb in the United Kingdom has shown that when a pesticide is needed, use of the lower rates commercially recommended for controlling root maggots can inhibit the development of enhanced biodegradation even if the soil has been conditioned by previous treatments (104).

Appropriate agronomic practices can greatly reduce the need for pesticides. Corn-soybean rotations have long been known as a panacea for controlling corn rootworm infestations. The prevalence of crop-rotation systems in the United Kingdom may explain why few pest-control problems have been associated with enhanced biodegradation of pesticides there (102).

Proper calibration of insecticide application equipment is critical for corn rootworm control. Because soil rapidly attenuates bioactivity by sorptive processes as well as by biodegradation, the initial amount of pesticide applied should conform with the manufacturer's recommendations. Yet 85% of farmers tested in a Canadian study applied less than the recommended amount of insecticide to cornfields, probably because of improper calibration practices (21).

In adhering to principles of IPM, application of corn rootworm insecticides closer to the time when larvae are actively feeding can be beneficial. Application of carbofuran during field cultivation protected corn roots significantly better than application at planting time (98).

Appropriate rotation of insecticides of different chemical classes has been studied as a technique to allay the development of insecticide resistance (36). It may also be a strategy for coping with enhanced biodegradation of carbamates. Work has not been published to support or negate the benefits of insecticide rotation in corn. Rotations of fensulfothion and carbofuran failed to prevent the enhancement of biodegradation of either compound in rutabaga fields (77). Alternating the use of carbofuran or fensulfothion with the use of chlorfenvinphos or terbufos has been suggested to reduce the possibility of enhanced biodegradation (79). A 3- or 4-yr crop rotation with use of insecticides only once every 3–4 yr was also suggested for allaying microbial enrichment. Efficacy of the thiocarbamate herbicides EPTC and butylate was improved by use only in alternate years (83, 117).

## Technological Strategies

Technological strategies require alterations in formulation chemistry or structural chemistry of the insecticide. The alternatives include extenders and inhibitors, new formulation technology, and directed chemistry.

Extenders and inhibitors are additives to pesticide formulations that improve residual bioactivity by preventing or at least slowing down biodegradation. Inhibitors, which are well known from enzymology, increase chemical persistence by inhibition of specific degradative enzymes. Extenders increase chemical persistence by generalized adverse effects on soil microbial populations. These compounds are not used in currently registered soil insecticide formulations. One commercial formulation of EPTC herbicide, however, contains dietholate, a parathion analog. This compound slows the biodegradation rate of EPTC in soils adapted for enhanced biodegradation of thiocarbamates (20, 43, 68). Unfortunately, after repeated applications the formulation containing the inhibitor eventually lost its efficacy (44, 83). The inhibitory effects of methyl carbamate insecticides on chlorpropham herbicide degradation are well known (50, 53), and extended dodder control was observed when chlorpropham was mixed with carbaryl (18). Several fungicides and fumigants extended insecticide persistence, presumbly by producing adverse effects on normal microbial communities (31, 102).

If the availability of an insecticide to microorganisms could be reduced without affecting uptake by the insect, then a longer effective residual life would result. Alternative formulation technologies, such as controlled release, may be feasible for extending the biological activity of soil insecticides (99). Controlled-release formulations of carbofuran persisted unchanged for 2 mo in soils adapted for enhanced biodegradation of carbofuran, but field applications did not result in improved corn rootworm control (A. S. Felsot & E. Levine, unpublished).

Insecticide chemistry needs to be directed toward specific soil pests to minimize the amount of chemical needed for control. At recommended rates of application, insecticide concentrations approximate 6 ppm in the upper 4 inches of soil. If concentrations below 0.1 ppm could be highly active against root-feeding insect pests, then when 98% of the insecticide had degraded, bioactivity would still suffice to control feeding damage. Furthermore, biodegradation in soil slows down significantly at such low insecticide concentrations.

The concept of a threshold concentration necessary to stimulate microbial enrichment suggests that significant lowering of initial rates of insecticide application could help alleviate degradation problems (4). The inability of low carbofuran concentrations (16) and low-rate field applications of aldicarb (104) to induce microbial enrichment supports this hypothesis.

Directed syntheses may yield chemical structures that are resistant to mineralizing activity of soil microflora, but are still cometabolized so that environmental pollution is minimized. For example, the commercially used thiocarbamate herbicides, which are readily susceptible to enhanced biodegradation, have alkyl chains attached to a thiocarbamyl functional group. Cycloate, however, a thiocarbamate that has a cyclohexane ring, does not seem to condition soil for its accelerated biodegradation (44).

The operational strategies are the most desirable for long-term management of biodegradation problems. They are more compatible with environmental goals and require less research and development than technological strategies, and with the exception of chemical class rotations, they can be implemented immediately. The most promising technological strategies may be new formulation technologies and directed chemistry, but unfortuantely these will be the most expensive.

## POSITIVE ASPECTS OF ENHANCED BIODEGRADATION

Given the widespread concern about nonpoint pollution of surface water and groundwater by pesticide residues, it is ironic that up until now I have presented a negative image of enhanced biodegradation in agroecosystems. I prefer to end on an upbeat note about this fascinating and potentially useful natural phenomenon. The study of microbial adaptation to pesticides and other environmental contaminants is leading to opportunities to harness microorganisms for cleanup of polluted soil and water. For example, pesticide wastes are commonly produced as the result of spills and equipment-rinsing procedures. Researchers are beginning to test the feasibility of using adapted bacterial inocula and bacterial enzymes to degrade concentrated pesticide wastes quickly (9, 49, 100). These efforts will accelerate as more is learned about the biochemical ecology of pesticide biodegradation in the soil. Molecular biology will be very useful for understanding the complex interactions between pesticides and the soil microflora.

## CONCLUSIONS

A number of insecticides used for control of soil insect pests can condition soil for microbial enrichment and subsequently enhanced biodegradation. The accelerated loss of chemicals has been linked to crop protection failures. The strongest associations have been made with carbofuran for control of corn rootworm and cabbage maggot and with diazinon for control of brown rice planthopper. The enhanced biodegradation of corn rootworm soil insecticides has serious implicatons in corn agroecosystems because the acreage treated is very large. However, the argument that many of the treatments for corn rootworm control may not be economical suggests that corn rootworm populations are below the economic injury level in many fields that are treated. Prophylactic treatments may exacerbate the enhancement of biodegradation, but in the absence of an economically damaging insect population, enhanced biodegradation may not even be noticed. Thus, the true impact of enhanced biodegradation in agroecosystems remains obscure. Not all soil insecticides can induce microbial enrichment; however, the chemical characteristics that stimulate adaptations are unknown. The best prescription for avoiding problems is enthusiastic adoption of ecologically based management principles. On a positive note, the study of enhanced biodegradation of insecticides will lead to biologically based methods for cleaning up chemical wastes.

*Literature Cited*

1. Abou-Assaf, N., Coats, J. R., Gray, M. E., Tollefson, J. J. 1986. Degradation of isofenphos in cornfields with conservation tillage practices. *J. Environ. Sci. Health. Part B* 21:425–46
2. Ahmad, N., Walgenbach, D. D., Sutter, G. R. 1979. Degradation rates of technical carbofuran and a granular formulation in four soils with known insecticide use history. *Bull. Environ. Contam. Toxicol.* 23:572–74
3. Alexander, M. 1981. Biodegradation of chemicals of environmental concern. *Science* 211:132–38
4. Alexander, M. 1985. Biodegradation of organic chemicals. *Environ. Sci. Technol.* 18:106–11
5. Audus, L. J. 1949. The biological de-

toxication of 2:4-dichlorphenoxyacetic acid in soil. *Plant Soil* 2:31–36

6. Audus, L. J. 1964. Herbicide behaviour in the soil. II. Interactions with soil microorganisms. In *The Physiology and Biochemistry of Herbicides,* ed. L. J. Audus, pp. 163–206. New York: Academic

7. Ball, H. J. 1977. Larval and adult control recommendations and insecticide resistance data for corn rootworms in Nebraska (1948–76). *Rep. No. 3, Agric. Exp. Stn., Univ. Nebr., Lincoln.* 20 pp.

8. Ball, H. J., Weekman, G. T. 1962. Insecticide resistance in the adult western corn rootworm in Nebraska. *J. Econ. Entomol.* 55:439–41

9. Barles, R. W., Daughton, C. G., Hsieh, D. P. H. 1979. Accelerated parathion degradation in soil inoculated with acclimated bacteria under field conditions. *Arch. Environ. Contam. Toxicol.* 8:647–60

10. Bewick, D. W., Hill, I. R., Pluckrose, J., Stevens, J. E. B., Weissler, M. S. 1986. The role of laboratory and field studies, using radiolabelled materials, in the investigation and mobility of tefluthrin in soil. In *Pests and Diseases. Proc. Br. Crop Prot. Conf., Brighton, UK,* pp. 459–68. Croydon, UK: Br. Crop Prot. Counc.

11. Bigger, J. H. 1963. Corn rootworm resistance to chlorinated hydrocarbon insecticides in Illinois. *J. Econ. Entomol.* 56:118–19

12. Camper, N. D., Fleming, M. M., Skipper, H. D. 1987. Biodegradation of carbofuran in pretreated and non-pretreated soils. *Bull. Environ. Contam. Toxicol.* 39:571–78

13. Carson, R. 1962. *Silent Spring.* Greenwich, Conn: Fawcett. 304 pp.

14. Chapman, R. A., Harris, C. R. 1982. Persistence of isofenphos and isazophos in a mineral and organic soil. *J. Environ. Sci. Health Part B* 17:355–61

15. Chapman, R. A., Harris, C. R., Harris, C. 1986. The effect of formulation and moisture level on the persistence of carbofuran in a soil containing biological systems adapted to its degradation. *J. Environ. Sci. Health Part B* 21:57–66

16. Chapman, R. A., Harris, C. R., Harris, C. 1986. Observations on the effect of soil type, treatment intensity, insecticide formulation, temperature and moisture on the adaptation and subsequent activity of biological agents associated with carbofuran degradation in soil. *J. Environ. Sci. Health Part B* 21:125–41

17. Chapman, R. A., Harris, C. R., Moy, P., Henning, K. 1986. Biodegradation of pesticides in soil: rapid degradation of isofenphos in a clay loam after a previous treatment. *J. Environ. Sci. Health Part B* 21:269–76

18. Dawson, J. H. 1984. Effect of carbaryl and PCMC on dodder (*Cuscuta* spp.) control with chlorpropham. *Weed Sci.* 32:290–92

19. Derbyshire, M. K., Karns, J. S., Kearney, P. C., Nelson, J. O. 1987. Purification and characterization of an *N*-methylcarbamate pesticide hydrolyzing enzyme. *J. Agric. Food Chem.* 35:871–77

20. Dowler, C. C., Marti, L. R., Kvien, C. S., Skipper, H. D., Gooden, D. T., Zublena, J. P. 1987. Accelerated degradation potential of selected herbicides in the southeastern United States. *Weed Technol.* 1:350–58

21. Ellis, C. R. 1982. A survey of granular application equipment and insecticide rates used for control of corn rootworms (Coleoptera: Chrysomelidae) in southern Ontario. *Proc. Entomol. Soc. Ont.* 113:29–34

22. Felsot, A. 1984. Persistence of isofenphos (Amaze) soil insecticide under laboratory and field conditions and tentative identification of a stable oxygen analog metabolite by gas chromatography. *J. Environ. Sci. Health Part B* 19:13–27

23. Felsot, A. 1985. Factors affecting the bioactivity of soil insecticides. *Proc. 37th Ill. Custom Spray Oper. Train. Sch.,* pp. 134–38. Urbana-Champaign, Ill: Univ. Ill.

24. Felsot, A., Maddox, J. V., Bruce, W. 1981. Enhanced microbial degradation of carbofuran in soils with histories of Furadan use. *Bull. Environ. Contam. Toxicol.* 26:781–88

25. Felsot, A. S. 1981. Factors affecting the performance of Furadan against corn rootworms. *Proc. 33rd Ill. Custom Spray Oper. Train. Sch.,* pp. 79–80. Urbana-Champaign, Ill: Univ. Ill.

26. Felsot, A. S. 1984. *Coping with enhanced biodegradation of agricultural chemicals.* Presented at Natl. Meet. Am. Chem. Soc., 187th, St. Louis

27. Felsot, A. S. 1986. Effect of conditioning soil for enhanced biodegradation of carbofuan on dissipation of other methyl carbamate insecticides. *Abstr. 6th Int. Congr. Pestic. Chem., Ottawa,* Abstr. 6B–08. Ottawa: IUPAC

28. Felsot, A. S., Baughman, T. A., Kuhlman, D. 1988. Monitoring corn rootworm (*Diabrotica* spp.) larvae for resistance to soil insecticides. *Abstr. 43rd Annu. Meet. North. Cent. Branch En-*

*tomol. Soc. Am., Denver, Colo.*, Abstr. 210

29. Felsot, A. S., Steffey, K. L., Levine, E., Wilson, J. G., 1985. Carbofuran persistence in soil and adult corn rootworm (Coleoptera: Chrysomelidae) susceptibility: relationship to the control of rootfeeding damage by larvae. *J. Econ. Entomol.* 78:45–52

30. Felsot, A. S., Wilson, J. G., Kuhlman, D. E., Steffey, K. L. 1982. Rapid dissipation of carbofuran as a limiting factor in corn rootworm (Coleoptera: Chrysomelidae) control in fields with histories of continuous carbofuran use. *J. Econ. Entomol.* 75:1098–103

31. Ferris, I. G., Lichtenstein, E. P. 1980. Interactions between agricultural chemicals and soil microflora and their effects on the degradation of [$^{14}$C]parathion in a cranberry soil. *J. Agric. Food Chem.* 28:1011–19

32. Forrest, M., Lord, K. A., Walker, N., Woodville, H. C. 1981. The influence of soil treatments on the bacterial degradation of diazinon and other organophosphorus insecticides. *Environ. Pollut. Ser. A* 24:93–104

33. Fournier, J. C., Codaccioni, P., Soulas, G., Repiquet, C. 1981. Soil adaptation to 2,4-D degradation in relation to the application rates and the metabolic behaviour of the degrading microflora. *Chemosphere* 10:977–84

34. Fryer, J. D., Smith, P. D., Hance, R. J. 1980. Field experiments to investigate long-term effects of repeated applications of MCPA, triallate, simazine and linuron: II. Crop performance and residues 1969–78. *Weed Res.* 20:103–10

35. Gauger, W. K., MacDonald, J. M., Agrian, N. R., Matthees, D. P., Walgenbach, D. D. 1986. Characterization of a streptomycete growing on organophosphate and carbamate insecticides. *Arch. Environ. Contam. Toxicol.* 15:137–41

36. Georghiou, G. P. 1980. Insecticide resistance and prospects for its management. *Residue Rev.* 76:131–45

37. Gorder, G. W., Dahm, P. A., Tolefson, J. J. 1982. Carbofuran persistence in cornfield soils. *J. Econ. Entomol.* 75:637–42

38. Gorder, G. W., Tollefson, J. J., Dahm, P. A. 1980. Carbofuran residue analysis and control of corn rootworm larval damage. *Iowa State J. Res.* 55:25–33

39. Greenhalgh, R., Belanger, A. 1981. Persistence and uptake of carbofuran in a humic mesisol and the effects of drying and storing soil samples on residue levels. *J. Agric. Food Chem.* 29:231–35

40. Harris, C. R., Chapman, R. A., Harris, C., Tu, C. M. 1984. Biodegradation of pesticides in soil: rapid induction of carbamate degrading factors after carbofuran treatment . *J. Environ. Sci. Health Part B* 19:1–11

41. Harris, C. R., Chapman, R. A., Morris, R. F., Stevenson, A. B. 1988. Enhanced soil microbial degradation of carbofuran and fensulfothion—a factor contributing to the decline in effectiveness of some soil insect control programs in Canada. *J. Environ. Sci. Health Part B* In press

42. Harvey, R. G. 1987. Herbicide dissipation from soils with different herbicide use histories. *Weed Sci.* 35:583–89

43. Harvey, R. G., Dekker, J. H., Fawcett, R. S., Roeth, F. W., Wilson, R. G. 1987. Enhanced biodegradation of herbicides in soil and effects on weed control. *Weed Technol.* 1:341–49

44. Harvey, R. G., McNevin, G. R., Albright, J. W., Kozak, M. 1986. Wild proso millet *(Panicum miliaceum)* control with thiocarbamate herbicides on previously treated soils. *Weed Sci.* 34:773–80

45. Hill, I. R., Wright, S. J. L., eds. 1978. *Pesticide Microbiology.* New York: Academic 844 pp.

46. Horng, L. C., Kaufman, D. D. 1987. Accelerated biodegradation of several organophosphate insecticides. *Abstr. Pap. 193rd Natl. Meet. Am. Chem. Soc., Denver, Colo.* Washington, DC: Am. Chem. Soc.

47. Hsu, T.-S., Bartha, R. 1979. Accelerated mineralization of two organophosphate insecticides in the rhizosphere. *Appl. Environ. Microbiol.* 37:36–41

48. Karns, J. S., Mulbry, W. W., Nelson, J. O., Kearney, P. C. 1986. Metabolism of carbofuran by a pure bacterial culture. *Pestic. Biochem. Physiol.* 25:211–17

49. Karns, J. S., Muldoon, M. T., Mulbry, W. W., Derbyshire, M. K., Kearney, P. C. 1987. Use of microorganisms and microbial systems in the degradation of pesticides. *ACS Symp. Ser.* 334:156–70

50. Kaufman, D. D., Blake, J., Miller, D. E. 1971. Methylcarbamates affect acylanilide herbicide residues in soil. *J. Agric. Food Chem.* 19:204–6

51. Kaufman, D. D., Edwards, D. F. 1983. Pesticide/microbe interaction effects on persistence of pesticides in soil. In *Pesticide Chemistry: Human Welfare and the Environment*, ed. J. Miyamoto, P. C. Kearney, 3:177–82. Oxford, UK: Pergamon

52. Kaufman, D. D., Katan, Y., Edwards, D. F., Jordan, E. G. 1983. Microbial

adaptation and metabolism of pesticides. In *Agricultural Chemicals of the Future. BARC (Beltsville Agric. Res. Cent.) Symp. No. 8*, ed. J. L. Hilton, pp. 437–51. Totowa, NJ: Rowman/Allanheld

53. Kaufman, D. D., Kearney, P. C., Von Endt, D. W., Miller, D. E. 1970. Methylcarbamate inhibition of phenylcarbamate metabolism in soil. *J. Agric. Food Chem.* 18:513–19

54. Kearney, P. C., Kellog, S. T. 1985. Microbial adaptation to pesticides. *Pure Appl. Chem.* 57:390–403

55. Kirkland, K., Fryer, J. D. 1966. Pretreatment of soil with MCPA as a factor affecting persistence of a subsequent application. *Proc. 8th Br. Weed Control Conf., Brighton, UK,* 2:616–21. Croydon, UK: Br. Weed Control Counc.

56. Kuhlman, D. E. 1974. Results of 1973 corn rootworm control in demonstration plots. *Proc. 26th Ill. Custom Spray Oper. Train. Sch.,* pp. 56–59. Urbana-Champaign, Ill: Univ. Ill.

57. Kuhlman, D. E., Steffey, K. L. 1981. Putting corn rootworm problems and control into perspective. *Proc. 33rd Ill. Custom Spray Oper. Train. Sch.,* pp. 81–87. Urbana-Champaign, Ill: Univ. Ill.

58. Kuhlman, D. E., Wedberg, J. 1977. Corn rootworm problems and solutions in Illinois. *Proc. 29th Ill. Custom Spray Oper. Train. Sch.,* pp. 81–89. Urbana-Champaign, Ill: Univ. Ill.

59. Lal, R., ed. 1984. *Insecticide Microbiology.* New York: Springer-Verlag. 268 pp.

60. Larkin, M. J., Day, M. J. 1986. The metabolism of carbaryl by three bacterial isolates, *Pseudomonas* spp. (NCIB 12042 & 12043) and *Rhodococcus* sp. (NCIB 12038) from garden soil. *J. Appl. Bacteriol.* 60:233–42

61. Laveglia, J., Dahm, P. A. 1977. Degradation of organophosphorus and carbamate insecticides in the soil and by soil microorganisms. *Ann. Rev. Entomol.* 22:483–513

62. Lee, A. 1984. EPTC (*S*-Ethyl *N,N*-dipropylthiocarbamate)–degrading microorganisms isolated from a soil previously exposed to EPTC. *Soil Biol. Biochem.* 16:529–31

63. Munnecke, D. M. 1977. Properties of an immobilized pesticide-hydrolyzing enzyme. *Appl. Environ. Microbiol.* 33:503–7

64. Munnecke, D. M. 1980. Enzymatic detoxification of waste organophosphate pesticides. *J. Agric. Food Chem.* 28:105–11

65. Nelson, L. M. 1982. Biologically-induced hydrolysis of parathion in soil:

isolation of hydrolyzing bacteria. *Soil Biol. Biochem.* 14:219–22

66. Newman, A. S., Thomas, J. R., Walker, R. L. 1952. Disappearance of 2,4-dichlorophenoxyacetic acid from soil. *Soil Sci. Soc. Am. Proc.* 16:21–24

67. Niemczyk, H. D., Chapman, R. A. 1987. Evidence of enhanced degradation of isofenphos in turfgrass thatch and soil. *J. Econ. Entomol.* 80:880–82

68. Obrigawitch, T., Roeth, F. W., Martin, A. R., Wilson, R. G. Jr. 1982. Addition of R-33865 to EPTC for extended herbicide activity. *Weed Sci.* 30:417–22

69. Racke, K. D., Coats, J. R. 1987. Enhanced degradation of isofenphos by soil microorganisms. *J. Agric. Food Chem.* 35:94–99

70. Racke, K. D., Coats, J. R. 1988. Comparative degradation of organophosphorus insecticides in soil: specificity of enhanced microbial degradation. *J. Agric. Food Chem.* 36:193–99

71. Racke, K. D., Coats, J. R. 1988. Enhanced degradation and the comparative fate of carbamate insecticides in soil. *J. Agric. Food Chem.* 36:1067–72

72. Racke, K. D., Lichtenstein, E. P. 1987. Effects of agricultural practices on the binding and fate of $^{14}$C-parathion in soil. *J. Environ. Sci. Health Part B* 22(1):1–14

73. Rajagopal, B. S., Brahmaprakash, G. P., Reddy, B. R., Singh, U. D., Sethunathan, N. 1984. Effect and persistence of selected carbamate pesticides in soil. *Residue Rev.* 93:1–199

74. Rajagopal, B. S., Panda, S., Sethunathan, N. 1986. Accelerated degradation of carbaryl and carbofuran in a flooded soil pretreated with hydrolysis products, 1-naphthol and carbofuran phenol. *Bull. Environ. Contam. Toxicol.* 36:827–32

75. Rajaram, K. P., Rao, Y. R., Sethunathan, N. 1978. Inhibition of biological hydrolysis of parathion in rice straw-amended flooded soil and its reversal by nitrogen compounds and aerobic conditions. *Pestic. Sci.* 9:155–60

76. Ramanand, K., Panda, S., Sharmila, M., Adhya, T. K., Sethunathan, N. 1988. Development and acclimatization of carbofuran-degrading soil enrichment cultures at different temperatures. *J. Agric. Food Chem.* 36:200–5

77. Read, D. C. 1983. Enhanced microbial degradation of carbofuran and fensulfothion after repeated applications to acid mineral soil. *Agric. Ecosyst. Environ.* 10:37–46

78. Read, D. C. 1986. Accelerated microbial breakdown of carbofuran in soil

from previously treated fields. *Agric. Ecosyst. Environ.* 15:51–61

79. Read, D. C. 1986. Influence of weather conditions and microorganisms on persistence of insecticides to control root maggots (Diptera: Anthomyiidae) in rutabagas. *Agric. Ecosyst. Environ.* 16:165–73

80. Read, D. C. 1987. Greatly accelerated microbial degradation of aldicarb in retreated field soil, in flooded soil, and in water. *J. Econ. Enotomol.* 80:156–63

81. Rhode, W. A., Johnson, A. W., Dowler, C. C., Glaze, N. C. 1980. Influence of climate and cropping patterns on the efficacy of ethoprop, methyl bromide, and DD-MENCS for control of root-knot nematodes. *J. Nematol.* 12:33–39

82. Rodriguez, L. D., Dorough, H. W. 1977. Degradation of carbaryl by soil microorganisms. *Arch. Environ. Contam. Toxicol.* 6:47–56

83. Rudyanski, W. J., Fawcett, R. S., McAllister, R. S. 1987. Effect of prior pesticide use on thiocarbamate herbicide persistence and giant foxtail *(Setaria faberi)* control. *Weed Sci.* 35:68–74

84. Serdar, C. M., Gibson, D. T., Munnecke, D. M., Lancaster, J. H. 1982. Plasmid involvement in parathion hydrolysis by *Pseudomonas diminuta. Appl. Environ. Microbiol.* 44:246–49

85. Sethunathan, N. 1971. Biodegradation of diazinon in paddy fields as a cause of its inefficiency for controlling brown planthoppers in rice fields. *PANS* 17:18–19

86. Sethunathan, N. 1973. Degradation of parathion in flooded acid soils. *J. Agric. Food Chem.* 21:602–4

87. Sethunathan, N., Caballa, S., Pathak, M. D. 1971. Absorption and translocation of diazinon by rice plants from submerged soils and paddy water and the persistence of residues in plant tissues. *J. Econ. Entomol.* 64:571–76

88. Sethunathan, N., Pathak, M. D. 1972. Increased biological hydrolysis of diazinon after repeated application in rice paddies. *J. Agric. Food Chem.* 20:586–89

89. Sethunathan, N., Rajaram, K. P., Siddaramappa, R. 1975. Persistence and microbial degradation of parathion in Indian rice soils under flooded conditions. In *Origin and Fate of Chemical Residues in Food, Agriculture and Fisheries,* pp. 9–18. Vienna: Int. Atomic Energy Agency. 189 pp.

90. Sethunathan, N., Yoshida, T. 1973. A *Flavobacterium* sp. that degrades diazinon and parathion. *Can. J. Microbiol.* 19:873–75

91. Siddaramappa, R., Tirol, A. C., Seiber, J. N., Heinrichs, E. A., Watanabe, I. 1978. The degradation of carbofuran in paddy water and flooded soil of untreated and retreated rice fields. *J. Environ. Sci. Health Part B* 13:369–80

92. Siddaramappa, R., Tirol, A., Watanabe, I. 1979. Persistence in soil and absorption and movement of carbofuran in rice plants. *J. Pestic. Sci.* 4:473–79

93. Simon-Sylvestre, G., Fournier, J. C. 1979. Effects of pesticides on the soil microflora. *Adv. Agron.* 31:1–92

94. Skipper, H. D., Murdock, E. C., Gooden, D. T., Zublena, J. P., Amakiri, M. A. 1986. Enhanced herbicide biodegradation in South Carolina soils previously treated with butylate. *Weed Sci.* 34:558–63

95. Smelt, J. H., Crum, S. J. H, Teunissen, W., Leistra, M. 1987. Accelerated transformation of aldicarb, oxamyl and ethoprophos after repeated soil treatments. *Crop Prot.* 6:295–303

96. Somasundaram, L., Racke, K. D., Coats, J. R. 1987. Effect of manuring on the persistence and degradation of soil insecticides. *Bull. Environ. Contam. Toxicol.* 39:579–86

97. Spain, J. C., Pritchard, P. H., Bourquin, A. W. 1980. Effects of adaptation on biodegradation rates in sediment/water cores from estuarine and freshwater environments. *Appl. Environ. Microbiol.* 40:726–34

98. Steffey, K., Kuhlman, D., Kinney, K. 1987. Corn rootworm larval control—research and management in Illinois. *Proc. Ill. Agric. Pestic. Conf., Urbana-Champaign,* pp. 50–62. Urbana-Champaign: Univ. Ill.

99. Stokes, R. A., Coppedge, J. R., Bull, D. L., Ridgway, R. L. 1973. Use of selected plastics in controlled release granular formulations of aldicarb and dimethoate. *J. Agric. Food Chem.* 21: 103–8

100. Sudhakar-Barik, Munnecke, D. M. 1982. Enzymatic hydrolysis of concentrated diazinon in soil. *Bull. Environ. Contam. Toxicol.* 29:235–39

101. Sudhakar-Barik, Waid, P. A., Ramakrishna, C., Sethunathan, N. 1979. A change in the degradation pathway of parathion after repeated applications to flooded soil. *J. Agric. Food Chem.* 27:1391–92

102. Suett, D.L. 1986. Accelerated degradation of carbofuran in previously treated field soils in the United Kingdom. *Crop Prot.* 5:165–69

103. Suett, D. L. 1987. Influence of treatment of soil with carbofuran on the sub-

sequent performance of insecticides against cabbage root fly *(Delia radicum)* and carrot fly *(Psila rosae)*. *Crop Prot.* 6:371–78

104. Suett, D. L., Jukes, A. A. 1988. Accelerated degradation of aldicarb and its oxidation products in previously treated soils. *Crop Prot.* In press

105. Sutter, G. R. 1982. Comparative toxicity of insecticides for corn rootworm (Coleoptera: Chrysomelidae) larvae in a soil bioassay. *J. Econ. Entomol.* 75: 489–91

106. Tam, A. C., Behki, R. M., Khan, S. U. 1987. Isolation and characterization of an *s* - ethyl-*N*, *N*-dipropylthiocarbamate-degrading *Arthrobacter* strain and evidence for plasmid-associated *s*-ethyl-*N,N*-dipropylthiocarbamate degradation. *Appl. Environ. Microbiol.* 53:1088–93

107. Tollefson, J. J. 1986. Why soil insecticidal immunity? *Solutions* 30:48–55

108. Torstensson, L. 1980. Role of microorganisms in decomposition. In *Interactions Between Herbicides and the Soil*, ed. R. J. Hance, pp. 159–78. New York: Academic. 349 pp.

109. Turpin, F. T., York, A. C. 1981. Insect management and the pesticide syndrome. *Environ. Entomol.* 10:567–72

110. Venkateswarlu, K., Gowda, T. K. S., Sethunathan, N. 1977. Persistence and biodegradation of carbofuran in flooded soils. *J. Agric. Food Chem.* 25:533–36

111. Venkateswarlu, K., Sethunathan, N. 1978. Degradation of carbofuran in rice

soils as influenced by repeated applications and exposure to aerobic conditions following anaerobiosis. *J. Agric. Food Chem.* 26:1148–51

112. Walker, A., Suett, D. L. 1986. Enhanced degradation of pesticides in soil: a potential problem for continued pest, disease and weed control. *Aspects Appl. Biol.* 12:95–103

113. Walker, N. 1982. Interactions of pesticides with soil microorganisms. In *Advances in Agricultural Microbiology*, ed. N. S. Subba-Rao, pp. 377–95. Boston: Butterworth. 704 pp.

114. Wiggins, B. A., Jones, S. H., Alexander, M. 1987. Explanations for the acclimation period preceding the mineralization of organic chemicals in aquatic environments. *Appl. Environ. Microbiol.* 53:791–96

115. Wilde, G., Mize, T. 1984. Enhanced microbial degradation of systemic pesticides in soil and its effect on chinch bug *Blissus leucopterus leucopterus* (Say) (Heteroptera: Lygaeidae) and greenbug *Schizaphis graminum rondani* (Homoptera: Aphididae) control in seedling sorghum. *Environ. Entomol.* 13:1079–82

116. Williams, I. H., Pepin, H. S., Brown, M. J. 1976. Degradation of carbofuran by soil microorganisms. *Bull. Environ. Contam. Toxicol.* 15:244–49

117. Zublena, J. P., Gooden, D. T., Skipper, H. D. 1987. Influence of crop-chemical rotations on enhanced biodegradation. *Agron. Abstr. Am. Soc. Agron.* 1987:195

*Ann. Rev. Entomol. 1989. 34:477–501*

# STRUCTURE AND FUNCTION OF THE DEUTOCEREBRUM IN INSECTS[1]

*Uwe Homberg, Thomas A. Christensen, and John G. Hildebrand*

ARL Division of Neurobiology, University of Arizona, Tucson, Arizona 85721

## PERSPECTIVES AND OVERVIEW

The deutocerebrum is usually regarded as a preoral neuromere subserving an antennal head segment (81). Most authors agree that it consists of two distinct neuropils (Figure 1): the antennal lobe (AL) and the antennal mechanosensory and motor center (AMMC), also called the dorsal lobe (10, 17, 35, 39). We adhere to this classical definition, but it should be noted that a different, extended definition of the deutocerebrum has recently been suggested by Strausfeld and coworkers (105, 106).

Both areas of the deutocerebrum receive primary sensory fibers from receptor cells in the antenna. Most and possibly all axons of olfactory receptor

---

[1]Abbreviations: AC, anterior cell group of antennal lobe; AL, antennal lobe; aL, α-lobe of the mushroom body; AMMC, antennal mechanosensory and motor center; bL, β-lobe of the mushroom body; Ca, calyces of the mushroom body; CF, centrifugal neuron; DACT, dorsal antenno-cerebral tract; DMACT, dorso-medial antenno-cerebral tract; IACT, inner antenno-cerebral tract; ILP, inferior lateral protocerebrum; IMP, inferior medial protocerebrum; LC, lateral cell group of antennal lobe; LCI, large dorsal cell cluster of the lateral cell group; LCII, postero-ventral cell cluster of the lateral cell group; LH, lateral horn; LN, local neuron; LPO, labial-palp pit organ; MACT, middle antenno-cerebral tract; MC, medial cell group of antennal lobe; MGC, macroglomerular complex; OACT, outer antenno-cerebral tract; P, pedunculus; PD, projection neuron in dorsal antenno-cerebral tract; PDM, projection neuron in dorso-medial antenno-cerebral tract; PIa,b,c, projection neuron in inner antenno-cerebral tract, types a,b,c; PM, projection neuron in middle antenno-cerebral tract; POa,b,c,d, projection neuron in outer antenno-cerebral tract, types a,b,c,d; SOG, suboesophageal ganglion; SB, sensilla basiconica; ST, sensilla trichodea.

0066-4170/89/0101–0477$02.00

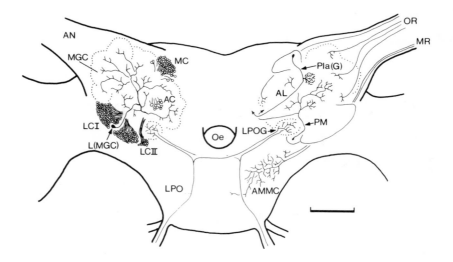

*Figure 1*   Frontal diagram of the brain of *Manduca sexta*, showing the anatomical organization of the antennal lobes. Olfactory receptor cells (*OR*) in the antennal flagellum send axons through the antennal nerve (*AN*) into the glomeruli in the AL of the deutocerebrum. Axons of mechanosensory receptor cells (*MR*) in the antenna bypass the AL and project to the antennal mechanosensory and motor center (*AMMC*) of the deutocerebrum. Receptor cells of the labial-palp pit organ (*LPO*) have bilateral projections into an identified glomerulus in the postero-ventral AL (*LPOG*). Somata of AL neurons reside in three cell groups: a medial group (*MC*), an anterior group (*AC*), and a lateral group, which consists of two cell clusters (*LCI, LCII*). The AL arborizations of three representative AL neurons are shown schematically. The local cell [*L(MGC)*] innervates many glomeruli, including the male-specific macroglomerular complex (*MGC*). The projection neuron of the inner antenno-cerebral tract [*PIa(G)*] innervates a single ordinary (non-MGC) glomerulus, and the projection neuron of the middle antenno-cerebral tract (*PM*) has its soma in LCII and multiglomerular AL arborizations. *Oe*, oesophagus. Scale bar: 200 μm.

cells in the third antennal segment, the flagellum, have terminal arborizations in the AL. The AMMC, on the other hand, receives axons from mechanosensory neurons in the two basal antennal segments, the scape and pedicel. The AMMC is furthermore innervated by dendrites of antennal-muscle motoneurons. Whereas the structure of antennal sensilla and the physiology of the underlying receptor cells have been studied in numerous species (for reviews see 1, 21, 53, 54, 56, 76, 100, 101, 111, 116, 117), the neuronal organization of the deutocerebrum has been investigated in depth for only a few species of insects. The AL has been studied in certain Lepidoptera, including the moths *Manduca sexta* (24, 38, 43), *Bombyx mori* (55, 64, 66, 78), and *Antheraea polyphemus* (7, 64); in several species of cockroaches, most prominently *Periplaneta americana* (8, 10, 11, 85); in certain Hymenoptera, including honey bees (3, 41, 73) and ants (32, 69); and in a few species of flies (80, 82, 83, 102, 103). This brief review describes the functional organization of the AL and its inputs, processing of sensory information, and connections to higher brain centers. Because information about the AMMC is very limited, only a fragmentary overview of its neuronal organization and connections can be given.

# ORGANIZATION OF THE DEUTOCEREBRUM

## Antennal Mechanosensory and Motor Center

The two basal segments of the antenna, the scape and pedicel, apparently house exclusively mechanosensory receptor cells associated with fields of bristles [Böhm's organ (14)], chordotonal organs [Johnston's organ (52); Janet's organ (48)], and other sensilla. The central projections of these receptor neurons are concentrated in the AMMC (Figure 1), but arborizations usually extend into the protocerebrum, the ipsilateral and contralateral suboesophageal ganglion, and even thoracic ganglia [*M. sexta* (39; S. Camazine, J. Hildebrand, U. Homberg & M. Stengl, unpublished observations), *Calliphora erythrocephala* (12, 105), *Locusta migratoria* and *Schistocerca gregaria* (16, 31)]. Because the neuropil of the AMMC is usually fused with surrounding areas of the protocerebrum and suboesophageal ganglion, its boundaries cannot easily be defined. Evidence from Lepidoptera suggests that some flagellar receptor cells, possibly also mechanosensory afferents, have axons that bypass the AL and also terminate in the AMMC (19, 39, 64). Mechanosensory fibers entering the brain through the tegumentary nerves also send side branches into the AMMC (2, 4, 45, 110).

Little is known about interneurons that innervate the AMMC. In flies and other insects as well, descending neurons from the brain have arborizations in the AMMC; these neurons possibly receive direct mechanosensory input (5, 33, 105).

Motoneurons that innervate the antennal muscles have their dendritic fields in the AMMC, but it is not known if they make monosynaptic connections with mechanosensory receptor cells (39, 63).

## Antennal Lobe

Besides primary afferents, the neuropil of the AL comprises neurites of three classes of central neurons (Table 1; Figures 1 and 2): local interneurons (LNs), projection neurons (PNs), and centrifugal neurons. The LNs are amacrine cells with aborizations confined to the AL (24, 28, 71, 97), while the PNs (also called output or principal neurons) have dendrites in the AL and axons that project into the protocerebrum. Centrifugal neurons have dendritic arborizations outside the AL (e.g. in the protocerebrum) and send axons into the AL.

**Table 1**  Cellular elements in the antennal lobe of *Manduca sexta*[a]

| Cell type | Former nomenclatures[b] 1 | 2 | Location of somata | Cells per AL | AL arborizations[c] | Arborizations outside AL | Putative transmitters[d] | Homologies (References) |
|---|---|---|---|---|---|---|---|---|
| **Sensory cells** | | | | | | | | |
| Olfactory (pheromone) receptors | | | Antennal flagellum, ST | 86,000 (male only) | u in MGC | | ACh | (Reviewed in 101) |
| Olfactory receptors | | | Antennal flagellum, ST, SB | 167,000[e] | u in G | | ACh | (Reviewed in 1, 21, 56, 116, 117) |
| Olfactory and/or $CO_2$ receptors (?) | | | LPO | 2,000 | u in LPOG | | | *Antheraea* (58), *Bombyx* (58), *Rhodogastria* (13), *Pieris* (65) |
| **Local interneurons** | | | | | | | | |
| L(MGC) | L(M1:MGC) LIIb | L(M2:MGC) LIIa | LCI | 360 | m | | GABA, peptides | |
| L(G) | L(M1:G) LIb | L(M2:G) LIa | LCI | | m | | GABA, peptides | *Drosophila* (15), *Formica* (32), *Apis* (73), *Musca* (104), *Locusta* (27, 28), *Periplaneta* (27, 28), *Bombyx* (55) |
| **Centrifugal neurons** | | | | | | | | |
| Unique cells | | | Protocerebrum | < 100 | m | Olfactory foci: LH, superior protocerebrum, ILP, IMP, lateral accessory lobe | | *Periplaneta* (27) |
| | | | LCI | 1 | m | LH, superior protocerebrum, central body | Serotonin | |

Projection neurons

| Type | PI | P | LC | No. | | Projection targets | Transmitter | Species (refs) |
|---|---|---|---|---|---|---|---|---|
| Pla(MGC1) | PI(U1:MGC) | PIII | LCI, MC | 35–40 (male only) | u[f] in MGC | Calyces, ILP, LH | | Periplaneta, (18), Bombyx (55) |
| Pla(MGC2) | PI(U2:MGC) | PIII | | | u in MGC | | | |
| Pla(G) | PI(U1:G) PI(U2:G) | PII PI | AC, LCI, MC | 360 | u[g] | Calyces, LH | | Apis (41, 73), Acheta (90), Drosophila (15), Periplaneta (27), Formica (32), Calliphora (102) |
| PIb | | | LCI (?) | | | Superior protocerebrum | | Bombyx (55) |
| PIc | P2 | | LCI (?) | | m | Calyces, LH | | Bombyx (55) |
| PM | | | LCII | 120 | m, u[h] | LH, inferior protocerebrum | GABA | |
| POa | P3 | | LCI | | m, u[h] | ILP, LH | | |
| POb | P3 | | LCI | | m | LH | | |
| POc | P3 | | LCI | | m | Calyces, LH | ACh, peptides | |
| POd | P3 | | LCI | 340 | m, u | Accessory calyx, AMMC, SOG, inferior protocerebrum, optic foci | | Apis (73), Bombyx (55) |
| PD | | | IMP | ~30 | m | LH, ILP | GABA, peptides | |
| PDM | | | SOG | ~10 | u[i] | Calyces, LH | Peptides | |

[a] Data from References 23, 24, 43, 57, 59, 71, 86; R. Kanzaki, unpublished. Abbreviations as in text; see footnote 1.

[b] Column 1 from References 23, 24; column 2 from Reference 71.

[c] u, uniglomerular; m, multiglomerular; G, ordinary glomerulus; LPOG, identified glomerulus in postero-ventral region of AL.

[d] ACh, acetylcholine.

[e] Numbers from male antenna.

[f] Dendritic tree restricted to dorso-lateral portion of MGC.

[g] Dendritic tree often has small side branches toward secondary glomeruli.

[h] Uniglomerular if dendritic tree is in MGC.

[i] Neurons innervate one glomerulus in both antennal lobes.

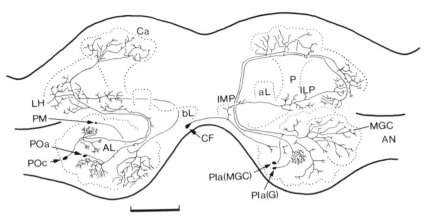

*Figure 2*  Horizontal diagram of the brain of *Manduca sexta,* illustrating typical *AL* projection neurons and a centrifugal neuron (*CF*). PIa projection neurons in the inner antenno-cerebral tract usually have uniglomerular AL arborizations. Those that innervate an ordinary glomerulus [*PIa(G)*] project to the calyces (*Ca*) of the mushroom body and the lateral horn (*LH*) of the ipsilateral protocerebrum. *PIa(MGC)* neurons, which innervate the macroglomerular complex (*MGC*), project to the calyces and a circumscribed area in the inferior lateral protocerebrum (*ILP*). Projection neurons in the middle antenno-cerebral tract (*PM*) have multiglomerular AL branches and extensive axonal projections in the protocerebrum, often including the lateral horn and the ILP. Two types of projection neurons in the outer antenno-cerebral tract are shown (*POa, POc*). Both have multiglomerular AL dendrites. Axons from POa cells have small side branches as they pass to terminal arborizations in the ILP. POc neurons innervate the lateral horn and subsequently the calyces. A centrifugal cell (*CF*) is shown with dendrites in the lateral horn, ILP and inferior medial protocerebrum (*IMP*), and wide-field terminals in the AL. *aL,* α-lobe; *bL,* β-lobe; *P,* pedunculus. Scale bar: 200 μm.

LNs and most PNs have their somata in cell groups at the periphery of the AL (20, 27, 28, 43, 55, 71). In *M. sexta* three such cell groups can be distinguished: a large lateral group with two distinguishable clusters of neurons (LCI, LCII; see Table 1), a smaller medial group (MC), and an even smaller anterior group (AC). Some PNs and most centrifugal neurons in *M. sexta* have their somata in the protocerebrum (8, 43; Figure 3).

Several fiber tracts connect the AL with the protocerebrum, the suboesophageal ganglion, and, especially in Diptera, the contralateral AL. The most prominent connection to the protocerebrum is the inner antenno-cerebral tract (IACT, also known as the tractus olfactorio-globularis or median antennoglomerular tract). It has been described in numerous species (60; for reviews see 17, 35, 104). A smaller tract, the middle antenno-cerebral tract (MACT), is known in *M. sexta* (43, 47) and *Apis mellifera* (3, 60, 74). It is probably homologous with the posterior branch of the tractus olfactorio-globularis in cockroaches and crickets (28, 49, 50, 114). A third tract, the

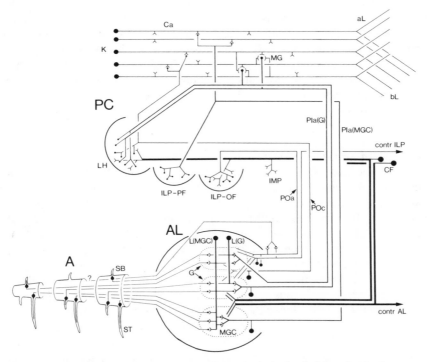

*Figure 3*  Schematic diagram of olfactory pathways in the brain of adult male *Manduca sexta*. Axons of pheromone receptor cells in sensilla trichodea (*ST*) on the antenna (*A*) terminate in the MGC of the AL. Sexually isomorphic olfactory receptor cells in sensilla basiconica (*SB*) project into ordinary glomeruli (*G*). Antennal receptor neurons synapse on two types of local interneurons, one with dendritic branches in many glomeruli, including the macroglomerular complex [*L(MGC)*], and the other with branches only in ordinary glomeruli [*L(G)*]. AL projection neurons make synaptic contact with L cells and project to olfactory foci in the protocerebrum (*PC*). Four major projection areas are the calyces (*Ca*) of the mushroom body, the lateral horn (*LH*) of the protocerebrum, olfactory foci in the inferior lateral protocerebrum (*ILP-OF*), and a pheromone focus in the inferior lateral protocerebrum (*ILP-PF*). Projection neurons in the outer antenno-cerebral tract (*POa*) presumably receive additional sensory input outside the AL from antennal receptor cells of unknown function. Some POa fibers project to the contralateral inferior protocerebrum (*contr ILP*). In the calyces of the mushroom body, projection neurons make contact with Kenyon cells (*K*) that send axon collaterals into α- and β-lobes (*aL, bL*). Projection neurons in the inner antenno-cerebral tract innervating ordinary glomeruli [*PIa(G)*] have highly divergent synaptic contacts in the calyces (microglomeruli, *MG*). PIa neurons innervating the macroglomerular complex [*PIa(MGC)*] and projection neurons in the outer antenno-cerebral tract (*POc*), on the other hand, probably have simple synapses with Kenyon cells. Centrifugal cells (*CF*) have extensive arborizations in the protocerebrum, often in the lateral horn and inferior lateral and medial protocerebrum (*ILP, IMP*). They have multiglomerular projections in one or both ALs. *contr AL*, contralateral AL.

outer antenno-cerebral tract (OACT), is known in *M. sexta* (43, 47) and *A. mellifera* (3, 60, 74) and is probably also present in flies (34, 80). Two smaller tracts, the dorsal antenno-cerebral tract (DACT) and the dorso-medial antenno-cerebral tract (DMACT), have been described in *M. sexta* (43, 47) and may have homologies in bees (3, 74). In *M. sexta* and probably other species as well, there are axonal connections to the suboesophageal ganglion [antenno-suboesophageal tract (43)]. In several insects, and most prominently in flies, commissures connect the right and left ALs (3, 49–51, 74, 80, 88, 104).

GLOMERULAR ORGANIZATION OF THE ANTENNAL LOBE    The neuropil of the AL includes distinct compartments called glomeruli. These are condensed, spheroidal regions of neuropil housing synapses between receptor axons and central neurons. A layer of glial processes separates the individual glomeruli (109), and glial cells may play an important role in the formation of glomeruli (77). The number of glomeruli seems to be species specific (85), ranging from fewer than 10 (e.g. in *Aedes aegypti*) to about 200 (e.g. in *Formica pratensis*). About 1000 very small glomerulus-like compartments have been counted in *L. migratoria* (28), but it is not clear that these structures are homologous to the glomeruli in other species. In general, glomeruli may be distributed throughout the AL, as in *P. americana* (28), or they may be arrayed around a central region of coarse neuropil formed largely by neurites of AL neurons, as in *M. sexta* (24, 38). In some species the glomeruli have been mapped and can be identified individually, on the basis of size, shape, and relative position (3, 20, 84, 85, 103).

The structure of most AL glomeruli resembles that of kidney glomeruli. A central, inner core, consisting largely of neurites of AL neurons, is surrounded by a cortical cup of receptor-axon terminals (3, 12, 47, 64, 69, 79). Certain glomeruli in *M. sexta, A. mellifera,* and probably other species are innervated throughout by receptor terminals and can therefore easily be identified (3; U. Homberg, unpublished observations).

ULTRASTRUCTURE    Synapses in the AL are largely or entirely restricted to the glomeruli (12, 95, 96, 109). In *M. sexta,* synapses are usually divergent. Most common are dyads in which one presynaptic element is apposed to two postsynaptic elements; some of these dyads also participate in serial synapses. Similar findings have been reported for other species (12, 69, 96). Intracellular staining and degeneration studies have revealed that neurites of AL neurons are both pre- and postsynaptic in the AL glomeruli (12, 109).

SEXUAL DIMORPHISM    The antennae and ALs, and probably higher brain centers as well, are sexually dimorphic in several orders. In Lepidoptera and

Dictyoptera the male antenna bears a large number of male-specific sensilla (sensilla trichodea in moths). These sensilla enclose olfactory receptor cells that are specialized to detect one or more of the female's sex pheromones, and they are usually not present in the female antenna (for reviews see 21, 101). The axons of sex-pheromone receptors terminate in an enlarged glomerulus, a macroglomerulus or macroglomerular complex (MGC), which is not present in female ALs (8, 19, 24, 64, 94). The MGC is also innervated by sexually dimorphic AL neurons (18, 23, 24, 27, 43, 55, 71, 93), as discussed below. The non–sexually dimorphic glomeruli are referred to as ordinary glomeruli throughout the text.

# CELLULAR COMPONENTS OF THE ANTENNAL LOBE

## Primary Afferents

The structure of antennal sensilla and the physiology of primary afferents that innervate them have been studied in numerous insect species and reviewed extensively (1, 21, 53, 56, 92, 101, 116, 117). In *M. sexta,* olfactory receptor cells reside in sensilla trichodea and sensilla basiconica (Table 1) and possibly other types of sensilla (86). The axons of these receptor cells project into single glomeruli of the ipsilateral AL (12, 19, 89, 103). In flies, however, many primary afferents project to a glomerulus in the ipsilateral AL and, via a commissure, to the corresponding glomerulus in the contralateral AL (103). All evidence so far suggests a functional rather than a spatial representation of the flagellum in the AL. Stocker et al (102a, 103) have suggested that certain types of sensilla, regardless of their location on the antenna, are represented in certain glomeruli. Furthermore, pheromone-sensitive afferents apparently project only to the MGC (8, 19, 64, 94; I. Harrow & T. Christensen, unpublished observations).

In addition to flagellar receptor neurons, the ALs receive projections from receptor cells in a labial-palp pit organ (LPO) (Table 1, Figure 1; 13, 58, 65). These sensory cells respond to carbon dioxide and perhaps other volatiles (13, 36). They have unilateral or bilateral projections in an identified glomerulus in the AL (LPO-glomerulus) that apparently is not innervated by antennal afferents (Figure 1). A similar area in *P. americana,* the lobus glomerulatus, also receives primary-afferent input from labial-palp chemoreceptors (8). These findings support the hypothesis of a functional topography of primary afferents in the AL.

## Local Interneurons

Amacrine cells have been described in the ALs of Hymenoptera (32, 73), Diptera (104), Orthoptera (28, 97), and Lepidoptera (47, 55, 71). These LNs innervate several (and sometimes all) glomeruli of the AL. In *M. sexta* the somata of LNs reside probably exclusively in cluster LCI of the lateral group of neurons (Table 1; Figure 1; 43). LNs that innervate the MGC in male *M. sexta* [designated L(MGC); see Table 1] are sexually dimorphic (43, 71). Biochemical, immunocytochemical, and pharmacological studies suggest that the majority of the approximately 360 LNs in *M. sexta* are GABAergic and inhibitory (47, 61, 72, 113). Immunocytochemical studies furthermore suggest that a variety of neuropeptides might be cotransmitters with GABA in certain subsets of LNs (42, 44, 62). Little is known about the remainder of the LNs, but some of them are believed to be excitatory interneurons (see below).

## Projection Neurons

In *M. sexta* and probably other species as well, several areas in the protocerebrum are innervated by PNs (43). These olfactory foci include the calyces of the mushroom body, the lateral horn of the protocerebrum, and an olfactory focus and a sex-pheromone focus in the inferior lateral protocerebrum (Figure 3). Several morphologically distinct types of PNs connect the AL with these olfactory foci (Figures 1–3; Table 1). These neurons have been assigned names according to the fiber tracts in which their axons project (Table 1): PIa, b, and c in the IACT; PM in the MACT; POa, b, c, and d in the OACT; PD in the DACT; and PDM in the DMACT.

PIa neurons have been described in considerable detail in several species (15, 18, 23–25, 27, 41, 43, 55, 71, 73, 90). These neurons usually have uniglomerular arborizations in the AL and axonal projections to the calyces of the mushroom body and the lateral horn of the protocerebrum (43, 105). The AL of male *M. sexta* contains about 30 male-specific PIa(MGC) neurons, which have no counterpart in females (Table 1; Figures 2 and 3). The projections of PIa(MGC) neurons in the calyces differ from those of PIa(G) neurons; the PIa(MGC) axons innervate only certain regions in the calyces and lack the varicose terminals typical of PIa(G) axons (43). The terminal arborizations of PIa(MGC) fibers define a sex-pheromone focus in the inferior lateral protocerebrum and overlap little with the projections of PIa(G) fibers in the lateral horn (Figures 2 and 3; 43). Similar observations have been reported for *P. americana* (8, 11). These observations, along with physiological findings (see below), strengthen the idea that the male-specific pathway for detection and processing of sex-pheromonal information constitutes a distinct olfactory subsystem.

With few exceptions, other types of PNs have so far been described only in *M. sexta* (40, 43). The following description therefore refers largely to this species (Table 1).

Two additional types of neurons have axons in the IACT: PIb neurons project not to the calyces but to the superior protocerebrum surrounding the pedunculus near its emergence from the calyces, and PIc neurons have multiglomerular AL arborizations (43). PNs in the MACT and OACT usually have multiglomerular AL arborizations, but those innervating the MGC have no ramifications in other glomeruli. PM neurons have extensive arborizations in the lateral protocerebrum; their somata have been found exclusively in cell cluster LCII (Figure 1; Table 1). Most PM neurons are also GABA immunoreactive (47). The somata of PO neurons, which project in the OACT, reside in cell cluster LCI. Four distinct morphological types have been found. POa neurons have numerous small side branches along their axons, most prominently in a bridge of neuropil between the AL and the protocerebrum, the AL isthmus, where antennal sensory fibers of unknown modality terminate (Figures 1 and 3). POa fibers terminate in an olfactory focus in the inferior lateral protocerebrum (Figures 2 and 3). POb neurons are characterized by unusually large, bulbous fiber specializations in the lateral horn and apparently high levels of intracellular acetylcholinesterase activity (46; S. Hoskins & U. Homberg, unpublished information) as well as immunoreactivity with peptide antisera (42, 62). Some POa and POb neurons connect the AL with the contralateral protocerebrum (Figure 3) and seem to be the only projection neurons to do so. POc neurons project first to the lateral horn and then into the calyces; similar neurons have also been found in bees (73) and silk moths (55). Finally, POd neurons have extensive dendritic fields in the AL, AMMC, optic foci, and tritocerebrum and project into a part of the calyces that is not innervated by other PNs.

The somata of PD neurons, which project in the DACT, are in the protocerebrum, while those of PDM neurons are in the suboesophageal ganglion. PDM neurons innervate corresponding ventral glomeruli in both ALs and send an axon via the ipsilateral DMACT to the calyces and lateral horn (57; R. Kanzaki, E. Arbas, N. Strausfeld & J. Hildebrand, in preparation). PD neurons send their primary neurites via a commissure into the contralateral AL, and then their axons continue via the DACT into the contralateral horn of the protocerebrum (43, 47; R. Kanzaki, E. Arbas, N. J. Strausfeld & J. Hildebrand, in preparation). PD neurons exhibit GABA immunoreactivity, and a subset is immunoreactive with peptide antisera (47; U. Homberg, T. Kingan & J. Hildebrand, in preparation).

Primary neurites of PD neurons in *M. sexta* generally form a pseudo-commissure between both ALs since each individual neuron arborizes in only one AL. However, neurons that connect both ALs have been described in Diptera (15, 102, 104). These neurons, which have extensive arborizations throughout both ALs, seem to be rare in other species (43, 74) and may correspond to the bilateral primary-afferent projections, which are apparently unique to Diptera.

## Centrifugal Neurons

While PNs are believed to have dendrites in the AL and axons projecting out of the AL, a smaller number of neurons appear to have dendrites in other parts of the brain and to send axons into the AL. Some examples of these neurons have been described in *P. americana* (8, 27), *M. sexta* (43; Table 1; Figure 2), and *Musca domestica* (104). Most of these cells have somata outside the AL, but in *M. sexta* a serotonin-immunoreactive neuron with its soma in the AL appears to have dendritic arborizations (weakly immunoreactive) in the ipsilateral horn of the protocerebrum and putatively axonal terminals in several areas of the protocerebrum and the contralateral AL (59; U. Homberg, unpublished information). Like this neuron, many centrifugal neurons seem to have feedback characteristics, with dendritic arborizations at least partly in olfactory foci in the protocerebrum. Centrifugal neurons usually have multi-glomerular ramificatons in one or both ALs.

## FUNCTIONS OF THE DEUTOCEREBRUM

Anatomical and physiological data suggest that a functional rather than a spatial representation of sensory inputs is mapped into the deutocerebrum. There appear to be largely separate areas for mechanosensory (AMMC) and for olfactory (AL) processing, and the AL is further topographically divided into areas designated for processing particular olfactory inputs (e.g. sex pheromones, plant and food odors, carbon dioxide). A similar organization may apply to the protocerebral projections. As we have seen, the pheromone-specific olfactory subsystem in males is clearly set apart from the general olfactory pathway in the AL, calyces, and lateral protocerebrum (8, 11, 43). While there is some evidence for multimodal integration in AL neurons (Table 2; also see below), the remainder of this discussion focuses mainly on the functions of the deutocerebrum as they relate to the processing of behaviorally relevant odors.

## Odor Quality Coding

Does the deutocerebrum contribute to an insect's ability to discriminate one odor from another? Only a decade ago, the answer to this question would have been "probably not." Anatomical studies suggested that the AL neuropil was no more than a convergence and relay station for antennal sensory axons (32, 114), and earlier electrophysiological studies tended to support this idea (9). Extracellular recordings from AL neurons in several species usually revealed simple excitatory responses to olfactory stimuli (7, 97, 107). In other studies, inhibitory responses were sometimes recorded (9, 115), but they were difficult to interpret. A confounding feature of these extracellular recordings was that if background activity was low or absent (e.g. 112), it was impos-

**Table 2**  Physiological profiles of morphologically identified cell types in antennal lobes

| Cell Type | Modality[a] | Antennal stimuli[b] | General response[c] | References |
|---|---|---|---|---|
| **Local interneurons** | | | | |
| *Manduca sexta* | | | | |
| 1.  L(M1:G) | O | Plant odors | + | 71 |
| 2.  L(M2:G) | O | Plant odors | + | 71 |
| 3.  L(M1:G, MGC) | O | (a)  Plant odors | + | 71 |
| | | (b)  Pheromones | + | |
| 4.  L(M2:G, MGC) | O | (a)  Plant odors | + | 71 |
| | | (b)  Pheromones | −/+/− | |
| *Periplaneta americana* | | | | |
| 1.  Unnamed local neurons | O | Food odors | + | 97 |
| 2.  IOsph | O | Pheromones | + | 27[d] |
| (MGC arbors) | | | | |
| 3.  IIO + M | O | (a)  Food odors | + | 27[d] |
| (MGC arbors) | | (b)  Pheromones | + | |
| | M | Air puff | + | |
| *Bombyx mori* | | | | |
| 1.  Unnamed local neurons | O | Bombykol | − | 55[e] |
| (MGC arbors) | M | Air puff | + | |
| 2.  Unnamed local neurons | O | Bombykol | + (on-off) | 55 |
| (no MGC arbors) | M | Air puff | + | |
| 3.  Unnamed local neurons | O | Bombykol | + | 55 |
| (no MGC arbors) | M | Air puff | − | |
| **Projection neurons** | | | | |
| *Antheraea pernyi, Antheraea* | | | | |
| *polyphemus* | | | | |
| 1.  Projection-like neurons | O | (a)  Pheromones | + | 7 |
| | | (b)  Plant odors | NR | |
| *Manduca sexta* | | | | |
| 1.  PIa(G) | O | Plant odors | + or − | 71 |
| 2.  PIa(G) | O | Plant odors | − | R. Kanzaki, unpublished |
| | M | Air puff | − | |
| 3.  PIc | M | (a)  Air-ipsi · | − | R. Kanzaki, |
| | | (b)  Air-contra | + | unpublished |
| 4.  PD | O· | Pheromones | + | R. Kanzaki, unpublished |
| 5.  PDM | O | Plant odors | − | R. Kanzaki, unpublished |
| | M | Air puff | + | |
| 6.  PIa(MGC1) | O | Bombykal | NR | 23 |
| (arbors confined to dorso- | | C15 | + | |
| lateral MGC) | | Blend | + | |

**Table 2**  (*continued*)

| Cell Type | Modality[a] | Antennal stimuli[b] | General response[c] | References |
|---|---|---|---|---|
| 7.  PIa(MGC2) | O | Bombykal | + | 23 |
| (arbors throughout entire | | C15 | NR | |
| MGC) | | Blend | + | |
| 8.  PIa(MGC2) | O | Bombykal | + | 23 |
| | | C15 | + | |
| | | Blend | + | |
| 9.  PIa(MGC2) | O | Bombykal | + | 23 |
| | | C15 | − | |
| | | Blend | −/+/− | |
| 10. PIa(MGC2) | O | Bombykal | − | 23 |
| | | C15 | + | |
| | | Blend | −/+/− | |
| 11. PIa(MGC2) | O | Bombykal | − | T. Christen- |
| | | C15 | − | sen, un- |
| | | Blend | − | published |
| *Bombyx mori* | | | | |
| 1.  Type Ia | O | Bombykol | + | 55[e] |
| (MGC arbors) | M | Air puff | − | |
| 2.  Type Ib | O | Bombykol | + | 55 |
| (MGC arbors) | M | Air puff | + | |
| *Periplaneta americana* | | | | |
| 1.  IOf (no MGC arbors) | O | Food odors | + | 27, 97 |
| 2.  IOsph (MGC arbors) | O | (*a*) Food odors | + | 27 |
| | | (*b*) Pheromones | + | |
| 3.  IIO + M | O | Food odors | + | 27[d] |
| (no MGC arbors) | M | Air puff | + | |
| 4.  Unnamed projection neurons | O | Periplanone B | + | 18 |
| (MGC arbors) | | Fraction A | + | |
| | | Blend | + | |
| 5.  Unnamed projection neurons | O | Periplanone B | + | 18 |
| (MGC arbors) | | Fraction A | NR | |
| | | Blend | + | |
| 6.  Unnamed projection neurons | O | (*a*) Periplanone B | NR | 18 |
| (arbors confined to dorso- | | Fraction A | + | |
| lateral area of MGC) | | Blend | + | |
| | | (*b*) Food odors | + | |
| 7.  Larval group A | O | (*a*) Periplanone B | NR | 89 |
| (arbors confined to dorsal area | | Fraction A | + | 89 |
| of MGC) | | Fraction D | + | |
| | | (*b*) Food odors | + | |
| 8.  Larval group B | O | (*a*) Periplanone B | + | 89 |
| (arbors throughout developing | | Fraction A | + | |
| MGC) | | Fraction D | NR | |
| | | (*b*) Food odors | + | |

**Table 2**  *(continued)*

| Cell Type | Modality[a] | Antennal stimuli[b] | General response[c] | References |
|---|---|---|---|---|
| *Apis mellifera* | | | | |
| 1.  IACT-I | O | Flower odor | − | 41 |
| | M | Air puff | − | |
| | H | Touch with water | + | |
| 2.  IACT-IIa | O | Flower odor | − | 41 |
| | M | Air puff | + | |
| | H | Touch with water | + | |
| 3.  IACT-IIb | O | Flower odor | + | 41 |
| | M | Air puff | − | |
| | H | Touch with water | + | |
| 4.  IACT-IIIa | O | Flower odor | NR | 41 |
| | M | Air puff | NR | |
| | H | Touch with water | + | |
| 5.  IACT-IIIb | O | Flower odor | NR | 41 |
| | M | Air puff | + | |
| | H | Touch with water | + | |
| 6.  IACT-IV | O | Flower odor | + | 41 |
| | M | Air puff | NR | |
| | H | Touch with water | + | |
| 7.  IACT-V | O | Flower odor | NR | 41 |
| | M | Air puff | + | |
| | H | Touch with water | − | |
| *Acheta domesticus* | | | | |
| 1.  Extrinsic mushroom body cells | O | Food odor | + | 91 |
| | M | Air puff | + | |
| | A | Sound | + (on-off) | |

[a]O, olfactory; M, mechanosensory; H, hygrosensory; A, acoustic.
[b]Blend, natural extract of female pheromone glands.
[c]+, excitation (sometimes with rebound inhibition); + (on-off), excitation at stimulus onset and offset; −, inhibition (sometimes with rebound excitation); −/+/−, triphasic response; NR, no response.
[d]See also Reference 9 for a multimodal neuron responding to temperature changes.
[e]See also Reference 78 for a study of unidentified deutocerebral units.

sible to tell if a neuron was inhibited or simply unresponsive to a given stimulus. It was clear, however, that the ALs in many insects were the sites of massive sensory convergence (Table 1), and that one likely function of these areas of the brain was to amplify the incoming signals and improve the signal-to-noise ratio in the system.

By far the best-studied example of quality coding in the ALs is in the sex pheromone–detecting subsystem in many male insects, in which thousands of pheromone receptors in each antenna project into the ipsilateral MGC (Table 1; Figure 3; reviewed in 8, 10, 11, 24, 66, 70, 76, 111). As indicated in Table 2, every LN or PN with arborizations in the MGC responds to pheromonal stimulation of the ipsilateral antenna. Moreover, in some species it has been shown that the sensory axons from other types of sensilla form a specific termination pattern in several of the ordinary glomeruli (73, 103). More recently, this observation has been extended to include nonantennal sensilla in the labial-palp pit organ (see above). These data further support the hypothesis that each glomerulus is functionally specialized and that any given complex of odors can be represented in the AL by the pattern of activity in a specific array of glomeruli. There does not appear to be a strict spatial representation of the antenna in the AL, but directional information is apparently preserved in many species by the unilaterality of central processing of antennal inputs. Even in Diptera, in which antennal inputs project bilaterally in the brain, the sensory projections into a small number of glomeruli are unilateral (103).

## Odor Quantity (Concentration) Coding

Some AL neurons in several species exhibit dose-dependent responses to olfactory stimuli (7, 18, 23, 55) according to a number of physiological criteria: the total number of action potentials evoked by each concentration of odorant, the mean rate of spiking, the instantaneous rate of spiking, or the response latency. Other AL neurons fire only after a certain threshold is reached and do not respond very differently to increasing stimulus concentrations. Thus some odor discrimination may occur independently of odor intensity. (See Reference 30 for a discussion on how a signal representing the ratio of the responses to two odors can permit discrimination of the two stimuli independently of concentration.)

## Odor Discrimination at the Single-Cell Level

In the pheromone-processing subsystem there is a physiologically diverse population of PNs that connect the MGC with higher-order neurons in the protocerebrum (see Table 2). Some PIa(MGC) neurons in *M. sexta* can discriminate one pheromone from another because they receive input only from one of the two classes of sex-pheromone receptors in the male antenna. One receptor class is selectively stimulated by bombykal, the most abundant pheromone in the natural blend (99). The second receptor class is stimulated by $E,Z$-11,13-pentadecadienal (C15) (K.-E. Kaissling, J. Hildebrand & J. Tumlinson, in preparation), which is an effective mimic of a second pheromone as shown with behavioral assays (J. Tumlinson, personal communication). If the antenna is stimulated with single pheromones, many PIa(MGC) neurons respond with a delayed (presumably polysynaptic) excitation to the presentation of bombykal alone (Figure 4, response *1*). Others show similar responses to C15 alone (response *3*). We therefore postulate that two independent populations of excitatory interneurons reside in the MGC, each receiving input from one class of receptors or the other. Because other PIa(MGC) cells are excited by both bombykal and C15 (response *2*), however, it is also possible that some excitatory interneurons receive input from both classes of receptors (not illustrated in Figure 4). Clearly, still other PIa(MGC) neurons integrate information from two separate interneuronal populations, since the polarity of the response to each pheromone is different. For example, in response 4 a neuron is inhibited by bombykal, is excited by C15, and responds to a blend of the two with a mixed triphasic response. While the details of the complex synaptic circuitry within the MGC are still unknown, there is strong pharmacological and immunocytochemical evidence that the inhibitory postsynaptic potential (IPSP) at the onset of this mixed response is due to a GABAergic synaptic mechanism and that many LNs in the AL are GABAergic (22, 47, 113). Through these polysynaptic interactions, a single PIa(MGC) neuron of this type can distinguish three different environmental situations: *(a)* the presence of bombykal alone, *(b)* the presence of C15 alone, and *(c)* the presence of the combination of the two or, more precisely, the natural pheromone blend. Recently, behavioral evidence has shown that a male moth uses information from the entire pheromone blend at all stages of his flight sequence in searching for a calling female (67, 68), but single pheromones can nevertheless influence specific behaviors in the complete sequence (108). The physiological diversity among PIa(MGC) neurons may therefore reflect a need for these insects to process information about blends and single pheromones differently. Moreover, PIa(MGC) neurons that receive mixed inputs (Figure 4, response 4) can track discontinuities in the pheromonal plume, which are known to be important in maintaining the male's upwind progress toward the source of pheromones (see references in 25).

Still other PIa(MGC) neurons are strictly inhibited by both pheromones in *M. sexta* (Figure 4, response 5). Cells that are inhibited by either bombykal or C15, but not by both, have not yet been found.

## Odor Quality Convergence

Another level of complexity within the AL neuropil is exemplified by PNs that, in addition to responding to sex pheromones, receive information about other behaviorally relevant odors. Such neurons in the American cockroach [analogous to the PIa(MGC) cells in *M. sexta*] are strongly excited or inhibited by various food odors (18, 27). These responses are thought to result from lateral interactions mediated through the multiglomerular LNs, since the MGC gets input only from pheromone receptors (8). The possibility that nonpheromonal odors directly stimulate pheromone receptor neurons has also been ruled out in *P. americana* through single-sensillum recordings (87). PIa(MGC) neurons in *M. sexta,* in contrast, do not respond strongly to tobacco or other host-plant odors (23), but there is a possibility that the relevant stimuli have not yet been tested.

No clear example of a distinct response to pheromones in a PIa(G) neuron

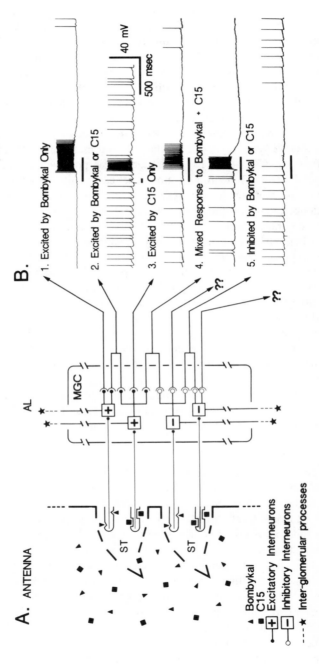

*Figure 4* A. Hypothetical "wiring" diagram showing the flow of pheromonal information in the olfactory system of male *Manduca sexta*, from the antenna to the output from the AL. For simplicity, only two of the ~80,000 sensilla trichodea (ST) are illustrated. The antenna is stimulated with equal amounts of bombykal and C15. This mixture has been shown to mimic the natural pheromone blend extracted from female lure glands (23). The smaller ordinary glomeruli are not illustrated. *B*. Physiological responses recorded from MGC output [Pla(MGC)] neurons indicate that information from bombykal- and C15-specific receptor neurons remains separated in some output lines from the AL, whereas it is combined in others. Activity from the ordinary glomeruli may influence the activity of Pla(MGC)s through interglomerular interactions.

has been reported in any species (Table 2). This would suggest that information from other glomeruli may flow into the MGC but not from the MGC to other glomeruli. Recent anatomical evidence from *M. sexta* (see above) also supports the idea that the pheromone-processing subsystem is largely separate from the rest of the olfactory system.

## Modality Convergence

While the primary function of the AL is to process olfactory information, some AL neurons clearly code multimodal inputs (Table 2), indicating a still higher level of complexity in this neuropil. Some cells respond not only to olfactory stimuli, but also to antennal stimulation with a puff of air (gentle mechanosensory stimulus) (27, 41, 55, 91). Some AL neurons also respond to temperature (10), humidity (41), and acoustic (91) stimuli. Many such responses are qualitatively and quantitatively different from responses to olfactory stimuli. Some of these responses may be mediated through centrifugal neurons that have various connections to other areas of sensory convergence in the protocerebrum (Figure 3; Table 1).

## Olfactory Memory

The mushroom bodies, in addition to participating in a wide variety of behavior patterns, have long been thought to be involved in olfactory learning and memory, but the ALs may contribute to memory formation as well (26). In bees trained in a one-trial conditioning paradigm, olfactory memory was impaired if the ALs were cooled by inserting tiny metal probes into the head, but the effect was more pronounced after the alpha lobes or calyces of the mushroom bodies were cooled. These results suggest that the mushroom bodies and perhaps the ALs are involved in memory consolidation, but we do not yet know if they are the sites of memory storage. More recent behavioral tests on structural brain mutants of *D. melanogaster* have suggested that the calycal projections are necessary for olfactory memory formation, while the terminals in the lateral protocerebrum could be important for olfactory discrimination (29, 37).

## Similarities to the Vertebrate Olfactory Bulb

The conspicuous glomerular organization in the insect AL is a common feature of the olfactory systems of other invertebrates such as crustaceans (6) and also of the vertebrate olfactory bulb. The latter system also contains processes from primary afferent axons as well as fibers from the other three major cell classes [the local interneurons (granule, periglomerular, short-axon cells), the principal (projection) neurons (mitral, tufted cells), and the centrifugal neurons] (75, 98). There is some evidence for topographical mapping of sensory afferents into the olfactory bulb, and some vertebrate glomeruli may have a functional specialization similar to the role of the MGC in insects (reviewed in 75). Both systems are characterized by a tremendous sensory convergence onto olfactory interneurons, complex synaptic interactions within the olfactory neuropil, and nonrandom projections from the olfactory neuropil to higher centers (75, 98).

## EPILOG

Explorations of the deutocerebrum of the insect brain have only begun. Further work is motivated by a need to understand how antennal sensory information, which is initially processed in the deutocerebrum, influences the behavior of insects. Moreover, the deutocerebrum is an experimentally favorable preparation for studies that may yield key insights about central processing of olfactory information in vertebrates and invertebrates alike. The main challenge of such research is to explain how the molecular information represented in the response spectra and patterns of primary olfactory receptor cells is mapped into central neural space, encoded by neurons in the brain, and integrated with other modalities, ultimately affecting behavior.

ACKNOWLEDGMENTS

We wish to thank J. Lawrence and D. Olson for help in preparing the manuscript and Dr. N. Antinoro for photographic assistance. Research in the authors' laboratory has been supported by grants from NIH (currently AI-23253 and NS-23405), NSF, the Army Research Office, and Monsanto Company.

*Literature Cited*

1. Altner, H., Prillinger, L. 1980. Ultrastructure of invertebrate chemo-, thermo-, and hygroreceptors and its functional significance. *Int. Rev. Cytol.* 67:69–139
2. Arbas, E. A., Hildebrand, J. G. 1986. Wind-sensitive head hairs: a mechanosensory pathway activating flight in the hawkmoth *Manduca sexta. Soc. Neurosci. Abstr.* 12:857
3. Arnold, G., Masson, C., Budharugsa, S. 1985. Comparative study of the antennal lobes and their afferent pathways in the worker bee and the drone *(Apis mellifera). Cell Tissue Res.* 242:593–605
4. Aubele, E., Klemm, N. 1977. Origin, destination, and mapping of tritocerebral neurons of locust. *Cell Tissue Res.* 178:199–219
5. Bacon, J., Strausfeld, N. J. 1986. The dipteran giant fiber pathway: neurons and signals. *J. Comp. Physiol. A* 158:529–48
6. Blaustein, D., Derby, C. E., Beall, A. C. 1987. The structure of chemosensory

centers in the brain of spiny lobsters and crayfish. In *Olfaction and Taste IX. Ann. NY Acad. Sci.* 510:180–83
7. Boeckh, J., Boeckh, V. 1979. Threshold and odor specificity of pheromone-sensitive neurons in the deutocerebrum of *Antheraea pernyi* and *A. polyphemus. J. Comp. Physiol.* 132:235–42
8. Boeckh, J., Ernst, K. D. 1987. Contribution of single unit analysis in insects to an understanding of olfactory function. *J. Comp. Physiol.* 161:549–65
9. Boeckh, J., Ernst, K. D., Sass, H., Waldow, U. 1975. Coding of odor quality in the insect olfactory pathway. In *Olfaction and Taste V,* ed. D. A. Denton, J. P. Coghlan, pp. 239–45. New York: Academic. 460 pp.
10. Boeckh, J., Ernst, K. D., Sass, H., Waldow, U. 1984. Anatomical and physiological characteristics of individual neurons in the central antennal pathway of insects. *J. Insect Physiol.* 30:15–26
11. Boeckh, J., Ernst, K. D., Selsam, P.

1987. Neurophysiology and neuroanatomy of the olfactory pathway in the cockroach. In *Olfaction and Taste IX. Ann. NY Acad. Sci.* 510:39–43

12. Boeckh, J., Sandri, C., Akert, K. 1970. Sensorische Eingänge und synaptische Verbindungen im Zentralnervensystem von Insekten. Experimentelle Degeneration in der antennalen Sinnesbahn von Fliegen und Schaben. *Z. Zellforsch. Mikrosk. Anat.* 103:429–46

13. Bogner, F., Boppré, M., Ernst, K. D., Boeckh, J. 1986. $CO_2$ sensitive receptors on labial palps of *Rhodogastria* moths (Lepidoptera: Arctiidae); physiology, fine structure and central projection. *J. Comp. Physiol.* 158:741–49

14. Böhm, L. K. 1911. Die antennalen Sinnesorgane der Lepidopteren. *Arb. Zool. Inst. Univ. Wien* 19:219–46

15. Borst, A. 1984. Untersuchungen zur zentralnervösen Verarbeitung olfaktorischer Reize bei *Drosophila melanogaster*. PhD dissertation. Univ. Würzburg, West Germany. 97 pp.

16. Bräunig, P., Pflüger, H. J., Hustert, R. 1983. The specificity of antennal nervous projections of locust mechanoreceptors. *J. Comp. Neurol.* 218:197–207

17. Bullock, T., Horridge, G. A. 1965. *Structure and Function in the Nervous System of Invertebrates.* San Francisco/London: Freeman. 1719 pp.

18. Burrows, M., Boeckh, J., Esslen, J. 1982. Physiological and morphological properties of interneurones in the deutocerebrum of male cockroaches, which respond to female pheromone. *J. Comp. Physiol.* 145:447–57

19. Camazine, S. M., Hildebrand, J. G. 1979. Central projections of antennal sensory neurons in mature and developing *Manduca sexta*. *Soc. Neurosci. Abstr.* 5:155

20. Chambille, I., Rospars, J. P. 1985. Neurons and identified glomeruli of antennal lobes during postembryonic development in the cockroach *Blaberus craniifer* Burm. (*Dictyoptera: Blaberidae*). *Int. J. Insect Morphol. Embryol.* 14:203–26

21. Chapman, R. F. 1982. Chemoreception. The significance of receptor numbers. *Adv. Insect Physiol.* 16:247–356

22. Christensen, T. A., Hildebrand, J. G. 1986. Synaptic relationships between local and projection neurons in the antennal lobe of the sphinx moth *Manduca sexta*. *Soc. Neurosci. Abstr.* 12:857

23. Christensen, T. A., Hildebrand, J. G. 1987. Male-specific, sex pheromone-selective projection neurons in the antennal lobes of the moth *Manduca sexta*. *J. Comp. Physiol. A* 160:553–69

24. Christensen, T. A., Hildebrand, J. G. 1987. Functions, organization, and physiology of the olfactory pathways in the lepidopteran brain. In *Arthropod Brain: Its Evolution, Development, Structure and Functions*, ed. A. P. Gupta, pp. 457–84. New York: Wiley. 588 pp.

25. Christensen, T. A., Hildebrand, J. G. 1988. Frequency coding by central olfactory neurons in the sphinx moth *Manduca sexta*. *Chem. Senses* 13:123–130

26. Erber, J., Homberg, U., Gronenberg, W. 1987. Functional roles of the mushroom bodies in insects. See Ref. 24, pp. 485–511

27. Ernst, K. D., Boeckh, J. 1983. A neuroanatomical study on the organization of the central antennal pathways in insects. III. Neuroanatomical characterization of physiologically defined response types of deutocerebral neurons in *Periplaneta americana*. *Cell Tissue Res.* 229:1–22

28. Ernst, K. D., Boeckh, J., Boeckh, V. 1977. A neuroanatomical study of the organization of the central antennal pathways in insects. II. Deutocerebral connections in *Locusta migratoria* and *Periplaneta americana*. *Cell Tissue Res.* 176:285–308

29. Fischbach, K. F., Heisenberg, M. 1984. Neurogenetics and behavior in insects. *J. Exp. Biol.* 112:65–93

30. Getz, W. M., Chapman, R. F. 1987. An odor discrimination model with application to kin recognition in social insects. *Int. J. Neurosci.* 32:963–78

31. Gewecke, M. 1979. Central projection of antennal afferents for the flight motor in *Locusta migratoria* (Orthoptera, Acrididae). *Entomol. Gen.* 5:317–20

32. Goll, W. 1967. Strukturuntersuchungen am Gehirn von *Formica*. *Z. Morphol. Ökol. Tiere* 59:143–210

33. Griss, C., Rowell, C. H. F. 1986. Three descending interneurons reporting deviation from course in the locust. I. Anatomy. *J. Comp. Physiol. A* 158:765–74

34. Groth, U. 1971. Vergleichende Untersuchungen über die Topographie und Histologie des Gehirns der Dipteren. *Zool. Jahrb. Abt. Anat. Ontog. Tiere* 88:203–319

35. Hanström, B. 1928. *Vergleichende Anatomie des Nervensystems der wirbellosen Tiere*. Berlin: Springer. 628 pp.

36. Harrow, I. D., Quartararo, P., Kent, K. S., Hildebrand, J. G. 1983. Central projections and possible chemosensory

function of neurons in a sensory organ on the labial palp of *Manduca sexta*. *Soc. Neurosci. Abstr.* 9:216

37. Heisenberg, M., Borst, A., Wagner, S., Byers, D. 1985. *Drosophila* mushroom body mutants are deficient in olfactory learning. *J. Neurogenet.* 2:1–30

38. Hildebrand, J. G. 1985. Metamorphosis of the insect nervous system. Influences of the periphery on the postembryonic development of the antennal sensory pathway in the brain of *Manduca sexta*. In *Model Neural Networks and Behavior*, ed. A. I. Selverston, pp. 129–48. New York: Plenum. 548 pp.

39. Hildebrand, J. G., Matsumoto, J. G., Camazine, S. M., Tolbert, L. P., Blank, S., et al. 1980. Organization and physiology of antennal centers in the brain of the moth *Manduca sexta*. In *Insect Neurobiology and Pesticide Action (Neurotox '79)*, pp. 375–82. London: Soc. Chem. Ind. 517 pp.

40. Hildebrand, J. G., Montague, R. A. 1985. Functional organization of olfactory pathways in the central nervous system of *Manduca sexta*. In *Mechanisms in Insect Olfaction*, ed. T. L. Payne, M. C. Birch, C. E. J. Kennedy, pp. 279–85. Oxford, UK: Clarendon. 364 pp.

41. Homberg, U. 1984. Processing of antennal information in extrinsic mushroom body neurons of the bee brain. *J. Comp. Physiol. A* 154:825–36

42. Homberg, U., Kingan, T. G., Hildebrand, J. G. 1987. Gastrin/CCK-like peptides in the brain of the tobacco hawkmoth *Manduca sexta*. *Soc. Neurosci. Abstr.* 13:235

43. Homberg, U., Montague, R. A., Hildebrand, J. G. 1988. Anatomy of antennocerebral pathways in the brain of the sphinx moth *Manduca sexta*. *Cell Tissue Res.* In press

44. Homberg, U., Montague, R. A., Hoskins, S. G., Kent, K. S., Hildebrand, J. G. 1986. Immunocytochemical mapping of transmitter candidates in the antennal lobes of *Manduca sexta*. *Soc. Neurosci. Abstr.* 12:949

45. Honegger, H. W. 1977. Interommatidial hair receptor axons extending into the ventral nerve cord in the cricket *Gryllus campestris*. *Cell Tissue Res.* 182:281–85

46. Hoskins, S. G., Hildebrand, J. G. 1983. Neurotransmitter histochemistry of neurons in the antennal lobes of *Manduca sexta*. *Soc. Neurosci. Abstr.* 9:216

47. Hoskins, S. G., Homberg, U., Kingan, T. G., Christensen, T. A., Hildebrand, J. G. 1986. Immunocytochemistry of

GABA in the antennal lobes of the sphinx moth *Manduca sexta*. *Cell Tissue Res.* 244:243–52

48. Janet, C. 1911. Sur l'existence d'un organe chordotonal et d'une vésicule pulsatile antennaires chez l'abeille et sur la morphologie de la tête de cette espèce. *C. R. Acad. Sci.* 152:110–12

49. Jawlowski, H. 1948. Studies on the insect brain. *Ann. Univ. Mariae Curie-Sklodowska Sect. C* 3:1–37

50. Jawlowski, H. 1954. Über die Struktur des Gehirnes bei Saltatoria. *Ann. Univ. Mariae Curie-Sklodowska Sect. C* 8:403–34

51. Jawlowski, H. 1957. Nerve tracts in bee (*Apis mellifica*) running from the light and antennal organs to the brain. *Ann. Univ. Mariae Curie-Sklodowska Sect. C* 12:307–23

52. Johnston, C. 1855. Auditory apparatus of *Culex* mosquito. *Q. J. Microsc. Sci.* 3:97–102

53. Kaissling, K.-E. 1971. Insect olfaction. In *Handbook of Sensory Physiology*, Vol. 4, Part 1, ed. L. M. Beidler, pp. 351–432. Berlin: Springer-Verlag. 518 pp.

54. Kaissling, K.-E. 1986. Chemo-electrical transduction in insect olfactory receptors. *Ann. Rev. Neurosci.* 9:121–45

55. Kanzaki, R., Shibuya, T. 1986. Identification of the deutocerebral neurons responding to the sexual pheromone in the male silkworm moth brain. *Zool. Sci.* 3:409–18

56. Keil, T. A., Steinbrecht, R. A. 1984. Mechanosensitive and olfactory sensilla of insects. In *Insect Ultrastructure*, ed. R. C. King, H. Akai, 2:477–516. New York: Plenum

57. Kent, K. S. 1985. Metamorphosis of the antennal center and the influence of sensory innervation on the formation of glomeruli in the hawkmoth *Manduca sexta*. PhD dissertation. Harvard Univ., Cambridge, Mass. 209 pp.

58. Kent, K. S., Harrow, I. D., Quartararo, P., Hildebrand, J. G. 1986. An accessory olfactory pathway in Lepidoptera: the labial pit organ and its central projections in *Manduca sexta* and certain other sphinx moths and silk moths. *Cell Tissue Res.* 245:237–45

59. Kent, K. S., Hoskins, S. G., Hildebrand, J. G. 1987. A novel serotonin-immunoreactive neuron in the antennal lobe of the sphinx moth *Manduca sexta* persists throughout postembryonic life. *J. Neurobiol.* 18:451–65

60. Kenyon, F. C. 1896. The brain of the bee. A preliminary contribution to the

morphology of the nervous system of the Arthropoda. *J. Comp. Neurol.* 6:133–210

61. Kingan, T. G., Hildebrand, J. G. 1985. γ-Aminobutyric acid in the central nervous system of metamorphosing and mature *Manduca sexta*. *Insect Biochem.* 15:667–75

62. Kingan, T. G., Homberg, U., Hildebrand, J. G. 1987. The distribution and characterization of FMRFamide-like peptides in the brain of the moth, *Manduca sexta*. *Soc. Neurosci. Abstr.* 13:234

63. Klepsch, U. 1986. Anatomische Charakterisierung antennaler Motoneurone im Gehirn von Grillen (*Gryllus campestris* und *Gryllus bimaculatus*). Diploma thesis. Tech. Univ. München, West Germany

64. Koontz, M. A., Schneider, D. 1987. Sexual dimorphism in neuronal projections from the antennae of silk moths (*Bombyx mori, Antheraea polyphemus*) and the gypsy moth (*Lymantria dispar*). *Cell Tissue Res.* 249:39–50

65. Lee, J. K., Altner, H. 1986. Primary sensory projections of the labial palp-pit organ of *Pieris rapae* L. (Lepidoptera, Pieridae). *Int. J. Insect Morphol. Embryol.* 15:439–48

66. Light, D. M. 1986. Central integration of sensory signals: an exploration of processing of pheromonal and multimodal information in lepidopteran brains. In *Mechanisms in Insect Olfaction*, ed. T. Payne, M. C. Birch, C. E. J. Kennedy, pp. 287–301. London: Oxford Univ. Press. 364 pp.

67. Linn, C. E., Campbell, M. G., Roelofs, W. L. 1986. Male moth sensitivity to multicomponent pheromones: critical role of female-released blend in determining the functional role of components and active space of the pheromone. *J. Chem. Ecol.* 12:659–68

68. Linn, C. E., Campbell, M. G., Roelofs, W. L. 1987. Pheromone components and active spaces: What do moths smell and where do they smell it? *Science* 237:650–52

69. Masson, C. 1972. Le système antennaire chez les fourmis. I. Histologie et ultrastructure du deutocérébron. Etude comparée chez *Camponotus vagus* (Formicinae) et *Mesoponera caffraria* (Ponerinae). *Z. Zellforsch. Mikrosk. Anat.* 134:31–64

70. Masson, C. 1977. Central olfactory pathways and plasticity of responses to odorous stimuli in insects. In *Olfaction and Taste VI*, ed. J. Le Magnen, P.

MacLeod, pp. 305–14. London: Inf. Retr.

71. Matsumoto, S. G., Hildebrand, J. G. 1981. Olfactory mechanisms in the moth *Manduca sexta*: response characteristics and morphology of central neurons in the antennal lobes. *Proc. R. Soc. London Ser. B* 213:249–77

72. Maxwell, G. D., Tait, J. F., Hildebrand, J. G. 1978. Regional synthesis of neurotransmitter candidates in the CNS of the moth *Manduca sexta*. *Comp. Biochem. Physiol. C* 61:109–19

73. Mobbs, P. G. 1982. The brain of the honeybee *Apis mellifera*. I. The connections and spatial organisation of the mushroom bodies. *Philos. Trans. R. Soc. London Ser. B* 298:309–54

74. Mobbs, P. G. 1985. Brain structure. In *Comprehensive Insect Physiology, Biochemistry, and Pharmacology*, Vol. 5, *Nervous System: Structure and Motor Function*, ed. G. A. Kerkut, L. I. Gilbert, pp. 299–370. Oxford: Pergamon. 646 pp.

75. Mori, K. 1987. Membrane and synaptic properties of identified neurons in the olfactory bulb. *Prog. Neurobiol. Oxford* 29:275–320

76. Mustaparta, H. 1984. Olfaction. In *Chemical Ecology of Insects*, ed. W. J. Bell, R. T. Cardé, pp. 37–70. Sunderland, Mass: Sinauer. 524 pp.

77. Oland, L. A., Tolbert, L. P. 1987. Glial patterns during early development of antennal lobes of *Manduca sexta*: a comparison between normal lobes and lobes deprived of antennal axons. *J. Comp. Neurol.* 255:196–207

78. Olberg, R. M. 1983. Interneurons sensitive to female pheromone in the deutocerebrum of the male silkworm moth *Bombyx mori*. *Physiol. Entomol.* 8:419–28

79. Pareto, A. 1972. Die zentrale Verteilung der Fühlerafferenz bei Arbeiterinnen der Honigbiene *Apis mellifera* L. *Z. Zellforsch. Mikrosk. Anat.* 131:109–40

80. Power, M. E. 1946. The antennal centers and their connections with the brain of *Drosophila melanogaster*. *J. Comp. Neurol.* 85:485–517

81. Rempel, J. G. 1975. The evolution of the insect head: the endless dispute. *Quaest. Entomol.* 11:7–25

82. Rodriguez, V. 1988. Spatial coding of olfactory information in the antennal lobe of *Drosophila melanogaster*. *Brain Res.* 453:299–307

83. Rodriguez, V., Buchner, E. 1984. [³H]2-deoxyglucose mapping of odor-induced neuronal activity in the antennal

lobes of *Drosophila melanogaster*. *Brain Res.* 324:374–78

84. Rospars, J. P. 1983. Invariance and sex-specific variations of the glomerular organization in the antennal lobes of a moth, *Mamestra brassicae*, and a butterfly, *Pieris brassicae*. *J. Comp. Neurol.* 220:80–96

85. Rospars, J. P. 1988. Structure and development of the insect antennodeutocerebral system. *Int. J. Insect Morphol. Embryol.* 17:243–94

86. Sanes, J. R., Hildebrand, J. G. 1976. Origin and morphogenesis of sensory neurons in an insect antenna. *Dev. Biol.* 51:300–19

87. Sass, H. 1983. Production, release and effectiveness of two female sex pheromone components of *Periplaneta americana*. *J. Comp. Physiol.* 152:309–17

88. Schäfer, S., Bicker, G. 1986. Distribution of GABA-like immunoreactivity in the brain of the honeybee. *J. Comp. Neurol.* 246:287–300

89. Schaller-Selzer, L. 1984. Physiology and morphology of the larval pheromone–sensitive neurones in the olfactory lobe of the cockroach, *Periplaneta americana*. *J. Insect Physiol.* 30:537–46

90. Schildberger, K. 1983. Local interneurons associated with the mushroom bodies and the central body in the brain of *Acheta domesticus*. *Cell Tissue Res.* 230:573–86

91. Schildberger, K. 1984. Multimodal interneurons in the cricket brain: properties of identified extrinsic mushroom body cells. *J. Comp. Physiol. A* 154:71–79

92. Schneider, D., Steinbrecht, R. A. 1968. Checklist of insect olfactory sensilla. *Symp. Zool. Soc. London* 23:279–97

93. Schneiderman, A. M. 1984. Postembryonic development of a sexually dimorphic sensory pathway in the central nervous system of the sphinx moth, *Manduca sexta*. PhD dissertation. Harvard Univ., Cambridge, Mass. 262 pp.

94. Schneiderman, A. M., Matsumoto, S. G., Hildebrand, J. G. 1982. Transsexually grafted antennae influence development of sexually dimorphic neurons in moth brain. *Nature* 298:844–46

95. Schürmann, F. W., Wechsler, W. 1969. Elektronenmikroskopische Untersuchung im Antennallobus des deutocerebrum der Wanderheuschrecke *Locusta migratoria*. *Z. Zellforsch. Mikrosk. Anat.* 95:223–48

96. Schürmann, F. W., Wechsler, W. 1970. Synapsen im Antennenhügel von *Locusta migratoria* (Orthoptera, Insecta). *Z. Zellforsch. Mikrosk. Anat.* 108:563–81

97. Selzer, R. 1979. Morphological and physiological identification of food odour specific neurons in the deutocerebrum of *Periplaneta americana*. *J. Comp. Physiol.* 134:159–63

98. Shepherd, G. M. 1972. Synaptic organization of the mammalian olfactory bulb. *Physiol. Rev.* 52:864–917

99. Starratt, A. M., Dahm, K. H., Allen, N., Hildebrand, J. G., Payne, T. L., Röller, H. 1978. Bombykal, a sex pheromone of the sphinx moth, *Manduca sexta*. *Z. Naturforsch. Teil C* 34:9–12

100. Steinbrecht, R. A. 1984. Chemo-, hygro-, and thermoreceptors. In *Biology of the Integument I*, ed. J. Breiter-Hahn, A. G. Matoltsy, K. S. Richards, pp. 523–53. Berlin: Springer. 84 pp.

101. Steinbrecht, R. A. 1987. Functional morphology of pheromone-sensitive sensilla. In *Pheromone Biochemistry*, ed. G. D. Prestwich, G. J. Blomquist, pp. 353–84. New York: Academic. 565 pp.

102. Stengl, M. 1985. Elektrophysiologische Ableitungen im Deutocerebrum von *Calliphora erythrocephala*. Diploma thesis. Univ. Würzburg, West Germany. 92 pp.

102a. Stocker, R. F., Gendre, N. 1988. Peripheral and central nervous effects of *Lozenge 3*: a *Drosophila* mutant lacking basiconic antennal sensilla. *Dev. Biol.* 127:12–24

103. Stocker, R. F., Singh, R. N., Schorderet, M., Siddiqi, O. 1983. Projection patterns of different types of antennal sensilla in the antennal glomeruli of *Drosophila melanogaster*. *Cell Tissue Res.* 232:237–48

104. Strausfeld, N. J. 1976. *Atlas of an Insect Brain*. Berlin: Springer. 214 pp.

105. Strausfeld, N. J., Bacon, J. P. 1983. Multimodal convergence in the central nervous system of dipterous insects. *Fortschr. Zool.* 28:47–76

106. Strausfeld, N. J., Bassemir, U., Singh, R. N., Bacon, J. P. 1984. Organizational principles of outputs from dipteran brains. *J. Insect Physiol.* 30:73–93

107. Suzuki, H., Tateda, H. 1974. An electrophysiological study of olfactory interneurons in the brain of the honey-bee. *J. Insect Physiol.* 20:2287–99

108. Teal, P.E.A., Tumlinson, J. H., Heath, R. R. 1986. Chemical and behavioral analyses of volatile sex components released by calling *Heliothis virescens* (F.) females (Lepidoptera: Noctuidae). *J. Chem. Ecol.* 12:107–26

109. Tolbert, L. P., Hildebrand, J. G. 1981. Organization and synaptic ultrastructure of glomeruli in the antennal lobes of the moth *Manduca sexta:* a study using thin sections and freeze-fracture. *Proc. R. Soc. London Ser. B* 213:279–301

110. Tyrer, N. M., Bacon, J. P., Davies, C. A. 1979. Sensory projections from the wind-sensitive head hairs of the locust *Schistocerca gregaria. Cell Tissue Res.* 203:79–92

111. Visser, J. H. 1986. Host odor perception in phytophagous insects. *Ann. Rev. Entomol.* 31:121–44

112. Waldow, U. 1977. CNS units in cockroach *(Periplaneta americana)*: specificity of response to pheromones and other odor stimuli. *J. Comp. Physiol.* 116:1–17

113. Waldrop, B., Christensen, T. A., Hildebrand, J. G. 1987. GABA-mediated synaptic inhibition of projection neurons in the antennal lobes of the sphinx moth, *Manduca sexta. J. Comp. Physiol. A* 161:23–32

114. Weiss, M. J. 1974. Neuronal connections and the function of the corpora pedunculata in the brain of the American cockroach, *Periplaneta americana* (L.). *J. Morphol.* 142:21–70

115. Yamada M. 1971. A search for odour encoding in the olfactory lobe. *J. Physiol. London* 214:127–43

116. Zacharuk, R. Y. 1980. Ultrastructure and function of insect chemosensilla. *Ann. Rev. Entomol.* 25:27–47

117. Zacharuk, R. Y. 1985. Antennae and sensilla. See Ref. 74, pp. 1–69

Ann. Rev. Entomol. 1989. 34:503–29

# LEAFHOPPER AND PLANTHOPPER TRANSMISSION OF PLANT VIRUSES

*Lowell R. Nault*

Department of Entomology, Ohio Agricultural Research and Development Center,
The Ohio State University, Wooster, Ohio 44691

*El Desouky Ammar*

Department of Plant Pathology, Ohio Agricultural Research and Development Center,
The Ohio State University, Wooster, Ohio 44691

## INTRODUCTION

The hoppers, as we refer collectively to the leafhoppers, planthoppers, and their Auchenorrhyncha relatives, were the first insects proven to be vectors of plant viruses. In 1883 a Japanese rice grower transmitted the rice dwarf virus using the leafhopper *Recelia dorsalis* (documented in 46). The first report of planthopper transmission of a plant virus was 39 years later, when Kunkel (85) reported that *Peregrinus maidis* was the vector of maize mosaic virus. Starting in the 1930s, reports documenting the role of leafhoppers and planthoppers as vectors of viruses and viruslike pathogens increased substantially (84, 113, 114). Until 1967, all plant pathogens transmitted by hoppers were thought to be viruses, including the frequently studied pathogens that cause aster yellows, corn stunt, and potato witches'-broom. The pathogens responsible for these yellows diseases were overlooked by investigators in electron microscopic studies of diseased plants and vectors (130) until workers in Japan discovered that phloem-limited, mycoplasmalike organisms (MLO), not viruses, were most probably the causal agents (38). Certain of these MLOs have been well characterized and classified as spiroplasmas (145). Other plant diseases formerly thought to be caused by hopper-transmitted viruses have subsequently been shown to be associated with phloem-limited, rickettsialike organisms (33) or xylem-limited bacteria (122).

0066-4170/89/0101-0503$02.00

In this review we focus on hopper-transmitted viruses but occasionally mention transmission by other insects. Aspects of virus and prokaryote transmission by hoppers were reviewed earlier (33, 84, 96, 111, 123, 133). In addition to concentrating on recently published studies, this review recognizes the recent family-level classification of viruses erected by the Plant Virus Subcommittee of the International Committee for Taxonomy of Viruses (ICTV) (100; R. I. Hamilton, personal communication). Furthermore, our consideration of virus-vector relationships is guided by recent studies in phylogeny, evolution, and molecular biology.

## CONCEPTS AND TERMINOLOGY

Fifty years have passed since Watson & Roberts (143) proposed a system to classify insect-transmitted plant viruses. They classified viruses into two groups, nonpersistent and persistent, to recognize differences in the length of time that vectors, once having acquired virus, are capable of inoculating plants. Sylvester (136) later improved the system by recognizing semipersistent viruses, whose persistence in vectors is intermediate between nonpersistent and persistent, and by applying more precise criteria for classifying the three groups. Basing terminology on the empiric characteristic of persistence has as its advantage experimental criteria that are easy to test. Furthermore, the terminology conveys the important transmission property of persistence to those interested in virus epidemiology and control. A shortcoming of the system, however, is that the terms fail to describe the mechanisms of transmission or routes of virus transport in the vector. Because such mechanisms or routes are known or strongly suspected for at least one virus from each of the ICTV plant virus taxa covered in this review, we organize our discussion according to a classification system that recognizes three groups of hopper-transmitted viruses: foregut-borne, circulative, and propagative.

Kennedy et al (81) used the term stylet-borne to describe the nonpersistent relationship for certain aphid-transmitted viruses. The term was based on a series of studies by Bradley (23), who treated the stylets of viruliferous aphids with formalin or UV radiation to render aphids less or non-inoculative. The data were interpreted as evidence that transmissible virus was carried on the stylets, specifically near the tips of the maxillary stylets (24). Harris (61) argued, however, that such treatments could affect aphid sensory receptors associated with stylet probing and feeding behavior and that this effect, rather than any inactivation of transmissible virus, could account for Bradley's results. An ingestion-egestion mechanism for transmission of both nonpersistently and semipersistently transmitted viruses was proposed (61, 62) whereby virus enters the foregut, attaches to the lining of the anterior portion of the alimentary canal, and is inoculated when the vector egests while probing a plant. Use of electron microscopy has not unequivocally provided evidence that supports either the stylet-borne or the ingestion-egestion mechanism for nonpersistent transmission of plant viruses by aphids (66). In a recent study, Berger & Pirone (18) used autoradiography to locate antigens of two potyviruses on the maxillary stylets and portions of the foregut of aphid vectors, leaving open the question of whether the stylets, foregut, or both are the sites of retention of transmissible virus.

Unlike for the nonpersistently transmitted viruses, electron microscopy has been useful in locating viruslike particles attached to the cuticular linings of the anterior alimentary canal of aphids and leafhoppers carrying semipersistently transmitted viruses (13, 31, 66). Thus, we introduce the term *foregut borne* in this review to describe the mode of transmission of these viruses. Foregut-borne viruses have no latent period in the vector, cannot be recovered from vector hemolymph, and cannot be transmitted after injection into the vector's hemocoel. Infectivity is retained only a few days and is lost after ecdysis.

Kennedy et al (81) used Black's term *circulative* for those viruses that are ingested, pass through the gut wall into the hemolymph, and then pass to the salivary glands to be discharged with the salivary secretions. Circulative viruses can be recovered experimentally from the vector's hemolymph, can be transmitted after injection into the vector's hemocoel, are not lost after ecdysis, and may be transmitted for weeks, sometimes for the life of the vector (137). The latent period, the time it takes for ingested virus to be incorporated into salivary secretions, may be several hours or days and is temperature dependent. The ability of vectors to transmit circulative viruses often declines with time but can be restored by allowing vectors to reacquire virus from infected plants. No circulative virus is transovarially transmitted from infective females to their progeny.

*Propagative* viruses are similar to circulative viruses in several transmission properties, but they multiply in their vectors. Such multiplication can be demonstrated by transmission studies in which virus is passed serially from insect to insect (either by needle injection or transovarial transmission) until the dilution attained in the final inoculative insects exceeds the dilution endpoint of the initial inoculum (144). Multiplication is also indicated by electron microscopic observation of aggregated virus particles and sites of viral assembly in vector cells (14, 128). Probably the best evidence supporting multiplication is obtained by quantitative serology that shows an increased virus titer in vectors after they are removed from a virus source (40, 56, 109). Propagative viruses usually require longer mean latent periods (one or more weeks) than circulative viruses (137). Some propagative viruses are transovarially transmitted from viruliferous females to their progeny (36, 137). Some authors (62, 94) include viruses that multiply (propagative) and those that do not (nonpropagative) within the circulative group. We prefer not to use the terms "circulative-nonpropagative" or "circulative-propagative," as they may imply an evolutionary relationship between circulative and propagative viruses that has not been established. Sylvester (137) has similarly considered these terms separately in his review of aphid-transmitted viruses.

## VECTOR TAXA

The homopteran suborder Auchenorrhyncha is composed of five families or superfamilies; the phylogenetic relationships of the cicadas (Cicadidae), spittlebugs (Cercopidae), planthoppers (Fulgoroidea), leafhoppers (Cicadellidae), and treehoppers (Membracoidea), as well as groups from the suborder Sternorrhyncha, have been summarized (107). Not all taxa within the Auchenorrhyncha transmit plant viruses. Tonkyn & Whitcomb (140) categorized the Auchenorrhyncha (as well as the aphids and their Sternorrhyncha relatives) into three distinct feeding guilds, those that are primarily phloem, xylem, or mesophyll feeders. Most leafhoppers, planthoppers, and treehoppers that are virus vectors are phloem feeders. All species of cicadas and spittlebugs are thought to be xylem feeders (140); none have been reported as virus vectors.

Among the approximately 60 recognized Cicadellidae subfamilies, eight have species that are vectors of plant pathogens (113, 114), but only two of these subfamilies have species that are virus vectors, the Agalliinae and Deltocephalinae. The Agalliinae vector species have herbaceous dicot plant hosts, whereas the Deltocephalinae vector species have primarily monocot hosts. That no leafhopper-borne viruses of woody dicots have been reported is surprising because many mollicutes infecting these hosts are transmitted by leafhoppers (113, 114). Of the approximately 15,000 described leafhopper species from about 2000 genera (114), only 49 species from 21 genera have been reported as virus vectors (36, 110, 113, 114).

Among the 20 Fulgoroidea families, only the Delphacidae have species confirmed as virus vectors. From other planthopper families, the Cixiidae are mollicute vectors and the Flatidae are vectors of plant pathogenic bacteria (116). The Delphacidae have monocot hosts, primarily the Gramineae (84, 116); thus, not surprisingly all viruses transmitted by these planthoppers have graminaceous hosts. Most cause important diseases in maize, rice, wheat, and other cereal crops. Of the more than 1100 described delphacid species from 137 genera (116), 27 species from 13 genera have been implicated as virus vectors. The treehoppers are represented by a single vector species (131).

## TRANSMISSION OF HOPPER-BORNE PLANT VIRUSES

The Homoptera have species that are vectors of viruses from 16 of the 34 plant virus families or groups currently recognized by the ICTV. The hoppers rank second to the aphids and their Sternorrhyncha relatives in the number of viruses transmitted and in the number of plant virus groups represented by these viruses (Table 1). Despite the greater number of plant viruses transmitted by aphids, Sinha (133) noted that the hoppers apparently have a more intimate and complex relationship with the viruses they transmit. Following is a discussion of transmission of hopper-borne plant viruses organized according to modes of transmission and virus taxonomy.

## Foregut-Borne Viruses

MAIZE CHLOROTIC DWARF VIRUS GROUP    Maize chlorotic dwarf virus (MCDV) has been designated as the type member of the group (54) with the rice tungro spherical virus (RTSV) as a probable member (Table 1). Particles of MCDV and RTSV are isometric, are approximately 30–33 nm in diameter, and contain single-stranded (ss) RNA. These viruses are semipersistently transmitted by leafhoppers in the subfamily Deltocephalinae. Neither virus is mechanically transmissible. Examination of phloem and vascular parenchyma of MCDV- and RTSV-infected plants reveals dense, granular inclusions that contain viruslike particles (13, 63, 148). Associated with the rice tungro disease is a second semipersistently transmitted leafhopper-borne virus, the rice tungro bacilliform virus (RTBV). Its particles measure 30–35 × 160–220 nm and are also found in the phloem of their hosts (25a, 71). The ICTV has not placed RTBV in a group, but Francki et al (45) have suggested that this virus be included in a group with the cacao swollen shoot virus.

The field vector *Graminella nigrifrons* can acquire MCDV from infected plants or inoculate it to healthy plants in 15 min (34), whereas *Nephotettix virescens* requires 30 min to acquire or inoculate RTSV (77). Longer acquisition and inoculation access periods increase transmission efficiency for both viruses (34, 77, 90, 112). The minimum time for leafhoppers to acquire and inoculate these viruses is assumed to be the time it takes for leafhoppers to penetrate to phloem cells, where most feeding and virus occur. The mean persistence of virus in leafhopper vectors at 25°C or higher is less than 24 hr, but at lower temperatures virus is retained for several days (90; L. R. Nault, unpublished). Nymphs and female or male adults are vectors; nymphs lose transmissible virus upon molting (34, 90, 112).

RTBV depends on RTSV for its transmission by leafhoppers (69, 72). RTBV can be transmitted from plants infected with both viruses or by leafhoppers from singly infected plants if leafhoppers are given access first to plants infected with RTSV and then to plants infected with RTBV, but not if the order of exposure is reversed. Leafhoppers that had acquired RTSV and were then held until they had lost their ability to transmit virus retained the ability to acquire and transmit RTBV (25). Leafhoppers given access to RTSV-infected plants and later fed RTSV antiserum lost most of their ability to transmit RTSV, but they could still acquire and transmit RTBV (70). These results strongly suggest that a factor from RTSV-infected plants, but not the RTSV virions, assists in RTBV transmission.

**Table 1**  Auchenorrhyncha-borne plant viruses and vector genera grouped according to mode of transmission, virus taxonomy, and vector family

| Transmission mode and virus group | Vector family | Virus | Vector genera | Ref. |
|---|---|---|---|---|
| **Foregut-borne** | | | | |
| Maize chlorotic dwarf virus | Cicadellidae | Maize chlorotic dwarf | *Graminella*, 6 others | 54, 110 |
| | | Rice tungro spherical | *Nephotettix, Recelia* | |
| Unclassified | Cicadellidae | Rice tungro bacilliform | *Nephotettix* | |
| **Circulative** | | | | |
| Geminivirus | Cicadellidae | Maize streak | *Cicadulina* | 65 |
| | | Beet curly top | *Circulifer* | |
| | | Chloris striate mosaic | *Nesoclutha* | |
| | | *Paspalum* striate mosaic | *Nesoclutha* | |
| | | Tobacco yellow dwarf | *Orosius* | |
| | | Wheat dwarf | *Psammotettix* | |
| | Membracidae | Tomato pseudo–curly top | *Micrutalis* | |
| **Propagative** | | | | |
| Marafivirus | Cicadellidae | Maize rayado fino | *Dalbulus*, 3 others | 50 |
| | | Bermuda grass etched-line | *Aconurella* | |
| | | Oat blue dwarf | *Macrosteles* | |
| Rhabdoviridae | Cicadellidae | Cereal chlorotic mottle | *Nesoclutha, Cicadulina* | 79, 80, 92, 147 |
| | | Oat striate mosaic | *Graminella* | |
| | | Potato yellow dwarf[a] | *Aceratagallia,*[b] *Agallia,*[b] *Agalliopsis*[b] | |
| | | Rice transitory yellowing | *Nephotettix* | |
| | | Sorghum stunt mosaic | *Graminella* | |
| | | Wheat striate mosaic | *Endria, Elymana* | |
| | | Winter wheat mosaic[a] | *Psammotettix* | |
| | Delphacidae | Barley yellow striate mosaic[a] | *Laodelphax* | |
| | | *Colocasia* bobone disease | *Tarophagus* | |
| | | *Cynodon* chlorotic streak | *Toya* | |
| | | *Digitaria* striate | *Sogatella* | |
| | | Finger millet mosaic | *Sogatella, Peregrinus* | |
| | | Iranian maize mosaic | *Ribautodelphax* | |
| | | Maize mosaic | *Peregrinus* | |
| | | Maize sterile stunt | *Sogatella, Peregrinus* | |
| | | Northern cereal mosaic | *Laodelphax, Muellerianella, Ribautodelphax, Unkanodes* | |
| | | Wheat chlorotic streak[a] | *Laodelphax* | |
| | | Wheat rosette stunt | *Laodelphax* | |

**Table 1**   *(Continued)*

| Transmission mode and virus group | Vector family | Virus | Vector genera | Ref. |
|---|---|---|---|---|
| Tenuivirus | Delphacidae | Rice stripe[a] | *Laodelphax, Terthron, Unkanodes* | 55 |
| | | Maize stripe[a] | *Peregrinus* | |
| | | Rice grassy stunt | *Nilaparvata* | |
| | | Rice hoja blanca[a] | *Sogatodes* | |
| | | European wheat striate mosaic[a] | *Javesella* | |
| Reoviridae | | | | |
| *Phytoreovirus* | Cicadellidae | Wound tumor (clover)[a] | *Agallia,*[b] *Agalliopsis,*[b] *Aceratagallia*[b] | 35 |
| | | Rice dwarf[a] | *Nephotettix, Recelia* | |
| | | Rice gall dwarf[a] | *Nephotettix, Recelia* | |
| *Fijivirus* | Delphacidae | Fiji disease[a] | *Perkinsiella* | 27, 27a, 35 |
| | | *Arrhenatherum* blue dwarf | *Javesella, Dicranotropis* (?) | |
| | | Cereal tillering disease | *Laodelphax, Dicranotropis* | |
| | | *Lolium* enation disease | *Javesella* | |
| | | Maize rough dwarf[a] | *Laodelphax, Delphacodes, Javesella, Sogatella* | |
| | | Oat sterile dwarf[a] | *Javesella, Dicranotropis, Ribautodelphax* | |
| | | Pangola stunt | *Sogatella* | |
| | | Rice black streaked dwarf | *Laodelphax, Unkanodes* | |
| | | Rice ragged stunt | *Nilaparvata* | |
| | | *Echinocloa* ragged stunt | *Sogatella* | |

[a]Transovarial transmission has been reported in some vectors.
[b]Genera from subfamily Agalliinae; all other cicadellids are from subfamily Deltocephalinae.

A similar situation pertains to MCDV. Purified MCDV particles fed to *G. nigrifrons* through Parafilm® membranes were not transmitted by leafhoppers. However, when leafhoppers were fed first on plants infected with a mild MCDV isolate and then on membranes containing purified virus of a severe MCDV isolate, both isolates were transmitted (75). Reversing the feeding sequence resulted in transmission of the mild but not the severe MCDV isolate. The most parsimonious explanation for these results with RTSV, RTBV, and MCDV is that a virus-induced helper component is involved in leafhopper transmission, possibly similar to the helper components required for the aphid transmission of potyviruses and caulimoviruses (121). The helper components for potyviruses have been identified as virus-coded proteins (139). Helper components have not been isolated yet for MCDV or RTSV.

The mode of transmission for MCDV may be similar to that suggested earlier for anthriscus yellows virus (66), which is semipersistently transmitted by aphids. MCDV-like particles (VLPs) embedded in a dense matrix have been observed attached to portions of the foregut of MCDV-exposed (31) or MCDV-inoculative (13) *G. nigrifrons*. Embedded VLPs were observed attached to the cuticular lining of the pharynx, cibarium (sucking pump), and precibarium (E. D. Ammar & L. R. Nault, unpublished) (Figures 1 and 2). Some authors have suggested that the helper component is the embedding matrix for VLPs in both plants and insects, although other roles in transmission have been suggested for helper component (66, 75).

We speculate that for foregut-borne viruses, leafhoppers and aphids ingest virus and helper component from the phloem of infected plants and that helper component with embedded virus specifically attaches to the pharynx and possibly other sites in the foregut. The binding of helper component and virus is reversible, and with time, virus detaches gradually and is finally lost. Apparently, virus detaches faster than helper component, which would account for the ability of RTSV-exposed leafhoppers to acquire and transmit RTBV after the ability to transmit RTSV is lost (25). Most virus with helper component probably passes into the midgut and is not transmitted. Nevertheless, some virus must pass back from its retention sites in the anterior alimentary canal to be injected into plant tissue, perhaps as insects are discriminating between acceptable and nonacceptable feeding sites with precibarial sensilla (Figures 1*b* and 2*a*) (15, 64, 103).

An understanding of the mode of transmission of foregut-borne viruses may lead to an explanation for vector specificity. RTSV is transmitted by four *Nephotettix* species and *Recelia dorsalis* but not by six other more distantly related Deltocephalinae species (77). In an extensive study of 25 grass-feeding leafhopper species from 13 genera in the subfamily Deltocephalinae, Nault & Madden (110) correlated the phylogenetic relatedness of leafhoppers with the ability to transmit MCDV. Efficient vector species were those that could use maize as a developmental host and were from the tribes Deltocephalini and advanced Euscelini. Leafhoppers from the primitive Euscelini and distantly related Macrostelini were poor vectors or nonvectors. We speculate

—————————————————————————————————————→

*Figure 1* (*a*) Sagittal view of the head of the leafhopper *Graminella nigrifrons* showing the anterior alimentary canal, salivary system, and surrounding structures. (*b*) Details of the precibarium and salivary syringe areas. *B*, brain; *cb*, cibarium; *cdm*, cibarial dilator muscle; *d*, cibarial diaphragm; *ds*, distal sensilla; *e*, esophagus; *ev*, esophageal valve; *fc*, food canal; *Lm*, labium; *Lr*, labrum; *MG*, midgut; *p*, precibarium; *Ph*, pharynx; *pr*, piston retractor muscle; *ps*, proximal sensilla; *pv*, precibarial valve; *pvm*, precibarial valve muscle; *S*, stylets; *sc*, salivary canal; *sd*, salivary duct; *seg*, subesophageal ganglion; *SG*, salivary gland; *ss*, salivary syringe; *tb*, tentorial bar (E. D. Ammar, unpublished).

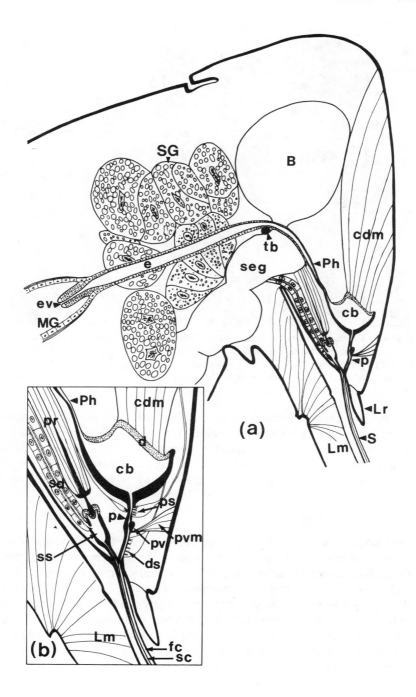

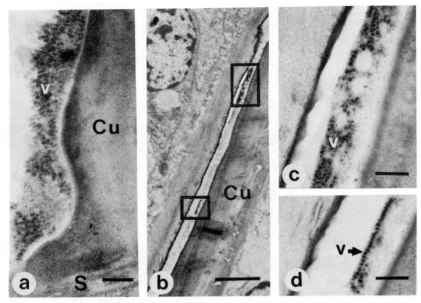

*Figure 2*    Electron micrographs of maize chlorotic dwarf virus–like particles (*v*), embedded in a dense matrix, in inoculative *Graminella nigrifrons*. (*a*) Particles attached to the cuticular lining (*Cu*) of the precibarium. (*b, c, d*) Particles attached to parts of the pharynx. Boxed areas in *b* are enlarged in *c* and *d*. *S*, precibarial sensillum. See Figure 1 for identification of these structures in the leafhopper's head. Scale bars in *a*, *c*, and *d* equal 200 nm, and in *b*, 2 μm.

that the physical and chemical properties of the cuticular lining of the foreguts of related vector leafhopper species are more similar to each other than to those of distantly related, nonvector species and that attachment and/or detachment of helper component and virus depend on these cuticular properties. However, the possible role of the salivary secretions in the vector specificity of foregut-borne viruses needs to be explored.

## Circulative Viruses

GEMINIVIRUSES    Geminiviruses have isometric particles approximately 16–18 nm in diameter that usually occur in pairs, hence the name of the group. Geminiviruses are unique in that they contain single-stranded, circular DNA molecules. The definitive and provisional members of the group can be defined by host relationships, symptoms produced in hosts, and vector taxa (44, 65). Members of one subgroup, the beet curly top virus (CTV) and tobacco yellow dwarf virus, cause stunting, yellowing, and leaf curling in dicots and are leafhopper borne. Tomato pseudo–curly top virus is similar to members of this subgroup except that the vector is a treehopper, *Micrutalis mallifera* (131). Members of a second subgroup, chloris streak mosaic virus (CSMV), maize streak virus (MSV), wheat dwarf virus, and *Paspalum* striate mosaic virus, cause striate mosaics in Gramineae and are also transmitted by leafhoppers. Two other subgroups cause diseases in dicots and are transmitted by whiteflies (Aleyrodidae). The existence of an aphid-transmitted geminivirus is also suspected (65). Several of the whitefly-transmitted geminiviruses are antigenically related and are transmitted by *Bemisia tabaci*. In contrast, the leafhopper-borne geminiviruses have vectors from several genera (Table 1) and are antigenically unrelated or distantly related. Furthermore, the whitefly-transmitted geminiviruses have bipartite genomes, that is, each geminate pair contains different genome segments, and both segments ($DNA_1$ and $DNA_2$) are necessary for infection of plants.

The three leafhopper-transmitted geminiviruses studied so far, CTV, MSV and CSMV, appear monopartite, i.e. each geminate pair contains two copies of the same genome (65). The nucleotide sequence of CTV reveals coding regions that are nearly homologous to $DNA_1$ of several whitefly-borne geminiviruses (135). The exception is the region that codes for coat protein; it more closely resembles that of other leafhopper-borne geminiviruses. Matthews (101) has suggested that differences between whitefly- and leafhopper-transmitted geminiviruses are sufficient to establish genera-level taxa for each group. Regardless of vector taxa, all geminiviruses have a circulative relationship with their vectors and are persistently transmitted. No evidence indicates that geminiviruses are propagative (65).

Acquisition threshold times for geminiviruses by their leafhopper vectors vary from a few seconds to one hour; longer feeding periods result in higher transmission rates and increased persistence in vectors (57, 65). The principle sites of virus acquisition by leafhoppers from plants are limited by the viruses' distribution; CTV is limited to the phloem, whereas CSMV and MSV invade the mesophyll and phloem (12, 44). The latent period of geminiviruses in leafhoppers ranges from 4 to 19 hr (57, 65). Presumably, penetration of the gut wall by virus and transport of virus into the salivary glands occur during the latent period, but details for the route of transport have not been established. Harrison (65) suggested that the route may be similar to that of the better-studied luteoviruses in their aphid vectors (52). Luteoviruses pass through the hindgut wall into the hemolymph of vector and nonvector aphid species, but only in vectors does virus penetrate beyond the basal plasmalemma of the accessory salivary glands. Comparing luteovirus transmission by aphids with geminivirus transmission by leafhoppers may be risky, however, because the morphology of the guts and salivary glands of these two homopteran families differ markedly (5, 6, 52). Perhaps these morphological and associated physiological differences will explain why the salivary glands of aphids regulate luteovirus transmission specificity (52), whereas the gut wall of leafhoppers apparently serves this function for MSV transmission (65).

Each leafhopper-transmitted geminivirus has a single vector species or several vector species from a single genus (Table 1). Two *Circulifer* species are reported as CTV vectors (65) and several *Cicadulina* species as vectors for MSV (127). Eight deltocephaline (non-*Cicadulina*) species have failed to transmit MSV under experimental conditions (58). Nonvectors and poor *Cicadulina* vectors of MSV acquire and retain smaller amounts of virus than the efficient vector *Cicadulina mbila* (22).

## Propagative Viruses

MARAFIVIRUSES   Maize rayado fino virus (MRFV) is the type member of this group (49, 50), which includes the serologically related oat blue dwarf virus (OBDV) (16, 17) and Bermuda grass etched-line virus (93) (Table 1). Marafiviruses have spherical particles 28–33 nm in diameter and a ss-RNA genome (49, 50). Virus particles are most frequently observed in the phloem and vascular parenchyma of their graminaceous hosts. The marafiviruses are not mechanically or seed transmitted.

Thresholds for acquisition and inoculation of MRFV and OBDV by leafhoppers range from several minutes to several hours, with longer feeding periods resulting in higher transmission levels (16, 49, 50). When a marafivirus is acquired from plants, a latent period of 7 days or longer is needed before leafhoppers become inoculative. Injection of virus into the hemocoel reduced the latent period to 1–3 days and increased the transmission rate (16, 108). Leafhoppers are better MRFV vectors if virus is acquired during nymphal stages rather than during the adult stage (108). Marafiviruses multiply in their vectors but are not transovarially transmitted (17, 49, 56).

Only 10–34% of *Dalbulus maidis* hoppers exposed to MRFV-infected plants transmitted virus, although nearly 80% contained virus as determined serologically (49, 50). Transmission of MRFV by *D. maidis* is under genetic control. Transmission rate was increased severalfold by selective breeding, but this enhanced ability dropped to normal levels after a few generations of random mating (49, 50, 108). Gamez & León (49, 50) proposed that MRFV vectors are recessive homozygotes for rare alleles that are rapidly diluted in outcrosses. Six of seven *Dalbulus* species are experimental MRFV vectors (108; R. Gomez-Luengo & L. R. Nault, unpublished). Nevertheless, *D. maidis* is the most efficient vector, which supports the view that the triad of MRFV, *D. maidis,* and maize has a long coevolutionary history (49, 50).

RHABDOVIRUSES    Of the taxonomic virus groups recognized by the ICTV, only the Rhabdoviridae and Reoviridae include viruses that can infect either vertebrates or invertebrates and plants (100). Plant rhabdoviruses have complex particles, bacilliform or bullet shaped, measuring 45–100 nm in diameter and 100–430 nm in length (44, 80). The particles are enveloped and contain four to six structural proteins and a single molecule of negative-sense ss-RNA. Most plant rhabdoviruses have a host range limited to the Gramineae, but some infect dicots (36). Eighteen rhabdoviruses have hopper vectors (Table 1), whereas others have aphid, lacebug, or mite vectors (80). Some investigators have proposed that plant rhabdoviruses be divided into two subgroups according to the properties of their proteins, the kinetics of their transcriptase activities, and the site of maturation of their particles in plant cells (44, 100). Such a division of the rhabdoviruses does not partition them by vector groups or families. Each hopper-transmitted virus is vectored either by cicadellids or delphacids, but not by both. A possible exception is finger millet mosaic virus, which was reported to be vectored by a delphacid and two cicadellid species; this unusual finding needs confirmation, however (36).

Acquisition thresholds of less than 1 min for wheat striate mosaic virus (WSMV) and 5–15 min for rice transitory yellowing vius (RTYV) probably reflect the more general distribution of WSMV in mesophyll and phloem cells (87) and the restricted distribution of RTYV in phloem cells and occasionally parenchyma cells bordering the vascular bundles (29, 32). Inoculation thresholds are less than 15 min for these two viruses (62). The latent period for RTYV in the vector is 3–66 days (28). The latent period for WSMV is shorter in efficient vectors than in inefficient vectors (62, 134). Nymphs of *Nephotettix cincticeps* are more efficient RTYV vectors than adults, and males are slightly more efficient than females (28).

A classic example of vector specificity for plant rhabdoviruses is provided by the two recognized strains of potato yellow dwarf virus (PYDV). One strain is transmitted by *Aceratagallia sanguinolenta* and other *Aceratagallia* species, but not by *Agallia constricta,* whereas the second strain is transmitted by *A. constricta* but not by the former species. (19). The molecular weights of three structural proteins differ for the two PYDV strains (2, 42). Adam & Hsu (2) reported that differences in the G (glycosylated) protein might be related to the selective transmission of these strains. The G protein, which protrudes from the virion envelope, functions in attachment of animal rhabdoviruses to host recognition sites on the plasma membrane during the early stages of infection (80). Recent findings have indicated that the same is probably true for PYDV in insect cell cultures (48). In addition to strain and inter-species specificity in vector transmission of PYDV, intraspecies specificity has been reported as well (19). Also, nontransmissible (exvectorial) isolates of PYDV have been selected by repeated transfer of virus by mechanical inoculation to plants (19).

Another example of vector specificity is the hopper transmission of Iranian maize mosaic virus [Shiraz maize rhabdovirus (80)], by *Ribautodelphax notabilis* in Iran (79). This virus is related serologically to one but not other maize mosaic virus (MMV) isolates. *Peregrinus maidis,* the only known vector of MMV in other parts of the world (44), inefficiently transmits Iranian MMV when it acquires virus from plants but transmits it efficiently when virus is injected into the hemocoel (E. D. Ammar, R. Gomez-Luengo & D. T. Gordon, unpublished). This evidence suggests that MMV-vector specificity is associated with gut barriers.

Experimental evidence of virus multiplication in hopper vectors has been reported for several rhabdoviruses (36, 80); some of these viruses are transovarially transmitted in their hopper vectors (Table 1). Some authors have suggested that most rhabdoviruses assemble and accumulate at similar sites (either perinuclear or cytoplasmic) in plant and vector cells (44, 80). An exception seems to be RTYV virions, which are primarily found in perinuclear spaces in plant cells (29) but accumulate mostly in cytoplasmic vacuoles of the leafhopper salivary gland cells (30). Recent ultrastructural studies on MMV in maize plants and in various organs of *P. maidis* have revealed differences in budding sites that may explain how this rhabdovirus is transmitted by its vector. In plant cells and in most vector tissues examined, MMV particles bud on nuclear and cytoplasmic membranes and accumulate in perinuclear spaces and in dilated cisternae connected with endoplasmic reticulum (14, 102). In secretory cells of the principal salivary glands, however, MMV particles bud mainly on the plasma membranes and accumulate in intercellular and extracellular spaces; these spaces are apparently connected with the extracellular vacuoles and canaliculi that lead to the salivary ductules and ducts (6, 14). A similar difference in the budding sites of the rabies virus in fox brain and salivary gland cells has been reported; plasma membrane budding in salivary gland cells is considered essential for bite transmission of this virus (106). Based on the above studies, Ammar (7) suggested an intercellular and extracellular route for MMV transport in the principal salivary glands of *P. maidis* that allows efficient discharge of virus with the saliva during feeding. A similar route has been suggested for corn stunt spiroplasma in the salivary glands of its leafhopper vector (97).

Replication of plant rhabdoviruses, particularly PYDV, has also been studied in cell cultures. In vector cell monolayers (VCM), Hsu et al (74) found no differences in susceptibility of two cell lines from *A. constricta* and *A. sanguinolenta* to their respective PYDV strains. Also, identical transcription and translation of viral proteins were found in both cell lines inoculated with the *A. sanguinolenta*–transmitted strain (124). In a cell line from the nonvector *Dalbulus elimatus*, however, the *A. constricta*–transmitted strain did not multiply, whereas the *A. sanguinolenta*–transmitted strain did, but much less efficiently than in the cell line of a vector species. The infection pathways of PYDV have recently been studied using cell lines from *A. sanguinolenta* (47). Inhibition of infection by lysosometric substances suggests an endocytotic pathway with fusion inside cells, although direct fusion of virus with the outer cell membrane was not ruled out.

TENUIVIRUSES    The tenuiviruses have RNA genomes, are associated with fine, variously configured filaments 3–8 nm in diameter, and produce large amounts of noncapsid protein in infected plants (41, 55, 73, 141). Tenuiviruses infect the Gramineae, are persistently transmitted by delphacid planthoppers (Table 1), and are not mechanically or seed transmitted. The recently reported maize yellow stripe virus apparently possesses several tenuivirus characteristics, including fine filaments and noncapsid protein, but it is transmitted by a cicadellid leafhopper, *Cicadulina chinai* (8, 9; E. D. Ammar & R. E. Gingery, unpublished).

Acquisition thresholds for tenuiviruses range from 10 min to several hours, whereas inoculation thresholds range from 30 sec to nearly an hour (55). The distribution of tenuiviruses in plant cells is not known, but maize stripe virus (MStpV) inclusions occur in most leaf tissues (10). Latent periods in the vector range from 3 to 36 days; most are between 7 and 21 days. Retention periods up to 84 days postacquisition have been reported for tenuiviruses in their vectors (3, 11). Females have been reported as better vectors than males for rice stripe virus (RSV) (55), and nymphs as better vectors than adults for MStpV (142). The latter report can be explained by recent data (E. D. Ammar & L. R. Nault, unpublished) that reveal a shorter latent period of MStpV in *P. maidis* nymphs than in adults. At 2 wk postacquisition nymphs were better vectors than adults, but at 4 wk postacquisition no transmission differences were found between nymphs and adults, males and females, or brachypterous and macropterous adults. A report that macropterous adults of *Nilaparvata lugens* cannot transmit rice grassy stunt virus (RGSV) (51) was contradicted later by Mathew & Basu (98), who reported equal transmission by both wing forms, nymphs and adults, and males and females. Transmission efficiency of European wheat striate mosaic virus (EWSMV) by several *Javesella pellucida* populations ranged from 0 to 89%, and that by several *J. dubia* populations ranged from 0 to 21% (3). Transovarial transmission to a high proportion of progeny has been reported for most tenuiviruses (Table 1).

Evidence that tenuiviruses multiply in their vectors has been summarized (109). Serological (ELISA) assays indicated that MStpV invades most organs of its *P. maidis* vectors. Virus was first detected in the midgut and thereafter in all other organs assayed, including reproductive organs of females and all but the testes in males. MStpV was detected in the salivary glands of all inoculative *P. maidis,* but not all individuals whose salivary glands were positive for virus were transmitters. Although MStpV produces large quantities of a noncapsid protein in plants, this protein could not be detected in infected insects (41), which suggests that the replication strategy for MStpV differs in plants and insects. The function of the noncapsid protein is unknown, but this protein might have a role in planthopper transmission. This possibility is suggested by studies that show that crude extracts but not highly purified preparations of several tenuiviruses are transmitted by vectors after needle injection (55). These crude extracts are probably contaminated with noncapsid protein, which may assist in planthopper transmission of tenuiviruses.

REOVIRUSES    Plant reoviruses have icosahedral, double-shelled particles that are 65–75 nm in diameter and contain 10 or 12 segments of genomic double-stranded (ds) RNA. Members of this group are transmitted by either leafhoppers or planthoppers; apparently none is transmitted by other insect groups or by seed (35, 44). The group is divided into two genera (Table 1). The genus *Phytoreovirus* includes three viruses that have 12 ds-RNA segments, have no spikes on either particle shell, and are transmitted by leafhoppers. The genus *Fijivirus* includes 10 viruses that have 10 ds-RNA segments, have spikes on the outer and inner shells, and are transmitted by delphacid planthoppers (43, 44). Rice ragged stunt virus (RRSV) was previously placed outside the fijiviruses because its particles appeared to lack an outer shell (44). However, intact double-shelled virions of RRSV have recently been found in leaf-dip preparations as well as in thin sections of leaves and planthopper vectors (C.-C. Chen, M. J. Chen, R.-J. Chiu & H. T. Hsu, personal communication). Chen et al (27a) have also reported similar findings for the serologically related *Echinochloa* ragged stunt virus, which is transmitted by a different planthopper genus (Table 1). Plant hosts of reoviruses are monocots, except for wound tumor virus (WTV), which has been isolated in nature only from leafhoppers but has a wide host range among 20 dicot families (35).

The acquisition and inoculation thresholds for reoviruses range from a few minutes for rice dwarf virus (RDV) to several hours for most others (35). The relatively rapid acquisition and inoculation times for RDV are probably associated with its occurrence in the mesophyll of infected plants, as compared with other reoviruses that are phloem-restricted (44). Latent periods range from 2 days for RRSV in *N. lugens* to 2 mo for pangola stunt virus in *Sogatella furcifera,* with most latent periods between 7 and 14 days (35). The vector usually retains inoculativity for life, frequently with intermittent transmission (35, 78). Transovarial transmission rates were reported as 1.8–

100% for all phytoreoviruses but only 0.2–17.0% for a few fijiviruses (Table 1). Generally, young nymphs are more efficient vectors of reoviruses than older nymphs or adults (35, 105). As an extreme example, Fiji disease virus (FDV) is transmitted if virus is acquired by first-instar nymphs but not by other instars or adults (76). Sinha (132) suggested that susceptibility and/or permeability of the gut to WTV decreases as the vector ages.

Ultrastructural and immunological studies have provided additional information concerning transmission mechanisms. Accumulations of WTV particles are high in various organs of A. constricta, the most efficient WTV vector, but low (with no virus observed in salivary glands) in Agalliopsis novella, an inefficient vector (59, 127a). Furthermore, ultrastructural studies have revealed that although WTV accumulates in almost all organs, relatively few particles are observed in the salivary glands (129). In contrast, RDV is observed in high concentrations in all organs, including the salivary glands of N. cincticeps (127a). These differences in occurrences of virus particles in salivary glands of their vectors may help explain why horizontal (vector-to-plant) but not why vertical (transovarial) transmission rates of WTV are lower than those of RDV (35). Low transmission rates may explain why WTV has never been found in plants in nature whereas RDV causes a widespread disease of rice.

The cytopathology and maturation sites of reoviruses in vertebrate, plant, and insect cells are apparently similar. Intracytoplasmic viroplasms, which are similar to the "virus factories" containing mature and immature progeny virions in cells infected with vertebrate reoviruses (149), occur in insect or plant cells infected with plant reoviruses (44, 128). Autoradiography and immunoelectron microscopy studies have indicated that the viral genome and viral proteins accumulate at viroplasms in RDV-infected VCMs of N. cincticeps (128). WTV antigens in A. constricta are detected first in the filter chamber, then in the hemolymph and several organs, and later in the anterior lobe of the salivary glands (133). The relative concentration of WTV rises steeply, increasing 1000-fold 30 days postacquisition, after which virus titer declines to one tenth of its peak concentration by day 45 (125). Caciagli & Casetta (26) reported that maximum inoculativity of maize rough dwarf virus (MRDV) by Laodelphax striatellus occurs 30 days after acquisition and remains constant thereafter. Nevertheless, virus antigens increase exponentially in the vector up to 45 days after acquisition.

The ability to infect VCMs with plant reoviruses is useful in studying infection and transmission mechanisms (1, 20). Multiplication of WTV in VCMs, coupled with results of infectivity assays, shows that virus increases exponentially between 6 and 12 hr postinoculation, with a doubling time of 58 min at 30°C (83). WTV readily infected cell lines from two vector species, A. constricta and A. novella, but only with difficulty infected the cell lines of a nonvector, *A. sanguinolenta*. WTV did not infect a cell line from *D. elimatus,* a nonagallian, nonvector species (20, 74). Omura et al (119) used VCMs to assay rice gall dwarf virus (RGDV) in *N. cincticeps* and found that virus titer in whole insects remained high up to 40 days postacquisition. This titer does not explain why transmission efficiency decreases with advancing age of viruliferous insects (78). Specific assays of virus in salivary glands rather than whole insects might provide a better understanding of this phenomenon. Whitcomb & Black (146) showed that a WTV clone that had been maintained exclusively in the host plant by vegetative propagation for several years lost its insect transmissibility; a similar finding was reported for RDV (82). Results of assays on VCM revealed a gradual decrease in the infectivity of WTV as the number of passages in plants increased (91). Electrophoretic analyses of the genomes from several WTV isolates showed that the transition to the exvectorial state is always accompanied by deletion mutations in one or more of four ds-RNA segments (126). Nuss (115) suggested that the gene products of segments S2 and S5 are needed for leafhopper transmission of WTV. The products of these two segments comprise the outer capsid (shell) of the virus, which indicates that these capsid proteins may be involved in the recognition of host cells, in virus penetration into cells, or both (1). Because removal of the outer protein coat by protease treatment apparently causes no loss of infectivity to VCM, Nuss (115) suggested that the products of S2 and S5 might perform multiple functions in the replication cycle of WTV in the vector.

## ORIGIN AND EVOLUTION OF AUCHENORRHYNCHA–BORNE PLANT VIRUSES

The reoviruses, regardless of whether they infect plants, vertebrates, or invertebrates, share many common features and probably evolved from a common ancestor (44, 67, 99). It can be convincingly argued that reoviruses, including those that infect plants, have an insect origin. Results of recent studies on two hopper-infecting reoviruses, the leafhopper A virus (LAV) and *Peregrinus maidis* virus (PMV), suggest this possibility. Maize wallaby ear disease (60) was thought to be caused by a plant virus until researchers showed that virus extracted from plants is LAV, that LAV does not multiply in maize, and that the wallaby ear disease is probably caused by an insect toxin (21, 117). LAV is structurally similar to the planthopper-borne fijiviruses (67) and is vertically transmitted from infected females to approximately 20% of their progeny. Moreover, LAV can be transmitted horizontally from infected to noninfected leafhoppers that feed on the same plant tissues (118). A similar example of an insect virus for which a plant serves as a transitory reservoir is the aphid-infecting *Rhopalosiphum padi* virus (RPV) (37). RPV is vertically transmitted to 28% of offspring but maintained in nearly 90% of the population by supplemental horizontal transmission in plants (53).

The second hopper-infecting reovirus, PMV, has 12 ds-RNA segments, as do the leafhopper-transmitted phytoreoviruses, but the virion morphology and cytopathology in insects resemble those of the planthopper-borne fijiviruses. PMV is unrelated serologically to LAV, FDV, or RGDV (39). Although horizontal transmission through plants has not been demonstrated for PMV, particles of similar morphology were reported to occur in the phloem of maize fed upon by *P. maidis* (86). The discovery of LAV and perhaps of PMV in plant tissues illustrates the opportunities that some insect reoviruses might have had in the past in establishing plants as alternate hosts.

That plant-infecting reoviruses originated in insects rather than plants is also suggested by their lack of seed transmission (44); thus, plants are dead-end hosts. Most plant-infecting reoviruses can survive in their vectors, at least for several generations, by transovarial transmission. Purcell (123) has argued that unless propagative phytopathogens are transovarially transmitted to 100% of vector offspring, they cannot be maintained in vector populations without horizontal transmission. Because plant reoviruses and LAV are transovarially transmitted by hoppers to less than 100% of progeny, plants have become integral to their perpetuation. Insects are probably the better-adapted and older hosts for the plant-infecting reoviruses because these viruses have fewer pathogenic effects on the hopper vectors (133) than on the host plants. Apparently, infections of reoviruses in their vectors are persistent rather than acute. For example, in vector cell lines WTV produces noncytopathic infections that are characterized by a brief acute phase followed by a persistent phase in subsequent cell passages (120). Although WTV accumulates in several of its vector organs and induces some cytopathic effects, the life span and fecundity of infected leafhoppers are unaffected (95, 128).

Similarly, the simplest evolutionary scenario is that hopper-borne rhabdoviruses and tenuiviruses originated in insects rather than in plants. These viruses are not seed-borne in plants, are for the most part transovarially transmitted, and usually do not cause disease in their hopper vectors. Early reports that EWSMV is pathogenic to its vector *J. pellucida* can be explained by the effects of inbreeding on the vector (133). Ammar (3, 4) found that EWSMV has no effect on the longevity or reproduction of its vector when virus is acquired from plants and is only slightly pathogenic when virus is acquired transovarially. Another tenuivirus, MStpV, and a rhabdovirus, MMV, apparently have no effect on adult longevity in their common planthopper vector *P. maidis* (E. D. Ammar & L. R. Nault, unpublished). Aggregates of MMV particles found in *P. maidis* cells (14) were smaller than the massive, often crystalline aggregates found in plant cells (102). Thus, the vector regulates the rate of virus replication better than the plant, which in turn could explain the pathogenicity found in plants and not in vectors.

The tenuiviruses and rhabdoviruses may share a common ancestor. Maize yellow stripe virus, which shares several properties with tenuiviruses (E. D. Ammar & R. E. Gingery, unpublished), is associated with helically wound filaments (8) that resemble the helical nucleoprotein core of degraded rhabdovirus particles (80). A relationship between the two virus groups is further suggested by the interference of a rhabdovirus with the infection of plants and vectors by a tenuivirus. The viruses MMV and MStpV interfere with each other in plants, and MMV strongly interferes with the transmission of MStpV by *P. maidis* (11). There are no other substantiated reports of interference between presumed unrelated viruses in their vectors (138).

The origin of the marafiviruses is more difficult to propose, as they are neither seed borne nor transovarially passed. The lack of pathogenic effects of MRFVs on their vectors (50), as compared with the effects on plants, suggests an insect origin. Nevertheless, Gamez & León (49, 50) favor a plant origin for MRFV and maintain that the virus secondarily adapted to an insect vector (and host) as a mechanism of viral perennation from one maize crop to another.

A plant origin for the geminiviruses is easier to support than an insect origin, although like the marafiviruses, the geminiviruses depend on vectors for survival because they are not seed borne. The geminiviruses may have originated from seed-borne ancestors that later evolved a more efficient strategy for transmission by insects. To have had an insect origin, geminiviruses would have had to begin as transovarially transmitted, propagative viruses, then to become adapted to plants, first by using plants as reservoirs and later by using plants as propagative hosts. Then geminiviruses would have had to lose their original propagative capabilities in insects.

It is even more difficult to support an insect origin for the MCDV group; no vestige indicates that these viruses originated from insect viruses.

The antiquity of hopper-borne viruses may never be known, because unlike their vectors, plant viruses have not left behind a fossil record. Nevertheless, we speculate that the plant-infecting reoviruses date back to the divergence of the major taxonomic groups of the Homoptera approximately 180 million years ago (68). In the ancestral state the Reoviridae probably had 10 ds-RNA genome segments (44); the state with 12 segments appears to be apomorphic, having evolved in *Phytoreovirus* and also apparently in PMV. Similarly, the absence of spikes on capsids, the most common condition, is probably ancestral, and the spikes in *Fijivirus* and the insect cytoplasmic polyhedrosis virus are probably derived characters. The different traits in *Phytoreovirus* and *Fijivirus* may have evolved after the divergence of the Cicadellidae and Delphacidae and therefore remained isolated and distinct. The same cannot be said for the Rhabdoviridae. No consistent associations can be made for rhabdovirus traits and vector families. As new traits evolved, so too must have the adoption of new vector groups (107). This concept is consistent with the colonization hypothesis of Mitter & Brooks (104), which explains how related parasites occupy different hosts.

## CONCLUDING REMARKS

Our understanding of Auchenorrhyncha-transmitted plant viruses currently is based primarily on those that infect the Gramineae and are vectored by cicadellid leafhoppers and delphacid planthoppers (Table 1). This sample of hopper-borne viruses is biased by the preponderance of research done in temperate climates and by a common focus on the world's principal grain crops. The bias is understandable given that hopper-borne viruses are a major constraint in or potential threat to the production of rice in Asia, of wheat and other small grains in Europe, and of maize in Africa, Latin America, and the United States. Virtually no attention has been paid to the potential of tropical treehoppers and nondelphacid planthoppers as vectors. These phloem-feeding groups are almost certainly vectors of a number of viruses (and MLOs) that infect tree fruits and herbaceous crops. Although we expect these undiscovered viruses to be added to the lists of the six plant virus groups covered in this review, some viruses may serve as type members for new groups.

Technical advances in electron microscopy, serology, and cell culture contributed to rapid advances in the understanding of hopper-borne viruses and their transmission in the 1950s and 1960s. Paradoxically, the resolution of the etiology of the yellows diseases in 1967 diverted the attention of many hopper-vector researchers from viruses to the mollicutes. Consequently, the pace of research on transmission of hopper-borne viruses in the past 20 years has lagged behind that of studies on transmission of other plant and animal viruses. A recent resurgence in the study of hopper-borne viruses and their transmission is highlighted in this review. The new technologies, particularly the more sensitive serological methods and molecular probes, the methods of sequencing virus genomes, the techniques for studying gene products, and improved methods for insect and plant cell cultures will permit studies comparing replication strategies of propagative viruses in insects with those in plants. Much work is still needed to further our understanding of the mechanisms involved in vector specificity of foregut-borne, circulative, and propagative viruses in their hopper vectors.

ACKNOWLEDGMENTS

We thank G. Adam, C.-C. Chen, and B. W. Falk for sending us manuscripts prior to publication, and F. E. Gildow, R. E. Gingery, D. T. Gordon, J. K.

Knoke, A. H. Purcell, and J. L. Todd for reviewing the manuscript. E. D. Ammar is on leave from the Department of Economic Entomology, Faculty of Agriculture, Cairo University, Egypt. Salaries and research support were provided by state and federal funds appropriated to the Ohio State University–Ohio Agricultural Research and Development Center. This is journal article No. 100–88.

## Literature Cited

1. Adam, G. 1984. Plant virus studies in insect cell cultures. In *Vectors in Virus Biology*, ed. M. A. Mayo, K. A. Harrap, pp. 37–62. New York: Academic
2. Adam, G., Hsu, H. T. 1984. Comparison of structural proteins from two potato yellow dwarf viruses. *J. Gen. Virol.* 65:991–94
3. Ammar, E. D. 1975. Effect of European wheat striate mosaic, acquired by feeding on diseased plants, on the biology of its planthopper vector *Javesella pellucida. Ann. Appl. Biol.* 79:195–202
4. Ammar, E. D. 1975. Effect of European wheat striate mosaic, acquired transovarially, on the biology of its planthopper vector *Javesella pellucida. Ann. Appl. Biol.* 79:203–13
5. Ammar, E. D. 1985. Internal morphology and ultrastructure of leafhoppers and planthoppers. See Ref. 111, pp. 127–62
6. Ammar, E. D. 1986. Ultrastructure of the salivary glands of the planthopper *Peregrinus maidis* (Hom., Delphacidae). *Int. J. Insect Morphol. Embryol.* 15:417–28
7. Ammar, E. D. 1987. Ultrastructural studies on the planthopper, *Peregrinus maidis* (Ashmead), vector of maize mosaic and maize stripe viruses. In *Proc. 2nd Int. Workshop Leafhoppers Planthoppers Econo. Import., Provo, Utah*, ed. M. R. Wilson, L. R. Nault, pp. 83–92. London: Commonw. Inst. Entomol.
8. Ammar, E. D. 1987. Fine, helical, filamentous structures associated with maize yellow stripe, a leafhopper-borne disease agent from Egypt. *Phytopathology* 77:1732 (Abstr.)
9. Ammar, E. D., Abul-Ata, A. E., El-Sheikh, M. A.. Sewify, G. H. 1987. Incidence of virus and viruslike disease syndromes on maize and sugarcane in Middle and Lower Egypt. *Egypt. J. Phytopathol.* 19:97–107
10. Ammar, E. D., Gingery, R. E., Nault, L. R. 1985. Two types of inclusions in maize infected with maize stripe virus. *Phytopathology* 75:84–89
11. Ammar, E. D., Gingery, R. E., Nault, L. R. 1987. Interactions between maize mosaic and maize stripe viruses in their insect vector, *Peregrinus maidis*, and in maize. *Phytopathology* 77:1051–56
12. Ammar, E. D., Gordon, D. T. 1988. Ultrastructure, cytopathology and distribution in maize leaves of maize streak and maize chlorotic mottle viruses. *5th Int. Congr. Plant Pathol, Kyoto, Japan*, p. 68 (Abstr.)
13. Ammar, E. D., Gordon, D. T., Nault, L. R. 1987. Ultrastructure of maize chlorotic dwarf virus infected maize and viruliferous leafhopper vectors. *Phytopathology* 77:1743 (Abstr.)
14. Ammar, E. D., Nault, L. R. 1985. Assembly and accumulation sites of maize mosaic virus in its planthopper vector. *Intervirology* 24:33–41
15. Backus, E. A. 1985. Anatomical and sensory mechanisms of planthopper and leafhopper feeding behavior. See Ref. 111, pp. 163–94
16. Banttari, E. E., Zeyen, R. J. 1970. Transmission of oat blue dwarf virus by the aster leafhopper following natural acquisition or inoculation. *Phytopathology* 60:399–402
17. Banttari, E. E., Zeyen, R. J. 1976. Multiplication of oat blue dwarf virus in the aster leafhopper. *Phytopathology* 66: 896–900
18. Berger, P. H., Pirone, T. P. 1986. The effect of helper component on the uptake and localization of potyviruses in *Myzus persicae. Virology* 153:256–61
19. Black, L. M. 1970. *Potato yellow dwarf virus. Descriptions of Plant Viruses, No. 35.* Kew, Surrey, UK: Commonw. Mycol. Inst./Assoc. Appl. Biol. 4 pp.
20. Black, L. M. 1979. Vector cell monolayers and plant viruses. *Adv. Virus Res.* 25:191–270
21. Boccardo, G., Hatta, T., Francki, R. I. B., Grivell, C. J. 1980. Purification and some properties of reovirus-like particles from leafhoppers and their possible involvement in wallaby ear disease in maize. *Virology* 100:300–13

22. Boulton, M. I., Markham, P. G. 1986. The use of squash-blotting to detect plant pathogens in insect vectors. In *Developments and Applications in Virus Testing*, ed. R. A. C. Jones, L. Torrance, pp. 55–69. Wellesbourne, UK: Assoc. Appl. Biol.

23. Bradley, R. H. E. 1964. Aphid transmission of stylet-borne viruses. In *Plant Virology*, ed. M. K. Corbet, H. D. Sisler, pp. 148–74. Gainesville, Fla: Univ. Fla. Press

24. Bradley, R. H. E. 1966. Which of an aphid's stylets carry transmissible virus? *Virology* 29:396–401

25. Cabauatan, P. Q., Hibino, H. 1985. Transmission of rice tungro bacilliform and spherical viruses by *Nephotettix virescens* Distant. *Philipp. Phytopathol.* 21:103–9

25a. Cabauatan, P. Q., Hibino, H. 1988. Isolation, purification, and serology of rice tungro bacilliform and rice tungro spherical viruses. *Plant Dis.* 72:526–28

26. Caciagli, P., Casetta, A. 1986. Maize rough dwarf virus (Reoviridae) in its planthopper vector *Laodelphax striatellus* in relation to vector infectivity. *Ann. Appl. Biol.* 109:337–44

27. Chang, V. C. S. 1977. Transovarial transmission of the Fiji disease virus in *Perkinsiella saccharicida* Kirk. *Sugarcane Pathol. News* 18:22–23

27a. Chen, C.-C., Chen, M. J., Chiu, R.-J., Hsu, H. T. 1988. Morphological comparisons of *Echinocloa* ragged stunt and rice ragged stunt viruses by electron microscopy. *Phytopathology* In press

28. Chen, C.-C., Chiu, R.-J., 1980. Factors affecting transmission of rice transitory yellowing virus by green leafhoppers. *Plant Prot. Bull. Taiwan* 22:297–306

29. Chen, M. J., Shikata, E. 1971. Morphology and intracellular localization of rice transitory yellowing virus. *Virology* 46:786–96

30. Chen, M. J., Shikata, E. 1972. Electron microscopy and recovery of rice transitory yellowing virus from its leafhopper vector, *Nephotettix cincticeps*. *Virology* 47:483–86

31. Childress, S. A. 1980. *The fate of maize chlorotic dwarf virus (MCDV) in the black-faced leafhopper, Graminella nigrifrons (Forbes) (Homoptera, Cicadellidae)*. PhD thesis. Texas A & M Univ., College Station

32. Chiu, R.-J., Jean, J.-H., Chen, M.-H., Lo, T.-C. 1968. Transmission of transitory yellowing virus of rice by two leafhoppers. *Phytopathology* 58:740–47

33. Chiykowski, L. N. 1981. Epidemiology of diseases caused by leafhopper-borne pathogens. In *Plant Diseases and Vectors, Ecology and Epidemiology*, ed. K. Maramorosch, K. F. Harris, pp. 106–59. New York: Academic

34. Choudhury, M. M., Rosenkranz, E. 1983. Vector relationships of *Graminella nigrifrons* to maize chlorotic dwarf virus. *Phytopathology* 73:685–90

35. Conti, M. 1984. Epidemiology and vectors of plant reolike viruses. In *Current Topics in Vector Research*, ed. K. F. Harris, 2:111–39. New York: Praeger

36. Conti, M. 1985. Transmission of plant viruses by leafhoppers and planthoppers. See Ref. 111, pp. 289–307

37. D'Arcy, C. J., Barnett, P. A., Hewings, A. D., Goodman, R. M. 1981. Purification and characterization of a virus from the aphid *Rhopalosiphum padi*. *Virology* 112:346–49

38. Davis, R. E., Whitcomb, R. F. 1971. Mycoplasmas, rickettsiae, and chlamydiae: possible relation to yellows diseases and other disorders of plants and insects. *Ann. Rev. Phytopathol.* 9:119–54

39. Falk, B. W., Kim, K. S., Tsai, J. H. 1988. Electron microscopic and physicochemical analysis of a reo-like virus of the planthopper *Peregrinus maidis*. *Intervirology* 29:195–206

40. Falk, B. W., Tsai, J. H. 1985. Serological detection and evidence for multiplication of maize mosaic virus in the planthopper, *Peregrinus maidis*. *Phytopathology* 75:852–55

41. Falk, B. W., Tsai, J. H., Lommel, S. A. 1987. Differences in levels of detection for the maize stripe virus capsid and major non-capsid proteins in plant and insect hosts. *J. Gen. Virol.* 68:1801–11

42. Falk, B. W., Weathers, L. G. 1983. Comparison of potato yellow dwarf virus serotypes. *Phytopathology* 73:81–85

43. Francki, R. I. B., Boccardo, G. 1983. The plant Reoviridae. In *The Reoviridae*, ed. W. K. Joklik, pp. 505–63. New York: Plenum

44. Francki, R. I. B., Milne, R. G., Hatta, T. 1985. *Atlas of Plant Viruses*, Vol. 1, Boca Raton, Fla: CRC

45. Francki, R. I. B., Milne, R. G., Hatta, T. 1985. *Atlas of Plant Viruses*, Vol. 2, Boca Raton, Fla: CRC

46. Fukushi, T. 1969. Relationships between propagative rice viruses and their vectors. In *Viruses, Vectors and Vegetation*, ed. K. Maramorosch, pp. 279–301. New York/London: Wiley

47. Gaedigk-Nitschko, K., Stussi-Garaud,

C., Mundry, K.-W., Adam, G. 1987. Infection pathways of plant rhabdoviruses for insect vector cells. *Int. Congr. Virol. 7th, Edmonton, Canada,* p. 215 (Abstr.)

48. Gaedigk-Nitschko, K., Adam, G., Mundry, K.-W. 1988. Role of the spike protein from potato yellow dwarf virus during infection of vector cell monolayers. In *Invertebrate Cell Systems in Application,* Vol. 1, ed. J. Mitsuhashi. Boca Raton, Fla: CRC. In press

49. Gamez, R., León, P. 1985. Ecology and evolution of a neotropical leafhopper-virus-maize association. See Ref. 111, pp. 331–50

50. Gamez, R., León, P. 1988. Maize rayado fino and related viruses. In *The Plant Viruses,* ed. R. Koenig, 3:213–33. New York: Plenum

51. Ghosh, A., John, V. T., Rao, J. R. K. 1979. Studies on grassy stunt disease of rice in India. *Plant Dis. Rep.* 63:523–25

52. Gildow, F. E. 1987. Virus-membrane interactions involved in circulative transmission of luteoviruses by aphids. In *Current Topics in Vector Research,* ed. K. F. Harris, 4:93–120. New York: Springer-Verlag

53. Gildow, F. E., D'Arcy, C. J. 1988. Barley and oats as reservoirs for an aphid virus and the influence on barley yellow dwarf virus transmission. *Phytopathology* 78:811–16

54. Gingery, R. E. 1988. Maize chlorotic dwarf and related viruses. See Ref. 50, pp. 259–72

55. Gingery, R. E. 1988. The rice stripe virus group. In *The Plant Viruses,* ed. R. G. Milne, 4:297–329. New York: Plenum

56. Gingery, R. E., Gordon, D. T., Nault, L. R. 1982. Purification and properties of maize rayado fino virus from the United States. *Phytopathology* 72:1313–18

57. Goodman, R. M. 1981. Geminiviruses. In *Handbook of Plant Virus Infections and Comparative Diagnosis,* ed. E. Kurstak, pp. 883–910. New York: Elsevier

58. Graham, C. L. 1979. Inability of certain vectors in North America to transmit maize streak virus. *Environ. Entomol.* 8:228–30

59. Granados, R. R., Hirumi, H., Maramorosch, K. 1967. Electron microscopic evidence for wound-tumor virus accumulation in various organs of an inefficient leafhopper vector, *Agalliopsis novella. J. Invertebr. Pathol.* 9:147–59

60. Grylls, N. E. 1975. Leafhopper trans-

mission of a virus causing maize wallaby ear disease. *Ann. Appl. Biol.* 79:283–96

61. Harris, K. F. 1977. An ingestion-egestion hypothesis of noncirculative virus transmission. In *Aphids as Virus Vectors,* ed. K. F. Harris, K. Maramorosch, pp. 166–220. New York: Academic

62. Harris, K. F. 1979. Leafhoppers and aphids as biological vectors: vector-virus relationships. See Ref. 96, pp. 217–308

63. Harris, K. F., Childress, S. A. 1983. Cytology of maize chlorotic dwarf virus infection in corn. *Int. J. Trop. Plant Dis.* 1:135–40

64. Harris, K. F., Treur, B., Tsai, J., Toler, R. 1981. Observations on leafhopper ingestion-egestion behavior: its likely role in the transmission of noncirculative viruses and other plant pathogens. *J. Econ. Entomol.* 74:446–53

65. Harrison, B. D. 1985. Advances in geminivirus research. *Ann. Rev. Phytopathol.* 23:55–82

66. Harrison, B. D., Murant, A. F. 1984. Involvement of virus-coded proteins in transmission of plant viruses by vectors. See Ref. 1, pp. 1–36

67. Hatta, T., Francki, R. I. B. 1982. Similarity in the structure of cytoplasmic polyhedrosis virus, leafhopper A virus and Fiji disease virus particles. *Intervirology* 18:203–8

68. Hennig, W. 1981. *Insect Phylogeny.* New York: Wiley

69. Hibino, H. 1983. Relations of rice tungro bacilliform and rice tungro spherical viruses with their vector *Nephotettix virescens. Ann. Phytopathol. Soc. Jpn.* 49:545–53

70. Hibino, H., Cabauatan, P. Q. 1987. Infectivity neutralization of rice tungro-associated viruses acquired by vector leafhoppers. *Phytopathology* 77:473–76

71. Hibino, H., Roechan, M., Sudarisman, S. 1978. Association of two types of virus particles with penyakit habang (tungro disease) of rice in Indonesia. *Phytopathology* 68:1412–16

72. Hibino, H., Saleh, N., Roechan, M. 1979. Transmission of two kinds of rice tungro-associated viruses by insect vectors. *Phytopathology* 69:1266–68

73. Hibino, H., Usugi, T., Omura, T., Tsuchizaki, T., Shohara, K., et al. 1985. Rice grassy stunt virus, a plant-hopper-borne circular filament. *Phytopathology* 75:894–99

74. Hsu, H. T., McBeath, J. H., Black, L. M. 1977. The comparative susceptibilities of cultured vector and nonvector

leafhopper cells to three plant viruses. *Virology* 81:257–62

75. Hunt, R. E., Nault, L. R., Gingery, R. E. 1988. Evidence for infectivity of maize chlorotic dwarf virus and a helper component in its leafhopper transmission. *Phytopathology* 78:499–504

76. Hutchinson, P. B., Francki, R. I. B. 1973. *Sugarcane Fiji disease virus. Descriptions of Plant Viruses, No. 119.* Kew, Surry, UK: Commonw. Mycol. Inst./Assoc. Appl. Biol. 3 pp.

77. Inoue, H., Hirao, J. 1981. Transmission of rice waika virus by green rice leafhoppers, *Nephotettix* spp. (Hemiptera: Cicadellidae). *Bull. Kyushu Natl. Agric. Exp. Sta.* 21:509–52

78. Inoue, H., Omura, T. 1982. Transmission of rice gall dwarf virus by the green rice leafhopper. *Plant Dis.* 66:57–59

79. Izadpanah, K., Ahmadi, A. A., Parvin, S., Jafari, S. A. 1983. Transmission, particle size and additional hosts of the rhabdovirus causing maize mosaic in Shiraz, Iran. *Phytopathol. Z.* 107:283–88

80. Jackson, A. O., Francki, R. I. B., Zuidema, D. 1987. Biology, structure and replication of plant rhabdoviruses. In *The Rhabdoviruses*, ed. R. R. Wagner, pp. 427–507. New York: Plenum

81. Kennedy, J. S., Day, U. F., Eastop, V. F. 1962. *A Conspectus of Aphids as Vectors of Plant Viruses.* London: Commonw. Inst. Entomol.

82. Kimura, I. 1976. Loss of vector-transmissibility in an isolate of rice dwarf virus. *Ann. Phytopathol. Soc. Jpn.* 42:322–24 (In Japanese with English summary)

83. Kimura, I., Black, L. M. 1972. Growth of wound tumor virus in vector cell monolayers. *Virology* 48:852–54

84. Kisimoto, R. 1973. Leafhoppers and planthoppers. In *Viruses and Invertebrates*, ed. A. J. Gibbs, pp. 137–56. New York: Elsevier

85. Kunkel, L. O. 1922. Insect transmission of yellow stripe disease. *Hawaii. Plant. Rec.* 26:58–64

86. Lastra, R., Carballo, O. 1983. Maize virus diesease problems in Venezuela. *Proc. Int. Maize Virus Dis. Colloq. Workshop, Wooster, Ohio, 1982,* ed. D. T. Gordon, J. K. Knoke, L. R. Nault, R. M. Ritter, pp. 83–86. Wooster, Ohio: Ohio State Univ., Ohio Agric. Res. Dev. Cent.

87. Lee, P. E. 1967. Morphology of wheat striate mosaic virus and its localization in infected cells. *Virology* 33:84–94

88. Deleted in proof

89. Deleted in proof

90. Ling, K. C., Tiongco, E. R. 1979. Transmission of rice tungro at various temperatures: a transitory virus-vector interaction. See Ref. 96, pp. 349–66

91. Liu, H. Y., Kimura, I., Black, L. M. 1973. Specific infectivity of different wound tumor virus isolates. *Virology* 51:320–26

92. Lockhart, B. E. L. 1986. Occurrence of cereal chlorotic mottle virus in Northern Africa. *Plant Dis.* 70:912–15

93. Lockhart, B. E. L., Khaless, N., Lennon, A. M., Maatauoi, M. E. 1985. Properties of Bermuda grass etched-line virus, a new leafhopper-transmitted virus related to maize rayado fino and oat blue dwarf viruses. *Phytopathology* 75:1258–62

94. Maramorosch, K. 1964. Virus-vector relationships: vectors of circulative and propagative viruses. See Ref. 23, pp. 175–93

95. Maramorosch, K. 1975. Infection of arthropod vectors by plant pathogens. In *Invertebrate Immunity*, ed. K. Maramorosch, R. E. Shope, pp. 49–53. New York: Academic

96. Maramorosch, K., Harris, K. F. 1979. *Leafhopper Vectors and Plant Disease Agents.* New York: Academic

97. Markham, P. G. 1983. Spiroplasmas in leafhoppers: a review. *Yale J. Biol. Med.* 56:745–51

98. Mathew, S. K., Basu, A. N. 1986. Relationship of rice grassy stunt virus with its planthopper vector. *Curr. Sci.* 55:1245–47

99. Matthews, R. E. F. 1981. *Plant Virology.* New York: Academic

100. Matthews, R. E. F. 1982. Classification and nomenclature of viruses. *Intervirology* 17:1–199

101. Matthews, R. E. F. 1985. Viral taxonomy for the nonvirologist. *Ann. Rev. Microbiol.* 39:451–74

102. McDaniel, L. L., Ammar, E. D., Gordon, D. T. 1985. Assembly, morphology, and accumulation of a Hawaiian isolate of maize mosaic virus. *Phytopathology* 75:1167–72

103. McLean, D. L., Kinsey, M. G. 1984. The precibarial valve and its role in the feeding behavior of the pea aphid, *Acyrthosiphon pisum. Bull. Entomol. Soc. Am.* 30:26–31

104. Mitter, C., Brooks, D. R. 1983. Phylogenetic aspects of coevolution. In *Coevolution*, ed. D. J. Futuyma, M. Slatkin, pp. 65–98. Sunderland, Mass: Sinauer

105. Morinaka, T., Putta, M., Chettanachit, D., Parejarearn, A., Disthaporn, S., et al. 1982. Transmission of rice gall dwarf virus by cicadellid leafhoppers *Recilia dorsalis* and *Nephotettix nigropictus* in Thailand. *Plant Dis.* 66:703–4

106. Murphy, F., Harrison, A. 1980. Electron microscopy of the rhabdoviruses on animals. In *Rhabdoviruses*, ed. D. H. Bishop, 1:65–107. Boca Raton, Fla: CRC

107. Nault, L. R. 1987. Origin and evolution of Auchenorrhyncha-transmitted, plant-infecting viruses. See Ref. 7, pp. 131–49

108. Nault, L. R., Gingery, R. E., Gordon, D. T. 1980. Leafhopper transmission and host range of maize rayado fino virus. *Phytopathology* 70:709–12

109. Nault, L. R., Gordon, D. T. 1988. Multiplication of maize stripe virus in *Peregrinus maidis*. *Phytopathology* 78:991–95

110. Nault, L. R., Madden, L. V. 1988. Phylogenetic relatedness of maize chlorotic dwarf virus leafhopper vectors. *Phytopathology* 78:In press

111. Nault, L. R., Rodriguez, J. G. 1985. *The Leafhoppers and Planthoppers*. New York: Wiley

112. Nault, L. R., Styer, W. E., Knoke, J. K., Pitre, H. N. 1973. Semipersistent transmission of leafhopper-borne maize chlorotic dwarf virus. *J. Econ. Entomol.* 66:1271–73

113. Nielson, M. W. 1979. Taxonomic relationships of leafhopper vectors of plant pathogens. See Ref. 96, pp. 2–37

114. Nielson, M. W. 1985. Leafhopper systematics. See Ref. 111, pp. 11–39

115. Nuss, D. L. 1984. Molecular biology of wound tumor virus. *Adv. Virus Res.* 29:57–93

116. O'Brien, L. B., Wilson, S. W. 1985. Planthopper systematics and external morphology. See Ref. 111, pp. 61–102

117. Ofori, F. A., Francki, R. I. B. 1983. Evidence that maize wallaby ear disease is caused by an insect toxin. *Ann. Appl. Biol.* 103:185–89

118. Ofori, F. A., Francki, R. I. B. 1985. Transmission of leafhopper A virus, vertically through eggs and horizontally through maize in which it does not multiply. *Virology* 144:152–57

119. Omura, T., Kimura, I., Tsuchizaki, T., Saito, Y. 1988. Infection by rice gall dwarf virus of cultured monolayers of leafhopper cells. *J. Gen. Virol.* 69:429–32

120. Peterson, A. J., Nuss, D. L. 1986. Regulation of expression of the wound tumor virus genome in persistently infected vector cells is related to change in translational activity of viral transcripts. *J. Virol.* 59:195–202

121. Pirone, T. P. 1977. Accessory factors in nonpersistent virus transmission. See Ref. 61, pp. 221–35

122. Purcell, A. H. 1979. Leafhopper vectors of xylem-borne plant pathogens. See Ref. 96, pp. 603–25

123. Purcell, A. H. 1982. Evolution of the insect vector relationship. In *Phytopathogenic Prokaryotes* 1:121–56, ed. M. S. Mount, G. H. Lacey. New York: Academic

124. Rathgeber, J., Adam, G., Mundry, K.-W. 1988. Protein synthesis in vector cell monolayers after infection with potato yellow dwarf virus. In *Vertebrate Cell Systems in Application*, Vol. 1, ed. S. Mitsuhashi. Boca Raton, Fla: CRC. In press

125. Reddy, D. V. R., Black, L. M. 1966. Production of wound-tumor virus and wound-tumor soluble antigen in the insect vector. *Virology* 30:551–61

126. Reddy, D. V. R., Black, L. M. 1974. Deletion mutations of the genome segments of wound tumor virus. *Virology* 61:458–73

127. Rose, D. J. W. 1978. Epidemiology of maize streak disease. *Ann. Rev. Entomol.* 23:259–82

127a. Shikata, E. 1979. Cytopathological changes in leafhopper vectors of plant viruses. See Ref. 96, pp. 309–25

128. Shikata, E. 1979. Rice viruses and MLO's and leafhopper vectors. See Ref. 96, pp. 515–27

129. Shikata, E., Maramorosch, K. 1965. Electron microscopic evidence for the systemic invasion of an insect by a plant pathogenic virus. *Virology* 27:461–75

130. Shikata, E., Maramorosch, K. 1969. Electron microscopy of insect-borne viruses *in situ*. See Ref. 46, pp. 393–415

131. Simons, J. N., Coe, D. M. 1958. Transmission of psuedo-curly top virus in Florida by a treehopper. *Virology* 6:43–45

132. Sinha, R. C. 1967. Response of wound tumor virus infection in insects to vector age and temperature. *Virology* 31:746–48

133. Sinha, R. C. 1973. Viruses and leafhoppers. See Ref. 84, pp. 493–511

134. Slykhuis, J. T. 1963. Vector and host relations of North American wheat striate mosaic virus. *Can. J. Bot.* 41:1171–85

135. Stanley, J., Markham, P. G., Callis, R. J., Pinner, M. S. 1986. The nucleotide sequence of an infectious clone of the

geminivirus beet curly top virus. *Eur. Mol. Biol. Organ. J.* 5:1761–68

136. Sylvester, E. S. 1958. Aphid transmission of plant viruses. *Proc. 10th Int. Congr. Entomol.* 3:195–204

137. Sylvester, E. S. 1980. Circulative and propagative virus transmission by aphids. *Ann. Rev. Entomol.* 25:257–86

138. Sylvester, E. S. 1985. Multiple acquisition of viruses and vector-dependent prokaryotes: consequences on transmission. *Ann. Rev. Entomol.* 30:71–88

139. Thornbury, D. W., Hellmann, G. M., Rhoads, R. E., Pirone, T. P. 1985. Purification and characterization of potyvirus helper component. *Virology* 144:260–67

140. Tonkyn, D. W., Whitcomb, R. F. 1987. Feeding strategies and the guild concept among vascular feeding insects and microorganisms. See Ref. 52, pp. 179–99

141. Toriyama, S. 1983. *Rice stripe virus. Descriptions of Plant Viruses, No. 269.* Kew, Surrey, UK: Commonw. Mycol. Inst./Assoc. Appl. Biol. 5 pp.

142. Tsai, J. H., Zitter, T. A. 1982. Characteristics of maize stripe virus transmission by the corn delphacid. *J. Econ. Entomol.* 75:397–400

143. Watson, M. A., Roberts, F. M. 1939. A comparative study of the transmission of *Hyocyamus* virus 3, potato virus Y and cucumber virus 1 by the vectors *Myzus persicae* (Sulz), *M. circumflexus* (Buckton), and *Macrosiphum gei* (Koch). *Proc. R. Soc. London Ser. B* 127:543–76

144. Whitcomb, R. F. 1972. Transmission of viruses and mycoplasmas by auchenorrhynchous Homoptera. In *Principles and Techniques in Plant Virology,* ed. C. I. Kado, H. O. Agrawal, pp. 168–203. New York: Van Nostrand Reinhold

145. Whitcomb, R. F. 1981. The biology of spiroplasmas. *Ann. Rev. Entomol.* 26:397–425

146. Whitcomb, R. F., Black, L. M. 1969. Demonstration of exvectorial wound-tumor virus. *Ann. Rev. Phytopathol.* 7:86–87

147. Wiese, M. V. 1977. *Compendium of Wheat Diseases.* St. Paul, Minn: Am. Phytopathol. Soc.

148. Yamashita, S., Doi, Y., Yora, K. 1977. Some properties and intracellular appearance of rice waika virus. *Ann. Phytopathol. Soc. Jpn.* 43:278–90

149. Zarbl, H., Millward, S. 1983. The reovirus multiplication cycle. In *The Reoviridae,* ed. W. K. Joklik, pp. 107–90. New York: Plenum

*Ann. Rev. Entomol. 1989. 34:531–64*

# INSECT HERBIVORES AND PLANT POPULATION DYNAMICS

*Michael J. Crawley*

Department of Pure and Applied Biology, Imperial College, Silwood Park, Ascot, Berkshire, England, SL5 7PY

## PERSPECTIVES AND OVERVIEW

### Distinguishing Performance and Dynamics

It is one thing to show that herbivorous insects affect plant performance. It is an entirely different matter to demonstrate that insect herbivory affects plant population dynamics. There is a vast literature on insect pests of crop plants that shows how attack by defoliating, sucking, stem-mining, and gall-forming species can delay seed ripening, reduce seed production and individual seed weights, reduce the rates of shoot and root growth, increase the susceptibility of plants to disease, and reduce the competitive ability of plants relative to their unattacked neighbors. This literature tells us virtually nothing, however, about the importance of insects in natural communities, chiefly because we have so little information on the regulation of plant populations in the wild. For example, we do not know whether plant recruitment is seed limited, so we cannot predict whether there would be an increase in plant numbers if the simple experiment of sowing extra seeds were carried out. Information on the nature of population regulation is vital because if plant recruitment is not seed limited, then insects that reduce seed production will not have an important effect on plant population dynamics.

This review concentrates on material published from 1984 to 1987. Earlier material has been covered in other reviews (36, 64, 70, 94, 107, 121, 182, 206). The first part examines the impact of insect feeding on different aspects of plant performance. The second part examines two of the best sources of evidence on the impact of insect feeding on plant population dynamics: (*a*) the release of specialist insect herbivores against target weed species in classical

531

0066-4170/89/0101–0531$02.00

biological control projects and (b) the exclusion of insect herbivores from natural plant communities by the repeated application of chemical insecticides. The limitations and shortcomings of these bodies of evidence are also discussed.

## Background

Until recently, conventional wisdom suggested that because the world was green, it was not possible that insect herbivores could be food limited (82, 167). The dramatic outbreaks of herbivorous insects that sometimes followed the indiscriminate use of broad-spectrum insecticides (51) lent credibility to the view that populations of insect herbivores are regulated at low densities by the actions of various natural enemies, notably predators, parasitoids, and diseases (87). At these low densities, the argument went, it was unlikely that insects could have an important effect on plant population dynamics.

The flaws in these arguments are numerous and have been discussed elsewhere (36, 62, 162). They can be summarized in three themes: (a) The world is not always green. (b) All that is green is not edible. (c) What is edible is not necessarily of sufficiently high quality to allow increase of the herbivore population. The evidence from natural populations does not provide convincing evidence in favor of the widespread importance of regulation by natural enemies (107) except in a handful of insect groups (e.g. leafminers, gall-formers, and some other small, sedentary species). Even in these few groups, however, direct experimental demonstration of population regulation by natural enemies by means of enemy exclusion has rarely been attempted.

It is possible, of course, that the assertion that insects have little impact on plant dynamics is correct, but for the wrong reasons. For example, recent work has shown that many herbivorous insects remain at low densities because of limited access to high-quality food (1, 3, 31, 83, 147, 150, 163, 175, 190, 211a, 212). They are food limited, but by the quality rather than the quantity of host plants available. Also, food quality and natural-enemy impact may be connected in that herbivore populations feeding on low-quality host plants will tend to have sufficiently low rates of increase that their populations could, at least in principle, be regulated by predators and parasites (124). A sudden rise in food quality might therefore trigger a population outbreak by inflating the rate of increase beyond the point at which the herbivore escapes natural-enemy control (174).

In any event, if insect herbivores are regulated by restricted access to high-quality food, it is not likely that they will be sufficiently abundant to have a profound effect on plant performance. Thus, insects may be food limited without imposing a substantial resource drain on their host plants, and quality-limited herbivores are not likely to be influential in plant population dynamics. While, in theory, low food quality might be expected to lead to higher rather than lower rates of damage (as the insects attempt to compensate for low nutrient concentrations by feeding at a higher rate), in practice it appears that their capacity to compensate in this way is extremely limited (158).

Sometimes insect populations are limited by the amount of plant food available (54, 114, 202) and yet have little or no impact on the population density of their host plants. This asymmetry may come about because the growing season of the plant exceeds the feeding period of the insect, so that postdefoliation regrowth is possible (35, 86, 211). Alternatively, the plant might have a reserve of material that is physically inaccessible to the herbivore, and regrowth might occur from these reserves following defoliation. In yet other cases, the seed bank in the soil might be so large that even dramatic reductions in seed production would not lead to reduced seedling recruitment (41, 96, 186, 209).

The current view is that natural enemies probably maintain some herbivorous insects at population densities so low that their effect on plant dynamics would be immeasurably small (87). A further group of herbivorous insects (including many phloem feeders) are regulated by low food quality; their population densities are too low to have substantial effects on plant recruitment in most years (175). Low food quality and natural-enemy attack may interact with one another (124), so that a population may be enemy regulated in years when food quality is low, but may break free of enemy control in years of higher food quality. Other herbivorous insects are limited by the amount of food available to them and regularly defoliate their food plants (54). This does not mean that the insects necessarily have any important effect on plant dynamics, and the asymmetry observed in many field examples is such that plant recruitment is not herbivore limited even though herbivore populations are food limited (44, 47). A few herbivorous insects are known to have important effects on the population dynamics of their host plants [bark beetles (12), spruce budworm (14), beech bark insects (210a), and gypsy moth (201)]. Other herbivorous species, though not significant on their own, may have incremental effects that, along with effects of other insects, have substantial impact on plant dynamics (84, 225). Overall, however, it appears that plants have more impact on the population dynamics of insects than insects have on the population dynamics of plants.

## EFFECTS OF INSECT HERBIVORY ON PLANT PERFORMANCE

### Flowering

Insect feeding affects flower production both directly, by the destruction of flowers and flower buds, and indirectly, through various kinds of damage that reduce bud production or bud burst. Flower and bud feeding are widespread among the main groups of herbivorous insects and are prevalent in all higher plant groups. Among coniferous trees, for example, a 40% female flower loss between May and August was attributed to insect feeding (mainly spruce budworm), with 22% of the surviving fruit crop showing signs of feeding damage (57). Cone-feeding insect populations often respond numerically to cone abundance, so attack in one year is reasonably predicted from a knowledge of the size of the cone crop in the previous year, which suggests that these insects are food limited. Whether or not their feeding influences the rate of flowering in future years has yet to be established (61, 105).

In angiosperms, flower loss attributable to insect attack ranges from less than 10% to almost 100%. For example, 80% of the bushes of the composite shrub *Gutierrezia microcephala* that were heavily attacked by the grasshopper *Hesperotettix viridis* failed to produce any flowers at all (171). Galling by *Euura lasiolepis* caused the loss of 43% of the reproductive buds in *Salix lasiolepis* (191). Grass-feeding aphids can cause greatly reduced flowering; *Holcaphis holci* feeding on tillers of the rhizomatous grass *Holcus mollis* can reduce the probability that a floral meristem is produced, and, at high aphid infestations, can preclude flowering altogether (J. Packham, personal communication). Feeding on buds and flowers by mirid bugs can cause heavy loss of fruit production (159). Mirids feeding on the flower stalks of grasses can cause silver top in which an entire panicle of unfilled seeds is produced as a result of blockage of the phloem (213). Flower consumption on the primary umbels of *Heracleum lanatum* by the moth *Depressaria pastinacella* leads to increased production of seeds from secondary umbels and to a change from male to female function in flowers on tertiary umbels (93).

Attack by the agamic generation of the cynipid gall wasp *Andricus quercuscalicis* caused 100% female flower mortality on certain trees of *Quercus robur*, with average rates of loss ranging between 26 and 47% over a 6-yr period (37). The sexual generation attacked the male flowers of a second host tree, *Quercus cerris*, and destroyed between 2 and 8% of its pollen production. This insect appears to be limited by the number of acorns available to the agamic generation on *Q. robur* and to have little impact on male flower function in *Q. cerris* (80).

Indirect effects of flower feeding on the rate of pollination, caused by reduced pollinator attraction, have received little attention. Given the balance of evidence that fruit set tends to be resource limited rather than pollination limited, reduced pollination may have little impact on plant population dynamics (52, 81, 101, 103, 178, 203, 205). It may, however, restrict the pollen choice of the female parent, and hence the genetic makeup of the progeny (145).

Plant size can affect the vulnerability of the plant to reduction of seed production by flower-feeding insects. For example, the host-specific moth

*Heliodines nyctaginella* causes a disproportionately great flower loss of small *Mirabilis hirsuta* plants because large plants flower early and thereby escape attack by the peak insect population (116). In other species, larger plants may flower longer than small ones, and may thus suffer a lower rate of flower loss to insects (200).

The factors determining the rate of flower loss in any given year, and the degree to which losses in one year influence the abundance of the insects and hence flower losses in subsequent years, have received little attention (152). Where density dependence in flower destruction has been investigated, either it has not been found (103) or flower attack rates have been found to be inversely density dependent (47, 56, 80). These results are consistent with the hypothesis that a fixed (or independently determined) number of insects attacks a variable number of flowers (where the number of flowers fluctuates in response to independent, external factors). Predator satiation therefore occurs in years of high flower production, but high rates of loss occur when low numbers of flowers happen to coincide with high insect densities.

## Fruit Production

Feeding on ripening fruits and on seeds prior to dispersal accounts for massive losses in reproductive potential in many plant species. In others, the fruits are so well protected against insect attack (by a thick armor plating or an array of chemical toxins) that predispersal seed predation is minimal. Among conifers, fruit losses to insects of about 20% are commonplace (57, 61, 74, 142, 151, 198), but these may be substantially lower than losses attributable to mammals and birds (42, 199).

Angiosperms also lose many of their undispersed seeds to insect herbivores. Sucking insects such as aphids can greatly reduce the rate of fruit filling and can sometimes curtail fruit production altogether. Sheppard (196) sprayed the umbel rays of *Heracleum sphondylium* to exclude hogweed aphid, *Carvariella pastinacae*, and found that seed production and seed size both increased substantially. On the most heavily infested umbels there was no seed production at all. The aphid *Staticobium staticis* prevented seed production by sea lavender in those parts of a salt marsh where aphid densities were highest (68). Dramatic losses of pecan nut yield were attributable to leaf-feeding aphids (228). Attack on ripening fruits by mirid buds causes a more or less linear decline in seed production as insect numbers increase (159). In a long-term experiment on the broom *Cytisus scoparius*, Waloff & Richards (215) excluded herbivorous insects over an 11-yr period. Fruit production of the sprayed bushes increased an average of 2.5 fold chiefly as a result of the exclusion of the aphid *Acyrthosiphon pisum spartii* and two psyllids of the genus *Arytaina*.

A considerable amount of work has been carried out on the guild of herbivorous insects that attack the flowerheads (capitula) of thistles and knapweeds. These plants are natives of southern Europe, and a number of them have become serious weeds when introduced into continental or Mediterranean climates elsewhere. Research has been aimed at discovering potential agents for biological weed control. There have been some notable successes, such as the introduction of the weevil *Rhinocyllus conicus* into Canada from France in 1968 to control the nodding thistle, *Carduus nutans* (231). Note that while the insect was able to control *C. nutans* in Canada, where the plant is alien, the same insect does not reduce the recruitment of the thistle in its native southern France, despite its inflicting over 90% seed losses. In addition, after sowing 1000 thistle seeds per $m^2$ no increase was observed in the number of thistle rosettes. Recruitment in this case appears not to be seed limited, though whether it is microsite limited or regulated by postdispersal seed predation or early seedling mortality is not yet clear (A. Sheppard, personal communication).

Chewing caterpillars can take a heavy toll of the capitula of some Compositae. Perhaps the best known example is the cinnabar moth, *Tyria jacobaeae*, which frequently strips all of the flower heads from ragwort plants on which egg batches are laid. Late-instar larvae, dispersing from defoliated plants, can inflict up to 90% loss of flower heads on nearby plants that escaped the attentions of ovipositing females. As with *C. nutans*, however, sowing 1000 seeds per $m^2$ did not increase the rate of recruitment of ragwort rosettes (46). Even though the insects cause large reductions in fruit survival, they do not appear to reduce the population density of the plants, at least in mesic grasslands (44) or the Breckland heaths of East Anglia (53). These dynamics contrast with those of the same plant species in more arid, open habitats on coastal sand dunes, where recruitment of new rosettes is reduced following years of heavy defoliation by the cinnabar moth (211).

In contrast to the two studies of Compositae described above, sowing 1000 hogweed seeds per $m^2$ did lead to increased recruitment. Thus ray-feeding aphids may well have reduced the number of hogweed rosettes that became established in the following spring. Plots sown with extra seed in 1982 (196) were still supporting conspicuously higher plant densities in 1988 (M. J. Crawley, unpublished results).

Other effects of insect herbivory on plant reproduction include reductions in the numbers of seeds per fruit (118) and delays in the timing of fruit maturation (140, 210). Ant-attended plants lost 33% of their fruits to insect herbivores, with the loss rising to 66% when the ants were experimentally excluded (100).

Insect herbivory frequently leads to reduced seed size (46, 93, 208), with potential effects on seedling competitiveness (see below). Insect herbivores may also influence the evolution of seed size, both by exhibiting feeding preferences for seeds of particular sizes (143, 164) and by inflicting a higher death rate on some sizes of seeds than on others. For example, the acorn weevil *Curculio glandium* kills a significantly greater proportion of small acorns than of larger acorns from the same tree (67).

Some insects cause such inconspicuous damage that assessment and attribution of seed losses is remarkably difficult. Heteroptera such as *Lygus* spp. can cause seed losses of more than 50% in umbelliferous vegetable crops such as carrot, celery, dill, fennel, parsley, and parsnip by killing the embryo without disturbing other parts of the seed (65).

Like flower losses, fruit losses appear more often to be inversely density dependent than directly density dependent. Thus, larger fruit crops suffer a lower loss rate than small ones (16, 152, 180, 200), presumably because insect abundance and fruit density are not closely coupled or are coupled with a substantial time lag. Inverse density dependence even occurs within multi-seeded fruits. The probability of death of an individual seed declines as the number of seeds per fruit increases (95) even though insects may show a preference for larger fruits (118, 141). Insect feeding may also increase the rate of abortion of partially filled fruits, though whether this causes a net reduction in the total seed crop is not clear (117, 203, 205).

Insect attack on fruits can reduce the attractiveness of fruits to seed dispersers (34, 138). As with so many other aspects of plant performance, however, no link has yet been established between reduced dispersal efficiency and reduced plant recruitment.

## Postdispersal Seed Mortality

Detailed field studies on the fate of seeds have been published for umbellifers (93, 196, 208), oaks (67, 80, 148), thistles (210b, 211b, 231), crucifers (59, 134), leguminous trees (6, 106), desert annuals (48–50), and many other plants. Attention has been paid to the selection of seeds of different sizes (20, 221) and to the growth and competitive ability of seedlings from seeds of different sizes (28, 46, 143).

Of the insect groups, most attention has been paid to beetles and ants (chiefly the desert harvester ants) (20, 27, 48, 137, 160). These studies are notable for the fact that some attempt has been made to demonstrate the effects of seed removal on plant recruitment (2, 9, 27, 49, 97, 119, 155, 165, 178, 184, 229). Where exclusion experiments have been carried out in the field, it has typically been found that seed predation by vertebrates is more important than insect granivory in affecting plant population dynamics (42, 49, 50). In other cases, despite predator satiation, plant recruitment is limited by rare, pulsed conditions (often of rainfall) (219). The view that dispersal of seeds by ants can target the seeds to locally improved soil conditions around ant nests (185), although an attractive hypothesis, has received little support from critical field studies (9, 183).

Once buried, seeds become exceptionally difficult to study, and virtually no work has been carried out to determine the loss of seeds that can be attributed to soil-dwelling insect herbivores (186).

## Seedling Mortality

Molluscan and vertebrate herbivores are traditionally regarded as the major seedling predators (36, 108), but several insect groups are important seedling feeders. These insects either kill the seedlings outright or seriously reduce their growth and competitive ability (153). Sawflies are pests of seedlings in gardens, but their role in natural vegetation is not clear. Mirid and pentatomid bugs can damage seedlings (76). Psyllids reduced the growth rate of *Ceanothus* seedlings in California, impairing their competitiveness with *Adenostoma* seedlings. In this case, exclusion of small mammals had a greater effect on seedling performance than insect exclusion (154). Frit flies *(Oscinella spp.)* are severe pests of spring-sown cereals and short-term grass leys, where the young larvae attack seedlings before the four-leaf stage. Attack later in the season, once the plant has tillered, is rarely lethal (108, 161). Repeated defoliation of seedlings of *Q. robur* by lepidopteran caterpillars spinning down from mature trees overhead may be one of the main factors preventing recruitment of oaks beneath the canopy of parent trees (195). However, vertebrate herbivores such as sheep and rabbits are usually more likely to prevent regeneration of oaks from seed (41, 173, 181).

Failure of buried seed to produce an emergent seedling is usually attributed to germination failure, simply as a matter of convenience. How many of these losses result from attack by soil insects on very young seedlings, prior to their emergence above soil level, is unknown. Large-scale, frequent, and destructive sampling would be required to answer this question.

## Defoliation

The rates of defoliation attributable to insect herbivores are traditionally considered to lie within the range of 5–10% of leaf area per year (15, 37, 79, 139). Average rates of defoliation may be higher in certain Australian eucalyptus forests (average about 15%), and within Australia defoliation rates increase with altitude (73). Critics have argued, however, that these percentages represent a serious underestimate of defoliation rates (135, 166) because they fail to account for leaf turnover during the growing season (i.e. for the appearance and disappearance of leaves between one sampling period and the next). It is certainly true that far too many estimates of defoliation have been based on a single sample of leaves, taken toward the end of the growing period once insect feeding was assumed to have ceased. Such studies have provided no estimates of the numbers of leaves that were consumed completely, leaving no trace, or of the effect of partial damage on the rate of premature abscission of leaves. In a detailed study of herbivory and leaf turnover in water lilies, for example, it was discovered that feeding by adult and larval *Pyrrhalta* beetles reduced average leaf longevity from 45 to 17 days. This degree of leaf turnover would have led to severe underestimation of primary production had a simple method such as the use of peak leaf standing crop been employed (214).

Studies involving careful, frequent measurements on large samples of individually marked leaves have been rare. In work on cohorts of individually marked birch leaves *(Betula pendula)*, Zakaria (230) measured every leaf every day and found that 11% of leaves produced from April to mid July fell prematurely or were consumed entirely, so that no trace of their existence remained by the date that a single damage sample would have been taken. Between mid July and August a further 6% were lost. Premature losses of leaves varied from tree to tree between 0 and 35%. Of the total annual leaf production, 13–35% of leaves were produced after the first spring flush was complete and would not have been detected if a single cohort of leaves had been tagged in spring.

Other studies have discovered even higher rates of within-season leaf loss, and it is clear that considerable care must be exercised to ensure that leaf turnover rates are known. For example, a number of studies on evergreen and semievergreen trees have shown that attack by herbivorous insects (notably by leafminers) can cause a substantial rise in the rate of premature leaf fall (24, 111, 176, 224). Attack by leaf-mining weevils causing premature leaf fall over successive years may eventually kill the host tree (102).

Refoliation following leaf loss to herbivorous insects is commonplace, especially in woody angiosperms (90, 91). It is important, therefore, in studies of leaf loss that these regrowth leaves be distinguished from primary leaves. Not only might failure to distinguish the classes of leaves lead to underestimates of leaf consumption, but since insect performance is often different on diets of regrowth and primary foliage (45, 88, 223), it might also lead to confusion over the cause of differences between the survival and fecundity of early and late insect cohorts.

## Growth and Reproduction

Plant growth rate, plant shape, and the allocation of resources to seed production are all influenced to a greater or lesser extent by insect herbivory. Even relatively low levels of insect herbivory can, if sustained, have a significant effect on the growth of young trees [e.g. *Acer pseudoplatanus* (218)]. Chronic levels of insect herbivory modified both plant shape and sex expression in pinon pine, *Pinus edulis* (222). In some cases, insect herbivores reduce growth without causing any reduction in sexual reproduction. For example, Karban (113) found that the cicada *Magicicada septendecim* caused a 30% reduction in ring width in *Quercus ilicifolia* oak wood without causing any reduction in acorn production. In other work on *Q. robur* (37), however, exclusion of insects using insecticides had no effect on ring width, but caused a marked increase in acorn production. Clearly there have been too few studies to allow any generalizations about the order in which different components of plant performance are impaired under attack by insect herbivores.

While insect attack is often the cause of reduced plant growth rate (91, 112), the rate of plant growth may itself influence both the probability and the consequences of insect attack (60, 120, 204, 217). For instance, stressed plants may be more susceptible to insect herbivory (see section on competitive ability, below). Similarly, individual plants that have slow growth rates by reason of their age (204) or the soil conditions in their particular microhabitat (125) might be more prone to insect attack or may suffer greater loss of performance with each individual insect they support.

The effect of insect feeding on plant performance is typically described by plotting a damage function. This is a graph of plant growth or seed production against insect population density. The shape of the damage function provides information on the nature and extent of plant compensation (36). Most insect-plant relationships are adequately described by simple, linear damage functions (7, 169). A smaller number are better described by curvilinear functions, which suggest plant compensation for low levels of insect feeding (18, 156, 159). Some studies have distinguished explicitly between effects on primary growth and on the extent of regrowth following defoliation (45, 93, 105, 157). Some insect species perform better on a diet of regrowth foliage (87, 177, 223), while other species perform less well (71, 126).

The current controversy as to whether and under what circumstances herbivory could ever benefit individual plants (8, 39, 146) rests largely on semantic issues. While there have been a few reports of increased plant performance following insect herbivory (i.e. a humped damage function) (17, 147, 168), I have found no detailed experimental study that has reported greater recruitment from plants attacked by insect herbivores than from unattacked control plants growing under the same competitive circumstances. One study currently in progress aims to test whether attack by the weevil *Curculio glandium* increases the fitness of *Q. robur*. Weevil attack causes death of the acorn in about 90% of cases. Intact acorns, however, suffer almost 100% predation from various vertebrate herbivores including rabbits, mice, squirrels, jays, and wood pigeons. Most of these animals reject weevily acorns, so the probability of an acorn producing a seedling might be higher for a weeviled than for an unweeviled acorn (67).

Clearly, the potential for regrowth and compensation depends critically upon the timing of insect attack (114, 156, 202). In general, the earlier the attack, the greater the possibility for regrowth. For example, an identical 3%

loss of final cabbage yield was caused by 25% defoliation early in the growing season and by as little as 5% defoliation at the time of heading (227). Late-season attack leaves no time for regrowth and may also cause grazed plants to enter the unfavorable season in a more vulnerable condition (e.g. more likely to succumb to frost or drought) (85).

A certain amount of insect damage inevitably escapes recognition because of its obscure effects and the small size of the individual animals. In such cases losses are wrongly attributed to other factors such as drought, fungi, frost, or deficiencies in soil nutrients that produce similar, nondistinctive symptoms. These misattributions of damage are especially likely when small sucking insects such as froghoppers and aphids attack plants such as grasses that are not usually thought to suffer severe invertebrate herbivory (25). In fact, insect herbivory may be profoundly important in the population dynamics of grasses (29, 30, 92).

## Competitive Ability

Indirect effects of feeding by insect herbivores, leading to reduced plant vigor and impaired competitive ability, are thought to be paramount in affecting plant population dynamics, especially by practitioners of biological weed control (78, 85, 193, 216). Defoliating insects, phloem-feeding aphids, whitefly or scale insects, various sap-feeding Heteroptera, and stem-boring flies, moths, and beetles have all been charged with reducing the vigor of their host plants sufficiently that the plants were unable to compete with more vigorous, unattacked, neighboring plants (references below). Few critical manipulative experiments have been carried out on this important topic (32, 33), however, and it is easy to overlook other, less obvious factors that might be the direct cause of the plants' decline. For instance, there have been numerous reports of proliferation of insect herbivores on plants stressed by air pollution (58, 66), drought (11, 35, 122, 130, 220), soil acidity or impeded drainage (13, 125, 134), shade (131, 136), soil nutrient deficiency (123, 192), resistant plant genotypes (127), previous defoliation (71, 84, 126, 170, 177, 187, 194), or previous low plant growth rate due to intense plant competition (60, 120, 204, 217). It is not clear to what extent, if any, the insect attack exacerbated the loss in vigor that the plants would have suffered in any case. Recent field studies using controlled release of air pollutants ($SO_2$) have, however, pointed to extra yield losses attributable to enhanced feeding by cereal and conifer aphids (66).

Several reports have suggested that differential rates of insect attack can relax or even reverse the competitive relationships between plant species in natural communities (69, 128, 175, 179, 226). For example, the grass aphid *Holcaphis holci* reduced the rate of tillering of its host plant *Holcus mollis,* and this reduction led to increases in the abundance of the low-growing herb

*Galium saxatile* in years when (or in places where) the aphid was abundant. In low-aphid years, the grass advanced at the expense of the herb (J. Packham, personal communication). Defoliation by insect herbivores reduced the competitive ability of certain shrubs in the Chilean matorral, effectively excluding them from plant communities (75). Defoliation by the chrysomelid beetle *Gastrophysa viridula* altered the outcome of competition between *Rumex crispus* and *Rumex obtusifolius* (10, 33).

These effects are by no means universal, and insect exclusion has been shown to have no measurable effect on interspecific plant competition in a number of studies where broad-spectrum insecticides have been applied to seminatural plant communities (see Table 4). Analysis of insecticide application experiments is necessarily rather crude, given the number of plant species present and the variety and trophic positions of insects excluded. Thus it is possible that there are responses in plant growth to reduced insect feeding, but they cancel one another out. This topic is discussed in more detail below.

## Death Rate of Established Plants

It is unusual for insect herbivores to cause the death of established perennial plants. Increased mortality has resulted from bark beetle attacks on mature coniferous trees (12, 188), attack by spruce budworm on balsam fir, *Abies balsamea* (14), and attack by gypsy moths on deciduous hardwood trees (201). In the last case, plant death rates are little affected unless individual trees suffer more than 50% defoliation for three or more consecutive years (90). Most deciduous hardwoods have substantial powers of regrowth following complete defoliation, and several species can withstand repeated defoliation even within the same growing season (e.g. *Q. robur*) (43).

Increased plant death rates as a result of insect herbivory have led to successful biocontrol of the ragwort *Senecio jacobaea* by the cinnabar moth, *T. jacobaeae*, in eastern Canada (86). Plants that have been severely defoliated by cinnabar moths are killed by the characteristically early frosts in this part of Canada. In other regions frosts occur later in the autumn, and ragwort regrowth appears after cinnabar moths have pupated. Plants that have been allowed more protracted regrowth prior to the first frosts do not suffer significantly higher death rates than undefoliated plants, and weed control is much less successful. When levels of plant competition experienced by the arid-land shrub *Gutierrezia microcephala* were reduced experimentally by removing neighboring individuals, the death rate of plants decreased from 47 to 22% on plots exposed to intense herbivory by the grasshopper *Hesperotettix viridis*. Ungrazed plants, protected with cages, showed no response to reduced competition; their death rate was 8% whether or not their neighbors were removed (172).

# EFFECTS ON PLANT POPULATION DYNAMICS

Despite the mass of evidence on the effects of herbivorous insects on plant performance, there is woefully little data on the effects of insect herbivory on plant population dynamics. Without a knowledge of the key factors acting at each stage in the plant's life cycle and which (if any) of these factors act in a density-dependent manner, it is impossible to predict how insect feeding will affect plant abundance.

## Evidence From Biological Weed Control

Releases of imported insects for the biological control of weeds represent some of the most impressive, large-scale ecological field experiments that have ever been carried out. Even though these attempts were not planned as experiments and tend to be neither replicated nor randomized (38, 40), it is important that we use the information they provide on insect-plant interactions to the full. The material presented in this section was assembled as part of the Silwood Project on the Biological Control of Weeds (see acknowledgments), which augmented Julien's catalog (110) of published weed-biocontrol results with firsthand, unpublished accounts. One aim of this work was to document control attempts that failed, either because the agent failed to establish or because, having established successfully, it failed to make any impact on the abundance of the target weed (78, 85, 193, 216).

INSECT TAXONOMY    The most frequently repeated successes in biological control are listed in Table 1, and the most successful individual cases are given in Table 2. It is difficult to generalize about the taxonomy of the insects involved. The *Opuntia* cacti, for example, have been controlled by species as different as a relatively large, chewing insect, *Cactoblastis cactorum,* and a tiny sucking species, the cochineal insect *Dactylopius ceylonicus.* Cochineal insects have been responsible for successful control much more frequently than moths (8:3 in Table 1; 7:1 in Table 2). Cochineal insects might even have been effective in the classic textbook example of cactus control in Queensland, Australia, had *C. cactorum* not been introduced before *D. ceylonicus* became firmly established. Out of the top 11 weed control cases, seven involved leaf or cladode suckers, three involved leaf-chewing insects, and only one involved a seed head–feeding insect *(Rhinocyllus conicus).* Leafminers, stem borers, and root feeders are not represented among these most successful cases (Table 2).

On host plants other than *Opuntia* and *Lantana* species, the insect group responsible for the largest number of conspicuous successes is the beetles. Weevils have controlled plants as different as the aquatic fern *Salvinia molesta,* the waterweed *Eichhornia crassipes,* and terrestrial weeds such as

**Table 1**  The most frequently repeated successes of biological weed control up to 1980[a]

| Rank | Plant | Insect | S[b] | E[c] | R[d] | %S[e] | %E[f] | Insect life history |
|---|---|---|---|---|---|---|---|---|
| 1 | Lantana camara | Teleonemia scrupulosa | 9 | 26 | 29 | 31 | 90 | Leaf feeding tingid |
| | | Uroplata girardi | 5 | 11 | 17 | 29 | 65 | Chrysomelid beetle |
| | | Hypena strigata | 3 | 6 | 7 | 43 | 86 | Noctuid moth |
| | | Three others | 4 | 14 | 24 | 17 | 58 | Various |
| 2 | Hypericum perforatum | Chrysolina quadrigemina | 4 | 7 | 8 | 50 | 88 | Chrysomelid beetle |
| | | Chrysolina hyperici | 3 | 6 | 10 | 30 | 60 | Chrysomelid beetle |
| | | Zeuxidiplosis giardi | 2 | 5 | 6 | 33 | 83 | Cecidomyiid midge |
| 3 | Eichhornia crassipes | Neochetina eichhorniae | 5 | 6 | 10 | 50 | 60 | Curculionid beetle |
| | | Sameodes albigutalis | 2 | 3 | 5 | 40 | 60 | Pyralid moth |
| | | Neochetina bruchi | 1 | 2 | 3 | 33 | 66 | Curculionid beetle |
| 4 | Opuntia vulgaris | Dactylopius ceylonicus | 6 | 7 | 9 | 66 | 78 | Cochineal insect |
| | | Cactoblastis cactorum | 1 | 20 | 22 | 5 | 91 | Pyralid moth |
| 5 | Cordia macrostachya | Metrogaleruca obscura | 3 | 3 | 3 | 100 | 100 | Chrysomelid beetle |
| | | Eurytoma attiva | 2 | 3 | 3 | 66 | 100 | Eurytomid wasp |
| 6 | Centaurea diffusa | Urophora affinis | 4 | 6 | 6 | 66 | 100 | Tephritid fly |
| 7 | Opuntia dillenii | Dactylopius opuntiae | 2 | 16 | 23 | 9 | 70 | Cochineal insect |
| | | Cactoblastis cactorum | 2 | 20 | 22 | 9 | 91 | Pyralid moth |
| 8 | Alternanthera phi-loxeroides | Agasicles hygrophila | 2 | 2 | 2 | 100 | 100 | Chrysomelid beetle |
| | | Vogtia malloi | 2 | 2 | 2 | 100 | 100 | Pyralid moth |
| 9 | Tribulus cistoides | Microlarinus lypriformis | 3 | 10 | 10 | 30 | 100 | Curculionid beetle |
| | | Microlarinus lareynii | 1 | 3 | 5 | 20 | 60 | Curculionid beetle |
| 10 | Linaria vulgaris | Gymnaetron antirrhini | 2 | 3 | 3 | 66 | 100 | Curculionid beetle |
| | | Brachypterolus pulicarius | 1 | 2 | 2 | 50 | 100 | Curculionid beetle |

[a]From Reference 110 and the Silwood Project on Weed Biocontrol.
[b]S = Number of releases leading to successful control.
[c]E = Number of releases leading to established insect populations.
[d]R = Total number of releases up to 1980.
[e]%S = Percentage of releases leading to successful control.
[f]%E = Percentage of releases leading to established insect populations.

**Table 2**  The most successful individual biocontrol releases, ranked by the degree of control, the area over which control was obtained, and the initial severity of the weed infestation[a]

| Rank | Plant | Insect | Location, year |
|------|-------|--------|----------------|
| 1 | *Opuntia vulgaris* | *Dactylopius ceylonicus* | India, 1795 |
| 2 | *O. vulgaris* | *D. ceylonicus* | Sri Lanka, 1795 |
| 3 | *Opuntia dillenii* | *Dactylopius opuntiae* | India, 1926 |
| 4 | *Carduus nutans* | *Rhinocyllus conicus* | Canada, 1968 |
| 5 | *Hypericum perforatum* | *Chrysolina quadrigemina* | USA, 1946 |
| 6 | *O. vulgaris* | *D. ceylonicus* | South Africa, 1913 |
| 7 | *Opuntia elatior* | *D. opuntiae* | Indonesia, 1935 |
| 8 | *Opuntia* sp. | *D. opuntiae* | Madagascar, 1923 |
| 9 | *Opuntia tuna* | *D. opuntiae* | Madagascar, 1928 |
| 10 | *Opuntia triacantha* | *Cactoblastis cactorum* | Nevis, West Indies, 1957 |
| 11 | *Salvinia molesta* | *Cyrtobagous salviniae* | Australia & New Guinea, 1980 |

[a]Maximum scores represent complete control of weeds that had previously formed impenetrable monocultures over large areas. Low scores represent local or patchy control of weeds that had low cover prior to release of the control agent. All the insect species in the table were recorded as having brought about "complete" control of the weed in question. Changing the definition of "successful control" alters the ranking slightly but does not have a marked effect on the weed species included as having been controlled.

the spiny-fruited *Tribulus terrestris* and the thistle *C. nutans*. Chrysomelids have controlled *Hypericum perforatum, Alternanthera philoxeroides,* and *Cordia macrostachya* (Table 1). Other insect groups previously thought to have little or no potential in weed biocontrol have produced some spectacular results in recent introductions. For example, the pteromalid gall wasp *Trichilogaster acaciaelongifoliae* was introduced into South Africa from Australia in an attempt to control the invasive Australian tree *Acacia longifolia*. The insects were considered so sensitive and delicate on first release that doubts were raised about whether they could even be successfully established. Since 1982, however, they have increased to such densities that tree branches are now weighed down by galls, and seed production has been reduced by over 95% (55).

Taxonomic mistakes have an interesting history in weed biocontrol. A number of failures in cactus control have been attributed to failure in identification of the host plant that resulted in the collection and release of insects from the wrong native strain. Ironically, other cases have led to successful control, apparently because of the small degree of taxonomic mismatching. Hokkanen & Pimentel (99) suggested that successful control would be more likely if the insect were not too closely adapted to the host plant, basing their argument on the notion that prolonged coevolution leads to

**Table 3**  Most frequently repeated failures of biological weed control (zero success rate)

| Rank | Plant | Insect | E[a] | R[b] | %E[c] | Plant life history |
|---|---|---|---|---|---|---|
| 1 | *Lantana camara* | *Diastema tigris* | 1 | 11 | 9 | Verbenaceae: perennial, straggling shrub with prickly stems, spread by seed, but regrows vigorously after cutting |
| | | *Leptobyrsa decora* | 4 | 11 | 36 | |
| | | *Ophiomyia lantanae* | 13 | 20 | 65 | |
| | | *Teleonemia elata* | 0 | 5 | 0 | |
| | | 17 others | 13 | 34 | 38 | |
| 2 | *Hypericum perforatum* | *Agrilus hyperici* | 2 | 7 | 29 | Hypericaceae: protohemicryptophyte, rhizomatous perennial |
| | | *Chrysolina varians* | 0 | 3 | 0 | |
| | | 4 others | 2 | 5 | 40 | |
| 3 | *Salvinia molesta* | *Cyrtobagous singularis* | 1 | 5 | 20 | Polypodiaceae: annual or perennial floating fern, spread by fragmentation |
| | | *Paulinia acuminata* | 7 | 10 | 70 | |
| | | *Samea multiplicalis* | 1 | 3 | 33 | |
| 4 | *Cirsium arvense* | *Altica carduorum* | 0 | 4 | 0 | Compositae: geophyte with root buds, rhizomatous perennial spread by seed and by rootstock fragments |
| | | *Urophora cardui* | 2 | 5 | 40 | |
| | | *Ceutorhynchus litura* | 2 | 2 | 100 | |
| | | *Lema cyanella* | 0 | 1 | 0 | |
| 5 | *Opuntia littoralis* | *Chelinidea vittiger* | 3 | 6 | 50 | Cactaceae: spread by seed and cladodes |
| | | *Melitara prodenialis* | 0 | 9 | 0 | |
| | | 4 others | 27 | 42 | 64 | |
| 6 | *Cyperus rotundus* | *Athesapeuta cyperi* | 1 | 7 | 14 | Cyperaceae: rhizomatous perennial spread by seed and tubers from rootstock |
| | | *Bactra minima* | 0 | 4 | 0 | |
| | | *Bactra venosana* | 2 | 5 | 40 | |
| 7 | *Opuntia ficus-indica* | *Archlagocheirus funestus* | 4 | 4 | 100 | Cactaceae: spread by seed and cladodes |
| | | *Melitara doddalis* | 0 | 3 | 0 | |
| | | *Melitara prodenialis* | 0 | 9 | 0 | |
| | | 3 others | 6 | 8 | 75 | |
| 8 | *Chromolaena odorata* | *Apion brunneonigrum* | 0 | 5 | 0 | Compositae: shortlived, scrambling perennial spread by seed |
| | | *Pareuchaetes pseudoinsulata* | 1 | 6 | 17 | |
| 9 | *Senecio jacobaea* | *Tyria jacobaeae* | 4 | 9 | 44 | Compositae: biennial or perennial, rosette-forming hemicryptophyte spread by seed and root fragments |
| | | *Hylemyia seneciella* | 2 | 4 | 50 | |
| | | *Longitarsus jacobaeae*[d] | 3 | 3 | 100 | |

**Table 3**    *(Continued)*

| Rank | Plant | Insect | E[a] | R[b] | %E[c] | Plant life history |
|------|-------|--------|------|------|-------|--------------------|
| 10 | *Euphorbia* × *pseudovirgata* | *Chamaesphecia tenthrediniformis* | 0 | 3 | 0 | Euphorbiaceae: rhizomatous peren- nial protohemicryp- tophyte |
| | | *Hyles euphorbiae* | 3 | 4 | 75 | |
| | | *Oberea erythro- cephala* | 2 | 2 | 100 | |
| | | *Chamaesphecia empiformis* | 0 | 2 | 0 | |

[a]E = Number of releases leading to establishment.
[b]R = Number of release attempts.
[c]%E = Percentage of releases leading to establishment.
[d]Successful control of *S. jacobaea* by *L. jacobaeae* has recently been achieved in Oregon (P. McEvoy, personal communication).

reduced virulence. Although this argument has been refuted by ecologists working with animal parasites [who point out that natural selection can lead to increases or decreases in virulence, depending upon the precise circumstances (38)], little work has been carried out with plants. A striking counterexample to Hokkanen & Pimentel's hypothesis is provided by the successful control of *Salvinia molesta*. At first both the fern and the insect were misidentified, and attempts at weed control brought little success (Table 3). Once the plant had been correctly identified, however, insect collection was carried out in its correct native range, and in due course weevils were collected that brought about spectacularly successful control of the weed in both Australia and Papua New Guinea. It later transpired that the successful insect was not the species it was thought to be, but a local endemic, new to science (207). Here, a level of control that had proved impossible with the more distantly related form was achieved using the more closely adapted species.

The most frequently repeated failures of biological weed control (Table 3) fall into two clear categories. The first two plants also top the list of repeated successes (Table 1); they are weeds whose control is possible but relatively unpredictable (and in the case of *Lantana camara*, control often involves several insect species), so that there have been many attempts, and many of these have led to failure. In the second category are plant species that for one reason or another are more difficult to control. All of the cases listed in Table 3 had success ratings of zero, either through failure of the insect to establish or because the established population had no impact on weed abundance. The table may tell us more about the optimism of biocontrol workers than about the ecology of the insect herbivores employed.

PLANT TAXONOMY    Given that most targets for biocontrol are nonnatives, it might be expected that the taxonomy of target weeds would reflect the invasive tendencies of different plant families (40). Some families, such as Compositae, are abundant in both native and alien floras and are not significantly overrepresented as aliens. Some families produce proportionately more invasive aliens than others (e.g. Labiatae, Cruciferae, and Caryophyllaceae). Some families are significantly underrepresented as aliens (e.g. Cyperaceae, Ericaceae, Orchidaceae, Juncaceae, and Polypodiaceae). Interestingly, biocontrol is rarely targeted against weeds from the overrepresented families of aliens, perhaps because so many of their species are annuals and therefore tend to be less serious weeds. In addition, annual plants may be less likely to succumb to biocontrol using insect herbivores (78, 84).

Leaving aside the *Opuntia* cacti and *Lantana camera* (which together account for over 40% of all weed biocontrol releases), most biocontrol projects have been targeted against weeds from Compositae (e.g. *Centaurea, Carduus, Senecio, Eupatorium,* and *Chromolaena* species). All the plants controlled successfully were perennials, and most were weeds of open habitats such as badly managed, semiarid pasturelands where there was little competition from native perennials. Another open habitat in which conspicuous success was achieved using insect biocontrol was lentic freshwater, where plants such as alligator weed *(A. philoxeroides)*, water hyacinth *(E. crassipes)*, and floating fern *(S. molesta)* have been dramatically reduced in abundance.

Repeated attempts to control weeds from certain plant families have led only to repeated failures (Table 3). Despite numerous attempts using a variety of insect species, nutsedge *(Cyperus rotundus)* and creeping thistle *(Cirsium arvense)* have never been successfully controlled, even though many of the insects have successfully established quite large populations. Indeed, the weevil *Apion ulicis,* introduced to control gorse *(Ulex europaeus)* in New Zealand, has become one of that country's most abundant insects, frequently destroying over 90% o .he seed crop (96). It fails as a control agent partly because gorse has such a large and long-lived seed bank and partly because the weevil has no deleterious effect on the growth or survival of the parent plant. In other cases, control agents have become established, but remain scarce as a result of attack by generalist natural enemies, inclement weather, disturbance, and other disasters (19). Insects that feed on the outside of plants are especially vulnerable to predation of their eggs by ants, and this is a frequent cause of failure of biocontrol releases in tropical environments (38).

The difficulty experienced in controlling certain pasture weeds (e.g. *Euphorbia* × *pseudovirgata)* may result partly from their low quality as food for insects and partly from their production of copious latex, which might limit the intrinsic rate of increase of herbivorous insects to such low levels that no depression in weed abundance is achieved (see above). Food quality had a key role in the successful control of the water fern *S. molesta* by the weevil *Cyrtobagous salviniae*. Unless nitrogen fertilizer was applied to plants in the initial release cages, the population of beetles did not build up rapidly enough to reduce the abundance of the fern. Insect population growth following fertilization of the plants was spectacular and brought about complete control of the plants over an area much larger than that to which fertilizer was initially applied (189).

INSECT DEMOGRAPHY    The demographic attributes of successful biocontrol agents have been reviewed elsewhere (38, 40). Small insect body size and high intrinsic rate of increase are associated with both an increased probability of establishment following release and an increased likelihood of significant weed control following successful establishment. Certain traits that were thought to be of possible importance to successful control, such as lack of obligatory diapause and high powers of dispersal, proved not to be of general significance.

In natural communities, large, mobile polyphagous insect herbivores tend to have more impact on plant dynamics than small, sedentary specialist species because the former insects are maintained at higher population densities by the abundance of less preferred, relatively low-quality food plants. Thus these insects tend to affect the population dynamics of less common plants of relatively high palatability, which may be driven locally to extinction (36). Specialist insects, especially sucking species like aphids, can have important effects on community dynamics, not directly by eating their preferred species to extinction, but by reducing the competitive ability of grazed plants relative to their herbivore-free neighbors.

PLANT LIFE HISTORY    Certain plants appear to be particularly difficult to control using insect herbivores (Table 3). Traits associated with this resistance include (*a*) a long growing period (especially a protracted growing period after a univoltine insect has entered its dormant or nonfeeding stage); (*b*) reserves of carbohydrates and proteins that are inaccessible to the herbivore (as in species with thick underground rhizomes or tough, woody stems); (*c*) high powers of regrowth following defoliation or stem destruction (i.e. growth that is not meristem limited); (*d*) ability to produce replacement crops of fruit following defloration; (*e*) low food quality (e.g. low tissue nitrogen concentration or high tannin content); and (*f*) large seed banks and protracted dormancy.

The duration of the period between insect release and achievement of successful biocontrol differs among plants with different life histories. Weeds of arid lands that are long-lived, achieve large adult size, spread mainly by vegetative means, possess high powers of regrowth, and represent low-quality food for insects have frequently been controlled, but the insects tend to spread slowly following release. In contrast, successful control of smaller, short-lived perennial weeds of temperate habitats tends to occur more rapidly, with the insects spreading quickly following release. This second group of plants tends to reproduce by seed, to have lower powers of regrowth, and to represent higher-quality food for insects (containing higher tissue nitrogen concentrations and lower concentrations of quantitative defensive chemicals such as tannins and lignin) (38).

CAVEATS    Weed biocontrol projects have provided some of the best field data on plant-herbivore dynamics. Releases are carried out on a spatial scale that could never be managed in experiments by individual ecologists. Unfortunately, what biocontrol projects gain in scale, they often lose in rigor. Very few releases are either replicated or randomized. Sometimes releases are made with minimal prerelease study and virtually no postrelease follow-up. Again, the data base is dominated by repeated attempts to control particular plant species: (a) weeds that have been successfully controlled elsewhere in the past (e.g. *Opuntia* and *Lantana* species), and (b) weeds that are unsuitable for control by other means (e.g. plants infesting marginal grazing land where cost precludes any other means of control). This taxonomic bias can be seen from the fact that of 627 documented cases up to 1980, 152 involved *Lantana camara* 117 involved various *Opuntia* species. Current biocontrol practice is aimed at improving the scientific value of these release experiments by developing protocols that will allow detailed prerelease and followup studies and, where possible, by setting up control plots in addition to replicated release sites, where weed dynamics can be monitored (J. K. Waage, personal communication).

An important question concerns the relevance of weed biocontrol data to an understanding of plant-herbivore dynamics in native vegetation. There are several concerns: (a) The weeds are generally alien plants growing in plant communities that are often quite different from those in which they evolved. (b) The insects are also alien, imported especially for the purposes of control. (c) The insects have been freed from their native natural enemies by careful screening prior to release. (d) The range of genetic variability in both plant and insect populations may be lower than in native communities as a result of the small size of the initial introductions. (e) The habitats in which biocontrol is practiced are often highly disturbed (e.g. overgrazed semiarid rangelands).

## Evidence From Insecticide Exclusion Experiments

The best evidence for the impact of insect herbivory on plant population dynamics would come from experiments in which insects were removed from large, replicated plots using minimally disruptive exclusion techniques (per-haps removal of the insects by hand) and in which the subsequent plant dynamics were compared with other plots that supported normal and en-hanced densities of insects. Unfortunately, hand removal of insects is pro-hibitively labor intensive in most cases, and we must make do with the removal of insects using chemical insecticides or exclusion cages. Both of these techniques have potentially serious shortcomings (discussed below), but they represent better evidence on the importance of insect herbivory in plant dynamics than can ever be obtained from purely observational studies.

Studies in which insecticides have been applied to natural and seminatural vegetation with the object of measuring the impact of insect herbivory on plant distribution and abundance are listed in Table 4. Studies of insecticide application have not been included if the object was merely to measure the impact of insect feeding on plant performance (e.g. growth, seed yields, or survivorship of individual crop plants). The table shows an approximately equal number of cases in which insect feeding did and did not affect plant populations. However, negative results may well have been underrecorded, especially in the early years.

Cantlon's (26) demonstration of a dramatic increase in the abundance of the hemiparasitic woodland herb *Melampyrum lineare* following the eradication of the katydid *Atalanticus testaceous* using insecticides was for many years the only report suggesting that a native insect population might maintain a natural plant population at low density. It is unfortunate that Waloff & Richards (215) in their classic study on broom (see above) did not monitor the recruitment of young plants on the sprayed and unsprayed plots; instead, seedlings were pulled up to maintain the original plant density. Insect exclu-sion did, however, significantly increase the longevity of the plants and dramatically increase their fecundity. Studies on the exclusion of insects (mainly flies of the genus *Oscinella*) from grasslands have shown that by reducing the vigor of the pasture grasses on which they feed, the flies allowed the ingress of less competitive, weedy species (30, 92). Thus the quality of sown pasture declined less rapidly when insects were excluded by insecticide application (29). Other studies have shown that varying the cutting frequency of sprayed grasslands can alter the importance of leaf- and planthoppers in reducing grass yields. Plots cut once lost 49% of their dry weight to insect feeding, while plots cut frequently to simulate close grazing lost only 4% (89). Insecticide application led to increases in grass dry weight yields for 12 of 14 cultivars tested by Byers & Jung (25). It is noteworthy that the grass species that produced the greatest yields on the insecticide-treated plots (*Festuca arundinacea* var. Fawn; 2.2 t/ha) ranked only eighth of 14 on the unsprayed plots. The species that ranked first on the unsprayed plots (*Poa pratensis* var. Kenblue) was ranked only fifth on the insecticide-treated plots.

**Table 4**  Experimental exclusion of insects from plant communities using insecticides

| Plant | Insect | Exclusion effects[a] | Remarks | Ref. |
|---|---|---|---|---|
| Grasses | Wireworms | Y | Reduced weed densities | 72 |
| *Melampyrum* sp. | Katydid | Y | Small scale | 26 |
| *Convolvulus* sp. | Various | N | Phytotoxic insecticide | 197 |
| *Calluna vulgaris* | *Strophingia ericae* | N | Exclusion during a psyllid population peak | 98 |
| *Lolium perenne* | Frit fly | Y | Increased competitiveness of dominant grass | 30 |
| *Cytisus scoparius* | Various | Y | Reduced plant mortality; recruitment not measured; not replicated | 215 |
| Forage grasses | Various | Y | Insects altered relative performance of 14 taxa | 25 |
| *Abies balsamea* | Spruce budworm | Y | Selective mortality | 14 |
| *Raphanus* sp. | Various | Y | Not replicated | 21 |
| *Haplopappus squarrosus* | Flower and seed feeders | Y | Seed limited | 132 |
| *Haplopappus venetus* | Flower and seed feeders | N | Microsite limited | 133 |
| *Ceanothus* sp., *Adenostoma* sp. | Various | N | Vertebrate herbivores affected seedling survival | 153 |
| *Solidago canadensis* | Three chrysomelids | Y | Maintenance of goldenrod dominance | 144 |
| *Quercus robur* | Various | N | More seed but not more seedlings | 37 |
| Annuals of secondary succession | Various | | Matched experiment in Iowa and England | b |
| First year in US | | N | | |
| Second year in US | | N | | |
| First year in UK | | Y | | |
| Second year in UK | | N | | |
| *Cardamine pratensis* | Various | N | More seeds but not more seedlings | 59 |
| *Heracleum sphondylium* | Ray aphids | Y | Seed limited | 196 |
| *Calluna vulgaris* | Various | N | No effect until increase in flowering in third year | c |
| *Holcus/Galium* sp. | *Holcaphis holci* | Y | Bedstraw declines on uninfested plots | d |
| *Medicago lupulina* | Weevils | Y | Only on sheep grazed plots | 77 |

**Table 4** *(Continued)*

| Plant | Insect | Exclusion effects[a] | Remarks | Ref. |
|---|---|---|---|---|
| *Vicia* spp. | Various | Y | Increased seedling densities | 22 |
| *Trifolium pratense* | Various | Y | Increased survivorship but no seedling recruitment | e |
| *Senecio jacobaea* | Cinnabar moth | N | Not seed limited | 44 |
| Grassland seed bank | Various | N | Pronounced effect of rabbit exclusion | 42 |
| *Poa pratensis* | Various | N | Build-up of thatch | 5 |

[a]Y denotes significant effects of insect exclusion on plant abundance. N, no significant effects on plant dynamics reported.
[b]S. Hendrix & V. K. Brown, personal communication.
[c]V. C. Brown & S. McNeill, personal communication.
[d]J. Packham, personal communication.
[e]A. Gange, personal communication.

Reanalysis of these data shows that there is no significant rank correlation between the yields of the treated and untreated plots (Spearman's $r = 0.235$; $p \gg 0.05$), which suggests that insect exclusion had a substantial effect in altering the relative performance of these grasses. The species were grown as monocultures, however, so these results cannot be interpreted as direct evidence of changes in interspecific competitive ability following insect exclusion.

After shrubs of *Haplopappus squarrosus* and *Haplopappus venetus* were sprayed with insecticide, seed production was increased in both species; plant recruitment, however, only increased in *H. squarrosus* (132). Evidently, recruitment was seed limited in *H. squarrosus* but not in *H. venetus,* so insect herbivory had no impact on the population dynamics of *H. venetus* despite its causing substantially reduced fecundity (133).

Some insecticide experiments have shown effects on plant abundance that subsequently turned out to be due to direct phytotoxic effects of the pesticide on the dominant species that allowed competitive release of previously suppressed plants (197). Subsequent workers have been more careful to choose chemicals that have no direct stimulatory or phytotoxic effects (references in Table 4). There are very few reports of stimulatory effects on plant growth, even with systemic organophosphorus insecticides, when insecticides are applied at recommended rates (109).

Long-term insecticide studies aimed at determining the effect of insect exclusion on plant recruitment have only been initiated recently, and few results have been published. Elsewhere (42) I have given details of a number of these studies in progress; here I simply report that exclusion of insects sometimes has no measurable effect on plant recruitment (e.g. in heather,

*Calluna vulgaris,* and in communities of arable weeds), sometimes modifies the competitive balance between different plant species (as when the grass aphid *H. holci* reduced the competitive ability of its host plant *H. mollis* and allowed an increase in the abundance of the herb *G. saxatile*), and occasionally has more pronounced effects on plant community dynamics (references in Table 4). In the few cases in which insect and vertebrate herbivores have been excluded singly and in combination, the vertebrate herbivores have always proved to have the greater impact on plant dynamics (41, 77, 153).

CAVEATS    There are several potential pitfalls in insecticide application experiments: (*a*) The chemical may be phytotoxic, perhaps differentially so, giving the erroneous impression of competitor release following insect exclusion. (*b*) The chemical may have some stimulatory effect on plant growth, either through hormonal action or fertilizer effects. (*c*) The chemical may actually increase the abundance of herbivorous insects if predatory insects suffer differentially high rates of mortality under insecticide treatment (4, 5). (*d*) The chemical may deter herbivorous animals other than insects (e.g. mollusks or small mammals) so that responses in plant abundance cannot be attributed directly to a reduction in insect herbivory. (*e*) The effects on the soil fauna and microflora are generally unknown (e.g. the chemical may kill mycorrhiza-feeding insects, thereby increasing the efficiency of mineral uptake). (*f*) Because the hypothesis being tested is that insect herbivores influence plant dynamics, responses due to these other processes might be attributed erroneously to insect herbivory.

Despite all these shortcomings, the technique holds tremendous potential for demonstrating the impact of insect feeding on plant population dynamics under field conditions. The more precisely the insecticide can be targeted to eliminate selected species of insects on particular species of plants (e.g. by applying the insecticide to individual leaves of the target plant species with a fine brush, with minimal disturbance to neighboring plants), the more intelligible the results will be. When all the plants in a community are sprayed, it is impossible to tease apart the direct negative effects of herbivory and the indirect negative effects of competitor release of a plant's neighbors. Similarly, improved insecticide controls need to be employed to convince skeptics that the observed effects really are due to reduced insect herbivory rather than to one or more of the potential side effects listed above. For example, hand picking of insects from plants can provide a useful control when associated with insecticide-spray and blank-spray controls. Although the technique is extremely laborious, work on ragwort (44) and oak (43) has affirmed that hand picking gives results that are indistinguishable from those obtained using insecticide treatment. More such studies are required.

# CONCLUSIONS

Insect herbivores can affect every aspect of plant performance, and there is a massive literature from pest control showing effects on plant growth, form, seed production, development rate, and survivorship. In contrast, there is an almost complete dearth of information on the role of insects in plant population dynamics.

Our knowledge of the effects of insect herbivores in natural vegetation is restricted to a small (but growing) number of studies using insecticidal exclusion in natural vegetation. These studies suggest that insects do influence plant population dynamics in some cases, but their effects are often subtle and are typically less pronounced than the effects of other kinds of herbivores [notably vertebrate herbivores (42)]. In several cases no effect on plant population dynamics has been observed despite substantial reductions in the numbers of herbivorous insects feeding on the plants (Table 4).

The different kinds of insect herbivory affect plant population dynamics in different ways. For example, postdispersal seed-feeding insects like ants can reduce plant recruitment in years or in places where recruitment is seed limited (e.g. in deserts or semiarid open woodland). However, following mast crops of seed or in habitats where plant recruitment is microsite limited, seed-feeding insects typically have no effect on the numbers of plants that become established. Predispersal seed predators have been used with considerable success in biological weed control (e.g. the weevil *Rhinocyllus conicus* against nodding thistle *Carduus nutans*). They have also been shown to limit plant recruitment in some natural communities (132) but not in others (133). Certain foliage-feeding and bark-dwelling forest insects can cause the premature death of trees over large areas, with profound effects on forest succession (14, 210a). Other forest insects have no measurable effect on tree seedling recruitment, despite substantial effects on tree growth or seed production. Insect feeding is much more likely to increase the death rate of established plants when these plants are growing in dense stands and are subject to intense inter- or intraspecific competition. We know virtually nothing about the effect of root- and mycorrhiza-feeding insects. Sucking insects can reduce seed production, seed size, and seedling recruitment. Perhaps the most important effect of sucking insects is in reducing vegetative growth rate, and hence in reducing competitive ability. Since the majority of aphids and hoppers are highly specific in their choice of food plants, this kind of insect herbivory provides a possible mechanism for the maintenance of plant species richness, prohibiting competitive exclusion by selectively reducing the competitive ability of the dominant plant species.

While we know what kind of manipulative field experiments need to be carried out to understand the effects of insect herbivory on plant population dynamics, there are only a handful of published studies that address this question directly. Many more studies are currently under way, but the number of habitats investigated and the replication of studies within given habitats need to be increased before we can begin to generalize about the importance of insect herbivores in the population dynamics of plants. On present evidence, it appears that insect herbivores have a measurable impact on plant abundance in about half the cases studied. In contrast, exclusion experiments have almost always suggested a significant role for vertebrate herbivory in plant dynamics. In short, it appears that plants have substantially more effect on herbivore population dynamics than insect herbivores have on plant population dynamics.

ACKNOWLEDGMENTS

The Silwood Project on Weed Biocontrol was the brainchild of Cliff Moran and Jeff Waage. Those most closely associated with the data collection were Matthew Cock, David Greathead, Peter Harris, Mike Hassell, John Lawton, George McGavin, Dieter Schroeder, and Vicky Taylor.

*Literature Cited*

1. Abrahamson, W. G., McCrea, K. D. 1986. Nutrient and biomass allocation in *Solidago altissima:* effects of two stem gallmakers, fertilization, and ramet isolation. *Oecologia* 68:174–80
2. Andersen, A. N. 1987. Effects of seed predation by ants on seedling densities at a woodland site in SE Australia. *Oikos* 48:171–74
3. Andow, D. A. 1984. Microsite of the green rice leafhopper, *Nephrotettix cincticeps* (Homoptera: Cicadellidae), on rice: plant nitrogen and leafhopper density. *Res. Popul. Ecol.* 26:313–29
4. Annecke, D. P., Karny, M., Burger, W. A. 1969. Improved biological control of the prickly pear, *Opuntia megacantha* Salm-Dyck, in South Africa through the use of an insecticide. *Phytophylactica* 1:9–13
5. Arnold, T. F., Potter, D. A. 1987. Impact of a high maintenance lawn-care program on nontarget invertebrates in Kentucky bluegrass turf. *Environ. Entomol.* 16:100–5
6. Auld, T. D. 1983. Seed predation in native legumes of south-eastern Australia. *Aust. J. Ecol.* 8:367–76
7. Bailey, W. C., Pedigo, L. P. 1986. Damage and yield loss induced by stalk borer (Lepidoptera: Noctuidae) in field corn. *J. Econ. Entomol.* 79:233–37
8. Belsky, A. J. 1986. Does herbivory

benefit plants? A review of the evidence. *Am. Nat.* 127:870–92
9. Bennett, A., Krebs, J. 1987. Seed dispersal by ants. *Trends Ecol. Evol.* 2:291–92
10. Bentley, S., Whittaker, J. B. 1979. Effects of grazing by a chrysomelid beetle, *Gastrophysa viridula,* on competition between *Rumex obtusifolius* and *Rumex crispus. J. Ecol.* 69:79–90
11. Bernays, E. A., Lewis, A. C. 1986. The effect of wilting on palatability of plants to *Schistocerca gregaria,* the desert locust. *Oecologia* 70:132–35
12. Berryman, A. A., Dennis, B., Raffa, K. F., Stenseth, N. C. 1985. Evolution of optimal group attack, with particular reference to bark beetles (Coleoptera: Scolytidae). *Ecology* 66:898–903
13. Bink, F. A. 1986. Acid stress in *Rumex hydrolapathum* (Polygonaceae) and its influence on the phytophage *Lycaena dispar* (Lepidoptera; Lycaenidae). *Oecologia* 70:447–451
14. Blais, J. R. 1983. Trends in the frequency, extent, and severity of spruce budworm outbreaks in eastern Canada. *Can. J. Forest Res.* 13:539–47
15. Blanton, C. M., Ewel, J. J. 1985. Leaf-cutting ant herbivory in successional and agricultural tropical ecosystems. *Ecology* 66:861–69
16. Borowicz, V. A., Juliano, S. A. 1986.

Inverse density-dependent parasitism of *Cornus amomum* fruit by *Rhagoletis cornivora*. *Ecology* 67:639–43

17. Boscher, J. 1979. Modified reproduction strategy of leek *Allium porrum* in response to a phytophagous insect, *Acrolepiopsis assectella*. *Oikos* 33:451–56

18. Breen, J. P., Teestes, G. L. 1986. Relationships of yellow sugarcane aphid (Homoptera: Aphididae) density to sorghum damage. *J. Econ. Entomol.* 79:1106–10

19. Briese, D. T. 1986. Factors affecting the establishment and survival of *Anaitis efformata* (Lepidoptera: Geometridae) introduced into Australia for the biological control of St. John's wort, *Hypericum perforatum*. II. Field trials. *J. Appl. Ecol.* 23:821–39

20. Brown, J. H., Reichman, O. J., Davidson, D. W. 1979. Granivory in desert ecosystems. *Ann. Rev. Ecol. Syst.* 10:201–27

21. Brown, V. K. 1982. The phytophagous insect community and its impact on early successional habitats. In *Proc. 5th Int. Symp. Insect-Plant Relat., Wageningen, the Netherlands*, ed. J. H. Visser, A. K. Minks, pp. 205–13. Wageningen, the Netherlands: Pudoc

22. Brown, V. K., Gange, A. C., Evans, I. M., Storr, A. L. 1987. The effect of insect herbivory on the growth and reproduction of two annual *Vicia* species at different stages in plant succession. *J. Ecol.* 75:1173–89

23. Bulow-Olsen, A. 1984. Diplochory in *Viola:* a possible relation between seed dispersal and soil seed bank. *Am. Midl. Nat.* 112:251–60

24. Bultman, T. L., Faeth, S. H. 1986. Selective oviposition by a leaf miner in response to temporal variation in abscission. *Oecologia* 69:117–20

25. Byers, R. A., Jung, G. A. 1979. Insect populations on forage grasses: effects of nitrogen fertilizer and insecticides. *Environ. Entomol.* 8:11–18

26. Cantlon, J. E. 1969. The stability of natural populations and their sensitivity to technology. In *Diversity and Stability in Ecological Systems*, ed. G. M. Woodwell, H. H. Smith, pp. 197–205. New York: Brookhaven Natl. Lab.

27. Carroll, C. R., Risch, S. J. 1984. The dynamics of seed harvesting in early successional communities by a tropical ant, *Solenopsis geminata*. *Oecologia* 61:388–92

28. Cideciyan, M. A., Malloch, A. J. C. 1982. Effects of seed size on the germination, growth and competitive ability of *Rumex crispus* and *Rumex obtusifolius*. *J. Ecol.* 70:227–32

29. Clements, R. O., Gilbey, J., Bentley, B. R., French, N., Cragg, I. A. 1985. Effects of pesticide combinations on the herbage yield of permanent pasture in England and Wales. In *Tests of Agrochemicals and Cultivars*, 6:126–27 Suppl. *Ann. Appl. Biol.* 106.

30. Clements, R. O., Henderson, I. F. 1979. Insects as a cause of botanical change in swards. *J. Br. Grassland Soc. Occas. Symp.* 10:157–60

31. Coley, P. D., Bryant, J. P., Chapin, F. S. 1985. Resource availability and plant antiherbivore defense. *Science* 230:895–99

32. Cottam, D. A. 1985. Frequency-dependent grazing by slugs and grasshoppers. *J. Ecol.* 73:925–33

33. Cottam, D. A., Whittaker, J. B., Malloch, A. J. C. 1986. The effects of chrysomelid beetle grazing and plant competition on the growth of *Rumex obtusifolius*. *Oecologia* 70:452–56

34. Courtney, S. P., Manzur, M. I. 1985. Fruiting and fitness in *Crataegus monogyna:* the effects of frugivores and seed predators. *Oikos* 44:398–406

35. Cox, C. S., McEvoy, P. B. 1983. Effect of summer moisture stress on the capacity of tansy ragwort *(Senecio jacobaea)* to compensate for defoliation by cinnabar moth *(Tyria jacobaeae)*. *J. Appl. Ecol.* 20:225–35

36. Crawley, M. J. 1983. *Herbivory. The Dynamics of Animal Plant Interactions*. Oxford: Blackwell Sci.

37. Crawley, M. J. 1985. Reduction of oak fecundity by low density herbivore populations. *Nature* 314:163–64

38. Crawley, M. J. 1986. The population biology of invaders. *Philos. Trans. R. Soc. London Ser. B* 314:711–31

39. Crawley, M. J. 1987. Benevolent herbivores? *Trends Ecol. Evol.* 2:167–68

40. Crawley, M. J. 1987. What makes a community invasible? In *Colonization, Succession and Stability*, ed. A. J. Gray, M. J. Crawley, P. J. Edwards, pp. 429–53. Oxford: Blackwell Sci.

41. Crawley, M. J. 1989. Rabbit grazing, plant competition and seedling recruitment in acid grassland. *J. Ecol.* In press

42. Crawley, M. J. 1988. The relative importance of vertebrate and invertebrate herbivores in plant population dynamics. In *Focus on Plant-Animal Interactions*, ed. E. A. Bernays. New York: CRC. In press

43. Crawley, M. J., Akhteruzzaman, M. 1988. Individual variation in the phenology of oak trees and its consequences for

herbivorous insects. *Funct. Ecol.* 2:409–15

44. Crawley, M. J., Islam, Z., Nachapong, M., Ohgushi, M., Pattrasudhi, R., et al. 1989. The population dynamics of cinnabar moth in mesic grasslands. *J. Anim. Ecol.* In press

45. Crawley, M. J., Nachapong, M. 1984. Facultative defences and specialist herbivores? Cinnabar moth *(Tyria jacobaeae)* on the regrowth foliage of ragwort *(Senecio jacobaea)*. *Ecol. Entomol.* 9:389–93

46. Crawley, M. J., Nachapong, M. 1985. The establishment of seedlings from primary and regrowth seeds of ragwort *(Senecio jacobaea)*. *J. Ecol.* 73:255–61

47. Crawley, M. J., Pattrasudhi, R. 1988. Interspecific competition between insect herbivores: asymmetric competition between cinnabar moth and the ragwort seed-head fly. *Ecol. Entomol.* 13:243–49

48. Davidson, D. W. 1985. An experimental study of diffuse competition in harvester ants. *Am. Nat.* 125:500–6

49. Davidson, D. W., Inouye, R. S., Brown, J. H. 1984. Granivory in a desert ecosystem: experimental evidence for indirect facilitation of ants by rodents. *Ecology* 65:1780–86

50. Davidson, D. W., Samson, D. A., Inouye, R. S. 1985. Granivory in the Chihuahuan Desert: interactions within and between trophic levels. *Ecology* 66:486–502

51. Debach, P. 1974. *Biological Control by Natural Enemies.* Cambridge, UK: Cambridge Univ. Press

52. Delph, L. F. 1986. Factors regulating fruit and seed production in the desert annual *Lesquerella gordonii. Oecologia* 69:471–76

53. Dempster, J. P. 1982. The ecology of the cinnabar moth *Tyria jacobaeae* L. (Lepidoptera, Arctiidae). *Adv. Ecol. Res.* 12:1–36

54. Dempster, J. P. 1983. The natural control of populations of butterflies and moths. *Biol. Rev.* 58:461–81

55. Dennill, G. B. 1985. The effect of the gall wasp *Trichilogaster acaciaelongifoliae* (Hymenoptera: Pteromalidae) on reproductive potential and vegetative growth of the weed *Acacia longifolia. Agric. Ecosyst. Environ.* 14:53–61

56. De Steven, D. 1983. Reproductive consequences of insect seed predation in *Hamamelis virginiana. Ecology* 64:89–98

57. Dewey, J. E. 1986. Western spruce budworm impact on Douglas-fir cone production. *Gen. Tech. Rep. Interm. Res. Stn.* 203:243–45

58. Dohmen, G. P., McNeill, S., Bell, J. N. B. 1984, Air pollution increases *Aphis fabae* pest potential. *Nature* 307:52–53

59. Duggan, A. E. 1988. *The population biology of orange tip butterfly and its host plant,* Cardamine pratensis. PhD thesis. Univ. London

60. Dunn, J. P., Kimmerer, T. W., Nordin, G. L. 1986. The role of host tree condition in attack of white oaks by the two-lined chestnut borer, *Agrilus bilineatus* (Weber) (Coleoptera: Buprestidae). *Oecologia* 70:596–600

61. Ehnstroem, B. 1985. Insect damage in Swedish forests since 1970: summary. *Sver. Skogvaardsforb. Tidskr.* 2:11–20

62. Ehrlich, P. R., Birch, L. C. 1967. The balance of nature and population control. *Am. Nat.* 101:97–107

63. Deleted in proof

64. Faeth, S. H. 1987. Community structure and folivorous insect outbreaks: the roles of vertical and horizontal interactions. In *Insect Outbreaks,* ed. P. Borbosat, J. C. Shultz. New York: Academic. In press

65. Flemion, F. 1958. Penetration and destruction of plant tissues during feeding by *Lygus lineolaris* P. de B. *Proc. 10th Int. Congr. Entomol., Montreal, 1956* 3:475–78

66. Flueckiger, W., Braun, S. 1986. Effect of air pollutants on insects and hostplant/insect relationships. In *How are the Effects of Air Pollutants on Agricultural Crops Influenced by the Interaction With Other Limiting Factors?,* pp. 79–91. Brussels: Comm. Eur. Commun.

67. Forrester, G. 1989. *The population ecology of acorn weevils and their influence on natural regeneration of oak.* PhD thesis. Univ. London

68. Foster, W. A. 1984. The distribution of the sea lavender aphid *Staticobium staticis* on a marine saltmarsh and its effects on host plant fitness. *Oikos* 42:97–104

69. Fowler, N. L., Rausher, M. D. 1985. Joint effects of competitors and herbivores on growth and reproduction in *Aristolochia reticulata. Ecology* 66:1580–87

70. Fowler, S. V., Lawton, J. H. 1985. Rapidly induced defences and talking trees: the devil's advocate position. *Am. Nat.* 126:181–95

71. Fowler, S. V., MacGarvin, M. 1986. The effects of leaf damage on the performance of insect herbivores on birch, *Betula pubescens. J. Anim. Ecol.* 55:565–73

72. Fox, C. J. S. 1958. Some effects of in-

secticides on the wireworms and vegetation of grassland in Nova Scotia. *Proc. 10th Int. Congr. Entomol., Montreal, 1956* 3:297–300

73. Fox, L. R., Morrow, P. A. 1983. Estimates of damage by herbivorous insects on *Eucalyptus* trees. *Aust. J. Ecol.* 8:139–47

74. Frank, C. J., Jenkins, M. J. 1987. Impact of the western spruce budworm (Lepidoptera: Tortricidae) on buds, developing cones, and seeds of Douglas-fir in west central Idaho. *Environ. Entomol.* 16:304–8

75. Fuentes, E. R., Etchegaray, J., Aljaro, M. E., Montenegro, G. 1981. Shrub defoliation by matorral insects. In *Ecosystems of the World*, Vol. 11, *Mediterranean-Type Shrublands*, ed. F. di Castri, D. Goodall, R. Specht, pp. 345–59. New York: Elsevier

76. Fye, R. E. 1984. Damage to vegetable and forage seedlings by overwintering *Lygus hesperus* (Heteroptera: Miridae) adults. *J. Econ. Entomol.* 77:1141–43

77. Gibson, C. W. D., Brown, V. K., Jepsen, M. 1987. Relationships between the effects of insect herbivory and sheep grazing on seasonal changes in an early successional plant community. *Oecologia* 71:245–53

78. Goeden, R. D., Kok, L. T. 1986. Comments on a proposed "new" approach for selecting agents for the biological control of weeds. *Can. Entomol.* 118:51–58

79. Gradwell, G. 1974. The effect of defoliators on tree growth. In *The British Oak*, ed. M. G. Morris, F. H. Perring, pp. 182–93. Faringdon, UK: Classey

80. Hails, R. 1988. *Population ecology of the gall wasp Andricus quercuscalicis.* PhD thesis. Univ. London

81. Hainsworth, F. R., Wolf, L. L., Mercier, T. 1984. Pollination and predispersal seed predation: net effects on reproduction and inflorescence characteristics in *Ipomopsis aggregata. Oecologia* 63:405–9

82. Hairston, N. G., Smith, F. E., Slobodkin, L. B. 1960. Community structure, population control and competition. *Am. Nat.* 94:421–25

83. Hargrove, W. W., Crossley, D. A., T. R. Seastedt, T. R. 1984. Shifts in insect herbivory in the canopy of black locust, *Robinia pseudoacacia*, after fertilization. *Oikos* 43:322–28

84. Harris, P. 1981. Stress as a strategy in the biological control of weeds. In *Biological Control in Crop Protection*, ed. G. C. Papavizas, pp. 333–40. Totowa, NJ: Allanheld, Osmun

85. Harris, P. 1986. Biological control of weeds. *Fortschr. Zool.* 32:123–38

86. Harris, P., Thompson, L. S., Wilkinson, A.T.S., Neary, M.E. 1978. Reproductive biology of tansy ragwort, climate and biological control by the cinnabar moth in Canada. In *Proc. 4th Int. Symp. Biol. Control Weeds*, ed. T. E. Freeman, pp. 163–73. Gainsville, Fl: CEPIFAS

87. Hassell, M. P., Anderson, R. M. 1984. Host susceptibility as a component in host-parasitoid systems. *J. Anim. Ecol.* 53:611–21

88. Haukioja, E., Neuvonen, S. 1985. Induced long-term resistance of birch foliage against defoliators: defensive or accidental? *Ecology* 66:1303–8

89. Hawkins, J. A., Wilson, C. L., Mondart, C. L., Nelson, B. D., Farlow, R. A., Schilling, P. E. 1979. Leafhoppers and plant hoppers in coastal bermudagrass: effect on yield and quality, and control by harvest frequency. *J. Econ. Entomol.* 72:101–4

90. Heichel, G. H., Turner, N. C. 1976. Phenology and leaf growth of defoliated hardwood trees. In *Perspectives in Forest Entomology*, ed. J. F. Anderson, H. K. Kaya, pp. 31–40. New York: Academic

91. Heichel, G. H., Turner, N. C. 1984. Branch growth and leaf numbers of red maple (*Acer rubrum* L.) and red oak (*Quercus rubra* L.): response to defoliation. *Oecologia* 62:1–6

92. Henderson, I. F., Clements, R. O. 1979. Differential susceptibility to pest damage in agricultural grasses. *J. Agric. Sci.* 93:465–72

93. Hendrix, S. D. 1984. Relations of *Heracleum lanatum* to floral herbivory by *Depressaria pastinacella. Ecology* 65:191–97

94. Hendrix, S. D. 1988. Herbivory and its impact on plant reproduction. in *Reproductive Strategies in Plants*, ed. J. Lovett-Doust, L. Lovett-Doust. New York: Oxford Univ. Press. In press

95. Herrera, C. M. 1984. Selective pressures on fruit seediness: differential predation of fly larvae on the fruits of *Berberis hisanica. Oikos* 42:166–70

96. Hill, R. L. 1982. *The phytophagous fauna of gorse* (Ulex europeus *L.*) *and host plant quality.* PhD thesis. Univ. London

97. Hobbs, R. J. 1985. Harvester ant foraging and plant species distribution in annual grassland. *Oecologia* 67:519–23

98. Hodkinson, I. D. 1973. The population dynamics and host plant interactions of *Strophingia ericae* (Curt.) (Homoptera:

Psylloidea). *J. Anim. Ecol.* 42:565–83

99. Hokkanen, H., Pimentel, D. 1984. New approach for selecting biological control agents. *Can. Entomol.* 116:1109–21

100. Horvitz, C. C., Schemske, D. W. 1984. Effects of ants and an ant-tended herbivore on seed production of a neotropical herb. *Ecology* 65:1369–78

101. Horvitz, C. C., Schemske, D. W. 1988. A test of the pollinator limitation hypothesis for a neotropical herb. *Ecology* 69:200–6

102. Hosking, G. P., Hutcheson, J. A. 1986. Hard beech *(Nothofagus truncata)* decline on the Mamaku Plateau, North Island, New Zealand. *NZ J. Bot.* 24:263–69

103. Howe, H. F., Westley, L. C. 1986. Ecology of pollination and seed dispersal. In *Plant Ecology,* ed. M. J. Crawley, pp. 185–215. Oxford: Blackwell Sci.

104. Inouye, D. W. 1982. The consequences of herbivory: a mixed blessing for *Jurinea mollis* (Asteraceae). *Oikos* 39:269–72

105. Islam, Z., Crawley, M. J. 1983. Compensation and regrowth in ragwort *(Senecio jacobaea)* attacked by cinnabar moth *(Tyria jacobaeae)*. *J. Ecol.* 71:829–43

106. Janzen, D. H. 1985. *Spondias mombin* is culturally deprived in megafauna-free forest. *J. Trop. Ecol.* 1:131–55

107. Jermy, T. 1984. Evolution of insect/host plant relationships. *Am. Nat.* 124:609–30

108. Jones, F.G.W., Jones, M.G. 1964. *Pests of Field Crops*. London: Arnold

109. Jones, V. P., Toscano, N. C., Johnson, M. W., Welter, S. C., Youngman, R. R. 1986. Pesticide effects on plant physiology: integration into a pest management program. *Bull. Entomol. Soc. Am.* 32:103–9

110. Julien, M. H., ed. 1987. *Biological Control of Weeds: A World Catalogue of Agents and their Target Weeds*. Wallingford, UK: CAB Int. 2nd ed.

111. Kahn, D. M., Cornell, H. V. 1983. Early leaf abscission and folivores: comments and considerations. *Am. Nat.* 122:428–32

112. Kappel, F., Proctor, J. T. A. 1986. Simulated spotted tentiform leafminer injury and its influence on growth and fruiting of apple trees. *J. Am. Soc. Hortic. Sci.* 111:64–69

113. Karban, R. 1980. Periodical cicada nymphs impose periodical oak tree wood accumulation. *Nature* 287:326–27

114. Karban, R. 1986. Interspecific competition between folivorous insects on *Erigeron glaucus*. *Ecology* 67:1063–72

115. Karel, A. K., Mghogho, R. M. K. 1985. Effects of insecticides and plant populations on the insect pests and yield of common bean *(Phaseolus vulgaris* L.). *J. Econ. Entomol.* 78:917–21

116. Kinsman, S., Platt, W. J. 1984. The impact of a herbivore upon *Mirabilis hirsuta,* a fugitive prairie plant. *Oecologia* 65:2–6

117. Kirkland, R. L., Goeden, R. D. 1978. Biology of *Microlarinus lareynii* (Col.: Curculionidae) on puncturevine in southern California. *Ann. Entomol. Soc. Am.* 71:13–18

118. Kirkland, R. L., Goeden, R. D. 1978. An insecticidal-check study of the biological control of puncturevine *(Tribulus terrestris)* by imported weevils, *Microlarinus lareynii* and *M. lypriformis* (Col.: Curculionidae). *Environ. Entomol.* 7:349–54

119. Kjellsson, G. 1985. Seed fate in a population of *Carex pilulifera* L. I. Seed dispersal and ant-seed mutualism. II. Seed predation and its consequences for dispersal and seed bank. *Oecologia* 67:416–23, 424–29

120. Ko, J. H., Morimoto, K. 1985. Loss of tree vigor and role of boring insects in red pine stands heavily infested by the pine needle gall midge in Korea. *Esakia* 23:151–58

121. Kulman, H. M. 1971. Effects of insect defoliation on growth and mortality of trees. *Ann. Rev. Entomol.* 16:289–324

122. Landsberg, J. 1985. Drought and dieback of rural eucalypts. *Aust. J. Ecol.* 10:87–90

123. Larsson, S., Wiren, A., Lundgren, L., Ericsson, T. 1985. Effects of light and nutrient stress on leaf phenolic chemistry in *Salix dasyclados* and susceptibility to *Galerucella lineola* (Coleoptera). *Oikos* 45:205–10

124. Lawton, J. H., McNeill, S. 1979. Between the devil and the deep blue sea: on the problems of being a herbivore. In *Population Dynamics,* ed. R. M. Anderson, B. D. Turner, L. R. Taylor, pp. 223–44. Oxford: Blackwell Sci.

125. Leather, S. R., Barbour, D. A. 1987. Associations between soil type, lodgepole pine *(Pinus contorta)* provenance, and the abundance of the pine beauty moth, *Panolis flammea. J. Appl. Ecol.* 24:945–51

126. Leather, S. R., Watt, A. D., Forrest, G. I. 1987. Insect-induced chemical changes in youthing lodgepole pine *(Pinus contorta)*: the effects of previous defoliation on oviposition, growth and

survival of the pine beauty moth, *Panolis flammea. Ecol. Entomol.* 12:275–81

127. Lechowicz, M. J., Jobin, L. 1983. Estimating the susceptibility of tree species to attack by the gypsy moth, *Lymantria dispar. Ecol. Entomol.* 8:171–83

128. Lee, T. D., Bazzaz, F. A. 1980. Effects of defoliation and competition on growth and reproduction in the annual plant *Abutilon theophrasti. J. Ecol.* 68:813–21

129. Deleted in proof

130. Lewis, A. C. 1984. Plant quality and grasshopper feeding: effects of sunflower condition on preference and performance in *Melanoplus differentialis. Ecology* 65:836–43

131. Lincoln, D. E., Mooney, H. A. 1984. Herbivory on *Diplacus aurantiacus* shrubs in sun and shade. *Oecologia* 64:173–76

132. Louda, S. M. 1982. Limitation of the recruitment of the shrub *Haplopappus squarrosus* (Asteraceae) by flower- and seed-feeding insects. *J. Ecol.* 70:43–53

133. Louda, S. M. 1983. Seed predation and seedling mortality in the recruitment of a shrub, *Haplopappus venetus* (Asteraceae), along a climatic gradient. *Ecology* 64: 511–21

134. Louda, S. M. 1986. Insect herbivory in response to root-cutting and flooding stress on a native crucifer under field conditions. *Acta Oecol.* 7:37–53

135. Lowman, M. D. 1985. Temporal and spatial variability in insect grazing of the canopies of five Australian rainforest tree species. *Aust. J. Ecol.* 10:7–24

136. Maiorana, V. C. 1981. Herbivory in sun and shade. *Biol. J. Linn. Soc.* 15:151–56

137. Majer, J. D., Lamont, B. B. 1985. Removal of seed of *Grevillea pteridifolia* (Proteaceae) by ants. *Aust. J. Bot.* 33:611–18

138. Manzur, M. I., Courtney, S. P. 1984. Influence of insect damage in fruits of hawthorn on bird foraging and seed dispersal. *Oikos* 43:265–70

139. Marquis, R. J. 1984. Leaf herbivores decrease fitness of a tropical plant. *Science* 226:537–39

140. Marshall, D. L., Levin, D. A., Fowler, N. L. 1985. Plasticity in yield components in response to fruit predation and date of fruit initiation in three species of *Sesbania* (Leguminosae). *J. Ecol.* 73:71–81

141. Marshall, D. L., Levin, D. A., Fowler, N. L. 1986. Plasticity of yield components in response to stress in *Sesbania macrocarpa* and *Sesbania vesicaria* (Leguminosae). *Am. Nat.* 127:508–21

142. Mattson, W. J. 1986. Competition for food between two principal cone insects of red pine, *Pinus resinosa. Environ. Entomol.* 15:88–92

143. Mazer, S. J. 1988. The quantitative genetics of life history and fitness components in *Raphanus raphanistrum* L. (Brassicaceae): ecological and evolutionary consequences of seed-weight variation. *Am. Nat.* 130:891–914

144. McBrien, H., Harmsen, R., Crowder, A. 1983. A case of insect grazing affecting plant succession. *Ecology* 64:1035–39

145. McDade, L. A., Davidar, P. 1984. Determinants of fruit and seed set in *Pavonia dasypetala* (Malvaceae). *Oecologia* 64:61–67

146. McNaughton, S. J. 1986. On plants and herbivores. *Am. Nat.* 128:765–70

147. McNeill, S., Southwood, T.R.E. 1978. The role of nitrogen in the development of insect/plant relationships. In *Biochemical Aspects of Plant and Animal Coevolution,* ed. J. B. Harborne, pp. 77–98. London: Academic

148. Deleted in proof

149. Deleted in proof

150. Meyer, G. A., Montgomery, M. E. 1987. Relationships between leaf age and the food quality of cottonwood foliage for the gypsy moth, *Lymantria dispar. Oecologia* 72:527–32

151. Miller, G. E. 1986. Insects and conifer seed production in the Inland Mountain West: a review. *Gen. Tech. Rep. Interm. Res. Stn.* 203:225–37

152. Miller, G. E., Hedin, A. F., Ruth, D. S. 1984. Damage by two Douglas-fir cone and seed insects: correlation with cone crop size. *J. Entomol. Soc. BC* 81:46–50

153. Mills, J. N. 1983. Herbivory and seedling establishment in post-fire southern California chaparral. *Oecologia* 60:267–70

154. Mills, J. N. 1984. Effects of feeding by mealybugs (*Planococcus citri,* Homoptera: Pseudococcidae) on the growth of *Colliguaya odorifera* seedlings. *Oecologia* 64:142–44

155. Mittelbach, G. G., Gross, K. L. 1984. Experimental studies of seed predation in old-fields. *Oecologia* 65:7–13

156. Miyashita, T. 1985. Estimation of the economic injury level in the rice leafroller, *Cnaphalocrocis medinalis* Guenee (Lepidoptera, Pyralidae). I. Relation between yield loss and injury of rice leaves at heading or in the grain filling period. *Jpn. J. Appl. Entomol. Zool.* 29:73–76

157. Moore, L. M., Wilson, L. F. 1986. Impact of the poplar gall saperda, *Saperda*

*inornata* (Coleoptera: Cerambycidae) on a hybrid *Populus* plantation in Michigan. *Great Lakes Entomol.* 19:163–67

158. Moran, N., Hamilton, W. D. 1980. Low nutritive quality as defence against herbivores. *J. Theor. Biol.* 86:247–54

159. Morrill, W. L., Ditterline, R. L., Winstead, C. 1984. Effects of *Lygus borealis* Kelton (Hemiptera: Miridae) and *Adelphocoris lineolatus* (Goeze) (Hemiptera: Miridae) feeding on sainfoin production. *J. Econ. Entomol.* 77:966–68

160. Morton, S. R. 1985. Granivory in arid regions: comparison of Australia with North and South America. *Ecology* 66:1859–66

161. Mowat, D. J., Jess, S. 1984. Oviposition by frit fly, *Oscinella frit*, on young ryegrass seedlings. *J. Appl. Ecol.* 21:915–20

162. Murdoch, W. W. 1966. Community structure, population control, and competition—a critique. *Am. Nat.* 100:219–26

163. Myers, J. H. 1985. Effect of physiological condition of the host plant on the ovipositional choice of the cabbage white butterfly, *Pieris rapae. J. Anim. Ecol.* 54:193–204

164. Nelson, D. M., Johnson, C. D. 1983. Stabilizing selection on seed size in *Astragalus* (Leguminosae) due to differential predation and differential germination. *J. Kans. Entomol. Soc.* 56:169–74

165. O'Dowd, D. J., Gill, A. M. 1984. Predator satiation and site alteration following fire: mass reproduction of alpine ash *(Eucalyptus delegatensis)* in southeastern Australia. *Ecology* 65:1052–66

166. Ohmart, C. P. 1984. Is insect defoliation in eucalypt forests greater than that in other temperate forests? *Aust. J. Ecol.* 9:413–18

167. Oksanen, L. 1983. Trophic exploitation and arctic phytomass patterns. *Am. Nat.* 122:45–52

168. Paige, K. N., Whitham, T. G. 1987. Overcompensation in response to mammalian herbivory: the advantage of being eaten. *Am. Nat.* 129:419–28

169. Pantoja, A., Smith, C. M., Robinson, J. F. 1986. Effects of the fall armyworm (Lepidoptera: Noctuidae) on rice yields. *J. Econ. Entomol.* 79:1324–29

170. Parker, M. A. 1984. Local food depletion and the foraging behaviour of a specialist grasshopper, *Hesperotettix viridis. Ecology* 65:824–35

171. Parker, M. A. 1985. Size dependent herbivore attack and the demography of an arid grassland shrub. *Ecology* 66:850–60

172. Parker, M. A., Salzman, A. G. 1985. Herbivore exclosure and competitor removal: effects on juvenile survivorship and growth in the shrub *Gutierrezia microcephala. J. Ecol.* 73:903–13

173. Pigott, C. D. 1985. Selective damage to tree-seedlings by bank voles *(Clethrionomys glareolus). Oecologia* 67:367–71

174. Port, G. R., Thompson, J. R. 1980. Outbreaks of insect herbivores on plants along motorways in the United Kingdom. *J. Appl. Ecol.* 17:649–56

175. Prestidge, R. A., McNeill, S. 1981. The role of nitrogen in the ecology of grassland Auchenorrhyncha. In *Nitrogen as an Ecological Factor*, ed. J. A. Lee, S. McNeill, I. H. Rorison, pp. 257–81. Oxford: Blackwell Sci.

176. Pritchard, I. M., James, R. 1984. Leaf mines: their effect on leaf longevity. *Oecologia* 64:132–39

177. Pullin, A. S. 1987. Changes in leaf quality following clipping and regrowth of *Urtica dioica*, and consequences for a specialist insect herbivore, *Aglais urticae. Oikos* 49:39–45

178. Queller, D. C. 1985. Proximate and ultimate causes of low fruit production in *Asclepias exaltata. Oikos* 44:373–81

179. Rai, J. P. N., Tripathi, R. S. 1985. Effect of herbivory by the slug, *Mariaella dussumieri*, and certain insects on growth and competitive success of two sympatric annual weeds. *Agric. Ecosyst. Environ.* 13:125–37

180. Randall, M. G. M. 1986. The predation of predispersed *Juncus squarrosus* seeds by *Coleophora alticolella* (Lepidoptera) larvae over a range of altitudes in northern England. *Oecologia* 69:460–65

181. Rawes, M. 1981. Further results of excluding sheep from high level grasslands in the north Pennines. *J. Ecol.* 69:651–69

182. Rhoades, D. F. 1985. Offensive-defensive interactions between herbivores and plants: their relevance in herbivore population dynamics and ecological theory. *Am. Nat.* 125:205–38

183. Rice, B., Westoby, M. 1986. Evidence against the hypothesis that ant-dispersed seeds reach nutrient-enriched microsites. *Ecology* 67:1270–74

184. Risch, S. J., Carroll, C. R. 1986. Effects of seed predation by a tropical ant on competition among weeds. *Ecology* 67:1319–27

185. Rissing, S. W. 1986. Indirect effects of granivory by harvester ants: plant spe-

cies composition and reproductive increase near ant nests. *Oecologia* 68:231–34

186. Roberts, H. A. 1986. Seed persistence in soil and seasonal emergence in plant species from different habitats. *J. Appl. Ecol.* 23:639–56

187. Roland, J., Myers, J. H. 1987. Improved insect performance from host-plant defoliation: winter moth on oak and apple. *Ecol. Entomol.* 12:409–14

188. Romme, W. H., Knight, D. H., Yavitt, J. B. 1986. Mountain pine beetle outbreaks in the Rocky Mountains: regulators of primary productivity? *Am. Nat.* 127:484–94

189. Room, P. M., Thomas, P. A. 1985. Nitrogen and establishment of a beetle for biological control of the floating weed *Salvinia* in Papua New Guinea. *J. Appl. Ecol.* 22:139–56

190. Rossiter, M., Schultz, J. C., Baldwin, I. T. 1988. Relationships among defoliation, red oak phenolics, and gypsy moth growth and reproduction. *Ecology* 69:267–77

191. Sacchi, C. F., Price, P. W., Craig, T. P., Itami, J. K. 1988. Impact of the stem galler, *Euura lasiolepis*, on sexual reproduction in the arroyo willow, *Salix lasiolepis. Ecology* In press

192. Salama, H. S., El-Sharif, A. F., Megahed, M. 1985. Soil nutrients affecting the population density of *Parlatoria zizyphus* (Lucus) and *Icerya purchasi* Mask. (Homopt., Coccoidea) on citrus seedlings. *Z. Angew. Entomol.* 99:471–76

193. Schroeder, D., Goeden, R. G. 1986. The search for arthropod natural enemies of introduced weeds for biocontrol—in theory and practice. *Biocontrol News Inf.* 7:147–55

194. Schultz, J. C. 1988. Plant responses induced by herbivory. *Trends Ecol. Evol.* 3:45–49

195. Shaw, M. W. 1974. The reproductive characteristics of oak. See Ref. 79, pp. 162–81

196. Sheppard, A. W. 1987. *Competition between the insect herbivores of hogweed* Heracleum sphondylium. PhD thesis. Univ. London

197. Shure, D. J. 1971. Insecticide effects on early succession in an old field ecosystem. *Ecology* 52:271–79

198. Skrzypczynska, M. 1985. Gall midge (Cecidomyiidae, Diptera) pests in seeds and cones of coniferous trees in Poland. *Z. Angew. Entomol.* 100:448–50

199. Smith, C. C. 1975. The coevolution of plants and seed predators. In *Coevolu-tion of Animals and Plants,* ed. L. E. Gilbert, P. H. Raven, pp. 53–77. Austin, Texas: Univ. Texas Press

200. Solomon, B. P. 1981. Response of a host-specific herbivore to resource density, relative abundance and phenology. *Ecology* 62:1205–14

201. Stalter, R., Serrao, J. 1983. The impact of defoliation by gypsy moths on the oak forest at Greenbrook Sanctuary, New Jersey. *Bull. Torrey Bot. Club* 110:526–29

202. Stamp, N. E. 1984. Effect of defoliation by checkerspot caterpillars *(Euphydryas phaeton)* and sawfly larvae (*Macrophya nigra* and *Tenthredo grandis*) on their host plants (*Chelone* spp.). *Oecologia* 63:275–80

203. Stephenson, A. G. 1981. Flower and fruit abortion: proximate causes and ultimate function. *Ann. Rev. Ecol. Syst.* 12:253–81

204. Stuart, J. D. 1984. Hazard rating of lodgepole pine stands to mountain pine beetle outbreaks in central Oregon. *Can. J. For. Res.* 14:666–71

205. Sutherland, S. 1986. Patterns of fruit set: what controls fruit-flower ratio in plants? *Evolution* 40:117–28

206. Tallamy, D. W. 1985. Squash beetle feeding behaviour: an adaptation against induced cucurbit defences. *Ecology* 66:1574–79

207. Thomas, P. A., Room, P. M. 1986. Taxonomy and control of *Salvinia molesta. Nature* 320:581–84

208. Thompson, J. N. 1985. Postdispersal seed predation in *Lomatium* spp. (Umbelliferae): variation among individuals and species. *Ecology* 66:1608–16

209. Thompson, K. 1986. Small-scale heterogeneity in the seed bank of an acidic grassland. *J. Ecol.* 74:733–38

210. Trumble, J. T., Ting, I. P., Bates, L. 1985. Analysis of physiological, growth, and yield responses of celery to *Liriomyza trifolii. Entomol. Exp. Appl.* 38:15–21

210a. Twery, M. J., Patterson, W. A. 1984. Variations in beech bark disease and its effects on species composition and structure of northern hardwood stands in central New England. *Can. J. For. Res.* 14:565–74

210b. van der Meijden, E., de Jong, T. J., Klinkhamer, P. G. L., Kooi, R. E. 1985. Temporal and spatial dynamics in populations of biennial plants. In *Structure and Functioning of Plant Populations,* ed. J. Haeck, J. W. Woldendorp, 2:91–103. The Hague: North-Holland

211. van der Meijden, E., van der Waals-Kooi, R. E. 1979. The population ecology of *Senecio jacobaea* in a sand dune system. I. Reproductive strategy and the biennial habit. *J. Ecol.* 67:131–53

211a. van Emden, H. F., Bashford, M. A. 1971. The performance of *Brevicoryne brassicae* and *Myzus persicae* in relation to plant age and leaf amino acids. *Entomol. Exp. Appl.* 14:349–60

211b. van Leeuwen, B. H. 1983. The consequences of predation in the population biology of the monocarpic species *Cirsium palustre* and *Cirsium vulgare*. *Oecologia* 58:178–87

212. Via, S. 1986. Genetic covariance between oviposition preference and larval performance in an insect herbivore. *Evolution* 40:778–85

213. Wagner, F., Ehrhardt, R. 1961. Untersuchungen am Stickanal der Graswanze *Miris dolobratus* L., der Urheberin der totalen Weissahrigkeit der Rotschwingels *(Festuca rubra)*. *Z. Pflanzenkr. Pflanzenschutz* 68:615–20

214. Wallace, J. B., O'Hop, J. 1985. Life on a fast pad: waterlily leaf beetle impact on water lilies. *Ecology* 66:1534–44

215. Waloff, N., Richards, O. W. 1977. The effect of insect fauna on growth, mortality and natality of broom, *Sarothamnus scoparius*. *J. Appl. Ecol.* 14:787–98

216. Wapshere, A. J. 1985. Effectiveness of biological control agents for weeds: present quandaries. *Agric. Ecosyst. Environ.* 13:261–80

217. Waring, R. H., Pitman, G. B. 1985. Modifying lodgepole pine stands to change susceptibility to mountain pine beetle attack. *Ecology* 66:889–97

218. Warrington, S., Whittaker, J. B. 1985. An experimental field study of different levels of insect herbivory induced by *Formica rufa* predation on sycamore *(Acer pseudoplatanus)*. I. Lepidoptera larvae. II. Aphidoidea. *J. Appl. Ecol.* 22:775–96

219. Wellington, A. R., Noble, I. R. 1985. Seed dynamics and factors limiting recruitment of the mallee *Eucalyptus incrassata* in semi-arid southeastern Australia. *J. Ecol.* 73:657–66

220. White T.C.R. 1984. The abundance of insect herbivores in relation to the availability of nitrogen in stressed food plants. *Oecologia* 63:90–105

221. Whitford, W. G. 1978. Foraging by seed-harvesting ants. In *Production Ecology of Ants and Termites*, ed. M. V. Brian, pp. 107–10. Cambridge, UK: Cambridge Univ. Press

222. Whitham, T. G., Mopper, S. 1985. Chronic herbivory: impacts on architecture and sex expression of pinyon pine. *Science* 228:1089–91

223. Wilcox, A., Crawley, M. J. 1988. The effects of host plant defoliation and fertilizer application on larval growth and oviposition behaviour in cinnabar moth. *Oecologia* 76:283–87

224. Williams, A. G., Whitham, T. G. 1986. Premature leaf abscission: an induced plant defence against gall aphids. *Ecology* 67:1619–27

225. Wilson, M, C., Stewart, J. K., Vail, H. D. 1979. Full season impact of the alfalfa weevil, meadow spittlebug and potato leafhopper in an alfalfa field. *J. Econ. Entomol.* 72:830–34

226. Windle, P. N., Franz, E. H. 1979. The effects of insect parasitism on plant competition: greenbugs and barley. *Ecology* 60:521–29

227. Wit, A. K. H. 1985. The relation between partial defoliation during the preheading stages of spring cabbage and yield, as a method to assess the quantitative damage induced by leaf mining insects. *Z. Angew. Entomol.* 100:96–100

228. Wood, B. W., Tedders, W. L., Dutcher, J. D. 1987. Energy drain by three pecan aphid species (Homoptera: Aphididae) and their influence on in-shell pecan production. *Environ. Entomol.* 16:1045–56

229. Wood, D. M., Andersen, M. C. 1988. Predispersal seed predation in *Aster ledophyllus:* does it matter? Experimental evidence from Mount St. Helens. *Ecology* In press

230. Zakaria, S. 1989. *The influence of previous insect feeding on the rate of damage of birch tree leaves*. PhD thesis. Univ. London

231. Zwolfer, H., Harris, P. 1984. Biology and host specificity of *Rhinocyllus conicus* (Froel.) (Col., Curculionidae), a successful agent for biocontrol of the thistle, *Carduus nutans* L. *Z. Angew. Entomol.* 97:36–62

# SUBJECT INDEX

## A

Abate, 415
*Acerartagallia sanguinolenta*, 515, 520
Acetylcholine, 80-81, 85, 88, 91, 103
*Acheta domesticus*, 105, 108
Acoustic
  sounding and remote sensing, 255
  strategies of *Nezara viridula*, 282
Acquisition of plant viruses, 507, 515, 517
*Acrosternum hilare*, 280
Action potential, 79, 81
*Acyrthosiphon pisum*, 535
Adaptation and learning, 317-18
*Aedeomyia*, 403
*Aedes*
  *aegypti*, 402-3, 406, 409-15, 484
  *albopictus*, 404
  *atropalpus*, 407-9
  *caspius*, 409
  *sollicitans*, 407
  *taeniorhynchus*, 403, 407
  *togoi*, 408-9, 413
  *triseriatus*, 403, 406-8, 415
  *vexans*, 404-5, 407
Aerial
  photography, 248-50, 259
  videography, 250
*Afroxylocopa*, 173
*Agallia constricta*, 515-17, 519-20
*Agalliopsis novella*, 519-20
*Agasicles hygrophila*, 543
*Agelenopsis*, 236
Aggregation by *Nezara viridula*, 275, 279-80
*Agrilus hyperici*, 546
*Agrotis munda*, 257
Air pollution and herbivory, 541
Aldicarb, 59, 298. 459, 465, 468, 470
Aldrin, 455
*Aleochara*
  *bilineata*, 127
  *bipustulata*, 127
*Aleurocanthus woglumi*, 261
Allelochemicals and *Heliothis*, 21
Allethrin, 77-96
*Alloeorhynchus*, 392

*Altica carduorum*, 546
*Amblyopone*, 194
  *pallipes*, 203
Amidation, 362
γ-Aminobutyric acid, 103
  receptor complex, 78, 80, 86-91
*Anaphes ovijentatus*, 390
*Anaptus major*, 387-88, 390
*Andricus quercuscalicis*, 534
Anesthesia by carbon dioxide, 98, 102-10
*Anopheles*, 403
  *albimanus*, 413
  *gambiae*, 352, 407
  *stephensi*, 408, 415
Anopheline resistance, 298
Antennal receptor cells, 483
*Antheraea*, 480
  *pernyi*, 489
  *polyphemus*, 478, 489
*Anthonomus grandis*, 60, 262, 285
Anthrax, 172-73
*Anticarsia gemmatalis*, 280
Antigens and scabies, 153-56
Ants
  and carbon dioxide, 99-100
  and carpenter bees, 169, 171
  and competition, 427
  and deutocerebrum, 478
  and foraging strategies, 191-210
  and guilds, 432
  and learning, 330
  and *Nezara viridula*, 275, 283-84
  and plant dynamics, 536-37, 555
  and remote sensing, 248
  and weed control, 548
*Apanteles*
  See *Cotesia*
*Aphaereta pallipes*, 127-28, 131
*Aphanotus brevicornis*, 172
Aphids, 218-19
  and *Brassica* crops, 213
  and fungi, 391
  and insecticides, 461
  and Nabidae, 392
  and oilseed crops, 221
  and plant dynamics, 534-35, 541, 549, 552, 554-55
  and plant viruses, 505-7, 511, 513, 515-16, 522
  and remote sensing, 258, 261

*Apion*
  *brunneonigrum*, 546
  *ulicis*, 548
*Apis mellifera*, 105, 482, 484, 491
  and remote sensing, 258
Apparency, 342
Apple maggot flies, 318, 323, 338, 340
*Aptus mirmicoides*, 385, 389
*Arachnocoris albomaculatus*, 393
*Archlagocheirus funestus*, 546
*Archytas marmoratus*, 56, 62-63
Armyworm, 217-19, 257
  and remote sensing, 263
Associative learning, 321-26
*Atalanticus testaceous*, 551
*Athalia proxima*, 214
*Athesapeuta cyperi*, 546
*Athrycia cinerea*, 389
ATP and pyrethroids, 88-89, 91
*Atrytonopsis*
  *edwardsi*, 237
  *ovinia*, 237
*Atta*, 199
  *cephalotes*, 195
*Attagenus*, 172
Attraction by carbon dioxide, 97, 99-100, 105
*Autochton*, 238
*Autographa californica*, 352
Aversion learning, 325-26, 328, 334, 339

## B

*Bacillus thuringiensis*, 57, 64-66, 69, 373-76
  endotoxin, 374, 376-80
*Bactra*
  *minima*, 546
  *venosana*, 546
Baculovirus
  genomes, 353-54
*Bagrada cruciferarum*, 214
Bark beetles, 533, 542
*Battus*, 319-20, 324, 329, 331, 336, 338
  *philenor*, 335
*Beauveria brassiana*, 57
Bees
  and carbon dioxide, 97, 99, 102, 105-7
  and deutocerebrum, 478

**565**

# CUMULATIVE INDEXES

## CONTRIBUTING AUTHORS, VOLUMES 25–34

**573**

# CHAPTER TITLES, VOLUMES 25–34

# Annual Reviews Inc.

## A NONPROFIT SCIENTIFIC PUBLISHER

4139 El Camino Way
P.O. Box 10139
Palo Alto, CA 94303-0897 • USA

Annual Reviews Inc. publications may be ordered directly from our office by mail, Telex, or use our Toll Free Telephone line (for orders paid by credit card or purchase order*, and customer service calls only); through booksellers and subscription agents, worldwide; and through participating professional societies. Prices subject to change without notice. ARI Federal I.D. #94-1156476

- **Individuals:** Prepayment required on new accounts by check or money order (in U.S. dollars, check drawn on U.S. bank) or charge to credit card—American Express, VISA, MasterCard.
- **Institutional buyers:** Please include purchase order number.
- **Students:** $10.00 discount from retail price, per volume. Prepayment required. Proof of student status must be provided (photocopy of student I.D. or signature of department secretary is acceptable). Students must send orders direct to Annual Reviews. Orders received through bookstores and institutions requesting student rates will be returned. You may order at the Student Rate for a maximum of 3 years.
- **Professional Society Members:** Members of professional societies that have a contractual arrangement with Annual Reviews may order books through their society at a reduced rate. Check with your society for information.
- **Toll Free Telephone orders:** Call 1-800-523-8635 (except from California) for orders paid by credit card or purchase order and customer service calls only. California customers and all other business calls use 415-493-4400 (not toll free). Hours: 8:00 AM to 4:00 PM, Monday-Friday, Pacific Time. **Written confirmation** is required on purchase orders from universities before shipment.
- **Telex: 910-290-0275**

**Regular orders:** Please list the volumes you wish to order by volume number.
**Standing orders:** New volume in the series will be sent to you automatically each year upon publication. Cancellation may be made at any time. Please indicate volume number to begin standing order.
**Prepublication orders:** Volumes not yet published will be shipped in month and year indicated.
**California orders:** Add applicable sales tax.
**Postage paid** (4th class bookrate/surface mail) **by Annual Reviews Inc.** Airmail postage or UPS, extra.

| ANNUAL REVIEWS SERIES | | Prices Postpaid per volume USA & Canada/elsewhere | Regular Order Please send: | Standing Order Begin with: |
| :--- | :--- | :--- | :--- | :--- |
| | | | Vol. number | Vol. number |
| **Annual Review of ANTHROPOLOGY** | | | | |
| Vols. 1-14 | (1972-1985) | $27.00/$30.00 | | |
| Vols. 15-16 | (1986-1987) | $31.00/$34.00 | | |
| Vol. 17 | (1988) | $35.00/$39.00 | | |
| Vol. 18 | (avail. Oct. 1989) | $35.00/$39.00 | Vol(s). _____ | Vol. _____ |
| **Annual Review of ASTRONOMY AND ASTROPHYSICS** | | | | |
| Vols. 1, 4-14, 16-20 | (1963, 1966-1976, 1978-1982) | $27.00/$30.00 | | |
| Vols. 21-25 | (1983-1987) | $44.00/$47.00 | | |
| Vol. 26 | (1988) | $47.00/$51.00 | | |
| Vol. 27 | (avail. Sept. 1989) | $47.00/$51.00 | Vol(s). _____ | Vol. _____ |
| **Annual Review of BIOCHEMISTRY** | | | | |
| Vols. 30-34, 36-54 | (1961-1965, 1967-1985) | $29.00/$32.00 | | |
| Vols. 55-56 | (1986-1987) | $33.00/$36.00 | | |
| Vol. 57 | (1988) | $35.00/$39.00 | | |
| Vol. 58 | (avail. July 1989) | $35.00/$39.00 | Vol(s). _____ | Vol. _____ |
| **Annual Review of BIOPHYSICS AND BIOPHYSICAL CHEMISTRY** | | | | |
| Vols. 1-11 | (1972-1982) | $27.00/$30.00 | | |
| Vols. 12-16 | (1983-1987) | $47.00/$50.00 | | |
| Vol. 17 | (1988) | $49.00/$53.00 | | |
| Vol. 18 | (avail. June 1989) | $49.00/$53.00 | Vol(s). _____ | Vol. _____ |
| **Annual Review of CELL BIOLOGY** | | | | |
| Vol. 1 | (1985) | $27.00/$30.00 | | |
| Vols. 2-3 | (1986-1987) | $31.00/$34.00 | | |
| Vol. 4 | (1988) | $35.00/$39.00 | | |
| Vol. 5 | (avail. Nov. 1989) | $35.00/$39.00 | Vol(s). _____ | Vol. _____ |

| ANNUAL REVIEWS SERIES | Prices Postpaid per volume USA & Canada/elsewhere | Regular Order Please send: Vol. number | Standing Order Begin with: Vol. number |
|---|---|---|---|
| **Annual Review of COMPUTER SCIENCE** | | | |
| Vols. 1-2 (1986-1987) | $39.00/$42.00 | | |
| Vol. 3 (1988) | $45.00/$49.00 | | |
| Vol. 4 (avail. Nov. 1989) | $45.00/$49.00 | Vol(s). _____ | Vol. _____ |
| **Annual Review of EARTH AND PLANETARY SCIENCES** | | | |
| Vols. 1-10 (1973-1982) | $27.00/$30.00 | | |
| Vols. 11-15 (1983-1987) | $44.00/$47.00 | | |
| Vol. 16 (1988) | $49.00/$53.00 | | |
| Vol. 17 (avail. May 1989) | $49.00/$53.00 | Vol(s). _____ | Vol. _____ |
| **Annual Review of ECOLOGY AND SYSTEMATICS** | | | |
| Vols. 2-16 (1971-1985) | $27.00/$30.00 | | |
| Vols. 17-18 (1986-1987) | $31.00/$34.00 | | |
| Vol. 19 (1988) | $34.00/$38.00 | | |
| Vol. 20 (avail. Nov. 1989) | $34.00/$38.00 | Vol(s). _____ | Vol. _____ |
| **Annual Review of ENERGY** | | | |
| Vols. 1-7 (1976-1982) | $27.00/$30.00 | | |
| Vols. 8-12 (1983-1987) | $56.00/$59.00 | | |
| Vol. 13 (1988) | $58.00/$62.00 | | |
| Vol. 14 (avail. Oct. 1989) | $58.00/$62.00 | Vol(s). _____ | Vol. _____ |
| **Annual Review of ENTOMOLOGY** | | | |
| Vols. 10-16, 18 (1965-1971, 1973) | | | |
| 20-30 (1975-1985) | $27.00/$30.00 | | |
| Vols. 31-32 (1986-1987) | $31.00/$34.00 | | |
| Vol. 33 (1988) | $34.00/$38.00 | | |
| Vol. 34 (avail. Jan. 1989) | $34.00/$38.00 | Vol(s). _____ | Vol. _____ |
| **Annual Review of FLUID MECHANICS** | | | |
| Vols. 1-4, 7-17 (1969-1972, 1975-1985) | $28.00/$31.00 | | |
| Vols. 18-19 (1986-1987) | $32.00/$35.00 | | |
| Vol. 20 (1988) | $34.00/$38.00 | | |
| Vol. 21 (avail. Jan. 1989) | $34.00/$38.00 | Vol(s). _____ | Vol. _____ |
| **Annual Review of GENETICS** | | | |
| Vols. 1-19 (1967-1985) | $27.00/$30.00 | | |
| Vols. 20-21 (1986-1987) | $31.00/$34.00 | | |
| Vol. 22 (1988) | $34.00/$38.00 | | |
| Vol. 23 (avail. Dec. 1989) | $34.00/$38.00 | Vol(s). _____ | Vol. _____ |
| **Annual Review of IMMUNOLOGY** | | | |
| Vols. 1-3 (1983-1985) | $27.00/$30.00 | | |
| Vols. 4-5 (1986-1987) | $31.00/$34.00 | | |
| Vol. 6 (1988) | $34.00/$38.00 | | |
| Vol. 7 (avail. April 1989) | $34.00/$38.00 | Vol(s). _____ | Vol. _____ |
| **Annual Review of MATERIALS SCIENCE** | | | |
| Vols. 1, 3-12 (1971, 1973-1982) | $27.00/$30.00 | | |
| Vols. 13-17 (1983-1987) | $64.00/$67.00 | | |
| Vol. 18 (1988) | $66.00/$70.00 | | |
| Vol. 19 (avail. Aug. 1989) | $66.00/$70.00 | Vol(s). _____ | Vol. _____ |
| **Annual Review of MEDICINE** | | | |
| Vols. 9, 11-15 (1958, 1960-1964) | | | |
| 17-36 (1966-1985) | $27.00/$30.00 | | |
| Vols. 37-38 (1986-1987) | $31.00/$34.00 | | |
| Vol. 39 (1988) | $34.00/$38.00 | | |
| Vol. 40 (avail. April 1989) | $34.00/$38.00 | Vol(s). _____ | Vol. _____ |